Interdisciplinary Biotechnological Advances

This series is an authoritative resource that explains recent advances and emerging directions in biotechnology, reflecting the forefront of research clearly and reliably, without excessive hype. Each volume is written by authors with excellent reputations and acknowledged expertise in the topic under discussion. The volumes span the entire field from an interdisciplinary perspective, covering everything from biotechnology principles and methods to applications in areas including genetic engineering, transgenic plants and animals, environmental problems, genomics, proteomics, diagnosis of disease, gene therapy, and biomedicine. The significance of these applications for the achievement of UN Sustainable Development Goals is highlighted. The series will be highly relevant for Master's and PhD students in Biotechnology, Nanochemistry, Biochemical Engineering, and Microbiology, medical students, academic and industrial researchers, agricultural scientists, farmers, clinicians, industry personnel, and entrepreneurs.

Jen-Tsung Chen · Masudulla Khan ·
Aiman Parveen · Jayanta Kumar Patra
Editors

Molecular and Biotechnological Tools for Plant Disease Management

Springer

Editors
Jen-Tsung Chen
Department of Life Sciences
National University of Kaohsiung
Kaohsiung, Taiwan

Aiman Parveen
Department of Botany
Mayfair College
Moradabad, India

Masudulla Khan
Aligarh Muslim University
Aligarh, India

Jayanta Kumar Patra
Dongguk University
Goyang, Korea (Republic of)

ISSN 2730-7069 ISSN 2730-7077 (electronic)
Interdisciplinary Biotechnological Advances
ISBN 978-981-97-7509-5 ISBN 978-981-97-7510-1 (eBook)
https://doi.org/10.1007/978-981-97-7510-1

This Springer imprint is published by the registered company Springer Nature Singapore Pte Ltd.
The registered company address is: 152 Beach Road, #21-01/04 Gateway East, Singapore 189721, Singapore

Preface

Each year, agricultural production has huge losses caused by diverse plant diseases all over the world. Therefore, there is a high demand to develop effective management systems for reducing such damage to crops from pests and pathogens. In the scenario of global climate change, and increasing pollutants, the issue of plant diseases can be even more serious under stressful environments due to the retard of growth, development, and defense mechanisms of crops. Among the approaches, molecular biology together with biotechnology can lead to a range of tools for managing plant diseases and are expected to increase agricultural production to support global food security and the vision of sustainable agriculture. This book provides edge knowledge of ways based on molecular strategies for disease diagnostics and breeding of tolerant crops covering the most significant pests and pathogens.

This book presents polymerase chain reaction (PCR)-based technologies for plant disease detection and identification, including real-time PCR, multiplex PCR, nested PCR, and PCR-restriction fragment polymorphism (RFLP). Moreover, upgraded tools for plant disease diagnostics were provided in detail, such as loop-mediated amplification (LAMP), nucleic acid sequence-based amplification (NASBA), and next-generation sequencing (NGS), which enable high-throughput ways of strategies in disease management. For in-field phenotyping, this book provides spectroscopy-based methods combined with unmanned aerial vehicles (UAVs) that can identify the type of diseases quickly and non-invasively. For years, biosensor technologies have been successfully applied in the field of plant-pathogen detection; this book highlights the promising methods such as antibody-based biosensors, lateral flow immunoassays (LFIA), enzyme-linked immunosorbent assay (ELISA), DNA-based biosensors, optical biosensors, electronic nose, and acoustic sensors, for constructing the next-generation of plant disease diagnostics.

In recent years, computer science and computational biology have had great breakthroughs and led to powerful bioinformatics approaches in plant pathology and may develop revolutionizing systems for plant disease management in the future. This book discusses a series of subtopics in artificial intelligence (AI), machine learning, deep learning, and the Internet of Things (IoT). Additionally, popular databases for the research of plant pathology were comprehensively summarized.

In facing pests and pathogens, plants have evolved immunity and possess defense mechanisms to combat diseases. For decades, scientists have conducted genomics and further multiple omics for releasing the genetic potential of plant immunity in benefiting programs of crop breeding. An emerging tool, genome editing, has enabled the precise breeding of crops with novel traits, and one of the leading systems is based on clustered regularly interspaced short palindromic repeats (CRISPR) with Cas9 protein. This book summarizes current achievements in genome editing for the development of biotic stress tolerant plants and proposes molecular strategies using knockout, knockdown, insertion, and replacement of target genes. Also, the method for multiplex genome editing was illustrated with its current applications. The ways of using genome editing for achieving plant pathogenic tolerance were discussed in subtopics such as the identification and characterization of susceptibility genes, uncovering plant-pathogen interactions, and targeting key disease-resistant genes. Future directions of genome editing in crop breeding and ethical considerations in agricultural production were comprehensively discussed.

This book *Biotechnological and Molecular Applications in Plant Disease Management* presents an overview of the latest advances and the role of plant biotechnology and molecular biology in disease management. Molecular techniques have accelerated the detection and identification of pests and pathogens in plants. Identification of resistance genes, characterization of desired resistance genes, and transfer of resistance gene(s) into plants have been performed by applying molecular techniques. The respective chapters explore emerging areas of plant biotechnology such as RNAi technology, genetic engineering, GWAS, CRISPR, and nanobiotechnology, and their applications in plant disease management. As a result, the chapters have been created to ensure that consistency and coherence are maintained throughout and that the researchers have access to the most information possible on this subject. This book includes the frontiers of epigenetics and epigenomics approaches for the management of plant diseases. One of the leading tools is based on RNA interference (RNAi) for mediating gene silencing to regulate plant biotic stress responses and enable crops to gain disease resistance against bacteria, fungi, viruses, nematodes, and insects. Several methods were discussed for delivering RNAi, including spraying, agroinfiltration, micro-bombardment, virus-induced gene silencing (VIGS), and nanotechnology-mediated systems. Also, the approach of plant microRNAs (miRNAs) was presented in detail including their biogenesis, fundamental mode of action, and the current applications against plant pathogens such as bacteria, fungi, viruses, and nematodes.

As an ideal reference for students, researchers, teachers, and professors, this book can inspire new ideas for future experimental designs in plant science, plant pathology, crop breeding, and plant disease management. In the end, the book editors appreciate all contributors for their excellent chapters and the publisher for their instruction and help. All the editors like to express their heartfelt gratitude to all whoever has provided their support and suggestions for this book. Without the unending help, prayers, and encouragement of their elders and youth in both happy and sad times, the editors could not have finished this endeavor. We hope that our attempts to guide readers toward a higher level of plant science will be successful.

Kaohsiung, Taiwan — Jen-Tsung Chen
Aligarh, India — Masudulla Khan
Moradabad, India — Aiman Parveen
Goyang, Korea (Republic of) — Jayanta Kumar Patra

Contents

Recent Strategies in the Management of Bacterial Diseases for Cereals

Salma Benchlih, Oumaima Benaissa, Ilham Dehbi, Kamal Aberkani, Zineb Belabess, Essaid Ait Barka, and Rachid Lahlali

Abstract Cereal crops are a vital element of the human diet. However, they can be affected by a wide range of pathogenic bacteria, resulting in considerable annual losses and economic consequences worldwide. Plant disease detection is essential for managing plant diseases to ensure high productivity and sustainable agriculture. Recent methods have been developed based on molecular biology techniques, notably high-throughput sequencing, and nanobiotechnology applied to plants using biosensors. Genome editing has proved useful for protecting cereal crops against pathogen infection. Plant growth-promoting rhizobacteria (PGPR) have also demonstrated their potential as an effective alternative to conventional chemical plant preservation techniques. They act through a variety of processes to improve plant development and yield while preventing pathogen infection. This chapter highlights the importance of cereal crops and the bacterial diseases associated with them, with particular emphasis on the Xanthomonas genus. It illustrates recent detection techniques for cereal crops. We also discuss the implementation of PGPRs as an affordable and environmentally friendly means of controlling pathogenic bacteria, describing the processes developed by biocontrol agents (BCAs) and highlighting their direct/

S. Benchlih · O. Benaissa · I. Dehbi · R. Lahlali (✉)
Phytopathology Unit, Department of Plant Protection, Ecole Nationale d'Agriculture de Meknès, Km 10, Rte Haj Kaddour, BP S/40, 50001 Meknes, Morocco
e-mail: rlahlali@enameknes.ac.ma

S. Benchlih · E. Ait Barka
Unité de Recherche Résistance Induite Et Bio-Protection Des Plantes-EA 4707-USC INRAE1488, Université de Reims Champagne-Ardenne, 51100 Reims, France

S. Benchlih · K. Aberkani
Faculté Poly-Disciplinaire de Nador, University Mohammed Premier, 60000 Oujda, Morocco

Z. Belabess
Plant Protection Laboratory, Regional Center of Agricultural Research of Meknes, National Institute of Agricultural Research, Km 13, Route Haj Kaddour, BP.578, 50001 Meknes, Morocco

R. Lahlali
Plant Pathology Laboratory, AgroBioSciences, College of Sustainable Agriculture and Environmental Sci-12 Ences, Mohammed VI Polytechnic University, Lot 660, Hay Moulay Rachid, 43150 Ben Guerir, Morocco

J. Chen et al. (eds.), *Molecular and Biotechnological Tools for Plant Disease Management*, Interdisciplinary Biotechnological Advances,
https://doi.org/10.1007/978-981-97-7510-1_1

indirect modes of action designed to protect cereal crops effectively and sustainably. We are also exploring genome-editing techniques to develop high-yielding, stress-tolerant plants.

Keywords Cereal crops · Bacterial pathogens · Biosensors · High-throughput sequencing · Plant growth promoting rhizobacteria (PGPR) · Genome editing · Sustainable protection

1 Introduction

Cereals are a fundamental component of the human diet, covering the caloric needs of the world's consumers (Food and Agriculture Organization (FAO) Data 2018). Maize, rice and wheat represent the three main cereal crops with the highest yields, widely regarded as the world's principal staple crops, with a high market value and great economic importance (Talhelm 2020). By 2050, the world's population could exceed 9 billion, representing a major challenge for agricultural systems, which must provide adequate food to satisfy the rapidly growing world population, with a requirement to increase food production to meet consumers' demands (Food and Agriculture Organization of the United Nations 2019). Yet, despite all the efforts, the number of severely malnourished and hungry people worldwide is still rising (Prosekov and Ivanova 2018; Kumar et al. 2017). In the same context, plant pathogens are a serious obstacle in the agricultural system, affecting 10–20% of global production with serious epidemics and considerable yield losses worldwide (Strange and Scott 2005). These pathogens include over 200 phytobacteria, capable of causing serious diseases in all the world's agricultural ecosystems (Sundin et al. 2016). Bacterial diseases are regarded as one of the most common and complicated difficulties to overcome in agricultural production, as they spread rapidly and are difficult to manage (Marin et al. 2019). *Xanthomonas* genus comprises the most harmful phytopathogenic bacteria of cereal crops, widespread throughout the world and responsible for diseases in several economically developed countries (Rodriguez-r et al. 2012).

Routine use of pesticides and chemical fertilizers has been widely adopted as a method of managing crop disease damage, to eliminate or limit the severity of disease phenotypes. However, it has seriously disrupted human and animal health, and it is becoming increasingly clear that long-term use of these agrochemicals can have several negative side-effects (Sundin et al. 2016; Toribio et al. 2021). Moreover, their use is hampered by the development of microbial resistance (Hawkins et al. 2019). As a result of chemical issues, extensive research has been conducted into finding alternative solutions to pathogenic infections (Newitt et al. 2019). The deployment of plant growth-promoting rhizobacteria (PGPR) has enjoyed notable success in cereal crops, as they have proved to be a valuable means of avoiding synthetic chemicals and represent a cost-effective, environmentally friendly approach to crop maintenance. They have therefore received considerable attention worldwide

(Glick 2020; Benchlih et al. 2023). PGPRs are characterized by their ability to stimulate plant development, boost crop yields, and mitigate pathogen attacks through a series of complex direct and indirect mechanisms (Singh et al. 2022; Ngalimat et al. 2021). Genome-editing technologies are also of great interest for the functional characterization of plant genes and the genetic selection of crops. They enable efficient exploration of molecular understanding, targeting different genes to obtain the desired plants, thereby developing high-yielding, and stress-tolerant plants (Ansari et al. 2020; Mohamed 2022).

This chapter illustrates the importance of cereal crops and associated bacterial diseases, focusing on the genus *Xanthomonas*. We have emphasized that detection is one of the primary means of combating plant pathogens, based on recent molecular techniques. In addition, we also present recent management strategies, including the use of PGPR and genome-editing technologies, enabling effective control against bacterial pathogens.

2 Bacterial Diseases of Cereal Crops

Cereal crops provide 70% of consumers' calorie requirements (Food and Agriculture Organization (FAO) Data 2018). Cereals are grown on 734.32 million hectares worldwide, producing 2,980.2 million tonnes (FAO Stat 2019). Interestingly, the three major cereal crops with the largest production are corn, rice, and wheat, which provide more than 42% of the calories consumed by the world's population (Matres et al. 2021). Due to their accessibility and affordability, they are the main components of the human diet and constitute an important source of nutrition in developing countries. In this respect, wheat and rice are the principal cereal species grown mainly for food (Talhelm 2020).

Corn is now the world's leading cereal crop in production terms. Over the last ten years, corn has overtaken rice and wheat as the world's leading crop (Serna-saldivar 2019). Due to the successful creation of both conventional and genetically modified genotypes with high yields and the improved adaptability of maize to a variety of circumstances (Serna-saldivar 2019; Shimada et al. 2021). Yet these valuable crops are seriously threatened by plant pathogens that destroy them annually, affecting 10–20% of global production, leading to severe yield losses valued in the billions of dollars, and depriving 800 individuals of an adequate diet. The pathogens include more than two hundred species of phytobacteria, capable of causing serious disease in all the world's agricultural ecosystems (Strange and Scott 2005; Benchlih et al. 2023). Indeed, bacterial infections are one of the most frequent and difficult agricultural challenges in the world. This is due to their wide distribution, the associated control challenges and, above all, the severity of the damage caused to affected plantations (Sundin et al. 2016; Marin et al. 2019). Moreover, they represent a real obstacle to the sustainable productivity of cereal crops, specifically those belonging to the *Xanthomonas* genus (Ngalimat et al. 2021).

2.1 *Corn (*Zea mays *L.)*

Corn is widely regarded as the world's leading staple grain, with an annual production of over one billion tonnes (Erenstein et al. 2022). Corn has a high market value and great economic importance, not only for human consumption but also for livestock feed, as an important component in industrial products, and as a source of biofuels (Food and Agriculture Organization (FAO) Data 2018). Therefore, Corn plays a diversified and active role in the world's food and nutrition systems (Shimada et al. 2021). However, the effects of global climate change represent a major challenge for corn production worldwide, with dramatic consequences for the plant's growth, development, and cultivation (Asseng et al. 2019). In addition to abiotic stress factors, plant pathogens are a real threat to corn productivity (Cairns et al. 2012). In particular, the pathogenic bacterium *Xanthomonas vasicola* pv. *Vasculorum* is responsible for bacterial leaf streak (BLS), a major disease that has recently spread to maize throughout North and South America (Fig. 1) (Lang et al. 2017). It is a universal bacterial disease that has spread since the first symptoms were identified in grain corn and popcorn (Coser et al. 2018). The first symptoms observed initially appeared on leaves in the form of fine linear lesions on the interveinal areas, ranging in color from translucent yellow to brown. In highly susceptible corn varieties, more severe symptoms appear as long, aggregated lesions that extend deep into the leaf blade and are dark in color (Coser et al. 2018; Hartman 2018). the Gram-positive coryneform bacterium *Clavibacter nebraskensis* can affect corn plants during all stages of their development, causing bacterial wilt and leaf blight, resulting in considerable economic losses in North America (Osdaghi et al. 2022). As a result, several international plant protection organizations have classified *C. nebraskensis* as a high-risk quarantine disease and strict controls are therefore implemented worldwide (EPPO code: CORBMI 2004). The bacterium was first identified in 1969. Since then, it has become a dangerous and deadly bacterial disease, presenting a wide variety of symptoms (Ikley et al. 2015). If infection occurs early, it leads to wilting and mortality of plants of susceptible genotypes and serves as a source of future infection (Eggenberger et al. 2015). Indeed, leaf blight is the most common symptom observed on older plants, characterized by oval, parallel, massive, beige to gray lesions on the leaves (Osdaghi et al. 2022; Ikley et al. 2015).

2.2 *Rice (*Oryza sativa*)*

Rice is a widespread food in Africa and Asia. Asia provides almost 90% of the world's rice production. (Firdaus et al. 2020). According to estimates, the world currently produces 755 million tonnes of rice, with a total of 167 million hectares under cultivation (Doni et al. 2022). However, every year, millions of hectares of rice are severely damaged by various microbial diseases, particularly those caused by bacteria, with severe epidemics resulting in crop losses of over 60% (Wubneh and

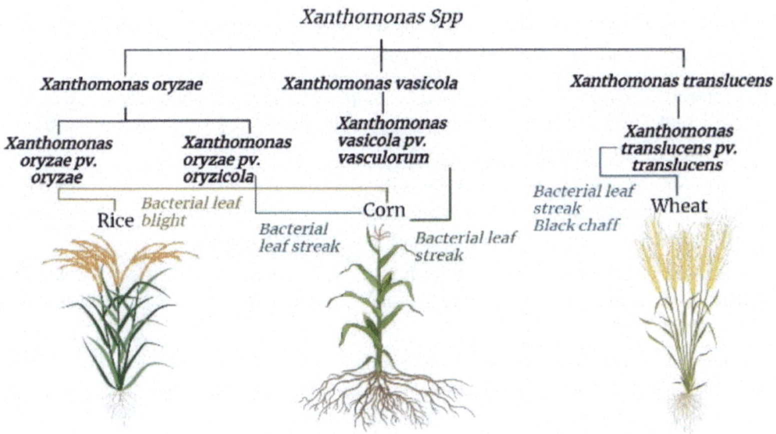

Fig. 1 The various *Xanthomonas* strains attacking the three main types of cereals

Bayu 2016; Wonni 2016). A wide range of pathogenic bacteria infects rice plants (Table 1). The genera *Xanthomonas, Burkholderia, Pseudomonas, Pantoea, Erwinia, Acidovorax, Dickeya, and Enterobacter* are the best known and include numerous species of pathogenic bacteria widely recognized as harmful to rice (Nandakumar et al. 2009; Niño-Liu et al. 2006). *Xanthomonas oryzae* pv. *oryzae (Xoo)* is the most destructive infection of rice production in South Asian countries, causing bacterial leaf blight (BLB) (Fig. 1). Once this pathogenic bacterium penetrates the host through the hydathodes or wounds, it can propagate into the xylem vessels, dispersing the infection worldwide (Oryza et al. 2022).In addition, *Burkholderia* and *Pantoea* bacterial species are among the main rice pathogens (Doni et al. 2019).

*2.3 Wheat (*Triticum aestivum*)*

Wheat is considered the largest agricultural system more than any other crop in the world, with over 220 million hectares of arable land used to cultivate wheat, with a total production of 750 million tonnes in 2018 (Food and Agriculture Organization (FAO) Data 2018). It is the principal grain used in human nutrition (Shewry 2009), Over one-third of all people use wheat as a staple meal, which accounts for 20% of daily caloric intake (Kolpakova et al. 2007). Every year, around 25–30% of wheat yield losses are linked to abiotic and biotic stresses caused by various wheat pests. Bacteria reduce annual wheat production by 10–40% when severe infections occur early in the plant's growth stage. (Table 1) (Tambong 2022). The pathogens include *Clavibacter michiganensis subsp. Tessellarius*, are responsible for a leaf disease characterized by a mosaic of small yellow lesions resembling infections by viral pathogens. The small leaf lesions may cluster together to form streaks (Duveiller

Table 1 Bacterial pathogens of cereal crops and associated symptoms

Crops	Diseases	Pathogens	Symptoms	References
Corn	Bacterial leaf streak	*Xanthomonas vasicola* pv. *vasculorum*	Development of yellow–brown lesions with wavy edges along the veins on the leaves	Coser et al. (2018)
	Leaf wilt, systemic wilt	*Clavibacter nebraskensis*	Leaf wilt Plant death Formation of necrotic lesions	Osdaghi et al. (2022)
Rice	Bacterial leaf blight	*Xanthomonas oryzae* pv. *Oryzae*	Leaf blight, wilt Kresek Yellow leaf or pale yellow	Ngalimat et al. (2021)
	Bacterial leaf streak	*Xanthomonas oryzae* pv. *Oryzicola*	Leaf death Discoloration and death of ovary, stamens, and seeds Brownish glumes Seeds do not reach maturity	
	Bacterial panicle blight	*Burkholderia glumae*	Seedling rot Grain rot in rice nurseries Bleaching rice seedlings	Cho et al. (2007)
	Bacterial seedling rot	*Burkholderia gladioli*	Plant rot and scorch Production of the toxic materials toxoflavin and tropolone resulting in rot	Maeda et al. (2006)
	Seedling blight	*Burkholderia plantarii*	Production of the toxic materials thioflavin and tropolone resulting in blight in rice seedlings	
		Pseudomonas plantarii	Basal chlorosis and leaf wilting Root development becomes slower	Saha et al. (2016)
	Bacterial brown stripe	*Pseudomonas syringae* pv. *Panici* *P. avenae*	Inhibition of germination	
	Halo blight	*P. syringae* pv. *Oryzae*	Light green to yellowish-brown circular tumors on leaf blades A clear halo surrounding the lesions	
	Sheath brown rot	*Pseudomonas fuscovaginae*	Discoloration of the leaf sheath Leaf desiccation Grains on infected panicles are discoloured, deformed or empty	
	Grain rot	*Pseudomonas glumae*	Brown, soft, water-soaked rot on seedlings Wilting or rotten leaves, grains become shrunken and pale green Brown margin between infected and healthy parts of the grain, slight rotting of the sheath or collar of the flag leaf	
	Rice foot rot	*Dickeya zeae*	Foot-rot	Bez et al. (2021)

(continued)

Table 1 (continued)

Crops	Diseases	Pathogens	Symptoms	References
	Bacterial palea browning	*Enterobacter cloacae*	Yellowing to pale browning, stunting, withering, drying, and death	Verma et al. (2019)
		Erwnia herbicola	Production of light-brown, water-soaked lesions on the lemma or palea Immature seeds lighter in color at harvest	Saha et al. (2016)
	Bacterial foot rot	*Erwinia chrysanthemi*	Brown rotting of leaf sheaths Leaf yellowing and wilting rotting of ligules, nodes, culms, and crowns	
Wheat	The bacterial mosaic pathogen	*Clavibacter michiganensis subsp. Tessellarius*	Leaf discoloration, mosaics, stunted plants, chlorosis and leaf lesions, and yield loss	Kleczewski et al. (2020)
	Bacteria blight	*Pseudomonas syringae* pv. *syringae*	Grain yield losses	Valencia-bot and Cisneros-l (2012)
	Bacterial leaf streak Black chaff	*Xanthomonas translucens* pv. *Translucens*	Water-soaked lesions that rapidly elongate along the parallel vernation of the leaf, drying out and necrotizing Loss of yield is due to leaf streaking Black flakes refer to partial or full blackening of the stalks	Tambong (2022)

et al. 1997). *Pseudomonas syringae* pv. *syringae*, is widely dispersed in soil and coexists with other bacterial species on wheat leaves. This pathogenic bacterium is responsible for wheat leaf blight and develops in response to particular environmental factors (Valencia-bot and Cisneros-l 2012). For example, under conditions of high humidity and rain, small water-soaked spots have been observed to develop into large, identifiable lesions, spots, or streaks, as well as other symptoms (Table 1) (Bockus et al. 2010). In addition, the pathogenic bacterium *Xanthomonas translucens* is responsible for both bacterial leaf streak and black bullet disease (Fig. 1), considered the most important bacterial disease of wheat and a potential biological threat to wheat growers worldwide. Every infected plant shows symptoms on its leaves and/or ears. Necrotic streaks immersed in water at the start of the disease then turn into transparent lesions (Duveiller et al. 1997).

3 Detection and Diagnosis of Pathogenic Bacteria for Cereals

In environmental samples such as water, soil, and plant surfaces, detection remains a very useful tool for confirming the presence of the micro-organisms in question. Diagnosis, on the other hand, involves determining the nature of the problem and its main cause (Louws et al. 1999). Consequently, identifying and diagnosing an unknown pathogen generally involves classifying it in a taxonomically recognized category according to its characteristics. Whatever characteristic is chosen, the diagnosis of a pathogen requires a high degree of sensitivity, specificity, and rapidity Research has focused on the identification of pathogenic fungi and viruses that affect plants (Awasthi 2015; Hariharan and Prasannath 2021). However, studies on cereal pathogenic bacteria are still in their infancy. This chapter aims to detect and diagnose the specific phytopathogenic bacteria of cereal crops that most disrupt yields, the main ones being *Xanthomonas oryzae, Xanthomonas translusens, Pseudomonas syringae, Clavibacter nebraskinsis, Burkholderia glumeae, Burkholderia gladioli, Dickeya, Erwinia,* and *Pantoa* genus (Mansfield et al. 2012; Fan et al. 2017; Tambong et al. 2021a; Peng et al. 2022; McNally et al. 2016; Lee et al. 2018; Uematsu et al. 2015; Prokić et al. 2020). This is achieved through traditional microbiological techniques and new molecular microbiology tools. We will also focus on biotechnology based on advanced nanodiagnostic and practical collaborative devices for on-site diagnosis of bacterial pathogens in order to develop and implement new disease management strategies (Tewari and Sharma 2019; López et al. 2003).

3.1 High-Throughput Sequencing and Molecular Approaches

The identification of different species requires the use of molecular methods, including sequencing, polymerase chain reaction (PCR), multilocus sequence analysis (MLSA), and loop-mediated isothermal amplification (LAMP) (Peng et al. 2022; Roach et al. 2018). The use of real-time PCR and PCR has transformed the acceptance of molecular diagnostic methods. Indeed, traditional diagnostic procedures require prior knowledge of the conditions in which the pathogen develops, whereas modern PCR-based approaches are more advantageous for the identification of bacterial pathogens. Therefore, because they are more reliable, sensitive and rapid, DNA-based pathogen diagnostic techniques are becoming increasingly popular (Tewari and Sharma 2019; Srinivasa et al. 2012). Understanding the epidemiology of diseases requires accurate identification of genera and species. MLSA as described by Maiden et al. (1998) is initially based on sequence typing, and used as a molecular typing method for characterizing pathogenic bacteria with variations in several housekeeping genes (Maiden 2006), and internal gene segments were assigned as allelic segments, this technology is mainly based on the concept of multilocus enzyme electrophoresis (Glaeser and Kämpfer 2015; Selander et al. 1990). In the loop-mediated isothermal amplification (LAMP) reaction, gene amplification is accomplished by repeating two types of extension reactions that occur across loop regions using inner and outer primer pairs, the rationale for which is based on the generation of large quantities of DNA amplicons with complementing sequences and alternating and repeating structures (Notomi et al. 2015).

Polymerase chain reaction (PCR)-based diagnostics were created to distinguish several species of *Xanthomonas* spp. with four primer sets; Xvv 3 to Xvv8 generated to specifically amplify *X. vasicola* pv. *vasculorum* from isolated DNA (Fan et al. 2017). Using the housekeeping gene, atpD, and the 16S rRNA gene, infected macerated tissues or heat-killed cells were employed for the detection of *Xvv,* and it is also detected by (LAMP) and TaqMan qPCR (Stulberg et al. 2020) (Table 2). Quantitative reverse transcription polymerase chain reaction (qRT-PCR) analysis by Fan et al. (2017) showed that mRNA levels of the hrp genes that code for the hypersensitive response and pathogenicity of *Xoo* and *X. oryzae* pv. *oryzicola*, as well as the regulatory genes hrpG and hrpX, were inhibited by treatment with phenolic compounds while suppressing expression of the type III secretion system (T3SS) (Fan et al. 2017). To detect the difference between *X. translucens* pv. *undulosa and X. translucens* pv. *translucens*, the bacterial leaf streak pathogen of wheat, DNA samples were tested by LAMP and SYBR Green-based qPCR using the F11/R11/P11 probe, at the SCB06003 locus (Sarkes et al. 2022). In addition, duplex qPCR was also performed using genomic DNA from 1000 bacterial cells on samples of infected barley grain, wheat and wheat leaves (Fu et al. 2023). Furthermore, maize leaf blight is caused by the bacterium *Clavibacter michiganensis* subsp. *nebraskensis* (Osdaghi et al. 2022). It was detected by CMN_01184 as a new genetic target, and used in conventional PCR and SYBR green-based quantitative PCR assays (McNally et al. 2016). An EvaGreen droplet digital PCR (dPCR) assay was developed based on SYBR Green real-time quantitative PCR for the detection and quantification of the two pathogens *Burkholderia glumae* and *Burkholderia gladioli* responsible for bacterial panicle of rice (Zhang et al. 2022). Additionally, strains of *B. glumae* strains were molecularly identified using phylogenetic analysis of 16S rRNA and gyrB genes (Lee et al. 2018; Gowda et al. 2022; Ramachandran et al. 2021). Besides, *Pantoea* spp strains obtained from rice leaves or seeds from African countries were identified by multiplex PCR in planta and in in vitro cultures with a detection threshold of 10^4 CFU/ml for heated cells and 0, 5 ng/l for total genomic DNA, which enabled three species to be distinguished, including strains of unresolved *Pantoea, Pantoea ananatis, Pantoea stewartii* and *Pantoea agglomerans* (Kini et al. 2021). Seven distinct genetic profiles and analysis of the recA gene sequence of strains recovered from rotting maize belonging to *Dickeya zeae* were discovered using ERIC-PCR analysis, indicating the existence of genetic diversity in the population of this disease in Serbia (Prokić et al. 2020). Substantial genetic variability with syrB genes from bacterial isolates of *Pseudomonas syringae* pv. *syringae* causing rice bacterial brown leaf spot obtained from different localities in the Chinese province of Heilongjiang were validated by biochemical tests with the LOPAT test and multilocus sequence analysis (MLSA) as well as repetitive PCR using BOX, ERIC and REP primers (Peng et al. 2022) (Table 2).

3.2 Nanodiagnostic Approaches

Nanotechnologies offer new instruments for the molecular detection of biotic and abiotic stress and the rapid detection of plant pathogens, using biosensors to monitor soil quality and plant vitality. This is an emerging technical and scientific innovation with the potential to change the agricultural industry (Hartman 2018). In addition, the biology, yield and nutritional value of crops are better understood, thanks to nanobiotechnology (Osdaghi et al. 2022). This chapter describes the use of nanotechnology in plant pathology and the development of biosensors capable of identifying the main bacterial diseases affecting cereal crops.

3.2.1 Nanoparticles (NPs)

The primary mechanisms behind the antimicrobial effects of most nanoparticles (NPs) can be attributed to three factors: photocatalysis, physical damage to bacterial cell walls, and the release of toxic metal ions. Among the NPs exhibiting photocatalytic activity, the majority are metal oxides like CuO and ZnO (Fig. 2) (Li et al. 2012), as well as metals like Ag (Choi and Hu 2008). Reactive oxygen species (ROS) are produced throughout photocatalysis in the form of hydrogen peroxide, hydroxyl radicals, and peroxide (Manke et al. 2013). These are toxic to bacteria because they have the potential to damage cellular building blocks such ADN, lipids, and proteins, leading to bacteriocidal and bacteriostatic effects (Storz and Imlay 1999). The creation of NPs from plant extracts represents a new substitute for traditional chemical or physical processes. Several techniques such as FTIR, XRD, UV-DRS, and TEM have been used to identify synthesized green nanoparticles (Fig. 2) (Doni et al. 2022).

- **ZnONPs**

ZnONPs produced by Olea europapaea had a significantly different antibacterial effect on *Xanthomonas oryzae* pv. *oryzae* strain GZ 0003 than ZnONPs produced by *Matricaria chamomilla* and *Lycopersicon esculentum*, with a zone of inhibition of 2, 2 cm to 16, 0 mg/ml. Similar to this, ZnO nanoparticles had a substantial impact on the Xoo GZ 0003 strain's bacterial cell memory, swimming motility, biofilm formation, and bacterial proliferation (Ogunyemi et al. 2019). Another study examined the crystalline and spherical form of zinc oxide nanoparticles (ZnONPs) with diameters ranging from 21 to 35 nm in a native *Bacillus cereus* RNT6 strain, which was taxonomically recognized by examination of the 16S rRNA gene (Wonni 2016), and reduced the cell number (OD600) of both pathogens in broth culture by 68.1% and

Table 2 PCR variants and the targets used to detect bacterial pathogens for cereals

Bacterial pathogens	Diseases caused	Target	Molecular technique	References
Xanthomonas vasicola pv. *vasculorum*	Bacterial leaf streak of Corn	Six genes (atpD, dnaK, gyrB, fusA, lepA, and rpoD)	Conventional PCR TaqMan qPCR	Lang et al. (2017); Stulberg et al. (2020)
X.oryzae pv. *Oryzae (Xoo)* *X. oryzae* pv. *Oryzicola (Xoc)*	Leaf blight and leaf streak diseases	Bacterial type III secretion systems (T3SSs) hrpG and hrpX regulatory genes	qRT-PCR	Fan et al. (2017)
Xanthomonas translucens pv. *undulosa*	Wheat bacterial leaf streak (BLS)	Whole genome	SYBR green-based qPCR and (LAMP) assay	Sarkes et al. (2021)
Xanthomonas translucens pv. *translucens and* pv. *undulosa*	Cereal bacterial leaf streak	Genomic DNA	Duplex quantitative PCR	Fu et al. (2023) Tambong et al. (2021b)
Clavibacter michiganensis subsp. *nebraskensis*	Goss's leaf blight and wilt of maize	CMN_01184 gene Ribosomal protein gene (rpsB, rpsG, rpmF)	Conventional PCR and SYBR green–based (qPCR) assays	McNally et al. (2016) Osdaghi et al. (2022)
Burkholderia glumae (BPB), B. gladioli	Bacterial panicle blight disease of rice	16S rRNA and gyrB genes	Droplet Digital PCR Assay	Zhang et al. (2022) Gowda et al. (2022) Ramachandran et al. (2021)
Burkholderia glumae, Burkholderia gladioli	Panicle blight and seed rot disease	16S rRNA gene	Monoplex PCR, duplex PCR	Lee et al. (2018)
Pantoea spp.	Bacterial leaf blight of rice	16S rRNA, atpD, gyrB, infB and rpoB genes Genomic DNA	Multiplex PCR	Kini et al. (2021)
Pantoea stewartii subsp. *stewartii*	Stewart's wilt	CpsD, pstS–glmS regions	Conventionnel PCR with LAMP	Uematsu et al. (2015)
Dickeya zeae	Maize stalk soft-rot	recA gene	ERIC-PCR analysis	Prokić et al. (2020)
Pseudomonas syringae pv. *syringae*	Rice bacterial leaf spot	16S rRNA sequences rpoD, gyrB, and gltA genes	BOX-PCR, ERIC-PCR	Peng et al. (2022)

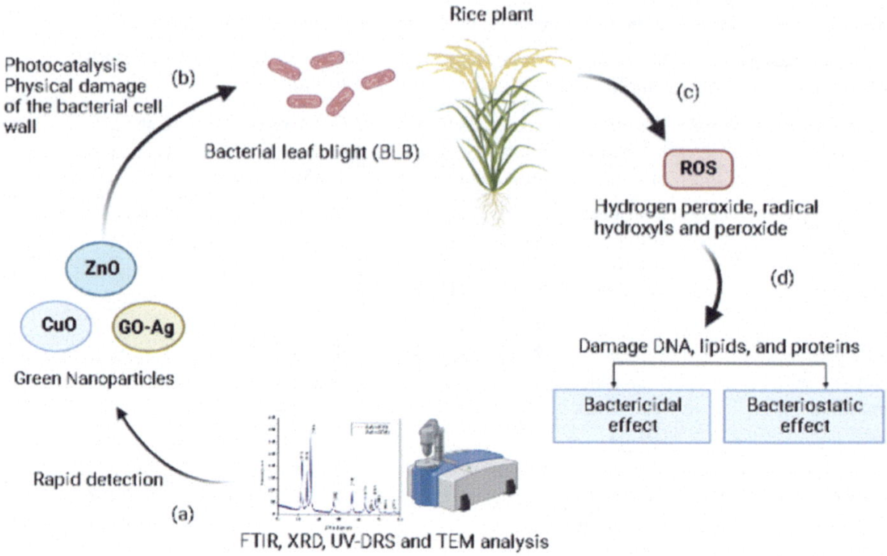

Fig. 2 Schematic illustration of the modes of action of nanobioparticles on a cereal plant infected with pathogenic bacteria: **a** detection of nanoparticles (ZnO, CuO, Ag) by FTIR, XRD, UV-DRS, and TEM analysis. **b** Photocatalysis is the physical damage caused when nanoparticles attach to the bacterial cell wall. **c** Release of reactive oxygen species (ROS) in the form, hydroxyl radicals and peroxide. **d** Destruction of bacterial DNA, lipids and proteins with a bacteriostatic and/or bactericidal effect

71.2%, respectively, when used at 50 g mL^{-1} against *B. gladioli* and *B. glumae.* Zones of inhibition were 2.18 cm and 2.83 cm respectively (Ahmed et al. 2021).

- **AgNPs**

The plant pathogen *Xanthomonas oryzae* pv. *oryzae* (*Xoo*) was combated using bioengineering agent nanoparticles (Ag NPs) made from a susceptible rice variety identified as Taichung Native-1. The nanoparticles were examined using transmission electron microscopy, zeta potential, dynamic light scattering and UV–visible absorption spectroscopy (UV–vis). The antibacterial efficacy of Ag NPs was assessed by in vitro and in vivo tests under stress conditions. Results showed that Ag NPs (10 μg/mL) effectively inhibit *Xoo* growth and colony formation, and are the most potent antibacterial agents (Namburi et al. 2021). A composite of graphene oxide and silver nanoparticles (GO-Ag) was created, and its antibacterial efficacy against *Xoo* was also studied. According to the results, the GO-Ag composite showed a nearly four-fold improvement over pure silver NPs. The concentration of 2.5 g mL^{-1} was low enough to make bacteria inactive. The delayed release of silver ions by the GO-Ag composite compromised cell integrity, causing leakage of intercellular contents. This combination reduced DNA replication and stimulated the generation of reactive oxygen species at the molecular level. In addition, the phytotoxicity test revealed that the GO-Ag composite as prepared was less harmful to plants than silver NPs, making it a suitable option for crop disease prevention (Liang et al. 2017) (Table 3). The three plant AgNPs, Al-AgNPs, Sm-AgNPs, and Tm-AgNPs, proved capable of preventing Xoo from forming biofilms and swarming, and showed the highest bacteriostatic effects against Xoo at 20 g/mL with inhibition zones of 16.5 mm, 14.5 mm and 12.4 mm, respectively (Tian et al. 2022). Interestingly, biorecognition uses antibodies directed against *Xoo* bacterial cells and their conjugation with gold nanoparticles and graphene quantum dots from which the interaction gave rise to an immunoreaction, which transfers energy to the quenching mechanism, based on fluorescence in the 100–105 CFU mL^{-1} range, with variable intensity related to the logarithm of Xoo cells. A specificity test against pathogens responsible for other plant diseases revealed good detection for *Xoo* (Awaludin et al. 2020).

- **Cu_2O and SnO_2**

A high antibacterial action has been demonstrated by nano Cu_2 made from O-Cu/alginate against *Xoo*. The development of sustainable agricultural production can be facilitated by the use of nano Cu_2O-Cu/alginate as fungicides to replace harmful agrochemicals (Shewry 2009; Ngoc et al. 2021). Furthermore, the results showed that the *Xoo* bacterium, was completely suppressed by nano Cu_2O-Cu/alginate containing 30 mg L^{-1} of copper ions (Kolpakova et al. 2007; Tambong et al. 2021b). Biosynthesized SnO_2 nanoparticles and their silver-doped variant with an average crystal size of 2.6 nm were investigated for their bactericidal effect against *Xoo*, with the following results: DNA cleavage, cellular protein leakage, and stress-induced enzyme release, biofilm inhibition against xanthomonadine and exopolysaccharide formation (Elizabeth Sunny and Kumar Shanmugam 2021). A colloidal gold-based lateral flow immunochromatography strip has been developed for the rapid identification of Pantoea stewartii subsp. stewartii in maize grain samples. A pair of highly sensitive, targeted anti-Pss antibodies was created by animal immunization and cell fusion. The results indicated that in corn kernels and 0.01 M PBS, Pss LDL was 10^5 CFU/mL (Feng et al. 2015) (Table 3).

4 Genome Editing Techniques

Genome-editing technologies are of great interest for the functional characterization of plant genes and the genetic breeding of crops (Xie and Yang 2013), revolutionizing research into the sustainable protection of diverse crops. The main aim of genetic selection in cereal crops is both to improve food production and to guarantee nutritional security for a rapidly growing world population (Ansari et al. 2020; Tyagi et al. 2021).

A series of fascinating new techniques have emerged, such as regularly interspaced palindromic repeats (CRISPR) and genome-associated protein 9 (Cas9), genome-wide association sequences (GWAS), and transcription activator-like effector nucleases (TALEN). All these techniques contribute to overcoming cereal crop disease

Table 3 Nano-bioparticles developed for the detection of bacterial pathogens in cereals

Type of nanoparticles	Bacterial pathogen detected	Results	References
Chitosan nanoparticles (CSNPs), zinc oxide nanoparticles (ZnONPs)	*Xanthomonas oryzae* pv. *oryzae*	Leakage of nutrients and nucleic material, bacterial death The antibacterial effect with a zone of inhibition of 2.2 cm at 16.0 mg/ml	Ogunyemi et al. (2019); Abdallah et al. (2020)
Mycosynthesis of ZnO nanoparticles using trichoderma spp.	*Xanthomonas oryzae* pv. *oryzae*		Shobha et al. (2020)
Silver nanoparticles (Ag NP) were produced utilizing leaf extract of rice cultivar Taichung native-1	*Xanthomonas oryzae* pv. *oryzae*	Treatment with Ag NP (10 μg/mL) resulted in an increase in seedling vigour index At a significantly lower dosage, pots treated with Ag NP (15 g/mL) in vivo in a greenhouse displayed disease severity of 26.6% and disease decline over control of 49.2%	Namburi et al. (2021)
A graphene oxide/silver nanoparticle composite (GO–Ag)	*Xanthomonas oryzae* pv. *oryzae*	Compared to pure AgNPs, the GO-Ag composite showed an improvement of roughly four times. At a comparatively modest concentration of 2.5 g mL^{-1}, the bacteria were fully inactivated	Liang et al. (2017)
Silver nanoparticles using three plant extracts	*Xanthomonas oryzae* pv. *oryzae*	Al-AgNPs, Sm-AgNPs, and Tm-AgNPs prevent the growth of biofilms and the movement of swarms	Tian et al. (2022)
Fluorescence-based immunoassay	*Xanthomonas oryzae* pv. *oryzae*	Interaction of AntiXoo-GQDs, *Xoo* cells, and AntiXoo-AuNPs resulted in an immunoreaction that transferred energy to the fluorescence-based quenching mechanism, which turned off	Awaludin et al. (2020)
Nano Cu_2O-Cu/alginate		Using nano Cu_2O-Cu/alginate with 30 mg. L^{-1} of copper ions fully prevented *Xoo* from growing	Ngoc et al. (2021)
Ag-doped SnO_2 nanoparticles	*Xanthomonas oryzae*	Inhibitory effect of nanoparticles on xanthomonadin and exopolysaccharide	Elizabeth Sunny and Kumar Shanmugam (2021)

(continued)

Table 3 (continued)

Type of nanoparticles	Bacterial pathogen detected	Results	References
Gold nanoparticle-labeled mAb, immunochromatographic test strip	*Pantoea stewartii subsp. stewartii*	Rapid detection	Feng et al. (2015)
Zinc oxide nanoparticles from a native *Bacillus cereus* Strain RNT6	*Burkholderia glumae and B. gladioli*	Significant antibacterial activity at 50 $\mu g\ mL^{-1}$	Tian et al. (2022)

pathogens by generating a wide range of specific genomic modifications, based on strategies that combine targeted mutagenesis and editing for biotechnological agricultural purposes (Mohamed 2022; Sedeek et al. 2019).

4.1 CRISPR/Cas9

Genetic modification of susceptibility (S) genes has long been used to ensure disease tolerance in a wide range of high-value crops, as a cost-effective strategy with minimal environmental impact (Liu et al. 2011). Recently, the most powerful and simplest genome-editing tool is CRISPR/Cas9, which comprises two arms: the CRISPR array is a sequence comprising 50-bp repeats separated by single spacers of the same length, and the Cas9 protein is a CRISPR-linked nuclease that contributes to foreign DNA degradation events (Tyagi et al. 2021). This technique has been successfully used in the most valuable crops, including cereals, due to its high acceptability, reduced time required and improved targeting (Islam 2018). In addition, genome-wide CRISPR/Cas9 screens enable new traits to be discovered, expanded, and accelerated in target crops essential for food security (Ansari et al. 2020; Sedeek et al. 2019), offering biotechnologists the prospect of creating a sustainable, high-performance agronomic system to improve yields (Islam 2018). Moreover, CRISPR/Cas9 genome-editing technology has provided a new opportunity to rapidly develop disease-resistant crop varieties (Tyagi et al. 2021; Islam 2018). To date, however, there are few reports on the deployment of CRISPR/Cas9 technology to improve resistance to bacterial diseases (Ansari et al. 2020; Zhou et al. 2015).

With rice, CRISPR/Cas9 has been used to boost resistance against bacterial blight (Zhou et al. 2015). Indeed, the effectiveness of this resistance approach relies on a better understanding of the molecular pathogen–host plant interaction (Mohamed 2022). The pathogenic bacterium (*Xoo)* affects rice primarily by repressing the host immune mechanisms through its type III secretion system (T3SS), which includes transcription activator effectors (TALs). Upon invasion by *Xoo*, TAL effectors are introduced into rice cells via the T3SS, inducing specific defense processes and activating transcription of the corresponding host genes through recognition and linking to specific effector binding elements (EBEs) in their promoter regions. (Ansari et al. 2020; Mohamed 2022; Zhang et al. 2021). The effector protein PthXo2X results from the expression of a principal virulence TAL effector gene in *Xoo*, which in turn increases susceptibility (S) genes expression in the host plant, notably the OsSWEET13 gene (Liu et al. 2011; Zhou et al. 2015). This OsSWEET13 susceptibility gene codes for a sucrose transporter involved in the interaction between the pathogenic bacterium and rice plant, potentially contributing to the remobilization of carbon sources to nutrient-poor xylem vessels, where *Xoo* colonization is favored (Chen et al. 2010). Consequently, CRISPR/Cas9 was implemented to create a null mutation in OsSWEET13. leading to efficient and extensive resistance against infection by *Xoo* (Zhou et al. 2015). Accordingly, the enhancement of host susceptibility is regarded as the most potent and successful means of achieving disease resistance in rice.

4.2 Genome-Wide Association Studies (GWAS)

In recent years, remarkable advances in sequencing technologies and computational tools, together with genome-wide association studies (GWAS), have provided a powerful and feasible strategy for identifying the natural variation behind complex quantitative traits in crop plants (Zhang et al. 2021). The characterization of the complex molecular landscape of intergenomic interactions between plants and pathogens could significantly enrich our comprehension of coevolutionary pathways and the adaptive capacity of microbial species in plant communities (Bartoli and Roux 2017).

Bacterial leaf streak (BLS)is one of the world's most serious and damaging rice diseases, resulting in substantial yield losses worldwide (Ry 2006). The selection of highly resistant rice varieties has become an essential and relevant objective in the fight against this disease (Mohamed 2022). Determining resistance genes/quantitative trait loci (QTL) and the functional context of BLS resistance is therefore crucial in rice breeding programs aimed at improving rice varieties. Interestingly, eleven QTLs for broad-spectrum BLS susceptibility were recently detected using GWAS (Bossa-castro et al. 2018). Furthermore, the research of Sattayachiti et al. (2020) reported that the genome-wide association study carried out on a series of various rice selections, allowing to identify 12 QTL, most notably qBLS5.1 and qBLS2.3, which were identified as potent QTL for wide-spectrum resistance to *Xoc*. The results suggest that the xa5 gene is a powerful wide-spectrum resistance factor and could be an important source of BLS resistance in rice breeding systems (Sattayachiti et al. 2020).

4.3 *Transcription Activator-Like Effector Nucleases (TALEN)*

TALEs have demonstrated their potential to regulate the activity of susceptible genes in rice. In this context, although over 40 genes responsible for tolerance to bacterial blight in rice have been discovered, only a limited number have been used to select resistant varieties (Han et al. 2020). These include recessive resistance genes such as xa13 and xa25, whose resistance depends on TAL effectors not detecting their promoters to activate them (Koebnik et al. 2015). The xa5 gene is valuable for rice breeding, as it is responsible for encoding an essential transcription factor (TFIIAγ), which confers broad resistance to the majority of *Xoo* strains. It was shown that the TALENs technique was used to modify the dominant allele of the xa5 gene and resulted in a series of Xa5/TFIIAγ5 mutant genes with several rice origins, thus improving rice resistance to *Xoo* (Hwang et al. 2020).

5 The Use of PGPR

As biological control is an integral part of effective plant disease management, the use of plant growth-promoting rhizobacteria (PGPR) has received considerable attention worldwide (Khan and Asadullah 2019), as an environmentally-friendly and low-cost disease control strategy. It is an ideal substitute for agrochemical methods and a genuine antibacterial solution, helping to improve crop yields, promote biodiversity and preserve soil fertility, thereby reducing the prevalence and severity of plant diseases (Benchlih et al. 2023). In terms of sustainability, their use is recommended to meet society's expectations for a healthier and more sustainable food supply (Marin et al. 2019). Indeed, these rhizobacteria could potentially improve plant development and performance via direct/indirect processes (Doni et al. 2022). Direct mechanisms employed by PGPR, either by helping to assimilate nutrients via nitrogen fixation, phosphorus solubilization, and the biosynthesis of siderophores, or by regulating plant growth via the synthesis of phytohormones (auxins, gibberellins, cytokinins) (Fig. 3) (Singh et al. 2022). In addition, PGPRs indirectly enhance plant growth by suppressing damaging pathogens and providing a healthy environment for plants, through antibiosis, competition with pathogens for nutrients and ecological niches, and induced systemic resistance (ISR) in host plants, which enhances their innate response mechanisms and provides effective protection against plant pathogens (Behera et al. 2021; Pieterse et al. 2014). Research into PGPRs has intensified and numerous experimental investigations have been performed on various crops, including maize, wheat and rice (Zahid et al. 2015; Sen and Chandrasekhar 2014; Abbasi 2015), improving the health and yield of cereal crops with minimal agrochemical inputs.

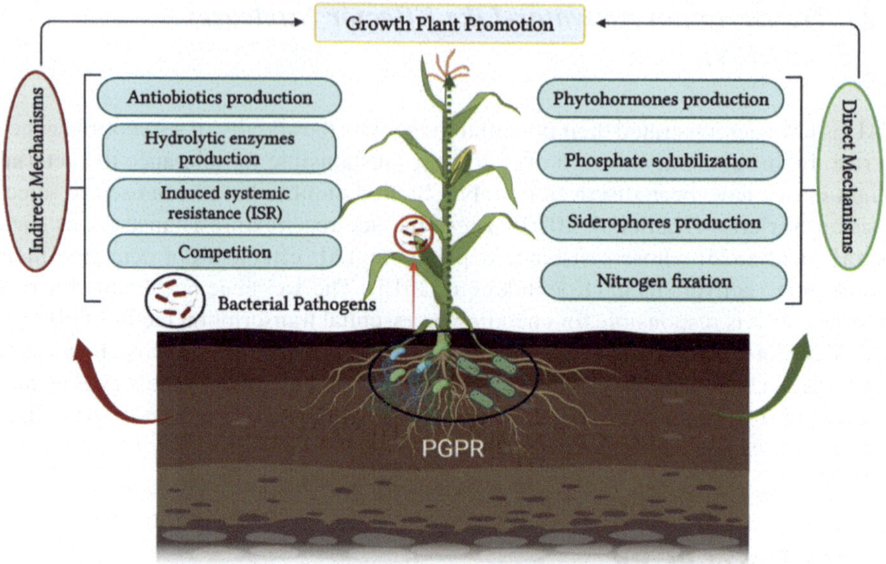

Fig. 3 Schematic illustration of PGPR modes of action to promote plant growth promotion of cereal crops

5.1 Direct PGPR Mechanisms to Promote Cereal Crops

The beneficial effects of PGPR on crop development and productivity have been widely reported, and biofertilization is considered as one of the interesting uses of PGPRs for cereal crops, exhibiting a key role in increasing nutrient uptake by host plants, through nitrogen fixation, improving nutrient availability in the rhizosphere and increasing root surface area (Fig. 3) (Santos et al. 2020). As numerous research studies have demonstrated that inoculating maize plants with PGPR can improve growth and grain yield, while also reducing fertilizer N (Table 4) (Salvo et al. 2018; Kuan et al. 2016). Interestingly, Breedt et al. (2017) has proved that the application of *Paenibacillus alvei*, *Bacillus. safensis, Bacillus. pumilus*, and *Brevundimonas vesicularis* resulted in yield increases varying from 24 to 34%. In addition, Cassan et al. (2009) reported that the inoculants *Azospirillum brasilense* and *Bradyrhizobium japonicum* can enhance seed germination as well as early development of maize seedlings. Besides, phosphorus (P), another essential mineral nutrient for plants, is required for almost all of the plants' major metabolic processes (Khan and Asadullah 2019). Some PGPRs are capable of solubilizing phosphates, such as *Bacillus*, *Burkholderia*, *Pseudomonas*, *Rhizobium*, or *Serratia* (Santos et al. 2020). In the same context, recent research studies have shown that inoculation with associated *Azospirillum brasilense*, *Bacillus subtilis*, or *Pseudomonas fluorescens* can boost phosphorus uptake capacity, helping to enhance maize development and yield (Table 4) (Pereira et al. 2020). Similarly, treatment with *Enterobacter cloacae* increased vegetative growth characteristics: root length, shoot length, as well as root and shoot biomass in comparison to the untreated control (Verma et al. 2019).

Various rhizobacteria are involved in the production of phytohormones, in particular indole-3-acetic acid (IAA), considered to be one of the key phytohormones that enhance plant nutrient supply and root parameters in cereal crops (Santos et al. 2020). In particular, recent investigation showed that Chryseobacterium sp. strain NGB-29 and *Flavobacterium* sp. strain NGB-31 are highly efficient IAA producers, promoting lateral and adventitious root formation of maize plants, compared to untreated control (Table 4) (Youseif 2018). Similarly, it was shown that treatment of rice plants with *Bacillus subtilis* subsp. and *Bacillus megatherium* showed a greater increase in vegetative parameters, which is attributed to their ability to boost nutrient uptake and the biosynthesis of phytohormones such as abscisic acid (ABA), (IAA), salicylic acid, gibberellic acid (El-Mageed et al. 2022). In the case of wheat, recent research investigations have shown that various rhizobacteria belonging *to Klebsiella, Enterobacter, Raoultella, Rhizobium, Agrobacterium, Bacillus,* and *Pseudomonas* could potentially fix nitrogen and solubilize phosphate/potassium (Table 4). In addition to their

Table 4 Direct mechanisms of PGPR to promote the growth of cereal crops

Cereal crops	Beneficial rhizobacteria	Factors promoting plant growth	References
Corn	*Bacillus. safensis* *Bacillus. pumilus* *Brevundimonas vesicularis*	PGPR inoculants increased corn's yield varying from 24 to 34%	Breedt et al. (2017)
	Azospirillum brasilense *Bradyrhizobium japonicum*	Nitrogen fixation Enhancement of seed germination as well as early development of maize seedlings	Cassán et al. (2009)
	Azospirillum brasilense *Bacillus subtilis* *Pseudomonas fluorescens*	Nitrogen fixation, phosphate solubilization, and leading to enhancement of maize development and yield	Pereira et al. (2020)
	Enterobacter cloacae	Phosphate solubilization Increased vegetative growth parameters: root length, shoot length, as well as root and shoot biomass	Verma et al. (2019)
	Chryseobacterium sp. strain NGB-29 *Flavobacterium* sp. Strain NGB-31	High efficient production of IAA, promoting lateral and adventitious root formation of maize plants	Youseif (2018)
Rice	*Bacillus subtilis* *Bacillus megatherium*	Enhancement of nutrient uptake and the biosynthesis of phytohormones including abscisic acid (ABA), (IAA), salicylic acid, gibberellic acid	El-Mageed et al. (2022)
Wheat	*Klebsiella, Enterobacter, Raoultella, Rhizobium, Agrobacterium, and Pseudomonas spp*	Nitrogen fixation and phosphate solubilization Biosynthesis of IAA	Wang et al. (2020); Taheri et al. (2022)

ability to synthesize IAA. A combination of growth-promoting bacteria with diverse functional properties effectively enhanced N/P/K uptake and growth of wheat plants (Wang et al. 2020; Taheri et al. 2023, 2022).

5.2 *Indirect Mechanisms*

5.2.1 Antibiosis

Biological control is an attractive and effective tool for combating bacterial diseases in cereal crops (Ngalimat et al. 2021). The leaf blight (BLB) is perceived as one of the most catastrophic rice diseases in the world, with high occurrence and severity in rice (Ngalimat et al. 2021; Saha et al. 2016). Members of *Bacillus* spp. are well known for their prevalence in the rhizosphere, and their use in the manufacture of biopesticides for crop preservation. Indeed, their success could be attributed to the production of a wide range of secondary metabolites (Table 5), especially surfactins lipopeptides, which have been shown to have effective antimicrobial activity (Marin et al. 2019; Glick 2020). As recent research has revealed the strong antagonistic effect of surfactin A C15, generated by *Bacillus velezensis* strain HN-2 against the rice pathogen *Xoo*, by severely disrupting the cell wall of the pathogenic bacterium, and the results have shown that these compounds could act as elicitors and be involved in triggering plant defense mechanisms (Jin et al. 2020). In addition, it was suggested that *B. subtilis A15 and B. methylotrophicus* H8 could synthesize several antimicrobial compounds, including bacillomycin, iturin and surfactin, with an antagonistic effect against *Xoo* in vitro. In addition, these strains proved effective against BLB disease in greenhouses (Elshakh et al. 2016). In addition, *Bacillus. amyloliquefaciens* is well known for its ability to produce two antibiotic molecules, difficidin, and bacilysin, both of which down-regulate the expression of genes influencing *Xanthomonas* virulence (Wu et al. 2015). Interestingly, *Bacillus* strains have been used as maize seed inoculants to protect against bacterial streaks induced by the bacterium *X. vasicola* pv. *vasculorum* (*Xvv*) (Table 5). In this context, recent research has shown that maize plants treated with *Bacillus pumilus* R190 and *B. subtilis* 6051 showed a reduction in disease gravity of 85% and 56% respectively in comparison with untreated plants, while also having the potential to trigger the plant's defense mechanisms against bacterial leaf streak (Bathke et al. 2022). With wheat, Taheri et al. (Taheri et al. 2023) demonstrated the activity of *B. subtilis* CB2, a seed endophyte, in the biocontrol of, due to its ability to produce surfactins and iturins, as well as hydrolytic enzymes like pectinase, cellulase, amylase. Implementing this technique under field conditions could provide an environmentally-friendly strategy for combating wheat plant diseases (Taheri et al. 2023). In the same context, it was revealed that the endophytic bacterium *Paenibacillus polymyxa* could suppress *X. translucens* under in *vitro* et *planta* conditions, by reducing the length of bacterial streak lesions on wheat leaves, and consequently decreasing the disease severity. The results showed the protease, amylase, pectinase, and cellulase production by the endophytic strain, which could therefore be a powerful antibacterial agent (Taheri et al. 2022).

Table 5 Indirect mechanisms of PGPR to manage Bacterial diseases of Cereal crops

Cereal crops	Phytopathogens	Beneficial rhizobacteria	Biocontrol mechanisms	References
Rice	*X. oryzae* pv. *Oryzae*	*B. subtilis A15,* *B.amyloliquefaciens D29* *B. methylotrophicus H8*	Antimicrobial compounds production (surfactins, iturins, bacillomycin) Induction of systemic resistance, through enhanced activity of defense related genes	Elshakh et al. (2016)
		Pseudomonas aeruginosa BRp3	Siderophores, rhamnolipids, phenazines production Stimulation of defense-related genes	Yasmin et al. (2017)
		B. amyloliquefaciens *A. spinulosporus*	Upregulation of defense related enzymes and acceleration of the SOD, PAL, PO, PPO activities	Jain et al. (2020)
		B. subtilis GBO3 *B. pumilus T4SE34*	Triggering defense mechanisms on rice plants by accumulation of antioxidant enzymes	Udayashankar et al. (2011)
Corn	*X. vasicola* pv. *vasculorum*	*Bacillus. pumilus R190* *B. subtilis 6051*	Reduced Bacterial leaf streak severity, through induction of systemic resistance on maize plants	Bathke et al. (2022)
Wheat	*X. translucens*	*B. subtilis CB2*	Production of surfactins and iturins, as well as hydrolytic enzymes like pectinase, cellulase, amylase	Taheri et al. (2023)
		Paenibacillus polymyxa	Reduced the length of bacterial streak lesions on wheat leaves Production of hydrolytic enzymes like cellulase, amylase, pectinase, and protease Increment on chlorophyll a, b, T, and carotenoids in plants treated	Taheri et al. (2022)

Like *Bacillus* spp, *Pseudomonas* spp is widely distributed in agricultural soils and possesses a variety of characteristics beneficial to cereal growth (Singh et al. 2022; Ngalimat et al. 2021). For example, strains of *Pseudomonas aeruginosa* have been used for rice seed treatment, notably the *P. aeruginosa* BRp3 strain, which was found to be potent in the biocontrol approach, because the findings suggest that the secondary metabolites produced by the BRp3 strain display an antagonistic effect against *X. oryzae* pv. *oryzae* and because of its potential role in stimulating the growth and production of Super Basmati rice (Yasmin et al. 2017).

5.2.2 Induced Systemic Resistance (ISR)

ISR is seen as an enhancement of plant's defense ability against various devastating pathogens, triggered by beneficial microorganisms present in the rhizosphere (Fig. 3) (Pieterse et al. 2014; Pérez-Montaño et al. 2014). Once the ISR is activated, the plant's antioxidant enzymes are produced, such as phenylalanine ammonia lyase (PAL), and peroxidase (PO). This enables plants to minimize the level of reactive oxygen species (ROS), the main cause of oxidative stress caused by infection by plant pathogens (Glick 2020; Jin et al. 2020). Moreover, the induction of pathogenesis-related genes is a key function in plant defense responses.

Besides the direct antagonistic character of *Bacillus* species against plant pathogens, numerous studies have demonstrated their effectiveness in stimulating plant defensive responses (Jin et al. 2020; Elshakh et al. 2016; Udayashankar et al. 2011). Indeed, a recent study showed that *Bacillus velezensis* HN-2 strain was able to activate the immune responses of rice plants, as research results revealed a significant increase in PAL activity in rice leaves when inoculated with this strain HN-22 (Jin et al. 2020). Interestingly, PAL is implicated in the biosynthesis of phenolic compounds, phytoalexins, and lignin, all of which are essential for plant disease resistance (Pieterse et al. 2014, 2009). It's worth mentioning that *Pr1a* gene levels also increased rapidly and significantly when rice was treated with surfactin A C15. This is generally regarded as a transcriptional marker of the SA-dependent induction pathway (Jin et al. 2020). In addition, the application of *B. amyloliquefaciens D29 and B. methylotrophicus H8* improved the activities of defense-related enzymes, enabling effective biocontrol of BLB disease in greenhouses (Elshakh et al. 2016). Similarly, seed treatment with *B. subtilis* GBO3 and *B. pumilus* T4SE34 strains boosted the accumulation of phenylalanine ammonia lyase, peroxidase and polyphenol oxidase in comparison with untreated control seedlings (Udayashankar et al. 2011).

6 Conclusion

The economic and social importance of cereal crops cannot be neglected, given the world's growing population and food production needs, as well as attacks by plant pathogens, particularly those caused by pathogenic bacteria. The development and improvement of rapid diagnostic tools, molecular techniques and disease detection represent a valuable approach to early intervention and targeted control measures. Over the last few decades, the development of fast and powerful new technologies has enabled agricultural biotechnology applications to make great strides, offering a wide range of options for better understanding the molecular functions of plants and breeding varieties. This information speeds up the identification of essential resistance characteristics. In addition, the application of precision farming techniques, including remote sensing and data analysis, has facilitated the precise application of disease management practices, optimizing resource use and minimizing environmental impact. Despite this progress, continued research and collaboration are essential to combat emerging bacterial diseases, develop innovative management strategies and ensure sustainable cereal production for food security.

References

Abbasi MK (2015) Isolation and characterization of rhizobacteria from wheat rhizosphere and their effect on plant growth promotion. 6:1–10. https://doi.org/10.3389/fmicb.2015.00198

Abdallah Y, Liu M, Ogunyemi SO et al (2020) Bioinspired green synthesis of chitosan and zinc oxide nanoparticles with strong antibacterial activity against rice pathogen *Xanthomonas oryzae* pv. *oryzae*. Mol 25:4795. https://doi.org/10.3390/MOLECULES25204795

Ahmed T, Wu Z, Jiang H et al (2021) Bioinspired green synthesis of zinc oxide nanoparticles from a native *Bacillus cereus* strain rnt6: characterization and antibacterial activity against rice panicle blight pathogens burkholderia glumae and *B. gladioli*. Nanomaterials 11. https://doi.org/10.3390/nano11040884

Ansari WA, Chandanshive SU, Bhatt V, Nadaf AB (2020) Genome editing in cereals: approaches, applications and challenges

Asseng S, Martre P, Maiorano A et al (2019) Climate change impact and adaptation for wheat protein. Glob Chang Biol 25:155–173. https://doi.org/10.1111/gcb.14481

Awaludin N, Abdullah J, Salam F et al (2020) Fluorescence-based immunoassay for the detection of *Xanthomonas oryzae* pv. *oryzae* in rice leaf. Anal Biochem 610:113876. https://doi.org/10.1016/J.AB.2020.113876

Awasthi LP (2015) Recent advances in the diagnosis and management of plant diseases

Bartoli C, Roux F (2017) Genome-wide association studies in plant pathosystems: toward an ecological genomics approach. 8. https://doi.org/10.3389/fpls.2017.00763

Bathke, Katie J, Charlene CJ, GYY (2022) Biological control of bacterial leaf streak of corn using systemic resistance-inducing Bacillus strains. Crop Prot 155:105932

Behera B, Das TK, Raj R et al (2021) Microbial consortia for sustaining productivity of non-legume crops: prospects and challenges. Agric Res 10:. https://doi.org/10.1007/s40003-020-00482-3

Benchlih S, Esmaeel Q, Aberkani K et al (2023) Modes of action of biocontrol agents and elicitors for sustainable protection against bacterial canker of tomato. Microorganisms 11:726. https://doi.org/10.3390/microorganisms11030726

Bez C, Esposito A, Thuy HD et al (2021) The rice foot rot pathogen *Dickeya zeae* alters the infield plant microbiome. https://doi.org/10.1111/1462-2920.15726

Bockus WW, Bowden RL, Morrill WL et al (2010) Compendium of wheat diseases and pests edited by. Am Phytopathol Soc 6

Bossa-castro AM, Tekete C, Raghavan C et al (2018) Allelic variation for broad-spectrum resistance and susceptibility to bacterial pathogens identified in a rice MAGIC population. 1559–1568. https://doi.org/10.1111/pbi.12895

Breedt G, Labuschagne N, Coutinho TA (2017) Seed treatment with selected plant growth-promoting rhizobacteria increases maize yield in the field. Ann Appl Biol 171:229–236. https://doi.org/10.1111/aab.12366

Cairns JE, Sonder K, Zaidi PH et al (2012) Maize production in a changing climate. impacts, adaptation, and mitigation strategies

Cassán F, Perrig D, Sgroy V et al (2009) *Azospirillum brasilense* Az39 and *Bradyrhizobium japonicum* E109, inoculated singly or in combination, promote seed germination and early seedling growth in corn (*Zea mays* L.) and soybean (*Glycine max* L.). Eur J Soil Biol 45:28–35. https://doi.org/10.1016/j.ejsobi.2008.08.005

Chen L, Hou B, Lalonde S et al (2010) Sugar transporters for intercellular exchange and nutrition of pathogens. Nature 468:527–532. https://doi.org/10.1038/nature09606

Cho H, Park S, Ryu C et al (2007) Interference of quorum sensing and virulence of the rice pathogen *Burkholderia glumae* by an engineered endophytic bacterium. https://doi.org/10.1111/j.1574-6941.2007.00280.x

Choi O, Hu Z (2008) Size dependent and reactive oxygen species related nanosilver toxicity to nitrifying bacteria. Environ Sci Technol 42:4583–4588. https://doi.org/10.1021/ES703238H

Coser TR, de Figueiredo CC, Jovanovic B et al (2018) Short-term buildup of carbon from a low-productivity pastureland to an agrisilviculture system in the *Brazilian savannah*. Agric Syst 166:184–195. https://doi.org/10.1016/j.agsy.2018.01.030

Di Salvo LP, Cellucci GC, Carlino ME, García de Salamone IE (2018) Plant growth-promoting rhizobacteria inoculation and nitrogen fertilization increase maize (*Zea mays* L.) grain yield and modified rhizosphere microbial communities. Appl Soil Ecol 126:113–120. https://doi.org/10.1016/j.apsoil.2018.02.010

Doni F, Suhaimi NSM, Mohamed Z et al (2019) Pantoea: a newly identified causative agent for leaf blight disease in rice. J Plant Dis Prot 126:491–494. https://doi.org/10.1007/s41348-019-00244-6

Doni F, Suhaimi NSM, Mispan MS et al (2022) Microbial contributions for rice production: from conventional crop management to the use of 'Omics' technologies. Int J Mol Sci 23. https://doi.org/10.3390/ijms23020737

dos Santos RM, Diaz PAE, Lobo LLB, Rigobelo EC (2020) Use of plant growth-promoting rhizobacteria in maize and sugarcane: characteristics and applications. Front Sustain Food Syst 4:1–15. https://doi.org/10.3389/fsufs.2020.00136

Duveiller E, Bragard C, Maraite H (1997) Bacterial leaf streak and black chaff caused by *Xanthomonas translucens*. In: Duveiller E, Fucikovskil L, Rudolph K (eds)

Eggenberger S, Pathology P, Diaz-arias MM (2015) Dissemination of Goss's wilt of corn and epiphytic *Clavibacter michiganensis* subsp. nebraskensis from Inoculum Point Sources. 100

Elizabeth Sunny N, Kumar Shanmugam V (2021) Anti-blight effect of green synthesized pure and Ag-doped tin oxide nanoparticles from *Averrhoa bilimbi* fruit extract towards *Xanthomonas oryzae*-the leaf blight pathogen of rice. Inorg Chem Commun 133:108866

El-Mageed A, Abd TA, Mageed E et al (2022) Plant growth—promoting rhizobacteria improve growth, morph—physiological responses, water productivity, and yield of rice plants under full and deficit drip irrigation. Rice. https://doi.org/10.1186/s12284-022-00564-6

Elshakh ASA, Anjum SI, Qiu W et al (2016) Controlling and defence-related mechanisms of bacillus strains against bacterial leaf blight of rice. 1–13. https://doi.org/10.1111/jph.12479

EPPO code: CORBMI (2004)

Erenstein O, Jaleta M, Sonder K et al (2022) Global maize production, consumption and trade: trends and R&D implications. Food Secur 14:1295–1319. https://doi.org/10.1007/s12571-022-01288-7

Fan SU, Tian F, Li J et al (2017) Identification of phenolic compounds that suppress the virulence of *Xanthomonas oryzae* on rice via the type III secretion system. Wiley Online Libr 18:555–568. https://doi.org/10.1111/mpp.12415

FAO STAT (2019) https://www.fao.org/faostat/fr/#data

Feng M, Kong D, Wang W et al (2015) Development of an immunochromatographic strip for rapid detection of *Pantoea stewartii* subsp. stewartii.mdpi.com

Firdaus RBR, Leong Tan M, Rahmat SR, Senevi Gunaratne M (2020) Paddy, rice and food security in Malaysia: a review of climate change impacts. Cogent Soc Sci 6. https://doi.org/10.1080/23311886.2020.1818373

Food and Agriculture Organization (FAO) Data (2018). https://faostat.fao.org/faostat/en/#data. Accessed 10 June 2023

Food and Agriculture Organization of the United Nations, International Fund for Agricultural Development, UNICEF, et al (2019) The state of food security and nutrition in the world : safeguarding against economic slowdowns and downturns

Fu H, Fleitas MC, Sarkes A et al (2023) Detection and differentiation of *Xanthomonas translucens* pathovars translucens and undulosa from wheat and barley by duplex quantitative PCR. biorxiv.org. https://doi.org/10.1101/2023.03.06.531419

Glaeser SP, Kämpfer P (2015) Multilocus sequence analysis (MLSA) in prokaryotic taxonomy. Syst Appl Microbiol 38:237–245. https://doi.org/10.1016/j.syapm.2015.03.007

Glick BR (2020) Introduction to plant growth-promoting bacteria. In: Beneficial plant-bacterial interactions. Springer, Cham. https://doi.org/10.1007/978-3-030-44368-9_1

Gowda HRA, Tripathi R, Tewari R, Vishunavat K (2022) Morphological and molecular characterization of *Burkholderia glumae* causing panicle blight of paddy. Physiol Mol Plant Pathol 117:101755. https://doi.org/10.1016/J.PMPP.2021.101755

Han J, Xia Z, Li P et al (2020) TALEN-based editing of TFIIAy5 changes rice response to *Xanthomonas oryzae* pv. *Oryzae*. 1–12. https://doi.org/10.1038/s41598-020-59052-w

Hariharan G, Prasannath K (2021) Recent Advances in molecular diagnostics of fungal plant pathogens: a mini review. 10:1–14. https://doi.org/10.3389/fcimb.2020.600234

Hartman TM (2018) Investigation of alternative hosts and agronomic factors affecting *Xanthomonas vasicola* pv. *vasculorum*; causal agent of bacterial leaf streak of corn. 1–115

Hawkins NJ, Bass C, Dixon A, Neve P (2019) The evolutionary origins of pesticide resistance. 94:135–155. https://doi.org/10.1111/brv.12440

Hwang IS, Oh EJ, Oh CS (2020) Transcriptional changes of plant defense-related genes in response to clavibacter infection in pepper and tomato. Plant Pathol J 36:450–458. https://doi.org/10.5423/PPJ.OA.07.2020.0124

Ikley JT, Wise KA, Johnson WG et al (2015) Annual Ryegrass (*Lolium multiflorum*), Johnsongrass (*Sorghum halepense*), and large crabgrass (*Digitaria sanguinalis*) are alternative hosts for *Clavibacter michiganensis* subsp. nebraskensis, causal agent of Goss's wilt of corn annual ryegrass. Weed Sci 63:901–909. https://doi.org/10.1614/WS-D-15-00028.1

Islam T (2018) Application of CRISPR/Cas9 genome editing technology for the improvement of crops cultivated in tropical climates: recent progress. Prospects Challenges 9:1–12. https://doi.org/10.3389/fpls.2018.00617

Jain A, Chatterjee A, Das S (2020) Synergistic consortium of beneficial microorganisms in rice rhizosphere promotes host defense to blight—causing *Xanthomonas oryzae* pv. *oryzae*. Planta 1–25. https://doi.org/10.1007/s00425-020-03515-x

Jin P, Wang Y, Tan Z (2020) Antibacterial activity and rice-induced resistance, mediated by C15surfactin A, in controlling rice disease caused by *Xanthomonas oryzae* pv. *oryzae*. Pestic Biochem Physiol 169:104669

Khan N, Asadullah, Bano A (2019) Plant growth promoting rhizobacteria for sustainable stress management

Kini K, Agnimonhan R, Dossa R et al (2021) Genomics-informed multiplex PCR scheme for rapid identification of rice-associated bacteria of the genus pantoea. Plant Dis 105. https://doi.org/10.1094/PDIS-07-20-1474-RE/ASSET/IMAGES/LARGE/PDIS-07-20-1474-REF4.JPEG

Kleczewski N, Chapara V, Bradley CA (2020) Occurrence of viruses and *Clavibacter michiganensis* in winter wheat in Illinois, 2009 to 2011 and 2019 to 2020. 21:317–320. https://doi.org/10.1094/PHP-07-20-0060-S

Koebnik R, Szurek B, Ghesqui A (2015) A knowledge-based molecular screen uncovers a broad-spectrum OsSWEET14 resistance allele to bacterial blight from wild rice. 694–703. https://doi.org/10.1111/tpj.13042

Kolpakova VV, Molchanova EN, Vasil AV, Chumikina LV (2007) Physicochemical properties of proteins from wheat grown under high-contrast climatic conditions. 43:382–390. https://doi.org/10.1134/S0003683807030180

Kuan KB, Othman R, Rahim KA, Shamsuddin ZH (2016) Plant growth-promoting rhizobacteria inoculation to enhance vegetative growth, nitrogen fixation and nitrogen remobilisation of maize under greenhouse conditions. PLoS ONE 11:1–19. https://doi.org/10.1371/journal.pone.0152478

Kumar A, Raghuwanshi BRMR, Singh V (2017) Co-inoculation with enterobacter and rhizobacteria on yield and nutrient uptake by wheat (*Triticum aestivum* L.) in the alluvial soil under indo-gangetic plain of India. J Plant Growth Regul 36:608–617. https://doi.org/10.1007/s00344-016-9663-5

Lang JM, Ducharme E, Caballero JI et al (2017) Detection and characterization of *Xanthomonas vasicola* pv. *vasculorum* (Cobb 1894) comb. nov. causing bacterial leaf streak of corn in the United States. Am Phytopath Soc 107:1312–1321. https://doi.org/10.1094/PHYTO-05-17-0168-R

Lee C, Lee HH, Mannaa M et al (2018) Genomics-based sensitive and specific novel primers for simultaneous detection of *Burkholderia glumae* and *Burkholderia gladioli* in rice seeds. Plant Pathol J 34:490–498. https://doi.org/10.5423/PPJ.OA.07.2018.0136

Li Y, Zhang W, Niu J, Chen Y (2012) Mechanism of photogenerated reactive oxygen species and correlation with the antibacterial properties of engineered metal-oxide nanoparticles. ACS Nano 6:5164–5173. https://doi.org/10.1021/NN300934K

Liang Y, Yang D, Cui J (2017) A graphene oxide/silver nanoparticle composite as a novel agricultural antibacterial agent against *Xanthomonas oryzae* pv. *oryzae* for crop disease management. New J Chem 41:13692–13699. https://doi.org/10.1039/C7NJ02942J

Liu Q, Yuan M, Zhou Y et al (2011) A paralog of the MtN3/saliva family recessively confers race-specific resistance to *Xanthomonas oryzae* in rice. Plant, Cell Environ 34:1958–1969. https://doi.org/10.1111/j.1365-3040.2011.02391.x

López M, Bertolini E, Olmos A et al (2003) Innovative tools for detection of plant pathogenic viruses and bacteria. Springer 6:233–243. https://doi.org/10.1007/s10123-003-0143-y

Louws FJ, Rademaker JLW, de Brujin FJ (1999). 81–125

Maeda Y, Shinohara H, Kiba A et al (2006) Phylogenetic study and multiplex PCR-based detection of *Burkholderia plantarii*, *Burkholderia glumae* and *Burkholderia gladioli* using gyrB and rpoD sequences Printed in Great Britain. 1031–1038. https://doi.org/10.1099/ijs.0.64184-0

Maiden MCJ (2006) Multilocus sequence typing of bacteria. https://doi.org/10.1146/annurev.micro.59.030804.121325

Maiden MCJ, Bygraves JA, Feil E, Morelli G, Russell JE, Urwin R, Zhang Q, Zhou J, Zurth K, Caugant DA, Feavers IM, Achtman M, Spratt BG (1998) Multilocus sequence typing: a portable approach to the identification of clones within populations of pathogenic microorganisms. 95:3140–3145. https://doi.org/10.1073/pnas.95.6.3140

Manke A, Wang L, Rojanasakul Y (2013) Mechanisms of nanoparticle-induced oxidative stress and toxicity. Biomed Res Int. https://doi.org/10.1155/2013/942916

Mansfield J, Genin S, Magori S et al (2012) Top 10 plant pathogenic bacteria in molecular plant pathology. Wiley Online Libr 13:614–629. https://doi.org/10.1111/J.1364-3703.2012.00804.X

Marin VR, Ferrarezi JH, Vieira G, Sass DC (2019) Recent advances in the biocontrol of *Xanthomonas* spp. World J Microbiol Biotechnol 35. https://doi.org/10.1007/s11274-019-2646-5

Matres JM, Hilscher J, Datta A et al (2021) Genome editing in cereal crops: an overview

McNally R, Ishimaru C, Phytopathology DM (2016) Undefined (2016) PCR-mediated detection and quantification of the Goss's wilt pathogen *Clavibacter michiganensis* subsp. nebraskensis via a novel gene target. Am Phytopath Soc 106:1465–1472. https://doi.org/10.1094/PHYTO-05-16-0190-R

Mohamed KAAHI (2022) Cereal diseases: nanobiotechnological approaches for diagnosis and management

Namburi KR, Kora AJ, Chetukuri A, Kota VSMK (2021) Biogenic silver nanoparticles as an antibacterial agent against bacterial leaf blight causing rice phytopathogen *Xanthomonas oryzae* pv. *oryzae*. Bioprocess Biosyst Eng 44:1975–1988. https://doi.org/10.1007/S00449-021-02579-7/METRICS

Nandakumar R, Shahjahan AKM, Yuan XL et al (2009) Burkholderia glumae and *B. gladioli* cause bacterial panicle blight in rice in the Southern United States. Plant Dis 93:896–905. https://doi.org/10.1094/PDIS-93-9-0896

Newitt JT, Prudence SMM, Hutchings MI, Worsley SF (2019) Biocontrol of cereal crop diseases using streptomycetes

Ngalimat MS, Hata EM, Zulperi D, Ismail SI (2021) Plant growth-promoting bacteria as an emerging tool to manage bacterial rice pathogens. 1–23

Ngoc DTB, Bui Duy D, Tuan LNA et al (2021) Effect of copper ions concentration on the particle size of alginate-stabilized Cu_2O-Cu nanocolloids and its antibacterial activity against rice bacterial leaf blight (*Xanthomonas oryzae* pv. *oryzae*). Adv Nat Sci Nanosci Nanotechnol 12:013001. https://doi.org/10.1088/2043-6262/ABEBD6

Niño-Liu DO, Ronald PC, Bogdanove AJ (2006) Xanthomonas oryzae pathovars: model pathogens of a model crop. Mol Plant Pathol 7:303–324. https://doi.org/10.1111/j.1364-3703.2006.00344.x

Notomi T, Mori Y, Tomita N, Kanda H (2015) Loop-mediated isothermal amplification (LAMP): principle, features, and future prospects. 53:1–5. https://doi.org/10.1007/s12275-015-4656-9

Ogunyemi SO, Abdallah Y, Zhang M et al (2019) Green synthesis of zinc oxide nanoparticles using different plant extracts and their antibacterial activity against *Xanthomonas oryzae* pv. *oryzae*. Artif Cells, Nanomed Biotechnol 47:341–352. https://doi.org/10.1080/21691401.2018.1557671

Oryza L, Ahmed T, Noman M et al (2022) Bioengineered chitosan-iron nanocomposite controls bacterial leaf blight disease by modulating plant defense response and nutritional status of rice (*Oryza sativa* L.). Nano Today 45:101547. https://doi.org/10.1016/j.nantod.2022.101547

Osdaghi E, Robertson AE, Jackson-Ziems TA et al (2022) Clavibacter nebraskensis causing Goss's wilt of maize: five decades of detaining the enemy in the New World. Mol Plant Pathol. https://doi.org/10.1111/MPP.13268

Peng L, Yang S, Zhang Y et al (2022) Characterization and genetic diversity of *Pseudomonas syringae* pv. *syringae* isolates associated with rice bacterial leaf spot in Heilongjiang, China. Biology (Basel) 11:720. https://doi.org/10.3390/BIOLOGY11050720/S1

Pereira NCM, Galindo FS, Gazola RPD et al (2020) Corn yield and phosphorus use efficiency response to phosphorus rates associated with plant growth promoting bacteria. Front Environ Sci 8:1–12. https://doi.org/10.3389/fenvs.2020.00040

Pérez-Montaño F, Alías-Villegas C, Bellogín RA et al (2014) Plant growth promotion in cereal and leguminous agricultural important plants: from microorganism capacities to crop production. Microbiol Res 169:325–336

Pieterse CMJ, Leon-Reyes A, Van Der Ent S, Van Wees SCM (2009) Networking by small-molecule hormones in plant immunity. Nat Chem Biol 5:308–316. https://doi.org/10.1038/nchembio.164

Pieterse CMJ, Zamioudis C, Berendsen RL et al (2014) Induced systemic resistance by beneficial microbes. Annu Rev Phytopathol 52:347–375. https://doi.org/10.1146/annurev-phyto-082712-102340

Prokić A, Zlatković N, Kuzmanović N et al (2020) Identification and characterization of *Dickeya zeae* strains associated with maize stalk soft-rot in northern Serbia. Eur J Plant Pathol 157:685–691. https://doi.org/10.1007/S10658-020-02019-4/METRICS

Prosekov AY, Ivanova SA (2018) Food security: the challenge of the present. Geoforum 91:73–77. https://doi.org/10.1016/j.geoforum.2018.02.030

Ramachandran K, Vijaya SI, Ahmad FN et al (2021) Characterization and identification of *Burkholderia glumae* as the causal pathogen of bacterial panicle blight of rice (*Oryza sativa* L.) in Malaysian rice granaries. J Gen Plant Pathol 87:164–169. https://doi.org/10.1007/S10327-021-00991-1

Roach R, Mann R, Gambley CG et al (2018) Identification of *Xanthomonas* species associated with bacterial leaf spot of tomato, capsicum and chilli crops in eastern Australia. Eur J Plant Pathol 150:595–608. https://doi.org/10.1007/S10658-017-1303-9

Rodriguez-r LM, Grajales A, Arrieta-ortiz ML et al (2012) Genomes-based phylogeny of the genus *Xanthomonas*

Ry SUMMA (2006) *Xanthomonas oryzae* pathovars: model pathogens of a model crop *Xanthomonas oryzae* pv. Oryzaeandx. Oryzae pv. oryzicola. 7:303–324. https://doi.org/10.1111/J.1364-3703.2006.00344.X

Saha S, Garg R, Biswas A, Rai AB (2016) Bacterial diseases of rice: an overview bacterial diseases of rice: an overview

Sarkes A, Yang Y, Dijanovic S et al (2022) Detection of *Xanthomonas translucens* pv. undulosa, pv. translucens, and pv. secalis by quantitative PCR. Plant Dis 106:2876–2883. https://doi.org/10.1094/PDIS-03-22-0574-RE

Sarkes A, Yang Y, Dijanovic S et al (2021) Detection of *Xanthomonas translucens* pv. *undulosa* from wheat by quantitative PCR. biorxiv.org. https://doi.org/10.1101/2021.06.10.447895

Sattayachiti W, Wanchana S, Arikit S, Patarapuwadol S, Vanavichit A, (2020) Genome-wide association analysis identifies resistance loci for bacterial leaf streak resistance in rice (*Oryza sativa* L.). 1–16

Sedeek KEM, Mahas A, Mahfouz M (2019) Plant genome engineering for targeted improvement of crop traits. 10:1–16. https://doi.org/10.3389/fpls.2019.00114

Selander RK, Beltran P, Smith NH et al (1990) Evolutionary genetic relationships of clones of salmonella serovars that cause human typhoid and other enteric fevers. 58:2262–2275

Sen S, Chandrasekhar CN (2014) Effect of PGPR on growth promotion of rice (*Oryza sativa* L.) under salt stress. 4:62–67

Serna-saldivar SO (2019) Corn history and culture. 1–18. https://doi.org/10.1016/B978-0-12-811971-6.00001-2

Shewry PR (2009) The HEALTHGRAIN programme opens new opportunities for improving wheat for nutrition and health. Nutr Bull 34:225–231. https://doi.org/10.1111/j.1467-3010.2009.01747.x

Shimada BS, Simon MV, Laís I et al (2021) Application technology in the use of correctives in corn culture. 43:114–120. https://doi.org/10.9734/JEAI/2021/v43i730719

Shobha B, Lakshmeesha TR, Ansari MA et al (2020) Mycosynthesis of ZnO nanoparticles using *Trichoderma* spp. isolated from rhizosphere soils and its synergistic antibacterial effect against *Xanthomonas oryzae* pv. *oryzae*. J Fungi 6:181. https://doi.org/10.3390/JOF6030181

Singh P, Singh RK, Zhou Y et al (2022) Unlocking the strength of plant growth promoting *Pseudomonas* in improving crop productivity in normal and challenging environments: a review. J Plant Interact 17:220–238. https://doi.org/10.1080/17429145.2022.2029963

Srinivasa C, Sharanaiah U, Shivamallu C (2012) Molecular detection of plant pathogenic bacteria using polymerase chain reaction single-strand conformation polymorphism. Acta Biochim Biophys Sin (Shanghai) 44:217–223. https://doi.org/10.1093/abbs/gmr129

Storz G, Imlay JA (1999) Oxidative stress. Curr Opin Microbiol 2:188–194. https://doi.org/10.1016/S1369-5274(99)80033-2

Strange RN, Scott PR (2005) Plant disease: a threat to global food security. Annu Rev Phytopathol 43:83–116. https://doi.org/10.1146/annurev.phyto.43.113004.133839

Stulberg MJ, Santillana G, Studholme DJ et al (2020) Genomics-informed molecular detection of *Xanthomonas vasicola* pv. vasculorum strains causing severe bacterial leaf streak of corn. 1174–1179. https://doi.org/10.1094/PHYTO-12-18-0453-R

Sundin GW, Castiblanco LF, Yuan X et al (2016) Bacterial disease management: challenges, experience, innovation and future prospects: challenges in bacterial molecular plant pathology. Mol Plant Pathol 17:1506–1518. https://doi.org/10.1111/mpp.12436

Taheri E, Tarighi S, Taheri P (2022) Characterization of root endophytic *Paenibacillus polymyxa* isolates with biocontrol activity against *Xanthomonas translucens* and *Fusarium graminearum*. Biol Control 174:105031. https://doi.org/10.1016/j.biocontrol.2022.105031

Taheri E, Tarighi S, Taheri P (2023) An endophytic bacterium with biocontrol activity against important wheat pathogens. Biol Control 183:105243. https://doi.org/10.1016/j.biocontrol.2023.105243

Talhelm T (2020) ScienceDirect emerging evidence of cultural differences linked to rice versus wheat agriculture. Curr Opin Psychol 32:81–88. https://doi.org/10.1016/j.copsyc.2019.06.031

Tambong JT (2022) We are IntechOpen , the world's leading publisher of Open Access books Built by scientists, for scientists TOP 1 % bacterial pathogens of wheat: identification, and taxonomy. In: Mahmood-ur-Rahman Ansari (ed) IntechOpen. Faisalabad, Pakistan, p 342

Tambong JT, Xu R, Gerdis S et al (2021) Molecular analysis of bacterial isolates from necrotic wheat leaf lesions caused by *Xanthomonas translucens*, and description of three putative novel species, *Sphingomonas albertensis* sp. nov., *Pseudomonas triticumensis* sp. nov. and *Pseudomonas foliumensis*. Front Microbiol 12. https://doi.org/10.3389/FMICB.2021.666689

Tambong JT, Xu R, Gerdis S et al (2021) Molecular analysis of bacterial isolates from necrotic wheat leaf lesions caused by *Xanthomonas translucens*, and description of three putative novel species, *Sphingomonas albertensis* sp. nov., *Pseudomonas triticumensis* sp. nov. and *Pseudomonas foliumensis*. Front Microbiol 12:666689. https://doi.org/10.3389/fmicb.2021.666689

Tewari S, Sharma S (2019) Molecular techniques for diagnosis of bacterial plant pathogens. Microb Divers Genomic Era 481–497. https://doi.org/10.1016/B978-0-12-814849-5.00027-7

Tian Y, Luo J, Wang H et al (2022) Bioinspired green synthesis of silver nanoparticles using three plant extracts and their antibacterial activity against rice bacterial leaf blight pathogen *Xanthomonas oryzae* pv. oryzae. Plants 11:2892. https://doi.org/10.3390/PLANTS11212892

Toribio AJ, Jurado MM, Suárez-Estrella F et al (2021) Application of sonicated extracts of cyanobacteria and microalgae for the mitigation of bacterial canker in tomato seedlings. J Appl Phycol 33:3817–3829. https://doi.org/10.1007/s10811-021-02599-6

Tyagi S, Kumar R, Kumar V et al (2021) Engineering disease resistant plants through CRISPR-Cas9 technology. https://doi.org/10.1080/21645698.2020.1831729

Udayashanlar AC, Nayaja SC, Reddy MS, Srinivas C (2011) Plant growth-promoting rhizobacteria mediate induced systemic resistance in rice against bacterial leaf blight caused by *Xanthomonas oryzae* pv. *oryza*. Biol Control 59:114–122

Uematsu H, Inoue Y, Ohto Y (2015) Detection of *Pantoea stewartii* from sweet corn leaves by loop-mediated isothermal amplification (LAMP). J Gen Plant Pathol 81:173–179. https://doi.org/10.1007/S10327-015-0580-4/METRICS

Valencia-bot AJ, Cisneros-l E (2012) A review of the studies and interactions of *Pseudomonas syringae* pathovars on wheat. https://doi.org/10.1155/2012/692350

Verma P, Agrawal N, Shahi SK (2019) Enterobacter cloacae strain PGLO9: potential source of maize growth promoting rhizobacteria *Enterobacter cloacae* strain PGLO9: potential source of maize growth promoting rhizobacteria

Wang J, Li R, Zhang H et al (2020) Beneficial bacteria activate nutrients and promote wheat growth under conditions of reduced fertilizer application. 1–12

Wonni I (2016) Evaluation of elite rice varieties unmasks new sources of bacterial blight and leaf streak resistance for Africa. Rice Res Open Access 4. https://doi.org/10.4172/2375-4338.1000162

Wu L, Wu H, Chen L et al (2015) Difficidin and bacilysin from Bacillus amyloliquefaciens FZB42 have antibacterial activity against *Xanthomonas oryzae* rice pathogens. Nat Publ Gr 1–9. https://doi.org/10.1038/srep12975

Wubneh WY, Bayu FA (2016) Assessment of diseases on rice (*Oriza sativa* L.) in major growing fields of Pawe district, Northwestern Ethiopia. World Sci News 42:13–23

Xie K, Yang Y (2013) RNA-guided genome editing in plants using a CRISPR—cas system. Mol Plant 6:1975–1983. https://doi.org/10.1093/mp/sst119

Yasmin S, Hafeez FY, Mirza MS et al (2017) Biocontrol of bacterial leaf blight of rice and profiling of secondary metabolites produced by rhizospheric *Pseudomonas aeruginosa* BRp3. 8. https://doi.org/10.3389/fmicb.2017.01895

Youseif SH (2018) Genetic diversity of plant growth promoting rhizobacteria and their effects on the growth of maize plants under greenhouse conditions. Ann Agric Sci 63:25–35

Zahid M, Kaleem Abbasi M, Hameed S, Rahim N (2015) Isolation and identification of indigenous plant growth promoting rhizobacteria from Himalayan region of Kashmir and their effect on improving growth and nutrient contents of maize (*Zea mays* L.). Front Microbiol 6:1–10. https://doi.org/10.3389/fmicb.2015.00207

Zhang F, Hu Z, Wu Z et al (2021) Reciprocal adaptation of rice and *Xanthomonas oryzae* pv. oryzae: cross-species 2D GWAS reveals the underlying genetics. Plant Cell 33:2538–2561. https://doi.org/10.1093/plcell/koab146

Zhang J, Luo J, Chen L et al (2022) Development of droplet digital PCR assay for detection of seed-borne *Burkholderia glumae* and *B. gladioli* causing bacterial panicle blight disease of rice. Microorganisms 10. https://doi.org/10.3390/MICROORGANISMS10061223

Zhou J, Peng Z, Long J, Sosso D, Liu BO, Joon-Seob Eom SH et al (2015) Gene targeting by the TAL effector PthXo2 reveals cryptic resistance gene for bacterial blight of rice. Plant J 82.4:632–643

Diagnostics and Detection Tools for Pathogens in Food Crops

Mohammed Taoussi, Mohammed Radi, Rachid Ezzouggari, Mohamed El Ahrach, Tourya Sagouti, Zineb Belabess, and Rachid Lahlali

Abstract For the past decades, the worldwide population has continued to grow rapidly, and ensuring a food supply has become a serious matter. Besides, agricultural production was considered a major pole of development and research oriented to better food quality, important yields, and above all, ensuring plant protection using common ancestral practices and products. Indeed, more attention was given to new methods that combine both minimum time consumption and efficiency. Unfortunately, as agriculture involves difficulties that can become more decisive, one of the

M. Taoussi · M. Radi · R. Ezzouggari · M. El Ahrach · R. Lahlali (✉)
Phytopathology Unit, Department of Plant Protection, Ecole Nationale d'Agriculture de Meknès, Km10, Rte Haj Kaddour, BP S/40, 50001 Meknes, Morocco
e-mail: rlahlali@enameknes.ac.ma

M. Taoussi · M. Radi
Laboratory of Environment and Valorization of Microbial and Plant Resources, Faculty of Sciences, Moulay Ismail University, PO Box 11201, Zitoune, Meknes, Morocco

R. Ezzouggari
Laboratory of Biotechnology, Conservation and Valorisation of Natural Resources (LBCVNR), Faculty of Sciences Dhar El Mehraz, Sidi Mohamed Ben Abdallah University, 30000 Fez, Morocco

M. El Ahrach
Laboratory of Functional Ecology and Engineering Environment, Department of Biology, Faculty of Sciences and Techniques, Sidi Mohamed Ben Abdellah University, Route d'Imouzzer, P.O. Box 2202, 30000 Fez, Morocco

T. Sagouti
Laboratoire de Virologie, Microbiologie Et Qualité/Ecotoxicologie Et Biodiversité, Faculté Des Sciences Et Techniques de Mohammedia, 20650 Mohammedia, Morocco

Z. Belabess
Plant Protection Laboratory, Regional Center of Agricultural Research of Meknes, National Institute of Agricultural Research, Km 13, Route Haj Kaddour, BP.578, 50001 Meknes, Morocco

R. Lahlali
Plant Pathology Laboratory, AgroBioSciences, College of Sustainable Agriculture and Environmental Sciences, Mohammed VI Polytechnic University, Lot 660, Hay Moulay Rachid, 43150 Ben Guerir, Morocco

J. Chen et al. (eds.), *Molecular and Biotechnological Tools for Plant Disease Management*, Interdisciplinary Biotechnological Advances,
https://doi.org/10.1007/978-981-97-7510-1_2

factors that hinder the agricultural system is food losses due to pathogen infections, which are the main cause of yield reductions resulting in a significant decrease in financial returns. To overcome this challenge, researchers in the plant pathology field take care of developing fast and accurate early detection methods, notably for emerging diseases or tough pathogens that reside more and spread fast and wide. To overcome this challenge, researchers in the plant pathology field take care to forge fast and accurate early detection methods, especially for novel diseases or persistent infections that survive alive for a long time and spread faster. Therefore, the use of recent molecular detection tools has enabled fast and real-time diagnosis of plant diseases, in addition to artificial sensors that are capable of simultaneously monitoring the plant's health status over time. In this chapter, an overview will be provided of the state-of-the-art, recent, and innovative tools applied in the early detection of emerging diseases. Moreover, a deep comparison between conventional and recent diagnostic methods for phytopathogens will be made. Interestingly, this review emphasizes the control strategies of plant diseases and the transition from less-effective practices and harmful chemicals to more fast and eco-friendly strategies resulting in reliable, productive, and sustainable agriculture.

Keywords Plant pathogens · Food crops · PCR · Biosensors · Sustainable agriculture

1 Introduction

Biotechnology is an applied science that involves the modification of genetic material and the proliferation of organisms using modern technology, producing enhanced and novel products with a variety of uses (Coletta et al. 2022). In agriculture, biotechnology tools such as molecular markers, genetic engineering, tissue culture, and molecular diagnosis have been developed to increase agricultural productivity and enhance pathologists' ability to improve crop resistance against plant diseases (James 2003). Additionally, from now on, it is essential for sustainable agriculture that biotechnology is used with conventional agricultural methods (Haggag 2008). Diagnosis of plant disease and accurate identification represent essential matters for resilient food security and the control of invasive pathogen transmission (Balodi et al. 2017).

Biotechnological tools greatly improve the procedure of detecting plant diseases with more precision. Several molecular methods are now available for plant disease detection, such as PCR isothermal amplification methods, DNA hybridization technology, and sequencing (McCartney et al. 2003; Barnes and Szabo 2007; Kang et al. 2010). Knowledge of the level of genetic and pathogenic variability present in a pathogen population, creation of disease-resistant cultivars, and evaluation of disease-resistance genes in various epidemiological regions are all necessary basics for the successful management of plant diseases. Plant diseases and pests are a significant threat to global food production, causing estimated losses ranging from

20 to 40% of total output. In the United States, plant diseases result in annual losses of over 20 billion dollars (Savary et al. 2019; Oerke 2006). These losses not only have economic consequences but also impact food security, leading to increased food prices, poor diets, and even civil unrest in vulnerable countries (Bebber et al. 2014). As the worldwide population is projected to attain 9.5 billion by 2075, a substantial increase in food production is necessary to meet demand, requiring advancements in agricultural research, the maximization of land resources for food production, and sustainable practices (Godfray et al. 2010). Despite these efforts, an estimated 805 million people still lack sufficient access to food (World Health Organization 2019). To address these challenges, plant health scientists are developing disease management strategies, including disease resistance, chemical and biocontrol agents, and microbiome enhancements (Savary et al. 2019). Plant pathology is a vital field of applied science that investigates how environmental factors and host circumstances affect diseases brought on by infectious microorganisms (Agrios 2005). Understanding the biology of plant-pathogen interactions is crucial to protecting food security and the environment (Baysal and Baştaş 2019), especially when plant diseases have become much more dangerous as a result of climate change and globalization emphasizing the need for innovative and sustainable detection and management methods (Bebber et al. 2013). But now advanced molecular tools, like CRISPR/Cas9 and next-generation sequencing, have granted new horizons into plant disease interaction and diagnostics supporting disease control strategies (Sperschneider et al. 2015). The latest features of technology applied in the plant pathology discipline aim to provide full details of plant diseases and develop effective disease management strategies (Savary et al. 2019).

This chapter offers an overview of the state-of-the-art on major diseases, including their impact on crops worldwide. Followed by a straight focus on novel techniques practiced in the diagnostics of plant diseases, the evolution from classic methods to what is now commonly used nowadays. Finally, particular attention is given to the current innovative techniques for monitoring emerging diseases and the accurate identification of the pathogens involved. Sections of this chapter will give a full idea of the development of the plant pathology field and its advanced techniques that encourage modern and sustainable agriculture, aiming to preserve global food security.

2 Principal Diseases and Impact on Crops Worldwide

Several major diseases affect crops around the world. These diseases can cause significant damage to agricultural production, resulting in yield losses and economic losses for farmers. The following are the main ones with their impacts on crops:

2.1 Viral Diseases

The whole world suffers from several challenges including climate change and demographic pressure, which cause several changes in agricultural practices and cropping systems that promote the appearance of destructive viral diseases (Jones and Naidu 2019). The estimated global economic loss is considered to be more than 30 billion US dollars per year due to pandemics and epidemics of viral diseases (Muniyappa 2014). Additionally, (47%) of pathogens causing re-emerging and emerging plant diseases globally belong to viruses (Anderson et al. 2004). Regarding, this situation, many causes have contributed to the rapid degradation of plants. First, the explosive growth of plant and plant products worldwide trade by multinational corporations is inserting harmful viral diseases into areas of the world where they were previously missing (Rodoni 2009; Jones 2021). Secondly, new harmful viral diseases appear at an accelerated rate due to the distancing of cultivated plants from their centers of domestication and toward faraway countries or continents where they are cultivated in monoculture (Jones 2020). Third, pandemics and epidemics of plant virus diseases are becoming more challenging to handle as a result of climatic instability due to global warming (Trebicki 2020). Some examples among the 13 main viral diseases are maize lethal necrosis caused by Machlomovirus (Redinbaugh and Stewart 2018), wheat yellow dwarf caused by Luteovirus and Polerovirus (Irwin and Thresh 1990), Citrus Tristeza caused by Closterovirus (Lee 2015), Banana bunchy top caused by Babuvirus (Moriones et al. 2017).

2.2 Bacterial Diseases

Plant pathogenic bacterial diseases have raised a lot of worry since they generate huge economic losses across all cultivated regions (Lamichhane 2015). The two primary categories of harmful bacteria are Gram-negative bacteria, which make up roughly 95% of bacteria, and Gram-positive bacteria, which make up less than 5% of bacteria. They differ mostly in structure and cell wall (Cui et al. 2021). Gram-negative bacteria of the genera Erwinia, *Pseudomonas*, and *Xanthomonas* are responsible for the major widespread severe impacts on plants, affecting approximately 400 plants like sugar cane, beans, cassava, cabbage, banana, citrus fruits, tomatoes, pepper, and rice are susceptible to their pathogenicity's (Ryan et al. 2011). The wide host diversity that the genus *Xanthomonas* can inhabit gives it yield losses in important crops such as tomatoes, peppers, and citrus fruits (Nakayinga et al. 2021) and causes a variety of diseases including bacterial spot, bacterial blight, and black rot (An et al. 2020).

The genus *Pseudomonas* is particularly one of the most prevalent diseases that affect the phyllosphere is *Pseudomonas syringae* (Xin et al. 2018). Given that it has been documented to affect more than 80 different plant species (Hirano and Upper 2000). *P. syringae* may grow as an epiphyte directly on the surface of plants. It multiplies in the apoplast after entering the plant through wounds or naturally occurring holes like stomata to spread disease. On tomatoes, a condition called bacterial speck develops necrotic lesions encircled by a yellow chlorotic halo (Cruz et al. 2010).

On the other hand, fire blight, caused by the phytopathogenic bacterium *Erwinia amylovora*, a member of the *Enterobacteriaceae* family represents a severe disease that affects plants belonging to the *Rosaceae* in most geographical production centers of the world (Cui et al. 2021). It is considered one of the most devastating apple diseases in the United States, with annual losses and control costs estimated at over $100 million (Norelli et al. 2003). There is also *A tumefaciens* most commonly known as one of the causes of the most serious plant diseases affecting various species of cultures around the world. *A. tumefaciens* is not a phytopathogen like the others but has a very rare characteristic, which is the capacity for genetic transformation (Hooykaas 2023).

2.3 Plant Parasitic Nematodes

PPNs (plant parasitic nematodes) are a highly severe issue since they can result in significant production losses in many commercially significant crops. Most NPPs are subsurface parasitic, which makes it possible for the harm to go undetected, especially on plant roots, which might lead people to overestimate how important these nematodes are (Abd-Elgawad 2020). Even though plant parasitic nematodes have existed forever, their losses have only recently been recognized and quantified. Recent estimates of these NPP-related losses have revealed a consistent rise in the numbers emerging from these analyses.; In advanced countries, these losses are thought to be 8.8% on average, but they can be as high as 14.6% in underdeveloped nations. However, the combination of the 21 most economically significant crops, which include 15 essential plant species, gives an idea of the annual losses brought on by nematodes that are considered a total of 173 billion dollars (McSorley et al. 1987; Gamalero and Glick 2020). Nematode losses are now higher in countries in development than in developed countries due to advances in NPP control methods in the latter (Abd-Elgawad and Askary 2015). Because figures for other crops have not been made, many experts think that total nematode losses are larger than these numbers (Berg and Taylor 2008).

In the APHIS U.S, it is cited that, in general, the most economically important phyto-nematodes are (PCN's) potato cyst nematodes represented by *Globodera pallida* and *Globodera rostochiensis* have worldwide distribution and impact savings of around 9% on potato crops (Gartner et al. 2021). Soybean cyst nematode or *Heterodera glycines* (SCN) (Tylka and Marett 2021), pine root-knot or *Bursaphelenchus xylophilus* is the responsible agent of pine wilt disease, false root-knot nematode of Columbia or *Meloidgyne fallax*, *Meloidogyne chitwoodi* is another root-knot nematode, *Meloidogyne enterolobii*, or root-knot nematode of Guava, *Pratylenchus fallax*, *Ditylenchus dipsaci*, or stem and bulb nematode, all of these species are cited in the APHIS Regulated Pest Table (Kantor et al. 2022).

2.4 Fungal Diseases

Phytopathogenic fungi are in their majority strict saprophytes and are the main causative agents of epidemics of cultivated plants, leading to persistent and substantial losses in crop yield every year all over the world (González-Fernández et al. 2010). These losses affect the world's top five crops in importance (wheat, rice, corn, potatoes, and soybeans). These crops would have fed 8.5% of the world's seven billion people in 2011 if these losses had been reduced (Fisher et al. 2012). The fungi that cause the greatest economic harm are: In the first place, *Magnaporthe oryzae* (Mo, anamorph *Pyricularia oryzae)* which represents the most commercially significant food crops are among the more than 50 grasses that are infected. Wheat, rice, and maize, in 85 countries, causing a loss of 10 to 35% of the harvest (Zhang et al. 2022), followed by *Botrytis cinerea*, which has a large host range of more than 200 species of plants, causing gray mold, it causes annual losses of 10 to 100 billion (Boddy 2016). In addition, most crops worldwide are susceptible to one or more species of *Colletotrichum* (*Ascomycota*), which cause anthracnose (deep sunken spots) and blight (tissue browning) of aerial tissues (Boddy 2016). One of the famous phenomena of fungal diseases worldwide is potato blight caused by *Phytophthora infestans* which was a subject of much attention in the mid-1800s, causing famine in Ireland, this disaster was attributed to the arrival of a virulent strain of the pathogen in Europe (Yuen 2021). Most fungi like the genera *Monascum, Alternaria, Cephalosporium, Fusarium, Penicillium, Aspergillus, Claviceps,* and *Drechslera,* mainly *Fusarium* and *Aspergillus* known to produce mycotoxins in food crops and trigger infections and among mycotoxins there is aflatoxin, fumonisins, trichothecenes, zearalenone, ochratoxin that are the most common ones (Gurikar et al. 2023).

2.5 Diseases Caused by Parasitic Plants

Parasitic plants have been little studied compared to the other diseases, they have been of the most difficult pests to control on crops globally. More than 4000 parasitic plants most hemiparasites with a percentage of 90% of this population settle in the host plant using unique organs they developed, which penetrate the vascular tissues of the host and supply them with the nutrition, water, and minerals they require. The haustorium is the name of this specific organ, and the remaining organisms are holoparasites (Cimmino et al. 2018; Erdogan 2022).

Facultative hemiparasites can live their whole life cycle without a host. Similar to obligatory parasites, which can be hemi- or holoparasites, obligate parasites require a host to survive and reproduce. Most facultative parasites may infect a variety of hosts. For instance, *Rhinanthus* attaches to more than 50 species of herbaceous plants and grasses. Obligate parasites are more likely to specialize on a single host plant species or a small host range (particularly holoparasites), such as *Striga* species have infected more than two-thirds of the 73 million hectares of cereals and legumes grown on the African continent, Only after a period of conditioning at the proper temperature do these species begin to sprout, at which point they are exposed to chemical cues like strigolactones, which are released by the host's roots to indicate the presence of mycorrhizal symbiotic arbuscular structures in the soil (Twyford 2018).

3 Detection and Diagnosis of Plant Pathogens

3.1 Conventional Characterization for Plant Pathogens

Classification systems for organisms were historically distinguished by observable characteristics, particularly fungi of medical importance and fungi found on plant parts and whole plants. These types often were poorly understood morphologically and the study of their morphology began to be clarified with the discovery of the light microscope. Thus, the approach based on phenotypic traits was born. Consequently, morphological criteria are the basis of the identification and classification of fungi. In particular the color, septation, and size of the spores, as well as the form of the spore-bearing structures. Up to the early decades of the twentieth century, classification systems highlighted these crucial elements of distinction more and more (Crous et al. 2015; Guarro et al. 1999).

The majority of pathogenic fungi only exhibit their vegetative phase during infections, which implies that they lack sporulation form and only exhibit hyphal components or other non-specific structures in the host tissue. Even though the color, form, and presence or absence of septa on these hyphae might help us identify them, a fungal culture is advised for a precise diagnosis. Growing isolates in appropriate culture media with the benefit of probes and species-specific antibodies, if necessary, can be very useful in such cases because it helps to distinguish their most characteristic features, and this procedure remains the most common in practice (Guarro et al. 1999).

Conidiogenesis (the process of conidial production) and conidial type are the most crucial qualities to watch for when identifying and classifying fungi (Cole and Samson 1979a). Frequently, a separate structure called a conidiophore is found that contains one or more conidiogenous cells (Hoog and Guarro 1995). Furthermore, in the absence of sporulation, a number of morphological features can still be considered for classification. Structures such as sclerotia, chlamydospores, etc., can be used for classification. The categorization of sexual sporulation can benefit greatly from a number of morphological characteristics, such as the fruiting body type and asci type, which are essential. For the identification of higher taxa, ascomata's form, color, and existence of an apical opening (ostiole) are also crucial (Guarro et al. 1999).

Transmission electron microscopy (TEM) corroborated the observations and recognized several details of particular taxonomic importance, such as cell wall stratification, which reveals significant differences between ascomycete and basidiomycete yeasts, Therefore, it is possible to classify conidia or yeast-like fungi as ascomycetes or basidiomycetes even in the absence of sexual structures (Kurtzman et al. 2011; Arx et al. 1982). In contrast, in the mid-1960s, scanning electron microscopy (SEM) allowed species identification by allowing for the first-time critical examination of differences in ascospore surface detail, and this breakthrough allowed the isolation of two previously highly similar plant diseases, most notably the rusts, and smuts. Overall, a significant contribution to understanding conidiogenesis patterns was made by the SEM (Hoog and Guarro 1995; Cole and Samson 1979b; Hyde et al. 2014).

Technology advancements have recently had an impact on morphological approaches such as digital taxonomy, and efficient statistical software, which allows reliable phenotypic studies, considering the use of computational techniques to create identification keys, which provide some solutions and raise the prospect of a revival in morphological research (Guarro et al. 1999). Ordinarily, basic visible signs following phyto-fungal infections have typically been the most helpful for conventional fungal disease diagnostic techniques, such as fungus propagules like conidia, sclerotia, or mycelia on the exterior surfaces of the flora, or symptoms of fungal diseases brought on by fungal pathogens after infection (Stehlíková et al. 2020). Therefore, the diagnosis of fungal plant diseases has shifted to molecular approaches that allow easy detection of fungal pathogens. Molecular tests can now easily compensate for the drawbacks of conventional approaches to fungus diagnosis (Hariharan and Prasannath 2021).

3.2 Accurate Diagnostic Tools for Plant Pathogens

3.2.1 Immunological Assays and Serological Methods

When cultivation ad hoc isn't possible, especially in the case of viruses, detection by serological assays is the key, and that is why it was developed in the first place (Alarcon et al. 1990; Caruso et al. 2002). Detection is now possible using polyclonal and monoclonal antisera and techniques such as serologically specific electron microscopy (SSEM) and enzyme-linked immunosorbent assay (ELISA), immunostrip assays, western blots, dot-blot immune-binding assays, and for more than a thousand other pathogens, bacteria, and fungi (Zarraonaindia et al. 2015). ELISA was first employed in the 1970s, and due to its great throughput potential, it is today regarded as the most used immunodiagnostic method depending on the organism, sample freshness, and titer, ELISA's sensitivity varies (Martinelli et al. 2015).

Concerning serological detection, it's a strategy that involves plant disease identification based on a color change of antibodies in the test. The immunoglobulin (Ig) proteins from an animal's body are the main component of the antibodies, which are produced in reaction to the presence of antigens, they can be proteins, complex carbohydrates polynucleotides, and lipopolysaccharides. Each antigen has its specific antibody that will bind specifically. Enzyme-linked immunosorbent assay, or ELISA, is the most often employed serological technique and was originally developed for the detection of plant viruses. It was then modified to meet the particular requirements of the detection test, including those for Direct ELISA (conjugated antibody-enzyme complex directly binds to antigen) and Indirect ELISA (antigen does not directly attach to antibody-enzyme conjugate) (Khurana 2006). The precision of ELISA has been improved with the utility of the recombinant antibodies that are commercially available, which makes the diagnosis more reliable (Lopez et al. 2001). Currently, ELISA is massively used for plant disease detection, including plant pathogenic bacteria and fungi, using polyclonal antisera and monoclonal (Balodi et al. 2017).

Numerous monoclonal and polyclonal antisera techniques, including ELISA, dot-blot immune-binding tests for Southern rice black-streaked dwarf disease (Wang et al. 2012), western blots, and serologically specific electron microscopy (SSEM), are used to detect phytoplasma diseases (Musetti et al. 2002) are also broadly used in the diagnosis of plant pathogens, few of them can be applicable at any level of the distribution or production system; for example, the lateral flow-ELISA is extensively used for *Phytophthora infestans, Ralstonia solanacearum,* the potyvirus group, tomato mosaic virus, more pathogens, etc. (Danks and Barker 2000). Likewise, for site-specific detection with tissue print-ELISA and other lateral flow devices (Balodi et al. 2017).

Other supplemental serological techniques have had a big influence on the revolution in plant disease detection and, at the same time, promoted the identification of many phytopathosystems, allowing precise quantification of pathogen participation in a diseased sample. Additionally, serological approaches can detect isolates with higher virulence or the ability to detect isolates with pesticide resistance. Thus, precise and measurable detection of pathogenic agents in impacted plant organs is fundamental to properly determining the required pesticide load before application (Balodi et al. 2017).

3.2.2 Molecular Methods

PCR-Based Methods

Real-Time PCR

Real-time PCR is undoubtedly based on the original PCR technology, but it is an upgraded approach that can measure DNA in addition to amplifying it, providing more details about the fungal infection. It can also provide information about how efficient a fungus control agent is (Mackay 2004; Bernreiter 2017). The PCR products may be tracked and evaluated continuously in closed tubes throughout the amplification step of RT-PCR by using either DNA intercalating stains like the Taq Manprobe which is a sequence-specific probe or also SYBR Green (Badali and Nabili 2013). These probes have two separate fluorescent dyes attached; one is a reporter dye that connects to the 5'end and the other is a quencher dye that attaches to the 3'end, preventing the emission of fluorescence due to their proximity. Reporter dye separates from quenching dye and begins to produce fluorescence as a result of Taq polymerase's exonuclease activity (Dasmahapatra and Mallet 2006). There were too many RT-PCR experiments performed to identify and measure plant diseases, such as the identification of *Phytophthora infestans*, which causes late blight in potatoes (Lees et al. 2012).

Multiplex PCR

By combining numerous primer pairs in one reaction mixture, multiplex PCR has shown to be a time and money saver, by concurrently focusing on several DNA sequences, it may identify multiple diseases. This technique is crucial for plants that become infected with multiple pathogens following amplification and detection in this method is done based on the molecular sizes of the pathogens on the agarose gels (Sint et al. 2012; Aslam et al. 2017).

A successful amplification requires that different sets of primers have corresponding annealing temperatures and relative concentrations, which calls for careful selection of the multiplex test primers (Zhao et al. 2014). *P. nicotinae, F. oxysporum, P. cactorum* and *B. cactivora* are responsible for important diseases in a broad range of plants, remarkably in the cactus industry using the multiplex PCR approach, this pathogenic fungus was identified in grafted cacti (Cho et al. 2016). Multiplex PCR studies are often successful, as demonstrated by the diagnostic test that used the TEF-1a gene regions and ITS-LSU to detect 12 fungi linked to fruit rot in cranberries. The fungal pathogens were *Phyllosticta vaccinia, Monilinia oxycocci, Physalospora vaccinia, Strasseria, Phomopsis vaccinia, Allantophomopsis cytisporea, Colletotrichum fiorinae, C. fructivorum, Coleophoma empetri, A. lycopodina, Phyllosticta elongate* and *Fusicoccum putrefaciens* (Pallás et al. 2018).

Nested PCR

Nested PCR, an end-point PCR modified version, is substantially more sensitive and selective for the identification of fungus than a single PCR (Hariharan and Prasannath 2021; Yeo and Wong 2002). This technique targets two successive cycles of PCR amplification using two sets of primer pairs. In the first round, an important portion of DNA is amplified using a single pair of non-specific primers, and the target for the second round is the amplified DNA sequence using the two internal primers (pathogen-specific primers) in the second round (Rahman et al. 2013; Bhat and Browne 2010). Through nested PCR, *P. granati,* the responsible agent of pomegranate crown rot or twig blight, yet, only when the sample utilized a low quantity of DNA considered by 10 pg (Yang et al. 2017). As well as the two specific *Phytophthora* species that cause orchid disease identified *P. palmivora* and *P. parasitica* were attainable to identify using nested PCR tests (Tsai et al. 2006).

PCR–RFLP

In this detection tool, DNA samples are typically digested by a panel of restriction enzymes as part of the restriction fragment polymorphism (RFLP) procedure, and the resultant patterns may then be tabulated and compared (Uijthof and Hoog 1995). Combined with PCR, the PCR–RFLP can quickly help in the differentiation between related species without the need for sequencing (Bernreiter 2017). This approach also aids in identifying species without the need for sub-culturing by amplifying the ITS region from removed plant tissue and then performing a restriction analysis. When the DNA is divided by size using conventional agarose gel electrophoresis, cutting the ITS segment with the restriction enzyme AluI produces a distinct pattern (Bernreiter 2017).

For instance, *Dothistroma septosporum* and *Dothistroma pini*, two closely similar species that cause red band needle blight disease in *Pinus*, were effectively distinguished using this approach (Barnes et al. 2004). PCR–RFLP test based on a single nucleotide variation in the TEF gene was also successfully applied in distinguishing Fumonisin-producing *Fusarium fujikuroi* strains from non-producing strains (Suga et al. 2014).

LAMP

Loop-mediated amplification (LAMP) is a more recent acid amplification method that has emerged as a helpful diagnostic tool for plant disease diagnosis, it is considered a more effective alternative to PCR (Le and Vu 2017). In essence, it is a DNA autocycling strand displacement amplification (Panno et al. 2020), where the targeted nucleic acid is amplified under isothermal conditions with great specificity for DNA amplification, which eliminates the requirement for a thermal cycler (Tsui et al. 2011) and it has ten times the accuracy and sensitivity of traditional PCR. There are two key stages: the first step, cycling amplification, followed by the elongation step (Ren et al. 2009).

For each LAMP test, two sets of internal primers—forward inner primer (FIP) and backward inner primer (BIP), one backward loop primer (B-Loop), and a second set of two outer primers—F3 and B3—are made to identify six different sequences on the targeted nucleic acid. The sense and anti-sense strands of the targeted DNA are represented by two distinct sequences in each FIP and BIP. With two extra loop primers—loop forward (LF) and loop backward (LB)—and high exponential and isothermal amplification, which takes 45–60 min at 60–65 C and produces 109–1010 fold target DNA, which is thought to be the optimal temperature range for Bst polymerase activity (Ren et al. 2009; Chander et al. 2014; Francois et al. 2011). It is not time-consuming, for example, in a LAMP assay sugar beet rust-causing fungus, *Uromyces betae*, was diagnosed by targeting the cytochrome b DNA sequence and it took only 30 min (Kaczmarek et al. 2019). In pine and conifer, pitch canker caused by the fungal pathogen *Fusarium circinatum* was diagnosed by quantitative LAMP (qLAMP) assays and conventional LAMP (cLAMP) and it seems that qLAMP tests were more specific than cLAMP tests when using LAMP probes targeting TEF-1a (Stehlíková et al. 2020).

NASBA

NASBA is a highly precise isothermal amplification technique that amplifies nucleic acids and is specifically intended for RNA detection, in which reverse transcriptase converts single-stranded RNA (ssRNA) to complementary DNA (cDNA). Hence the name of self-sustained sequence replication (3SR) (Hariharan and Prasannath 2021; Sergentet-Thévenot et al. 2008). The procedure takes place at a specific temperature of 41 °C, and three enzymes work together to generate more than 109 copies of the nucleic acid sequence in just 30 min. Reverse transcriptase, RNase H, and T7 RNA polymerase from the Avian myeloblastosis virus (Chang et al. 2012; Mazumdar et al. 2012). NASBA cancels the use of thermal cycler without any side effects on the amplification capacity, it is on a level with real-time based PCR tests or outperforms them (Chang et al. 2012; Loens et al. 2006) and at the end amplicons can be detected by probe-capture hybridization using electrochemiluminescence (Mazumdar et al. 2012; Heo et al. 2019). Real-time fluorescence detection, similar to tests for the sugarcane yellow leaf virus, can be used in the NASBA-based plant pathogen testing (Gonc 2002), strawberry vein banding virus (Klerks et al. 2004), apple stem pitting virus, *Arabis mosaic* virus, *Fusarium poae,* potato virus Y, *Ralstonia solanacearum* (Szemes and Schoen 2003), and apple stem pitting virus (Klerks et al. 2001).

An assay for finding Apple scar skin viroid (ASSVd) was developed using NASBA and real-time molecular beacon detection during the amplification process. An assay for detecting Apple scar skin viroid (ASSVd) was developed using nucleic acid sequence-based amplification (NASBA) and real-time detection during the amplification process using molecular beacons (Heo et al. 2019).

Next-Generation Sequencing

Next-generation sequencing (NGS) or also high throughput sequencing is a novel technique for diagnosis that has introduced inventive methods to detect phytopathogenic agents (Chalupowicz et al. 2019). It is based on numerous consecutive steps, including DNA isolation, library preparation, massively parallel sequencing, bioinformatics analysis, and variant/mutation identification and interpretation (Qin 2019). Some of the available advanced sequencing techniques in NGS include pyrosequencing, massively parallel signature sequencing, polony sequencing, and sequencing by oligonucleotide ligation detection (SOLID) (Rajesh and Jaya 2017).

Basically, a lot of RNA is converted into a cDNA library and either one or both ends have adaptors linked to them. In order to get short sequences from either one end, as in single-end sequencing, or both ends, as in pair-end sequencing, each fragment with or without amplification is then sequenced in a high-throughput manner. The reads depend on the DNA sequencing method employed and are typically between 30 and 400 bp (Mortazavi et al. 2008). Pyrosequencing, polony sequencing, massively parallel signature sequencing, and sequencing by oligonucleotide ligation detection (SOLID) are a few of the sequencing methods often employed in HTS (Rajesh and Jaya 2017). However, the most popular ones are Ion Torrent and (SOLID), and both are promoted by Illumina's HiSeq and Life Technologies (Wang and Peng 2014), this platform has established the benchmark for NGS and is the most broadly useful NGS platform for RNA-Seq. A desktop sequencer called MiSeq was also recently introduced (Kukurba and Montgomery 2015).

The sequencing platforms used by Applied Biosystems SOLiD, Illumina, and Ion Torrent all employ sequencing by synthesis. These systems can produce reads that are longer than those produced by the FLX Genome Sequencer, but they also cost more to operate and have greater error rates. Furthermore, compared to the FLX Genome Sequencer, the Illumina and Ion Torrent technologies enable quicker sequencing times and offer better throughput (Maliogka et al. 2018). For high throughput sequencing of plant pathogens, Illumina provides nine distinct sequencers. These sequencers can produce brief reads with a length of 50 to 300 nucleotides that range from 4 million to 20 billion. The method combines PCR with sequencing-by-synthesis on a flow cell with lanes. Sequencing can be done on entire DNA or RNA, on partially or fully purified viral particles, or siRNAs. It makes use of four nucleotides that have been tagged with various fluorescent dyes. Low-quality readings are filtered out (Maliogka et al. 2018). The Ion Personal Genome Machine (PGM), the first sequencer to be commercially released on the Ion Torrent platform, was unveiled by Life Technologies in 2010. The method works by measuring the concentration of hydrogen ions (H+), which are produced when a nucleotide joins a DNA strand. Sequencing chips are microprocessor chips with many micromachined wells that make up their structure. These chips can sequence RNA, viral particles, and siRNA using short reads of 200–400 nucleotides (Maliogka et al. 2018).

The MinION platform, created by Oxford Nanopore, is smaller than a smartphone and connects to a laptop or PC through USB. A motor protein is used in this nanopore sequencing system to push a single-stranded DNA or RNA molecule through a protein or solid-state nanopore, where an ionic current sets a voltage and produces nucleotide-dependent electronic current signals (Li et al. 2020; Mthethwa et al. 2021; Zhou et al. 2022; Liefting et al. 2021). There have been too many studies utilizing NGS to diagnose cucurbit downy mildew caused by the pathogen *Pseudoperonospora cubensis* based on RNA and DNA. Seven distinct areas in *P. cubensis* were identified during this approach thanks to comparative genomics utilizing RNA-Seq of closely related species *P. humuli*, which enables the creation of diagnostic markers (Withers et al. 2016). Stone and pome fruits are harmed both after harvest and before by the brown rot disease-causing fungus *Monilinia fructicola*. The *M. fructicola* Mfrc123 strain's genome was sequenced using a hybrid and hierarchical de novo association technique. Using third-generation sequencing systems from Pacific Biosciences (PacBio) and Illumina for short-run NGS (Miccolis Angelini et al. 2019). Another hybrid technique yielded 547 Mb as the overall genome size of *Hemileia vastarix* race XXXIII after sequencing the coffee rust fungus's genome using PacBio RS II and Illumina HiSeq systems (Porto et al. 2019).

CRISPR-Cas9 as a Genome Editing Tool

CRISPR-Cas9 is a method that runs based on the RNA-mediated sequence-specific adaptive immune system of the prokaryotes (Hu et al. 2019). CRISPR-Cas is not only a nuclease system; it is an inheritable prokaryotic immune system that produces sequence-specific adaptive immunity against viruses (bacteriophages), conjugative plasmids, and foreign DNAs (Grissa et al. 2008; Makarova et al. 2015; Marraffini and Sontheimer 2010; Ricroch et al. 2017).

The CRISPR RNA (crRNA) binds to the trans-activating crRNA (tracrRNA) in order to direct double-stranded cleavage in the target DNA by Cas proteins. On the other hand, three nucleotides are undoubtedly needed for DNA binding; this sequence, which is 50-NGG-30 and is recognized as an adjacent protospacer motif (PAM) recognition sequence (placed at the gRNA/DNA interface), is what is known as a PAM (Sorek et al. 2013). The CRISPR-Cas9 system may cause a double-strand break (DSB) in the genome which makes it a valuable genomic scissor for genome editing (Jinek et al. 2012). The DSB originated by CRISPR is repaired by two different systems, either a low fidelity non-homologs end joining (NHEJ) process or a more specific or homology-directed repair (HDR) system (Liu et al. 2019). Through the error-prone nature of NHEJ, the target sequence can be cut by either insertion or deletion (InDels), which causes loss of function and results in genetically modified plants. With the use of 20 nucleotides or less, synthetic guide RNAs (gRNAs) complementary to a target DNA have been developed to produce target site specificity (Jinek et al. 2012), and target different genes by a simple modification (Shan et al. 2014). Cloning the gRNA into the chosen vector follows gRNA synthesis and PAM sequence identification, and is then followed by transformation, screening, and validation (Lone et al. 2018; Pickar-Oliver and Gersbach 2019).

With the exception of a few live systems, CRISPR-Cas9 has so far been utilized to change the genomes of several plant species, from monocots to dicots. Model plants that have been subjected to genome editing include *Nicotiana benthamiana, Arabidopsis thaliana*, as well as cash crops including tomato, rice, potato wheat, cotton, maize, citrus, barley, grape, and soybean. Despite the ploidy complexity of their genomes (Shan et al. 2014; Ahmad et al. 2018; Brooks et al. 2014). CRISPR/Cas9 has demonstrated real success when used to modify single or several targeted genes in *Arabidopsis*, tomato, tobacco, sweet orange sorghum, and rice (Shan et al. 2014; Belhaj et al. 2013; Bortesi and Fischer 2015; Feng et al. 2013; Jiang et al. 2013) (Fig. 1).

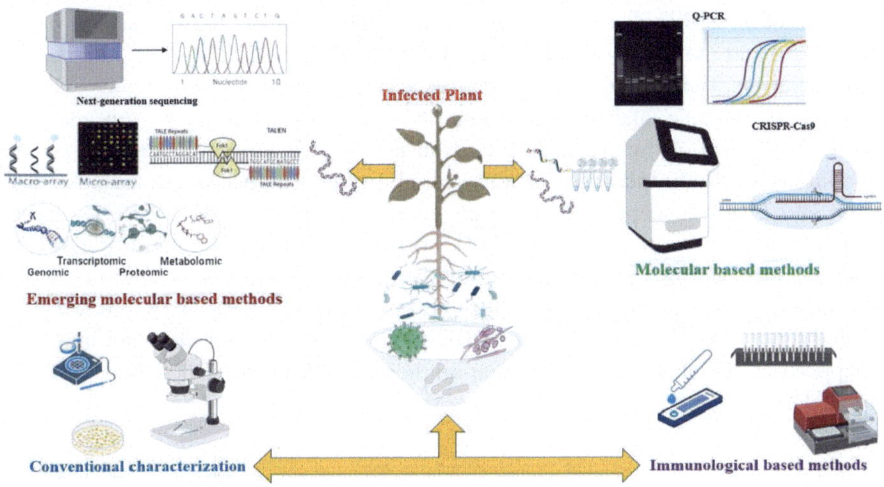

Fig. 1 Laboratory-based techniques for plant pathogen diagnostics from conventional to emerging methods

TALEN as a Genome Editing Tool

TALE technology is a recently discovered molecular technique in which TALE proteins function to find basic recognition codes in modular DNA (Boch et al. 2009; Moscou and Bogdanove 1501). It uses transcription factors extracted from the phytopathogenic bacteria *Xanthomonas* spp. They are known by having a DNA-binding domain of tandem repeats made of 33–35 almost identical amino acids (Moscou and Bogdanove 1501) without two repeat-variable residues (RVDs), they determine the TALE DNA-binding particularity (Deng et al. 2012). The procedure involves the translocation of TALE proteins into the nucleus by injecting them into plant cells through a type III secretion pathway (Moscou and Bogdanove 1501; Lange et al. 2013). Afterward, they activate or transactivate downstream genes by attaching the TALEs to the promoters of either executor R genes as avirulence factors or susceptibility genes as virulence factors respectively (Boch et al. 2009; Schornack et al. 2013). However, TALE nucleases (TALENs) are known as the same process, when there is a link between DNA binding domain and FokI cleavage domain (Becker 2012).

In their mode of action, double-strand breaks (DSBs) are induced at specific target positions within a eukaryotic genome, which activates natural DNA damage repair pathways by either introducing a small insertion/deletion (indel) at the breakpoint that gives frame-shifts that knockouts a targeted gene, or by non-homologous end joining (NHEJ). Alternately, a use a template-dependent process properly known as homology-directed repair (HDR), precisely inserts a homologous donor template at the desired locus (Chapman et al. 2012; Jankele and Svoboda 2014; Lieber 2010). Now it's already performed in the model plants, *Brachypodium* and *Arabidopsis*

(Christian et al. 2013; Zhang et al. 2013), rice (Zhang et al. 2013), barley (Wendt et al. 2013), and maize (Char et al. 2015; Liang et al. 2014).

In rice crops, the bacteria *X. oryzae* uses an injection of TALE proteins AvrXa7 that binds to the rice susceptibility gene OsSWEET14 promoter region and activates transcription of the sucrose efflux transporter. It is a specific strategy for the bacterial cells to help increase nutrient availability and induce the disease because the OsSWEET14 susceptibility gene encodes a member of the SWEET sucrose efflux transporter (Yuan and Wang 2013). Fortunately, the employment of TALEN-mediated gene editing was very useful; in this case, the editing was done in the rice bacterial blight S gene, specifically in the promoter of the EBE site; consequently, the binding of TALEs AvrXa7 to the EBE site could not be possible and that false its expression, which stops the propagation the causal bacteria (Boch et al. 2009; Li et al. 2012).

3.3 In-Field Detection of Plant Pathogens

3.3.1 Spectroscopy-Based Methods

The detection and identification of plant diseases are decisive for an optimal and targeted application of plant protection measures in agriculture (Thomas et al. 2018). Furthermore, the rapid advancement of agricultural technologies has increased the need for non-invasive plant disease detection techniques. That's why recent research has been conducted intensively on the application of spectroscopy to plant disease detection based on criteria such as speed, non-invasiveness, and specificity of a disease type (Khaled et al. 2018). The potential of spectroscopy applied in agriculture can offer valuable insights into the field of plant pathology. During host–pathogen interactions, several mechanisms occur that lead to changes in plant physiology, biochemistry, and morphology. These pathogen mechanisms alter the optical properties of crops, which is what spectroscopy perceives as a signal for plant pathogen detection (Cotrozzi 2022). On the VIS–NIR spectrum, symptomatic aspects and phyto-physiological changes of diseases and pests (pigment deterioration, cell deterioration, scab formation, etc.) reveal very precise spectral reflectance responses (Zhang et al. 2019).

Vis–NIR spectroscopy is based on wavelengths of the detected light that samples absorb, reflect, or transmit in the Vis and NIR ranges, which are associated with sample staining (Farber et al. 2019). As an example, Wahabzada et al. (2016) analyzed the relative spectra for three leaf diseases of barley (powdery mildew, leaf spot, and brown rust) and their respective biochemical characteristics in the visible and near-infrared (IR) regions. The study showed that the pigment's wavelength of powdery mildew damage is between 500 and 650 nm, and the most suitable wavelength for chlorosis caused by net blotch is between 500 and 580 nm, and certainly at 550 and 700 nm.

In addition, several bands in the shortwave infrared (SWIR) region (11,000–2500 nm) are especially sensitive to plant or soil water content, while ideally complementing conventional VIS–NIR sensors (400–1000 nm), which particularly indicates the propagation characteristics of plant diseases and pests (Zhang et al. 2014). Indeed, each host–pathogen interaction exhibits specific spatiotemporal evolutions that influence wavelength ranges during distinctive phases of pathogenesis (Mahlein et al. 2018). Thus, it can be argued that pathogens have different actions, resulting in different symptoms in the host, which is why the appropriate wavelengths for monitoring are different (Zhang et al. 2020). As a result, reflectance spectroscopy now represents one of the most widely used approaches for the detection, identification, and quantification of plant diseases at different scales (Cotrozzi 2022). Currently, many agricultural activities are carried out using drones, particularly in the fields of disease identification (Castro et al. 2015), plant health surveys (Garcia et al. 2020), and production assessment (Qi Yang 2019).

Unmanned aerial vehicles (UAVs) are aerial robots that fly without a human pilot on board, these devices can be controlled by remote control or perform several operations simultaneously by a computer running intelligent algorithms. The operation of these vehicles relies on various pieces of equipment such as cameras, sensors, motors, rotors, controllers, and so on (Bouguettaya et al. 2023). One simple method of identifying plant diseases is to use drones equipped with algorithms that provide automatic knowledge of the disease. The information accumulated through these algorithms contributes to both disease identification and severity (Abdu et al. 2020).

To identify plant diseases, drones detect physiological disorders generated by leaf reflectance in the near-infrared spectrum (Di Gennaro et al. 2016). Two types of agricultural UAV models can be used for monitoring and diagnosing plant diseases: fixed-wing and rotary-wing UAVs (Kim et al. 2019).

Di Gennaro and colleagues (2016) used a modified OktoXL Mikrocopter (HiSystems GmbH, Moomerland, Germany) to monitor grapevine leaf-band disease. In addition, soybean leaf spots and powdery mildew were detected using a popular quadcopter, the Phantom-3 (Shenzhen DJI Sciences and Technologies Ltd., Shenzhen, China) (Tetila et al. 2017). Xavier and colleagues (2019) used a multi-rotor drone to detect Rumaria leaf blight disease in cotton (Xavier et al. 2019). A hexacopter, DJI Matrice 600 pro (Shenzhen DJI Sciences and Technologies Ltd., Shenzhen, China), was utilized to detect and target bacterial spots on tomato leaves (Abdulridha et al. 2020). As a result, UAVs are effective and provide a wealth of details on crop conditions, helping to improve crop management. Nevertheless, there are still major prospects for diagnosing plant pathologies. For this, we need to combine the latest deep learning architectures and new technologies with UAV platforms to develop an effective solution for improving crop productivity (Bouguettaya et al. 2023).

3.4 *Biosensor Technologies for Plant Pathogen Detection*

3.4.1 Antibody-Based Biosensors

Pathogen detection with antibody-based sensors refers to biosensors that use antibodies as recognition elements to detect and identify pathogens. These sensors exploit the specific binding affinity between antibodies and their target pathogens to achieve highly sensitive and selective detection (Fang and Ramasamy 2015). The basic principle of antibody-based sensors is to immobilize antibodies on a sensor surface or incorporate them into a detection system. When a sample containing the target pathogen is introduced, the antibodies bind to the pathogen. Surface proteins or other biomarkers, trigger a measurable signal. Due to their many uses and strong antigen–antibody interactions, antibodies are affinity biorecognition components that have been in use for twenty years. Immunoglobulins (Ig) are the structure of antibodies (Schroeder and Cavacini 2010). Immunosensors are two types of biosensors that use the interaction between an antibody and an antigen or have an embedded antibody as a ligand. Labeled immunosensors are introduced, with the label being considerably detectable. Unlabeled immunosensors are designed to uniquely identify the antigen–antibody complex by evaluating the physical changes produced by complex formation. The antigen–antibody complex is sensitively assessed by label measurement. Only the quality of the antibody used and its storage conditions, which might vary greatly from one nation to another, serve as a restriction (Lim and Ahmed 2016).

Theoretically, immunoassays can detect bacterial, fungal, and viral pathogens since they are relevant to all plant diseases that express antigenic molecules (Mehetre et al. 2021). Numerous immunoassays are available for the detection of infections, with the majority of them being employed in a clinical environment. Some of these immunoassays include chemiluminescent immunoassays, fluorescence immunoassays, latex agglutination tests, western blot, etc. Below are listed the most popular immunological methods for identifying plant pathogens (Haselbeck et al. 2022).

Lateral Flow Immunoassays (LFIA)

Lateral flow immunoassay (LFIA) is a commonly used immunoassay for the detection of plant pathogens (López-Soriano et al. 2017). These tests use nitrocellulose membranes to detect the presence of an antigen under target capillary forces in a sample. Antibodies present in the sample bind to the membrane, and a visual chemical reaction is used to determine whether the antigen is present or not (López-Soriano et al. 2017; Singh et al. 2021).

The tests primarily aim to identify the target pathogen qualitatively, and they provide results by the visual observation of a colored line in the test region. However, if it is linked to a sensor, semi-quantification is possible (Posthuma-Trumpie et al. 2009). LFIAs are easy to use, portable, inexpensive, and can provide results in approximately 10 min, not including sample preparation (Singh et al. 2021). The speed of analysis and the ease of use in the field has prompted the creation of a number of LFIAs to identify plant pathogens. For example, a polyclonal LFIA has been developed for the detection of *Xanthomonas campestris* pv. *Musacearium* (Hodgetts et al. 2015).

Enzyme-Linked Immunosorbent Assay (ELISA)

It is considered the most popular immunological method. It was initially created in the 1970s and has since gained widespread acceptance as a reliable technique for microbial pathogen identification. Because it is a quick procedure that is adaptable to automation and high-throughput screening, it is commonly employed (Martinelli et al. 2015). Multiple formats Multiple ELISA formats have been developed including Indirect, Direct, Competitive, and Sandwich Direct, Indirect, Sandwich, and Competitive ELISA (Alhajj et al. 2023), which all share the same principle, namely the use of specific antibodies conjugated to an active substance. Antibodies that are specifically bound to an enzyme that can change a colorless substrate into a colorful result. The amount of the target antigen, and consequently the target pathogen, in the sample, determines how much color is produced (Richard et al. 2014).

Since its creation, ELISA tests have been widely used in agriculture to identify plant diseases. For instance, EPPO guidelines advise using ELISA testing to check for the presence of viruses on the grafting material of fruit trees (Boonham et al. 2014) and even to this day, new ELISA tests are being developed to detect plant pathogens. For instance, the detection of the plant pathogen *Xylella fastidiosa* by the development of a double antibody sandwich ELISA (DAS-ELISA) (Gorris et al. 2021).

3.4.2 DNA-Based Biosensors

The interaction of a molecular probe with a particular affinity for target analysis is the basis of biosensors (Murphy 2006). Due to stable chemistry, DNA probe-based biosensors offer the benefits of high-temperature tolerance, facile modification, and effective surface regeneration (Wang 2000). More significantly, targeted screening of DNA libraries can produce DNA probes with various affinities for substances (Canoura et al. 2021; Zhuo et al. 2017). The two forms of functional DNA strands, DNA aptamer, and DNAzyme, constitute the foundation of these biosensors.

The Use of Aptamers as an Alternative for Antibodies

Aptamers are short oligonucleotides (DNA or RNA) of 10 to 100 nucleotides, having strong affinity and a particular three-dimensional conformation for the target analyte/pathogen, which are alternatives to the use of antibodies (Zhao et al. 2018). These aptamers can be used similarly to antibodies that target particular antigens.

Because they are inexpensive, deterioration-resistant, small in size, and easily manufactured. Aptamers have great potential to take the place of antibodies (Toh et al. 2015). The use of aptasensors is further restrained because of their sensitivities to several factors such as temperature, pH, air temperature, water temperature, air temperature, etc. (Shahdordizadeh et al. 2017). Despite its still limited use in the detection of plant pathogens, aptamer-based selection has demonstrated its relevance in some studies, notably in the detection of soybean rust fungi and apple stem spit virus (Krivitsky et al. 2021).

DNAzyme-Based Biosensors

Functional oligonucleotides like DNAzyme can work as sensors that can transmit a detectable signal or as molecular recognition components in biosensors that give great specificity to their target analyte (Khan et al. 2021). They possess enzymatic capabilities that allow them to mediate a variety of chemical processes, including DNA and RNA cleavage, ligation, and phosphorylation (Morrison et al. 2018). However, certain substances, such as metal ions and proteins, are required for some DNAzymes to function. (Silverman 2005). Due to this characteristic, these DNAzymes can be utilized to design biosensors that can find these molecules. It has been demonstrated that many DNAzymes can differentiate between comparable stimuli, such as various bacterial strains, enabling the construction of specific biosensors with great specificity (Baum and Silverman 2008; Huang et al. 2013).

DNAzymes possess two distinctive properties: The capacity to recognize and cleave, which may be exploited to build recognition elements primarily directed at cofactors, and the numerous enzymatic turnover features, which are ideally suited for signal amplification (Gong et al. 2014).

Some of the uses of DNAzyme in plant pathogen detection, cucumber green mottle mosaic virus (CGMMV) is detected using a visual color change assay that uses cDNA made from total mRNA extracted from infected crop seedlings in combination with the restriction enzyme AcII, terminal deoxynucleotidyl transferase (TdT) activity, Mg^{2+}-dependent DNAzyme, and hemin/G-quadruplex DNAzyme (Wang et al. 2019a).

Similar to this work, an identification of the Watermelon Mosaic Virus (WMV) was done using electrochemical differential pulse voltammetric measurements with the help of the exonuclease and polymerase activity of T4 DNA polymerase, Mg^{2+} dependent DNAzyme, and hemin/GQuadruplex DNAzyme cascade reaction series (Wang et al. 2019b).

3.4.3 Optical Biosensors

The principle of these biosensors is based on the detection of optical changes that occur when biological molecules bind to their specific ligand. These changes may be evaluated by employing a light source to monitor the interaction of a target analyte and a ligand using an optical signal chemiluminescence or fluorescence or a color created directly by a bioreceptor and a biomarker or by a recognition procedure (Dyussembayev et al. 2021). It consists of an optical transmission medium, a biorecognition component that is immobilized, and a signal-detecting system. With the naked eye, the colorimetric optical signals may be seen. Optical sensors use a photodetector, such as a photodiode or a photomultiplier, to give quantitative measurements by converting an optical signal into a quantifiable electrical output (Vizzini et al. 2019). Among the optical biosensors developed for the detection of plant pathogens, fluorescence-based assays, colorimetric biosensors, and SPR surface plasmon resonance-based biosensors are the most common (Naresh and Lee 2021). The basic principle of fluorescence-based immunoassays is that antibodies or plant target molecules, which are labeled with fluorochrome or fluorophore molecules, generate light during biological recognition (Chen and Wang 2020). Plant diseases are detected using certain characteristics, such as plant growth, condition, signs, and leaf damage (Wang et al. 2019c).

The colorimetric biosensor is a paper-based sensor that is commonly used in laboratories for quick diagnosis and is extremely economical, simple to use, and very affordable. Probably one of the most widespread tools, which allows the user to immediately detect a small number of samples within 10–15 min through a color change, this analytical equipment has four separate buffers: the sample containing the analyte is deposited in the first buffer, which is comprised of cellulose; the second buffer is a glass fiber soaked in a bio conjugate solution; the third buffer is the detection buffer; and a control line is printed (Khater et al. 2019). Recent studies show that colloidal gold is most often used as a label in immunoassays (Koczula and Gallotta 2016). For example, In 2012, Drygin et al. created lateral flow immunoassays for a variety of plant diseases, most notably for the potato virus X (Drygin et al. 2012), *Fusarium* in Corn (Xu et al. 2019), and *Pantoea stewartii subsp. stewartii* (Pss) in corn too (Feng et al. 2015).

Surface plasmon waves are produced when polarized light illuminates a metal surface at the interface between two media with different refractive indices and are used by SPR-based biosensors to measure changes in refractive index brought on by molecular interaction on a metal surface. The SPR phenomenon causes a reduction in the intensity of the reflected light compared to the incident light at a certain angle known as the resonance angle, depending on the thickness of the layer on the metal's surface (Solaimuthu et al. 2020). A novel phenomenon known as localized surface plasmon resonance (LSPR) was created when the SPR phenomenon was applied to metal-based nanomaterials, such as gold and silver (Damborský et al. 2016).

SPR-based biosensors have been the subject of several studies. For instance, SPR-based, label-free biosensors have been developed to detect plant pathogens, including fungal agents such as *Fusarium culmorum* (Zezza et al. 2006), same case for *Phyththora infestans* and *Puccinia striiformis* (Skottrup et al. 2007), cowpea mosaic virus, tobacco mosaic virus and lettuce mosaic virus as plant viruses (Boltovets et al. 2002; Torrance et al. 2006).

3.4.4 E-nose

The electronic nose (electronic nose), (e-nose), also known as an artificial nose or an artificial olfactory system, is a device that mimics the olfactory senses of humans (Sun et al. 2019). The Electronic Nose System is an intelligent sensor-based tool created to recognize and classify complex odors using a variety of non-selective sensors (Jiarpinijnun et al. 2020). An inclusive EN system is usually derived from single or multiple sensors that can potentially detect different volatile metabolites emanating from the sample and a detection system from machine learning algorithms and data processing (Rahman et al. 2020). The sensor array consists of several gas sensors that have been coated with various odor compounds to be sensitive to various chemicals (Boeker 2014). Currently, several types of sensors are readily available, including metal oxide semiconductors (MOS) (Nunes et al. 2019), metal–oxide–semiconductor field-effect transistors (MOSFETs) (Loutfi et al. 2015), the piezoelectric crystal (PC) (Pohanka 2018), Quartz crystal microbalance (QCM) (Wang 2020), surface acoustic wave (SAW) (Kus et al. 2020) and electrochemical sensor (Xiao et al. 2019). These sensors can detect small changes in volatile metabolites in the sample, enabling a sensitive and early detection process (Luo et al. 2022). In addition, various recent works have proven that EN has the potential to detect and discriminate changes in volatile metabolites of different crop maturity stages, disease identification, quality assessment, evaluation of shelf life, and sensory evaluation, thus realizing a real-time monitoring system (Mohd Ali et al. 2020).

An e-nose system with ten different MOS sensors was used by M. Mohd Ali et al. 2020 (Chang et al. 2017) for the detection of phytopathogenic bacteria on papaya and chili plants, on the other hand, Jiarpinijnun et al. (2020) used this technology to detect *Aspergillus* fungal infection on brown jasmine rice grains as early as possible. In a different research, the e-nose system's potential for detecting citrus *Tristeza* virus infection in mandarins was assessed. Using an adaptive classification method, a classification accuracy of more than 97% has been attained to distinguish between healthy samples and contaminated samples (Hazarika et al. 2020).

3.4.5 Acoustic Sensors

Acoustic energy can be produced as a vibrating concurrent pressure wave and transmitted through gases, liquids, and solids. The lowest frequency of the acoustic spectrum is in the frequency range of 20 Hz related to infrasound. Up to or well above 20 kHz for ultrasound, which is frequently employed as a diagnostic and therapeutic tool in the medical sector. Through mechanical and thermal processes, this frequency range can interact with biological tissues (Speaks 2017). The main components of this system are an audio probe, responsible for sound acquisition, a low-power processor, which can run the detection algorithm to process the captured sound, a wireless communication interface, and a power supply, responsible for providing appropriate power to the bioacoustic sensor node (Rach et al. 2013). When plants are infected with pathogens such as bacteria, fungi, or pests, they often produce distinct acoustic emissions or vibrations. These sounds can be produced by the movement of air or fluids within plant tissues, as well as changes in the structural integrity of the plant. Since the early 1900s, acoustic pest detection has been reported but has only gained momentum in the last two decades (Mankin et al. 2011). Few studies carried out for the detection of microscopic pathogens Papadakis et al. developed in 2015 (Papadakis et al. 2015) a multiplex PCR for the acoustic detection of three bacteria *Pseudomonas syringae* pv., *Ralstonia solanacearum* and *Xanthomonas campestris* pv. Several other recent works have been reported on acoustic pest detection, including (i) an automated acoustic detection platform for the detection of cicadid pests (Quesada Gigas; HQ) in coffee plantations (Escola et al. 2019). On the other hand, this system reaches rates above 90% for the early detection of the red palm weevil caused by (RPW, *Rynchophorus ferrugineus*) (Rach et al. 2013) installed inside the tree trunk used to collect sound signals from a set of quarantined palms costing around $30 while exhibiting 96.41% accuracy (Hetzroni et al. 2016). In addition to the use of monitoring species-specific pests in soils, e.g. stridulations, Larval feeding sounds: short "clicks" and "pops" in the frequency range of 1–8 kHz, and longer lasting sounds around 3 kHz were recorded. Both manual and automated energy thresholds were used to identify pests, where a rate above 75% (Görres and Chesmore 2019).

Acoustic signature detection for pest monitoring is of plants even before visible disease symptoms appear, this enables early detection and intervention, which enables real-time, continuous, and inexpensive disease management that takes advantage of advanced microphone and signal processing technologies, this can be particularly useful in agricultural settings or nurseries, where early identification of diseased plants is crucial (Kashyap and Kumar 2021) (Fig. 2).

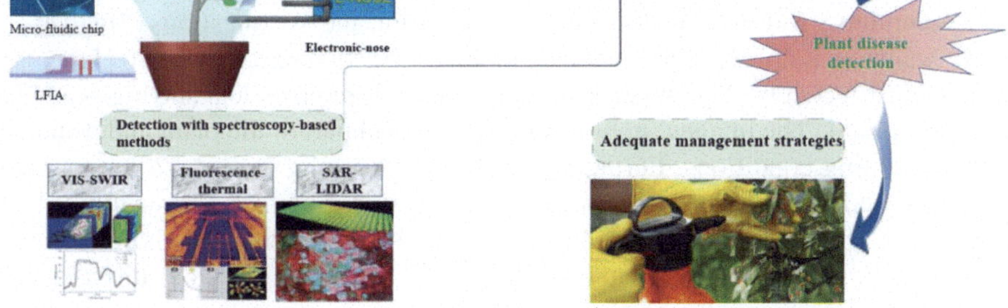

Fig. 2 In-field-based methods for plant disease detection, combining detection with spectroscopy and biosensing

3.5 *Information Technology in Plant Pathology*

Information technology (IT) has revolutionized the way that plant pathology is practiced. In the past, plant pathologists relied on traditional methods of disease detection and control, such as visual inspection and laboratory testing. However, IT has made it possible to collect and analyze large amounts of data in real time, which has led to the development of new and more effective methods of disease management (Islam et al. 2017).

One of the most important applications of IT in plant pathology is in-field monitoring. In-field monitoring involves the use of sensors and other devices to collect data on plant health conditions in real time. This data can then be used to identify and track plant diseases, as well as to assess the effectiveness of disease control measures (Kabir et al. 2021).

The combination of AI, ML, and the IoT is creating new opportunities for the early detection, diagnosis, and management of plant diseases. These technologies are helping to improve crop yields, reduce the use of pesticides, and protect food security (Kabir et al. 2021).

3.5.1 Artificial Intelligence

In earlier times, farmers consulted experts or used their own experience for the identification of diseases in their crops, but now artificial intelligence (AI) has become an important tool for the detection and recognition of plant diseases (Singh et al. 2021). AI sensors can detect and identify weeds, and the methodologies utilized for disease detection include segmentation of the affected part and classification of the diseases (Singh et al. 2021). AI can provide more accurate and precise results than traditional methods of disease detection and help in the early detection of plant diseases to prevent their spread, which can lead to more effective disease management (Kothari 2018).

For methodology, AI plant disease diagnosis tools work by using image recognition technology to identify diseases in plants. They provide a diagnosis of the disease and treatment recommendations, and some tools also provide real-time monitoring and a community forum for users to share information and ask questions (Singh et al. 2021; Anand et al. 2022; Sujawata and Chouhanb 2021; Kamdar et al. 2021).

To work effectively, AI plant disease diagnosis tools require clear and high-quality images of plants. The images should show the affected part of the plant and be taken in good lighting conditions. Furthermore, A study by Jung et al. in 2023 on deep learning-based disease detection in plants used images of diseased-healthy plant pairs to construct a disease detection model. The images were of high quality and showed the affected part of the plant (Jung et al. 2023). Thus, plant diseases are detected by using manual observation of the disease symptoms present on parts of plants, especially leaves. However, this method has significant limitations in terms of accuracy and speed because AI detection is often trained on idealized data sets, which may not reflect real-world conditions which can make it difficult to apply AI algorithms in real-world settings, where there may be variability in lighting, weather, or other factors (Anand et al. 2022).

An AI-driven program called Plantix is quickly becoming indispensable to farmers all around the world. By offering reservoirs of information on illnesses, insect infestations, and nutrient deficiencies—all by just uploading a photo of the relevant crop—it is primarily helping farmers and gardeners. The user is then given a choice between receiving a direct health diagnostic for the crop or receiving a comparable list of probable crop health issues (Svetskiy 2022).

An AI-powered application called PlantVillage offers a database of plant illnesses and associated symptoms. Users may upload plant pictures, and the software will offer diagnostic and treatment suggestions. Additionally, it offers a community forum where users may discuss plant problems and ask inquiries (Kreuze et al. 2022).

One of the main challenges in using AI for plant disease detection is the lack of data. AI algorithms require large amounts of data to learn and improve their accuracy. However, there is often a limited amount of data available for specific plant diseases, which can make it difficult to train AI models (Kamdar et al. 2021). Plant diseases can vary in their symptoms and severity, which can make it difficult for AI algorithms to accurately detect and diagnose them. This variability can be due to environmental factors, genetic variation, or other factors (Kamdar et al. 2021).

Another challenge in using AI for plant disease detection is limited access to technology. Not all farmers or agricultural workers have access to the necessary technology or resources to use AI tools effectively, in addition to the cost of implementing AI technology for plant disease detection can be a barrier for some farmers or agricultural workers as well as the cost of hardware, software, and training can be prohibitive for some users (Sujawata and Chouhanb 2021).

3.5.2 Machine Learning

Plant disease detection has made extensive use of machine learning techniques including feature extraction and classification (Shoaib et al. 2023). Modern machine learning techniques are being developed by researchers to enable quick data processing of large data sets. This not only minimizes human labor but also makes it possible to review data objectively. The potential to find likely coherent factors during plant-pathogen-environment interactions is particularly great for deep learning systems (Thomas et al. 2022).

Machine learning can help automate the process of plant disease detection and reduce the time and cost associated with manual inspection. It can also improve the accuracy of disease detection and enable early intervention to prevent the spread of disease. However, it is important to note that the effectiveness of machine learning algorithms can vary depending on the quality and quantity of data available for training and evaluation (Shoaib et al. 2023; Potgieter and Arif 2019). Machine learning algorithms can be used to analyze data collected from sensors and remote sensing devices to identify and track plant diseases (Mahlein 2016). Machine learning algorithms such as SVM, K-NN, and CNN have been used to distinguish diseased or non-diseased leaves (Harakannanavar et al. 2022).

The identification and categorization of plant diseases have been accomplished using deep learning algorithms (Potgieter and Arif 2019; Andrew et al. 2022). From digital images, these models can correctly diagnose and categorize plant diseases. It has demonstrated good outcomes in recognizing diseases and anomalies in plants, as well as in diagnosing plant diseases from digital photos. It has the potential to change how plant diseases are found, treated, and managed (Thakur et al. 2022).

The categorization of plant diseases from digital photos has been already done using machine learning (ML) methods such as the c4.5 classifier, tree bagger, and linear support vector machines (Barbedo 2021). Researchers have used a public dataset of diseased and healthy plant leaves collected under controlled conditions to train a deep convolutional neural network for image-based plant disease detection (Sperschneider 2019).

Researchers have used convolution neural networks (CNN) to classify potato diseases (Jackulin and Murugavalli 2022), by using a dataset of images of healthy and diseased potato leaves to train the models (Dhaka et al. 2021).

While machine learning has shown great potential in plant disease detection, there are still some limitations that need to be addressed to improve the accuracy and reliability of the algorithms. One of the main limitations of using ML for plant disease detection is the limited dataset available for training the algorithms (Jackulin and Murugavalli 2022). The accuracy of the algorithms depends on the quality and quantity of the data used for training (Hasan et al. 2020). This means that the algorithms may not be able to detect diseases in plants that are not in this position or an uncontrolled environment (Mohanty et al. 2016), which makes machine learning classified as not a fully reliable method of identifying and preventing the spread of plant diseases depending on the collected data and algorithms (Shoaib et al. 2023).

3.5.3 The Internet of Things (IoT)

The Internet of Things (IoT) has the potential to revolutionize plant pathology by enabling the detection, analysis, and classification of diseases affecting crops without human intervention. It can be used to investigate plant diseases and nutrient levels using image analysis techniques (Suganya et al. 2019). It involves using sensors and cameras to capture images of plants, which are then analyzed using machine learning algorithms to detect and classify diseases. IoT can be used to develop automated disease detection systems that can identify the presence of disease in plants without human intervention (Nawaz et al. 2020). The sensor nodes help in collecting data on environmental factors such as humidity, temperature, and soil moisture, which are then analyzed to optimize crop growth and detect the presence of disease allowing farmers to take preventive measures before the disease spreads (Liu et al. 2022; Nayagam et al. 2023; Khan et al. 2020). However, for special cases like pest detection, the IoT devices require proximity sensors to emit a signal once an object is near plants

(Khan et al. 2020), the color sensors indicate the changes in leaves and stem colors that can be explained by the presence of a disease in the field (Perera 2021).

The use of IoT devices for plant disease detection requires a certain level of technical expertise, which may limit their use by farmers who lack the necessary skills (Nawaz et al. 2020) and the large amount of data generated by IoT devices for plant disease detection can be difficult to manage and analyze, which may limit their effectiveness (Perera 2021). Nevertheless, the reliability of IoT devices for plant disease detection can be affected by environmental factors such as weather conditions, which may affect the accuracy of the data collected (Yakkundimath et al. 2018).

A vivid example of IoT uses Shiva et al. in 2019 created using an Internet of thing-based system a remote sensing setup that takes into consideration the conditions where the plant grows and also the plant's illnesses, with an integrated sensor that asses soil and atmospheric humidity and moisture. Web cameras capture images, which are subsequently preprocessed to include edge detection, defect identification, image enhancement, scaling, and RGB to grayscale conversion. The features of citrus canker illnesses, such as skewness, entropy, energy, and kurtosis (damaged lemon crop), are then categorized with the use of SVM. Therefore, the program will spray by pumping the recommended nutrients into the affected leaves when it identifies the illness on the field. This study also included the lemon plant to show how it might be used for other crops in the future with positive results (Shiva et al. 2019).

4 Database Programs for Plant Pathogens

Plant pathogen database systems are critical in monitoring and mitigating the hazards associated with plant diseases. These advanced software tools offer comprehensive and structured archives of data on plant diseases, their characteristics, distribution, and management measures. These database projects, by harnessing the power of technology, enable researchers, plant health practitioners, and policymakers to access accurate and up-to-date information, facilitating effective decision-making and improving overall plant health management (Baysal and Baştaş 2019). In addition, plant pathogen database packages may offer capabilities such as interactive mapping tools that visually illustrate disease distribution patterns across different locations. This spatial information aids in the tracking of disease action and possible introduction, allowing for early discovery and intervention to avert broad outbreaks (Buja et al. 2021).

4.1 CABI/Plantwise

CABI is an international development organization that works to improve people's lives through a better understanding of plant and animal health. Plantwise is a global program led by CABI that helps farmers lose less of what they grow to plant health problems (Otieno 2016).

In particular in developing nations, the Plantwise program has created plant clinics across Africa, Asia, and Latin America. This initiative is run by "plant doctors," who provide on-the-spot diagnosis and support through pest diagnoses and management recommendations, including biological control options (Barratt et al. 2018; Romney et al. 2013). The program serves as a great intermediary, assisting in the formation of networks and connections while balancing supply push and demand pull. One of the elements that ensure the program's continuance to meet new requirements and difficulties is experimenting and learning from experiences (Romney et al. 2013). The program could not function effectively without a perfect adherence to several key principles:

- Training agricultural advisors on visual pest diagnosis and how to give good advice based on integrated pest management (IPM) principles to farmers by serving as plant doctors,
- Establishing networks of plant clinics where farmers can consult the plant doctors on issues affecting the health of their crops and get timely and good actionable advice.
- Developing an access-controlled databank and associated tools that ease data capture, storage, and retrieval by plant doctors and other plant health stakeholders to help inform decisions and actions on plant health issues frequently raised by farmers in any country.
- Developing open-access information resources for easy reference to support the work of plant doctors and other intermediaries.
- Building the capacity of in-country experts in the development of extension material such as factsheets and Pest Management Decision Guides (Otieno et al. 2020) (Fig. 3).

Plantwise also works to build the capacity of its partners to provide plant health advice. This includes training in plant identification, diagnosis, and management. Plantwise also provides its partners with access to a range of resources, including diagnostic tools, crop management guides, and pesticide registers (Powell 2017). In fact, it has helped to improve the livelihoods of millions of farmers around the world. In 2018, the platform helped farmers to save an estimated $1.5 billion in crop losses. It has also helped to reduce the use of pesticides and other chemicals by farmers. Plantwise has been recognized for its work by a number of organizations, including the World Food Prize, the CGIAR Systemwide Program on Collective Action for Nutrition, and the International Plant Protection Convention (Sivapragasam and Loke 2017).

For example, in Bangladesh, Plantwise helped farmers manage the papaya mealybug, a pest that was causing significant damage to papaya crops. Plantwise provided farmers with information on the pest and their management, including the use of biological control agents, and helped establish a network of plant clinics to provide farmers with access to plant health information and diagnosis of plant health issues (Otieno et al. 2021). Another example in Honduras, Plantwise helped farmers

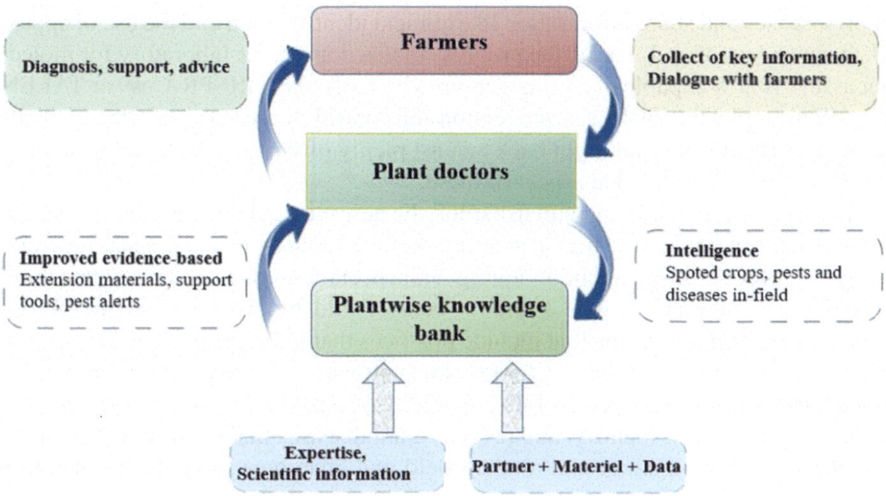

Fig. 3 Information flow in the plantwise program

manage the coffee berry borer, a pest that was causing significant damage to coffee crops. Plantwise provided farmers with information on the pest and its management, including the use of pheromone traps, and helped establish a network of plant clinics to provide farmers with access to plant health information and diagnosis of plant health issues (Scott et al. 2021).

5 Conclusion and Perspectives

Food security became a public issue mostly because of all the agricultural losses brought on by plant diseases by the development of the agricultural sector throughout the world, especially when resilient agriculture is now on promotion and the movement toward organic consumption has gained a respectable audience. Plant hosts can be easily threatened by their interaction with pathogens. Some of them are specific to their host and some have a large-scale host and can invade a large spectrum of crops, this threat has been present since the beginning of time. Plants can resist biotic invasion naturally but most of them can't, and this is where research is responsible to overcome plant disease through strategic management but firstly by detecting it in the early stages which avoids a lot of damage and yield losses when crops are in biotic danger.

Yet, monitoring plant disease in the field has gained a lot of concern recently, coupled with the use of high-tech features like UAV, and AI, which makes tracking pathogens easier. As well as international platforms like CABI-Plantwise that help report diseases and share information like clinical identification and the use of appropriate control manner. Luckily, the progress that is done in the laboratory for molecular assays is now capable of editing genomes by tools like CRISPR-Cas9 or TALEN technology applicable in disease prevention and control by creating genetically modified organisms (GMO) that fight back against plenty of viruses, bacteria, or fungi or by suppressing the virulence aspect of these.

Future perspectives should aim to exploit the new molecular techniques for detection and make them more available in the market and industry by creating commercial kits for instinct detection or collaborating with special labs that can investigate more in further pathologies, as well as more focus should be given on integrated pest management, with programs that include practices that make crops more resistant to plant bio-aggressors like breeding for resistant cultivars, cultural practices, biological control. In addition, more accessibility should be present in the use of recent devices implemented to track, identify, and prevent plant diseases that occur in major crops to make agriculture a smart and modern field and guarantee food safety for future populations.

References

Abd-Elgawad MM (2020) Plant-parasitic nematodes and their biocontrol agents: current status and future vistas. Management of phytonematodes: recent advances and future challenges. https://doi.org/10.1007/978-981-15-4087-5

Abd-Elgawad MMM, Askary TH (2015) Impact of phytonematodes on agriculture economy. Biocontrol agents of phytonematodes, pp 3–49. https://doi.org/10.1079/9781780643755.0003

Abdu AM, Mokji MM, Sheikh UU (2020) Automatic vegetable disease identi fi cation approach using individual lesion features. Comput Electron Agric 176:105660. https://doi.org/10.1016/j.compag.2020.105660

Abdulridha J, Ampatzidis Y, Charan S, Pamela K (2020) Detection of target spot and bacterial spot diseases in tomato using UAV—based and benchtop—based hyperspectral imaging techniques. Precis Agric 0123456789. https://doi.org/10.1007/s11119-019-09703-4

Agrios GN (2005) Plant pathology academic press. San Diego, USA

Ahmad S, Wei X, Sheng Z, Hu P, Tang S (2018) CRISPR/Cas9 for development of disease resistance in plants: recent progress, limitations and future prospects. Briefings Funct Genomics 19(1):26–39. https://doi.org/10.1093/bfgp/elz041

Alarcon B, Lopez MM, Cambra M, Gorris MT, Guerri J (1990) Differentiation of *Erwinia carotovora* subsp. carotovora and *Erwinia carotovora* subsp. atroseptica isolated from potato by Western blot and subsequent indirect ELISA. J Appl Bacteriol 69(1):17–24. https://doi.org/10.1111/j.1365-2672.1990.tb02906.x

Alhajj M, Zubair M, Farhana A (2023) Enzyme linked immunosorbent assay. StatPearls. https://www.ncbi.nlm.nih.gov/books/NBK555922/

An S et al (2020) Society of nematologists crop loss assessment committee correspondence should be addressed: jitang@gxu.edu.cn. The main route of Xanthomonas species (spp.) transmission is via contaminated. FEMS Microbiol Rev 44(1):1–32

Anand R, Mishra RK, Khan R (2022) Plant diseases detection using artificial intelligence. Appl Mach Learn Agric 173–190

Anderson PK, Cunningham AA, Patel NG, Morales FJ, Epstein PR, Daszak P (2004) Emerging infectious diseases of plants: pathogen pollution, climate change and agrotechnology drivers. Trends Ecol Evol 19(10):535–544. https://doi.org/10.1016/j.tree.2004.07.021

Andrew J, Eunice J, Popescu DE, Chowdary MK, Hemanth J (2022) Deep learning-based leaf disease detection in crops using images for agricultural applications. 1–19

Aslam S, Tahir A, Aslam MF, Alam MW, Shedayi AA, Sadia S (2017) Recent advances in molecular techniques for the identification of phytopathogenic fungi—a mini review. J Plant Interact 12(1):493–504. https://doi.org/10.1080/17429145.2017.1397205

Badali H, Nabili M (2013) Molecular tools in medical mycology; where we are! Jundishapur J Microbiol 6(1):1–3. https://doi.org/10.5812/jjm.8566

Balodi R, Bisht S, Ghatak A, Rao KH (2017) Plant disease diagnosis: technological advancements and challenges. Indian Phytopathol 70(3):275–281. https://doi.org/10.24838/ip.2017.v70.i3.72487

Barbedo JGA (2021) Deep learning applied to plant pathology: the problem of data representativeness. Trop Plant Pathol 0123456789. https://doi.org/10.1007/s40858-021-00459-9

Barnes I, Crous PW, Wingfield BD, Wingfield MJ (2022) Multigene phylogenies reveal that red band needle blight of Pinus is caused by two distinct species of Dothistroma, *D. septosporum* and *D. pini*. Stud Mycol 50(2):551–565. Accessed: 29 Aug 2022. [Online]. Available: http://www.eppo.org/QUARANTINE/QP_fungi.htm

Barnes CW, Szabo LJ (2007) Detection and identification of four common rust pathogens of cereals and grasses using real-time polymerase chain reaction. Phytopathology 97(6):717–727. https://doi.org/10.1094/PHYTO-97-6-0717

Barratt BIP, Moran VC, Bigler F, van Lenteren JC (2018) The status of biological control and recommendations for improving uptake for the future. Biocontrol 63(1):155–167. https://doi.org/10.1007/s10526-017-9831-y

Baum DA, Silverman SK (2008) Deoxyribozymes: useful DNA catalysts in vitro and in vivo. Cell Mol Life Sci 65(14):2156–2174. https://doi.org/10.1007/s00018-008-8029-y

Baysal Ö, Baştaş KK (2019) Proteomic tools in plant pathology for sustainable agriculture

Bebber DP, Ramotowski MAT, Gurr SJ (2013) Crop pests and pathogens move polewards in a warming world. Nat Clim Chang 3(9):1–4. https://doi.org/10.1038/nclimate1990

Bebber DP, Holmes T, Gurr SJ (2014) The global spread of crop pests and pathogens. Glob Ecol Biogeogr 23(12):1398–1407. https://doi.org/10.1111/geb.12214

Becker M (2012) Method of the year 2011. Nat Methods 9:1–1

Belhaj K, Chaparro-Garcia A, Kamoun S, Nekrasov V (2013) Plant genome editing made easy: targeted mutagenesis in model and crop plants using the CRISPR/Cas system. Plant Methods 9(1). https://doi.org/10.1186/1746-4811-9-39

Berg RH, Taylor C (2008) Cell biology of plant nematode parasitism, vol 15. Springer Sci Bus Media 18

Bernreiter A (2017) Artículo científico molecular diagnostics to identify fungal plant pathogens-a review of current methods. [Online]. Available: www.ncbi.nlm.nih.gov/genbank/

Bhat RG, Browne GT (2010) Specific detection of *Phytophthora cactorum* in diseased strawberry plants using nested polymerase chain reaction. Plant Pathol 59(1):121–129. https://doi.org/10.1111/j.1365-3059.2009.02147.x

Boch J et al (2009) Breaking the code of DNA binding specificity of TAL-type III effectors. Science 326(5959):1509–1512. https://doi.org/10.1126/science.1178811

Boddy L (2016) Pathogens of autotrophs. Academic Press, Fungi, pp 245–292

Boeker P (2014) On 'Electronic Nose' methodology. Sens Actuators, B Chem 204:2–17. https://doi.org/10.1016/j.snb.2014.07.087

Boltovets PM, Boyko VR, Kostikov IY, Dyachenko NS, Snopok BA, Shirshov YM (2002) Simple method for plant virus detection: effect of antibody immobilization technique. J Virol Methods 105(1):141–146. https://doi.org/10.1016/S0166-0934(02)00098-8

Boonham N et al (2014) Methods in virus diagnostics: from ELISA to next generation sequencing. Virus Res 186:20–31. https://doi.org/10.1016/j.virusres.2013.12.007

Bortesi L, Fischer R (2015) The CRISPR/Cas9 system for plant genome editing and beyond. Biotechnol Adv 33(1):41–52. https://doi.org/10.1016/j.biotechadv.2014.12.006

Bouguettaya A, Zarzour H, Kechida A, Taberkit AM (2023) A survey on deep learning-based identification of plant and crop diseases from UAV-based aerial images. Cluster Comput 26(2):1297–1317. https://doi.org/10.1007/s10586-022-03627-x

Brooks C, Nekrasov V, Lipppman ZB, Van Eck J (2014) Efficient gene editing in tomato in the first generation using the clustered regularly interspaced short palindromic repeats/CRISPR-associated9 system. Plant Physiol 166(3):1292–1297. https://doi.org/10.1104/pp.114.247577

Buja I et al (2021) "Advances in plant disease detection and monitoring: from traditional assays to in-field diagnostics. Sensors 21(6):1–22. https://doi.org/10.3390/s21062129

Canoura J, Yu H, Alkhamis O, Roncancio D, Farhana R, Xiao Y (2021) Accelerating Post-SELEX aptamer engineering using exonuclease digestion. J Am Chem Soc 143(2):805–816. https://doi.org/10.1021/jacs.0c09559

Caruso P, Gorris MT, Cambra M, Palomo JL, Collar J, López MM (2002) Enrichment double-antibody sandwich indirect enzyme-linked immunosorbent assay that uses a specific monoclonal antibody for sensitive detection of *Ralstonia solanacearum* in asymptomatic potato tubers. Appl Environ Microbiol 68(7):3634–3638. https://doi.org/10.1128/AEM.68.7.3634-3638.2002

Chalupowicz L et al (2019) Diagnosis of plant diseases using the nanopore sequencing platform. Plant Pathol 68(2):229–238. https://doi.org/10.1111/ppa.12957

Chander Y et al (2014) A novel thermostable polymerase for RNA and DNA loop-mediated isothermal amplification (LAMP). Front Microbiol 5:395. https://doi.org/10.3389/FMICB.2014.00395/BIBTEX

Chang CC, Chen CC, Wei SC, Lu HH, Liang YH, Lin CW (2012) Diagnostic devices for isothermal nucleic acid amplification. Sensors (Switzerland) 12(6):8319–8337. https://doi.org/10.3390/s120608319

Chang Z et al (2017) Bacterial infection potato tuber soft rot disease detection based on electronic nose. Open Life Sci 12(1):379–385. https://doi.org/10.1515/biol-2017-0044

Chapman JR, Taylor MRG, Boulton, SJ (2012) Playing the end game: DNA double-strand break repair pathway choice. Mol Cell 47(4):497–510. https://doi.org/10.1016/j.molcel.2012.07.029

Char SN et al (2015) Heritable site-specific mutagenesis using TALENs in maize. Plant Biotechnol J 13(7):1002–1010. https://doi.org/10.1111/pbi.12344

Chen C, Wang J (2020) Optical biosensors: an exhaustive and comprehensive review. Analyst 145(5):1605–1628

Cho HJ, Hong SW, Kim HJ, Kwak YS (2016) Development of a multiplex PCR method to detect fungal pathogens for quarantine on exported cacti. Plant Pathol J 32(1):53–57. https://doi.org/10.5423/PPJ.NT.09.2015.0184

Christian M, Qi Y, Zhang Y, Voytas DF (2013) Targeted mutagenesis of *Arabidopsis thaliana* using engineered TAL effector nucleases. G3 Genes Genomes Genet 3(9):1697–1705. https://doi.org/10.1534/g3.113.007104

Cimmino A, Masi M, Rubiales D, Evidente A, Fernández-Aparicio M (2018) Allelopathy for parasitic plant management. Nat Prod Commun 13(3):289–294. https://doi.org/10.1177/1934578x1801300307

Cole GT, Samson RA (1979) Patterns of development in conidial fungi. https://www.cabdirect.org/cabdirect/abstract/19791356541. Accessed 8 Aug 2022

Cole GT, Samson RA (1979) Patterns of development in conidial fungi. https://www.cabdirect.org/cabdirect/abstract/19791356541. Accessed 8 Aug 2022

Coletta A, Bartolini N, Maselli G, Kehs A, McCloskey P, Hughes DP (2022) Optimal deployment in crowdsensing for plant disease diagnosis in developing countries. IEEE Internet Things J 9(9):6359–6373. https://doi.org/10.1109/JIOT.2020.3002332

Cotrozzi L (2022) Spectroscopic detection of forest diseases: a review (1970–2020). J for Res 33(1):21–38. https://doi.org/10.1007/s11676-021-01378-w

Crous PW, Hawksworth DL, Wingfield MJ (2015) Identifying and naming plant-pathogenic fungi: past, present, and future. Annu Rev Phytopathol 53:247–267. https://doi.org/10.1146/annurev-phyto-080614-120245

Cruz L, Cruz J, Eloy M, Oliveira H, Vaz H, Tenreiro R (2010) First report of bacterial speck of tomato caused by *Pseudomonas syringae* pv. tomato race 1 in Portugal. Plant Dis 94(12):1504–1504

Cui Z, Huntley RB, Zeng Q, Steven B (2021) Temporal and spatial dynamics in the apple flower microbiome in the presence of the phytopathogen *Erwinia amylovora*. ISME J 15(1):318–329. https://doi.org/10.1038/s41396-020-00784-y

Damborský P, Švitel J, Katrlík J (2016) Optical biosensors. Essays Biochem 60(1):91–100. https://doi.org/10.1042/EBC20150010

Danks C, Barker I (2000) On-site detection of plant pathogens using lateral-flow devices. EPPO Bull 30(3–4):421–426. https://doi.org/10.1111/j.1365-2338.2000.tb00922.x

Dasmahapatra KK, Mallet J (2006) Taxonomy: DNA barcodes: recent successes and future prospects. Heredity 97(4):254–255. https://doi.org/10.1038/sj.hdy.6800858

De Castro AI, Ehsani R, Ploetz RC, Crane JH (2015) Detection of laurel wilt disease in avocado using low altitude aerial imaging. PLoS ONE 10(4):1–13. https://doi.org/10.1371/journal.pone.0124642

de Lange O et al (2013) Breaking the DNA-binding code of *Ralstonia solanacearum* TAL effectors provides new possibilities to generate plant resistance genes against bacterial wilt disease. New Phytol 199(3):773–786. https://doi.org/10.1111/nph.12324

De Miccolis Angelini RM, Romanazzi G, Pollastro S, Rotolo C, Faretra F, Landi L (2019) New high-quality draft genome of the brown rot fungal pathogen *Monilinia fructicola*. Genome Biol Evol 11(10):2850–2855. https://doi.org/10.1093/gbe/evz207

Deng D et al (2012) Structural basis for sequence-specific recognition of DNA by TAL effectors. [Online]. Available: www.sciencemag.org

Dhaka VS, Meena SV, Rani G, Sinwar D, Ijaz MF, Wo M (2021) A survey of deep convolutional neural networks applied for prediction of plant leaf diseases

Di Gennaro SF et al (2016) Unmanned Aerial Vehicle (UAV)-based remote sensing to monitor grapevine leaf stripe disease within a vineyard affected by esca complex Published by: Firenze University Press on behalf of the Mediterranean Phytopathological Union Stable. Phytopathol Mediterr 55(2):262–275

Drygin YF et al (2012) Highly sensitive field test lateral flow immunodiagnostics of PVX infection. Appl Microbiol Biotechnol 93(1):179–189. https://doi.org/10.1007/s00253-011-3522-x

Dyussembayev P, Sambasivam K, Bar I, Brownlie JC, Shiddiky MJ, Ford R (2021) Biosensor technologies for early detection and quantification of plant pathogens. Front Chem 9:636245

Erdogan P (2022) Parasitic plants in agriculture and management. Parasit Plants

Escola JPL, Guido RC, da Silva IN, Cardoso AM, Maccagnan DHB, Dezotti AK (2019) Automated acoustic detection of a cicadid pest in coffee plantations. Comput Electron Agric 169:105215. https://doi.org/10.1016/j.compag.2020.105215

Fang Y, Ramasamy RP (2015) Current and prospective methods for plant disease detection. Biosensors 5(3):537–561. https://doi.org/10.3390/bios5030537

Farber C, Mahnke M, Sanchez L, Kurouski D (2019) Trends in analytical chemistry advanced spectroscopic techniques for plant disease diagnostics. A review. Trends Anal Chem 118:43–49. https://doi.org/10.1016/j.trac.2019.05.022

Feng Z et al (2013) Efficient genome editing in plants using a CRISPR/Cas system. Cell Res 23(10):1229–1232. https://doi.org/10.1038/cr.2013.114

Feng M, Kong D, Wang W, Liu L, Song S, Xu C (2015) Development of an immunochromatographic strip for rapid detection of *Pantoea stewartii* subsp. stewartii. Sensors (Switzerland) 15(2):4291–4301. https://doi.org/10.3390/s150204291

Fisher MC, Henk DA, Briggs LC, Brownstein CJ, Madoff JS, McCraw SL, Gurr SJ (2012) Emerging fungal threats to animal, plant and ecosystem health. Nature 484:186–194

Francois P et al (2011) Robustness of a lo̊op-mediated isothermal ampli¢cation reaction for diagnostic applications. https://doi.org/10.1111/j.1574-695X.2011.00785.x

Gamalero E, Glick BR (2020) The use of plant growth-promoting bacteria to prevent nematode damage to plants. Biology (Basel) 9(11):1–13. https://doi.org/10.3390/biology9110381

Garcia H, Flores H, Ascencio-Hernández A, Khalil-Gardezi R, Tijerina-Chávez O, Mancilla L, Pena VM (2020) Corn grain yield estimation from vegetation indices, canopy cover, plant density, and a neural network using multispectral and RGB images acquired with unmanned aerial vehicles. Agric 10(7):1–24

Gartner U et al (2021) Resisting potato cyst nematodes with resistance. Front Plant Sci 12:1–18. https://doi.org/10.3389/fpls.2021.661194

Godfray HCJ et al (2010) Food security: the challenge of feeding 9 billion people. Science 327(5967):812–818

Gonc MC (2022) The use of molecular beacons combined with NASBA for the sensitive detection of Sugarcane yellow leaf virus, pp 401–407

Gong L et al (2014) DNAzyme-based biosensors and nanodevices. Chem Commun. https://doi.org/10.1039/C4CC06855F

González-Fernández R, Prats E, Jorrín-Novo JV (2010) Proteomics of plant pathogenic fungi. J Biomed Biotechnol

Görres CM, Chesmore D (2019) Active sound production of scarab beetle larvae opens up new possibilities for species-specific pest monitoring in soils. Sci Rep 9(1):1–9. https://doi.org/10.1038/s41598-019-46121-y

Gorris MT, Sanz A, Peñalver J, López MM, Colomer M, Marco-Noales E (2021) Detection and diagnosis of *Xylella fastidiosa* by specific monoclonal antibodies. Agronomy 11(1):1–14. https://doi.org/10.3390/agronomy11010048

Grissa I, Bouchon P, Pourcel C, Vergnaud G (2008) On-line resources for bacterial micro-evolution studies using MLVA or CRISPR typing. Biochimie 90(4):660–668. https://doi.org/10.1016/j.biochi.2007.07.014

Guarro J, Gené J, Stchigel AM (1999) Developments in fungal taxonomy. [Online]. Available: https://journals.asm.org/journal/cmr

Gurikar C, Shivaprasad DP, Sabillón L, Nanje Gowda NA, Siliveru K (2023) Impact of mycotoxins and their metabolites associated with food grains. Grain Oil Sci Technol 6(1):1–9. https://doi.org/10.1016/j.gaost.2022.10.001

Haggag WM (2008) Original article biotechnological aspects of plant resistant for fungal diseases management. Plant Breed 2(1):1–18

Harakannanavar SS, Rudagi JM, Puranikmath VI, Siddiqua A (2022) Plant leaf disease detection using computer vision and machine learning algorithms. 3:305–310. https://doi.org/10.1016/j.gltp.2022.03.016

Hariharan G, Prasannath K (2021) Recent advances in molecular diagnostics of fungal plant pathogens: a mini review. Front Cell Infect Microbiol. 10. Front Media 11. https://doi.org/10.3389/fcimb.2020.600234

Hasan RI, Yusuf SM, Alzubaidi L (2020) Review of the state of the art of deep learning for plant diseases: a broad analysis and discussion. 1–25

Haselbeck AH, Im J, Prifti K, Marks F, Holm M, Zellweger RM (2022) Serology as a tool to assess infectious disease landscapes and guide public health policy. Pathogens 11(7):1–15. https://doi.org/10.3390/pathogens11070732

Hazarika S, Choudhury R, Saikia S, Sarma U (2020) Pathogen detection in Khasi mandarin orange using serological and electronic nose diagnostic technique. Int J Innov Technol Explor Eng 4(9):2981–2985. https://doi.org/10.35940/ijitee.d2077.029420

Heo S, Kim HR, Lee HJ (2019) Development of a quantitative real-time nucleic acid sequence based amplification (NASBA) assay for early detection of apple scar skin viroid. Plant Pathol J 35(2):164–171. https://doi.org/10.5423/PPJ.OA.10.2018.0206

Hetzroni A, Soroker V, Cohen Y (2016) Toward practical acoustic red palm weevil detection. Comput Electron Agric 124:100–106. https://doi.org/10.1016/j.compag.2016.03.018

Hirano SS, Upper CD (2000) Bacteria in the leaf ecosystem with emphasis on *Pseudomonas syringae*—a pathogen, ice nucleus, and epiphyte. Microbiol Mol Biol Rev 64(3):624–653. https://doi.org/10.1128/mmbr.64.3.624-653.2000

Hodgetts J et al (2015) Development of a lateral flow device for in-field detection and evaluation of PCR-based diagnostic methods for Xanthomonas campestris pv. musacearum, the causal agent of banana xanthomonas wilt. Plant Pathol 64(3):559–567. https://doi.org/10.1111/ppa.12289

de Hoog GS, Guarro J (1995) Atlas of clinical fungi. https://www.cabdirect.org/cabdirect/abstract/19951202596. Accessed 08 Aug 2022

Hooykaas PJJ (2023) The Ti Plasmid, driver of agrobacterium pathogenesis. Phytopathology 113(4):594–604. https://doi.org/10.1094/PHYTO-11-22-0432-IA

Hu N et al (2019) Rapid and user-friendly open-source CRISPR/Cas9 system for single- or multi-site editing of tomato genome. Hortic Res 6(1). https://doi.org/10.1038/s41438-018-0082-6

Huang PJJ, Liu M, Liu J (2013) Functional nucleic acids for detecting bacteria. Rev Anal Chem 32(1):77–89. https://doi.org/10.1515/revac-2012-0027

Hyde KD et al (2014) One stop shop: backbones trees for important phytopathogenic genera: I. Fungal Divers 67(1):21–125. https://doi.org/10.1007/s13225-014-0298-1

Irwin ME, Thresh JM (1990) Epidemiology of barley yellow dwarf: a study in ecological complexity. Annu Rev Phytopathol 28:393–424. https://doi.org/10.1146/annurev.py.28.090190.002141

Islam M, Dinh A, Khan W, Bhowmik P (2017) Detection of potato diseases using image segmentation and multiclass support vector machine. In: 2017 IEEE 30th Canadian conference on electrical and computer engineering (CCECE), pp 8–11

Jackulin C, Murugavalli S (2022) Measurement: sensors A comprehensive review on detection of plant disease using machine learning and deep learning approaches. Meas Sens 24:100441. https://doi.org/10.1016/j.measen.2022.100441

James C (2003) Preview: global status of commercialized transgenic crops: 2003

Jankele R, Svoboda P (2014) TAL effectors: tools for DNA targeting. Brief Funct Genomics 13(5):409–419. https://doi.org/10.1093/bfgp/elu013

Jiang W, Zhou H, Bi H, Fromm M, Yang B, Weeks DP (2013) Demonstration of CRISPR/Cas9/sgRNA-mediated targeted gene modification in Arabidopsis, tobacco, sorghum and rice. Nucleic Acids Res 41(20). https://doi.org/10.1093/nar/gkt780

Jiarpinijnun A, Osako K, Siripatrawan U (2020) Visualization of volatomic profiles for early detection of fungal infection on storage Jasmine brown rice using electronic nose coupled with chemometrics. Meas J Int Meas Confed 157:107561. https://doi.org/10.1016/j.measurement.2020.107561

Jinek M, Chylinski K, Fonfara I, Hauer M, Doudna JA, Charpentier E (2012) A programmable dual-RNA-guided DNA endonuclease in adaptive bacterial immunity. [Online]. Available: http://science.sciencemag.org/

Jones RAC (2020) Disease pandemics and major epidemics arising from new encounters between indigenous viruses and introduced crops. Viruses 12(12):1388

Jones RAC (2021) Global plant virus disease pandemics and epidemics. Plants 10(2):1–41. https://doi.org/10.3390/plants10020233

Jones RAC, Naidu RA (2019) Global dimensions of plant virus diseases: current status and future perspectives. Annu Rev Virol 6:387–409

Jung M et al (2023) Construction of deep learning-based disease detection model in plants. Sci Rep 13(1):1–13. https://doi.org/10.1038/s41598-023-34549-2

Kabir MM, Ohi AQ, Mridha MF (2021) A multi-plant disease diagnosis method using convolutional neural network. 24. https://doi.org/10.1007/978-3-030-65927-1_2

Kaczmarek AM, King KM, West JS, Stevens M, Sparkes D, Dickinson MJ (2019) A loop-mediated isothermal amplification (LAMP) assay for rapid and specific detection of airborne inoculum of uromyces betae (Sugar Beet Rust). Plant Dis 103(3):417–421. https://doi.org/10.1094/PDIS-02-18-0337-RE

Kamdar JH, Jasani MD, Jasani JD, Praba J, Georrge JJ (2021) 11 Artificial intelligence for plant disease detection: past, present, and future. Internet Things Mach Learn Agric 223–238. https://doi.org/10.1515/9783110691276-011

Kang S et al (2010) The promise and pitfalls of sequence-based identification of plant-pathogenic fungi and oomycetes. Phytopathology 100(8):732–737. https://doi.org/10.1094/PHYTO-100-8-0732

Kantor M, Handoo Z, Kantor C (2022) Top ten most important US-regulated and emerging plant-parasitic nematodes. Horticulturae 8(3):208

Kashyap B, Kumar R (2021) Sensing methodologies in agriculture for monitoring biotic stress in plants due to pathogens and pests. Inventions 6(2). https://doi.org/10.3390/INVENTIONS6020029

Khaled AY, Aziz SA, Bejo SK (2018) Early detection of diseases in plant tissue using spectroscopy—applications and limitations. Appl Spectrosc Rev 53(1):36–64. https://doi.org/10.1080/05704928.2017.1352510

Khan S, Burciu B, Filipe CDM, Li Y, Dellinger K, Didar TF (2021) DNAzyme-based biosensors: immobilization strategies, applications, and future prospective. ACS Nano 15(9):13943–13969. https://doi.org/10.1021/acsnano.1c04327

Khan FA, Ibrahim AA, Zeki AM (2020) Environmental monitoring and disease detection of plants in smart greenhouse using internet of things environmental monitoring and disease detection of plants in smart greenhouse using internet of things

Khater M, de la Escosura-Muñiz A, Quesada-González D, Merkoçi A (2019) Electrochemical detection of plant virus using gold nanoparticle-modified electrodes. Anal Chim Acta 1046:123–131. https://doi.org/10.1016/j.aca.2018.09.031

Khurana SMP (2006) Detection of plant pathogens : development and applications. 59(1):1–15

Kim J, Kim S, Ju C, Son HIL, Member S (2019) Unmanned aerial vehicles in agriculture : a review of perspective of platform, control, and applications. IEEE Access 7

Klerks MM, Leone G, Lindner JL, Schoen CD, Van Den Heuvel JFJM (2001) Rapid and sensitive detection of apple stem pitting virus in apple trees through RNA amplification and probing with fluorescent molecular beacons. 91(11):1085–1091

Klerks MM, Schoen CD, Thompson JR, Jelkmann W, Spak J (2004) Real-time NASBA for detection of Strawberry vein banding virus, pp 213–221

Koczula KM, Gallotta A (2016) Lateral flow assays. Essays Biochem 60(1):111–120. https://doi.org/10.1042/EBC20150012

Kothari J (2018) Plant disease identification using artificial intelligence: machine learning approach. Artic Int J Innov Res Comput Commun Eng 7(11):11082–11085. [Online]. Available: https://papers.ssrn.com/sol3/papers.cfm?abstract_id=3729753

Kreuze J et al (2022) Innovative digital technologies to monitor and control pest and disease threats in root, tuber, and banana (RT&B) cropping systems: progress and prospects. https://doi.org/10.1007/978-3-030-92022-7_15

Krivitsky V, Granot E, Avidor Y, Borberg E, Voegele RT, Patolsky F (2021) Rapid collection and aptamer-based sensitive electrochemical detection of soybean rust fungi airborne urediniospores. ACS Sensors 6(3):1187–1198. https://doi.org/10.1021/acssensors.0c02452

Kukurba KR, Montgomery SB (2015) RNA sequencing and analysis. Cold Spring Harb Protoc 2015(11):pdb.top084970. https://doi.org/10.1101/PDB.TOP084970

Kurtzman CP, Fell JW, Boekhout T (2011) XanthophyllomycesGolubev

Kus F, Altinkok C, Zayim E, Erdemir S, Tasaltin C, Gurol I (2021) Surface acoustic wave (SAW) sensor for volatile organic compounds (VOCs) detection with calix[4]arene functionalized Gold nanorods (AuNRs) and silver nanocubes (AgNCs). Sens Actuators B Chem 330:129402. https://doi.org/10.1016/j.snb.2020.129402

Lamichhane JR (2015) Bacterial diseases of crops: elucidation of the factors that lead to differences between field and experimental infections, vol 134. Elsevier Ltd. https://doi.org/10.1016/bs.agron.2015.06.006

Le DT, Vu NT (2017) Progress of loop-mediated isothermal amplification technique in molecular diagnosis of plant diseases. Appl Biol Chem 60(2):169–180. https://doi.org/10.1007/s13765-017-0267-y

Lee RF (2015) Control of virus diseases of citrus, 1st edn, vol 91, no. 1. Elsevier Inc. (2015). https://doi.org/10.1016/bs.aivir.2014.10.002

Lees AK, Sullivan L, Lynott JS, Cullen DW (2012) Development of a quantitative real-time PCR assay for Phytophthora infestans and its applicability to leaf, tuber and soil samples. Plant Pathol 61(5):867–876. https://doi.org/10.1111/j.1365-3059.2011.02574.x

Li Z, Yu T, Paul R, Fan J, Yang Y, Wei Q (2020) Agricultural nanodiagnostics for plant diseases: recent advances and challenges. Nanoscale Adv. 2(8):3083–3094. https://doi.org/10.1039/c9na00724e

Li T, Liu B, Spalding MH, Weeks DP, Yang B (2012) High-efficiency TALEN-based gene editing producesdisease-resistant rice

Liang Z, Zhang K, Chen K, Gao C (2014) Targeted mutagenesis in Zea mays using TALENs and the CRISPR/Cas system. J Genet Genomics 41(2):63–68. https://doi.org/10.1016/j.jgg.2013.12.001

Lieber MR (2010) The mechanism of double-strand DNA break repair by the nonhomologous DNA end-joining pathway. Ann Rev Biochem 79:181–211. https://doi.org/10.1146/annurev.biochem.052308.093131

Liefting LW, Waite DW, Thompson JR (2021) Application of oxford nanopore technology to plant virus detection. Viruses 13(8). https://doi.org/10.3390/v13081424

Lim SA, Ahmed MU (2016) Electrochemical immunosensors and their recent nanomaterial-based signal amplification strategies: a review. RSC Adv 6(30):24995–25014. https://doi.org/10.1039/c6ra00333h

Liu Z et al (2022) Internet of things (IoT) and machine learning model of plant disease prediction—blister blight for tea plant. IEEE Access 10:44934–44944. https://doi.org/10.1109/ACCESS.2022.3169147

Liu M et al (2019) Methodologies for improving HDR efficiency. Front Genet 10. https://doi.org/10.3389/fgene.2018.00691

Loens K et al (2006) Detection of rhinoviruses by tissue culture and two independent amplification techniques, nucleic acid sequence-based amplification and reverse transcription-PCR, in children with acute respiratory infections during a winter season. J Clin Microbiol 44(1):166–171. https://doi.org/10.1128/JCM.44.1.166-171.2006

Lone BA, Karna SKL, Ahmad F, Shahi N, Pokharel YR (2018) CRISPR/Cas9 system: a bacterial tailor for genomic engineering. Genet Res Int 2018. https://doi.org/10.1155/2018/3797214

López-Soriano P et al (2017) Lateral flow immunoassay for on-site detection of *Xanthomonas arboricola* pv. *Pruni* in symptomatic field samples. PLoS ONE 12(4):1–13. https://doi.org/10.1371/journal.pone.0176201

Loutfi A, Coradeschi S, Mani GK, Shankar P, Rayappan JBB (2015) Electronic noses for food quality: a review. J Food Eng 144:103–111. https://doi.org/10.1016/j.jfoodeng.2014.07.019

Luo CX, Schnabel G, Hu M, De Cal A (2022) Global distribution and management of peach diseases. Phytopathol Res 4(1). https://doi.org/10.1186/s42483-022-00134-0

Mackay IM (2004) Real-time PCR in the microbiology laboratory. Clin Microbiol Infect 10(3):190–212. https://doi.org/10.1111/j.1198-743X.2004.00722.x

Mahlein A-K (2016) Present and future trends in plant disease detection. Plant Dis 100(2):1–11. https://doi.org/10.1007/s13398-014-0173-7.2

Mahlein A, Kuska MT, Behmann J, Polder G, Walter A (2018) Hyperspectral sensors and imaging technologies in phytopathology: state of the art. Annu Rev Phytopathol 56:535–558

Makarova KS et al (2015) An updated evolutionary classification of CRISPR-Cas systems. Nat Rev Microbiol 13(11):722–736. https://doi.org/10.1038/nrmicro3569

Maliogka VI et al (2018) Recent advances on detection and characterization of fruit tree viruses using high-throughput sequencing technologies. Viruses 10(8):1–23. https://doi.org/10.3390/v10080436

Mankin RW, Hagstrum DW, Smith MT, Roda AL, Kairo MTK (2011) Perspective and promise: a century of insect acoustic detection and monitoring. Am Entomol 57(1):30–44. https://doi.org/10.1093/ae/57.1.30

Marraffini LA, Sontheimer EJ (2010) CRISPR interference: RNA-directed adaptive immunity in bacteria and archaea. Nat Rev Genet 11(3):181–190. https://doi.org/10.1038/nrg2749

Martinelli F et al (2015) Advanced methods of plant disease detection. A review. Agron Sustain Develop 35(1):1–25. https://doi.org/10.1007/s13593-014-0246-1

Mazumdar RM, Chowdhury A, Shahnewaj Bin Mannan K (2012) Nucleic acid sequence based amplification (NASBA)-prospects and applications

McCartney HA, Foster SJ, Fraaije BA, Ward E (2003) Molecular diagnostics for fungal plant pathogens. Pest Manage Sci 59(2):129–142. https://doi.org/10.1002/ps.575

McSorley R et al (1987) Bibliography of estimated crop losses in the united states due to plant-parasitic nematodes. J Nematol 19(1):6–12 [Online]. Available: http://www.pubmedcentral.nih.gov/articlerender.fcgi?artid=2618703&tool=pmcentrez&rendertype=abstract

Mehetre GT et al (2021) Current developments and challenges in plant viral diagnostics: a systematic review. Viruses 13(3):1–31. https://doi.org/10.3390/v13030412

Lopez MM et al (2001) Strategies for improving serological and molecular detection of plant pathogenic bacteria. Plant Pathogenic Bacteria

Mohanty SP, Hughes DP, Salathé M (2016) Using deep learning for image-based plant disease detection. Front Plant Sci 7(September):1–10. https://doi.org/10.3389/fpls.2016.01419

Mohd Ali M, Hashim N, Abd Aziz S, Lasekan O (2020) Principles and recent advances in electronic nose for quality inspection of agricultural and food products. Trends Food Sci Technol 99:1–10. https://doi.org/10.1016/j.tifs.2020.02.028

Moriones E, Praveen S, Chakraborty S (2017) Tomato leaf curl new delhi virus: an emerging virus complex threatening vegetable and fiber crops. Viruses 9(10). https://doi.org/10.3390/v9100264

Morrison D, Rothenbroker M, Li Y (2018) DNAzymes: selected for applications. Small Methods 2(3):1–12. https://doi.org/10.1002/SMTD.201700319

Mortazavi A, Williams BA, McCue K, Schaeffer L, Wold B (2008) Mapping and quantifying mammalian transcriptomes by RNA-Seq. Nat Methods 5(7):621–628. https://doi.org/10.1038/nmeth.1226

Moscou MJ, Bogdanove AJ (2009) A simple cipher governs DNA recognition by TAL effectors. Science 326(5959):1501. https://doi.org/10.1126/science.1178817

Mthethwa NP, Amoah ID, Reddy P, Bux F, Kumari S (2021) A review on application of next-generation sequencing methods for profiling of protozoan parasites in water: current methodologies, challenges, and perspectives. J Microbiol Methods 187:106269. https://doi.org/10.1016/j.mimet.2021.106269

Muniyappa V (2016) K. Subramanya Sastry and Thomas A. Zitter: Plant virus and viroid diseases in the tropics, volume 2: epidemiology and management (Springer, 2014, 489 pp, ISBN 978–94–007–7819–1). Virus Disease 27(1):106–106. https://doi.org/10.1007/s13337-015-0297-1

Murphy L (2006) Biosensors and bioelectrochemistry. Curr Opin Chem Biol 10(2):177–184. https://doi.org/10.1016/j.cbpa.2006.02.023

Musetti R, Loi N, Carraro L, Ermacora P (2002) Application of immunoelectron microscopy techniques in the diagnosis of phytoplasma diseases. Microsc Res Tech 56(6):462–464. https://doi.org/10.1002/jemt.10061

Nakayinga R, Makumi A, Tumuhaise V, Tinzaara W (2021) *Xanthomonas bacteriophages*: a review of their biology and biocontrol applications in agriculture. BMC Microbiol 21(1):1–20. https://doi.org/10.1186/s12866-021-02351-7

Naresh V, Lee N (2021) A review on biosensors and recent development of nanostructured materials-enabled biosensors. Sensors 21(4):1109. https://doi.org/10.1016/j.tibtech.2019.02.005

Nawaz MA et al (2020) Plant disease detection using internet of thing (IoT). 11(1):505–509

Nayagam MG, Vijayalakshmi B, Somasundaram K, Mukunthan MA, Yogaraja CA, Partheeban P (2023) Control of pests and diseases in plants using IOT Technology. Meas Sens 26:100713. https://doi.org/10.1016/j.measen.2023.100713

Norelli JL, Jones AL, Aldwinckle HS (2003) Fire blight management in the twenty-first century: using new technologies that enhance host resistance in apple. Plant Dis 87(7):756–765. https://doi.org/10.1094/PDIS.2003.87.7.756

Nunes D, Pimentel A, Gonçalves A, Pereira S, Branquinho P, Barquinha R, Martins R (2019) Metal oxide nanostructures for sensor applications. Semicond Sci Technol 34(4):043001. [Online]. Available: https://doi.org/10.1088/2053-1583/abe778

Oerke EC (2006) Crop losses to pests. J Agric Sci 144(1):31–43. https://doi.org/10.1017/S0021859605005708

Otieno W (2016) Plantwise: an innovative approach to reduce crop losses by sharing plant health knowledge

Otieno W, Ochilo W, Migiro L, Jenner W, Kuhlmann U (2020) Tools for pest and disease management by stakeholders: a case study on plantwise. 151–174. https://doi.org/10.19103/as.2020.0080.06

Otieno W, Ochilo W, Migiro L (2021) Tools for pest and disease management by stakeholders : a case study on plantwise

Pallás V, Sánchez-Navarro JA, James D (2018) Recent advances on the multiplex molecular detection of plant viruses and viroids. Front Microbiol 9. https://doi.org/10.3389/fmicb.2018.02087

Panno S et al (2020) Plants loop mediated isothermal amplification: principles and applications in plant virology. Plants. https://doi.org/10.3390/plants9040461

Papadakis G, Skandalis N, Dimopoulou A, Glynos P, Gizeli E (2015) Bacteria murmur: application of an acoustic biosensor for plant pathogen detection. PLoS ONE 10(7):1–11. https://doi.org/10.1371/journal.pone.0132773

Perera L (2021) A smart solution for plant disease detection based on IoT a smart solution for plant disease detection based on IoT

Pickar-Oliver A, Gersbach CA (2019) The next generation of CRISPR–Cas technologies and applications. Nat Rev Mol Cell Biol 20(8):490–507. https://doi.org/10.1038/s41580-019-0131-5

Pohanka M (2018) Overview of piezoelectric biosensors, immunosensors and DNA sensors and their applications. Materials (Basel) 11(3). https://doi.org/10.3390/ma11030448

Porto BN et al (2019) Genome sequencing and transcript analysis of *Hemileia vastatrix* reveal expression dynamics of candidate effectors dependent on host compatibility. PLoS ONE 14(4):e0215598. https://doi.org/10.1371/JOURNAL.PONE.0215598

Posthuma-Trumpie GA, Korf J, Van Amerongen A (2009) Lateral flow (immuno)assay: its strengths, weaknesses, opportunities and threats. A literature survey. Anal Bioanal Chem 393(2):569–582. https://doi.org/10.1007/s00216-008-2287-2

Potgieter MHSJ, Arif KM (2019) Plant disease detection and classification by deep learning. Plants 32–34

Powell A (2017) CABI's innovative use of technology, data, and knowledge transfer to reduce crop losses in the developing world. Food Energy Secur 6(3):94–97. https://doi.org/10.1002/fes3.113

Qi Yang LSLL (2019) 2019 IEEE international geoscience and remote sensing symposium: proceedings

Qin D (2019) Next-generation sequencing and its clinical application. Cancer Biol Med 16(1):4–10. https://doi.org/10.20892/j.issn.2095-3941.2018.0055

Rach MM, Gomis HM, Granado OL, Malumbres MP, Campoy AM, Martín JJS (2013) On the design of a bioacoustic sensor for the early detection of the red palm weevil. Sensors (Switzerland) 13(2):1706–1729. https://doi.org/10.3390/s130201706

Rahman M, Uddin M, Sultana R, Moue A, Setu M (2013) Polymerase chain reaction (PCR): a short review

Rahman MS et al (2020) Selected sensor technology innovation in food quality and safety. https://doi.org/10.1007/978-3-030-47166-8_4

Rajesh T, Jaya M (2017) Next-generation sequencing methods. Curr Dev Biotechnol Bioeng Funct Genomics Metab Eng 143–158. https://doi.org/10.1016/B978-0-444-63667-6.00007-9

Redinbaugh MG, Stewart LR (2018) Maize lethal necrosis: an emerging, synergistic viral disease. Annu Rev Virol 5(July):301–322. https://doi.org/10.1146/annurev-virology-092917-043413

Ren CH, Hu CQ, Luo P, Wang QB (2009) Sensitive and rapid identification of *Vibrio vulnificus* by loop-mediated isothermal amplification. Microbiol Res 164(5):514–521. https://doi.org/10.1016/J.MICRES.2008.05.002

Richard A, Kaslow, Stanberry LR, Le Duc JW (2014) Viral infections of humans: epidemiology and control. Viral Infect Humans Epidemiol Control 1–1215. https://doi.org/10.1007/978-1-4899-7448-8

Ricroch A, Clairand P, Harwood W (2017) Use of CRISPR systems in plant genome editing: toward new opportunities in agriculture. Emerg Topics Life Sci 1(2):169–182. https://doi.org/10.1042/ETLS20170085

Rodoni B (2009) The role of plant biosecurity in preventing and controlling emerging plant virus disease epidemics. Virus Res 141(2):150–157

Romney D, Day R, Faheem M, Finegold C, Lamontagne-Godwin J, Negussie E (2013) Plantwise: putting innovation systems principles into practice. Agric Dev 18:27–31

Ryan RP et al (2011) Pathogenomics of Xanthomonas: understanding bacterium-plant interactions. Nat Rev Microbiol 9(5):344–355. https://doi.org/10.1038/nrmicro2558

Savary S, Willocquet L, Pethybridge SJ, Esker P, Mcroberts N, Nelson A (2019) The global burden of pathogens and pests on major food crops. Nat Ecol Evol 3(3):430–439. https://doi.org/10.1038/s41559-018-0793-y

Schornack S, Moscou MJ, Ward ER, Horvath DM (2013) Engineering plant disease resistance based on TAL effectors. Annu Rev Phytopathol 51:383–406. https://doi.org/10.1146/annurev-phyto-082712-102255

Schroeder HW, Cavacini L (2010) Structure and function of immunoglobulins. J Allergy Clin Immunol 125(2 SUPPL. 2):S41–S52. https://doi.org/10.1016/j.jaci.2009.09.046

Scott P, Strange R, Korsten L, Gullino ML (2021) Plant pathology in the 21st century plant diseases and food security in the 21st century. [Online]. Available: www.isppweb

Sergentet-Thévenot D, Montet M, Vernozy-Rozand C (2008) Challenges to developing nucleic acid sequence based amplification technology for the detection of microbial pathogens in food. Undefined

Shahdordizadeh M, Taghdisi SM, Ansari N, Alebooye Langroodi F, Abnous K, Ramezani M (2017) Aptamer based biosensors for detection of *Staphylococcus aureus*. Sens Actuators B Chem 241:619–635. https://doi.org/10.1016/j.snb.2016.10.088

Shan Q, Wang Y, Li J, Gao C (2014) Genome editing in rice and wheat using the CRISPR/Cas system. Nat Protoc 9(10):2395–2410. https://doi.org/10.1038/nprot.2014.157

Shiva RH, Ganesh H, Chinnayan DR (2019) IOT based leaf disease detection and fertilizer recommendation. Int J Innov Technol Explor Eng 9(2):132–136. https://doi.org/10.35940/ijitee.a5228.129219

Shoaib M, Shah B, Ei-sappagh S, Hussain T, Ali F (2023) An advanced deep learning models-based plant disease detection: a review of recent research. 1–22. https://doi.org/10.3389/fpls.2023.1158933

Silverman SK (2005) In vitro selection, characterization, and application of deoxyribozymes that cleave RNA. Nucleic Acids Res 33(19):6151–6163. https://doi.org/10.1093/nar/gki930

Singh A et al (2021) Challenges and opportunities in machine-augmented plant stress phenotyping. Trends Plant Sci 26(1):53–69. https://doi.org/10.1016/j.tplants.2020.07.010

Singh T, Kumar K, Bedi SS (2021) A review on artificial intelligence techniques for disease recognition in plants. IOP Conf Ser Mater Sci Eng 1022(1). https://doi.org/10.1088/1757-899X/1022/1/012032

Sint D, Raso L, Traugott M (2012) Advances in multiplex PCR: balancing primer efficiencies and improving detection success. Methods Ecol Evol 3(5):898–905. https://doi.org/10.1111/j.2041-210X.2012.00215.x

Sivapragasam A, Loke WH (2017) Plantwise-a center for agriculture and bioscience international (CABI)–led extension program for farmers. Mysore J Agric Sci 212–218

Skottrup P et al (2007) Detection of fungal spores using a generic surface plasmon resonance immunoassay. Biosens Bioelectron 22(11):2724–2729. https://doi.org/10.1016/j.bios.2006.11.017

Solaimuthu A, Vijayan AN, Murali P, Korrapati PS (2020) Nano-biosensors and their relevance in tissue engineering. Curr Opin Biomed Eng 13:84–93. https://doi.org/10.1016/j.cobme.2019.12.005

Sorek R, Lawrence CM, Wiedenheft B (2013) CRISPR-mediated adaptive immune systems in bacteria and archaea. Annu Rev Biochem 82:237–266. https://doi.org/10.1146/annurev-biochem-072911-172315

Speaks CE (2017) Introduction to sound: acoustics for the hearing and speech sciences. Plural Publishing

Sperschneider J, Dodds PN, Gardiner DM, Manners JM, Singh KB, Taylor JM (2015) Advances and challenges in computational prediction of effectors from plant pathogenic fungi. PLoS Pathog 11(5):1–8. https://doi.org/10.1371/journal.ppat.1004806

Sperschneider J (2019) Tansley insight Machine learning in plant—pathogen interactions: empowering biological predictions from field scale to genome scale. https://doi.org/10.1111/nph.15771

Stehlíková D, Luchi N, Aglietti C, Pepori AL, Diez JJ, Santini A (2020) Real-time loop-mediated isothermal amplification assay for rapid detection of *Fusarium circinatum*. Biotechniques 69(1):11–17. https://doi.org/10.2144/BTN-2019-0168

Suga H et al (2014) A single nucleotide polymorphism in the translation elongation factor 1α gene correlates with the ability to produce fumonisin in Japanese *Fusarium fujikuroi*. Fungal Biol 118(4):402–412. https://doi.org/10.1016/J.FUNBIO.2014.02.005

Suganya E, Sountharrajan S, Shandilya SK, Karthiga M (2019) Chapter 5—IoT in agriculture investigation on plant diseases and nutrient level using image analysis techniques. Elsevier Inc. https://doi.org/10.1016/B978-0-12-817356-5.00007-3

Sujawata GS, Chouhanb JS (2021) Application of artificial intelligence in detection of diseases in plants: a survey. Turk J Comput Math Educ 12(3):3301–3305. https://doi.org/10.17762/turcomat.v12i3.1581

Sun Y, Wang J, Sun L, Cheng S, Xiao Q (2019) Evaluation of E-nose data analyses for discrimination of tea plants with different damage types. J Plant Dis Prot 126(1):29–38. https://doi.org/10.1007/s41348-018-0193-1

Svetskiy AV (2022) Application of artificial intelligence in agriculture. Сельское Хозяйство 3:1–12. https://doi.org/10.7256/2453-8809.2022.3.39469

Szemes M, Schoen CD (2003) Design of molecular beacons for AmpliDet RNA assay—characterization of binding stability and probe specificity. 315:189–201. https://doi.org/10.1016/S0003-2697(02)00692-9

Tetila EC, Machado BB, Alessandro N, Belete DS, Guimarães DA, Pistori H (2017) Identification of soybean foliar diseases using unmanned aerial vehicle images. IEEE Geosci Remote Sens Lett 1–5

Thakur PS, Khanna P, Sheorey T, Ojha A (2022) Trends in vision-based machine learning techniques for plant disease identification: a systematic review. Expert Syst Appl 208:118117. https://doi.org/10.1016/j.eswa.2022.118117

Thomas S et al (2018) Benefits of hyperspectral imaging for plant disease detection and plant protection: a technical perspective. J Plant Dis Prot. https://doi.org/10.1007/s41348-017-0124-6

Thomas M, René K, Ina HJH, Kaitlin G, Anna MG, Stefan B (2022) Digital plant pathology: a foundation and guide to modern agriculture. J Plant Dis Prot 129(3):457–468. https://doi.org/10.1007/s41348-022-00600-z

Toh SY, Citartan M, Gopinath SCB, Tang TH (2015) Aptamers as a replacement for antibodies in enzyme-linked immunosorbent assay. Biosens Bioelectron 64:392–403. https://doi.org/10.1016/j.bios.2014.09.026

Torrance L, Ziegler A, Pittman H, Paterson M, Toth R, Eggleston I (2006) Oriented immobilisation of engineered single-chain antibodies to develop biosensors for virus detection. J Virol Methods 134(1–2):164–170. https://doi.org/10.1016/j.jviromet.2005.12.012

Trebicki P (2020) Climate change and plant virus epidemiology. Virus Res 286:198059

Tsai H-L, Huang L-C, Ann P-J, Liou R-F (2006) Detection of orchid Phytophthora disease by nestedPCR. Bot Stud 47

Tsui CKM et al (2011) Molecular techniques for pathogen identification and fungus detection in the environment

Twyford AD (2018) Parasitic plants. Curr Biol 28(16):R857–R859. https://doi.org/10.1016/j.cub.2018.06.030

Tylka GL, Marett CC (2021) Known distribution of the soybean cyst nematode, *Heterodera glycines*, in the United States and Canada in 2020. Plant Heal Prog 22(1):72–74. https://doi.org/10.1094/PHP-10-20-0094-BR

Uijthof JMJ, de Hoog GS (1995) PCR-ribotyping of type isolates of currently accepted Exophiala and Phaeococcomyces species. Antonie van Leeuwenhoek 68(1):35–42. https://doi.org/10.1007/BF00873290

Vizzini P, Braidot M, Vidic J, Manzano M (2019) Electrochemical and optical biosensors for the detection of campylobacter and listeria: an update look. Micromachines 10(8). https://doi.org/10.3390/mi10080500

von Arx JA, van der Walt JP, Liebenberg NVDM (1982) The classification of taphrina and other fungi with yeast-like cultural states. Mycologia 74(2):285–296. https://doi.org/10.1080/00275514.1982.12021503

Wahabzada M, Mahlein A, Bauckhage C, Steiner U (2016) Plant phenotyping using probabilistic topic models : uncovering the hyperspectral language of plants. Nat Publ Gr 1–11. https://doi.org/10.1038/srep22482

Wang J (2000) From DNA biosensors to gene chips. Nucleic Acids Res 28(16):3011–3016

Wang Z et al (2012) The development and application of a dot-ELISA assay for diagnosis of southern rice black-streaked dwarf disease in the field. Viruses 4(1):167–183. https://doi.org/10.3390/v4010167

Wang X et al (2019c) A living plant cell-based biosensor for real-time monitoring invisible damage of plant cells under heavy metal stress. Sci Total Environ 697:134097. https://doi.org/10.1016/j.scitotenv.2019.134097

Wang L (2020) Metal-organic frameworks for QCM-based gas sensors: a review. Sens Actuators A Phys 307:111984. https://doi.org/10.1016/j.sna.2020.111984

Wang D, Peng Y (2014) Assessing the impact of dominant sequencing-based gene expression profiling techniques (SGEPTs) on phytopathogenic fungi. [Online]. Available: http://epg.science.cmu.ac.th/ejournal/

Wang Y, Liu J, Zhou H (2019a) Visual detection of cucumber green mottle mosaic virus based on terminal deoxynucleotidyl transferase coupled with DNAzymes amplification. https://doi.org/10.3390/s19061298

Wang Y, Li B, Liu J, Zhou H (2019) T4 DNA polymerase-assisted upgrade of a nicking/polymerization amplification strategy for ultrasensitive electrochemical detection of Watermelon mosaic virus. 2915–2924

Wendt T et al (2013) TAL effector nucleases induce mutations at a pre-selected location in the genome of primary barley transformants. Plant Mol Biol 83(3):279–285. https://doi.org/10.1007/s11103-013-0078-4

Withers S, Gongora-Castillo E, Gent D, Thomas A, Ojiambo PS, Quesada-Ocampo LM (2016) Using next-generation sequencing to develop molecular diagnostics for pseudoperonospora cubensis, the cucurbit downy mildew pathogen. Phytopathology 106(10):1105–1116. https://doi.org/10.1094/PHYTO-10-15-0260-FI

World Health Organization (2019) The state of food security and nutrition in the world 2019: safeguarding against economic slowdowns and downturns. Food & Agriculture Org

Xavier TWF et al (2019) Identification of ramularia leaf blight cotton disease infection levels by multispectral, multiscale UAV imagery. Drones 3(2)

Xiao T, Huang J, Wang D, Meng T, Yang X (2020) Au and Au-based nanomaterials: synthesis and recent progress in electrochemical sensor applications. Talanta 206:120210. https://doi.org/10.1016/j.talanta.2019.120210

Xin XF, Kvitko B, He SY (2018) Pseudomonas syringae: what it takes to be a pathogen. Nat Rev Microbiol 16(5):316–328. https://doi.org/10.1038/nrmicro.2018.17

Xu S, Zhang G, Fang B, Xiong Q, Duan H, Lai W (2019) Lateral flow immunoassay based on polydopamine-coated gold nanoparticles for the sensitive detection of zearalenone in maize. ACS Appl Mater Interfaces 11(34):31283–31290. https://doi.org/10.1021/acsami.9b08789

Yakkundimath R, Saunshi G, Kamatar V (2018) Plant disease detection using IoT. 8(9)

Yang K, Lee I, Nam S (2017) Development of a rapid detection method for Peronospora destructor using loop-mediated isothermal amplification (LAMP). HortScience 52(9S):S1–S497. https://doi.org/10.21273/HORTSCI.52.9S.S1

Yeo SF, Wong B (2002) Current status of nonculture methods for diagnosis of invasive fungal infections. Clin Microbiol Rev 15(3):465–484. https://doi.org/10.1128/CMR.15.3.465-484.2002

Yuan M, Wang S (2013) Rice MtN3/saliva/SWEET family genes and their homologs in cellular organisms. Mol Plant 6(3):665–674. https://doi.org/10.1093/mp/sst035

Yuen J (2021) Pathogens which threaten food security: *Phytophthora infestans*, the potato late blight pathogen. Food Secur 13(2):247–253. https://doi.org/10.1007/s12571-021-01141-3

Zarraonaindia I et al (2015) The soil microbiome influences grapevine-associated microbiota. MBio 6(2). https://doi.org/10.1128/MBIO.02527-14/ASSET/ABDBB043-6786-4B93-9405-1548161CDD1D/ASSETS/GRAPHIC/MBO0021522390004.JPEG

Zezza F, Pascale M, Mulè G, Visconti A (2006) Detection of *Fusarium culmorum* in wheat by a surface plasmon resonance-based DNA sensor. J Microbiol Methods 66(3):529–537. https://doi.org/10.1016/j.mimet.2006.02.003

Zhang N, Yang G, Pan Y, Yang X, Chen L (2020) A review of advanced technologies and development for hyperspectral-based plant disease detection in the past three decades. Remote Sens 12(19):3188

Zhang J, Pu R, Yuan L, Huang W, Nie C, Yang G (2014) Integrating remotely sensed and meteorological observations to forecast wheat powdery mildew at a regional scale. IEEE J Sel Top Appl Earth Obs Remote Sens 7(11):4328–4339

Zhang H, Islam T, de Liu W (2022) Integrated pest management programme for cereal blast fungus *Magnaporthe oryzae*. J Integr Agric 21(12):3420–3433. https://doi.org/10.1016/j.jia.2022.08.056

Zhang Y et al (2013) Rapid and efficient gene modification in rice and brachypodium using TALENs. Mol Plant 6(4):1365–1368. https://doi.org/10.1093/mp/sss162

Zhang J et al (2019) Monitoring plant diseases and pests through remote sensing technology: a review. Comput Electron Agric 165. https://doi.org/10.1016/j.compag.2019.104943

Zhao X, Lin CW, Wang J, Oh DH (2014) Advances in rapid detection methods for foodborne pathogens. J Microbiol Biotechnol 24(3):297–312. https://doi.org/10.4014/jmb.1310.10013

Zhao S, Wang S, Zhang S, Liu J, Dong Y (2018) State of the art: lateral flow assay (LFA) biosensor for on-site rapid detection. Chinese Chem Lett 29(11):1567–1577. https://doi.org/10.1016/j.cclet.2017.12.008

Zhou Y et al (2022) Application of nanopore sequencing in the detection of foodborne microorganisms. Nanomaterials 12(9). https://doi.org/10.3390/nano12091534

Zhuo Z et al (2017) Recent advances in SELEX technology and aptamer applications in biomedicine. Int J Mol Sci 18(10):1–19. https://doi.org/10.3390/ijms18102142

Molecular Diagnostics and Management of Phyto-Parasitic Nematodes

Salah-Eddine Laasli, Najwa Seddiqi Kallali, Ikram Legrifi, Jihane Kenfaoui, Khadija Goura, Fouad Mokrini, Mustapha Barakate, Abdelfattah A. Dababat, and Rachid Lahlali

Abstract Phyto-parasitic nematodes (PPNs) are a significant threat to global agriculture due to their ability to cause substantial crop losses. Effective management of PPNs requires the development and implementation of accurate diagnostic tools and efficient management strategies. Molecular diagnostic tools have emerged as a promising approach to rapidly and accurately identify PPNs and gain insights into their genetic diversity and virulence. In this chapter, the current state of molecular diagnostic tools for PPNs and their application in the management of these pests is reviewed. An overview of the various molecular techniques used for the detection of PPNs, including polymerase chain reaction (PCR)-based techniques, DNA sequencing, and isothermal amplification techniques is provided. The advantages and limitations of each of these tools and case studies demonstrating their application for the identification of specific PPN species are also highlighted. Next, the

S.-E. Laasli · N. S. Kallali · I. Legrifi · J. Kenfaoui · K. Goura · R. Lahlali (✉)
Department of Plant and Environment Protection, Ecole Nationale d'Agriculture de Meknes, Km. 10, Route Haj Kaddour, B.P. S/40, 50001 Meknès, Morocco
e-mail: rlahlali@enameknes.ac.ma

S.-E. Laasli
Crop Protection Unit, Agronomic and Veterinary Institute, IAV-Morocco, Rabat, Morocco

N. S. Kallali · K. Goura
Laboratory of Biotechnology and Valorization of Phyto-Resources, Faculty of Sciences, Moulay Ismail University of Meknes, Avenue Zitoun, 50001 Meknes, Morocco

I. Legrifi · J. Kenfaoui
Laboratory of Functional Ecology and Environmental Engineering, Department of Biology, FST-Fez, Sidi Mohamed Ben Abdellah University, BP 2202, 30050 Fez, Morocco

F. Mokrini
Biotechnology Unit, Regional Center of Agricultural Research, INRA-Morocco, Rabat, Morocco

M. Barakate · R. Lahlali
Plant Pathology Laboratory, AgroBioSciences, Mohammed VI Polytechnic University, Lot 660, Hay Moulay Rachid, 43150 Ben Guerir, Morocco

A. A. Dababat
International Maize and Wheat Improvement Center (CIMMYT), P.O. Box 39, Emek, Ankara 06511, Turkey

J. Chen et al. (eds.), *Molecular and Biotechnological Tools for Plant Disease Management*, Interdisciplinary Biotechnological Advances,
https://doi.org/10.1007/978-981-97-7510-1_3

various molecular techniques used for the management of PPNs, including RNA interference (RNAi), genome editing, and gene silencing are thoroughly discussed. Finally, the challenges and considerations (regulatory and ethical) associated with the development and implementation of molecular diagnostic and management tools for PPNs are highlighted. Molecular diagnostic and management tools have emerged as powerful approaches for the effective management of PPNs. Continued development and improvement of these tools will be critical for reducing the impact of PPNs on global agriculture. The integration of molecular techniques with traditional management strategies has the potential to provide more efficient and sustainable solutions to PPN management.

Keywords Diagnostic tools · Genome editing · Nematodes · PCR-based techniques · RNA interference

1 Introduction

Phyto-parasitic nematodes (PPNs) encompass a diverse array of pathogens that infiltrate a wide crop range, including vegetables, fruits, and cereals (Tileubayeva et al. 2021; Treonis et al. 2018). These minuscule pests engage in intricate interactions with their host plants, impeding nutrient uptake, inciting physiological disorders, and facilitating secondary infections by other pathogens (Aleuy and Kutz 2020; Bernard et al. 2017). Consequently, they inflict substantial harm on crop yield and quality, causing extensive and significant economic losses for farmers (Bozbuga et al. 2018).

The timely detection and accurate identification of PPNs play a crucial role in implementing effective management strategies (Kantor et al. 2022). Early identification empowers growers to enact appropriate control measures before substantial damage occurs (Ahuja and Somvanshi 2021). Furthermore, precise species identification is vital since different PPN species exhibit variations in host range, pathogenicity, and susceptibility to control methods. Erroneous identification may result in ineffective or unnecessary control measures, wastage of resources, and potential exacerbation of the issue (Carneiro et al. 2017).

Although traditional methods of PPN detection and identification, such as visual observation, symptomatology, and soil sampling combined with microscopic examination, exist, they possess limitations in terms of accuracy, time consumption, and sensitivity (de Oliveira et al. 2011). Consequently, molecular diagnostic tools have emerged as promising alternatives (Blok 2005). These tools offer rapid and precise identification of PPNs, facilitating a better understanding of their genetic diversity and virulence; they have revolutionized the field of nematology (Sivasubramaniam et al. 2020). Polymerase chain reaction (PCR)-based techniques, including conventional PCR, nested PCR, and real-time PCR, have been widely employed to detect and quantify PPNs in various substrates such as soil, plant tissues, and nematode cysts (Braun-kiewnick and Kiewnick 2018; Keçici et al. 2022). These techniques utilize specific primers that target conserved regions of the nematode genome, allowing target DNA amplification, which can be subsequently detected using gel electrophoresis or fluorescence-based methods (Subbotin et al. 2018).

DNA sequencing, particularly through the utilization of high-throughput sequencing technologies, has facilitated the accurate identification and characterization of PPNs (Palomares-Rius et al. 2017). By sequencing specific gene regions, such as ribosomal DNA (rDNA) or mitochondrial DNA (mtDNA), researchers can compare obtained sequences with existing databases, enabling species-level identification and phylogenetic classification (Sultana et al. 2013; Rybarczyk-Mydłowska et al. 2012). This approach has significantly enhanced our understanding of PPN diversity and their relationships with different host plants (Palomares-Rius et al. 2017).

Molecular techniques present considerable potential for PPN management (Abd-Elgawad 2024). RNA interference (RNAi), a natural mechanism for gene silencing, has been harnessed as a targeted approach to suppress essential genes in PPNs (Lilley et al. 2012). By introducing double-stranded RNA molecules specific to nematode genes, RNAi disrupts vital physiological processes, leading to the inhibition of nematode development and a subsequent reduction in their pathogenicity and reproductive capabilities (Banerjee et al. 2017). Moreover, genome editing techniques, exemplified by CRISPR-Cas9, enable precise and focused manipulation of the nematode genome (Huang et al. 2023). This powerful tool allows for the modification or disruption of specific genes involved in nematode parasitism, reproduction, or survival (Lok 2019). By selectively targeting genes that are crucial for nematode fitness, it becomes possible to impair their ability to infect plants or limit their reproductive potential, thus mitigating the detrimental impact of PPNs (Sivasubramaniam et al. 2020). Furthermore, gene silencing approaches, including RNAi and small RNA-based technologies, effectively downregulate key genes associated with nematode parasitism (Whangbo and Hunter 2008). By delivering small interfering RNAs (siRNAs) or short hairpin RNAs (shRNAs) that specifically target nematode genes, gene silencing disrupts nematode development, impedes their ability to establish feeding sites, and interferes with their capacity to evade plant defenses (Price et al. 2023).

In this chapter, we provide an overview of the importance of early detection and accurate identification of PPNs, highlighting the essential role played by molecular diagnostic tools in achieving these goals. We discuss the application of molecular techniques to PPN management, focusing on RNA interference (RNAi), genome editing, and gene silencing as promising approaches. In addition, we highlight the limitations and impacts on non-target organisms and the environment associated with each technique.

2 Molecular Diagnostic Tools for PPNs

Swift and accurate detection of PPNs is crucial for the successful implementation of effective management strategies. Traditional methods of PPN diagnosis (Yeates 2009), such as visually examining symptoms could be an exhaustive and labor-intensive approach (Shao et al. 2023). Additionally. These methods often require skilled taxonomists to accurately identify PPN species. Morphological identification can be particularly challenging because closely related species may have similar traits (Shao et al. 2023; Munawar et al. 2023; Almasco et al. 2021). Molecular diagnostic tools have revolutionized the nematode field by providing fast, reliable, and efficient approaches for PPN processing (Goverse and Mitchum 2022). These tools rely on genetic material such as DNA or RNA to identify the PPN type. Advances in molecular biology have paved the way for many technologies that transcend the limitations of traditional methods (Ibrahim et al. 2019a; Montarry et al. 2021; Anjam et al. 2022; Abd-Elgawad 2022).

In this section, we delve into the execution of molecular diagnostic tools for PPNs, entailing PCR-based techniques, DNA sequencing, DNA barcoding, and isothermal amplification techniques. We scrutinize the advantages and constraints of these tools and showcase case studies that highlight their effectiveness in discerning and overseeing specific PPN species. The fusion of these molecular tools into PPN management strategies harbors the immense potential for sustainable and scrupulous control measures.

2.1 PCR-Based Techniques

Polymerase chain reaction (PCR) has revolutionized the field of molecular diagnostics and is widely used for the detection and diagnosis of PPN. PCR-based methods have many advantages, including sensitivity, specificity, and versatility, making them useful for the study and management of PPNs (Saeki et al. 2003; Sudha et al. 2017; Berry et al. 2007).

2.1.1 Conventional PCR

Conventional PCR is a technique that uses the principle of DNA amplification to detect and identify plant parasitic nematodes. PPN is a microscopic roundworm that infects plant roots and causes significant damage to important crops and plants. Association between conventional PCR and PPN in the presence of a gene signature in the nematode genome (Irdani et al. 2008; He et al. 2021a). The principle of conventional PCR involves the amplification of specific DNA sequences using a pair of primers that flank the target region of interest. In the case of PPNs, the primers target the conserved regions within the nematode genome, such as the ribosomal DNA (rDNA) sequences. These regions exhibit a high degree of sequence conservation across different nematode species, while still possessing sufficient genetic variation to distinguish between taxa.

To make PCR samples for PPN detection, DNA is first extracted from existing nematodes (soil, tissue, or other samples). The extracted DNA is then used as a template for PCR reactions with specific primers designed to amplify the conserved region of the nematode DNA. During the PCR amplification cycle, the DNA sample is denatured to separate the two DNA strands, and the primers are annealed to their complementary sequences. DNA polymerase then synthesizes a new DNA strand that extends from the primer and replicates the target DNA region. As the PCR cycles progress, the DNA of the target PPN is amplified, increasing the copy number of the target DNA.

The amplified DNA product is then analyzed using gel electrophoresis, where the DNA fragments are separated by size and visualized under ultraviolet light. By comparing the size of the amplified DNA sequence with the weighted marker, the presence and identity of certain types of PPNs can be determined (Hassan et al. 2022; Sayler et al. 2012; Huang et al. 2017a; Chowdhury and Yan 2021; Carneiro et al. 2017). The relationship between conventional PCR and PPNs is rooted in the principle that the unique DNA sequences within the nematode genome can be amplified using specific primers. These primers ensure efficient PPN characterization, substituting the morphological identification methods that often require laborious and time-consuming microscopic examination (Berry et al. 2007; Sayler et al. 2012). This technique has greatly contributed to the field of PPN research and management demonstrating its immense value. This technique's accuracy, the ability to detect even minute amounts of genetic material, cost-effectiveness, and compatibility with other methods have established it as a crucial component in molecular diagnostics for plant parasitic nematodes (Sayler et al. 2012; Kawanobe and Toyota 2018; Peng et al. 2017).

Conventional PCR has stood as a bedrock technique in PPN research, bestowing a myriad of possibilities that add to the cognition, spotting, and handling of these economically significant pests. This technique empowers the swift and precise discernment of PPN species. Through the concoction of primers tailor-made to hone in on enshrined segments of the nematode genome, researchers can magnify DNA extracted from nematodes retrieved from soil or plant tissues. The ensuing PCR products can subsequently beheld on an agarose gel, facilitating the distinction of PPN species grounded on their distinctive DNA banding arrangements. This technique is pivotal for comprehending the multiplicity and dispersal of PPNs in diverse agricultural systems (Carneiro et al. 2017).

The diagnosis of plant diseases caused by nematodes is artfully achieved through the utilization of Conventional PCR. By extracting the genetic material from ailing plant tissues and submitting it to the scrutiny of PCR analysis, diligent researchers can unveil the existence of pernicious PPNs Heavily damaging crops. This prompt and precise diagnosis facilitates the prompt implementation of control measures, thereby curtailing the potential for grievous crop losses (Subbotin et al. 2005). Moreover, the surveillance of nematode resistance against control measures, such as chemical nematicides or resistant crop varieties is impressively empowered (Atkins et al. 2003).

2.1.2 Real-Time PCR

Real-time PCR, also known as quantitative PCR (qPCR), a promising technique used to identify, track, and quantify plant-parasitic nematodes, proceeds through a series of steps. These steps include sample preparation, primer fabrication, fluorescent probe selection, real-time PCR instrument setup, and data translation (Sayler et al. 2012; Adams 2020).

To get started, DNA must be procured from the nematode specimen. This task entails employing a range of DNA extraction techniques, contingent upon the nature of the sample (e.g., nematode cysts, eggs, or individual nematodes). Subsequently, the extracted DNA is meticulously purified to eradicate any contaminating elements that might impede the PCR reaction. Next, specific primers are typically designed and intended to intensify a distinct DNA sequence corresponding to the nematode species under diagnosis. These primers, comprising concise DNA sequences, adhere to complementary regions of the targeted DNA. In the realm of real-time PCR, a duo of primers is employed, one for each DNA strand. qPCR employs luminous probes to meticulously scrutinize the amplification process as it unfolds in real-time (Arya et al. 2005; He et al. 2021; Žibrat et al. 2021; Toyota et al. 2008). There are two types of luminescent probes commonly used. The TaqMan probe is a hydrolysis probe with a fixed luminescent dye (reporter gene) at the 5′ end and a quencher compound at the 3′ end. During the PCR reaction, the probe is immobilized on the target DNA sequence between the forward and reverse primers. When the DNA is amplified, DNA polymerase separates the probe by removing the emitted dye material. These outcomes in strong detected fluorescence and measured by real-time PCR technology (Ma et al. 2011; Sapkota et al. 2016; Mokrini et al. 2013). On the other hand, SYBR Green is a DNA-binding pigment that intercalates into the double-stranded DNA during the amplification process. As the DNA undergoes amplification, the SYBR Green dye fastens to the freshly synthesized DNA, yielding an augmentation in fluorescence. The intensity of fluorescence is measured at each cycle, providing real-time information on the amplification process (Baidoo and Yan 2021; Berry et al. 2008; Arora et al. 2020).

qPCR necessitates a specialized contraption known as a thermocycler or a qPCR contraption. This contraption is armed with fluorescence detection capacities and precise temperature mastery. The contraption navigates through a chain of temperature transformations, encompassing denaturation, annealing, and extension, to magnify the target DNA. During the qPCR reaction, the contraption ceaselessly scrutinizes the fluorescence signal at each cycle. The fluorescence intensity is gauged and documented, producing a real-time amplification curve. The cycle threshold (Ct) value, denoting the cycle number at which the fluorescence signal surpasses a predetermined threshold, is employed for DNA quantification and qualification in a proper organism. qPCR data analysis also encompasses the utilization of controls to safeguard the accuracy and dependability of the findings (Zhou et al. 2019; Lehnert et al. 2018; Green and Sambrook 2018; Mendoza-Gallegos et al. 2018).

Furthermore, this technique empowers the meticulous quantification of PPN populations in diverse samples, like soil, plant tissues, or nematode extraction suspensions. By employing precise primers and fluorescent probes that target enduring DNA regions of PPNs, researchers can scrutinize the amplification of nematode DNA in real-time throughout the PCR reaction. This grants the ability to ascertain the initial nematode burden in a sample and estimate the size of the population, thereby facilitating studies on nematode dynamics, disease progression, and treatment efficacy (Arora et al. 2020; Christoforou et al. 2014a; Yan et al. 2013). In addition, it endows a remarkably perceptive and expeditious approach to the premature detection and monitoring of PPN infestations. By uncovering minuscule quantities of nematode DNA in samples, such as asymptomatic plants or soil, real-time PCR enables the establishment of early warning systems and timely implementation of management strategies. This proves particularly invaluable for quarantine and biosecurity purposes, as it enables the identification and containment of PPN outbreaks before they unleash substantial devastation (Ma et al. 2011; Mokrini et al. 2013; Toyota 2021).

2.1.3 Multiplex PCR

Multiplex PCR wields immense power as a molecular diagnostic technique, empowering the simultaneous magnification of numerous target DNA sequences through the utilization of multiple primer sets within a solitary reaction. This technique has garnered noteworthy fascination in the realm of plant pathology, encompassing the exploration of PPNs, owing to its capacity to promptly and precisely discern multiple PPN species within a single sample (Devran et al. 2018; Hu et al. 2011).

The principle behind Multiplex PCR is to conceive and integrate multiple sets of primers, each tailored to a distinct PPN species, into a single PCR reaction. These primers are devised to pinpoint peculiar genomic regions within the PPN genomes, enabling the selective magnification of DNA process fragments from diverse PPN species in only reaction tubes. To enhance multiplexing process, every primer set is adorned with a specific fluorophore or detection system that engenders a discernible signal. These fluorophores emit light of disparate wavelengths, which can be detected and distinguished by employing suitable detection systems, such as capillary electrophoresis or fluorescent DNA analyzers. Throughout amplification, the primer sets adhere to their corresponding target DNA sequences within the PPN genomes, and DNA amplification materializes concurrently. The resulting DNA fragments, with divergent lengths corresponding to each PPN species, can be segregated and visualized based on their sizes utilizing gel electrophoresis or capillary electrophoresis. The prevalence of specific DNA fragments signifies the occurrence of corresponding PPN species in the sample (Hu et al. 2011; Madani et al. 2008).

To dazzle and discriminate the magnified DNA fragments derived from various PPN species, fluorophores are frequently employed to label the primers. Each primer ensemble is labeled with a particular fluorophore, radiating a luminosity of a unique wavelength that can be perceived and distinguished with the aid of suitable detection systems (Marek et al. 2010; Dauphinais et al. 2018). Primers in Multiplex PCR often sport fluorophores that emit light of distinct wavelengths, like FAM (6-carboxyfluorescein), HEX (hexachloro-fluorescein), TAMRA (tetramethylrhodamine), or JOE (6-carboxy-4′,5′-dichloro-2′,7′-dimethoxy-fluorescein). These fluorophores are handpicked based on available detection systems and compatibility with the chosen detection method. For the fixation, the linker molecule deftly fastens the fluorophores to the 5′ end of the forward or reverse primers (Nazarenko et al. 2002). This ensures that the fluorophores don't meddle with the primer's binding affinity or amplification efficiency. The crucial step is to validate the Multiplex PCR assay and thwart any cross-reactions or false signals from the labeled primers and fluorophores. This involves testing the assay with known PPN species samples and comparing the observed results against the expected amplicon sizes and fluorescence patterns (Hu et al. 2011; Dauphinais et al. 2018).

Multiplex PCR has transformed PPN investigation by proffering a potent instrument for concurrently detecting, pinpointing, and delineating diverse nematode species. This technique amplifies our comprehension of PPN taxonomy, population kinetics, host particularity, and resistance mechanisms. The utilization of multiplex PCR in PPN research contributes to the formulation of efficacious nematode management strategies, the cultivation of resilient crop varieties, and the establishment of stalwart quarantine measures to safeguard agricultural systems (Edwards et al. 1994).

A recent study conducted by Chen et al. (2022) emphasized the adoption and development of a qPCR assay for the detection of *Meloidogyne enterolobii* in soil samples from banana plantations. However, *M. enterolobii,* a voraciously predatory root-knot nematode, unleashes catastrophic devastation upon banana plantations. Swift identification of this nematode is crucial for the streamlined control and confinement of the affliction. In this study, a qPCR assay was realized to unveil the existence of *M. enterolobii* in soil samples extracted from banana plantations. In effect, soil samples were obtained from the flourishing banana plantations of Hainan, China. DNA extraction from the soil samples was facilitated through a commercial kit. A series of primers and a TaqMan probe were intricately fashioned, drawing inspiration from the internal transcribed spacer (ITS) region of *M. enterolobii.* The qPCR assay underwent meticulous honing to amplify sensitivity and specificity, employing synthetic DNA templates and *M. enterolobii* genomic DNA. The validation of the assay involved the utilization of soil samples infiltrated with conspicuous quantities of *M. enterolobii* eggs. The results of the qPCR assay showcased its prowess in unearthing the DNA of *M. enterolobii* in soil samples, with an astonishing detection threshold of 10 eggs per gram of soil. The assay exhibited unparalleled specificity, free from any cross-reactivity with other plant-parasitic nematodes frequently stumbled upon in banana plantations. The validation experiment successfully discerned *M. enterolobii* in soil samples tainted with a mere 10 eggs per gram of soil.

2.2 *DNA Sequencing*

DNA sequencing, a scientific technique of paramount importance, uncovers the complex arrangement of nucleotides within a DNA molecule. This revolutionary method has sparked a revolution in the field of genetics, permeating various fields of research, including the study of PPNs. The latter, minuscule stealthily infiltrate plants, damaging crops, inflicting substantial harm. Gaining insight into the genetic composition of these PPNs proves indispensable in formulating efficacious strategies to curtail their dissemination and mitigate the havoc they wreak. DNA sequencing emerges as an important tool in unraveling the genetic information harbored by these organisms. A panoply of DNA sequencing techniques exists, each bearing its unique advantages and limitations. In this discourse, we will explore two frequently employed methods: Sanger sequencing and Next-Generation Sequencing (NGS) (Kikuchi et al. 2017; Sultana et al. 2013; Shendure et al. 2011; Garrido-Cardenas et al. 2017; Mohammadi 2022; Dey 2018).

2.2.1 Sanger Sequencing

Sanger sequencing, commonly called Chain termination sequencing, has been an important method in the domain of molecular biology and genetics research for numerous decades. This technique rests upon the integration of dideoxynucleotides (ddNTPs) during the DNA synthesis process (Dey 2018; Valencia et al. 2013). In Sanger sequencing, the DNA template of interest submits an initial amplification using the polymerase chain reaction (PCR) with specially designed primers that include the target region. The PCR products then submit a sequencing reaction, where a combination of ordinary deoxynucleotides (dNTPs), fluorescently labeled dideoxynucleotides (ddNTPs), DNA polymerase, and sequencing primers are introduced. As the sequencing reaction unfolds, the DNA polymerase randomly incorporates ddNTPs instead of the corresponding dNTPs, leading to chain termination and resulting in an amalgamation of DNA fragments of assorted lengths. The fragments are cleaved asunder, their sizes are distinguished through capillary gel electrophoresis, and the fragments marked with fluorescent labels are revealed through the illumination of lasers. The outcome is an electropherogram that unveils the DNA sequence, with the peaks illustrating the succession of nucleotides in the desired expanse (Besnard et al. 2014; Martin et al. 2008; Wylie et al. 2004; Vossenberg et al. 2022).

2.2.2 NGS: Next-Generation Sequencing

Next-Generation Sequencing (NGS), as its name implies, wields a potent technology that embodies the subsequent phase in the metamorphosis of DNA sequencing. In contrast to the Sanger sequencing technique, which sequences a single DNA fragment at a time, NGS is an immensely efficient technology with the capacity to sequence millions of DNA fragments at once. It has revolutionized genomics research, facilitating expeditious and economical sequencing of entire genomes (Abbasi and Masoumi 2020; Mahesh Dahake 2021). The NGS's modus operandi has a sequence of steps:

At first the extraction of DNA from the specimen of interest. The extracted DNA is subsequently fragmented into tinier fragments. These fragments are affixed to distinctive adapters on each terminus, which are employed in subsequent phases of the sequencing procedure (Besnard et al. 2014; Yang and Kim 2018). The adapted DNA fragments are escalated in a process recognized as clonal amplification. This can be accomplished either in solution (as observed in the instance of Illumina platforms) or on a solid surface like beads (as seen in Ion Torrent platforms). The outcome is a library of DNA template clusters, each encompassing numerous copies of an unparalleled DNA fragment from the original sample (Besnard et al. 2014; Fantini et al. 2017; Parkinson et al. 2012).

The primed library is loaded onto a sequencing platform. The sequencing process relies on fabricating a fresh DNA strand that complements the template strand. With each assimilated nucleotide, a distinctive signal is engendered and perceived by the sequencing machine. The specific signal nature hinges on the sequencing platform: for instance, Illumina platforms employ fluorescent signals, whereas Ion Torrent platforms apprehend pH fluctuations (Besnard et al. 2014; Mahesh Dahake 2021; Yang and Kim 2018). The signals stemming from the sequencing process are transmuted into a string of nucleotide bases (A, T, C, G), thus yielding crude sequence scans. These scans are subsequently arranged and amalgamated employing intricate bioinformatics tools to forge an all-encompassing genome sequence. Further scrutiny can detect genes, mutations, and sundry genomic traits (Pabinger et al. 2014; Korneliussen et al. 2014; Horner et al. 2009; Scholz et al. 2012).

The NGS bestows a sweeping and all-encompassing lens to scrutinize the existence of plant parasitic nematodes, allowing for profound revelations into their intricate biology, ecological dynamics, and entanglements with their host plants (Mahesh Dahake 2021). The important specific applications of NGS to unfurl in the realm of plant parasitic nematodes. NGS holds the power to unravel the entirety of plant parasitic nematodes' genetic blueprint. This technology can uncover priceless insights into the genetic constitution of these organisms, potential virulence factors, and genes intertwined with host specificity and resistance to nematicides (Danchin and Perfus-Barbeoch 2009; Abad et al. 2008). Moreover, transcriptome analysis or RNA sequencing is also empowered, opening windows for researchers to fathom the genes that come alive during diverse stages of the nematode life cycle or amid varying environmental circumstances. It proves especially valuable in scrutinizing gene expression transformations throughout host infection (Jacob and Mitreva 2011; Siddique et al. 2022; Maboreke et al. 2016).

Interestingly, scientists can delve into the intricate realm of nematodes thriving within soil samples (the nematode metagenome). This process unravels the diversity and profusion of plant parasitic nematodes in various agricultural environments, ultimately facilitating the creation of more potent strategies for pest management

(Li et al. 2014; Zakeel and Safeena 2020; Tian et al. 2015; Gendron et al. 2023). Furthermore, the advent of NGS technology empowers the exploration of epigenetic modifications within nematodes, which possess the ability to influence gene expression without altering the underlying DNA sequence. This endeavor provides invaluable insights into how these modifications may contribute to the nematode's remarkable adaptability to diverse hosts and environments (Chatterjee et al. 2018; Hewezi et al. 2017; Pratx et al. 2018; Gao et al. 2014).

2.2.3 Sanger and NGS Sequencing

Studying PPNs can be an arduous task, so it's crucial to explore the many facets of Sanger sequencing and NGS. Sanger sequencing, with its precision and focused sequencing, proves itself adept at scrutinizing specific genes and validating familiar mutations. Meanwhile, NGS beckons with its ability to embark on through high-throughput sequencing, unraveling genetic mysteries with de novo sequencing and conducting sweeping genomic analysis to unearth cryptic genetic variations and structural anomalies (Fridman et al. 2021; Pareek et al. 2011). While researchers should take into account their research goals, financial constraints, and available resources in choosing between these sequencing techniques, it's worth noting that a fusion of both methodologies might be the most potent strategy to adopt.

2.3 DNA Barcoding

DNA barcoding is a molecular technique that employs specific DNA sequences in specific genome regions for species identification and classification. It has been suggested as a good diagnostic method for many diseases, especially in taxa that do not have a morphological diagnosis. DNA barcoding has been investigated for various nematode groups, including marine nematodes, and root-knot nematodes. The COI (mitochondrial cytochrome c oxidase subunit I) gene was discovered as a barcode marker in *C. elegans*. DNA barcoding is a pertinent diagnosis tool, in important condition to maintain linkage of DNA barcoding, morphological and topological location sequence, as exemplified in many misidentified types of GenBank (Köhler 2007; Li et al. 2015; Mwamula et al. 2020).

Standardized regions of the genome are chosen to employ as a barcoding DNA platform. The COI gene is frequently utilized, but alternative regions might be applicable contingent on the taxonomic group under scrutiny. DNA samples are procured from the specimens of interest. This may entail extracting DNA from tissue samples, whole organisms, or environmental samples. The DNA barcode region is boosted through PCR approaches. This encompasses utilizing specific primers that zero in on the DNA barcode region and amplifying it to generate numerous copies of the DNA fragment. The amplified DNA fragment is sequenced using Sanger sequencing NGS technologies. This will yield a DNA sequence that can be employed for species identification and classification. DNA barcoding has revolutionized the identification of parasitic plant nematodes, providing a fast and accurate method for species determination. The technique offers significant advantages over traditional morphological approaches, allowing the identification of PPNs at different life stages and from complex samples. The following table presents an overview of some species of PPNs identified by DNA barcoding (Li et al. 2015; Mwamula et al. 2020; Powers 2004; Summerbell et al. 2005).

DNA barcoding has metamorphosed the PPN identification aspects, offering sundry advantages such as expeditious and high-throughput identification, premature detection, and standardized protocols. Nevertheless, it is crucial to contemplate the restraints linked with intraspecific variability, primer biases, hybridization, and meager reference databases. By recognizing these restraints and harnessing the advantages, DNA barcoding can substantially augment our comprehension of PPNs and bolster efficacious management strategies for these economically paramount plant pathogens.

Janssen et al. (2017) demonstrated the power of DNA barcoding in a study aimed at identifying different species in *Pratylenchus penetrans* populations. To find morphological and species-specific links, the ever-changing polygenic phylogeny of the nematode populations was deftly reconstructed using mitochondrial and ribosomal gene sequences. This study also revealed a high cryptic species diversity number embodied in the *Pratylenchus* genus, implying that standalone morphological-based identification may fail for taxonomically confusing groups. The fusion of molecular and morphological data has yielded remarkable advancements in the identification and categorization of species within the *Penetrans* group. A groundbreaking manifestation of this amalgamation lies in the use of a distinctive barcode via a COI gene. By showcasing the pivotal synergy between molecular and morphological analyses, this research underscores the paramount significance of DNA barcoding in the precise delineation and classification of nematodes. Moreover, it accentuates the vast potential of this method as an influential instrument for safeguarding biodiversity, fortifying biosecurity, and implementing effective pest control measures.

2.4 Isothermal Amplification Techniques

Isothermal amplification techniques (IATs) have gained attention as derivatives of PCR for PPN detection due to their advantages in terms of simplicity, rapidity, and robustness under isothermal conditions (Bodulev and Sakharov 2020). Unlike PCR, which requires temperature cycling, isothermal amplification methods allow for DNA amplification at a constant temperature, simplifying the workflow and reducing the need for sophisticated equipment (Boonbanjong et al. 2022).

The loop-mediated isothermal amplification (LAMP) has been the most established IAT approach for efficient PPN detection (Table 1). Based on the principle of low-temperature (60–65 °C) DNA amplification, LAMP uses specially designed primers to identify multiple target sequences in the nematode genome (Ahuja and Somvanshi 2021).

In LAMP, a comprehensive set of six oligonucleotide primers is employed. This set comprises two outer primers referred to as F3 and B3, two inner primers known as BIP and FIP, and two loop primers denoted as LF and LB (Tomita et al. 2008; Gadkar et al. 2018). These primers possess the ability to recognize multiple target sequences within the PPN genome, ensuring a high level of specificity and amplification efficiency (Ahuja and Somvanshi 2021). The inner primers consist of two important regions complementary to each other and opt to bind to the DNA in-target. The inclusion of loop primers is vital for initiating and expediting the amplification process, leading to a twofold increase in speed (Qianru et al. 2018). The loop primers have an important part in facilitating the formation of stem-loop DNA structures, which serve as templates for subsequent DNA synthesis and amplification

Table 1 LAMP technique utilization cases to phyto-parasitic nematode detection

Genus name	Nematode species	Target region	Host	References
Meloidogyne	*M. incognita, M. javanica, M. arenaria,* and *M. hapla*	ITS of rDNA	Tomato	Niu et al. (2011)
	M. mali	TS-5.8S rDNA		Zhou et al. (2017)
	M. enterolobii	5S rDNA-IGS2		Niu et al. (2012)
	M. chitwoodi and M. fallax	IGS2-18S		Zhang and Gleason (2018)
	M. partityla	ITS-5.8S rDNA	Mature pecan trees	Waliullah et al. (2020)
Bursaphelenchus	*B. xylophilus*	ITS-rDNA	Pine	Kikuchi et al. (2009)
	B. xylophilus	Pectate lyase-3		Kang et al. (2015)
	B. xylophilus	ITS-rDNA	*Pinus armandii* var	Kanetani et al. (2011)
	B. cocophilus	D2-D3 of rDNA	Coconut and oil palm trees	Ide et al. (2017)
Tylenchulus	*T. semipenetrans*	ITS1	Citrus rhizosphere soil	Song et al. (2017)
	T. semipenetrans	ITS-rDNA	Citrus orchards	Lin et al. (2016)
Ditylenchus	*D. destructor*	28S rRNA	Potato	Deng et al. (2019)
Radopholus	*R. similis*	D2-D3 of rDNA	Anthurium	Peng et al. (2012)
Anguina	*A. agrostis*	ITS rDNA	–	Yu et al. (2020)
	A. wevelli	ITS rDNA	–	Yu et al. (2018)

(Ding et al. 2019). The LAMP reaction progresses through several steps, including strand displacement, DNA synthesis, and DNA annealing. The continuous displacement of the synthesized DNA strands ensures a rapid and exponential amplification of the target DNA (Shirshikov and Bespyatykh 2022).

Recombinase-based amplification (RPA) is also a promising technique for isothermal amplification (Daher et al. 2016; Ju et al. 2019). Its potential in detecting PPN has been demonstrated across various sample types, including soil, plant tissues, and nematode cysts (Yao et al. 2021). RPA utilizes the principles of strand displacement and recombinase-mediated recombination to achieve the amplification of target DNA sequences under isothermal conditions. Key components involved in RPA include a recombinase enzyme, a single-stranded DNA-binding protein (SSB), a strand-displacing DNA polymerase, and specific primers (Lobato and Sullivan 2017).

The recombinase enzyme plays a vital role in initiating the amplification process by facilitating the recognition and binding of the DNA template, resulting in the formation of a presynaptic filament (Kojima et al. 2021). Designed to specifically complement the desired DNA sequence, specific primers anneal to the presynaptic filament with high specificity. The strand-displacing DNA polymerase then displaces and extends the annealed DNA strands in a strand-displacement manner, utilizing the displaced strands as templates for further amplification. This process of DNA synthesis leads to the extension of the 3′ ends of the displaced strands, ultimately generating double-stranded DNA products (Syeda et al. 2014). The efficiency of this DNA synthesis step enables rapid and exponential amplification of the target DNA. In RPA, the amplification process is cyclic, allowing for repeated rounds of primer binding, strand displacement, and DNA synthesis. This continuous amplification significantly increases the amount of target DNA within a short period, showcasing the efficiency of RPA as an amplification technique (Daher et al. 2016).

The applications of isothermal amplification techniques in the detection of PPNs have demonstrated great potential for various aspects of PPN management. An exemplary case is the utilization of LAMP in detecting the root-knot nematode *Meloidogyne* spp. (Niu et al. 2012; Zhang and Gleason 2018). Specific LAMP assays have been developed to target genomic regions of *Meloidogyne* spp., enabling rapid and accurate identification of this economically significant nematode genus. As mentioned by Niu et al. (2011), a LAMP assay targeting the internal transcribed spacer (ITS) region of ribosomal DNA was successfully employed to specifically detect *M. incognita, M. arenaria, M. javanica*, and *M. hapla* in infected plant roots and soil samples. The LAMP assay exhibited high sensitivity, capable of detecting as few as one second-stage juvenile per reaction. The results obtained from the LAMP assay were comparable to conventional PCR but with added advantages of simplicity, speed, and visual result interpretation. This case study highlights the potential of LAMP as a powerful tool for on-site detection and surveillance of *Meloidogyne* spp., assisting in the management of this destructive PPN (Niu et al. 2011).

Another case study involves the application of RPA in the detection of the cyst nematode *Globodera pallida*, which causes significant damage to potato crops (Wang et al. 2022). In a study by Yao et al. (2021), an RPA assay targeting a specific DNA sequence derived from the mitochondrial genome of *G. pallida* was developed. The

RPA assay exhibited high sensitivity and specificity, detecting *G. pallida* DNA in soil samples with as few as 10 juveniles. Additionally, the RPA assay demonstrated tolerance to inhibitory substances commonly found in soil, such as humic acid, suggesting its robustness in complex sample matrices. The results obtained from the RPA assay were consistent with conventional PCR but with the advantage of rapid detection within 30 min. This case study showcases the potential of RPA as a valuable tool for the rapid and sensitive detection of *G. pallida* in soil samples, facilitating early detection and targeted control measures (Yao et al. 2021).

3 Molecular Techniques for PPN Management

Molecular techniques have undergone significant advancements in recent years and have strong parts in the management of PPNs in agricultural systems. These techniques encompass extensive tool streams and approaches that effectively detect and characterize PPN species, as well as the elucidation of their genetic and molecular mechanisms. The release of new technologies (e.g. NGS) has revolutionized PPN-based research by unlocking comprehensive patterns of their transcriptomes, metagenomes, and genomes (Castagnone-Sereno et al. 2013; Filipiak et al. 2021).

The omics approaches facilitated by high-throughput technologies have ensured the discovery of novel targets for PPN management. For instance, researchers have identified genes involved in nematode parasitism and host response, providing potential targets for RNA interference (RNAi)-based control strategies (Pentimone et al. 2020). RNAi, a gene silencing mechanism has shown promise in suppressing PPN populations by delivering specific RNA strands that target essential genes or regulatory pathways (Atkinson et al. 2012), while the genome editing approach has demonstrated efficacy against various PPN species and holds potential as a sustainable and environmentally friendly method for nematode control.

3.1 Gene Silencing

3.1.1 RNA Interference (RNAi)

RNA interference (RNAi) is a naturally occurring, highly conserved mechanism of gene regulation that has been harnessed as a powerful tool for gene silencing in various organisms, including nematodes (Iqbal et al. 2016). The discovery of RNAi has not only revolutionized our understanding of gene regulation but has also provided a powerful technological tool for structural and functional genomic studies and the development of novel therapeutic strategies (Long et al. 2009).

Advancements in the application of RNAi for nematode management have been greatly facilitated by the identification of key target genes crucial to the biology and parasitic behavior of PPNs. These target genes encompass various aspects of nematode life, including development, reproduction, and parasitism (Tamilarasan and Rajam 2013; Rosso et al. 2009). By selectively silencing these genes, researchers have successfully induced a diverse range of phenotypic effects in PPNs (e.g. *M. incognita*, *Heterodera glycines*, and *G. pallida*), such as developmental arrest, reduced fertility, impaired host invasion, and enhanced host resistance (Dalzell et al. 2011; Chen et al. 2015; Kumar et al. 2022; Li et al. 2023) (Table 2). For instance, by suppressing the CLE-like genes in *M. incognita*, researchers observed a significant decline in nematode reproduction and an increase in host plant resistance (Kumar et al. 2022). Additionally, the introduction of non-nematode-derived double-stranded RNAs (dsRNAs) targeting conserved genes in *G. pallida* and *M. incognita* resulted in noteworthy phenotypic alterations, including reduced motility, abnormal development, and impaired infectivity (Dalzell et al. 2011; Li et al. 2023). Another study targeting a C-type lectin gene in *H. glycines* through RNAi-mediated silencing led to a substantial decrease in nematode infection and reproduction in soybean plants (Jacob 2009).

3.1.2 RNAi Mechanisms in Managing PPNs

RNAi has been seen as a vital technological tool in plant protection against PPNs, with its mechanism playing an integral role in the process. This mechanism involves the PTGS (post-transcriptional gene silencing) of specific genes in the nematodes, which effectively disrupts their ability to function properly or survive. RNAi operates by generating siRNAs (small interfering RNAs) that match the sequence of targeted nematode genes. These siRNAs then orientate RISC (RNA-induced silencing complex) to the mRNA, where it binds and prevents the mRNA from being translated into protein, thereby silencing the gene. The effectiveness of RNAi in controlling plant-parasitic nematodes (PPNs) is well-documented, with several studies showing its potential for crop protection against these destructive pests (Sharma et al. 2023).

The RNAi potential as an eminent control aspect against PPNs has been further explored through the implementation of host-based silencing platforms. For instance, HIGS (host-induced gene silencing) involves the expression of specific RNAi triggers in host plants, inducing gene silencing in the nematodes that consume the plant tissue. This approach has shown significant potential in enhancing plant resistance to nematodes, with the RNAi mechanism effectively reducing nematode infection and reproduction. In addition, research is being conducted to identify key genes in nematodes that can be targeted for silencing, with some success being seen in the silencing of genes participated in nematode reproduction, and parasitism. RNAi response in PPNs has been well-documented, with a similar mechanism also observed in plants (Gangwar and Neeraj 2023). RNAi autonomously functions in mammalian cells, spreading completely to neighboring cell materials and various tissues in both plants and nematodes. Researchers have identified an RNAi-deficient (sid) locus in *C. elegans*, which plays a key role in transmitting the silencing signal between cells. This

Table 2 Target genes and phenotypic effects of RNAi in phyto-parasitic nematodes

Target gene	Nematode species	Gene function	Phenotypic effect of RNAi	References
CLE-like genes	*Meloidogyne incognita*	Regulate cell differentiation and migration	Disrupted development of feeding structures	Kumar et al. (2022)
C-type lectin	*Heterodera glycines*	Involved in pathogenesis and immune evasion	Reduced nematode reproduction and infectivity	Jacob (2009)
Mitogen-activated protein kinase (MAPK)	*Heterodera avenae*	Key signaling molecule in stress response	Decreased hatching and reduced pathogenicity	Chen et al. (2015)
Chitinase	*Radopholus similis*	Degrades chitin in the nematode cuticle	Impaired molting and reduced population density	Urwin et al. (2002)
Glutathione S-transferase (GST)	*Bursaphelenchus xylophilus*	Detoxifies xenobiotics and protects against oxidative stress	Increased susceptibility to oxidative stress and reduced reproduction	Li et al. (2023)
Neprilysin-like protein	*Pratylenchus penetrans*	Involved in host-parasite interaction	Reduced nematode invasion and damage to host roots	Roderick et al. (2016)
Peroxiredoxin	*Heterodera schachtii*	Protects against oxidative stress	Impaired survival and reduced pathogenicity	Perry and Maule (2004)
Aquaporin	*Rotylenchulus reniformis*	Regulates water movement in the nematode	Reduced nematode motility and impaired development	Habash et al. (2017)
Serine protease inhibitor (SPI)	*Bursaphelenchus xylophilus*	Inhibits host defense responses	Enhanced host defense and reduced nematode fitness	Redding et al. (2018)

(continued)

Table 2 (continued)

Target gene	Nematode species	Gene function	Phenotypic effect of RNAi	References
Heat shock protein 90 (HSP90)	*Meloidogyne javanica*	Chaperone protein involved in stress response	Decreased hatching, delayed development, and reduced pathogenicity	Wen et al. (2023)
Expansin-like protein	*Heterodera avenae*	Facilitates cell wall extension and loosening	Altered feeding behavior and reduced nematode reproduction	Lourenço-Tessutti et al. (2015)
RNA-dependent RNA polymerase (RdRP)	*Globodera rostochiensis*	Involved in RNAi and antiviral defense	Reduced nematode growth and reproduction	Liu et al. (2016)
Fatty acid- and retinol-binding protein (FAR)	*Meloidogyne javanica*	Involved in nutrient acquisition and lipid metabolism	Reduced nematode survival and development	Sharma et al. (2022)
Pectate lyase	*Meloidogyne graminicola*	Degrades pectin in the plant cell wall	Impaired migration and reduced population density	Iberkleid et al. (2013)
Superoxide dismutase (SOD)	*Ditylenchus destructor*	Scavenges reactive oxygen species	Increased susceptibility to oxidative stress and reduced survival	Chen et al. (2021)
Mitogen-activated protein kinase (MAPK)	*Bursaphelenchus xylophilus*	Key signaling molecule in stress response	Reduced nematode motility and altered feeding behavior	Solo et al. (2021)
Cysteine protease	*Heterodera schachtii*	Involved in the degradation of host proteins	Impaired nematode feeding and reduced pathogenicity	Jacob (2009)

(continued)

Table 2 (continued)

Target gene	Nematode species	Gene function	Phenotypic effect of RNAi	References
Acetylcholinesterase (AChE)	*Radopholus similis*	Breaks down acetylcholine neurotransmitter	Impaired nematode movement and reduced population density	Modesto et al. (2022)
NADH dehydrogenase	*Globodera pallida*	Part of the mitochondrial respiratory chain	Decreased energy production and impaired development	Li et al. (2023)
Vacuolar ATPase	*Heterodera glycines*	Involved in acidification of intracellular compartments	Impaired nematode digestion and reduced reproduction	Rosana et al. (2016)
Excretory-secretory (ES) protein	*Radopholus similis*	Modulates host immune response	Reduced nematode pathogenicity and altered host defense	Marmonier et al. (2019)
Prolyl endopeptidase	*Heterodera glycines*	Involved in protein digestion and processing	Impaired nematode feeding and reduced reproduction	Austin et al. (2019)
Tyrosine phosphatase	*Bursaphelenchus xylophilus*	Regulates phosphorylation and signaling	Altered nematode development and reduced pathogenicity	Huang et al. (2017)
Choline kinase	*Pratylenchus penetrans*	Involved in phospholipid biosynthesis	Reduced nematode reproduction and impaired development	Noon et al. (2015)
Collagenase	*Radopholus similis*	Degrades collagen in host tissues	Impaired nematode migration and reduced pathogenicity	Dutta et al. (2015)

(continued)

Table 2 (continued)

Target gene	Nematode species	Gene function	Phenotypic effect of RNAi	References
Calcium channel	*Meloidogyne hapla*	Mediates calcium influx and signaling	Disrupted nematode development and reduced reproduction	Neupane et al. (2019)
Pyruvate dehydrogenase complex	*Globodera rostochiensis*	Involved in energy production	Reduced nematode mobility and impaired reproduction	Cui et al. (2018)
Actin-binding protein	*Heterodera glycines*	Regulates cytoskeletal dynamics	Impaired nematode movement and reduced pathogenicity	Tian et al. (2016)
Histone methyltransferase	*Meloidogyne incognita*	Modifies histones and regulates gene expression	Altered nematode development and reduced reproduction	Dalzell et al. (2011)
Proteasome subunit	*Bursaphelenchus xylophilus*	Part of the proteasome complex for protein degradation	Impaired nematode growth and reduced reproduction	Chung et al. (2014)
Neuropeptide Y-like receptor	*Heterodera avenae*	Regulates neuropeptide signaling	Disrupted nematode behavior and reduced reproduction	Martin et al. (2009)
Nucleotide excision repair protein	*Radopholus similis*	Repairs DNA damage	Reduced nematode survival and impaired reproduction	Zhang et al. (2015)
Serotonin receptor	*Meloidogyne graminicola*	Regulates serotonin signaling	Altered nematode behavior and reduced reproduction	Hada et al. (2021)

locus encodes a protein, SID1, that features 11 transmembrane domains, indicating it might work as an import–export channel or potentially as a potential receptor (Chen and Rechavi 2022). In most PPNs, their slightly alkaline gut lumen hinders the quick digestion of nucleic acids. There are three ways to induce RNAi in these nematodes: feeding them bacteria that produce dsRNA, orally administering dsRNA from a solution, or through microinjection (Menezes et al. 2022). Once inside the nematodes, the dsRNA spreads systemically, leading to in-cell gene silencing that possesses an identical sequence material of mRNA. If PPNs are involved, RNAi phenotypes can be inherited for at least two generations, a behavior that could help control evolution (Menezes et al. 2022). There is evidence that mutations in some *C. elegans* genes contribute to RNAi susceptibility in different tissues. This may be a good outcome for plant nematodes because the integration of *rrf-3* with target genes can result in efficient RNAi.

Despite this potential, a major challenge remains: The small size of parasite-infected nematodes poses a formidable challenge for dsRNA microinjection. These nematodes generally do not drink water until they enter the plant (Dalzell et al. 2011; Bairwa et al. 2022). Even better is the recent discovery of the success of using FITC (fluorescent isothiocyanate) in dsRNA solutions as a perspective indicator to monitor health and select individuals for future experiments (Zhao et al. 2022).

In a remarkable twist, scientists conducted RNAi experiments targeting specific transcripts of cysteine proteinases in PPNs, which unexpectedly altered their sexual fate, favoring the production of males at a few post-infection days. Astonishingly, this manipulation did not decrease the overall number of PPNs successfully infecting plants. Further investigations delved into the effects of subjecting *H. glycines* to dsRNA corresponding to a gene of unknown function, which not only influenced sexual destiny but also resulted in a significant 41% reduction in the number of retrieved nematodes (Sharma et al. 2013). Intriguingly, dsRNA-based treatment targeting the MSP (main sperm protein) protein did not exhibit any discernible impact on nematode development or sexual destiny. However, a recent analysis revealed a notable decrease in mRNA and MSP transcripts upon the attainment of sexual maturity by males (Joshi et al. 2020). These findings highlight the intricate interplay between gene expression, sexual differentiation, and parasitic behavior, providing fascinating insights into the complex mechanisms underlying nematode biology and potential avenues for future research.

In more recent research, an efficient FITC uptake was observed in *M. incognita* when soaked, with more than 90% of individuals consuming the dye material (Hada et al. 2023). To examine the impact of targeting the dual oxidase enzyme, researchers fed *M. incognita* juveniles with dsRNA derived from the gene encoding this enzyme. The results revealed a notable decrease in the size and number of female nematodes implemented at 14 and 35 days post-infection. Moreover, overall egg production declined by approximately 70% compared to the control group (Hada et al. 2023). Another noteworthy achievement in RNAi experiments involved a chitin synthase gene in *M. artiellia*. Researchers successfully induced RNAi by soaking the developing eggs of the nematode in dsRNA targeting this gene (Fanelli et al. 2005; Vasanthrao et al. 2023). The phenotypic effects resulting from the RNAi treatment persisted for up to five weeks after the initial exposure. Interestingly, dsRNA timing delivery did not appear to be quite important for the effectiveness of the RNAi effect, as demonstrated by the delayed expression of a specific gene called MSP. This finding holds significant implications for the potential application of RNAi *in planta*, suggesting that precise timing may not be essential for inducing the desired RNAi effect.

3.1.3 *In Planta* Potential of Gene Silencing

The use of RNAi technology in combatting phyto-parasitic nematodes depends largely on the in-plant expression and consumption of a dsRNA that aligns with a crucial parasite gene (Sindhu et al. 2009). Even though this consumption could occur during the invasion phase, it's believed that going after key genes after feeding is more beneficial (Senthil-Kumar and Mysore 2011). The end goal is to initiate an RNAi phenotype that could cause significant harm or even death to the parasite (Liu et al. 2021).

Several factors need to be considered in this process. A partial effect might hinder the quality of plant resistance against nematodes. It's worth noting that the RNAi process is different in mammals as compared to nematodes (Liu et al. 2021; Knox et al. 2007). Both plants and nematodes have RNA polymerases amplifying siRNA, triggering a systemic RNAi response (Geldhof et al. 2006). Therefore, simply feeding nematodes with appropriate dsRNA can activate a cells-based RNA effect (Bakhetia et al. 2005). Generally, dsRNA occurred in a specific niche where the targeted gene resides, directly leading to specific phenotypic aspects of RNAi. In that manner, the dsRNA destiny in a plant lacking a target gene remains a mystery (Zhang et al. 2012).

A significant question to answer is whether a total knockout can be achieved for PPNs (Eves-van den Akker et al. 2014). The ingestion of siRNA or dsRNA by implemented PPNs requires their traversal through the feeding tube (Bakhetia et al. 2005) (Fig. 1). This tube functions as a filter at a molecular level, and it's not clear whether all plant nematodes can effectively consume dsRNA (Zhang et al. 2012). In *H. schachtii*, the feeding tube allows subunits of 20 kDa but blocks those of 40 kDa (Dalzell et al. 2010). *G. pallida* limits the uptake of GFP, while some nematode species (e.g. *M. incognita*) consume it without any problems (Kakrana et al. 2017). Therefore, these nematodes with no feeding tube production are the preferred organisms for initial testing (Gheysen and Vanholme 2007).

The linear structure of A-type RNAs has an estimated maximum angle of 26 Å, regardless of their length. The linear form is more common than the one in which single-stranded RNA can be unwrapped as a base pair duplex (Dubreuil et al. 2009). Therefore, this type of RNA can be withdrawn from the feeding tube for a long time (Dutta et al. 2015). However, siRNAs can potentially be isolated in the condition of a specific protein complex (RISC) formation before uptake. In that sense, some siRNAs can be recruited while not bound to the RISC complex (Dubreuil et al. 2009).

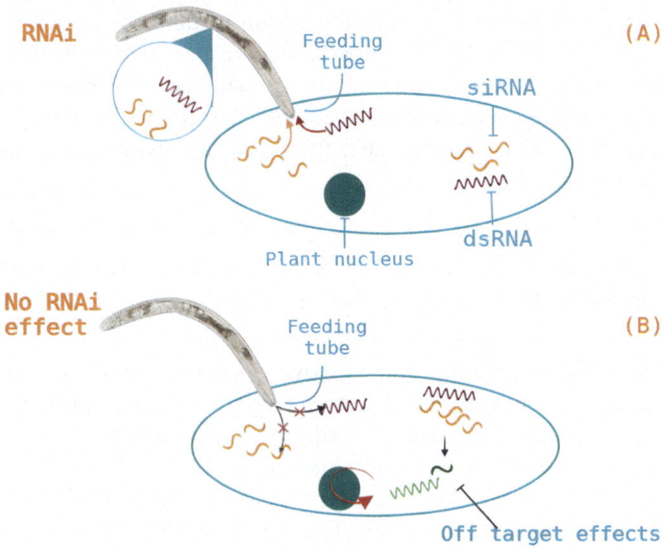

Fig. 1 Ingestion patterns of RNAi in phyto-parasitic nematodes. **a** Normal siRNA and dsRNA ingestion; **b** no ingestion causing off-target effects. Adapted from Bakhetia et al. (2005)

3.1.4 Efficacy, Durability, and Biosafety

It is generally believed that the efficacy of RNAi hinges on the nematode's capacity to counterbalance the loss of a target gene's functionality. Investigations involving *C. elegans* have revealed that RNAi can trigger lethal outcomes, implying that analogous genes may serve as top-priority targets for similar examinations with PPNs (Tamilarasan and Rajam 2013). The integration of several transgenes into a solitary construct has been proven to fortify resistance (Dutta et al. 2015), similar to the concurrent expression of dsRNA from various unrelated genes (Fuller et al. 2008). This implies that a multifaceted dsRNA that targets two or more plant parasitic nematode genes could accomplish superior efficacy. The RNAi operation is based on homology, which means a single construct could hypothetically diminish the expression of diverse genes (Marques et al. 2022). This paves the way for the creation of enduring resistance against plant parasitic nematodes by multiple gene processing (Joshi et al. 2022).

From a biosafety standpoint, RNAi-mediated resistance boasts various benefits. For example, dsRNA cannot generate a functional protein, which should limit unintended effects. This presents a considerable advantage over other proteins with transgenic nature, such as *Bacillus thuringiensis*' (endotoxins (Bt)), presently utilized to manage pests (mainly insects) in certain crops. Moreover, dsRNA is not a novel component in the human diet, as it's frequently consumed in virus-infected crop plants. This hints that an RNAi-based plant defense could be a biosecure method, although additional research is required for confirmation. The employment of predominantly active promoters occurring within feeding cells could provide an extra safety layer, particularly for migratory endoparasites such as *R. similis* which do not stimulate feeding cells (Joshi et al. 2022). The use of dsRNA that doesn't impact other invertebrates when consumed could also safeguard non-target species. Nonetheless, it's imperative to verify that the dispensed dsRNA does not trigger off-target effects on any plant tissues.

3.2 Genome Editing

Recent advancements in genome editing have opened up new possibilities for targeting PPNs and developing resistant cultivars. By directly inserting isolated genes into the genomes of desired organisms, breeders now have a potentially sustainable tool to effectively manage PPNs (Sivasubramaniam et al. 2020) For instance, the key to potential host plant resistance lies in accurately classifying the species of root-knot nematodes (RKNs) since resistance is effective only against specific races or species (Bali et al. 2021). In tomato, various resistance gene platforms to different plant pathogens have already been transferred successfully, resulting in reduced disease incidence (Shilpa et al. 2022).

Genetic engineering has made significant progress thanks to the arrival of CRISPR/Cas9, a revolutionary technology in plant breeding research. This advanced tool has greatly transformed both fundamental and applied studies in this field (Sivasubramaniam et al. 2020). CRISPR/Cas9 functions by combining tracrRNA and crRNA chimera type (sgRNA), which directs the Cas9 nuclease to create a site-specific double-strand break (DSB) in the target DNA sequence (Culp et al. 2020).

The field of genome editing is rapidly gaining momentum, and engineered nucleases play a crucial role in streamlining the process. Among the primary nucleases used for this technology are effector-based nucleases and CRISPR systems. These two techniques are widely employed in modern applications due to their effectiveness (Lo et al. 2013). With ongoing advancements, gene pyramiding and trait stacking in various plants, such as soybean, hold immense potential for achieving durable resistance. A dedicated CRISPR/Cas9 genomic platform for soybean has been established, promising to facilitate future studies and overcome associated challenges (Kang 2016) (Fig. 2). Researchers have identified several key genes in various PPN species that serve as potential targets for CRISPR-Cas9 intervention. Among these, the CLE genes found in *Meloidogyne* spp. have been highlighted in nematode-feeding site formation within host plants (Vieira and Gleason 2019). Disrupting these signal peptides may impair the establishment of feeding sites, thereby reducing nematode virulence. In *Heterodera* spp., heat shock proteins (HSP genes) are critical for the nematode's response to environmental stress. By targeting these genes, researchers aim to impair the nematode's ability to cope with adverse conditions, making it more susceptible to environmental challenges.

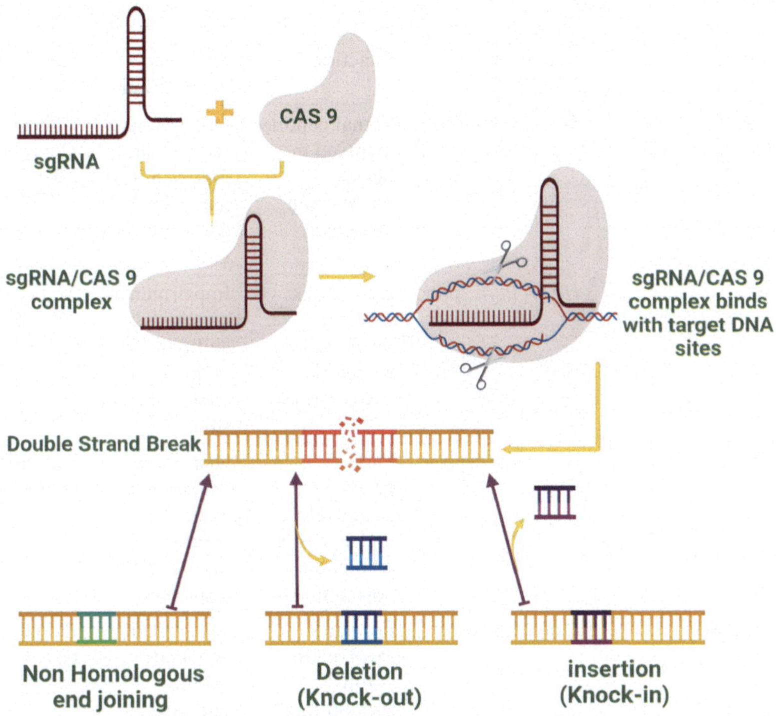

Fig. 2 Mechanism of gene editing by CRISPR-Cas9 technique. Adapted from Negi et al. (2022)

Another crucial target is the effector genes secreted by *Globodera* spp. These parasitic proteins manipulate the host plant's immune responses, facilitating nematode infestation. By using CRISPR-Cas9 to interfere with effector gene expression, researchers seek to thwart the nematode's ability to exploit plant defenses (Karibasappa et al. 2021). Chitinase genes, present in *Radopholus* spp., are essential for nematode cuticle formation. Disruption of these genes could compromise the integrity of the nematode cuticle, leading to reduced survival and infectivity (Parmar et al. 2017). In *Pratylenchus* spp., GST (glutathione-S-transferase) genes have an important part in detoxifying plant-produced toxins, enhancing the nematode's immune system against plant defenses. By targeting GST genes, scientists aim to disrupt this detoxification mechanism, increasing the nematode's susceptibility to plant toxins (Rajput et al. 2021) (Table 3).

Additionally, researchers have identified cystatin genes in *Heterodera* spp. These genes encode protease inhibitors that interfere with nematode digestion. CRISPR-Cas9-based knockout of cystatin genes could inhibit the nematode's ability to break down host plant proteins for nutrition. Furthermore, a pivotal gene known as 16D10 in *Meloidogyne* spp. has been identified, which is essential for nematode parasitism. Targeting this gene with CRISPR-Cas9 technology results in impaired nematode infectivity and reduced pathogenicity (Banerjee et al. 2017).

Table 3 List of targeted genes of plant-parasitic nematodes by CRISPR-Cas9 approach

Targeted Gene	Nematode Species	Function	Potential Application	References
CLE genes	*Meloidogyne* spp.	Signal peptides involved in nematode feeding site formation	Disruption of feeding site establishment, reducing nematode virulence	Vieira and Gleason (2019)
HSP genes	*Heterodera* spp.	Heat shock proteins involved in nematode stress response	Impairment of nematode's ability to cope with environmental stress	You et al. (2022)
Effector genes	*Globodera* spp.	Parasitic proteins secreted by nematodes to manipulate host defenses	Disruption of nematode's ability to manipulate plant immune responses	Karibasappa et al. (2021)
Chitinase genes	*Radopholus* spp.	Enzymes involved in chitin metabolism, important for nematode cuticle formation	Impairment of nematode cuticle integrity, reducing survival and infectivity	Parmar et al. (2017)
Glutathione-S-transferase (GST) genes	*Pratylenchus* spp.	Detoxification enzymes involved in nematode defense against plant-produced toxins	Disruption of nematode's detoxification mechanisms, increasing susceptibility to plant defenses	Rajput et al. (2021)

(continued)

Table 3 (continued)

Targeted Gene	Nematode Species	Function	Potential Application	References
Cystatin genes	*Heterodera* spp.	Protease inhibitors that interfere with nematode digestion	Inhibition of nematode's ability to break down host plant proteins for nutrition	Fosu-Nyarko and Jones (2015)
Parasitism gene (16D10)	Meloidogyne spp.	Essential gene involved in nematode parasitism	Knockout of this gene results in impaired nematode infectivity and reduced pathogenicity	Banerjee et al. (2017)

3.3 Advantages and Limitations

The stream of gene silencing and genome editing technologies, such as RNA interference (RNAi) and CRISPR-Cas9, respectively, has revolutionized the field of plant parasitic nematode management. Firstly, gene silencing through RNAi provides a targeted approach to nematode control. By silencing essential nematode genes, it is possible to disrupt their reproduction cycle and reduce their infectivity to host plants (Pandey et al. 2022). This method is advantageous as it is highly specific, reducing the risk of off-target effects that can occur with traditional chemical pesticides. Secondly, genome editing technologies like CRISPR-Cas9 offer the potential for durable resistance to plant parasitic nematodes. By editing the host plant's genome to enhance its natural defense mechanisms or to disrupt the nematode's infectivity, it is possible to create varieties that are inherently resistant to nematode infection (O'Halloran 2019). This approach is advantageous as it can potentially potential solutions to nematode problems, decreasing the need for repeated pesticide applications.

Despite these advantages, there are also limitations following gene silencing and genome editing in managing plant parasitic nematodes. A major challenge with RNAi-based methods is the variability in effectiveness. Not all nematode species are equally susceptible to RNAi, and even within a single species, different individuals may respond differently to the same RNAi treatment (Dalzell et al. 2011). This variability can make it difficult to achieve consistent control across a population. In terms of genome editing, one of the main limitations is the potential for off-target effects. While CRISPR-Cas9 is generally considered to be highly specific, there is still a risk of unintended edits occurring, which could potentially have negative impacts on the host plant (Zhang and Gleason 2018) Furthermore, both methods face regulatory and public acceptance challenges. The use of genetic technologies in agriculture is a contentious issue, with concerns about potential environmental and health impacts. This can limit the adoption of these technologies, regardless of their potential benefits for nematode control (Wolt et al. 2016).

4 Challenges and Future Directions

Currently, the majority of laboratories throughout the world are using molecular approaches to identify plant-parasitic nematodes because the costs of reagents and equipment of some molecular techniques are affordable, moreover, there is an increasing curiosity about molecular taxonomy among young scientists (Oliveira 2011; Nega 2014). These techniques have been utilized in the field and are highly sensitive to detecting single nematodes from challenging materials such as species mixtures and soil samples (Randig et al. 2002; Yan et al. 2008). Nematode genome sequencing is generating an abundance of genetic data, providing the exploration of comparative genomes and the identification of target DNA regions that can be utilized as diagnostic markers (Carneiro et al. 2017).

Certainly, molecular diagnostics is a powerful tool for identifying plant parasitic nematodes, however, there are several challenges associated with this approach. Usually, the main challenges are the problems encountered during the optimization and validation of protocols and methods in the field of nematode genome sequencing. DNA extraction methods, sample conditions, the quantity of targeted DNA in a sample, contamination from other sources, and false positive and negative results, all of which need to be utilized with prudence to prevent compromising diagnosis (Rothman 2004).

Over several decades, the implementation of PCR-based approaches using distinct primers in PPN diagnostics has highlighted several issues (Rothman 2004). Firstly, PCR multiplies DNA from both live and dead organisms (MacMillan et al. 2006; Christoforou et al. 2014). Accordingly, this limits the application of the previous technique for estimating the effectiveness of pesticide treatments on parasite populations. To solve this problem, an alternative approach known as reverse transcriptase (RT)-PCR can be utilized. This technique includes extracting mRNA from live nematodes. The extracted mRNA is then transformed into complementary DNA using reverse transcriptase (RT), and this cDNA is subsequently amplified and quantified using PCR (Leal et al. 2013). By focusing on mRNA, which is only present in live organisms, this method allows for a more accurate estimation of the nematode population and the efficacy of pesticide applications (Perry and Subbotin 2007). Furthermore, Leal et al. (2013), have established a similar approach to RT-PCR called reverse transcription loop-mediated isothermal amplification (RT-LAMP), a test that detects an amount of mRNA as a survival indicator to selectively identify live Pine Wood Nematode (PWN) in wood. When compared to PCR, this method of diagnosis approach has been proven to be more precise, quicker, more cost-effective, and enables simpler visual detection.

The second limitation of molecular methods is the dependence on DNA-based techniques, which might result in false negatives when nematodes are not actively feeding on plant roots (Ye 2012). Although it is recommended to use a conserved sequence area for primer design in PCR diagnostics, it is also necessary to be aware of the possibility of mutations within this area in certain desired species. As a result, these changes may render the target sequences undetectable by the PCR assay (Perry and Subbotin 2007). This can be a particular issue in situations where nematodes are present in low quantity or are not actively causing visible damage to the plant (Subbotin 2021).

The third disadvantage of this approach, which is the opposite of the above-mentioned one, is the possibility of a false-positive result. Primer design constantly depends on available data about DNA sequences for the targeted species and closely associated nematodes (Borst et al. 2004). Meanwhile, analogous fragments might be found for a previously unstudied nematode as well as the species of interest. Moreover, during the preparation of samples, contamination may occur, as a result, the high sensitivity of PCR might give a false-positive reaction (Irdani et al. 2008). Therefore, all diagnostic procedures require rigorous guidelines for contamination avoidance in the PCR mixture process. These restrictions suggest that, initially, the PCR approach needs to be used with caution, and the advisor or researchers ought to be conscious of the unique characteristics of each approach. In addition, when in doubt, it is inevitably essential to validate the identity using various methods, including classic morphological criteria (Perry and Subbotin 2007).

Another factor to consider is the process of extraction, which highly relies on various indicators, namely the nematode stage of life (egg, juvenile, adult), the anticipated quantity of nematodes in a sample, the age and preparation of the sample (aged, fresh, frozen), the robustness of the amplification reaction (purified nucleic acid or crude nucleic acid), and the origin material being extracted (root extracts, soil extracts, plant macerated material) (Berry et al. 2007). Taking these factors into account ensures that adequate methods for extraction are used to establish high-quality nucleic acids for precise and accurate PCR diagnostics (Abd-Elgawad 2021). However, there is currently little knowledge concerning the extraction and quantification of nematode genomic nucleic acid.

Indeed, PPN morphological identification is puzzling due to their small size, inconsistent distribution in soil, and the difficulties in identifying their key traits utilizing a light microscope (Castagnone-sereno et al. 2011). Another comparable difficulty has been noted with the use of molecular diagnostics, the detection of positional homology being an important obstacle to their application in reconstructing phylogeny between taxa (Abebe et al. 2011).

The most common illustration of this limitation is the case of the *M. incognita* group (MIG), also known as tropical root-knot nematodes. For the latter, the traditional barcoding genes (e.g. 18S rRNA) might not furnish the requisite in species identification. This could be attributable to diverse factors, encompassing the existence of intragenomic variations, which signify genetic variations within the same individual or genome (Hugall et al. 1999). In such confusing situations, researchers actively seek alternative mitochondrial genes to serve as barcodes. Mitochondrial genes, with their variable traits, are frequently used in DNA barcoding to aid in species characterization (Janssen et al. 2016). Furthermore, Dorris et al. (1999), suggested that when DNA data are utilized alone, there is a possibility of encountering noise and artifacts, which may occur during manipulation, data collection, analysis, or interpretation. Additionally, collecting and analyzing molecular data requires a large amount of personal effort, and employing only DNA-similarity searches using BLAST might produce misleading findings. In the case of PPN, the presence of intra- or interspecific thresholds can be perplexing. Many different taxon species can have identical score features, whereas the exact species might be extremely distinct between taxa groups (Qing et al. 2020). Overall, morphological identification is sometimes challenging owing to the flexibility of phenotypes and the lack of key identification aspects, while the molecular aspect is extremely complex because of the poor database and complicated genetic background. In this regard, it has been suggested that morphological and molecular techniques are not supposed to be viewed as opposing but instead as complementary, to confirm the result of taxonomical identification and eventually lead to new insight (Subbotin et al. 2005).

Thereby, to strengthen the influence of molecular diagnostics and management measures, it is critical to understand and overcome implementation obstacles. Since these molecular approaches need a large laboratory establishment, highly trained employees, and specific equipment to accomplish a successful identification and diagnosis. These issues limit their direct application in field and point-of-care settings for the timely identification of PPNs. Moreover, it is crucial to accentuate the comprehension and expertise needed for the rational interpretation of molecular data, therefore reducing errors to insure precise findings (Ahuja et al. 2021). Especially given the fact of the incorporation of molecular diagnostics into existing management practices, which needs careful planning and coordination to improve both efficiency and efficacy (Ibrahim et al. 2019). Therefore, future directions for molecular diagnostics and PPN control should incorporate advancements in technology and tools. Investigation of new technologies, like CRISPR-based methods, has the potential to clarify the complexity of PPN genomes, discover important virulence components, and gain insight into resistance mechanisms (Rajput et al. 2021; El-Sappah et al. 2019).

Additionally, DNA barcoding and direct sequencing of nucleic acids employing next-generation sequencing technologies are becoming increasingly popular in the diagnosis of PPN (Braun-Kiewnick and Kiewnick 2018). Both approaches are expensive, but they can provide exact identification of nematodes while additionally improving the link between classical and molecular classification of PPN. The development of high-throughput and affordable diagnostic tools will improve accessibility and enable quick on-site detection of PPNs, allowing for early and informed management decisions (Blok 2005). This collection of sequence data, together with increases in computational ability and methodological advancements, now provides new options for the study of comparative genomics (Habash et al. 2017). In the coming years, there is bound to be an immense requirement for developing an assay based on new technologies that can be utilized in the field by specialists to observe pathogen invasions in soil or plants (Gasser et al. 2008). HTS is difficult to utilize, and its implementation in the field or farm may be challenging.

After all, developing sample collection, preservation, and storage procedures would increase the reliability and consistency of molecular diagnostic tools, allowing for more accurate predictions of PPN populations, distribution, and management strategies (Abd-Elgawad 2020). Furthermore, the data generated by robust developing technologies will need an increasing number of database developments in the future. As a result, sharing access to datasets and analytical pipelines in bioinformatics should be extensively promoted (Seesao et al. 2017). In summary, it is important to recognize that each method employed in plant-parasitic nematode management and diagnosis has merits and drawbacks. No solitary approach can provide a comprehensive solution alone. Therefore, the combination of various molecular and phenotypic tools, as well as available data from DNA sequencing are pivotal to enhancing the benefits of each method while minimizing their limitations and accomplishing sustainable management of plant-parasitic nematodes.

5 Conclusion

Molecular diagnostics techniques, such as PCR and NGS, have greatly enhanced our ability to accurately identify and characterize PPN species. By analyzing the DNA or RNA of nematodes, these techniques enable researchers to differentiate between species, determine their genetic diversity, and track their population dynamics. Furthermore, the application of NGS has allowed for the comprehensive exploration of nematode genomes and transcriptomes, shedding light on their complex biology and potential vulnerabilities.

Gene silencing technologies, particularly RNAi, have shown promise in controlling PPNs by disrupting essential genes in these pests. RNAi triggers the degradation of specific mRNA, effectively silencing the target gene and impairing nematode development and reproduction. The successful application of RNAi in reducing nematode infestation and crop damage has opened up new possibilities for sustainable and environmentally friendly PPN management strategies.

Looking ahead, the development and refinement of molecular techniques hold several prospects in PPN research and management. These include:

(i) Improved diagnostics: Advances in molecular diagnostics will continue to enhance our ability to accurately identify and characterize PPN species. The development of more efficient and cost-effective techniques, such as loop-mediated isothermal amplification (LAMP) or digital PCR, will allow for rapid and on-site detection of nematodes, aiding in surveillance and monitoring efforts.

(ii) Targeted control strategies: Further exploration of the nematode genome and transcriptome through NGS technologies will uncover new potential target genes for gene silencing. By identifying and disrupting essential genes involved in nematode virulence, reproduction, or behavior, we can develop specific and effective RNAi-based control strategies tailored to different PPN species.

(iii) Delivery systems optimization: Continued research into the optimization of delivery systems for dsRNA molecules will enhance the efficiency and practicality of gene silencing approaches. Investigating alternative delivery methods, such as nanoparticles or viral vectors, may offer more targeted and efficient delivery of dsRNA to nematodes, reducing off-target effects and increasing overall control efficacy.

(iv) Integration of molecular techniques with other control methods: The integration of molecular techniques with other management strategies, such as cultural practices, biological control agents, or chemical treatments, holds promise for developing integrated pest management (IPM) programs. Combining molecular diagnostics for accurate nematode identification with gene silencing approaches and other control methods can provide a multi-faceted and sustainable approach to PPN management.

In conclusion, the future of PPN research and management lies in the continued development and application of molecular techniques. The advancement of molecular diagnostics and gene silencing technologies will enable more accurate nematode identification, improved understanding of nematode biology, and the development of targeted and sustainable control strategies. By harnessing the power of molecular techniques, we can pave the way for effective nematode management, ultimately enhancing agricultural productivity and contributing to global food security.

Acknowledgements This study was financially supported by all author's institutes listed above.

References

Abad P, Gouzy J, Aury JM, Castagnone-Sereno P, Danchin EGJ, Deleury E, Perfus-Barbeoch L, Anthouard V, Artiguenave F, Blok VC et al (2008) Genome sequence of the metazoan plant-parasitic nematode meloidogyne incognita. Nat Biotechnol 26. https://doi.org/10.1038/nbt.1482

Abbasi S, Masoumi S (2020) Next-generation sequencing (NGS). Int J Adv Sci Technol 29. https://doi.org/10.1007/978-981-16-1037-0_23

Abd-Elgawad MMM (2020) Can rational sampling maximise isolation and fix distribution measure of entomopathogenic nematodes? Nematology 22:907–916. https://doi.org/10.1163/15685411-00003350

Abd-Elgawad MMM (2021) Optimizing sampling and extraction methods for plant-parasitic and entomopathogenic nematodes

Abd-Elgawad MMM (2022) Understanding molecular plant–nematode interactions to develop alternative approaches for nematode control. Plants 11

Abd-Elgawad MM (2024) Upgrading strategies for managing nematode pests on profitable crops. Plants 13(11):1558

Abebe E, Mekete T, Thomas WK (2011) A critique of current methods in nematode taxonomy. Afr J Biotechnol 10:312–323. https://doi.org/10.5897/AJB10.1473

Adams G (2020) A beginner's guide to RT-PCR, QPCR and RT-QPCR. Biochemist 42. https://doi.org/10.1042/bio20200034

Ahuja A, Somvanshi VS (2021) Diagnosis of plant-parasitic nematodes using loop-mediated isothermal amplification (LAMP): a review. Crop Prot 147:105459

Ahuja A, Kushwah J, Mathur C, Chauhan K, Dutta TK, Somvanshi VS (2021) Identification of galtox, a new protein toxin from photorhabdus bacterial symbionts of heterorhabditis nematodes. Toxicon 194:53–62. https://doi.org/10.1016/j.toxicon.2021.02.011

Aleuy OA, Kutz S (2020) Adaptations, life-history traits and ecological mechanisms of parasites to survive extremes and environmental unpredictability in the face of climate change. Int J Parasitol Parasites Wildl 12:308–317

Almasco A, Yabut AJ, Zamora MJ, Alvarez LV, Mapanao CP, Rendon AO, Zurbano LY, Bellere AD (2021) Morphological identification of plant parasitic nematodes found in roots and soil of pomelo (*Citrus maxima* (J. Burm.) Merr.). Int J Agric Technol 17

Anjam MS, Siddique S, Marhavy P (2022) RNA isolation from nematode-induced feeding sites in arabidopsis roots using laser capture microdissection. In: Methods in molecular biology, vol 2494

Arora D, Yan G, Baidoo R (2020) Developing a real-time PCR assay for direct detection and quantification of *Pratylenchus scribneri* in field soil. Nematology 22. https://doi.org/10.1163/15685411-00003336

Arya M, Shergill IS, Williamson M, Gommersall L, Arya N, Patel HRH (2005) Basic principles of real-time quantitative PCR. Expert Rev Mol Diagn 5

Atkins SD, Clark IM, Sosnowska D, Hirsch PR, Kerry BR (2003) Detection and quantification of *Plectosphaerella cucumerina*, a potential biological control agent of potato cyst nematodes, by using conventional PCR, real-time PCR, selective media, and baiting. Appl Environ Microbiol 69. https://doi.org/10.1128/AEM.69.8.4788-4793.2003

Atkinson HJ, Lilley CJ, Urwin PE (2012) Strategies for transgenic nematode control in developed and developing world crops. Curr Opin Biotechnol 23(2):251–256

Austin HW, McNeece BT, Sharma K, Niraula PM, Lawrence KS, Klink VP (2019) An expanded role of the SNARE-containing regulon as it relates to the defense process that Glycine max has to *Heterodera glycines*. J Plant Interact 14(1):276–283

Baidoo R, Yan G (2021) Developing a real-time PCR assay for direct identification and quantification of soybean cyst nematode, *Heterodera glycines*, in soil and its discrimination from sugar beet cyst nematode, *Heterodera Schachtii*. Plant Dis 105. https://doi.org/10.1094/PDIS-01-21-0129-RE

Bairwa A, Venkatasalam EP, Subhash S, Dipta B (2022) Management of potato cyst nematodes (Globodera Spp.) using biotechnological approaches. In: Genetic methods and tools for managing crop pests. Springer Nature Singapore, Singapore, pp 343–360

Bakhetia M, Charlton W, Atkinson HJ, McPherson MJ (2005) RNA interference of dual oxidase in the plant nematode *Meloidogyne incognita*. Mol Plant-Microbe Interact 18(10):1099–1106

Bali S, Zhang L, Franco J, Gleason C (2021) Biotechnological advances with applicability in potatoes for resistance against root-knot nematodes. Curr Opin Biotechnol 70:226–233

Banerjee S, Banerjee A, Gill SS, Gupta OP, Dahuja A, Jain PK, Sirohi A (2017) RNA interference: a novel source of resistance to combat plant parasitic nematodes. Front Plant Sci 8:834

Bernard GC, Egnin M, Bonsi C (2017) The impact of plant-parasitic nematodes on agriculture and methods of control. Nematol-Concepts Diag Control 121–151

Berry S, Fargette M, Morand S, Cadet P (2007) Reliability of PCR-based techniques for detection and discrimination of plant-parasitic nematodes of sugarcane. Nematology 9. https://doi.org/10.1163/156854107781487297

Berry SD, Fargette M, Spaull VW, Morand S, Cadet P (2008) Detection and quantification of root-knot nematode (*Meloidogyne javanica*), lesion nematode (*Pratylenchus zeae*) and dagger nematode (*Xiphinema elongatum*) parasites of sugarcane using real-time PCR. Mol Cell Probes 22. https://doi.org/10.1016/j.mcp.2008.01.003

Besnard G, Jühling F, Chapuis É, Zedane L, Lhuillier É, Mateille T, Bellafiore S (2014) Fast assembly of the mitochondrial genome of a plant parasitic nematode (*Meloidogyne graminicola*) using next generation sequencing. C R Biol 337. https://doi.org/10.1016/j.crvi.2014.03.003

Blok VC (2005) Achievements in and future prospects for molecular diagnostics of plant-parasitic nematodes. Can J Plant Path 27:176–185. https://doi.org/10.1080/07060660509507214

Bodulev OL, Sakharov IY (2020) Isothermal nucleic acid amplification techniques and their use in bioanalysis

Boonbanjong P, Treerattrakoon K, Waiwinya W, Pitikultham P, Japrung D (2022) Isothermal amplification technology for disease diagnosis. Biosensors (Basel) 12:677

Borst A, Box ATA, Fluit AC (2004) False-positive results and contamination in nucleic acid amplification assays: suggestions for a prevent and destroy strategy. Eur J Clin Microbiol Infect Dis 23:289–299

Bozbuga R, Lilley CJ, Knox JP, Urwin PE (2018) Host-specific signatures of the cell wall changes induced by the plant parasitic nematode, Meloidogyne Incognita. Sci Rep 8:17302

Braun-Kiewnick A, Kiewnick S (2018) Real-time PCR, a great tool for fast identification, sensitive detection and quantification of important plant-parasitic nematodes. Eur J Plant Pathol 152:271–283. https://doi.org/10.1007/s10658-018-1487-7

Carneiro RMDG, de Lima FS, Correia VR (2017) Methods and tools currently used for the identification of plant parasitic nematodes. Nematol Concepts Diag Control. https://doi.org/10.5772/intechopen.69403

Castagnone-Sereno P, Danchin EG, Perfus-Barbeoch L, Abad P (2013) Diversity and evolution of root-knot nematodes, genus Meloidogyne: new insights from the genomic era. Annu Rev Phytopathol 51:203–220

Castagnone-sereno P, Skantar A, Robertson L (2011) Genomics and molecular genetics of plant-nematode interactions. Genomics Mol Genet Plant-Nematode Interact 443–464. https://doi.org/10.1007/978-94-007-0434-3

Chatterjee N, Gim J, Choi J (2018) Epigenetic profiling to environmental stressors in model and non-model organisms: ecotoxicology perspective. Environ Health Toxicol 33. https://doi.org/10.5620/eht.e2018015

Chen X, Rechavi O (2022) Plant and animal small RNA communications between cells and organisms. Nat Rev Mol Cell Biol 23(3):185–203

Chen C, Liu S, Liu Q, Niu J, Liu P, Zhao J, Jian H (2015) An ANNEXIN-like protein from the cereal cyst nematode Heterodera avenae suppresses plant defense. PLoS ONE 10(4):e0122256

Chen J, Li Z, Lin B, Liao J, Zhuo K (2021) A Meloidogyne graminicola pectate lyase is involved in virulence and activation of host defense responses. Front Plant Sci 12:651627

Chen Y, Long H, Feng T, Pei Y, Sun Y, Zhang X (2022) Development of a novel primer–taqman probe set for diagnosis and quantification of meloidogyne enterolobii in soil using QPCR and droplet digital PCR assays. Int J Mol Sci 23. https://doi.org/10.3390/ijms231911185

Chowdhury IA, Yan G (2021) Development of real-time and conventional PCR assays for identifying a newly named species of root-lesion nematode (*Pratylenchus dakotaensis*) on soybean. Int J Mol Sci 22. https://doi.org/10.3390/ijms22115872

Christoforou M, Pantelides IS, Kanetis L, Ioannou N, Tsaltas D (2014) Rapid detection and quantification of viable potato cyst nematodes using QPCR in combination with propidium monoazide. Plant Pathol 63. https://doi.org/10.1111/ppa.12193

Chung HY, Kang MJ, Han HR, Sim JS, Oh BJ, Choi I, Hahn BS (2014) Expressed sequence tags (EST) analysis of a normalized full-length cDNA library from the pinewood nematode (*Bursaphelenchus xylophilus*). Afr J Biotechnol 13(33)

Cui JK, Peng H, Qiao F, Wang GF, Huang WK, Wu DQ, Peng D (2018) Characterization of putative effectors from the cereal cyst nematode *Heterodera avenae*. Phytopathology 108(2):264–274

Culp E, Richman C, Sharanya D, Jhaveri N, van den Berg W, Gupta BP (2020) Genome editing in the nematode *Caenorhabditis briggsae* using the CRISPR/Cas9 system. Biol Methods Protoc 5:bpaa003

Daher RK, Stewart G, Boissinot M, Bergeron MG (2016) Recombinase polymerase amplification for diagnostic applications. Clin Chem 62:947–958

Dalzell JJ, McMaster S, Fleming CC, Maule AG (2010) Short interfering RNA-mediated gene silencing in *Globodera pallida* and *Meloidogyne incognita* infective stage juveniles. Int J Parasitol 40(1):91–100

Dalzell JJ, McVeigh P, Warnock ND, Mitreva M, Bird DM, Abad P, Maule AG (2011) RNAi effector diversity in nematodes. PLoS Negl Trop Dis 5(6):e1176

Danchin EGJ, Perfus-Barbeoch L (2009) The genome sequence of meloidogyne incognita unveils mechanisms of adaptation to plant-parasitism in metazoa. Evol Biol

Dauphinais N, Vandal M, Gagnon AÈ, Bélair G, Véronneau PY, Mimee B (2018) Development and application of a multiplex QPCR method for the simultaneous detection and quantification of *Pratylenchus alleni* and *P. penetrans* in Quebec, Canada. Plant Dis 102. https://doi.org/10.1094/PDIS-08-17-1222-RE

de Oliveira CMG, Monteiro AR, Blok VC (2011) Morphological and molecular diagnostics for plant-parasitic nematodes : working together to get the identification done. Trop Plant Pathol 36:65–73

Deng M-H, Zhong L-Y, Kamolnetr O, Limpanont Y, Lv Z-Y (2019) Detection of helminths by loop-mediated isothermal amplification assay: a review of updated technology and future outlook. Infect Dis Poverty 8:1–22

Devran Z, Polat İ, Mıstanoğlu İ, Baysal Ö (2018) A novel multiplex PCR tool for simultaneous detection of three root-knot nematodes. Austr Plant Pathol 47. https://doi.org/10.1007/s13313-018-0570-z

Dey P (2018) Sanger sequencing and next-generation gene sequencing: basic principles and applications in pathology. In: Basic and advanced laboratory techniques in histopathology and cytology

Ding X, Xu Z, Yin K, Sfeir M, Liu C (2019) Dual-priming isothermal amplification (DAMP) for highly sensitive and specific molecular detection with ultralow nonspecific signals. Anal Chem 91:12852–12858

Dorris M, De Ley P, Blaxter M (1999) Molecular analysis of nematode diversity and the evolution of parasitism. Parasitol Today 15:188–193. https://doi.org/10.1016/S0169-4758(99)01439-8

Dubreuil G, Magliano M, Dubrana MP, Lozano J, Lecomte P, Favery B, Rosso MN (2009) Tobacco rattle virus mediates gene silencing in a plant parasitic root-knot nematode. J Exp Bot 60(14):4041–4050

Dutta TK, Banakar P, Rao U (2015) The status of RNAi-based transgenic research in plant nematology. Front Microbiol 5:760

Edwards MC, Gibbs RA (1994) Multiplex PCR: advantages, development, and applications. Genome Res 3. https://doi.org/10.1101/gr.3.4.S65

El-Sappah AH, Islam MM, El-Awady HH, Yan S, Qi S, Liu J, Cheng GT, Liang Y (2019) Tomato natural resistance genes in controlling the root-knot nematode. Genes (Basel) 10. https://doi.org/10.3390/genes10110925

Eves-van den Akker S, Lilley CJ, Ault JR, Ashcroft AE, Jones JT, Urwin PE (2014) The feeding tube of cyst nematodes: characterisation of protein exclusion. PLoS One 9(1):e87289

Fanelli E, Di Vito M, Jones JT, De Giorgi C (2005) Analysis of chitin synthase function in a plant parasitic nematode, *Meloidogyne artiellia*, using RNAi. Gene 349:87–95

Fantini M, Pandolfini L, Lisi S, Chirichella M, Arisi I, Terrigno M, Goracci M, Cremisi F, Cattaneo A (2017) Assessment of antibody library diversity through next generation sequencing and technical error compensation. PLoS One 12. https://doi.org/10.1371/journal.pone.0177574

Filipiak A, Malewski T, Matczyńska E, Tomalak M (2021) Molecular variation among virulent and avirulent strains of the quarantine nematode *Bursaphelenchus xylophilus*. Mol Genet Genomics 296(2):259–269

Fosu-Nyarko J, Jones MG (2015) Application of biotechnology for nematode control in crop plants. In: Advances in botanical research, vol 73. Academic Press, pp 339–376

Fridman H, Bormans C, Einhorn M, Au D, Bormans A, Porat Y, Sanchez LF, Manning B, Levy-Lahad E, Behar DM (2021) Performance comparison: exome sequencing as a single test replacing sanger sequencing. Mol Genet Genomics 296. https://doi.org/10.1007/s00438-021-01772-3

Fuller VL, Lilley CJ, Urwin PE (2008) Nematode resistance. New Phytol 180(1):27–44

Gadkar VJ, Goldfarb DM, Gantt S, Tilley PAG (2018) Real-time detection and monitoring of loop mediated amplification (LAMP) reaction using self-quenching and dequenching fluorogenic probes. Sci Rep 9:2–11. https://doi.org/10.1038/s41598-018-23930-1

Gangwar S, Neeraj AKSAS (2023) Collective effect of host delivered RNAi splicing factor and integrase gene on the reproductive potential of root-knot nematode (*Meloidogyne incognita*). Pharma Innov J 12(5):1240–1244

Gao F, Wang R, Liu M (2014) *Trichinella spiralis*, potential model nematode for epigenetics and its implication in metazoan parasitism. Front Physiol 2014

Garrido-Cardenas JA, Garcia-Maroto F, Alvarez-Bermejo JA, Manzano-Agugliaro F (2017) DNA sequencing sensors: an overview. Sensors (Switzerland) 17

Gasser RB, Bott NJ, Chilton NB, Hunt P, Beveridge I (2008) Toward practical, DNA-based diagnostic methods for parasitic nematodes of livestock—bionomic and biotechnological implications. 26:325–334. https://doi.org/10.1016/j.biotechadv.2008.03.003

Geldhof P, Visser A, Clark D, Saunders G et al (2006) RNA interference in parasitic helminths: current situation, potential pitfalls and future prospects. Parasitology 134(5):609–619

Gendron EM, Sevigny JL, Byiringiro I, Thomas WK, Powers TO, Porazinska DL (2023) Nematode mitochondrial metagenomics: a new tool for biodiversity analysis. Mol Ecol Resour. https://doi.org/10.1111/1755-0998.13761

Gheysen G, Vanholme B (2007) RNAi from plants to nematodes. Trends Biotechnol 25(3):89–92

Goverse A, Mitchum MG (2022) At the molecular plant–nematode interface: new players and emerging paradigms. Curr Opin Plant Biol 67

Green MR, Sambrook J (2018) Analysis and normalization of real-time polymerase chain reaction (PCR) experimental data. Cold Spring Harb Protoc 2018. https://doi.org/10.1101/pdb.top095000

Habash SS, Sobczak M, Siddique S, Voigt B, Elashry A, Grundler FM (2017) Identification and characterization of a putative protein disulfide isomerase (HsPDI) as an alleged effector of *Heterodera schachtii*. Sci Rep 7(1):13536

Hada A, Singh D, Satyanarayana KKVV, Chatterjee M, Phani V, Rao U (2021) Effect of fluensulfone on different functional genes of root-knot nematode. J Nematol 53(1):1–14

Hada A, Singh D, Banakar P, Papolu PK, Kassam R, Chatterjee M, Rao U (2023) Host-delivered RNAi-mediated silencing using fusion cassettes of different functional groups of genes precludes *Meloidogyne incognita* multiplication in *Nicotiana tabacum*. Plant Cell Rep 42(1):29–43

Hassan NA, Noor Badi FA, Mohd-Shaharuddin N, Wan Yusoff WS, Lim YAL, Chua KH, Sidi Omar SFN, Chang LY, Majid HA, Ngui R (2022) A conventional multiplex PCR for the detection of four common soil-transmitted nematodes in human feces: development and validation. Trop Biomed 39. https://doi.org/10.47665/tb.39.1.016

He Q, Wang D, Tang B, Wang J, Zhang D, Liu Y, Cheng F (2021) Rapid and sensitive detection of meloidogyne graminicola in soil using conventional PCR, loop-mediated isothermal amplification, and real-time PCR methods. Plant Dis 105, https://doi.org/10.1094/PDIS-06-20-1291-RE

Hewezi T, Lane T, Piya S, Rambani A, Rice JH, Staton M (2017) Cyst nematode parasitism induces dynamic changes in the root epigenome. Plant Physiol 174. https://doi.org/10.1104/pp.16.01948

Horner DS, Pavesi G, Castrignano' T, de Meo PDO, Liuni S, Sammeth M, Picardi E, Pesole G (2009) Bioinformatics approaches for genomics and post genomics applications of next-generation sequencing. Brief Bioinform 11. https://doi.org/10.1093/bib/bbp046

Hu MX, Zhuo K, Liao JL (2011) Multiplex PCR for the simultaneous identification and detection of *Meloidogyne incognita*, *M. enterolobii*, and *M. javanica* using DNA extracted directly from individual galls. Phytopathology 101. https://doi.org/10.1094/PHYTO-04-11-0095

Huang Q, Lin B, Cao Y, Zhang Y, Song H, Huang C, Sun T, Long C, Liao J, Zhuo K (2023) CRISPR/Cas9-mediated mutagenesis of the susceptibility gene OsHPP04 in rice confers enhanced resistance to rice root-knot nematode. Front Plant Sci 14:1134653

Huang D, Yan G, Skantar AM (2017) Development of real-time and conventional PCR assays for identifying stubby root nematode *Paratrichodorus allius*. Plant Dis 101. https://doi.org/10.1094/PDIS-10-16-1431-RE

Hugall A, Stanton J, Moritz C (1999) Reticulate evolution and the origins of ribosomal internal transcribed spacer diversity in apomictic meloidogyne. Mol Biol Evol 16:157–164. https://doi.org/10.1093/oxfordjournals.molbev.a026098

Iberkleid I, Vieira P, de Almeida Engler J, Firester K, Spiegel Y, Horowitz SB (2013) Fatty acid-and retinol-binding protein, Mj-FAR-1 induces tomato host susceptibility to root-knot nematodes. PLoS ONE 8(5):e64586

Ibrahim HMM, Ahmad EM, Martínez-Medina A, Aly MAM (2019) Effective approaches to study the plant-root knot nematode interaction. Plant Physiol Biochem 141

Ide T, Kanzaki N, Giraldo PPP, Giblin-Davis RM (2017) Loop-mediated isothermal amplification (lamp) for detection of the red ring nematode, *Bursaphelenchus cocophilus*. Nematology 19:559–565

Iqbal S, Fosu-Nyarko J, Jones MG (2016) Genomes of parasitic nematodes (*Meloidogyne hapla, Meloidogyne incognita, Ascaris suum* and *Brugia malayi*) have a reduced complement of small RNA interference pathway genes: knockdown can reduce host infectivity of *M. incognita*. Funct Integr Genomics 16:441–457

Irdani T, For I, Dna-based S, Pcr S, Dna R (2008) Molecular identification of some plant parasitic nematode speCIES. REDIA-J Zool XCI

Jacob J (2009) Plant-parasitic nematodes: from genomics to functional analysis of parasitism genes. Ghent University

Jacob J, Mitreva M (2011) Transcriptomes of plant-parasitic nematodes. In: Genomics and molecular genetics of plant-nematode interactions. Springer Netherlands, pp 119–138

Janssen T, Karssen G, Verhaeven M, Coyne D, Bert W (2016) Mitochondrial coding genome analysis of tropical root-knot nematodes (meloidogyne) supports haplotype based diagnostics and reveals evidence of recent reticulate evolution. Sci Rep 6:1–13. https://doi.org/10.1038/srep22591

Janssen T, Karssen G, Orlando V, Subbotin SA, Bert W (2017) Molecular characterization and species delimiting of plant-parasitic nematodes of the genus pratylenchus from the penetrans group (Nematoda: Pratylenchidae). Mol Phylogenet Evol 117. https://doi.org/10.1016/j.ympev.2017.07.027

Joshi I, Kumar A, Kohli D, Singh AK, Sirohi A, Subramaniam K, Jain PK (2020) Conferring root-knot nematode resistance via host-delivered RNAi-mediated silencing of four Mi-msp genes in Arabidopsis. Plant Sci 298:110592

Joshi I, Kohli D, Pal A, Chaudhury A, Sirohi A, Jain PK (2022) Host delivered-RNAi of effector genes for imparting resistance against root-knot and cyst nematodes in plants. Physiol Mol Plant Pathol 118:101802

Ju Y, Lin Y, Yang G, Wu H, Pan Y (2019) Development of recombinase polymerase amplification assay for rapid detection of *Meloidogyne incognita, M. javanica, M. arenaria,* and *M. enterolobii*. Eur J Plant Pathol 155:1155–1163

Kakrana A, Kumar A, Satheesh V, Abdin MZ, Subramaniam K, Bhattacharya RC, Jain PK (2017) Identification, validation and utilization of novel nematode-responsive root-specific promoters in Arabidopsis for inducing host-delivered RNAi mediated root-knot nematode resistance. Front Plant Sci 8:2049

Kanetani S, Kikuchi T, Akiba M, Nakamura K, Ikegame H, Tetsuka K (2011) Detection of *Bursaphelenchus xylophilus* from old discs of dead *Pinus armandii* var. amamiana trees using a new detection kit. For Pathol 41:387–391

Kang JS, Kim A, Han HR, Moon YS, Koh YH (2015) Development of two alternative loop-mediated isothermal amplification tools for detecting pathogenic pine wood nematodes. For Pathol 45:127–133

Kang J (2016) Application of CRISPR/Cas9-mediated genome editing for studying soybean resistance to soybean cyst nematode. Doctoral dissertation, University of Missouri-Columbia

Kantor M, Handoo Z, Kantor C, Carta L (2022) Top ten most important US-regulated and emerging plant-parasitic nematodes. Horticulturae 8:208

Karibasappa CS, Singh Y, Aravind T, Singh KP (2021) Concept of effectors and receptors in improving plant immunity. Emerg Trends Plant Pathol 475–497

Kawanobe M, Toyota K (2018) Application of a simple and high-throughput DNA extraction method to real-time PCR quantification of target plant-parasitic nematodes in nematode communities. Nematol Res (Japan J Nematol) 48. https://doi.org/10.3725/jjn.48.1

Keçici Aİ, Bozbuğa R, Öcal A, Yüksel E, Özer G, Yildiz Ş, Lahlali R, Slaats B, Dababat AA, İmren M (2022) Diversity and identification of plant-parasitic nematodes in wheat-growing ecosystems. Microorganisms 10:1534

Kikuchi T, Aikawa T, Oeda Y, Karim N, Kanzaki N (2009) A rapid and precise diagnostic method for detecting the pinewood nematode *Bursaphelenchus xylophilus* by loop-mediated isothermal amplification. Phytopathology 99:1365–1369

Kikuchi T, Eves-Van Den Akker S, Jones JT (2017) Genome evolution of plant-parasitic nematodes. Annu Rev Phytopathol 55. https://doi.org/10.1146/annurev-phyto-080516-035434

Knox DP, Geldhof P, Visser A, Britton C (2007) RNA interference in parasitic nematodes of animals: a reality check? Trends Parasitol 23(3):105–107

Köhler F (2007) From DNA taxonomy to barcoding—how a vague idea evolved into a biosystematic tool. Zoosystematics Evol 83. https://doi.org/10.1002/mmnz.200600025

Kojima K, Juma KM, Akagi S, Hayashi K, Takita T, O'Sullivan CK, Fujiwara S, Nakura Y, Yanagihara I, Yasukawa K (2021) Solvent engineering studies on recombinase polymerase amplification. J Biosci Bioeng 131:219–224

Korneliussen TS, Albrechtsen A, Nielsen R (2014) ANGSD: analysis of next generation sequencing data. BMC Bioinformatics 15. https://doi.org/10.1186/s12859-014-0356-4

Kumar A, Joshi I, Changwal C, Sirohi A, Jain PK (2022) Host-delivered RNAi-mediated silencing of the root-knot nematode (Meloidogyne incognita) effector genes, Mi-msp10 and Mi-msp23, confers resistance in Arabidopsis and impairs reproductive ability of the root-knot nematode. Planta 256(4):74

Leal I, Foord B, Allen E, Campion C, Rott M, Green M (2013) Development of two reverse transcription-PCR methods to detect living Pinewood nematode, *Bursaphelenchus xylophilus*, in wood. For Pathol 43:104–114. https://doi.org/10.1111/efp.12003

Lehnert M, Kipf E, Schlenker F, Borst N, Zengerle R, Von Stetten F (2018) Fluorescence signal-to-noise optimisation for real-time PCR using universal reporter oligonucleotides. Anal Methods 10. https://doi.org/10.1039/c8ay00812d

Li Y, Wu B, Hao X, Diao J, Cao J, Tan R, Ma L (2023) Functional analysis of 3 genes in xenobiotic detoxification pathway of *Bursaphelenchus xylophilus* against matrine. Pestic Biochem Physiol 190:105334

Li Y, Lawrence GW, Lu S, Balbalian C, Klink VP (2014) Quantitative field testing heterodera glycines from metagenomic DNA samples isolated directly from soil under agronomic production. PLoS One 9. https://doi.org/10.1371/journal.pone.0089887

Li X, Yang Y, Henry RJ, Rossetto M, Wang Y, Chen S (2015) Plant DNA barcoding: from gene to genome. Biol Rev Camb Philos Soc 90

Lilley CJ, Davies LJ, Urwin PE (2012) RNA interference in plant parasitic nematodes: a summary of the current status. Parasitology 139:630–640

Lin B, Wang H, Zhuo K, Liao J (2016) Loop-mediated isothermal amplification for the detection of *Tylenchulus semipenetrans* in soil. Plant Dis 100:877–883

Liu J, Peng H, Cui J, Huang W, Kong L, Clarke JL, Peng D (2016) Molecular characterization of a novel effector expansin-like protein from *Heterodera avenae* that induces cell death in *Nicotiana benthamiana*. Sci Rep 6(1):35677

Liu S, Geng S, Li A, Mao Y, Mao L (2021) RNAi technology for plant protection and its application in wheat. Abiotech 1–10

Lo TW, Pickle CS, Lin S, Ralston EJ, Gurling M, Schartner CM, Meyer BJ (2013) Precise and heritable genome editing in evolutionarily diverse nematodes using TALENs and CRISPR/Cas9 to engineer insertions and deletions. Genetics 195:331–348

Lobato IM, Sullivan CKO (2017) Recombinase polymerase amplification: basics, applications and recent advances. Trends Anal Chem 98:19–35. https://doi.org/10.1016/j.trac.2017.10.015
Lok JB (2019) CRISPR/Cas9 mutagenesis and expression of dominant mutant transgenes as functional genomic approaches in parasitic nematodes. Front Genet 10:656
Long CR, Tessanne KJ, Golding MC (2009) Applications of RNA interference-based gene silencing in animal agriculture. Reprod Fertil Dev 22(1):47–58
Lourenço-Tessutti IT, Souza Junior JDA, Martins-de-Sa D, Viana AAB, Carneiro RMDG, Togawa RC, Grossi-de-Sa MF (2015) Knock-down of heat-shock protein 90 and isocitrate lyase gene expression reduced root-knot nematode reproduction. Phytopathology 105(5):628–637
Ma Y, Xie H, Wang J, Lui C (2011) Detection of second-stage juveniles of anguina agrostis using TaqMan real-time PCR. Russ J Nematol 19
Maboreke HR, Feldhahn L, Bönn M, Tarkka MT, Buscot F, Herrmann S, Menzel R, Ruess L (2016) Transcriptome analysis in oak uncovers a strong impact of endogenous rhythmic growth on the interaction with plant-parasitic nematodes. BMC Genomics 17. https://doi.org/10.1186/s12864-016-2992-8
MacMillan K, Blok V, Young I, Crawford J, Wilson MJ (2006) Quantification of the slug parasitic nematode phasmarhabditis hermaphrodita from soil samples using real time QPCR. Int J Parasitol 36:1453–1461. https://doi.org/10.1016/j.ijpara.2006.08.005
Madani M, Ward LJ, De Boer SH (2008) Multiplex real-time polymerase chain reaction for identifying potato cyst nematodes, globodera pallida and globodera rostochiensis, and the tobacco cyst nematode, *Globodera tabacum*. Can J Plant Pathol 30. https://doi.org/10.1080/070606608 09507555
Mahesh Dahake AP (2021) Next generation sequencing: principle and applications. Int J Curr Microbiol Appl Sci 10. https://doi.org/10.20546/ijcmas.2021.1008.039
Marek M, Zouhar M, Douda O, Mazakova J, Rysanek P (2010) Bioinformatics-assisted characterization of the ITS1–5·8S-ITS2 segments of nuclear RRNA gene clusters, and its exploitation in molecular diagnostics of european crop-parasitic nematodes of the genus ditylenchus. Plant Pathol 59. https://doi.org/10.1111/j.1365-3059.2010.02322.x
Marmonier A, Perfus-Barbeoch L, Rancurel C, Boissinot S, Favery B, Demangeat G, Brault V (2019) In Vitro acquisition of specific small interfering RNAs inhibits the expression of some target genes in the plant ectoparasite Xiphinema index. Int J Mol Sci 20(13):3266
Marques CA, Ridgway M, Tinti M, Cassidy A, Horn D (2022) Genome-scale RNA interference profiling of *Trypanosoma brucei* cell cycle progression defects. Nat Commun 13(1):5326
Martin J, Abubucker S, Wylie T, Yin Y, Wang Z, Nematode MM, Update N (2008) Improvements enabling more efficient data mining and comparative nematode genomics. Nucleic Acids Res 2009:37. https://doi.org/10.1093/nar/gkn744
Martin J, Abubucker S, Wylie T, Yin Y, Wang Z, Mitreva M (2009) Improvements enabling more efficient data mining and comparative nematode genomics. Nucleic Acids Res 237(SUPPL. 1). https://doi.org/10.1093/nar/gkn744
Mendoza-Gallegos RA, Rios A, Garcia-Cordero JL (2018) An affordable and portable thermocycler for real-time PCR made of 3D-printed parts and off-the-shelf electronics. Anal Chem 90. https://doi.org/10.1021/acs.analchem.7b04843
Menezes PS, Yan Y, Yang Y, Mitter N, Mahony TJ, Mody KT (2022) RNAi-based biocontrol of pests to improve the productivity and welfare of livestock production. Appl Biosci 1(3):229–243
Modesto I, Mendes A, Carrasquinho I, Miguel CM (2022) Molecular defense response of pine trees (Pinus spp.) to the parasitic nematode *Bursaphelenchus xylophilus*. Cells 11(20):3208
Mohammadi MM, Bavi O (2022) DNA sequencing: an overview of solid-state and biological nanopore-based methods. Biophys Rev 14
Mokrini F, Waeyenberge L, Viaene N, Abbad Andaloussi F, Moens M (2013) Quantitative detection of the root-lesion nematode, pratylenchus penetrans, using QPCR. Eur J Plant Pathol 137. https://doi.org/10.1007/s10658-013-0252-1
Montarry J, Mimee B, Danchin EGJ, Koutsovoulos GD, Ste-Croix DT, Grenier E (2021) Recent advances in population genomics of plant-parasitic nematodes. Phytopathology 111

Munawar M, Rahman AU, Castillo P, Yevtushenko DP (2023) New records of ditylenchus species from Southern Alberta, Canada. Plants 12. https://doi.org/10.3390/plants12050998

Mwamula AO, Lee G, Kim YH, Kim YH, Lee KS, Lee DW (2020) Molecular phylogeny of several species of hoplolaimina (Nematoda: Tylenchida) associated with turfgrass in Korea, with comments on their morphology. Nematology 23

Nazarenko I, Lowe B, Darfler M, Ikonomi P, Schuster D, Rashtchian A (2002) Multiplex quantitative PCR using self-quenched primers labeled with a single fluorophore. Nucleic Acids Res 30. https://doi.org/10.1093/nar/30.9.e37

Nega A (2014) Review on nematode molecular diagnostics: from bands to barcodes. 4:129–154

Negi C, Vasistha NK, Singh D, Vyas P, Dhaliwal HS (2022) Application of CRISPR-mediated gene editing for crop improvement. Mol Biotechnol 64:1198–1217

Neupane S, Purintun JM, Mathew FM, Varenhorst AJ, Nepal MP (2019) Molecular basis of soybean resistance to soybean aphids and soybean cyst nematodes. Plants 8(10):374

Niu JH, Jian H, Guo QX, Chen CL, Wang XY, Liu Q, Guo YD (2012) Evaluation of loop-mediated isothermal amplification (LAMP) assays based on 5S RDNA-IGS2 regions for detecting *Meloidogyne enterolobii*. Plant Pathol 61:809–819

Niu J, Guo Q, Jian H, Chen C, Yang D, Liu Q, Guo Y (2011) Rapid detection of *Meloidogyne* spp. by LAMP assay in soil and roots. Crop Protect 30:1063–1069

Noon JB, Hewezi T, Maier TR, Simmons C, Wei JZ, Wu G, Baum TJ (2015) Eighteen new candidate effectors of the phytonematode *Heterodera glycines* produced specifically in the secretory esophageal gland cells during parasitism. Phytopathology 105(10):1362–1372

O'Halloran DM (2019) Genome aware CRISPR gRNA target prediction for parasitic nematodes. Mol Biochem Parasitol 227:25–28

Pabinger S, Dander A, Fischer M, Snajder R, Sperk M, Efremova M, Krabichler B, Speicher MR, Zschocke J, Trajanoski Z (2014) A survey of tools for variant analysis of next-generation genome sequencing data. Brief Bioinform 15. https://doi.org/10.1093/bib/bbs086

Palomares-Rius JE, Cantalapiedra-Navarrete C, Archidona-Yuste A, Subbotin SA, Castillo P (2017) The utility of MtDNA and RDNA for barcoding and phylogeny of plant-parasitic nematodes from longidoridae (Nematoda, Enoplea). Sci Rep 7:10905

Pandey P, Mysore KS, Senthil-Kumar M (2022) Recent advances in plant gene silencing methods. Plant Gene Silencing: Methods Protocols 1–22

Pareek CS, Smoczynski R, Tretyn A (2011) Sequencing technologies and genome sequencing. J Appl Genet 52

Parkinson NJ, Maslau S, Ferneyhough B, Zhang G, Gregory L, Buck D, Ragoussis J, Ponting CP, Fischer MD (2012) Preparation of high-quality next-generation sequencing libraries from picogram quantities of target DNA. Genome Res 22. https://doi.org/10.1101/gr.124016.111

Parmar N, Singh KH, Sharma D, Singh L, Kumar P, Nanjundan J, Thakur AK (2017) Genetic engineering strategies for biotic and abiotic stress tolerance and quality enhancement in horticultural crops: a comprehensive review. Biotech 7:1–35

Peng H, Peng D, Hu X, He X, Wang Q, Huang W, He W (2012) Loop-mediated isothermal amplification for rapid and precise detection of the burrowing nematode, *Radopholus Similis*, directly from diseased plant tissues. Nematology 14:977–986

Peng H, Long H, Huang W, Liu J, Cui J, Kong L, Hu X, Gu J, Peng D (2017) Rapid, simple and direct detection of meloidogyne hapla from infected root galls using loop-mediated isothermal amplification combined with FTA technology. Sci Rep 7. https://doi.org/10.1038/srep44853

Pentimone I, Colagiero M, Rosso LC, Ciancio A (2020) Omics applications: towards a sustainable protection of tomato. Appl Microbiol Biotechnol 104:4185–4195

Perry RN, Maule AG (2004) Physiological and biochemical basis of behaviour. In: Nematode behaviour. CABI Publishing, pp 197–238

Perry RN, Subbotin SA (2007) Molecular diagnostics of plant-parasitic nematodes. In: Biotechnology and plant disease management. ISBN 9781845932886

Powers T (2004) Nematode molecular diagnostics: from bands to barcodes. Annu Rev Phytopathol 42

Pratx L, Rancurel C, Da Rocha M, Danchin EGJ, Castagnone-Sereno P, Abad P, Perfus-Barbeoch L (2018) Genome-wide expert annotation of the epigenetic machinery of the plant-parasitic nematodes *Meloidogyne* spp., with a focus on the asexually reproducing species. BMC Genomics 19. https://doi.org/10.1186/s12864-018-4686-x

Price JA, Ali MF, Major LL, Smith TK, Jones JT (2023) An eggshell-localised annexin plays a key role in the coordination of the life cycle of a plant-parasitic nematode with its host. PLoS Pathog 19:e1011147

Qianru Y, Kelly JD, Beilei G (2018) Loop-mediated isothermal amplification for salmonella detection in food and feed: current applications and future directions. Foodborne Pathogens Dis 15. https://doi.org/10.1089/fpd.2018.2445

Qing X, Wang M, Karssen G, Bucki P, Bert W, Braun-Miyara S (2020) PPNID: a reference database and molecular identification pipeline for plant-parasitic nematodes. Bioinformatics 36:1052–1056. https://doi.org/10.1093/bioinformatics/btz707

Rajput M, Choudhary K, Kumar M, Vivekanand V, Chawade A, Ortiz R, Pareek N (1914) RNA interference and CRISPR/Cas gene editing for crop improvement: paradigm shift towards sustainable agriculture. Plants 2021:10

Rajput M, Choudhary K, Kumar M, Vivekanand V, Chawade A, Ortiz R, Pareek N (2021) RNA interference and crispr/Cas gene editing for crop improvement: paradigm shift towards sustainable agriculture. Plants 10:1–31. https://doi.org/10.3390/plants10091914

Randig O, Bongiovanni M, Carneiro RMDG, Castagnone-Sereno P (2002) Genetic diversity of root-knot nematodes from brazil and development of scar markers specific for the coffee-damaging species. Genome 45:862–870. https://doi.org/10.1139/g02-054

Redding NW, Agudelo P, Wells CE (2018) Multiple nodulation genes are up-regulated during establishment of Reniform nematode feeding sites in soybean. Phytopathology 108(2):275–291

Roderick H, Tripathi L, Poovarasan S (2016) Transgenic approaches to improve resistance to nematodes and weevils. Banana: genomics transgenic approaches for genetic improvement, pp 247–260

Rosana OB, Eapen SJ, Krishna PB (2016) Cloning and characterization of genes encoding two detoxifying enzymes, glutathione s-transferase and carboxylesterase, from burrowing nematode (*Radopholus similis*). Int J Parasitol Res 8(1):173–183

Rosso MN, Jones JT, Abad P (2009) RNAi and functional genomics in plant parasitic nematodes. Annu Rev Phytopathol 47:207–232

Rothman RE, Yang S (2004) PCR-based diagnostics for infectious diseases: uses, limitations, and future applications in acute-care settings. LANCET Infect Dis 96. https://doi.org/10.1016/j.coldregions.2013.09.003

Rybarczyk-Mydłowska K, Mooyman P, van Megen H, van den Elsen S, Vervoort M, Veenhuizen P, van Doorn J, Dees R, Karssen G, Bakker J (2012) Small subunit ribosomal DNA-based phylogenetic analysis of foliar nematodes (Aphelenchoides Spp.) and their quantitative detection in complex DNA backgrounds. Phytopathology 102:1153–1160

Saeki Y, Kawano E, Yamashita C, Akao S, Nagatomo Y (2003) Detection of plant parasitic nematodes, *Meloidogyne incognita* and *Pratylenchus coffeae* by multiplex PCR using specific primers. Soil Sci Plant Nutr 49. https://doi.org/10.1080/00380768.2003.10410010

Sapkota R, Skantar AM, Nicolaisen M (2016) A TaqMan real-time PCR assay for detection of meloidogyne hapla in root galls and in soil. Nematology 18. https://doi.org/10.1163/15685411-00002950

Sayler RJ, Walker C, Goggin F, Agudelo P, Kirkpatrick T (2012) Conventional PCR detection and real-time PCR quantification of reniform nematodes. Plant Dis 96. https://doi.org/10.1094/PDIS-12-11-1033-RE

Scholz MB, Lo CC, Chain PSG (2012) Next generation sequencing and bioinformatic bottlenecks: the current state of metagenomic data analysis. Curr Opin Biotechnol 23

Seesao Y, Gay M, Merlin S, Viscogliosi E, Aliouat-Denis CM, Audebert C (2017) A review of methods for nematode identification. J Microbiol Methods 138:37–49. https://doi.org/10.1016/j.mimet.2016.05.030

Senthil-Kumar M, Mysore KS (2011) Caveat of RNAi in plants: the off-target effect. In: RNAi and plant gene function analysis: methods and protocols, pp 13–25

Shao H, Zhang P, Peng D, Huang W, Kong L, Li C, Liu E, Peng H (2023) Current advances in the identification of plant nematode diseases: from lab assays to in-field diagnostics. Front Plant Sci 14

Sharma N, Prasad A, Prasad M (2022) Role of the Sw5 gene cluster in the fight against plant viruses. J Virol 96(5):e02084-e2121

Sharma S, Kumar A, Kumari N, Walia A (2023) RNAi as a tool to enhance crop yield and biotic stress management in the plants. Plant Cell Tissue Organ Cult 152(3):437–454

Sharma VK, Sanghera GS, Kashyap PL, Sharma BB, Chandel C (2013) RNA interference: a novel tool for plant disease management. Afr J Biotechnol 12(18)

Shendure JA, Porreca GJ, Church GM, Gardner AF, Hendrickson CL, Kieleczawa J, Slatko BE (2011) Overview of DNA sequencing strategies. Curr Protoc Mol Biol. https://doi.org/10.1002/0471142727.mb0701s96

Shilpa, Sharma P, Thakur V, Sharma A, Rana RS, Kumar P (2022) A status-quo review on management of root knot nematode in tomato. J Hortic Sci Biotechnol 97:403–416

Shirshikov FV, Bespyatykh JA (2022) Loop-mediated isothermal amplification : from theory to practice. Russ J Bioorg Chem 48:1159–1174. https://doi.org/10.1134/S106816202206022X

Siddique S, Radakovic ZS, Hiltl C, Pellegrin C, Baum TJ, Beasley H, Bent AF, Chitambo O, Chopra D, Danchin EGJ et al (2022) The genome and lifestage-specific transcriptomes of a plant-parasitic nematode and its host reveal susceptibility genes involved in trans-kingdom synthesis of vitamin B5. Nat Commun 13. https://doi.org/10.1038/s41467-022-33769-w

Sindhu AS, Maier TR, Mitchum MG, Hussey RS, Davis EL, Baum TJ (2009) Effective and specific in planta RNAi in cyst nematodes: expression interference of four parasitism genes reduces parasitic success. J Exp Bot 60(1):315–324

Sivasubramaniam N, Hariharan G, Zakeel MCM (2020) Sustainable management of plant-parasitic nematodes: an overview from conventional practices to modern techniques. In: Recent advances and future challenges, pp 353–399

Solo N, Kud J, Dandurand LM, Caplan A, Kuhl JC, Xiao F (2021) Characterization of superoxide dismutase from the potato cyst nematode, *Globodera pallida*. Phytopathology 111(11):2110–2117

Song Z-Q, Cheng J-E, Cheng F-X, Zhang D-Y, Liu Y (2017) Development and evaluation of loop-mediated isothermal amplification assay for rapid detection of *Tylenchulus semipenetrans* using DNA extracted from soil. Plant Pathol J 33:184

Subbotin SA, Toumi F, Elekçioğlu IH, Waeyenberge L, Maafi ZT (2018) DNA barcoding, phylogeny and phylogeography of the cyst nematode species of the avenae group from the genus heterodera (Tylenchida: Heteroderidae). Nematology 20:671–702

Subbotin SA (2021) Molecular identification of nematodes using polymerase chain reaction (PCR). Techniques for Work with Plant and Soil Nematodes. Wallingford UK CABI, pp 218–239

Subbotin SA, Madani M, Krall E, Sturhan D, Moens M (2005) Molecular diagnostics, taxonomy, and phylogeny of the stem nematode ditylenchus dipsaci species complex based on the sequences of the internal transcribed spacer-RDNA. Phytopathology 95. https://doi.org/10.1094/PHYTO-95-1308

Sudha SN, Ravindar G, Rao NP, Rajalingam G (2017) Molecular identification of *Rotylenchulus reniformis* (Nematoda) by using its region of RDNA. Int J Pure Appl Zool 5

Sultana T, Kim J, Lee SH, Han H, Kim S, Min GS, Nadler SA, Park JK (2013) Comparative analysis of complete mitochondrial genome sequences confirms independent origins of plant-parasitic nematodes. BMC Evol Biol 13. https://doi.org/10.1186/1471-2148-13-12

Summerbell RC, Lévesque CA, Seifert KA, Bovers M, Fell JW, Diaz MR, Boekhout T, De Hoog GS, Stalpers J, Crous PW (2005) Microcoding: the second step in DNA barcoding. Philos Trans Roy Soc B: Biol Sci 360. https://doi.org/10.1098/rstb.2005.1721

Syeda AH, Hawkins M, Mcglynn P, Daley JM, Gaines WA, Zelensky A, Kanaar R, Wyman C, Wyatt HDM, West SC et al (2014) Recombination and replication. Cold Spring Harb Perspect Biol 6:1–15

Tamilarasan S, Rajam MV (2013) Engineering crop plants for nematode resistance through host-derived RNA interference. Cell Dev Biol 2(2)

Tian B, Li J, Oakley TR, Todd TC, Trick HN (2016) Host-derived artificial microRNA as an alternative method to improve soybean resistance to soybean cyst nematode. Genes 7(12):122

Tian BY, Cao Y, Zhang KQ (2015) Metagenomic insights into communities, functions of endophytes, and their associates with infection by root-knot nematode, *Meloidogyne incognita*, in tomato roots. Sci Rep 5. https://doi.org/10.1038/srep17087

Tileubayeva Z, Avdeenko A, Avdeenko S, Stroiteleva N (2021) Plant-parasitic nematodes affecting vegetable crops in greenhouses. Saudi J Biol Sci 28:5428–5433. https://doi.org/10.1016/j.sjbs.2021.05.075

Tomita N, Mori Y, Kanda H, Notomi T (2008) Loop-mediated isothermal amplification (LAMP) of gene sequences and simple visual detection of products. Nat Protoc 3:877–882. https://doi.org/10.1038/nprot.2008.57

Toyota K (2021) Studies on ecology, diagnosis, and control of soilborne plant pathogens and plant parasitic nematodes: a synthesis. Soil Sci Plant Nutr 67

Toyota K, Shirakashi T, Sato E, Wada S, Min YY (2008) Development of a real-time PCR method for the potato-cyst nematode globodera rostochiensis and the root-knot nematode *Meloidogyne incognita*. Soil Sci Plant Nutr 54. https://doi.org/10.1111/j.1747-0765.2007.00212.x

Treonis AM, Unangst SK, Kepler RM, Buyer JS, Cavigelli MA, Mirsky SB, Maul JE (2018) Characterization of soil nematode communities in three cropping systems through morphological and DNA metabarcoding approaches. Sci Rep 8:1–12

Urwin PE, Lilley CJ, Atkinson HJ (2002) Ingestion of double-stranded RNA by preparasitic juvenile cyst nematodes leads to RNA interference. Mol Plant-Microbe Interact 15(8):747–752

Valencia CA, Pervaiz MA, Husami A, Qian Y, Zhang K (2013) Sanger sequencing principles, history, and landmarks

Vasanthrao T, Routhu GK, Nagaraju YKNRY (2023) Host induced gene silencing for biotic stress management. SR Edu Publications

Vieira P, Gleason C (2019) Plant-parasitic nematode effectors—insights into their diversity and new tools for their identification. Curr Opin Plant Biol 50:37–43

van de Vossenberg BTLH, van Kempen TL, van Duijnhoven AALAM, Warbroek T, van der Gouw LP, Apolônio Silva De Oliveira D, van Heese EYJ, Karssen G (2022) Development and validation of a high-throughput sequencing test for mitogenome and RDNA assembly and annotation, and its use in support of nematode identification of regulatory concern. Phyto Front TM. https://doi.org/10.1094/phytofr-04-22-0041-fi

Waliullah S, Bell J, Jagdale G, Stackhouse T, Hajihassani A, Brenneman T, Ali ME (2020) Rapid Detection of pecan root-knot nematode, *Meloidogyne partityla*, in laboratory and field conditions using loop-mediated isothermal amplification. PLoS ONE 15:e0228123

Wang X, Lei R, Peng H, Jiang R, Shao H, Ge J, Peng D (2022) Rapid diagnosis and visual detection of potato cyst nematode (*Globodera Rostochiensis*) using recombinase polymerase amplification combination with lateral flow assay method (RPA-LFA). Agronomy 12:2580

Wen TY, Wu XQ, Ye JR, Qiu YJ, Rui L, Zhang Y (2023) Two novel *Bursaphelenchus xylophilus* kunitz effector proteins using different infection and survival strategies to suppress immunity in pine. Phytopathology 113(3):539–548

Whangbo JS, Hunter CP (2008) Environmental RNA interference. Trends Genetics 24:297–305

Wolt JD, Wang K, Sashital D, Lawrence-Dill CJ (2016) Achieving plant CRISPR targeting that limits off-target effects. The Plant Genome 9: 2016–05

Wylie T, Martin JC, Dante M, Mitreva MD, Clifton SW, Chinwalla A, Waterston RH, Wilson RK, McCarter JP (2004) Nematode.Net: a tool for navigating sequences from parasitic and free-living nematodes. Nucleic Acids Res 32. https://doi.org/10.1093/nar/gkh010

Yan G, Smiley RW, Okubara PA, Skantar A, Easley SA, Sheedy JG, Thompson AL (2008) In DNA extracts from soil. Plant Dis 92:1480–1487

Yan G, Smiley RW, Okubara PA, Skantar AM, Reardon CL (2013) Developing a real-time PCR assay for detection and quantification of pratylenchus neglectus in soil. Plant Dis 97. https://doi.org/10.1094/PDIS-08-12-0729-RE

Yang M, Kim JW (2018) Principles of genetic counseling in the era of next-generation sequencing. Ann Lab Med 38

Yao K, Peng D, Jiang C, Zhao W, Li G, Huang W, Kong L, Gao H, Zheng J, Peng H (2021) Rapid and visual detection of heterodera schachtii using recombinase polymerase amplification combined with Cas12a-mediated technology. Int J Mol Sci 22:12577

Ye W (2012) Development of primetime-real-time PCR for species identification of soybean cyst nematode (*Heterodera glycines* Ichinohe, 1952) North Carolina. J Nematol 44(3):284

Yeates G (2009) Plant nematodes: methodology, morphology, systematics, biology and ecology. Nematology 11. https://doi.org/10.1163/138855409x412136

You H, Gordon CA, MacGregor SR, Cai P, McManus DP (2022) Potential of the CRISPR-Cas system for improved parasite diagnosis: CRISPR-Cas mediated diagnosis in parasitic infections. BioEssays 44:2100286

Yu L, Song S, Yu C, Qi L, Yu Z, Jiao B, Yang J (2018) A loop mediated isothermal amplification (LAMP) assay for rapid and reliable detection of anguina wevelli, a grass parasitic nematode. Eur J Plant Pathol 150:725–734

Yu L, Song S, Yu C, Jiao B, Tian Y, Li Y (2020) Rapid detection of anguina agrostis by loop-mediated isothermal amplification. Eur J Plant Pathol 156:819–825

Zakeel MCM, Safeena MIS (2020) Metagenomic insights into interactions between plant nematodes and endophytic microbiome. In: Management of phytonematodes: recent advances and future challenges

Zhang C, Xie H, Xu CL, Cheng X, Li KM, Li Y (2012) Differential expression of Rs-eng-1b in two populations of *Radopholus similis* (Tylenchida: Pratylecnchidae) and its relationship to pathogenicity. Eur J Plant Pathol 133:899–910

Zhang C, Xie H, Cheng X, Wang DW, Li Y, Xu CL, Huang X (2015) Molecular identification and functional characterization of the fatty acid-and retinoid-binding protein gene Rs-far-1 in the burrowing nematode *Radopholus similis* (Tylenchida: Pratylenchidae). PLoS ONE 10(3):e0118414

Zhang L, Gleason C (2018) Loop-mediated isothermal amplification for the diagnostic detection of *Meloidogyne chitwoodi* and *M. fallax*. Plant Dis 1–27. https://doi.org/10.1094/pdis-01-18-0093-re

Zhao Y, Zhou Q, Zou C, Zhang K, Huang X (2022) Repulsive response of Meloidogyne incognita induced by biocontrol bacteria and its effect on interspecific interactions. Front Microbiol 13:994941

Zhou Q, Cai Y, Gu J, Wang X, Chen J (2017) Rapid and sensitive detection of meloidogyne mali by loop-mediated isothermal amplification combined with a lateral flow dipstick. Eur J Plant Pathol 148:755–769

Zhou S, Gou T, Hu J, Wu W, Ding X, Fang W, Hu Z, Mu Y (2019) A highly integrated real-time digital PCR device for accurate DNA quantitative analysis. Biosens Bioelectron 128. https://doi.org/10.1016/j.bios.2018.12.055

Žibrat U, Gerič Stare B, Knapič M, Susič N, Lapajne J, Širca S (2021) Detection of root-knot nematode meloidogyne luci infestation of potato tubers using hyperspectral remote sensing and real-time PCR molecular methods. Remote Sens (Basel) 13. https://doi.org/10.3390/rs13101996

Advances in the Management of Bacterial Diseases of Vegetable Crops

Sabyasachi Banerjee

Abstract Global agricultural production is severely hampered and substantial yearly losses are sustained as a result of plant diseases resulting from bacterial infections. Consistently successful control of these diseases may be exceedingly challenging to achieve, especially when growers rely on highly disease-susceptible cultivars in response to market demands and when climatic circumstances favor the growth of pathogens. However, there is a broad list of bacterial disease concerns for which there are few effective treatment alternatives, including both new and emergent diseases (like potato's zebra chip) and old problems in new geographical locations (like kiwifruit's bacterial canker in New Zealand). Due to a growing human population and decreasing farmland, there's an urgent necessity for enhanced food production on a stable, reliable basis. The discovery and implementation of sustainable disease control strategies for bacterial infections is a crucial component in ensuring the long-term viability of agricultural production systems that consistently provide high yields. Additionally, due to the rising development of resistance to conventional bactericides, the discovery of innovative management techniques has come to the forefront. There are a variety of growing research emphases that will serve as a solid foundation for future bacterial disease management. These include fundamental research to discover essential pathogen targets for control, innovative approaches, and ways of administration. This chapter discusses a variety of innovative approaches now being utilized to manage microbial diseases in crops to maximize the growth of a sustainable and efficient food supply.

Keywords Bacterial diseases · Vegetables · Crops · Techniques · Agriculture · Biological control

---

S. Banerjee (✉)
Department of Pharmaceutical Chemistry, Gupta College of Technological Sciences, Ashram More, G.T. Road, Asansol, West Bengal 713301, India
e-mail: sabyasachibanerjee04@gmail.com

J. Chen et al. (eds.), *Molecular and Biotechnological Tools for Plant Disease Management*, Interdisciplinary Biotechnological Advances,
https://doi.org/10.1007/978-981-97-7510-1_4

1 Introduction

The presence of essential nutrients in vegetables is of paramount importance in the preservation of human health, as they serve as vital sources of energy for diverse metabolic processes occurring within the human body (Pujeri et al. 2015). India contributes for around 14% (approximately 200 million metric tons in the year 2022) of global vegetable production, making it the second-largest producer after China. Vegetables have a greater rate of return per area and time invested than any other crop. According to Pujeri et al. (2015), vegetable consumption will rise to 350 million tons by 2030 due to the nation's 1.8% annual population growth (Pujeri et al. 2015; Shahid et al. 2017). There is no such thing as a vegetarian diet since vegetables are an integral part of every cuisine on the planet. They provide the bulk of our daily vitamin and nutrient intake, making them essential to a healthy diet. Bioactive substances such as glucosinolates, phenolics, carotenoids, vitamins C and E, and selenium are abundant in plants (particularly Brassica vegetables like cabbage, broccoli, brussels sprouts, kale, and cauliflower). Due to their accessibility in local markets, low cost, and customer desire, these vegetables are among the most significant vegetables eaten worldwide (Herr 2010). Vegetable phytochemicals have been shown to protect against oxidative stress, lower cancer risk, boost the immune system, and block tumor development and malignant mutations (Herr 2010; Kestwal et al. 2011). Brassica glucosinolates may reduce inflammation and oxidative stress, according to certain research (Wu et al. 2004; Noyan-Ashraf et al. 2006). Despite its importance to human health, vegetable output varies greatly among nations. Vegetables cultivated in various agronomic locations suffer a significant yield loss owing to plant diseases. Artificial fertilizers and insecticides are among the many agrochemicals used worldwide to increase vegetable production, maintain crop quality, and curb crop losses caused by plant diseases (Ngowi et al. 2007). However, when these compounds are applied more than their allowed levels, they alter soil fertility and have a negative impact on vegetable yield. In light of the worldwide menace that phytopathogens pose to vegetable cultivation and the adverse consequences of agrochemicals employed to regulate these pathogens, researchers are seeking a secure and efficacious substitute for the administration of vegetable ailments. The utilization of microorganisms has resulted in a certain degree of alleviation from the exorbitant expenses associated with chemicals. Here, the effort is focused on learning the mechanisms behind how soil microorganisms inhibit diseases and identifying the best candidates for increasing crop yields without resorting to harmful pesticides (Kanjanamaneesathian 2015).

Numerous horticultural crops, such as vegetables, face serious threats from soil-borne plant infections such as fungi, nematodes, bacteria, and viruses. These pathogens provide a significant issue because their population in the soil is often increased by intensive farming systems with limited rotating crop techniques. In addition, pest organisms have a long lifespan and are unable to be removed from the ground. Therefore, succeeding crops planted in the same area might be further damaged by the disease complexity caused by the infection from various pathogens in the soil. When it comes to inflicting damage and financial losses on vegetables, fungi stand out as the most significant and widespread category of diseases (Salau et al. 2015). Most vegetable production loss may be attributed to fungal diseases like Rhizoctonia, Fusarium, Verticillium, etc., which cause wilt and powdery scab. Losses in agricultural output due to fungal infections may reach as high as 12% before harvest, and they can range from 10 to 30% after harvest in underdeveloped countries (Kim et al. 2003; Tripathi and Dubey 2008; Fatima et al. 2009). Cucumber production losses worldwide range from 70 to 100% owing to *Fusarium* spp. It is estimated that soilborne illnesses account for more than 70% of crop loss in India (Egel and Martyn 2007).

Pathogenic fungi thrive on vegetables because of their lower pH, higher water content, and nutritional content. Improper procedures during production, packing, storage, and transport can contribute to the spread of various diseases (Mari and Guizzardi 1998; Sharma 2006). Vegetable quality and yield may be impacted by the presence, growth, and colonization of fungus. Numerous fungi, such as *Aspergillus parasiticus*, *A. flavus*, *Fusarium* spp., and *Penicillium* spp., are recognized for causing severe illnesses in various kinds of vegetables. These fungi also produce mycotoxins that render the food hazardous for human consumption (Brewer et al. 2013). Mycotoxins contaminate more than 25% of the global vegetable supply, with over 300 fungal metabolites posing potential hazards to both human and animal health. Mycotoxins have been found to induce various types of harm, including but not limited to edema, toxic hepatitis, hemorrhage, hepatic carcinoma, and immunosuppression. These adverse effects are attributed to the genotoxicity, carcinogenicity, nephrotoxicity, teratogenicity, hepatotoxicity, and reproductive problems associated with mycotoxins (Makun 2013). Several fungi that are prevalent in India, such as *Phoma* spp., *Penicillium* spp., *Fusarium* spp., *A. parasiticus*, *A. flavus*, *A. tamarii*, *Cladosporium cladosporioides*, *A. candidus*, *A. nidulans*, *Aspergillus niger*, *Alternaria alternata*, and *Curvularia* spp., are known to cause significant damage to various types of vegetables (Reddy 2014). Soil is the preferred habitat for these fungi, which have been traced to infecting over 400 plant species, including vegetables, legumes, and grains. In the Gangetic Plains of West Bengal, it was observed that the reduction in yield in tomato crop caused by *A. solani* varied from 0.75 to 0.77 tonne per hectare with an increase of one percent in the severity of the disease over two years (Saha and Das 2012). The occurrence of black spot or leaf spot disease can be attributed to various species of *Alternaria*. This particular ailment is commonly observed on plants belonging to the Solanaceae family as well as the *Brassica* genus (Mamgain and Roychowdhury 2013). The disease in examining resulted in yield reductions ranging from 35 to 70%. Furthermore, the seeds that were affected by the disease frequently exhibited fungal growth both internally and externally (Srivastava et al. 2011). The soil-borne plant disease Fusarium *spp.* is another major factor in the drastic decline in vegetable yields. Infected plants may show symptoms such as root rot, wilting, root cortical deterioration, yellowing, rosette, and premature death due to a variety of *Fusarium* species. One of the greatest dangers to global tomato output is the *Phytophthora infestans*-caused disease known as late blight. *P. infestans* is capable of causing damage to the leaves and stem of tomato plants, as it disseminates via asexual sporangia that are transmitted through the air throughout the period of plant growth. Fruit rots in the field or after harvest, both due to the late blight. Diseases cause numerous challenges for vegetable crops, including slow-growing rate, a decline in crop quality, a decrease in productivity and yield, deformities in plant tissues or entire plants, a rise in production expenses, making veggies unappealing and unsellable, dying of the entire plant and lastly, cutting down farmers' or cultivators' profit margins (Shahid et al. 2017) (Table 1).

2 Managing Vegetable Crop Bacterial Diseases

2.1 Utilization of Copper and Antibiotics for Managing Bacterial Diseases and the Subsequent Development of Resistance

The management of bacterial diseases at the individual level using chemical methods has been influenced by various factors. These factors include the ability to reach the microbe present on plant surfaces, the presence of effective mechanisms of action, the susceptibility of the pathogen to specific chemicals, the economic importance of the affected crops, and the market demand for the industrial application of these chemicals. In contrast to fungicides, a limited number of chemical agents aimed at controlling bacterial diseases in plants have been commercially introduced. The primary factor contributing to this phenomenon is the fact that antibiotics are the most effective antibacterial agents currently available, and their development has primarily been geared toward clinical applications rather than plant agriculture. The early attempts at chemical regulation of bacterial disorders employed a comprehensive strategy, whereby a diverse array of accessible substances were tested toward a broad spectrum of maladies (Zaumeyer 1958). The findings of the study indicate that copper compounds and the antibiotic streptomycin have demonstrated significant efficacy as bactericidal spray treatments for managing bacterial ailments in plants. The compounds in question were first developed in the 1880s and 1950s, respectively. They have since gained significant popularity for their ability to selectively target bacterial pathogens, including *E. amylovora*, *Xanthomonas* spp., and *Pseudomonas* spp. (McManus et al. 2002). While bactericides seemed to be effective in managing diseases, the utilization of streptomycin with copper has been hindered via the emergence of resistance in the plant pathogen population (McManus et al. 2002; Cooksey 1990). Utilization of antibiotics with copper sprays on a large scale and for prolonged periods, or the application of a high frequency of such sprays within a single season, is associated with the emergence of resistance in pathogenic populations. Moreover, it is noteworthy that resistance has predominantly emerged through the acquisition of genetic elements that encode determinants of resistance. This phenomenon encompasses the horizontal transmissions of factors influencing resistance throughout agriculture ecological systems, which also affects non-target microbiota. The prevailing resistance to antibiotics turmoil that is impacting people

Table 1 List of common vegetable crop bacterial infections and their symptoms (Saxena et al. 2020)

Crops affected	Bacterial diseases	Symptoms	Dissemination facilitating factors
Potato, tomato; sweet potato, capsicum, carrots	*Erwinia* spp., *Pseudomonas* spp.	Wet, squishy, slimy rot affects vegetable heads, edible roots, curds, leaves, and stems. May have a disagreeable stink	Humid and warm conditions
Capsicum, tomato, chilli	Clavibacter michiganensis (cankers)	Stems discolor; Plants and seedlings may die; Stem and fruit cankers; Older plants feature one-sided yellowing and wilting leaves	High humidity and moderate temperature
Tomato, potato, eggplant, capsicum	*Ralstonia solanacearum*	Plants wilt, yellow, and stunt, but sometimes they just wilt and die quickly without any spotting or yellowing at all; vascular tissue looks brown and water-soaked, and a white ooze develops when pressure is applied to damaged tubers or stems	Extreme heat, saturated soil, and inadequate drainage. Hot, dry weather makes symptoms of an illness worse since it speeds up the process of wilting
Tomato	*Pseudomonas syringae*	Raised black dots on fruit and foliage encircled by a golden halo	Overhead sprinklers and high humidity
Capsicum, cucurbits, tomato	*Xanthomonas campestris*	Capsicum and tomato: Sticky patches that darken from brown to black on the leaves and stems; the fruit may develop a circular mark with a center scab Cucurbits: Small water-soaked or greasy patches appear first on the underside of leaves, followed by yellowing on the top side; light-brown ooze may be produced from water-soaked marks on fruit	windy weather and overhead irrigation
Peas	*Pseudomonas syringae*	Spots on leaves and stipules that absorb water and become black in cold weather or dark brown in warmer temperatures are called water spots Pods that have been waterlogged have sunken, dark brown areas	Wet and windy with a low temperature

worldwide is attributed to horizontal gene transfer, as proven by various research. This mechanism enables bacteria to acquire resistance determinants even from phylogenetically distinct species (Perry and Wright 2013; Smillie et al. 2011). The ubiquitous presence of similar antibiotic resistance determinants among bacterial species across diverse geographical regions is indicative of the interconnectedness of global ecosystems (Blair et al. 2015; Forsberg et al. 2012; Sundin 1996). The manifestation of plant pathogenic organisms and other bacterial colonies that are linked with plants in these interconnected ecological systems on a global scale was also observed upon the disclosure of the genetic factors that confer resistance to copper and streptomycin, which will be further elaborated on. The utilization of antibiotics is prevalent in both the animal and plant kingdoms and is also employed in animals to stimulate growth. The extensive dissemination of strAB genes across various bacterial genera inhabiting diverse environments is a clear indication of the remarkable potential and scope of horizontal gene transfer among bacterial colonies as a means of propagating ecology-friendly advantageous characteristics (Sundin 1996). Several other antibiotics were employed as substitutes for streptomycin, mainly for the reason of resistance or in certain pathosystems in which streptomycin's efficacy is limited. The utilization of Oxytetracycline has been observed in the United States and Mexico to manage fire blight (*E. amylovora*) in pome fruit trees, while in the United States, it has been used on nectarine and peach to regulate *Xanthomonas arboricola* pv. pruni growth which causes bacterial spots. Additionally, in Latin American countries, it has been employed on vegetable crops to target *Xanthomonas* spp. and *Pseudomonas* spp. (McManus et al. 2002). Latin American countries have utilized gentamicin for managing fire blight and diverse vegetable diseases, while in Israel, oxilinic acid is being employed for controlling fire blight (Shtienberg et al. 2001). A study reports the detection of gentamicin-resistant bacteria in lettuce from oxolinic acid and Costa Rica resistance in *E. amylovora* in Israel, which occurred one to three years after its introduction (Kleitman et al. 2005; Rodríguez et al. 2006). Kasugamycin was approved for application in the United States to control fire blight, particularly in orchards that harbor *E. amylovora* strains that are resistant to streptomycin. This antibiotic has been utilized in European Union nations for over thirty years, but its use was prohibited in 2007 following the ban on antibiotic application. In Japan, kasugamycin has been employed as a seed treatment for bacterial infections. Kasugamycin's applications are limited to plant agriculture and its spectrum of activity appears to be narrower than that of streptomycin. The utilization of antibiotics was hindered or prohibited in particular global regions due to the possible influence of antibiotic usage upon the transmission of antibiotic resistance to pathogenic microorganisms. Furthermore, based on the prevalence of horizontal gene transfer and the acquiring of streptomycin-resistant genes by pathogens of plants across various ecological circumstances, it can be inferred that the emergence of antibiotic resistance is an inevitable outcome following the introduction of a novel antibiotic. The primary veracity pertaining to the utilization of antibiotics for crop ailments is that any approach that favors developing resistance in the intended pathogen is not viable in the long run.

2.2 *Exploring Substitutes for Copper-Based Bactericides and Antibiotics in the Regulation of Bacterial Diseases*

The recognition that the utilization of copper and/or antibiotics to manage bacterial diseases is suboptimal or restricted by regulatory measures has spurred investigations and explorations into alternative approaches. This section briefly examines the implementation of biological control, the application of systemic acquired resistance (SAR), and inducers of the utilization of anti-bacterial peptides.

The concept of biological control entails the utilization of living organisms as a means to effectively regulate or manage populations of undesirable organisms such as weeds, pests, or other harmful entities. The approach to pest management under consideration involves the utilization of natural adversaries of the target organism to reduce its population, rather than using chemical pesticides or other synthetic methods. Biological control can be used in a variety of settings, including forestry, agriculture, and urban environments, and represents a viable and environmentally sustainable strategy for the control of pests. The term "biological control" encompasses the utilization of advantageous microorganisms, as well as by products or extracts derived from plants or animals, to mitigate plant diseases. Biological control agent (BCA) population management practices can be categorized generically as either inundated or augmentative. In the former, the BCA was employed in adequate numbers to curtail the disease, even in the absence of reproduction. In contrast, the latter involves introducing the BCA in sufficient numbers to establish a self-sustaining population capable of effectively suppressing the disease (Johnson 2010). In certain scenarios, antimicrobial molecules that are secreted by opposing microorganisms are utilized in lieu of the microorganisms themselves. The subject matter pertaining to biological control and its scientific foundations has been thoroughly examined, and our objective is not to provide a redundant summary of this area of study. The aim is to emphasize current endeavors in the field of biocontrol and elucidate novel developments that may impact forthcoming management endeavors. Systems that require management within a limited timeframe seem to offer the most promising opportunities for achieving success with biological control on a commercial scale. The utilization of *Agrobacterium radiobacter* strain K84 is a notable instance of the efficacy of biological control in managing bacterial disorders. This technique is used to prevent crown gall, a disease caused by *Agrobacterium tumefaciens*, from damaging the exposed surfaces of the stems of arboreal plants (Kerr 2016). The utilization of biological control methods has proven to be successful in managing the blossom blight stage of fire blight, specifically within the constrained blooming period of pear or apple trees. The efficacy of this system has exhibited diverse levels of achievement across various regions, a phenomenon that can be ascribed to disparities in environmental circumstances (Johnson 1998; Sundin et al. 2009).

Regrettably, the temporal range of infection is considerably extended for numerous bacterial diseases, as well as soil-borne pathogens and pathogens that cause fruit and leaf spot. This factor significantly diminishes the likelihood of achieving consistent efficacy with biological control measures. Challenges arise in such scenarios due to the requirement of maintaining ample populations of antagonistic agents to safeguard the host tissues at precise locations where pathogens establish colonization. The utilization of hypersensitive response and pathogenicity (hrp) mutants as antagonistic competitors has shown some success in these scenarios. However, the absence of commercial development of such strains suggests that there may be inherent limitations in the implementation of this strategy that could potentially hinder its effectiveness (Hert et al. 2009; Moss et al. 2007).

The deployment of certain biocontrol bacteria, such as *Pantoea agglomerans* and *Burkholderia* spp., has been restricted due to biosafety concerns, particularly their classification as opportunistic human pathogens. This has highlighted the necessity for the identification of new bio-control strains that are categorized as QPS/GRAS (qualified presumption of safety/generally regarded as safe). The utilization of bacteriophage as a means of managing bacterial diseases is a biological control technique that was the focus of scientific investigations at different intervals during the previous forty years (Jones et al. 2007). Bacteriophages are widely distributed in various ecosystems across the globe. Phages with the ability to induce lysis in plant-detrimental bacterial species can be obtained through the isolation process from soil or host plant tissue. The utilization of inundative application strategies is commonly observed in the context of disease management. The effectiveness of bacteriophages is primarily constrained by their environmental stability and the emergence of spontaneous mutations that confer resistance in the pathogenic bacteria they target. The renewed attention toward bacteriophage as a means of controlling disease is primarily observed in the field of clinical medicine, owing to the increasing challenge posed by antibiotic resistance. Similarly, with regard to bacterial diseases in plants, there is a growing interest in the utilization of phages. However, extensive field surveys and the refinement of formulations and spraying schedules are imperative to enable the advancement of efficacious implications. To achieve enhanced and enduring implementation in commercial systems, it is imperative to acquire a comprehensive comprehension of the potential antagonists' ecology, the ways they interact with plants, their favorable and unfavorable correlations with the plant metagenome of the microbiota, and their connections with pathogens. The utilization of functional genomic analyses seems to be a reliable approach to acquiring information regarding the present biocontrol agents that are considered the best in the field. However, it is imperative to combine this method with extensive and focused field endeavors to evaluate the conduct of these organisms in various environmental conditions and plant host variations (Massart et al. 2015).

2.2.1 Anti-Microbial Peptides

Anti-microbial peptides (AMPs) are short chains of amino acids, usually less than 50, which are synthesized by oomycetes, fungi, and bacteria. These peptides function as a mechanism of inter-microbial competition. Moreover, AMPs are produced by plants and animals as a component of their innate immune system in responding to microbe issues. A significant proportion of these peptides exhibit a (+)ve charge and have the ability to either disrupt cell membranes through insertion or hinder nucleic acid or protein synthesis upon cellular uptake (Brogden 2005). Research into the utilization of peptides for controlling plant bacterial diseases has been underway for over ten years. The principal techniques utilized encompass the administration of AMPs onto the exterior of plants, the exploitation of indigenous fungus or bacterial inhibitors that generate and discharge AMPs as a biological regulation, the expression of an animal defensin, like cecropin, in a genetically altered plant, or the application of an improved synthetically produced version of plant or animal defensins in a genetically modified context (Montesinos 2007). In specific scenarios, such as those involving specialty food crops, the application of AMPs onto agricultural plants may present advantages over the utilization of transgenic plants.

This phenomenon can be attributed to the derivation of certain highly efficient AMPs from amphibians and insects. Nonetheless, a crucial impediment to this procedure is the expense associated with the production of AMPs. The utilization of plants as "biofactories" for the synthesizing of AMPs, has the potential to be a financially viable method of production. This could ultimately aid in managing bacterial diseases through the usage of AMPs (Nadal et al. 2012; Parachin et al. 2012). Despite the productivity and maturity of the AMP field, the potential of AMPs for plant disease management is still a subject of ongoing research and development, primarily focused on the screening of new candidates (Breen et al. 2015).

2.2.2 Systemic Acquired Resistance, Induced Systemic Resistance, and Associated Tactics

The activation of SAR in plants can be achieved by administering a chemical inducer, such as acibenzolar-S-methyl (Ngowi et al. 2007), which is a product available in the market. This approach has been regarded as a significant accomplishment in the translation of fundamental research discoveries into practical disease control measures (Durrant 2004). In many pathosystems, including tomato bacterial wilt and canker, and fire blight, foliar applications of ASM have been used to prime plants for exposure to pathogen challenge (Maxson-Stein et al. 2002; Pradhanang et al. 2005; Sen et al. 2015). Soil drenches containing ASM and other SAR inducers have been shown to effectively manage citrus canker on immature grapefruit plants (Graham 2011). Furthermore, there is an ongoing assessment of SAR inducers as potential solutions for addressing emerging and consequential disease problems, such as Huanglongbing in Florida (Li et al. 2016). The effectiveness of ASM as a standalone material is generally inadequate. However, when ASM is integrated into a comprehensive program, it can lead to the attainment of the desired control levels. This integration also reduces the need for other materials like copper or antibiotics (Roberts et al. 2008).

Trichoderma spp., *Bacillus* spp., *Pseudomonas* spp., and other beneficial rhizosphere bacteria and fungi are used to stimulate plant defense systems and apply induced systemic resistance (ISR). Plant hormones play a key role in facilitating this process (Pieterse et al. 2014). The efficacy of ISR as a management strategy for various bacterial pathogens such as *R. solanacearum*, *Xanthomonas oryzae* pv. oryzae, and *P. syringae* has been demonstrated by Kloepper et al. (2004). Further strategies utilized to augment plant resistance against pathogenic bacteria, specifically *Ralstonia solanacearum*, involve the utilization of inorganic constituents, such as silicon. The potential of silicon to strengthen structural defenses against pathogens and activate resistance via hormone-controlled signaling pathways, akin to ISR, has been postulated and demonstrated by Ghareeb et al. (2011).

2.3 Managing Vegetable Crop Fungal Diseases

The safety of chemicals utilized for the management of phytopathogens is contingent upon their adherence to regulatory thresholds and proper application techniques. Pesticides, including fungicides, are widely utilized in the agricultural industry to safeguard crops, such as vegetables, against insect pests, both pre-and post-harvest, with the aim of enhancing food security. However, despite their benefits, pesticides have been known to have adverse health effects on consumers throughout the food chain (Aktar and Sengupta 2009; Damalas 2011). The application of pesticides has experienced a surge in recent times owing to its expeditious action and ease of application in comparison to alternative pest control measures, as noted by Gilden et al. (Gilden and Huffling 2010).

In India, the cultivation of horticultural crops is often plagued by significant pest infestations, resulting in the utilization of approximately 13–14% of pesticides during the production of vegetables (Agnihotri 1999). Despite their potential environmental hazards, protective fungicides continue to be utilized for managing fungal diseases among pesticides (Vaish and Sinha 2003).

The management of *Phomopsis vexans*, the causal agent of phomopsis leaf blight in brinjal, was documented in literature through the application of fungicides that are readily available in the market, including Carbendazim, Captan, Dithane M-45, and Antracol (Rohini et al. 2015). Among the four fungicides evaluated, carbendazim exhibited the greatest degree of disease protection. Therefore, it is recommended that this particular fungicide be employed as a means of managing phomopsis leaf blight in brinjal.

Singh and Agrawal (1999) found that carbendazim exhibited significant efficacy toward *P. vexans* in field conditions, leading to a disease incidence rate of only 4.3% and a fruit yield of 222.83 quintal/ha, which was the highest recorded (Singh and Agrawal 1999). Habib and colleagues (2007) conducted an assessment of four fungicides, namely Benlate, Topsin, Dithane M-45, and captan, in relation to their efficacy against the seed-borne mycoflora of eggplant. The study revealed that those fungicides exhibited detrimental effects on the fungal pathogens *A. alternata* and *F. solani* (Habib et al. 1925). Thippeswamy et al. discovered that carboxin and mancozeb were efficacious in mitigating phomopsis fruit rot in brinjal during field experiments (Thippeswamy et al. 2006). Beura et al. reported on the efficacy of six fungicides against *P. vexans*, the causal agent of phomopsis blight and fruit rot of brinjal. In field trials, it was observed that carbendazim at a concentration of 0.1% exhibited superior efficacy in managing fruit rot and twig blight diseases in brinjal when compared to the other five fungicides tested (Beura and Mahanta 2008). In their study, Islam and Meah employed various fungicides, including Vitavax and Bavistin, as well as botanicals such as allamanda and garlic, and bioagents like *Trichoderma harzianum* T22, *T. harzianum* CP, and Trichoderma *spp.* EP, for the purpose of seed treatment in brinjal. The efficacy of all treatments in managing eggplant seedling diseases in the nursery bed was observed (Islam and Meah 2011). Hossain and colleagues in the year 2013 conducted a study wherein they utilized four fungicides and three micronutrients to combat *P. vexans*, the causative agent of phomopsis blight and fruit rot disease in eggplant, under in vitro conditions. The researchers noted that the application of Bavistin at a concentration of 0.1% demonstrated a higher level of efficacy in inhibiting both spore-germinating and mycelial growing of *P. vexans*. On the other hand, while micronutrients did not demonstrate a substantial impact on this disease, they were found to be significantly more efficacious compared to the control group (Hossain et al. 2013). Sahar et al. conducted an investigation to evaluate the effectiveness of 4 systemic fungicides (Difenoconazole, Aliette, Nativo, and Topsin-M) and a bio-agent *Bacillus subtilis* in controlling *Fusarium oxysporum* f. sp. melongenae. The study was conducted in laboratory and greenhouse settings using the poison food technique (Sahar et al. 2013). Among the formulations tested, Topsin-B and -M. *B. subtilis* exhibited significant efficacy in reducing both the growth of mycelial and disease prevalence. A study performed by Muneeshwar et al. revealed that the effectiveness of 2 fungicides and a bio-agent in inhibiting the mycelial growth of *P. vexans*, which is responsible for spreading brinjal fruit rot and leaf blight disease, was inconsistent (Sharma and Razdan 2012). A subsequent investigation was carried out by Pani and Singh (2013) to manage phomopsis blight and fruit rot of brinjal, which is attributed to *P. vexans*. The field trial was conducted over two years, from 2011 to 2012. The researchers employed a seed treatment method utilizing carboxin + thiram at a rate of 2 gm/kg, in addition to a foliar utilization of copper oxychloride at a concentration of 0.3% within the field. Both treatments exhibited a significant reduction in the occurrence of seedling-blight, -mortality, and fruit rot infections, while also resulting in a substantial increase in the yield of brinjal (Pani and Singh 2013). The sporulation and growth of various fungi responsible for post-harvest fruit rot of chili were found to be completely inhibited by carbendazim and mancozeb (Suryawanshi 2001; Pan and Acharya 1995). According to Das and Mohanty, Dithane M-45 and Bavistin exhibited high efficacy in controlling anthracnose and fruit rot in chili (Das 1988). In another study, Rahman and Bhattiprolu employed the usage of fungicides and mycorrhizal fungi in treating seeds and soil to manage fungal disorders such as damping-off in brinjal, tomato, and chili during the nursery phase (Rahman and Bhattiprolu 2010). According to research by Sabalpara et al. various fungicides including thiophanate methyl (Topsin-M 45–75% wp), carbendazim (Bavistin 50% wp), propiconazole (Tilt 25% EC), mancozeb (Dithane-M 45–75% wp), carbendazim 25% + iprodione 25% (Quintal 50% wp), copper hydroxide (Kocide 75% wp), copper oxychloride (Blitox 50% wp), flusilazole (Nustar 40% EC), aureofungin (Aureofungin 46.15% SP), mancozeb 63% + carbendazim 12% (Sixer 75% wp), and fenamidone 10% + mancozeb 50% (Secure 60% WDG) were evaluated for their effectiveness against *P. vexans*, the causal agent of fruit rot in brinjal, using the poisoned food technique in PDA medium (Rahman and Bhattiprolu 2010). The application of Carbendazim and Thiophanate Methyl at varying concentrations of 250, 500, and 1000 ppm resulted in significant inhibition, ranging from 93 to 100%, of the pathogen's mycelial growth. In their study, Rahman et al. in the year 1988 conducted an evaluation of various fungicides for their efficacy against onion purple blotch. The findings of the study revealed that mancozeb exhibited the greatest efficacy in disease management. In previous studies, it has been observed that mancozeb, a nonsystemic fungicide, was able to inhibit *Alternaria porri* growth (Rahman et al. 1988). Specifically, Chethana et al. found that a concentration of 0.3% of mancozeb resulted in complete inhibition of growth (Chethana et al. 2011), while Mishra and Gupta reported that a concentration of 0.2% was effective against *A. porri* (Mishra 2012). Madhavi et al. conducted a study wherein they evaluated the efficacy of 5 fungicides, namely carbendazim, mancozeb, captafol, benlate, and blitox, against *A. porri*. The results indicated that mancozeb exhibited the highest effectiveness, followed by Blitox and benlate. The efficacy of fungicide concentrations ranging from 10 to 500 ppm in suppressing fungal growth was demonstrated (Madhavi et al. 2012). Rahman et al. (1988) reported on the effectiveness of various fungicides, including dicloran and mancozeb, in treating the purple blotch disease. Their findings indicated that mancozeb was the highest efficient fungicide for treating this disorder. In a subsequent study, the effectiveness of various systemic and non-systemic fungicides was assessed in vitro toward *A. porri*, the pathogen responsible for purple blotch disease in onions. The systemic fungicides tested included zineb (Indofil Z-78 75% WP), chlorothalonil (Kavach 75%WP), propineb (Antracol 70% WP), and mancozeb (Agastya M-45 75%WP), while the nonsystemic fungicides included difenoconazole (Score 25 EC), propiconazole (Tilt 25 EC), tebuconazole (Folicur 25 EC), and hexaconazole (Contaf 5 EC). The study assessed various nonsystemic fungicides for their efficacy against *A. porri* in vitro. Results indicated that mancozeb 75WP exhibited the highest inhibitory effect (89.51%) on the pathogen's mycelial growth, followed by chlorothalonil (84.94%), zineb (78.15%), and propineb 70%WP (86.30%) (Rahman et al. 1988). According to Priya et al., the mycelial growth of the pathogen was inhibited to the greatest extent by propiconazole 25 EC among the systemic fungicides tested. Difenoconazole 25 EC, hexaconazole 5 EC 70%, and tebuconazole 25 EC followed in order with inhibitory effects of 86.30%, 91.11%, and 80.86%, respectively (Priya et al. 2015).

2.4 *Bio-Management of Diseases Affecting Vegetables*

Fungal pathogens are a prominent cause of agricultural yield losses among various types of phytopathogens. Fungicides have been employed in agricultural practices worldwide to minimize crop losses and enhance crop production. The adverse environmental impacts associated with fungicides, as reported by Thabet et al. have compelled researchers to explore alternative methods for managing ailments (Thabet and Shendy 2016). Additionally, their detrimental effects on beneficial soil microorganisms (Solecki et al. 2005), the emergence of fungicide resistance among pathogens (Hobbelen et al. 2014), as well as the expenses and inadequate application/delivery methods available to farmers with limited resources, underscore the necessity of exploring alternative approaches to managing insect pests. In light of these considerations, there has been a recent shift in focus toward the utilization of certain biological resources that are deemed to be environmentally benign, cost-effective, and readily implementable (Satish et al. 2007; Taiga et al. 2008; Kotan et al. 2016). The utilization of biological materials, specifically non-pathogenic soil microbiota, for the purpose of managing plant pathogens, commonly referred to as the "biological control" approach, has presented a potential resolution to the costly and ecologically hazardous chemicals. In general, the concept of "biological control" pertains to the utilization of one or multiple organisms, either in combination or in isolation, for the purpose of eradicating harmful microorganisms. Plant growth-promoting rhizobacteria (PGPR) have been widely utilized as biological control agents to manage disease and improve crop yield, particularly in vegetables. PGPRs are a prevalent type of soil microflora (Kloepper 1978) and have been significantly studied for their effectiveness in controlling ailments (Beneduzi et al. 2012) and yield enhancement (Zaidi et al. 2015). Numerous PGPR species have been recognized as prevalent bio-control agents. These include Alcaligenes, Actinoplanes, Azotobacter, Bacillus, Enterobacter, Flavobacterium, Hafnia, Arthrobacter, Erwinia, Actinobacteria, Pasteuria, Pseudomonas, Rhizobium, Cellulomonas, Serratia, Bradyrhizobium, and Xanthomonas (Bjelić et al. 2015; Chen et al. 2016; Gunes et al. 2015; Zamoum et al. 2015; Panpatte et al. 2016; Kokalis-Burelle 2015; Ahmed et al. 2016). *Bacillus* and *Pseudomonas* spp. were extensively employed as bio-control agents for disease suppression respectively (Rahman et al. 2016; Mehrabi et al. 2016). Bacillus species are classified as Gram-positive bacteria and can endure unfavorable environmental circumstances by generating endospores. These microorganisms are particularly advantageous for the development of commercial formulations. *Pseudomonas* spp. have garnered significant interest as biocontrol agents due to their root-colonizing capabilities, catabolic versatility, and production of antifungal metabolites such as pyrrolnitrin, 2,4-diacetylphloroglucinol (DAPG), phenazines, and pyoluteorin. Extensive research has been conducted on Gram-negative bacteria due to their notable ability to produce a diverse array of antifungal compounds, which make them effective biocontrol agents. Bottiglieri and Keel have reported the production of DAPG (Bottiglieri et al. 2006), while Yan et al. have demonstrated the synthesis of pyoluteorin (Yan et al. 2007). The production of pyrrolnitrin was reported by Hammer et al. (1997), and Mavrodi et al. reported the production of phenazines (Mavrodi et al. 2006).

2.4.1 The Mechanism Underlying the Repression of Disease by Bio-Control Bacteria

Siderophores

Siderophores, which are lower molecular weight components that chelate iron, are commonly prepared by specific microorganisms and plants (Gupta et al. 2015). These compounds are also known to participate in nutrient competition and antibiosis (Shanmugaiah et al. 2015). The synthesis of siderophores, namely pseudobactin, and pyoverdine, has been recognized as a novel approach employed by PGPR to manage harmful phytopathogens that pose a threat to various crops, including vegetables (Hassen et al. 2016; Iqbal et al. 2012). Siderophores serve as a crucial mechanism for facilitating iron uptake by the producing organism in situations where iron availability is limited. The efficacy of the siderophore produced by *B. subtilis* CAS15 in managing Fusarium wilt (*F. oxysporum* Schl. f. sp. capsici) of pepper has been reported by Yu et al. (2011). Similarly, the siderophore produced by *Bacillus amyloliquefaciens* (DSBA-11 and -12) has demonstrated efficacy in the management of bacterial wilt (*Ralstonia solanacearum*) of tomato, as reported by Singh et al. (2016).

Antibiosis

The suppression of diseases is significantly influenced by antibiotics that are secreted by PGPR (Reddy 2014). Various antagonistic bacteria are known to produce antibiotics such as DAPG, butyrolactones (Lanteigne et al. 2012), oomycin A, kanosamine, pyoluteorin, xanthobaccin, zwittermicin A, pyrrolnitrin, oligomycin A, phenazine-1-carboxylic acid, and viscosinamide (Whipps 2001). Additionally, Ulloa-Ogaz et al. have reported that various antibiotics, including rhamnolipids, aerugine, phenazine-1-carboxamide, azomycin, pseudomonic acid, cepaciamide A, anti-tumor antibiotics caspofungins, FR901463, and anti-viral anti-biotic karalicin, exhibit antiviral, insecticidal, antimicrobial, antihelminthic, antioxidant, phytotoxic, and cytotoxic properties, as well as the capacity to enhance the growth of plants (Ulloa-Ogaz and Muñoz-Castellanos 2015). According to Raaijmakers et al., a considerable proportion of anti-biotics exhibits a broad-spectrum activity (Raaijmakers et al. 1999).

Parasitism or Lysis

The utilization of hydrolytic enzymes by pathogenic fungi to facilitate parasitism represents an essential mechanism in the biological control of fungal infections. According to Huang et al. and Xiao et al., certain strains of PGPR can secrete a significant amount of lytic enzymes, which exhibit antifungal properties (Huang et al. 2005; Xiao 2009). According to Ramyabharathi et al., *B. subtilis* strain EPCO 16 seemed to produce potent enzymes such as β-1,3-glucanase and chitinase, which exhibited strong inhibitory effects against *F. oxysporum* f. sp. lycopersici, a pathogen that affects tomato plants (Fig. 1) (Ramyabharathi et al. 2012). According to Dubey et al., BSK17, a strain of *B. subtilis*, has been observed to exhibit competence and antagonistic activity through the production of chitinase and β-1,3-glucanase (Dubey

Fig. 1 Leaf spot disease caused by *Septoria lycopersici* (obtained from Panno et al. 2021, distributed under Creative Commons Attribution 4.0 International License, which permits unrestricted use, distribution, and reproduction in any medium. http://creativecommons.org/licenses/by/4.0/) (Panno et al. 2021)

et al. 2014). According to the findings of Kumar et al., β-1,3-glucanase and chitinase represent a notable group of lytic enzymes responsible for the degradation of primary constituents of the fungal cell walls, including laminarin and chitin (Kumar and Dubey 2012). Chitinases are a significant type of hydrolytic enzyme because chitin, which is a linear polymer composed of *N*-acetyl-D-glucosamine, serves as a primary cell wall component in most phytopathogenic fungi. The antifungal properties of purified chitinases derived from *Serratia plymuthica* (Frankowski et al. 2001), *Trichoderma harzianum* (El-Katatny et al. 2000), and *Streptomyces* spp. (Gomes et al. 2001) were found to be significant. Several strains of plant growth-promoting rhizobacteria (PGPR) secrete various hydrolytic enzymes, including urease, amylase (Shrivastava and Kumar 2017), and catalase (Patel and Pratibha 2014), which are important groups of enzymes with antifungal activity. Furthermore, certain PGPRs that exhibit antagonistic behavior can produce various enzymes such as lipases, proteases (Pereg and McMillan 2015), chitosanases (Kumar et al. 2015), and glucanases (Figueroa-López et al. 2016) that modify chitin and chitosan. These enzymes are utilized to impede the growth of pathogens.

2.4.2 Instances of Some Common Vegetable Diseases Managed by Plant Growth-Promoting Rhizobacteria (PGPR)

Several commonly consumed vegetables, such as broccoli, cucumber, lettuce, onion, and spinach, are susceptible to various diseases, which will be elaborated as follows.

The cultivation of broccoli is often impacted by various biotic agents that lead to the development of diseases, resulting in significant reductions in yield (El-Mohamedy 2012). Pythium root is one of the options available. According to Abdelzaher (2003) and Tanina et al. (2004), rot is considered as the most severe disease. While fungicides have been utilized as a means of mitigating Pythium-related ailments, it is important to note that their implementation can have negative ramifications on the surrounding ecosystem (Abdelzaher 2003; Tanina et al. 2004). Presently, there is a growing focus on utilizing the biological control method, which entails employing antagonistic microorganisms such as bacteria and fungi, as a substitute for fungicides (Gravel et al. 2004). The present study employed *Pseudomonas fluorescens* as a biocontrol agent to manage the incidence of broccoli root rot disease instigated by the pathogens *Rhizoctonia* and *P. ultimum*. According to Slininger et al. (2000), *Pseudomonas fluorescens* 2–79 strain has the ability to synthesize 2-acetamidophenol (AAP) and Phenazine-1-carboxylic acid (PCA) (Slininger et al. 2000). According to Savchuk (2002), *Pseudomonas* spp. DF41 strain has been found to be highly proficient in inhibiting the mycelial growth of *S. sclerotiorum* and also in suppressing the germination of sclerotia and ascospores (Savchuk 2002). According to the findings of Fernando et al. (2007), the observed suppression of mycelial growth in *S. sclerotiorum* can be attributed to the existence of phenazine bio-synthetic genes and the production of PCA by PA-23 (Fernando et al. 2007). According to Georgakopoulos and colleagues' (2002) research, enhanced biocontrol in broccoli was attained through the application of *P. fluorescens* and *B. subtilis* via drenching or seed coating with peat carrier (Georgakopoulos et al. 2002). A pot experiment was undertaken to assess the efficacy of fungus and bacteria isolates in the management of Pythium root rot diseases. The experiment involved the use of soil that was artificially infested under greenhouse conditions. This study was conducted by El-Mohamedy (2012). The experiment involved the testing of two isolates of *T. harzianum* (B1 and G1), two isolates of *T. viride* (G1 and G3), and one isolate each of *P. fluorescens* (B3) and *B. subtilis* (B1). The findings of this investigation demonstrate a significant decrease in the occurrence of Pythium root rot in broccoli plants subsequent to the introduction of microbial inoculants. The combined application of bio-control agents through soil mixing and root dipping methods has been observed to be more effective in suppressing Pythium root rot incidence compared to individual treatments. The observed reductions were 84.4%, 88%, 80.2%, 77.2%, 75%, and 80% for *T. harzianum* B1 and G1, *T. viride* G1 and G3, *B. subtilis* B1, and *P. fluorescens* B3, respectively. The present study found that *T. harzianum* isolates B1 and G1, as well as *T. viride* isolates G1 and G3, exhibited greater inhibitory effects compared to other isolates of *T. viride* and *T. harzianum*. These isolates were observed to completely suppress the growth of the pathogen. The bacteria exhibiting antagonistic behavior demonstrated a noteworthy decrease in the proliferation of *P. ultimum*. The *B. subtilis* B3 isolate and *P. fluorescens* B3 isolate exhibited a greater inhibitory effect. The study's results led to the conclusion that, in general, the inhibitory effect on growth was greater for antagonistic fungi than for bacterial agents (El-Mohamedy 2012).

Cucumber plants are significantly affected by pathogen attacks worldwide. *Fusarium oxysporum* f. sp. cucumerinum is a pathogen that induces wilt disease and was identified as the most deleterious pathogen, according to Zhou et al. (Zhou et al. 2008). Currently, there is a lack of a dependable, economically feasible, and effective approach for the management of *Fusarium* wilt disease in cucumbers. Notwithstanding these challenges, certain measures such as the implementation of fungicidal agents have been employed in order to combat pathogenic invasions (Mishra et al. 2000). Mohamed et al. (2015) conducted a study in which they applied various biocontrol agents, such as *Bacillus pumilus*, *B. subtilis*, zinc oxide nanoparticles, clove and castor oils, as well as recommended rates of fungicide (cymoxanil + famoxadone), to decrease the toxicological effects of fungicides on crop residue. The experiment was conducted for a duration of 2 consecutive growing seasons in a controlled greenhouse environment, with a specific focus on the management of downy mildew disease in cucumber plants. The findings indicate a noteworthy rise in the height of plants, the number of fruits per plant, and the yield of fruits after all treatments in comparison to the control. The study found that *B. pumilus* and *B. subtilis* strains, specifically the EMs1 strain, exhibited noteworthy efficacy in mitigating the incidence of downy mildew in cucumber plants when used as bioagents. The findings of this research indicated that the application of treatment combinations can effectively mitigate the severity of downy mildew in cucumber plants, without any negative impact on their growth or development (Mohamed et al. 2015). Raupach and Kloepper (2000) conducted a study that discovered the effectiveness of PGPR strains *B. pumilus* (INR7 strain), *B. subtilis* (GB03 strain), and *Curtobacterium flaccumfaciens* (ME1 strain) in reducing the incidence of foliar diseases in cucumber. The research focused on the examination of two plant diseases, namely angular leaf spot, which is attributed to *Pseudomonas syringae*, and a concurrent invasion of anthracnose and angular leaf spot, which results from *Colletotrichum orbiculare*. Other researchers have also conducted similar endeavors targeting *P. cubensis*, yielding encouraging outcomes (Raupach 2000). Anand et al. (2007) reported that the cucumber pathogens were significantly reduced by *Bacillus* strains Z-X-3 and Z-X-10, *T. harzianum*, and *P. fluorescens*, resulting in a reduction of 46% (Anand et al. 2007). The research conducted by Rakh et al. in 2011 has confirmed the potential of *Pseudomonas monteilii* as a biocontrol agent for the management of stem rot disease in cucumbers. The utilization of bio-control agents such as *T. viride*, *Gliocladium virens*, and *T. harzianum* has proven to be effective in managing the pathogen, resulting in a reduction of up to 68% (Rakh et al. 2011). Additionally, the production of sclerotia has been decreased by 98% (Pant and Mukhopadhyay 2012; Dutta and Das 2002). On the contrary, lettuce has the potential to be contaminated by various microbes such as fungi, bacteria (*Salmonella* and *E. coli*), soil viruses, and nematodes. As a result, a comprehensive range of over seventy-five lettuce disorders with varying causes and etiologies have been documented (Raid 2004). While certain diseases have limited significance and distribution, others can result in significant yield and quality losses even under favorable conditions. Lettuce is known to be affected by various diseases such as Cercospora leaf spot, anthracnose (*Microdochium panattonianum*), damping-off, grey mold (*Botrytis cinerea*) (Fig. 2), downy mildew (*Bremia lactucae*), Sclerotinia rot (lettuce drop) (*S. minor* and *Sclerotinia sclerotiorum*), Septoria leaf spot, and bottom rot (*Rhizoctonia solani*). Currently, there is a lack of efficient and effective techniques for managing bacterial and fungal diseases. Fiume and Fiume (2005) conducted an assessment of the efficacy of biological control agents for the management of fungal diseases in lettuce, to minimize the dependency on fungicides. *Coniothyrium minitans*, a fungus with antagonistic properties, were employed to target and eliminate the sclerotia present in the soil (Ume and Ume 2005). El-Tarabily et al. (2000) conducted a study wherein they examined non-streptomycete actinomycetes ($N = 35$) and streptomycetes ($N = 94$) in vitro to determine their capability to repress the development of Sclerotinia minor, a pathogen that causes basal drop disease in lettuce. Within these various cultural groups, it was discovered that three distinct isolates (namely, *Micromonospora carbonacea*, *S. viridodiasticus*, and *Serratia marcescens*) exhibited the most potent suppressive qualities, as well as significant antifungal properties. Moreover, these isolates exhibited proficient colonization in both the rhizosphere of lettuce and roots. The in vitro cultivation of *S. minor* exhibited a significant reduction in growth, which can be attributed to the heightened secretion of chitinase and β-1,3-glucanase (El-Tarabily et al. 2000). Certain strains of *Bacillus* have been documented to possess the ability to manage fungal infections in lettuce. The study conducted by Chowdhury et al. (2013) reports that *B. amyloliquefaciens* (FZB42 strain) has the potential to decrease the severity of bottom rot disease in lettuce, which is caused by the soilborne pathogen *R. solani* (Chowdhury et al. 2013). The recognition of the relevance of employing biologic preparations as a substitute for fungicides in managing lettuce bottom rot has been established. Trichoderma has been recognized as a benign alternative among a range of bio-control agents, with no adverse effects on human health or the environment (Pinto and Cipriano 2014). Therefore, there has been a significant rise in the industrial manufacturing of Trichoderma in numerous countries in recent years, as reported by Verma et al. (2007). *Trichoderma* strains are recognized for their ability to synthesize diverse secondary metabolites, such as antibiotics, that exhibit antimicrobial properties. The synthesis of these compounds is contingent upon the particular strain, and they can pertain to a diverse range of non-volatile and volatile compound categories, including but not limited to water-soluble compounds, linear oligopeptides, and peptaibols (Vinale et al. 2008). Peptaibols have been identified as biomolecules that can inhibit the activity of β-glucan synthase in the host fungus. Furthermore, it has been discovered that they exhibit a synergistic effect with β-glucanases, thereby impeding the reconstitution of the fungal cell wall. Vinale et al. (2008) reported that the action of β-glucanases is facilitated by this phenomenon. According to Vinale et al. (2008), *T. viride* strains can generate antibiotics, including gliovirin, gliotoxin, and viridiol. These antibiotics exhibit antagonistic properties against various plant pathogens (Vinale et al. 2008). Maplestone et al. (1991) reported on a study investigating the efficacy of controlling bottom rot using a 0.5% *w/w* concentration of a single isolate of *T. harzianum*. The study involved planting lettuce on beds under plastic tunnels with a density of 1 kg per m^2 of *T. harzianum* (Maplestone et al. 1991).

The onion possesses a multitude of medicinal properties and has been utilized in medicine since ancient times. The insufficient output of onion in numerous countries has been linked to various factors, primarily the prevalence of phytopathogenic

Fig. 2 Grey mold on tomato petioles caused by *Botrytis cinerea* (obtained from Panno et al. 2021, distributed under Creative Commons Attribution 4.0 International License, which permits unrestricted use, distribution, and reproduction in any medium. http://creativecommons.org/licenses/by/4.0/) (Panno et al. 2021)

diseases that result in significant reductions in crop yield. Fungal diseases, including but not limited to purple blotch, leaf blight, basal rot, blight, downy mildew, and storage rots, are considered to be the most severe diseases among various types of diseases. Purple blotch is a highly detrimental ailment that is prevalent in nearly all regions where onions are cultivated, resulting in significant crop damage in field settings. Despite the extensive efforts of numerous researchers in studying this pathogen and its treatment, the persistence of the disease continues to pose a significant obstacle to onion cultivation. In general, the management of these diseases primarily involves the utilization of synthetic fungicides, as noted by Gade et al. (2011). Naguleswaran and colleagues (2014) conducted a field experiment and observed that the application of *Trichoderma viride* through foliar means, in conjunction with bulb treatment, resulted in enhanced yield and improved yield-related parameters, including bulb circumference, basal diameter, and mean number of bulbs per bunch (Naguleswaran et al. 2014). According to Khokhar et al. (2013), certain pathogens, such as *Penicillium* species, have been utilized as a biological control agent against the onion fungal pathogen *A. niger* due to their antagonistic properties. *Penicillium viridicatum* and *P. roqueforti* exhibited significant inhibitory effects on the growth of *A. niger*, with reductions of 60% and 66%, respectively (Khokhar et al. 2013). Apart from that, Malathi (2013) and Sudhasha et al. (2008) conducted a comparable experiment and found that the application of *T. harzianum* to seeds resulted in a significant decrease in the occurrence of basal rot in onions, both in pot and field settings. In a recent study, bio-control agents had been extracted from the rhizosphere soil of onion crops grown in various regions of Tamil Nadu, India. The effectiveness of these agents in controlling basal rot of onion was then assessed through in vitro experimentation. The study found that *Trichoderma* spp. isolates, particularly *T. harzianum* (TH 3 strain), exhibited the highest level of inhibition at 83%. Additionally, the Pf 12 strain of *Pseudomonas* significantly decreased the mycelial growth of *F. oxysporum* f. sp. cepae by 75%. Following the laboratory assessments, the efficacious biocontrol agents underwent additional evaluations in both greenhouse and field settings. The utilization of mixed biocontrol agents comprising of fungal and bacterial strains (Pf12 + Pf27 + TH3) resulted in a noteworthy reduction of the disease by 85% when compared to other combination treatments. The study determined that the utilization of biocontrol agents could potentially function as a substitute for chemical control methods in the mitigation of onion basal rot. This could result in the simultaneous improvement of onion growth and yield, as per the findings of Malathi in 2015 (Malathi 2013; Sudhasha et al. 2008). In a separate study, Sadik et al. (2013) identified *P. agglomerans* as the most efficient bio-control agent for managing bacterial ailments in onions caused by *P. ananatis*, *P. marginalis*, X. retroflexus, and *P. viridiflava.* The study reported a percentage inhibition of 25.6%, 24.8%, 14.4%, and 26.7%, respectively (Sadik et al. 2013).

Spinach is known to be susceptible to various pathogens worldwide. The primary factor responsible for this issue is the presence of downy mildew. Spinach is susceptible to various significant diseases, including Fusarium wilt caused by *F. oxysporum* f. sp. spinaciae, white rust disease caused by *Albugo occidentalis*, and root-knot nematode caused by *Meloidogyne* spp. The predominant foliar diseases include anthracnose and white rust caused by *Colletotrichum dematium.* The presence of these illnesses has the potential to diminish the overall quality and marketability of spinach. Nevertheless, the management of these diseases is primarily reliant on systemic and biological fungicides. The growing apprehension regarding the ecological consequences of chemical fungicides has led to a surge in the exploration of biocontrol approaches for regulating soil-borne phytopathogenic fungi. The efficacy of a blend of synthetic and biological fungicides was evaluated for its potential to control various spinach diseases, including Fusarium wilts. The combination of Methyl benzimidazole carbamate fungicides and biological fungicides has been observed to be highly efficacious in mitigating the occurrence of Fusarium wilt of spinach, as reported by Elmer and McGovern in 2004 and Everts et al. in 2014 (Elmer 2004; Everts et al. 2014). Jones (2000) and Everts et al. (2014) have demonstrated the suppressive effects of various demethylation inhibitor fungicides, including difenoconazole, tebuconazole, ipconazole, and prothioconazole, on Fusarium wilt in spinach and other leafy vegetables. The utilization of Serenade ASO (*B. subtilis*, Bayer Crop Science, Bethesda, NC, USA) as a biologic control agent for the suppression of white rust in spinach has been approved and registered in Canada (Everts et al. 2014; Jones 2000). Additionally, synthetic fungicides such as cyazofamid and fosetyl-AL (OMAFRA 2014) have also been registered for the same purpose. Various microorganisms such as Bacillus spp., Pseudomonas spp., Burkholderia spp., *Enterobacter cloacae*,

Penicillium oxalicum, *Streptomyces griseoviridis*, *Gliocladium catenulatum*, Trichoderma *spp.*, hypovirulent binucleate Rhizoctonia, and non-pathogenic Fusarium *spp.* have been reported to exhibit biological control of Fusarium wilt (Bell et al. 1998; Horinouchi et al. 2010; Larena et al. 2003; Larkin 1998; Lemanceau et al. 1993; Minuto et al. 1995; Nion and Toyota 2008; Rose et al. 2003). Numerous biological fungicides are presently accessible in Northern America for the management of soil-borne and foliar diseases in spinach, both in the greenhouse as well as field settings. Various organic products have been evaluated through field trials to ascertain their efficacy in controlling Fusarium wilt of spinach, as reported by Cummings et al. and Cummings (Cummings 2007; Cummings et al. 2009). A comparable investigation revealed that the utilization of the bio-fungicide Prestop (*G. catenulatum*, Verdera Oy, Espoo, Finland) through drenching resulted in a reduction of 37–56% in the occurrence of Fusarium wilt in comparison to the plants that were not treated. The application of soil drench, as observed by Cummings et al. (2009), increased pre-emergence damping-off of spinach, despite a reduction in wilt symptoms (Cummings et al. 2009). According to Rose et al. (2003), the employment of this particular biological control method resulted in a significant reduction of Fusarium wilt in greenhouse vegetables cultivated in soilless media (Rose et al. 2003) (Table 2).

3 Exploring the Potential of Innovation in the Management of Bacterial Diseases and Its Prospects

The previous sections have provided an account of the historical approach to managing bacterial plant diseases. The discourse reveals that management has predominantly employed two intervention approaches, specifically the administration of chemical or biological substances onto plant surfaces to reduce pathogenic communities or the usage of host resistance mechanisms to combat pathogens at the host interface. In the pursuit of novel approaches to disease management, endeavors may adopt a more precise and deliberate approach, such as the directed focus on pathogen virulence or the meticulous refinement of host plant genetic manipulation. As elaborated later in the text, the era of genetics and genomics holds promise for developing innovative management approaches that are believed to be ecologically sound and less susceptible to resistance development.

3.1 Nanoparticles

Nanoparticles (NPs) are defined as particles whose size falls within the range of 1–100 nm. Particles falling within this particular size range exhibit distinctive physical and chemical characteristics, such as a significant surface area-to-mass ratio, elevated reactivity, and exceptional interactions with biological systems (Zhang et al.

Table 2 Compilation of significant bio-control agents, along with their respective mode of action and the soilborne pathogens they target

Target pathogen	Bio-control agent	Mode of action	References
Sclerotinia trifoliorum, *S. sclerotiorum*	*Coniothyrium minitans*	Lysis mediated by β-1,3 glucanase and chitinase	Pertot et al. (2016)
Fusarium spp., *Pythium* spp., *Aspergillus flavus*, *Rhizoctonia solani*	*Bacillus amyloliquefaciens*, *B. subtilis*, *B. pumilus*, *B. firmus*	Dueling, antibiotic resistance, and induced resistance	Pertot et al. (2016); Shafi et al. (2017)
R. solani, *Verticillium dahliae*, and nematodes	*Purpureocillium lilacinum*	Parasitism	Lan et al. (2017)
Pythium, Cladosporium, Plicaria, Didymella, Rhizoctonia, Alternaria, Helminthosporium, Penicillium, Verticillium, Botrytis, Phytophthora, Fusarium species	*Gliocladium catenulatum*	By producing toxin	Pertot et al. (2016)
R. solani	Pseudomonas *spp.*	Synthesizing siderophores, and antibiotics	Pertot et al. (2016); Howell and Stipanovic (1980)
Heterobasidion annosum	*Phlebiopsis gigantea*	–	Pertot et al. (2016)
Geotrichum, Verticillium, Sclerotinia, Armillaria, Monosprascus, Aphanomyces, Phytomatotricum, Pythium, Phytophthora, Rhizoctonia, Fusarium species	Streptomyces spp.	Mycoparasitism	

2008). Certain properties exhibited by these substances render them highly effective as antimicrobial agents, while other properties make them optimal candidates for serving as carriers or delivery systems for other antimicrobial agents. The assessment of NPs to manage plant diseases has primarily concentrated on oomycete and fungal diseases (Servin et al. 2015; Banerjee et al. 2022). Nonetheless, encouraging outcomes have also been achieved in the context of bacterial diseases, as elaborated upon below.

The primary mechanisms underlying the antimicrobial properties of numerous types of nanoparticles (NPs) are attributed to three key factors: photocatalysis, the infliction of physical harm to the bacterial cell envelope, and the discharge of hazardous metal ions. Metal oxides, including TiO_2, CuO, Fe_3O_4, and ZnO, as well as metals like Ag, have been identified as the primary types of nanoparticles that exhibit photocatalytic activity (Li et al. 2012; Choi and Hu 2008). The process of photocatalysis results in the production of reactive oxygen species such as hydroxyl radicals, hydrogen peroxide, and peroxide (Manke et al. 2013). Reactive oxygen species exhibit toxicity toward bacteria by causing harm to cellular components such as lipids, DNA, and proteins, thereby leading to bactericidal and bacteriostatic consequences (Storz 1999). The utilization of a titanium dioxide (TiO_2) NP formulation in plant disease management has been found to induce photocatalysis, leading to notable anti-microbial properties toward the spot pathogen *Xanthomonas perforans* (Paret et al. 2013). The photocatalytic effect toward *X. perforans* was significantly increased by doping the TiO_2 NPs with Zn and Ag. In a study conducted by Paret et al. (2013), it was observed that the severity of bacterial spots in tomato plants infected with *X. perforans* was significantly reduced upon treatment with TiO_2/Zn NPs at a concentration of approximately 500 to 800 ppm. This reduction in severity was found to be superior to that observed in both untreated and copper control groups (Paret et al. 2013). Minimum inhibitory doses (83.3 and 108.3 ppm, respectively) for the anti-bacterial activity of Ag and ZnO nanoparticles against *E. amylovora* were found to be encouraging (Sundin et al. 2016). Certain types of nanoparticles possess the ability to exhibit antimicrobial properties by means of ion emission, specifically Zn^{2+}, Ag^{+}, and Cu^{2+}, which are known to have detrimental effects on bacterial cells. The antimicrobial effectiveness of Ag NPs was significantly enhanced by the introduction of Ag^{+} (Lok et al. 2007). The augmentation of antimicrobial properties of nanoparticles can be achieved through certain modifications and enhancements. One such approach involves the amalgamation of nanoparticles with a supporting material, which has demonstrated favorable antimicrobial outcomes in recent studies. According to Ocsoy et al. (2013), it has been observed that bare NPs have a tendency to aggregate, which results in a reduction of their antimicrobial efficacy. The utilization of Graphene oxide (GO) as a substrate has been observed in the growth of Ag NPs. This has resulted in a significant reduction in agglomeration, thereby improving the anti-microbial action toward *X. perforans* spot pathogen (Ocsoy et al. 2013). The present study conducted a glasshouse assay to evaluate the efficacy of Ag@dsDNA@GO NPs in controlling bacterial spots on infected tomato transplants. The results indicated that the treatment with 100 ppm of Ag@dsDNA@GO NPs exhibited a comparable level of control to the conventional copper treatment. Apart from exhibiting anti-microbial properties, NPs have been proposed as effective carriers for various other antimicrobial agents. Various delivery systems for nanoparticles have been researched for the purpose of delivering active ingredients. These systems include dendrimers, hydrogel, carbon nanotubes, liposomes, micelles, and nano- and micro-emulsions (De et al. 2014). Nanoemulsions have demonstrated the potential to deliver herbicides that are agriculturally significant (Lim et al. 2012). In addition, it was discovered that a nano-emulsion formulation enhanced the ampicillin antibiotic's permeability across the citrus cuticle and into the phloem by means of a foliar spray that specifically targeted Huanglongbing disease. The plant cuticle, consisting of wax, cutin, and pectin, serves as the primary obstacle to the infiltration of antimicrobial agents into plant tissues. The potential benefits for managing bacterial plant diseases are significant with the development of anti-microbial delivery methods that can penetrate the cuticle, as a multitude of plant-pathogenic bacteria target xylem and phloem tissues (Yang et al. 2015).

To summarize, nanoparticles exhibit numerous favorable characteristics that could render them highly effective as anti-microbial agents for treating bacterial infections in plants in the coming years. The utilization of anti-microbial NPs within the agricultural sector has the potential to substantially expand the range of available antimicrobial agents. Furthermore, a majority of NPs exhibit antimicrobial properties via diverse mechanisms, potentially mitigating the likelihood of resistance emergence. The utilization of NPs as carriers for pesticides presents novel prospects for improving the effectiveness of existing anti-microbial agents that aim to combat pathogens within plants. Notwithstanding, there exist certain constraints that require resolution, including but not limited to expenses, particle coalescence, material readiness, and effective delivery of the active ingredient. In addition, it is imperative to ascertain the environmental destiny of NPs prior to their widespread implementation. Furthermore, it's crucial to assess the likelihood of NPs persisting in food items due to their potential toxicity effects (Gaillet et al. 2015; Nel et al. 2006).

3.2 Targeting the T3SS

The Type III Secretion System (T3SS) is a structure resembling a needle that allows effector proteins to be transported directly from the bacterial cytoplasm to the cytoplasm of the host cell. According to Stavrinides et al., the host immune response is inhibited by these secreted effectors, which are essential in bacterial infection. Given that the T3SS plays a crucial role in the pathogenesis of numerous Gram-negative bacterial pathogens, it presents a clear and viable target for impeding pathogenesis in the development of innovative disease management approaches (Stavrinides et al. 2008). In a study conducted by Yang et al., it was discovered that *trans*-cinnamic acid (TCA) and *O*-coumaric acid (OCA), which are plant phenolic compounds, act as inducers of the T3SS genes in *Dickeya dadantii*. The biosynthetic precursors of salicylic acid (SA), a plant defense hormone, are TCA and OCA. Additionally, research has indicated that TCA and OCA are involved in plant defense (Montesano et al. 2005; Vidal et al. 1997). The aforementioned discovery implies that plant phenolic compounds may serve as valuable assets in the quest for identifying prospective T3SS inhibitors. In this investigation, many phenolic compounds were tested for their capacity to halt the transcription of hrpA, the gene responsible for encoding the T3SS pilus structural component in *D. dadantii*. This method was used to discover p-coumaric acid (PCA), a novel phenolic chemical, which acts as a suppressor of the T3SS (Li et al. 2009). In the biosynthesis of phenylpropanoids, PCA plays an intermediary role. In response to microbial harm, plants biosynthesize a family of secondary compounds called phenylpropanoids (Dixon and Paiva 1995; Feys and

Parker 2000). The potential inhibitory effect of PCA on T3SS was investigated and utilized for enhancing inhibition efficacy via compound modification. A library of PCA derivatives was screened, leading to the discovery of a component (*trans*-4-hydroxycinnamohydroxamic acid) with 8 times more detrimental impact against T3SS structural and regulatory genes compared to PCA (Li et al. 2015). The expression of T3SS genes in both *E. amylovora* and the multihost organism *Pseudomonas aeruginosa* has been shown to be modulated by diverse plant phenolic chemicals. The observed outcomes are analogous to the impacts of compounds belonging to the salicylidene acylhydrazide family, as they not only impede the transcription of genes associated with the T3SS but also diminish the synthesis of amylovoran exopolysaccharide in *E. amylovora* (Yang et al. 2014). Furthermore, a considerable number of the PCA derivatives that have been previously discussed exhibited minimal impact on bacterial proliferation in vitro. This observation suggests that the impact of these compounds on the emergence of resistance may not be as substantial as that of traditional bactericidal agents.

3.3 *Targeting Quorum Sensing*

The process of quorum sensing (Iqbal et al. 2012) is a mechanism of intercellular communication that regulates multiple characteristics in response to cell density. The aforementioned process is facilitated through the involvement of extracellular signaling molecules, namely diffusible signal factors (DSFs) and acyl homoserine lactones (AHLs) (Von Bodman et al. 2003; Rutherford and Bassler 2012). At least 9 plant pathogen genera use quorum sensing (Iqbal et al. 2012) to regulate crucial features in pathogen-vector and host–pathogen interactions (Helman and Chernin 2015). The exploitation of bacterial recognition of an extracellular signaling molecule through QS offers a potentially fruitful approach for manipulation, whereby the modulation of signal degradation or excessive production of the signaling molecule can be precisely targeted. Numerous bacteria have been found to possess AHL-degrading enzymes, which were utilized in disease management endeavors. AHL-degrading organisms have been applied via flooding or AHL-degrading enzymes were expressed in transgenic plants to accomplish this (Dong et al. 2001; Molina et al. 2003). Transgenic plants have been used in a variety of situations where the production of AHL molecules has impeded pathogenic attacks. The activation of QS-regulated traits at unsuitable timing could lead to dysfunctions in the coordination of pathogenesis, which has been referred to as "pathogen confusion" (Lindow et al. 2014). The identification of specific plants that demonstrate a reaction to QS molecules and then generate enzymes that inhibit quorum formation suggests that the host can potentially impact the bacterial population dynamics. This observation has been made by Bauer and Mathesius (2004) and Dudler and Eberl (2006), indicating that there's still a significant amount of knowledge to be gained before anti-QS strategies can effectively be used in the sustainable regulation of plant diseases.

3.4 Targeting Bacterial Biofilms

Many plant-pathogenic bacteria rely heavily on biofilms as a virulence factor because they provide superior defense against anti-microbial medications, environmental stressors, and host defense mechanisms. This has been extensively documented in the literature (Koczan et al. 2009; Rodrigues et al. 2008). Consequently, the investigation of inhibitors that can control bacterial infection by impeding biofilm formation has emerged as a compelling area of research. The process of biofilm formation generally encompasses three primary phases, namely initial attachment to the surface, the maturation of the biofilm, and the subsequent dispersal (Kostakioti and Hadjifrangiskou 2013). The primary focus of contemporary anti-biofilm inhibitors, exemplified by indole and *D*-amino acid derivatives, primarily revolves around the prevention of initial biofilm formation. This has led to a decrease in the severity of illness caused by animal and plant pathogens, as well as an increase in bacterial vulnerability to traditional antibiotics and copper (Brackman et al. 2011; Kolodkin-Gal et al. 2010; Li and Wang 2014; Worthington et al. 2012). The bacterial secondary messenger, known as Bis-(3′,5′)-cyclic dimeric guanosine monophosphate (c-di-GMP), plays a significant role in regulating various cellular behaviors and virulence in numerous bacterial species (Römling et al. 2013; Yuan et al. 2015).

According to Chua et al. (2015), recent research has indicated that a reduction in intra-cellular c-di-GMP levels can initiate the process of biofilm dispersal, which has been suggested as a potentially effective approach for eliminating biofilms. The compound 2-aminoimidazole (2AI) has demonstrated significant potential as it effectively inhibits the in vitro formation of biofilm by the bacterium *Xanthomonas euvesicatoria* (spot pathogen) (Chua et al. 2015). According to Worthington et al., the combination of 2AI and copper hydroxide exhibited superior efficacy in controlling bacterial spot infection in tomato crops during field experiments, as compared to the individual application of either material. The citrus canker pathogen *X. citri* ssp. citri can be inhibited in its biofilm formation by two additional compounds, namely 3-indoylacetonitrile and *D*-leucine (Worthington et al. 2012). In glasshouse assays conducted by Li and Wang (2014), it was observed that the application of these compounds either alone or in conjunction with copper sulfate resulted in a reduction of bacterial colonies and disease incidence on citrus leaves. Furthermore, it has been observed that the chemical compound N-acetylcysteine exhibits a reduction in adhesion and a hindrance in the formation of biofilm in the pathogenic microorganism *Xylella fastidiosa* that resides in the xylem (Worthington et al. 2012).

According to research by Muranaka et al. (2013), the minimum inhibitory concentration of N-acetylcysteine in the xylem of citrus plants was not met using fertigation or any other means. In contrast, the chemical was found to lessen the severity of symptoms and the number of germs in glasshouse studies (Muranaka et al. 2013). The creation of potent biofilm inhibitors has the potential to be a significant advancement, particularly in the control of foliar pathogens. The effectiveness of bactericides can be hindered by the formation of biofilms, making this development crucial

(Redondo et al. 2015). Therefore, it is not necessary to develop inhibitors as independent products. Instead, they could be utilized as complementary agents to augment the effectiveness of current bactericides.

3.5 Genome Editing Techniques to Develop Sustainable Resistance Mechanisms

The successful decoding of the TALE code pertaining to the binding of DNA at specific sites has paved the way for novel opportunities in the development of enhanced and efficacious techniques for genome manipulation. TALENs, the short form for transcription activator-like effector nucleases, represent a category of molecular instruments employed for the purpose of inducing double-strand breaks (DSBs) in DNA at specific loci. They have a highly specific sequence-directed mechanism thanks to the fusion of TALE-binding domains and the FokI nuclease cleavage domain (Christian et al. 2010; Miller et al. 2011). The application of DSBs can activate the inherent DNA repair mechanisms of an organism, including non-homologous end joining. This particular mechanism is known to be error-prone and can lead to the introduction of variable-length insertions or deletions at the site of the break. As a result, the target gene may experience frame shifts, ultimately leading to its loss of function. The second mechanism, as described by Joung and Sander (2013), pertains to homology-directed repair. This mechanism entails the integration of homologous DNA templates at the designated site, thereby facilitating the accurate addition or removal of nucleotides within a particular locus (Joung and Sander 2013). The efficiency of genome editing technology has been demonstrated in various plant organisms, including *Arabidopsis thaliana*, soybean, tobacco, corn, and rice, which are not only model plants but also economically significant crops (Li et al. 2012; Haun et al. 2014; Liang et al. 2014). Li et al. (2012) have reported a noteworthy instance of genome modification through the utilization of TALEN for the purpose of engineering rice cultivars that are resistant to bacterial blight (Li et al. 2012). The utilization of TALEN technology was implemented in this investigation to aim at the S gene Os11N3 (OsSWEET14), which is responsible for encoding a component of the SWEET sucrose efflux transporter family. This gene is stimulated by *X. oryzae* by a minimum of four distinct TALE proteins (PthXo3, TalC, Tal5, and AvrXa7) in order to redirect sugar transportation and promote pathogen's nourishment and expansion within the host (Streubel et al. 2013). Li and colleagues performed a targeted genome editing using sequence-specific TALENs to modify the effector binding element (Dudler and Eberl 2006) located in the promoter region of OsSWEET14. The S gene could not be activated because the promoter had become resistant to PthXo3 and AvrXa7. Nevertheless, the physiological role of the sucrose efflux mechanism in the plant remained unimpaired (Li et al. 2012).

The CRISPR/Cas9 system has gained significant utilization in genome editing as a recently developed technology. The CRISPRs, or Clustered Regularly Interspaced

Short Palindromic Repeats, are a type of DNA repeat that is commonly found in Archaea and approximately 40% of bacterial species. Bacteriophages or plasmids can be effectively neutralized through the acquisition of immunity by means of these entities (Horvath and Barrangou 2010). The aforementioned unique genetic markers are comprised of recurring palindromic sequences, which range from 21 to 47 base pairs in length. These sequences are interspersed with hypervariable spacers that display similarity to viral and plasmid sequences of exogenous origin, with a length range of 21 to 72 base pairs. The aforementioned arrays are frequently situated in close proximity to helper cas (CRISPR-associated) genes which are responsible for encoding helicases, nucleases, and polymerases. Upon transcription, spacer sequences produce diminutive CRISPR-RNA (crRNA) fragments which form a hybrid with a minute non-coding transactivating crRNA (tracrRNA). The utilization of a double-stranded RNA molecule serves as a means to specifically target foreign DNA sequences via complementarity. This process directs the Cas protein complex toward the identified sequences, ultimately resulting in the degradation of DNA through double-strand cleavage (Horvath and Barrangou 2010). The scientific community has documented the existence of three distinct categories, denoted as I, II, and III, of the CRISPR/Cas system. The cleavage of DNA in type II systems is attributed to the RNA-guided nuclease Cas9, which possesses nuclease domains. The combination of tracrRNA and crRNA into a singular small guide molecule, known as sgRNAs and measuring approximately twenty nucleotides, was shown to effectively introduce genome modifications with a high degree of target specificity (Jinek et al. 2012). The utilization of multiple sgRNAs in conjunction with Cas9 enables the concurrent targeting of multiple loci, owing to the compact nature of the sgRNAs.

The feature of multiplex genome editing presents a comparative advantage in relation to other genome editing technologies, such as TALENs. To avoid double-strand cleavage, researchers have used a Cas9 endonuclease variant to induce nickase activity. This approach has been employed to selectively target regulatory elements of a particular gene, resulting in the repression of its transcription. This mechanism is commonly referred to as CRISPR interference (CRISPRi) (Qi et al. 2013).

The utilization of CRISPR/Cas9 genome editing technology has demonstrated efficacy in the genetic modification of individual or multiple gene targets in various plant species, including tobacco, *A. thaliana*, sweet orange, and rice, as evidenced by studies conducted by Gao et al. (2015); Jia and Wang (2014); Shan et al. (2014). Furthermore, recent studies have demonstrated the effectiveness of this novel technology in inducing mutations within the TaMLO-A1 allele of *Triticum aestivum*, which encodes a protein responsible for suppressing the plant's immune response to powdery mildew. Wang et al. found that using CRISPR/Cas editing may successfully aid in the creation of long-lasting resistance, regardless of ploidy level. In particular, the aforementioned alteration led to protection against the infection (Wang et al. 2014).

4 Conclusion

Vegetables play a crucial role in everyday life because of the macro and micronutrients, vitamins, antioxidants, and proteins they provide. The widespread use of synthetic pesticides has hampered conventional methods of combating the pests that eat away vegetable crops. Alternative methods for disease control in vegetable crops have been developed due to the higher costs of synthetically produced pesticides coupled with the potential risks to human health and environmental damage. Since they are both cost-effective and eco-friendly, microbial biocontrol agents are emerging as a possible technique for managing diseases in vegetable crops. Many growers have found great success at a cheap cost employing microbial formulations for disease management, but we need a better understanding of the mechanics underlying disease suppression to fully make use of these tools. If additional farmers in other agronomic zones learn about and use this inexpensive method, it might have a significant impact on the vegetable farming industry.

There has been an explosion of study into the molecular biology and pathophysiology of plant-disease-causing bacteria in the last two decades. The present investigation has examined the factors that establish the conditions for interactions between hosts and pathogens. As a result, we have entered a novel epoch in the management of bacterial illnesses. There are various approaches to achieve this objective, such as targeting particular virulence factors or utilizing molecular breeding techniques to generate strains that exhibit broad-spectrum resistance over an extended period. New possibilities for disease control through host resistance have emerged as a result of a deeper knowledge of the role of type III effector proteins and their respective targets in a variety of plant hosts made possible by the growing dataset. Developing new sources of resistance to bacterial infections in plants no longer necessitates the introduction of foreign DNA material into the host genome, thanks to modern genome editing technologies like CRISPR/Cas9 and TALEN systems. Novel approaches and the ability to combine resistance characteristics offer promising opportunities for the development of multi-resistant varieties using more efficient and expedient methods than traditional breeding techniques. The management of bacterial plant diseases is currently undergoing a transitional phase. The present discourse pertains to techniques that are presently in the experimental phase. Consequently, farmers will persist in utilizing traditional techniques such as the application of bactericides and biological controls, as well as the cultivation of resistant plant varieties. The imperative to ensure food security and enhance crop productivity through genetic and genomic innovations in plant bacteriology research necessitates commensurate progress in addressing engineering and sociological issues, which are equally indispensable. Advancements in genetics and genomics can be considered as the second generation of bacterial disease management. The attainment of sustainable disease management through the third generation is anticipated to necessitate the adoption of innovative measures, which entail the efficacious execution of inventive management approaches and the endorsement of these approaches by consumers.

References

Abdelzaher HMA (2003) Biological control of root rot of cauliflower (caused by *Pythium ultimumvar.* ultimum) using selected antagonistic rhizospheric strains of *Bacillus subtilis*. New Zealand J Crop Horticult Sci 31(3):209–220

Agnihotri N (1999) Pesticides safety evaluation and monitoring. All India coordinated research project on pesticides residues, division of agricultural chemicals. IARI, New Delhi

Ahmed AQ, Javed N, Khan SA, Abbas H, Kamran MK (2016) Efficacy of rhizospheric organism *Rhizobium leguminosarum* against *Meloidogyne incognita* in soybean. 53:377–381

Aktar W, Sengupta D, Chowdhury AJ (2009) Impact of pesticides use in agriculture: their benefits and hazards. Interdiscip Toxicol 2(1):1

Anand P, Kunnumakkara AB, Newman RA, Aggarwal BB (2007) Bioavailability of curcumin: problems and promises. Mol Pharmaceutics 4(6):807–818

Banerjee S, Mazumder S, Chatterjee D, Bose S, Majee SB (2022) Nanotechnology for cargo delivery with a special emphasis on pesticide, herbicide, and fertilizer. Nano-enabled agrochemicals in agriculture. Elsevier, pp 105–144

Bauer WD, Mathesius U (2004) Plant responses to bacterial quorum sensing signals. Curr Opin Plant Biol 7(4):429–433

Bell AA, Hubbard JC, Liu L, Davis RM, Subbarao KV (1998) Effects of chitin and chitosan on the incidence and severity of Fusarium yellows of celery. Plant Dis 82(3):322–328

Beneduzi A, Ambrosini A, Passaglia LMP (2012) Plant growth-promoting rhizobacteria (PGPR): their potential as antagonists and biocontrol agents. Genet Mol Biol 35:1044–1051

Beura S, Mahanta I, Mohapatra KJ (2008) Economics and chemical control of Phomopsis twig blight and fruit rot of brinjal. 46(1):73–76

Bjelić DĐ, Marinković JB, Tintor BB, Tančić SL, Nastasić AM, Mrkovački NB (2015) Screening of Azotobacter isolates for PGP properties and antifungal activity. Zbornik Matice srpske za prirodne nauke 2015(129):65–72

Blair JM, Webber MA, Baylay AJ, Ogbolu DO, Piddock LJ (2015) Molecular mechanisms of antibiotic resistance. Nat Rev Microbiol 13(1):42–51

Bottiglieri M, Keel CE (2006) Characterization of PhlG, a hydrolase that specifically degrades the antifungal compound 2,4-diacetylphloroglucinol in the biocontrol agent Pseudomonas fluorescens CHA0. Appl Environ Microbiol 72(1):418–427

Brackman G, Cos P, Maes L, Nelis HJ, Coenye TJ (2011) Quorum sensing inhibitors increase the susceptibility of bacterial biofilms to antibiotics in vitro and in vivo. Antimicrobiol Agents Chemotherapy 55(6):2655–2661

Breen S, Solomon PS, Bedon F, Vincent DJ (2015) Surveying the potential of secreted antimicrobial peptides to enhance plant disease resistance. Front Plant Sci 6:900

Brewer JH, Thrasher JD, Hooper D (2013) Chronic illness associated with mold and mycotoxins: is naso-sinus fungal biofilm the culprit? Toxins 6(1):66–80

Brogden KA (2005) Antimicrobial peptides: pore formers or metabolic inhibitors in bacteria? Nat Rev Microbiol 3(3):238–250

Chen P, Peng Y, Chung W, Chung K, Huang H, Huang J et al (2016) Inhibition of *Penicillium digitatum* and citrus green mold by volatile compounds produced by *Enterobacter cloacae*. J Plant Pathol Microbiol 7(3)

Chethana B, Ganeshan G, Manjunath B (2011) Screening of genotypes and effect of fungicides against purple blotch of onion. J Agric Technol 7(5):1369–1374

Choi O, Hu ZJ (2008) Size dependent and reactive oxygen species related nanosilver toxicity to nitrifying bacteria. ACS Nano 42(12):4583–4588

Chowdhury SP, Dietel K, Rändler M, Schmid M, Junge H, Borriss R et al (2013) Effects of *Bacillus amyloliquefaciens* FZB42 on lettuce growth and health under pathogen pressure and its impact on the rhizosphere bacterial community. PloS One 8(7):e68818

Christian M, Cermak T, Doyle EL, Schmidt C, Zhang F, Hummel A et al (2010) Targeting DNA double-strand breaks with TAL effector nucleases. Genetics 186(2):757–761

Chua SL, Hultqvist LD, Yuan M, Rybtke M, Nielsen TE, Givskov M et al (2015) In vitro and in vivo generation and characterization of *Pseudomonas aeruginosa* biofilm–dispersed cells via c-di-GMP manipulation. Nat Protocol 10(8):1165–1180

Cooksey DA (1990) Genetics of bactericide resistance in plant pathogenic bacteria. 28(1):201–219

Cummings JA (2007) Evaluation of seed and drench treatments for management of damping-off and seedling blight pathogens of spinach for organic production. Washington State University

Cummings JA, Miles CA, du Toit LJ (2009) Greenhouse evaluation of seed and drench treatments for organic management of soilborne pathogens of spinach. Plant Dis 93(12):1281–1292

Damalas CA, Eleftherohorinos IG (2011) Pesticide exposure, safety issues, and risk assessment indicators. Int J Environ Res Pub Health 8(5):1402–1419

Das S, Mohanty K (1988) Management of die back/twig blight of chilly with fungicides. Indian J Plant Protect 16:109–111

De A, Bose R, Kumar A, Mozumdar S (2014) Targeted delivery of pesticides using biodegradable polymeric nanoparticles. Springer

Dixon RA, Paiva NL (1995) Stress-induced phenylpropanoid metabolism. Plant Cell 7(7):1085

Dong Y-H, Wang L-H, Xu J-L, Zhang H-B, Zhang X-F, Zhang L-H (2001) Quenching quorum-sensing-dependent bacterial infection by an *N*-acyl homoserine lactonase. Nature 411(6839):813–817

Dubey R, Khare S, Kumar P, Maheshwari D (2014) Combined effect of chemical fertilisers and rhizosphere-competent Bacillus subtilis BSK17 on yield of *Cicer arietinum*. Arch Phytopathol Plant Protect 47(19):2305–2318

Dudler R, Eberl L (2006) Interactions between bacteria and eukaryotes via small molecules. Curr Opin Biotechnol 17(3):268–273

Durrant WE, Dong X (2004) Systemic acquired resistance. Annu Rev Phytopathol 42:185–209

Dutta P, Das BC (2002) Management of collar rot of tomato by *Trichoderma* spp. and chemicals. Indian Phytopathol 55(2):235–237

Egel D, Martyn RJ (2007) Fusarium wilt of watermelon and other cucurbits. Plant Health Instruct 10:1094

El-Katatny M, Somitsch W, Robra K-H, El-Katatny M, Gübitz GM (2000) Production of chitinase and β-1, 3-glucanase by *Trichoderma harzianum* for control of the phytopathogenic fungus *Sclerotium rolfsii*. 38(3):173–180

Elmer WH, McGovern RJCP. Efficacy of integrating biologicals with fungicides for the suppression of *Fusarium wilt* of cyclamen. Crop Protect 23(10):909–914

El-Mohamedy RSR (2012) Biological control of Pythium root rot of broccoli plants under greenhouse conditions. J Agri Technol 8:1017–1028

El-Tarabily K, Soliman M, Nassar A, Al-Hassani H, Sivasithamparam K, McKenna F et al (2000) Biological control of Sclerotinia minor using a chitinolytic bacterium and actinomycetes. Plant Pathol 49(5):573–583

Everts KL, Egel DS, Langston D, Zhou X-G (2014) Chemical management of *Fusarium wilt* of watermelon. Crop Protect 66:114–119

Fatima N, Batool H, Sultana V, Ara J, Ehteshamul-Haque S (2009) Prevalence of post-harvest rot of vegetables and fruits in Karachi, Pakistan. Pak J Bot 41(6):3185–3190

Fernando W, Nakkeeran S, Zhang Y, Savchuk S (2007) Biological control of *Sclerotinia sclerotiorum* (Lib.) de Bary by Pseudomonas and Bacillus species on canola petals. Crop Protect 26(2):100–117

Feys BJ, Parker JE (2000) Interplay of signaling pathways in plant disease resistance. Trends Genet 16(10):449–455

Figueroa-López AM, Cordero-Ramírez JD, Martínez-Álvarez JC, López-Meyer M, Lizárraga-Sánchez GJ, Félix-Gastélum R et al (2016) Rhizospheric bacteria of maize with potential for biocontrol of *Fusarium verticillioides*. SpringerPlus 5(1):330

Forsberg KJ, Reyes A, Wang B, Selleck EM, Sommer MO, Dantas GJ (2012) The shared antibiotic resistome of soil bacteria and human pathogens. Science 337(6098):1107–1111

Frankowski J, Lorito M, Scala F, Schmid R, Berg G, Bahl H (2001) Purification and properties of two chitinolytic enzymes of *Serratia plymuthica* HRO-C48. Arch Microbiol 176:421–426

Gade R, Zote K, Mayee CD (2011) Integrated management of pigeon pea wilt using fungicide and bioagent

Gaillet S, Rouanet JM (2015) Silver nanoparticles: their potential toxic effects after oral exposure and underlying mechanisms–a review. Food Chem Toxicol 77:58–63

Gao J, Wang G, Ma S, Xie X, Wu X, Zhang X et al (2015) CRISPR/Cas9-mediated targeted mutagenesis in *Nicotiana tabacum*. Plant Mol Biol 87:99–110

Georgakopoulos D, Fiddaman P, Leifert C, Malathrakis NE (2002) Biological control of cucumber and sugar beet damping-off caused by *Pythium ultimum* with bacterial and fungal antagonists. J Appl Microbiol 92(6):1078–1086

Ghareeb H, Bozsó Z, Ott PG, Repenning C, Stahl F, Wydra K et al (2011) Transcriptome of silicon-induced resistance against *Ralstonia solanacearum* in the silicon non-accumulator tomato implicates priming effect. Physiol Mol Plant Pathol 75(3):83–89

Gilden RC, Huffling K, Sattler BJ (2010) Pesticides and health risks. Gynecol Neonatal Nursing 39(1):103–110

Gomes R, Semedo L, Soares R, Linhares L, Ulhoa C, Alviano C et al (2001) Purification of a thermostable endochitinase from Streptomyces RC1071 isolated from a cerrado soil and its antagonism against phytopathogenic fungi. J Appl Microbiol 90(4):653–661

Graham JH, Myers ME (2011) Soil application of SAR inducers imidacloprid, thiamethoxam, and acibenzolar-*S*-methyl for citrus canker control in young grapefruit trees. Plant Dis 95(6):725–728

Gravel V, Martinez C, Antoun H, Tweddell RJ (2004) Evaluation of antagonistic microorganisms as bio control agents of root rot (*Pythium ultimum*) of greenhouse tomatoes in rock wool. Can J Pathol 26:152

Gunes A, Karagoz K, Turan M, Kotan R, Yildirim E, Cakmakci R et al (2015) Fertilizer efficiency of some plant growth promoting rhizobacteria for plant growth. Res J Soil Biol 7(2):28–45

Gupta G, Parihar SS, Ahirwar NK, Snehi SK, Singh V (2015) Plant growth promoting rhizobacteria (PGPR): current and future prospects for development of sustainable agriculture. J Microb Biochem Technol 7(2):096–102

Habib A, Sahi ST, Ghazanfer M, Ali SJ (1925) Evaluation of some fungicides against seed born mycoflora of eggplant and their comparative efficacy regarding seed germination

Hammer PE, Hill DS, Lam ST, Van Pée K-H, Ligon JM (1997) Four genes from *Pseudomonas fluorescens* that encode the biosynthesis of pyrrolnitrin. Appl Environ Microbiol 63(6):2147–2154

Hassen AI, Bopape F, Sanger LK (2016) Microbial inoculants as agents of growth promotion and abiotic stress tolerance in plants. Microbial Inocultants Sustain Agric Prod 23–36

Haun W, Coffman A, Clasen BM, Demorest ZL, Lowy A, Ray E et al (2014) Improved soybean oil quality by targeted mutagenesis of the fatty acid desaturase 2 gene family. Plant Biotechnol J 12(7):934–940

Helman Y, Chernin L (2015) Silencing the mob: disrupting quorum sensing as a means to fight plant disease. Mol Plant Pathol 16(3):316–329

Herr I, Büchler MW (2010) Dietary constituents of broccoli and other cruciferous vegetables: implications for prevention and therapy of cancer. Cancer Treatment Rev 36(5):377–383

Hert A, Marutani M, Momol M, Roberts P, Olson S, Jones JJA, et al (2009) Suppression of the bacterial spot pathogen *Xanthomonas euvesicatoria* on tomato leaves by an attenuated mutant of *Xanthomonas perforans*. Appl Environ Microbiol 75(10):3323–3330

Hobbelen PH, Paveley ND, van den Bosch FJ (2014) The emergence of resistance to fungicides. PLoS One 9(3):e91910

Horinouchi H, Muslim A, Hyakumachi M (2010) Biocontrol of *Fusarium wilt* of spinach by the plant growth promoting fungus Fusarium equiseti GF183. J Plant Pathol 249–254

Horvath P, Barrangou RJS (2010) CRISPR/Cas, the immune system of bacteria and archaea. Science 327(5962):167–170

Hossain MI, Islam M, Uddin M, Arifuzzaman S, Hasan G (2013) Control of Phomopsis blight of egg plant through fertilizer and fungicide management. Int J Agric Res Innov Technol 3:66–72

Howell C, Stipanovic R (1980) Suppression of *Pythium ultimum*-induced damping-off of cotton seedlings by *Pseudomonas fluorescens* and its antibiotic, pyoluteorin. Phytopathology 70(8):712–715

Huang C-J, Wang T-K, Chung S-C, Chen C-Y (2005) Identification of an antifungal chitinase from a potential biocontrol agent, *Bacillus cereus* 28–9. BMB Rep 38(1):82–88

Iqbal MA, Khalid M, Shahzad SM, Ahmad M, Soleman N, Akhtar N (2012) Integrated use of *Rhizobium leguminosarum*, plant growth promoting rhizobacteria and enriched compost for improving growth, nodulation and yield of lentil (*Lens culinaris* Medik.). Chilean J Agric Res 72(1):104–110

Islam M, Meah MJ (2011) Association of Phomopsis vexans with eggplant (*Solanum melongena*) seeds, seedlings and its management. Agriculturists 9(1–2):8–17

Jia H, Wang N (2014) Targeted genome editing of sweet orange using Cas9/sgRNA. PLoS One 9(4):e93806

Jinek M, Chylinski K, Fonfara I, Hauer M, Doudna JA, Charpentier EJ (2012) A programmable dual-RNA–guided DNA endonuclease in adaptive bacterial immunity. Science 337(6096):816–821

Johnson K, Stockwell V (1998) Management of fire blight: a case study in microbial ecology. Annu Rev Phytopathol 36(1):227–248

Johnson KB (2010) Pathogen refuge: a key to understanding biological control. Annu Rev Phytopathol 48:141–160

Jones RK (2000) Assessments of Fusarium head blight of wheat and barley in response to fungicide treatment. Plant Dis 84(9):1021–1230

Jones JB, Jackson L, Balogh B, Obradovic A, Iriarte F, Momol M (2007) Bacteriophages for plant disease control. Annu Rev Phytopathol 45:245–262

Joung JK, Sander JD (2013) TALENs: a widely applicable technology for targeted genome editing. Nat Rev Mol Biol 14(1):49–55

Kanjanamaneesathian M (2015) Biological control of diseases of vegetables grown hydroponically in Thailand: challenge and opportunity. Recent Patents Biotechnol 9(3):214–222

Kerr A (2016) Biological control of crown gall. Aust Plant Pathol 45:15–18

Kestwal RM, Lin JC, Bagal-Kestwal D, Chiang BH (2011) Glucosinolates fortification of cruciferous sprouts by sulphur supplementation during cultivation to enhance anti-cancer activity. Food Chem 126(3):1164–1171

Khokhar I, Haider MS, Mukhtar I, Mushtaq S (2012) Biological control of *Aspergillus niger*, the cause of black-rot disease of Allium cepa L. (onion), by *Penicillium* species. J Agrobiol 29(1):23

Kim M-K, Choi G-J, Lee H-SJ (2003) Chemistry f. Fungicidal property of *Curcuma longa* L. rhizome-derived curcumin against phytopathogenic fungi in a greenhouse. J Agric Food Chem 51(6):1578–1581

Kleitman F, Shtienberg D, Blachinsky D, Oppenheim D, Zilberstaine M, Dror O et al (2005) *Erwinia amylovora* populations resistant to oxolinic acid in Israel: prevalence, persistence and fitness. Plant Pathol 54(2):108–115

Kloepper JW (1978) Plant growth-promoting rhizobacteria on radishes. In: Proceedings of the 4th internet conference on plant pathogenic bacter, station de pathologie vegetale et phytobacteriologie, INRA, Angers, France, pp 879–882

Kloepper JW, Ryu C-M, Zhang S (2004) Induced systemic resistance and promotion of plant growth by *Bacillus* spp. Phytopathology 94(11):1259–1266

Koczan JM, McGrath MJ, Zhao Y, Sundin GW (2009) Contribution of *Erwinia amylovora* exopolysaccharides amylovoran and levan to biofilm formation: implications in pathogenicity. Phytopathology 99(11):1237–1244

Kokalis-Burelle NJ (2015) Pasteuria penetrans for control of *Meloidogyne incognita* on tomato and cucumber, and *M. arenaria* on snapdragon. J Nematol 47(3):207

Kolodkin-Gal I, Romero D, Cao S, Clardy J, Kolter R, Losick R (2010) D-amino acids trigger biofilm disassembly. Science 328(5978):627–629

Kostakioti M, Hadjifrangiskou M, Hultgren SJ (2013) Bacterial biofilms: development, dispersal, and therapeutic strategies in the dawn of the postantibiotic era. Cold Spring Harbor Perspect Med 3(4):a010306

Kotan G, Kardaş F, Yokuş ÖA, Akyıldırım O, Saral H, Eren T, et al (2016) A novel determination of curcumin via Ru@ Au nanoparticle decorated nitrogen and sulfur-functionalized reduced graphene oxide nanomaterials. Anal Methods 8(2):401–408

Kumar P, Dubey R, Maheshwari DJ (2012) Bacillus strains isolated from rhizosphere showed plant growth promoting and antagonistic activity against phytopathogens. Microbiol Res 167(8):493–499

Kumar G, Desai S, Amalraj L, Reddy G (2015) Isolation of fluorescent *Pseudomonas* spp. from diverse agro-ecosystems of India and characterization of their PGPR traits. Bacteriol J 5(1):13–24

Lan X, Zhang J, Zong Z, Ma Q, Wang Y (2017) Evaluation of the biocontrol potential of *Purpureocillium lilacinum* QLP12 against *Verticillium dahliae* in eggplant. BioMed Res Int

Lanteigne C, Gadkar VJ, Wallon T, Novinscak A, Filion MJP (2012) Production of DAPG and HCN by *Pseudomonas* sp. LBUM300 contributes to the biological control of bacterial canker of tomato. Phytpathology 102(10):967–973

Larena I, Sabuquillo P, Melgarejo P, De Cal A (2003) Biocontrol of Fusarium and Verticillium wilt of tomato by *Penicillium oxalicum* under greenhouse and field conditions. 151(9):507–512

Larkin RP, Fravel DR (1998) Efficacy of various fungal and bacterial biocontrol organisms for control of *Fusarium wilt* of tomato. Plant Dis 82(9):1022–1028

Lemanceau P, Bakker PA, De Kogel WJ, Alabouvette C, Schippers BJA (1993) Antagonistic effect of nonpathogenic *Fusarium oxysporum* Fo47 and pseudobactin 358 upon pathogenic *Fusarium oxysporum* f. sp. dianthi. Appl Environ Microbiol 59(1):74–82

Li J, Wang NJP (2014) Foliar application of biofilm formation–inhibiting compounds enhances control of citrus canker caused by *Xanthomonas citri* subsp. citri. Am Phytopathol Soc 104(2):134–142

Li Y, Peng Q, Selimi D, Wang Q, Charkowski AO, Chen X et al (2009) The plant phenolic compound p-coumaric acid represses gene expression in the *Dickeya dadantii* type III secretion system. Appl Environ Microbiol 75(5):1223–1228

Li Y, Zhang W, Niu J, Chen YJ (2012) Mechanism of photogenerated reactive oxygen species and correlation with the antibacterial properties of engineered metal-oxide nanoparticles. ACS Nano 6(6):5164–5173

Li Y, Hutchins W, Wu X, Liang C, Zhang C, Yuan X et al (2015) Derivative of plant phenolic compound inhibits the type III secretion system of *Dickeya dadantii* via HrpX/HrpY two-component signal transduction and Rsm systems. Mol Plant Pathol 16(2):150–163

Li J, Trivedi P, Wang N (2016) Field evaluation of plant defense inducers for the control of citrus huanglongbing. Phytopathology 106(1):37–46

Liang Z, Zhang K, Chen K, Gao C (2014) Targeted mutagenesis in *Zea mays* using TALENs and the CRISPR/Cas system. J Genetics Genomics 41(2):63–68

Lim CJ, Basri M, Omar D, Rahman MBA, Salleh AB, Rahman MBA et al (2012) Physicochemical characterization and formation of glyphosate-laden nano-emulsion for herbicide formulation. Indus Crops Prod 6(1):607–613

Lindow S, Newman K, Chatterjee S, Baccari C, Iavarone AT, Ionescu M (2014) Production of Xylella fastidiosa diffusible signal factor in transgenic grape causes pathogen confusion and reduction in severity of Pierce's disease. Mol Plant Microbe Interact 27(3):244–254

Lok C-N, Ho C-M, Chen R, He Q-Y, Yu W-Y, Sun H et al (2007) Silver nanoparticles: partial oxidation and antibacterial activities. JBIC J Biol Inorgan Chem 12:527–534

Madhavi M, Kavitha A, Vijayalakshmi MJ (2012) Studies on *Alternaria porri* (Ellis) Ciferri pathogenic to onion (Allium cepa L.). Arch Appl Sci Res 4(1):1–9

Makun H (2013) Mycotoxin and food safety in developing countries. BoD–Books on Demand

Malathi S (2013) Production of cell wall degrading enzymes by *Fusarium oxysporum* f. sp. cepae causing basal rot of onion and its histopathological changes. 6(2):412–418

Mamgain A, Roychowdhury R, Tah J (2013) Alternaria pathogenicity and its strategic controls. Res J Biol 1:1–9

Manke A, Wang L, Rojanasakul YJ (2013) Mechanisms of nanoparticle-induced oxidative stress and toxicity

Maplestone P, Whipps J, Lynch JJP (1991) Effect of peat-bran inoculum of Trichoderma species on biological control of *Rhizoctonia solani* in lettuce. Plant Soil 136:257–263

Mari M, Guizzardi M (1998) The postharvest phase: emerging technologies for the control of fungal diseases. Phytoparasitica 26:59–66

Massart S, Perazzolli M, Höfte M, Pertot I, Jijakli MH (2015) Impact of the omic technologies for understanding the modes of action of biological control agents against plant pathogens. BioControl 60:725–746

Mavrodi DV, Blankenfeldt W, Thomashow LS (2006) Phenazine compounds in fluorescent *Pseudomonas* spp. biosynthesis and regulation. Annu Rev Phytopathol 44:417–445

Maxson-Stein K, He S-Y, Hammerschmidt R, Jones AL (2002) Effect of treating apple trees with acibenzolar-S-methyl on fire blight and expression of pathogenesis-related protein genes. Plant Dis 86(7):785–790

McManus PS, Stockwell VO, Sundin GW, Jones AL (2002) Antibiotic use in plant agriculture. Annu Rev Phytopathol 40(1):443–465

Mehrabi Z, McMillan VE, Clark IM, Canning G, Hammond-Kosack KE, Preston G et al (2016) *Pseudomonas* spp. diversity is negatively associated with suppression of the wheat take-all pathogen. Sci Rep 6(1):29905

Miller JC, Tan S, Qiao G, Barlow KA, Wang J, Xia DF et al (2011) A TALE nuclease architecture for efficient genome editing. Nat Biotechnol 29(2):143–148

Minuto A, Migheli Q, Garibaldi A (1995) Evaluation of antagonistic strains of *Fusarium* spp. in the biological and integrated control of Fusarium wilt of cyclamen. Crop Protect 14(3):221–226

Mishra R, Gupta RP (2012) In vitro evaluation of plant extracts, bio-agents and fungicides against purple blotch and Stemphylium blight of onion. J Medicinal Plants Res 6(48):5840–5843

Mishra A, Liu C, He B, Tan RJ (2000) Efficacy of isoalantolactone, a sesquiterpene lactone from *Inula racemosa*, as herbal fungitoxicant to control 'take-all' disease of wheat. 42(4):131–137

Mohamed A, Hamza A, Derbalah A (2015) Recent approaches for controlling downy mildew of cucumber under greenhouse conditions. Plant Protect Sci 52(1):1–9

Molina L, Constantinescu F, Michel L, Reimmann C, Duffy B, Défago G (2003) Degradation of pathogen quorum-sensing molecules by soil bacteria: a preventive and curative biological control mechanism. FEMS Microbiol Ecol 45(1):71–81

Montesano M, Brader G, Ponce De León PI, Palva ET (2005) Multiple defence signals induced by *Erwinia carotovora* ssp. *carotovora* elicitors in potato. Mol Plant Pathol 6(5):541–549

Montesinos EJ (2007) Antimicrobial peptides and plant disease control. FEMS Microbiol Lett 270(1):1–11

Moss W, Byrne J, Campbell H, Ji P, Bonas U, Jones J et al (2007) Biological control of bacterial spot of tomato using hrp mutants of *Xanthomonas campestris* pv. vesicatoria. Biol Control 41(2):199–206

Muranaka LS, Giorgiano TE, Takita MA, Forim MR, Silva LF, Coletta-Filho HD et al (2013) *N*-Acetylcysteine in agriculture, a novel use for an old molecule: focus on controlling the plant–pathogen *Xylella fastidiosa*. PLoS One 8(8):e72937

Nadal A, Montero M, Badosa E, Messeguer J, Montesinos L, Montesinos E et al (2012) Constitutive expression of transgenes encoding derivatives of the synthetic antimicrobial peptide BP100: impact on rice host plant fitness. BMC Plant Biol 12(1):1–22

Naguleswaran V, Pakeerathan K, Mikunthan G (2014) Biological control: a promising tool for bulb-rot and leaf twisting fungal diseases in red onion (*Allium cepa* L.) in Jaffna district. World Appl Sci J 31(6):1090–1095

Nel A, Xia T, Madler L, Li NJ (2006) Toxic potential of materials at the nanolevel. Science 311(5761):622–627

Ngowi A, Mbise T, Ijani A, London L, Ajayi OJ (2007) Pesticides use by smallholder farmers in vegetable production in Northern Tanzania. Crop Protect 26(11):1617

Nion YA, Toyota KJ (2008) Suppression of bacterial wilt and *Fusarium wilt* by a *Burkholderia nodosa* strain isolated from Kalimantan soils, Indonesia. Microbes Environ 23(2):134–141

Noyan-Ashraf MH, Wu L, Wang R, Juurlink BHJ (2006) Dietary approaches to positively influence fetal determinants of adult health. FASEB J 20(2):371–373

Ocsoy I, Paret ML, Ocsoy MA, Kunwar S, Chen T, You M, et al (2013) Nanotechnology in plant disease management: DNA-directed silver nanoparticles on graphene oxide as an antibacterial against *Xanthomonas perforans*. ACS Nano 7(10):8972–8980

Pan S, Acharya S (1995) Studies on the seed borne nature of Phomopsis vexans (Sacc. and Syd.) Harter. 39(3):193–198

Pani B, Singh D, Nanda S (2013) Chemical control and economics of phomopsis blight and fruit rot of brinjal in the eastern Ghat highland zone of Odisha. Int J Agric Environ Biotechnol 6(4):581–584

Panno S, Davino S, Caruso AG, Bertacca S, Crnogorac A, Mandić A et al (2021) A review of the most common and economically important diseases that undermine the cultivation of tomato crop in the mediterranean basin. Agronomy 11(11):2188

Panpatte DG, Jhala YK, Shelat HN, Vyas RV (2016) *Pseudomonas fluorescens*: a promising biocontrol agent and PGPR for sustainable agriculture. Microbial Inoculants Sustain Agric Productivity 257–270

Pant R, Mukhopadhyay AJ (2012) Integrated management of seed and seedling rot complex of soybean

Parachin NS, Mulder KC, Viana AAB, Dias SC, Franco OL (2012) Expression systems for heterologous production of antimicrobial peptides. Peptides 38(2):446–456

Paret ML, Vallad GE, Averett DR, Jones JB, Olson SM (2013) Photocatalysis: effect of light-activated nanoscale formulations of TiO_2 on *Xanthomonas perforans* and control of bacterial spot of tomato. Phytopathology 103(3):228–236

Patel T, Pratibha PB (2014) Study on rhizospheric microflora of wild and transgenic varieties of *Gossypium species* in monsoon. Res J Recent Sci 3(IVC-2014):42–51

Pereg L, McMillan M (2015) Scoping the potential uses of beneficial microorganisms for increasing productivity in cotton cropping systems. Soil Biol Biochem 80:349–358

Perry JA, Wright GD (2013) The antibiotic resistance "mobilome": searching for the link between environment and clinic. Front Microbioi 4:138

Pertot I, Alabouvette C, Esteve EH, Franca S (2016) Focus Group soil-borne diseases—the use of microbial biocontrol agents against soil—borne diseases. Agric Innov

Pieterse CM, Zamioudis C, Berendsen RL, Weller DM, Van Wees SC, Bakker P (2014) Induced systemic resistance by beneficial microbes. Annu Rev Phytopathol 52:347–375

Pinto ZV, Cipriano MAP, Santos AS, Pfenning LH, Patrício FRA (2014) Control of lettuce bottom rot by isolates of Trichoderma spp. Summa Phytopathol 40:141–146

Pradhanang P, Ji P, Momol M, Olson S, Mayfield J, Jones JJ (2005) Application of acibenzolar-S-methyl enhances host resistance in tomato against *Ralstonia solanacearum*. Plant Dis 89(9):989–993

Priya R, Sataraddi A, Darshan S (2015) Efficacy of non-systemic and systemic fungicides against purple blotch of onion (*Allium cepa* L.) caused by *Alternaria porri* (Ellis) Cif. Int J Recent Sci 6(9):6519–6521

Pujeri U, Pujar A, Hiremath S, Pujari K, Yadawe M (2015) Analysis of pesticide residues in vegetables in Vijayapur, Karnataka India. World J Pharm Pharm Sci 4(7):1743–1750

Qi LS, Larson MH, Gilbert LA, Doudna JA, Weissman JS, Arkin AP et al (2013) Repurposing CRISPR as an RNA-guided platform for sequence-specific control of gene expression. Cell 152(5):1173–1183

Raaijmakers JM, Bonsall RF, Weller DM (1999) Effect of population density of *Pseudomonas fluorescens* on production of 2, 4-diacetylphloroglucinol in the rhizosphere of wheat. Phytopathology 89(6):470–475

Rahman M, Bhattiprolu S (2010) Efficacy of fungicides and mycorrhizal fungi for the control of damping-off disease in nurseries of tomato, chilli and brinjal crops. 18(2)

Rahman M, Ahmed H, Mian I (1988) Efficacy of fungicides in controlling purple leaf blotch of onion. Bangladesh J Plant Path 4(1&2):71–76

Rahman MME, Hossain DM, Suzuki K, Shiiya A, Suzuki K, Dey TK et al (2016) Suppressive effects of *Bacillus* spp. on mycelia, apothecia and sclerotia formation of *Sclerotinia sclerotiorum* and potential as biological control of white mold on mustard. Aust Plant Pathol 45:103–117

Raid RN (2004) Lettuce diseases and their management. Disease Fruits Veg Diagnosis VVI Manage 121–47

Rakh R, Raut L, Dalvi S, Manwar AV (2011) Biological control of Sclerotium rolfsii, causing stem rot of groundnut by *Pseudomonas* cf. monteilii 9. Recent Res Sci Technol 3(3):26–34

Ramyabharathi S, Meena B, Raguchander T (2012) Induction of chitinase and β-1, 3-glucanase PR proteins in tomato through liquid formulated Bacillus subtilis EPCO 16 against Fusarium wilt. J Today's Biol Sci Res Rev 1:50–60

Raupach G, Kloepper JJ (2000) Biocontrol of cucumber diseases in the field by plant growth-promoting rhizobacteria with and without methyl bromide fumigation. Plant Dis 84(10):1073–1075

Reddy PP (2014) Plant growth promoting rhizobacteria for horticultural crop protection. Springer

Redondo C, Sena-Vélez M, Gell I, Ferragud E, Sabuquillo P, Graham J et al (2015) Influence of selected bactericides on biofilm formation and viability of *Xanthomonas citri* subsp. citri. Crop Protect 78:204–213

Roberts P, Momol M, Ritchie L, Olson S, Jones J, Balogh B (2008) Evaluation of spray programs containing famoxadone plus cymoxanil, acibenzolar-*S*-methyl, and Bacillus subtilis compared to copper sprays for management of bacterial spot on tomato. Crop Protect 27(12):1519–1526

Rodrigues CM, Takita MA, Coletta-Filho HD, Olivato JC, Caserta R, Machado MA et al (2008) Copper resistance of biofilm cells of the plant pathogen *Xylella fastidiosa*. Appl Microbiol Biotechnol 77:1145–1157

Rodríguez C, Lang L, Wang A, Altendorf K, García F, Lipski A et al (2006) Lettuce for human consumption collected in Costa Rica contains complex communities of culturable oxytetracycline-and gentamicin-resistant bacteria. Appl Environ Microbiol 72(9):5870–5876

Rohini A, Gowtham H, Niranjana SJ (2015) Evaluation of efficacy of fungicides against Phomopsis leaf blight of brinjal (*Solanum melongena* L.). Int J Agric Sci Res 5:45–50

Römling U, Galperin MY, Gomelsky MJM, Reviews MB (2013) Cyclic di-GMP: the first 25 years of a universal bacterial second messenger. Microbiol Mol Biol Rev 77(1):1–52

Rose S, Parker M, Punja ZJ (2003) Efficacy of biological and chemical treatments for control of Fusarium root and stem rot on greenhouse cucumber. Plant Dis 87(12):1462–1470

Rutherford ST, Bassler BL (2012) Bacterial quorum sensing: its role in virulence and possibilities for its control. Cold Spring 2(11):a012427

Sadik S, Mazouz H, Benbouazza ABA, Achbani EH (2015) Biological control of bacterial onion diseases using a bacterium, *Pantoea agglomerans* 2066–7. Int J Sci Res 4(1):103–111

Saha P, Das S (2012) Assessment of yield loss due to early blight (*Alternaria solani*) in tomato. Indian J Plant Protect 40(3):195–198

Sahar P, Sahi ST, Jabbar A, Rehman A, Riaz K, Hannan AJ (2013) Chemical and biological management of *Fusarium oxysporum* f. sp melongenae. Pak J Phytopathol 25(2):155–159

Salau I, Shehu K, Kasarawa A, Sambo S, Shahida AJ (2015) A fungi associated with post-harvest rot of commonly consumed fruits in Sokoto metropolis, Nigeria. J Adv Bot Zool 3(3):1–4

Satish S, Mohana D, Ranhavendra M, Raveesha KJ (2007) Antifungal activity of some plant extracts against important seed borne pathogens of *Aspergillus* sp. J Agric Technol 3(1):109–119

Savchuk S (2002) Evaluation of biological control control of Sclerotinia scleroiorum on Canola (*Brassica napus*) in the lab, in the greenhouse, and in the field. MSc. thesis, University of Manitoba

Saxena A, Rai D, Kalra A, Saxena MJ (2020) Bacterial diseases of crops: epidemiology, symptoms and management. Int J Curr Microbiol App Sci 9(6):1483–1499

Sen Y, van der Wolf J, Visser RG, van Heusden SJ (2015) Bacterial canker of tomato: current knowledge of detection, management, resistance, and interactions. Plant Dis 99(1):4–13

Servin A, Elmer W, Mukherjee A, De la Torre-Roche R, Hamdi H, White JC et al (2015) A review of the use of engineered nanomaterials to suppress plant disease and enhance crop yield. J Nanoparicle Res 17:1–21

Shafi J, Tian H, Ji M (2017) Bacillus species as versatile weapons for plant pathogens: a review. Biotechnol Biotechnol Equip 31(3):446–459

Shahid M, Zaidi A, Khan MS, Rizvi A, Saif S, Ahmed B (2017) Recent advances in management strategies of vegetable diseases. Microbial Strateg Veg Prod 197–226

Shan Q, Wang Y, Li J, Gao C (2014) Genome editing in rice and wheat using the CRISPR/Cas system. Nat Proto 9(10):2395–2410

Shanmugaiah V, Nithya K, Harikrishnan H, Jayaprakashvel M, Balasubramanian NJ (2015) Biocontrol mechanisms of siderophores against bacterial plant pathogens. Sustain Appro Control Plant Pathogenic Bacteria 167–190

Sharma N, Tripathi A (2006) Biotechnology. Fungitoxicity of the essential oil of *Citrus sinensis* on post-harvest pathogens. World J Microbiol Biotechnol 22:587–593

Sharma M, Razdan V, Rajik M (2012) In vitro evaluation of fungicides and biocontrol agents against Brinjal leaf blight and fruit rot pathogen *Phomopsis vexans* (Sacc. & Syd.) Harter. BIOINFOLET Quart J Life Sci 9(3):327–332

Shrivastava P, Kumar R, Yandigeri MS (2017) In vitro biocontrol activity of halotolerant *Streptomyces aureofaciens* K20: a potent antagonist against *Macrophomina phaseolina* (Tassi) Goid. Saudi J Biol Sci 24(1):192–199

Shtienberg D, Zilberstaine M, Oppenheim D, Herzog Z, Manulis S, Shwartz H et al (2001) Efficacy of oxolinic acid and other bactericides in suppression of *Erwinia amylovora* in pear orchards in Israel. Phytoparasitica 29(2):143

Singh A, Agrawal K (1999) Fungicidal control of phomopsis fruit rot of brinjal. 26(2)

Singh D, Yadav DK, Chaudhary G, Rana VS, Sharma RK (2016) Potential of *Bacillus amyloliquefaciens* for biocontrol of bacterial wilt of tomato incited by *Ralstonia solanacearum*. J Plant Pathol 7(327):1–6

Slininger P, Burkhead K, Schisler D, Bothast R (2000) Isolation, identification, and accumulation of 2-acetamidophenol in liquid cultures of the wheat take-all biocontrol agent *Pseudomonas fluorescens* 2–79. Appl Microbiol Biotechnol 54:376–381

Smillie CS, Smith MB, Friedman J, Cordero OX, David LA, Alm EJ (2011) Ecology drives a global network of gene exchange connecting the human microbiome. Nature 480(7376):241–244

Solecki R, Davies L, Dellarco V, Dewhurst I, Van Raaij M, Tritscher A et al (2005) Guidance on setting of acute reference dose (ARfD) for pesticides. Food Chem Toxicol 43(11):1569–1593

Srivastava M, Gupta S, Saxena A, Shittu LJA (2011) A review of occurrence of fungal pathogens on significant brassicaceous vegetable crops and their control measures. 3(2):70–79

Stavrinides J, McCann HC, Guttman DSJ (2008) Host–pathogen interplay and the evolution of bacterial effectors. Cell Microbiol 10(2):285–292

Storz G, Imlayt JAJ (1999) Oxidative stress. Curr Opin Microbiol 2(2):188–194

Streubel J, Pesce C, Hutin M, Koebnik R, Boch J, Szurek B (2013) Five phylogenetically close rice SWEET genes confer TAL effector-mediated susceptibility to *Xanthomonas oryzae* pv. *oryzae*. New Phytologist 200(3):808–819

Sudhasha S, Usharani S, Ravimycin T (2008) Surveillance of onion basal rot disease incidence caused by *Fusarium oxysporum* f. sp. cepae and varietal reaction under field condition. 3(2):369–337

Sundin G, Bender CJ (1996) Dissemination of the strA-strB streptomycin-resistance genes among commensal and pathogenic bacteria from humans, animals, and plants. Mol Ecol 5(1):133–143

Sundin GW, Werner NA, Yoder KS, Aldwinckle HS (2009) Field evaluation of biological control of fire blight in the eastern United States. Plant Dis 93(4):386–394

Sundin GW, Castiblanco LF, Yuan X, Zeng Q, Yang CH (2016) Bacterial disease management: challenges, experience, innovation and future prospects: challenges in bacterial molecular plant pathology. Mol Plant Pathol 17(9):1506–1518

Suryawanshi A, Deokar C (2001) Effect of fungicides on growth and sporulation of fungal pathogens causing fruit rot of chilli. 88(1/3):181–182

Taiga A, Suleiman M, Sule W, Olufolaji DB (2008) Comparative in vitro inhibitory effects of cold extracts of some fungicidal plants on *Fusarium oxysporium* mycelium. Afr J Biotechnol 7(18)

Tanina K, Tojo M, Date H, Nasu H, Kasuyama S (2004) Pythium rot of chingensai (*Brassica campestris* L. chinensis group) caused by *Pythium ultimum* var. ultimum and *Pythium aphanidermatum*. J Gen Plant Pathol 70:188–191

Thabet WM, Shendy AH, Gadalla SA (2016) Chronic exposure of insecticide and fungicide as indicator of health impact in some commonly consumed leafy vegetables: case study. Cogent Food Agric 2(1):1193926

Thippeswamy B, Krishnappa M, Chakravarthy C, Sathisha A, Jyothi S, Kumar KV (2006) Pathogenicity and management of phomopsis blight and leaf spot in brinjal caused by *Phomopsis vexans* and *Alternaria solani*. Indian Phytopathol 59(4):475–481

Tripathi P, Dubey N, Shukla AK (2008) Biotechnology. Use of some essential oils as post-harvest botanical fungicides in the management of grey mould of grapes caused by *Botrytis cinerea*. World J Microbiol Biotechnol 24:39–46

Ulloa-Ogaz A, Muñoz-Castellanos L, Nevárez-Moorillón GV (2015) Biocontrol of phytopathogens: antibiotic production as mechanism of control. Battle Against Microbial Pathogens: Basic Sci Technol Adv Edu Prog:305–309

Ume F, Ume G (2005) Biological control of botrytis gray mould and sclerotinia drop in lettuce

Vaish D, Sinha A (2003) Determination of tolerance in *Rhizoctonia solani*, *Trichoderma virens* and *Trichoderma* sp (isolate 20) to systemic fungicides. Indian J Plant Pathol 21(1–2):48–50

Verma M, Brar SK, Tyagi RD, Surampalli RY, Valero JR (2007) Industrial wastewaters and dewatered sludge: rich nutrient source for production and formulation of biocontrol agent, *Trichoderma viride*. World J Microbiol and Biotechnol 23:1695–1703

Vidal S, de León IP, Denecke J, Palva ET (1997) Salicylic acid and the plant pathogen *Erwinia carotovora* induce defense genes via antagonistic pathways. Plant J 11(1):115–123

Vinale F, Sivasithamparam K, Ghisalberti EL, Marra R, Woo SL, Lorito MJ et al (2008) Trichoderma–plant–pathogen interactions. Soil Biol Biochem 40(1):1–10

Von Bodman SB, Bauer WD, Coplin DL (2003) Quorum sensing in plant-pathogenic bacteria. Annu Rev Phytopathol 41(1):455–482

Wang Y, Cheng X, Shan Q, Zhang Y, Liu J, Gao C et al (2014) Simultaneous editing of three homoeoalleles in hexaploid bread wheat confers heritable resistance to powdery mildew. Nat Biotechnol 32(9):947–951

Whipps JM (2001) Microbial interactions and biocontrol in the rhizosphere. J Exp Bot 52(suppl_1):487–511

Worthington R, Rogers S, Huigens III R, Melander C, Ritchie D (2012) Foliar-applied small molecule that suppresses biofilm formation and enhances control of copper-resistant *Xanthomonas euvesicatoria* on pepper. Plant Dis 96(11):1638–1644

Wu L, Ashraf MHN, Facci M, Wang R, Paterson PG, Ferrie A et al (2004) Dietary approach to attenuate oxidative stress, hypertension, and inflammation in the cardiovascular system. Proc Natl Acad Sci 101(18):7094–7099

Xiao L, Xie C-C, Cai J, Lin Z-J, Chen Y-H (2009) Identification and characterization of a chitinase-produced Bacillus showing significant antifungal activity. Curr Microbiol 58:528–533

Yan A, Wang X, Zhang X, Xu YQ (2007) LysR family factor PltR positively regulates pyoluteorin production in a pathway-specific manner in *Pseudomonas* sp. M18. Sci China Ser C Life Sci 50(4):518
Yang F, Korban SS, Pusey PL, Elofsson M, Sundin GW, Zhao Y (2014) Small-molecule inhibitors suppress the expression of both type III secretion and amylovoran biosynthesis genes in *Erwinia amylovora*. Mol Plant Pathol 15(1):44–57
Yang C, Powell CA, Duan Y, Shatters R, Zhang MJ (2015) Antimicrobial nanoemulsion formulation with improved penetration of foliar spray through citrus leaf cuticles to control citrus huanglongbing. PLos One 10(7):e0133826
Yu X, Ai C, Xin L, Zhou G (2011) The siderophore-producing bacterium, Bacillus subtilis CAS15, has a biocontrol effect on Fusarium wilt and promotes the growth of pepper. Eur J Soil Biol 47(2):138–145
Yuan X, Khokhani D, Wu X, Yang F, Biener G, Koestler BJ et al (2015) Cross-talk between a regulatory small RNA, cyclic-di-GMP signalling and flagellar regulator FlhDC for virulence and bacterial behaviours. Environ Microbiol 17(11):4745–4763
Zaidi A, Ahmad E, Khan MS, Saif S, Rizvi A (2015) Role of plant growth promoting rhizobacteria in sustainable production of vegetables: current perspective. Scientia Horticulturae 193:231–239
Zamoum M, Goudjal Y, Sabaou N, Barakate M, Mathieu F, Zitouni A et al (2015) Biocontrol capacities and plant growth-promoting traits of endophytic actinobacteria isolated from native plants of *Algerian Sahara*. J Plant Dis Protect 122:215–223
Zaumeyer WJ (1958) Antibiotics in the control of plant diseases. 12(1):415–440
Zhang L, Gu F, Chan J, Wang A, Langer R, Farokhzad OJ et al (2008) Nanoparticles in medicine: therapeutic applications and developments. Clin Pharmacol Therap 83(5):761–769
Zhou X, Wu F, Wang X, Yuan Y (2008) Progresses in the mechanism of resistance to Fusarium wilt in cucumber (*Cucumis sativus* L.). J Northeast Agric Univ 15(3):1–6

Emerging Approaches Based on Genome-Wide Association Studies (GWAS) for Crop Disease Tolerance

Shikha Dubey and Arvind Mohanan

Abstract The trait of resistance or tolerance to specific pathogens is pivotal for crop resilience, mitigating losses without resorting to excessive chemical treatments. The identification of genes imparting resistance to a particular disease demands sophisticated mapping approaches. Genome-wide association studies (GWAS) involve scanning an organism's genome using hundreds of thousands of genetic markers to test the association of a specific genomic region with resistance to a pathogen. This is particularly beneficial for controlling soil-inhabiting and insect-transmitted pathogens that are challenging to manage with chemicals. GWAS can also unveil virulence genes in pathogens, although it lags a few steps behind. Compared to traditional linkage-based QTL mapping, GWAS offers advantages by significantly reducing the time required for fine mapping for resistance to a pathogen. It also aids in identifying whether complex epistatic networks govern the expression of the trait of interest. With the prevalent availability of sequence information for most crop species and a decrease in genotyping costs, GWAS has become the most preferred method for targeting specific resistance loci. The ensuing literature delves into the methodology of GWAS through case studies, outlines the advantages and disadvantages of GWAS, and explores its application in identifying pathogenicity factors in microbes.

Keywords Resistance/tolerance · GWAS · Pathogens · QTL mapping · Pathogenicity · Plant pathosystem

S. Dubey
Okra and Gourds Breeder, Bayer Crop Science, Vegetable Seeds, BRDC, Bangalore, Karnataka, India
e-mail: shikha.dubey1@bayer.com

A. Mohanan (✉)
Department of Plant Pathology, Sri Karan Narendra Agriculture University, Jobner, Rajasthan 303329, India
e-mail: arvindmohanan120894@gmail.com

J. Chen et al. (eds.), *Molecular and Biotechnological Tools for Plant Disease Management*, Interdisciplinary Biotechnological Advances,
https://doi.org/10.1007/978-981-97-7510-1_5

1 Introduction

In recent years, an array of technologies has played a pivotal role in achieving remarkable milestones. This surge in technology adoption across various sectors is attributed to the swift advancement in computer programs and algorithms, which not only reduce human intervention but also minimize errors in experiments. The field of plant breeding has also experienced significant benefits, both in terms of principle and application, thanks to technological advancements. Plant breeding depends on crossing the desired parents to obtain a specific gene combination in the offspring. While transferring a gene from a wild variety, a backcross approach is typically employed. In this method, the initial crosses between the donor and recipient parent are succeeded by multiple backcrosses to the recurrent parent, ultimately yielding the desired phenotype. The usage of marker-assisted backcrossing has shortened the period and enhanced the precision of the final product. Alternatively, genetic engineering allows for the insertion of a gene from one organism into another through recombinant DNA technology.

The success of these techniques hinges on accurate information about the location of the desired gene within an organism's genome. Consequently, quantitative trait locus (QTL) mapping approaches have been widely utilized in the past decades to map genes related to diseases and other agronomic traits. QTL mapping, also known as linkage mapping, is a statistical method that utilizes polymorphisms in markers linked to a gene to identify the causal variant(s) for a specific trait. It has been an effective method to dissect the regions of the genome that co-segregate with a given trait in the F_2 population or RILs. The key genes of the flowering time pathway in *Arabidopsis* have been located in this way (Kowalski et al. 1994; Clarke et al. 1995). However, QTL mapping has some limitations. First, it can only help to explore the allele diversity present in the parents of the mapping population (Borevitz and Nordborg 2003). Secondly, the amount of recombination that occurs during the creation of the mapping population (F_2, RIL, or DH) impacts the resolution of mapping. Several rounds of intercrossing during the establishment of RILs improve the mapping resolution (Balasubramanian et al. 2009). In order to improve the allelic diversity, multiple lines that are genetically diverse have been intercrossed to establish a multi-parent RIL (Kover et al. 2009; Huang et al. 2011).

Due to the quantum of time required for developing and maintaining mapping populations for QTL analysis and to solve the problem of low allelic diversity, genome-wide association studies (GWAS) have been used widely for the past few decades. In this approach, hundreds of thousands of markers are screened across the entire genome of numerous individuals to pinpoint the markers linked to the trait of interest. In GWAS, the fundamental methodology involves assessing the correlation between each genotyped marker and a scored phenotype of interest across a substantial number of individuals. This approach was first pioneered in human genetics almost a decade ago (Hirschhorn and Daly 2005). As new organisms are sequenced now and then, the total number of GWAS demonstrates a rising trend. The details regarding the number and kind of GWAS literature published to date across different species can be accessed from resources like GWAS Atlas (Tian et al. 2020). These repositories can function as valuable sources of knowledge before initiating an experiment. Another tool, EasyGWAS serves as a knowledge hub for computing, storing, sharing, annotating, and comparing GWAS, as outlined by Grimm et al. (2017). Additionally, it offers conveniences for replicating GWAS signals and improving the detection power in GWAS, as highlighted by Chen et al. (2024).

2 GWAS: The Technique

As previously noted, conducting a GWAS entails screening hundreds of individuals for a large number of markers across the genome. Before initiating the experiment, careful analysis of the type, quantity, and resources required is essential. The key considerations are detailed below and visually represented in Fig. 1:

1. **Plant material**: The plant material selected for a GWAS must consist of genotypes that exhibit both genetic and phenotypic diversity. These individuals could encompass a core set from a germplasm collection, advanced breeding lines, or advanced progeny from a multiparent intercross, and they should be sourced from geographically and genetically distinct origins. The selected lines must be homozygous or nearly homozygous. Sometimes genotypes that are morphologically diverse may not show sufficient polymorphism for genetic markers.

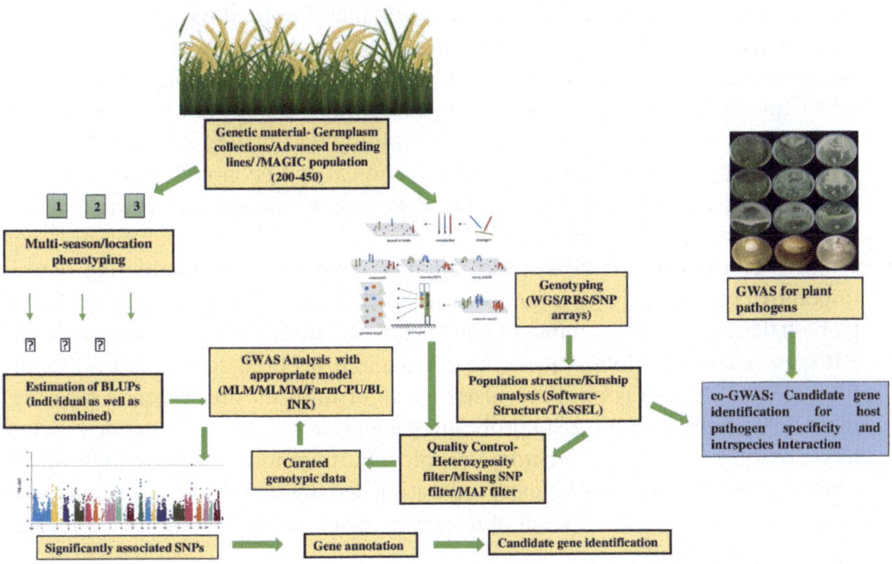

Fig. 1 Brief description of methodology of GWAS for plant disease tolerance

Therefore, a population analysis is conducted before GWAS to ensure that the GWAS panel employed is unstructured, thereby preventing false associations.

2. **Phenotyping**: The phenotyping of the GWAS panel should be highly precise and conducted with replications, preferably at multiple locations and/or seasons. A multi-environment GWAS enables the dissection of genotype x environment interactions, proving beneficial for complex traits such as yield (Eltaher et al. 2021). A deep learning image-based phenotyping was employed to conduct GWAS on sudden death syndrome of soybean [*Glycine max* L. (Merr.)] utilizing visual field disease severity ratings and images collected from 473 accessions (Rairdin et al. 2022). Appropriate experimental design methods should be applied in the field layout to minimize errors and enhance the precision of phenotyping.
3. **Genotyping**: Genotyping requires high-quality DNA that has been quantified properly. There are two objectives of genotyping. The first is to understand the population structure and the second is to detect linkage disequilibrium (LD) between markers and loci of interest. To study the population structure, a large number of adequately distributed markers are sufficient. However, the computational method used should be robust. Mixed models are recommended for this purpose to accommodate unmodeled factors (Sul et al. 2018). For the purpose of GWAS, three types of genotyping methods are most widely adopted: whole-genome resequencing (WGR), reduced representation sequencing (RRS), and single nucleotide polymorphism (SNP) arrays. WGR and RRS leverage next-generation sequencing (NGS) technologies and bioinformatics pipelines to align reads to a reference genome, enabling the identification of both SNPs and genotypes (Nielsen et al. 2011). SNP arrays use allele-specific oligonucleotide (ASO) probes (including target SNP loci and their flanking regions) fixed on a solid support (called a chip), which infer the genotype based on the hybridization signal with a complementary fragment (Pavan et al. 2020). Setting aside cost and technical considerations, the choice of genotyping method for a GWAS depends on the objectives of the study, the crop's ploidy and mating system, the scale, crop-specific genomic features, and the technical and economic aspects associated with each genotyping method. The details of the above genotyping methods can be referred from Pavan et al. (2020).
4. **Quality control of the SNP data**: Quality control of the genotyping data is imperative before utilizing it for downstream analysis. This process is conducted by both the genotyping companies and the investigator. For NGS methods (WGR, RRS, *etc.*), loci with low sequencing depth and low Phred quality scores (Q) are filtered out. SNP arrays utilize clustering algorithms on fluorescence measurement data obtained from ASO probes to categorize samples into genotype clusters, creating an allelic discrimination plot. Furthermore, an evaluation of QC scores gauges the quality of cluster separation and the signal-to-background ratio (Pavan et al. 2020). At the level of the investigator, the following QC procedure should be followed:

   (a) Filtering SNPs with low call rates

(b) Filtering for Minor Allele Frequencies (MAFs) between 1–5% to avoid false positives. It is recommended to first filter rare SNPs before considering the removal of individuals to prevent unnecessary exclusion. When dealing with binary traits such as disease resistance, it is crucial to ensure that there are no structured differences in call rates between the two groups to avoid introducing bias in association tests.
(c) Filtering of loci showing excessive heterozygosity (based on mating system).
(d) Filtering for duplicate individuals or those related by ancestry (pairwise identity-by-state (IBS) or identity-by-descent (IBD)). Parametric methods such as those used in popular software like STRUCTURE (Pritchard et al. 2000) or ADMIXTURE (Alexander et al. 2009) can be used for this purpose. Non-parametric methods, such as principal component analysis (PCA) and multidimensional scaling (MDS), can assist in identifying population structure. However, if outlier sub-populations or individuals exist, these methods may not effectively capture the variation attributable to population structure (Liu et al. 2013).

5. **Association mapping analysis**: Various models have been developed to examine the association between SNPs and measured traits. These models undergo continuous refinement over time to address factors not previously studied or to provide additional options for investigators. Models for GWAS include Efficient Mixed model Associations (EMMA), Multi Locus model (MLM), Multi Locus Mixed Model (MMLM) (Segura et al. 2012), fixed and random model circulating probability unification (Farm CPU) (Liu et al. 2016), Bayesian information and LD iteratively nested keyway (BLINK) (Huang et al. 2018).

3 Post GWAS Steps

Significant associations uncovered by GWAS can be classified based on p-value, the significance of the same SNPs across multiple environments or for related traits, and the degree of confounding with population structure (Spindel et al. 2018). Once promising SNPs are finalized, the genes underlying those peaks can be identified. In many instances, significant SNPs are likely to be near and exhibit strong LD with, true causal variants. If other genes are present near the true causal variant, it is best to manually examine those nearby genes for relevant functions as predicted by homology, gene networks, or other methods (Tibbs Cortes et al. 2021).

4 Application of GWAS in Plant Diseases

The escalating number and severity of plant diseases necessitate an approach that can swiftly identify host resistance, expediting breeding efforts. A meta-analysis performed on GWAS for plants and pathogens revealed a general pattern for plants, indicating that the genetic architecture of disease resistance is polygenic, dynamic during infection kinetics, and influenced by the environment. GWAS reduced the time of identification of QTLs and revealed both complex epistatic networks and a genetic architecture dependent upon the geographical scale (Demirjian et al. 2023). In that study, 162 articles published over the last 18 years were analyzed, encompassing 271 GWAS conducted on plant-pathogen systems. Among these, 143 articles focused on plants, while 9 articles specifically addressed pathogens. The potential of genome-wide association mapping for studying disease resistance loci was initially demonstrated successfully in fine mapping previously cloned genes linked to the presence/absence of a hypersensitive response in *Arabidopsis thaliana* (Aranzana et al. 2005). Since then, *A. thaliana* has been the subject of the highest number of GWAS studies among wild plants.

Disease resistance breeding is the incorporation of resistance genes in otherwise susceptible plants. Implementing this approach stands out as the most sustainable and efficient method for mitigating the adverse effects on crop yield and quality induced by major phytopathogens. The resistance to a pathogen may be quantitative or qualitative. Qualitative resistance is endowed by a single gene (either dominant or recessive), exhibiting race specificity and offering a high level of resistance. Conversely, quantitative resistance is non-specific to races and is controlled by a variable number of genes, ranging from few to many, which may function in an additive or partially dominant manner (Wisser et al. 2006). While qualitative resistance is relatively easier to breed for, plant breeders predominantly emphasize quantitative resistance because of its greater durability and the ability to offer protection against a wide spectrum of pathogens (Poland et al. 2009).

The theoretical foundation for qualitative resistance lies in the gene-for-gene hypothesis, introduced by Flor in 1971. According to this hypothesis, a resistant reaction in the host plant requires a dominant or semi-dominant *R* gene that corresponds to the avirulence (*Avr*) gene present in the pathogen. The interaction of the *R* gene of the host with the *Avr* gene of the pathogen stimulates the downstream signaling cascades triggering the secretion of antimicrobial compounds which in turn leads to resistance expression in the host. Meanwhile, if the *R* gene or the *Avr* gene gets altered or mutated, plant receptors fail to identify the *Avr* gene leading to a susceptible reaction or winning situation for the pathogen. The molecular mechanism underlying this host–pathogen interaction was first studied by Jones and Dangl (2006). A complex multiphase host–pathogen interplay occurs when a plant is invaded by a pathogen. The plant recognizes microbial-associated molecular patterns (MAMPs) like flagellin in bacteria, glucan in oomycetes, chitin in fungi, etc., using pattern-recognizing receptors (PRRs). The recognition triggers an innate immune response in the host plant, also termed basal line defense. In response to this, the virulent pathogen releases effector proteins to destroy the basal defense of the host. In turn, the host plant activates its second line of defense, i.e. resistance genes such as NBS-LRR and CC-NBS-LRR. The resistance genes recognize the pathogen-secreted effector proteins and destroy them. This immunity is known as the effector-triggered immunity. ETI is associated with a hypersensitive response causing localized cell death at the site of infection (Dodds and Rathjen 2010).

GWAS serves as a strong tool for identifying genes associated with both qualitative and quantitative resistance. The following section will explore marker-trait associations uncovered through GWAS for significant types of plant pathogens affecting crops.

Maize: Wang et al. (2012) undertook a GWAS to investigate head smut resistance in maize using the Illumina Maize SNP50 array. Their study, which involved a panel of 144 inbred lines, identified 18 previously unknown candidate genes associated with head smut resistance. These candidate genes were further categorized into three groups: resistance genes, disease response genes, and other genes potentially involved in plant disease resistance. Additionally, Omondi et al. (2023) conducted GWAS for gray leaf spot and northern corn leaf blight (NCLB). Focusing on NCLB, Rashid et al. (2020) carried out a high-resolution GWAS using three association mapping panels, predominantly composed of tropical lines adapted to various geographical regions. High-density SNPs from genotyping-by-sequencing (GBS) were used for genotyping after applying quality control. In the analysis, across the three mapping panels, a total of 22 SNPs were identified as being associated with NCLB, and haplotype regression revealed the presence of 17 significant haplotypes with a false discovery rate (FDR) of ≤ 0.05. Charcoal rot caused by *Macrophomina phaseolina* is a serious concern for maize cultivators. CIMMYT Asia conducted a GWAS utilizing an association mapping panel comprising 396 tropical-adapted lines. Disease severity was phenotyped at two hotspot locations in India. After filtering for quality control, a set of 2,96,497 high-quality SNPs was employed for genotyping. Among them, 19 SNPs were found to be associated with charcoal rot resistance, displaying *p*-values ranging from 5.88×10^{-6} to 4.8×10^{-5}. The validation of these variants was carried out using two $F_{2:3}$ populations. Two QTLs belonging to different resistant parents and having overlapping physical intervals, *qMSR6* and *qFMSR6* were present on chromosome 6 (Rashid et al. 2021). Maize rough dwarf stands as a significant viral disease affecting maize crops in China. Zhao et al. (2022) employed disease severity index (DSI) data from 292 maize inbred lines and an F_6 linkage population, collected across multiple environments over two years. The genotyping data was derived from maize SNP 50,000 chips, and GWAS was conducted using four analytical models. The results unveiled notable associations between resistance to maize rough dwarf disease and 22 SNPs situated on chromosomes 1, 3, 4, 6, 7, and 8, demonstrating statistical significance with a P-value < 0.0001. The SNPs located on chromosomes 3, 6, and 8 exhibited consistencies with the QTL regions identified through linkage mapping in an RIL population. Candidate genes identified by GWAS encompassed an LRR receptor-like serine/threonine-protein kinase (GRMZM2G141288) and a DRE-binding protein (GRMZM2G006745).

Rice: Rice blast poses a serious and persistent challenge across all rice-growing regions. The annual crop losses attributed to this disease could sustain nearly 60 million people (Zeigler et al. 1994). Currently, approximately 83 major blast-resistance genes have been identified, and of these, at least 24 genes have been successfully cloned (https://www.ricedata.cn/gene).

Blast resistance is commonly categorized into complete and partial resistance, as outlined by Wang et al. (1994). Complete resistance to blast disease is qualitative and specific to particular races. In contrast, partial resistance to blast is recognized as both quantitative and enduring. In a GWAS study on rice blast resistance, Wang et al. (2014) identified 30 Marker Trait Associations (MTA) involving 366 diverse accessions of indica rice. The strongest signal (*Chr11_6526998*, $P = 1.17 \times 10^{-17}$) was located within the *Pia*-blast resistance gene. An additional association signal (*Chr11_30606558*) was identified in the vicinity of the QTL *Pif*. In a study conducted by Kang et al. (2016), GWAS was performed on the rice diversity panel 1 (RDP1), genotyped with 700,000 SNPs and exposed to five different races of *Magnaporthe oryzae*. This analysis led to the identification of ninety blast resistance-associated loci, with 82 being novel regions, and 15 co-localizing with known Blast resistance loci. In a separate study, Li et al. (2019) discovered a total of 56 QTLs associated with rice blast resistance, utilizing 234 cultivars (RDP1) and three isolates of *M. oryzae* in Southern China. Among the identified QTLs, one localized with the resistance gene *Pik* locus, provided resistance against all three isolates. The genomic sequence analysis of the resistant cultivars led to the identification of a new *Pik* allele, designated as *Pikx*.

Wheat: The MTA for yellow rust, stripe rust and Fusarium head blight of wheat has been identified using GWAS (Gangurde et al. 2022). Kumar et al. (2020) conducted a genome-wide association scan in a panel comprising 483 spring wheat genotypes to pinpoint novel alleles associated with resistance to stripe rust, stem rust, and leaf rust. The genotyping of the panel was executed using a 35,000 SNP array, and the wheat genotypes were subjected to inoculation with 18 pathotypes of the three rusts under field conditions. Each rust type was evaluated at two locations for two successive crop seasons. The findings indicated a notable influence of population structure within the panel on the disease response. Employing a compressed MLM approach, 25 genomic regions were identified, showing associations with resistance to at least two rusts. Gene annotation showed several of these regions to be co-localized with resistance-associated genes viz., NBS-LRR, E3-ubiquitin protein ligase, ABC transporter protein, etc. Septoria tritici blotch (STB) is another foliar disease that is a menace to wheat production in Europe (Zakieh et al. 2023). A panel of 316 winter wheat breeding lines from a commercial breeding program was evaluated for STB resistance at the seedling stage under controlled conditions followed by GWAS analysis. Nine single and multi-locus GWAS models successfully identified 24 MTAs grouped into 20 QTLs associated with seedling-stage resistance to STB. Significantly, this study documented, for the first time, the presence of seven QTLs located on chromosomes 1B, 2A, 2B, 5B (two), 7A, and 7D.

Barley: Spot blotch, induced by *Cochliobolus sativus*, presents a substantial impediment to realizing the maximum yield potential of barley crops. To unravel the genetic foundation of this disease, a GWAS was performed at both seedling and adult plant stages, utilizing a diverse array of 336 barley genotypes. The genotyping process utilized a 9,000 SNP Illumina Infinium chip. Notably, nine QTLs associated with Spot Blotch resistance were identified at the seedling stage, while eight QTLs were uncovered at the adult stage. These findings shed light on the genetic factors that influence resistance, as elucidated by Gyawali et al. in 2018.

In a parallel study, Gutiérrez et al. (2015) employed a multi-environment multi-QTL mixed models approach to discern the genetic architecture of field-level resistance to leaf rust, stripe rust, and spot blotch in 360 Latin American barley accessions. This comprehensive investigation revealed significant MTAs for spot blotch (9 QTLs), leaf rust (6 QTLs), and stripe rust (7 QTLs). Both global and environment-specific QTL were detected, providing insights into the complex genetic underpinnings of resistance to these fungal diseases in barley.

Vegetable crops: In an investigation on powdery mildew disease resistance in pumpkins, GWAS augmented in the identification of the *AP2*-like ethylene-responsive transcription factor (*CmoAP2/ERF*) as a major contributor to disease resistance. During the GWAS validation, six SNPs were discerned within the coding sequence of this gene, differentiating between the resistant and susceptible parent (Alavilli et al. 2022). Liu et al. (2020) conducted a GWAS to identify novel genetic loci associated with downy mildew resistance in a cucumber core germplasm set. Eighteen downy mildew-associated loci were discovered, with sixteen of them coinciding with QTLs previously reported in other studies.

In a field evaluation for white rust resistance, a GWAS panel comprising 346 spinach USDA germplasm accessions underwent genotyping using 13,235 WGR generated nine SNPs, located on chromosomes 2, 3, 4, and 6 which were associated with white rust resistance (Shi et al. 2022).

A GWAS was conducted on a core set of 191 tomato genotypes, utilizing the 51,000 Axiom® tomato array to identify MTAs for bacterial wilt resistance. A total of eight MTAs were detected at a significance level of $p < 0.0005$, and two MTAs consistently appeared on chromosomes 4 and 12. These MTAs, along with their corresponding QTLs (*Bwr-4 and Bwr-12*), explained 8.36–18.28% of the total phenotypic variations in both disease assays (Nguyen et al. 2021).

Groundnut: Resistance to aflatoxin production is a prime focus in peanut breeding initiatives. In a study conducted by Yu et al. (2020), seeds from a Chinese peanut mini-core collection comprising 99 accessions were assessed for their susceptibility to aflatoxin production using a laboratory kernel inoculation assay. Simultaneously, the same accessions underwent genotyping through Restriction-site associated DNA sequencing (RAD-Seq) for a comprehensive GWAS. A total of 60 SNP markers linked to aflatoxin production were identified, collectively explaining 16.87–31.70% of phenotypic variation (PVE). Notably, *SNP02686* and *SNP19994* exhibited significant influence with PVE values of 31.70% and 28.91%, respectively. Further scrutiny pinpointed that SNPs within the coding sequence contributed to the variability in aflatoxin production capacity.

In a separate investigation, Oteng-Frimpong et al. (2023) employed six multi-locus GWAS models on BLUP data derived from 294 African groundnut germplasm. The screening encompassed tikka disease (Early and Late Leaf Spot) as well as image-based indices assessing leaf spot disease severity during the years 2020 and 2021. Additionally, 8,772 high-quality SNPs from 48,000 SNP array platforms were utilized. A total of 97 associated SNPs were identified, with 29 unique SNPs detected by at least two models across 16 chromosomes. The explained phenotypic variation ranged from 0.01 to 62.76% for one or more traits, shedding light on the genetic factors influencing disease resistance in the groundnut germplasm.

Soybean: Chang et al. (2016) conducted a GWAS using the USDA-ARS Soybean Germplasm Collection to investigate disease responses. Data on disease responses to one bacterial disease, five fungal diseases, two nematode-induced diseases, and three viral diseases were obtained from the public germplasm resources information network. Genotyping data from the public SNP dataset (*SoySNP50K*) were employed. The study identified significantly associated SNPs with disease ratings across the mentioned diseases. Subsequently, within the LD regions of these significant SNPs, there was an enrichment of LRR receptor-like kinases and NBS-LRR candidate resistance genes.

5 Greater Opportunities to Explore GWAS for Plant Pathogens

Evolutionary changes in the pathogen genomes play a pivotal role in causing epidemics in crops, allowing pathogens to evade the host plant's surveillance system (Bartoli et al. 2016). The identification of pathogenicity genes, such as effectors and genes involved in nutrition and cell wall degradation, is crucial for developing effective breeding strategies. However, the genetic and molecular mechanisms enabling pathogens to overcome host defense systems are not thoroughly investigated. Natural genetic variation in pathogens is less explored due to the preference for more feasible reverse genetic approaches based on mutant analysis or complementation (Sánchez-Vallet et al. 2018; Falush 2016).

A limited number of studies have explored the natural genetic variation in pathogens using GWAS, primarily because reverse genetic approaches are more practical. Demirjian et al. (2022) reported only 19 plant pathogen GWAS consisting of 1 bacterium and 18 fungi, with an average population size of approximately 100. In contrast, there have been 334 microbial GWAS reported on human pathogens up to 2019 (San et al. 2019).

The wheat pathogen *Zymoseptoria tritici* has been extensively studied. A global collection of 106 strains was subjected to GWAS to identify polymorphisms linked to virulence in two wheat varieties. The study identified 25 distinct genomic loci associated with the pathogen's reproductive success, suggesting host specialization. The gain in virulence was attributed to a segregating gene deletion polymorphism, highlighting the role of chromosomal structure turnover in driving host specialization (Hartmann et al. 2017). Subsequent studies on *AvrStb6*, the first functionally validated avirulence gene in *Z. tritici*, revealed its role in producing an avirulence phenotype on wheat cultivars carrying the *Stb6* resistance gene (Zhong et al. 2017). Diversity in the *AvrStb6* gene and the incidence of haplotypes allowing evasion of *Stb6*-mediated resistance in recently evolved *Z. tritici* populations were further investigated (Stephens et al. 2021). Surprisingly, the originally described avirulence isoform of *AvrStb6* was absent among modern *Z. tritici* isolates, indicating ongoing evolutionary changes.

The relatively low number of GWAS on plant pathogenic microbes can be attributed to several reasons. Limited genetic consortia have the mandate to establish large collections of whole-genome-sequenced pathogenic strains in crops and wild species (Karasov et al. 2018). Additionally, incomplete information on the population genetic processes of pathogens and opportunistic modes of reproduction for population expansion poses challenges. In asexually reproducing bacteria or fungi, low genetic diversity theoretically results in a long LD decay, affecting mapping resolution. This issue can be addressed by including a large number of strains from different locations and geographies. Another approach is co-GWAS, where host and pathogen genomes are studied simultaneously.

6 Integrated GWAS for Plant Pathogen: Co-GWAS

The co-Genome Wide Association Study (co-GWAS) serves as an extension of single-phenotype GWAS, incorporating genotypic information from both hosts and pathogens simultaneously. Leveraging clinical data and genome-scale SNP arrays, co-GWAS has proven effective in identifying clinically significant genomic variants associated with human diseases, including HIV infection, Hepatitis C, and pneumococcal meningitis (Dexter et al. 2023). This approach facilitates the exploration of interspecies interactions. In a study on the *Arabidopsis thaliana–Xanthomonas arboricola* pathosystem, Wang et al. (2018) utilized co-GWAS to pinpoint candidate genes responsible for host–pathogen specificity. The integration of whole-genome sequence data for interacting species allowed for a more detailed understanding of the genetic architecture of complex traits, particularly due to the shortened LD decay in the pathogen.

A similar strategy was employed to investigate the genetic systems governing the reciprocal adaptation of rice and its bacterial pathogen, *Xanthomonas oryzae* (*Xoo*) responsible for bacterial blight disease. The study utilized whole-genome sequencing and extensive phenotyping data from 701 rice accessions and 23 diverse *Xoo* strains. This comprehensive analysis identified 47 *Xoo* virulence-related genes and 318 rice quantitative resistance genes localized in 41 genomic regions. Genome-wide interactions between the identified virulence-related genes and quantitative resistance genes revealed the presence of well-known resistance genes and virulence genes, along with numerous previously uncharacterized ones (Zhang et al. 2021). Such studies contribute valuable insights into the co-evolution of plant–pathogen systems over time.

7 Perspectives and Future Prospects

Despite the considerable success achieved through GWAS, this method has certain drawbacks associated with population structure and rare alleles, leading to potentially misleading associations. To address these issues, two approaches are suggested. The first involves the development of novel statistical methods for exploring rare functional alleles (Listgarten et al. 2013; Kaakinen et al. 2017). Alternatively, artificial populations can be designed to balance allelic frequencies and control population structure (Romero Navarro et al. 2017; Wen et al. 2019). For a more in-depth discussion, readers are referred to other sources (Gibson 2012; Cockram and Mackay 2018).

GWAS often identifies loci with small effect sizes, explaining only a modest proportion of heritability (Manolio et al. 2009). Furthermore, non-causative loci may exhibit more significant associations than causative ones, a phenomenon known as synthetic associations. These arise due to linkage drag caused by LD between common markers and rare causative variants (Dickson et al. 2010; Liu and Yan 2019), contributing to the missing heritability problem in GWAS. Sometimes, trait variation results from multiple alleles within a gene, making haplotype or gene-based methodologies more suitable for identifying such situations. Thus, improving haplotype-based association methods to handle multiple alleles is recommended. Joint utilization of linkage and association mapping, either in the same step or with linkage mapping validating association results, enhances the robustness of findings (Wu et al. 2002; Omondi et al. 2023).

Advancements in technology for identifying disease resistance loci, coupled with open-domain resources and knowledge, have progressed rapidly in the past decade. Any new plant disease GWAS experiment should consider and build upon validated results from previous studies. After fine mapping and rigorous validation, the identified MTAs can be employed for marker-assisted selection and gene pyramiding, strengthening the application of experimental results in crop improvement. Establishing a repository categorizing MTAs by crop or pathogen will enable easier access to this information. Continuous improvements in NGS and bioinformatics tools hold promise for overcoming GWAS limitations and making results more accessible across diverse scientific disciplines. Once receptor genes in plant–pathogen interactions are identified through GWAS, the CRISPR-Cas system can be employed to knock out these genes, enhancing resistance levels, as demonstrated in rice (Huang et al. 2020a, b, c).

The presence of disease complexes poses a challenge during large-scale genotyping and phenotyping for GWAS. Developing efficient phenotyping screening protocols is crucial for achieving research objectives. Recent advances in statistical tools for multi-trait Genome-Wide Association mapping are expected to facilitate the identification of pleiotropic loci underlying multi-pathogen responses (Korte et al. 2012). Metagenomic resources can aid in describing multiple pathogen infections through the characterization of pathobiota, defined as the complex of microorganisms with the potential to cause disease in the plant host (Kamada et al. 2013).

In the case of viral pathogens, vectors tend to colonize the most susceptible genotype in the population. Choudhury et al. (2019) tested a wheat diversity panel for resistance to *Barley yellow dwarf virus* (BYDV) with both natural and artificial inoculation protocols, observing no clear correlations between resulting phenotypes. Therefore, suitable transmission methods should be employed for a homogeneous inoculum distribution. Visual scoring may be confounded by other physiological factors, necessitating supplementation with biochemical assays such as ELISA/PCR/q-RT-PCR for virus load detection (Pimenta et al. 2020). Similar to the emphasis on plant germplasm collections, the preservation and maintenance of strains of plant pathogens should be incentivized. The COVID-19 pandemic in 2020–21 highlighted how a pathogen can alter the course of human existence. Similarly, beneficial microorganisms, whether bacteria, fungi, or viruses, can prove useful in times of uncertainty. Encouragement should be provided for the development and maintenance of plant pathogen repositories, at least for research purposes.

8 Conclusion

Recent technological strides have transformed plant breeding, reducing human intervention and errors. Traditionally reliant on crossbreeding and genetic engineering, plant breeding benefits from marker-assisted backcrossing and recombinant DNA technology for efficient trait introgression. Accurate gene location within the genome is crucial for success. While traditional methods like QTL mapping have been effective, they have limitations, prompting the rise of GWAS. This technique screens numerous markers across diverse genomes to identify trait associations, yet challenges like population structure and rare functional alleles persist. The emerging field of co-GWAS simultaneously explores genotypic information from hosts and pathogens, proving effective in understanding genetic systems governing plant-pathogen interactions (Fig. 1). Overcoming GWAS challenges requires novel statistical methods, NGS, and advancements in technology. Continuous improvement of phenotyping protocols, coupled with maintaining pathogen strain collections, is vital. Integrating cutting-edge technologies, innovative methodologies, and collaborative efforts will shape the future of plant breeding and disease resistance research, contributing to global agricultural sustainability.

References

Alavilli H, Lee JJ, You CR et al (2022) GWAS reveals a novel candidate gene *CmoAP2/ERF* in pumpkin (*Cucurbita moschata*) involved in resistance to powdery mildew. Int J Mol Sci 23(12):6524

Alexander DH, Novembre J, Lange K (2009) Fast model-based estimation of ancestry in unrelated individuals. Genome Res 19:1655–1664

Aranzana MJ, Kim S, Zhao K et al (2005) Genome-wide association mapping in Arabidopsis identifies previously known flowering time and pathogen resistance genes. PLoS Genet 1:e60
Balasubramanian S, Schwartz C, Singh A et al (2009) QTL mapping in new *Arabidopsis thaliana* advanced intercross-recombinant inbred lines. PLoS ONE 4(2):e4318
Bartoli C, Roux F, Lamichhane JR (2016) Molecular mechanisms underlying the emergence of bacterial pathogens: an ecological perspective. Mol Plant Pathol 17(2):303–310
Borevitz JO, Nordborg M (2003) The impact of genomics on the study of natural variation in Arabidopsis. Plant Physiol 132(2):718–725
Chang HX, Lipka AE, Domier LL et al (2016) Characterization of disease resistance loci in the USDA soybean germplasm collection using genome-wide association studies. Phytopathology 106(10):1139–1151
Chen J, Tan C, Zhu M et al (2024) CropGS-Hub: a comprehensive database of genotype and phenotype resources for genomic prediction in major crops. Nucl Acids Res 52(5):1519-D1529
Choudhury S, Larkin P, Xu R et al (2019) Genome wide association study reveals novel qtl for *Barley yellow dwarf virus* resistance in wheat. BMC Genom 20:891
Clarke JH, Mithen R, Brown JK, Dean C (1995) QTL analysis of flowering time in *Arabidopsis thaliana*. Mol Gen Genet 248(3):278–286
Cockram J, Mackay I (2018) Genetic mapping populations for conducting high-resolution trait mapping in plants. In: Varshney RK, Pandey MK, Chitikineni A (ed) Plant genet molecular biology, pp 109–138
Demirjian C, Razavi N, Desaint H et al (2022) Study of natural diversity in response to a key pathogenicity regulator of *Ralstonia solanacearum* reveals new susceptibility genes in *Arabidopsis thaliana*. Mol Plant Pathol 23(3):321–338
Demirjian C, Vailleau F, Berthomé R et al (2023) Genome-wide association studies in plant pathosystems: success or failure? Trends Plant Sci 28(4):471–485
Dexter E, Fields PD, Ebert D (2023) Uncovering the genomic basis of infection through co-genomic sequencing of hosts and parasites. Mol Biol Evol 40(7):145
Dickson SP, Wang K, Krantz I et al (2010) Rare variants create synthetic genome-wide associations. PLoS Biol 8(1):1000294
Dodds PN, Rathjen JP (2010) Plant immunity: towards an integrated view of plant- pathogen interactions. Nat Rev Genet 11:539–548
Eltaher S, Baenziger PS, Belamkar V et al (2021) GWAS revealed effect of genotype × environment interactions for grain yield of Nebraska winter wheat. BMC Genom 22(1):1–14
Falush D (2016) Bacterial genomics: microbial GWAS coming of age. Nat Microbiol 1:16059
Flor HH (1971) Current status of the gene-for-gene concept. Annu Rev Phytopathol 9(1):275–296
Gangurde SS, Xavier A, Naik YD et al (2022) Two decades of association mapping: insights on disease resistance in major crops. Front Plant Sci 13:1064059
Gibson G (2012) Rare and common variants: twenty arguments. Nat Rev Genet 13(2):135–145
Grimm DG, Roqueiro D, Salomé PA et al (2017) EasyGWAS: a cloud-based platform for comparing the results of genome-wide association studies. Plant Cell 29:5–19
Gutiérrez L, Germán S, Pereyra S et al (2015) Multi-environment multi-QTL association mapping identifies disease resistance QTL in barley germplasm from Latin America. Theor Appl Genet 128:501–516
Gyawali S, Chao S, Vaish SS et al (2018) Genome wide association studies (GWAS) of spot blotch resistance at the seedling and the adult plant stages in a collection of spring barley. Mol Breed 38:1–14
Hartmann FE, Sánchez-Vallet A, McDonald BA et al (2017) A fungal wheat pathogen evolved host specialization by extensive chromosomal rearrangements. ISME J 11(5):1189–1204
Hirschhorn JN, Daly MJ (2005) Genome-wide association studies for common diseases and complex traits. Nat Rev Genet 6(2):95–108
Huang X, Paulo MJ, Boer M et al (2011) Analysis of natural allelic variation in Arabidopsis using a multiparent recombinant inbred line population. Proc Natl Acad Sci USA 108(11):4488–4493

Huang M, Liu X, Zhou Y et al (2018) BLINK: a package for the next level of genome-wide association studies with both individuals and markers in the millions. GigaScience 8:1–12
Huang S, Xin S, Xie G et al (2020a) Mutagenesis reveals that the rice *osmpt3* gene is an important osmotic regulatory factor. Crop J 8(3):465–479
Huang Y, Han Z, Cheng N et al (2020b) Minor effects of 11 *dof* family genes contribute to the missing heritability of heading date in rice (*Oryza sativa*). Front Plant Sci 10:1739
Huang YY, Liu XX, Xie Y et al (2020c) Identification of feronia-like receptor genes involved in rice-*Magnaporthe oryzae* interaction. Phytopathol Res 2(1):1–10
Jones JD, Dangl JL (2006) The plant immune system. Nature 444(7117):323
Kaakinen M, Mägi R, Fischer K et al (2017) MARV: a tool for genome-wide multi-phenotype analysis of rare variants. BMC Bioinform 18:110
Kamada N, Chen GY, Inohara N et al (2013) Control of pathogens and pathobionts by the gut microbiota. Nat Immunol 14:685–690
Kang H, Wang Y, Peng S et al (2016) Dissection of the genetic architecture of rice resistance to the blast fungus *Magnaporthe oryzae*. Mol Plant Pathol 17(6):959–972
Karasov TL, Almario J, Friedemann C et al (2018) *Arabidopsis thaliana* and Pseudomonas pathogens exhibit stable associations over evolutionary time scales. Cell Host Microbe 24:168–179
Korte A, Vilhjalmsson BJ, Segura V et al (2012) A mixed-model approach for genome-wide association studies of correlated traits in structured populations. Nat Genet 44:1066–1071
Kover PX, Valdar W, Trakalo J et al (2009) A multiparent advanced generation inter-cross to fine-map quantitative traits in *Arabidopsis thaliana*. PLoS Genet 5(7):e100055
Kowalski SP, Lan TH, Feldmann KA, Paterson AH (1994) QTL mapping of naturally-occurring variation in flowering time of *Arabidopsis thaliana*. Mol Gen Genet 245(5):548–555
Kumar D, Kumar A, Chhokar V et al (2020) Genome-wide association studies in diverse spring wheat panel for stripe, stem, and leaf rust resistance. Front Plant Sci 11:748
Li W, Chern M, Yin J et al (2019) Recent advances in broad-spectrum resistance to the rice blast disease. Curr Opin Plant Biol 50:114–120
Listgarten J, Lippert C, Heckerman D (2013) FaST-LMM-select for addressing confounding from spatial structure and rare variants. Nat Genet 45(5):470–471
Liu HJ, Yan J (2019) Crop genome-wide association study: a harvest of biological relevance. Plant J 97(1):8–18
Liu L, Zhang D, Liu H et al (2013) Robust methods for population stratification in genome wide association studies. BMC Bioinform 14:132
Liu X, Huang M, Fan B et al (2016) Iterative usage of fixed and random effect models for powerful and efficient genome-wide association studies. PLoS Genet 12:e1005767
Liu X, Lu H, Liu P (2020) Identification of novel loci and candidate genes for cucumber downy mildew resistance using GWAS. Plants 9(12):1659
Manolio TA, Collins FS, Cox NJ et al (2009) Finding the missing heritability of complex diseases. Nature 461(7265):747–753
Nguyen TT, Le NT, Sim SC (2021) Genome-wide association study and marker development for bacterial wilt resistance in tomato (*Solanum lycopersicum* L.). Sci Hortic 289:110418
Nielsen R, Paul JS, Albrechtsen A et al (2011) Genotype and SNP calling from next-generation sequencing data. Nat Rev Genet 12:443–451
Omondi DO, Dida MM, Berger DK et al (2023) Combination of linkage and association mapping with genomic prediction to infer QTL regions associated with gray leaf spot and northern corn leaf blight resistance in tropical maize. Front Genet 14:1282673
Oteng-Frimpong R, Karikari B, Sie EK (2023) Multi-locus genome-wide association studies reveal genomic regions and putative candidate genes associated with leaf spot diseases in African groundnut (*Arachis hypogaea* L.) germplasm. Front Plant Sci 13:1076744
Pavan S, Delvento C, Ricciardi L et al (2020) Recommendations for choosing the genotyping method and best practices for quality control in crop genome-wide association studies. Front Genet 11:447

Pimenta RJG, Aono AH, Villavicencio RCB et al (2020) Genome-wide association mapping of sugarcane yellow leaf virus resistance. Sci Rep 11(1):15730

Poland JA, Balint-Kurti PJ, Wisser RJ et al (2009) Shades of gray: the world of quantitative disease resistance. Trends Plant Sci 14(1):21–29

Pritchard JK, Stephens M, Donnelly P (2000) Inference of population structure using multilocus genotype data. Genetics 155:945–959

Rairdin A, Fotouhi F, Zhang J et al (2022) Deep learning-based phenotyping for genome wide association studies of sudden death syndrome in soybean. Front Plant Sci 13:966244

Rashid Z, Sofi M, Harlapur SI et al (2020) Genome-wide association studies in tropical maize germplasm reveal novel and known genomic regions for resistance to Northern corn leaf blight. Sci Rep 10:21949

Rashid Z, Kaur H, Babu V et al (2021) Identification and validation of genomic regions associated with charcoal rot resistance in tropical maize by genome-wide association and linkage mapping. Front Plant Sci 12:726767

Romero Navarro JA, Willcox M, Burgueño JN et al (2017) A study of allelic diversity underlying flowering-time adaptation in maize landraces. Nat Genet 49(3):476–480

San JE, Baichoo S, Kanzi A et al (2019) Current affairs of microbial genome-wide association studies: approaches, bottlenecks and analytical pitfalls. Front Microbiol 10:3119

Sánchez-Vallet A, Hartmann FE, Marcel TC et al (2018) Nature's genetic screens: using genome-wide association studies for effector discovery. Mol Plant Pathol 19(1):3

Segura V, Vilhjálmsson BJ, Platt A et al (2012) An efficient multi-locus mixed-model approach for genome-wide association studies in structured populations. Nat Genet 44:825–830

Shi A, Bhattarai G, Xiong H (2022) Genome-wide association study and genomic prediction of white rust resistance in USDA GRIN spinach germplasm. Hortic Res 9:uhac069

Spindel JE, Dahlberg J, Colgan M et al (2018) Association mapping by aerial drone reveals 213 genetic associations for Sorghum bicolor biomass traits under drought. BMC Genom 19:679

Stephens C, Ölmez F, Blyth H et al (2021) Remarkable recent changes in the genetic diversity of the avirulence gene *AvrStb6* in global populations of the wheat pathogen *Zymoseptoria tritici*. Mol Plant Pathol 22(9):1121–1133

Sul JH, Martin LS, Eskin E (2018) Population structure in genetic studies: Confounding factors and mixed models. PLoS Genet 14(12):e1007309

Tian D, Wang P, Tang B et al (2020) GWAS Atlas: a curated resource of genome-wide variant-trait associations in plants and animals. Nucleic Acids Res 48:927–932

Tibbs Cortes L, Zhang Z et al (2021) Status and prospects of genome-wide association studies in plants. Plant Genome 14:e20077

Wang GL, Mackill DJ, Bonman JM et al (1994) RFLP mapping of genes conferring complete and partial resistance to blast in a durably resistant rice cultivar. Genetics 136(4):1421–1434

Wang M, Yan J, Zhao J et al (2012) Genome-wide association study (GWAS) of resistance to head smut in maize. Plant Sci 196:125–131

Wang C, Yang Y, Yuan X et al (2014) Genome-wide association study of blast resistance in indica rice. BMC Plant Biol 14:1–11

Wang M, Roux F, Bartoli C et al (2018) Two-way mixed-effects methods for joint association analysis using both host and pathogen genomes. PNAS 115(24):5440–5449

Wen YJ, Zhang YW, Zhang J et al (2019) An efficient multi-locus mixed model framework for the detection of small and linked QTLs in F-2. Brief Bioinform 20(5):1913–1924

Wisser RJ, Balint-Kurti PJ, Nelson RJ (2006) The genetic architecture of disease resistance in maize: a synthesis of published studies. Phytopathology 96(2):120–129

Wu R, Ma CX, Casella G (2002) Joint linkage and linkage disequilibrium mapping of quantitative trait loci in natural populations. Genetics 160:779–792

Yu B, Jiang H, Pandey MK et al (2020) Identification of two novel peanut genotypes resistant to aflatoxin production and their SNP markers associated with resistance. Toxins 12(3):156

Zakieh M, Alemu A, Henriksson T et al (2023) Exploring GWAS and genomic prediction to improve *Septoria tritici* blotch resistance in wheat. Sci Rep 13(1):15651

Zeigler RS, Leong SA, Teng PS (1994) Rice blast disease. Int Rice Res Inst

Zhang F, Hu Z, Wu Z et al (2021) Reciprocal adaptation of rice and *Xanthomonas oryzae* pv. *oryzae*: cross-species 2D GWAS reveals the underlying genetics. Plant Cell 33(8):2538–2561

Zhao M, Liu S, Pei Y et al (2022) Identification of genetic loci associated with rough dwarf disease resistance in maize by integrating GWAS and linkage mapping. Plant Sci 315:111100

Zhong Z, Marcel TC, Hartmann FE et al (2017) A small secreted protein in *Zymoseptoria tritici* is responsible for avirulence on wheat cultivars carrying the *Stb6* resistance gene. New Phytol 214(2):619–631

Recent Diagnostics, Detection, and Monitoring Tools: Implications for Plant Pathogens and Their Management

Kuheli Deb, Saurav Paul, Rikita Bardhan, and Sudip Choudhury

Abstract Effective diagnostic, detection, and real-time monitoring tools have become the key components for the effective management of plant pathogens. Climate change, increased demands, and intensified global trade have made the timely detection and identification of diseases in crops to ensure global food productivity and agricultural sustainability. A number of advanced pathogen detection and monitoring techniques have been developed for various plant pathogens, and this field of research is continuously progressing. This chapter describes and reviews various techniques, both direct and indirect, for the identification of plant diseases used in agriculture. Laboratory-based techniques such as polymerase chain reaction (PCR), lateral flow immunoassays (LFIA), fluorescence *in-situ* hybridization (FISH), enzyme-linked immunosorbent assay (ELISA), gas chromatography-mass spectrometry (GC–MS) are discussed along with an overview of biosensors capable to exhibit selective bio-recognition. The application of the Internet of Things (IoT) and Information and Communication Technology (ICT) in agriculture for phytopathological management has been discussed.

Keywords Plant-pathogen · Detection · Management · Diagnostics

1 Introduction

Plant pathogens are organisms that can cause disease in plants. These can be viruses, bacteria, fungi, or nematodes. Plant diseases can cause significant damage to crops, leading to crop losses and economic losses for farmers. A staggering 20–40% of the world's crops are lost annually due to pests, as per the estimate of the Food and

K. Deb · S. Paul · R. Bardhan · S. Choudhury (✉)
Department of Chemistry, Assam University, Silchar, Assam, India
e-mail: sudipch1@gmail.com; sudip.choudhury@aus.ac.in

S. Choudhury
Centre for Soft Matter, Assam University, Silchar, Assam, India

J. Chen et al. (eds.), *Molecular and Biotechnological Tools for Plant Disease Management*, Interdisciplinary Biotechnological Advances,
https://doi.org/10.1007/978-981-97-7510-1_6

Agriculture Organization of the United Nations (FAO), which includes important food crops like rice, wheat, corn, potatoes, etc. (Buja et al. 2021; Oerke 2006). At the same time each year, plant pathogens cause worldwide crop losses of about 14% of global crop production which is estimated to be $220 billion worth (Agrios 2005; He and Creasey Krainer 2020; Savary et al. 2019; Prasanna et al. 2022; Paula and Marc 2013), where cereals, fruits, and vegetables are the most affected crops. In developing countries, where food insecurity is already a major problem, plant diseases can be devastating.

Recent outbreaks due to the bacterium *Candidatus liberobacter asiaticus* (Citrus greening disease) endemic caused orange fruit harvests reduction by ~20% from the year 2011 to 2014 in Thailand (Thoraneenitiyan et al. 2022; Ghosh et al. 2022). Plum pox virus, a devastating disease of stone fruits since 2010, *Hemileia vastatrix* fungus causing coffee rust, bacterium *Xanthomonas vasicola* pv. *maize* causing maize lethal necrosis, are some of the new and emerging diseases due to plant pathogens. At the same time, we must not forget the centuries-old pathogenic diseases as well, like potato brown rot disease caused by the fungus *Phytophthora infestans* which has the potential to cause disaster on production of this important crop.

To highlight the importance of efficient and sustainable plant health management, The United Nations designated 2020 as the International Year of Plant Health (IYPH). The initiative was extended to July 1, 2021, due to disruptions caused by the COVID-19 pandemic (FAO website https://www.fao.org/news/story/en/item/1402920/icode/). With the FAO anticipating a 10 billion global population by 2050, food production needs to rise by 60% to meet this demand (Ristaino et al. 2021; Gregory et al. 2009). Therefore, early detection and identification of plant pathogens become crucial for effective disease management, ensuring food security for a growing world. In recent years, there has been a growing focus on developing new diagnostics, detection, and monitoring tools for plant pathogens.

Plant pathologists, researchers, and experts have developed a diverse array of tools to monitor plant health. Such practices range from simple observation to the use of state-of-the-art molecular techniques. In the following sections, some of the important techniques used for the detection and management of plant pathogens are discussed.

2 Traditional Diagnostics Tools

Traditional diagnostics tools for plant pathogens include visual inspection, microscopic examination, culture-based methods, etc.

Table 1 Diagnostic information about different bands of electromagnetic radiation

Band	Wavelength	Diagnostic ability
Visible	400–700 nm	In the diseased canopy, reflectance increases in this spectral band due to chlorophyll loss and the presence of surface spores
Near Infra-Red (NIR)	700–1200 nm	Diseased canopies exhibit reflectance in shorter wavelengths with reduced intensity compared to healthy canopies which reflect at longer wavelengths in the NIR band
Short-wave Infra-Red (SWIR)	1200–2400 nm	Water absorbs in 1400, 1900, and 2400 nm. This band has relatively small direct diagnostic effects. However, reflectance in this band may give indirect evidence of stress
Thermal infrared (TIR)	8000–14,000 nm	This band gives information about the transpiration rate in plants. Reduced transpiration increases leaf temperature, which may indicate root diseases, xylem infection, foliar diseases, etc

2.1 Visual Inspection

This is the simplest and most common method for detecting plant diseases. The symptoms of plant diseases can be easily identified by trained professionals. However, visual inspection is not always reliable, as some plant diseases may not have visible symptoms.

A range of diverse responses can be exhibited by plants when under stress or affected by pathogens. Changes in the appearance of the leaf (shape, color, size), changes in density and morphology of the canopy, troubled photosynthesis, disturbances in transpiration, etc., are some of such responses (West et al. 2017), which provide significant visible changes. The differences in the appearance between vigorous and diseased plants in different wavelength windows also provide valuable information about plant health (West et al. 2017), as presented in Table 1.

2.2 Microscopic Examination

This method involves examining plant tissue under a microscope to identify the presence of pathogens. Microscopic examination is more reliable than visual inspection, but it requires specialized equipment and training.

2.3 Culture-Based Methods

This method involves growing the pathogen in a laboratory culture medium. The pathogen can then be identified by its morphology and biochemical properties. Culture-based methods are relatively slow and can be difficult to use for some pathogens.

3 Molecular Diagnostics Tools

Molecular diagnostics tools are based on the detection of nucleic acids (DNA or RNA) from pathogens. These tools are more sensitive and specific than traditional diagnostics tools, and they can be used to detect pathogens that are even difficult to culture. Some important molecular diagnostics tools for plant pathogens are discussed.

3.1 Lateral Flow Immunoassays (LFIA)

Lateral flow (LF) assays, also known as immunochromatographic strips, offer a simple and rapid method for detecting the presence or absence of a target molecule (analyte). The method is simple and rapid. In the assay, the analyte (in liquid form) is allowed to travel through a polymeric strip by capillary action. The strip is impregnated at various zones with specific probe molecules. As the analyte solution moves through various probe-containing zones, the probe molecules can interact with the analyte. Positive interactions are usually reported as specific color changes in the marked zone. Lateral flow strips are composed of four pads,

(i) *Sample pad*: The sample is dropped on this pad (made of cellulose)
(ii) *Conjugate pad*: Contains zone impregnated with the detector reagent (made of glass fiber). The conjugate pad is responsible for the uniform transfer of the detector reagent and test sample onto the detection pad.
(iii) *Detection pad*: The test line (TL) and control line (CL) are printed here (made of nitrocellulose)
(iv) *Adsorption pad*: It wicks off the excess liquid and prevents backflow of liquid (made of cellulose) (Koczula and Gallotta 2016)

Antibodies, aptamers, and DNA probes are generally applied recognition elements in lateral flow assay. To increase sensitivity or specific color response, the recognition elements are modified with AuNPs, magnetic nanoparticles, fluorescent probes, etc.

López-Soriano and coworkers (López-Soriano et al. 2017) developed an LF immunoassay prototype to detect *X. arboricola* pv. *Pruni*, the pathogen responsible for the bacterial spot disease of stone fruits and almonds. They attached polyclonal antibodies combined with carbon nanoparticles as the recognition element.

The resulting LFIA prototype exhibited excellent selectivity for *X. arboricola* pv. *Pruni with* LOD of 10^4 CFU ml^{-1}. In another work, Byzova and group designed LFIA for the detection of Grapevine leafroll-associated virus 3 (GLRaV-3) using antibodies immobilized on gold nanoparticles for target recognition (Byzova et al. 2018). Recently, Razo et al. (2021) developed LFIA to detect *Erwinia amylovora* which causes fire blight disease, using IgG antibodies as the recognition probe.

Selvarajan et al. (2020) reported the development of a lateral flow immunoassay (LFIA) test with high diagnostic sensitivity for the on-site detection of Banana bract mosaic virus (BBrMV), a serious pathogen against banana cultivation. The LFIA designed by them was in double-antibody sandwich format. The Conjugate pad contained Anti-BBrMV coat protein Immunoglobulins (IgG) conjugated to gold nanoparticles. The test line and control line were printed on nitrocellulose filter membrane with anti-BBrMV coat protein Immunoglobulins (IgG) and goat anti-rabbit IgG, respectively. The development of a pink band on the LFIA strip indicated a positive result.

In an interesting work, Razo and the group used a combination of magnetic nanoparticles (MNPs) and gold nanoparticles (GNPs) to enhance the sensitivity of LFIA. The conjugates of magnetic nanoparticles with antibodies realized an increase in the concentration of the analytes. At the same time, the presence of gold nanoparticles caused aggregation of antibodies conjugated MNPs thereby enhancing the visibility of the result. They successfully applied this double enhancement LFIA strategy on potato virus X. The magnetic nanoparticles (MNPs) conjugated with biotinylated antibodies specific to potato virus X (PVX) and gold nanoparticles (GNPs) conjugated with streptavidin were synthesized and used in this work, resulting in a very high sensitivity of 0.25 ng mL^{-1} for the target analyte (Razo et al. 2018).

Beyond single-pathogen detection, the LFIA technique can also be developed for the simultaneous detection of multiple pathogens. In a recent paper, duplex lateral flow immunochromatographic assay has been reported for the simultaneous detection of two sugarcane viruses under field conditions, viz. Sugarcane mosaic virus (SCMV) and Sugarcane streak mosaic virus (SCSMV) with LOD of 10^{-12} g mL^{-1}. They selected polyclonal antibodies (a mixture of antibodies) specific to the SCSMV and SCMV, and prepared separate conjugates with gold nanoparticles. The two separate AuNP-antibody conjugation pads were prepared and placed together in a single strip to achieve duplex detection (Thangavelu et al. 2022).

Apart from cases discussed so far, many more LFIA strips have been fabricated by researchers to detect various plant pathogens, like, *Villosiclava virens* causing Rice false smut (Fu et al. 2017), *Phytophthora ramorum* and *Phytophthora kernoviae* causing sudden oak death (Tomlinson et al. 2010), *Dickeya dianthicola* and *Dickeya solani* causing potato blackleg (Safenkova et al. 2017), *Ralstonia solanacearum* causative pathogen for wilt in potatoes (Panferov et al. 2016), etc.

3.2 Enzyme-Linked Immunosorbent Assay (ELISA)

This method employs antibodies as highly specific probes to detect the presence of specific proteins, characteristic of different pathogens. ELISA is a relatively simple and inexpensive method that can be used to detect a wide range of pathogens.

In an ELISA assay, the antigen specific to a pathogen is immobilized on a solid surface. This immobilization can be done either by directly grafting the antigen on the solid surface or by using a capture antibody which itself is already immobilized on the surface. The immobilized antigen is then complexed to a detection antibody. There are four main types of ELISA,

(i) Direct ELISA
(ii) Indirect ELISA
(iii) Sandwich ELISA
(iv) Competitive ELISA

ELISA has been successfully designed for detection of Plum pox virus responsible for stone fruit tree sharka disease (Guo et al. 2023), Maize chlorotic mottle virus causing chlorotic mottle in maize (Zhang et al. 2023a), Rice tungro virus (Henry Sum et al. 2017), *Ralstonia solanacearum* race 3 biovar 2 causing bacterial wilt in tomato, brown rot in potato, and Southern wilt of geranium (Behiry et al. 2018), etc.

3.3 Polymerase Chain Reaction (PCR)

Polymerase chain reaction (PCR) is a powerful laboratory technique used to rapidly generate millions of copies from a specific DNA segment. This amplification process involves repeatedly cycling through precise heating and cooling steps in a solution containing the target DNA, specific primers, and essential enzymes. The specific DNA segment to be amplified is defined by two short pieces of DNA called primers. The primers bind to the DNA at the beginning and end of the target segment, acting as starting points for a DNA polymerase enzyme. The enzyme DNA polymerase then uses the primers to synthesize new strands of DNA complementary to the target segment, effectively replicating the target segment. This process is repeated many times, resulting in the exponential amplification of the target DNA segment.

PCR is a very versatile technique with many applications. It can be used to:

- **Detect the presence of a specific DNA sequence in a sample**. PCR can be used to detect the DNA of a pathogen, such as a virus or bacteria.
- **Quantification of the amount of a specific DNA sequence in a sample**. This can be used to measure the amount of DNA in a sample.
- **Clone a DNA segment for further study**. By rapidly generating millions of copies of a specific DNA segment, PCR enables in-depth analysis and studies in more detail.

- **Create DNA fingerprints**. PCR can be used to create a unique DNA profile for an organism, which can be used for identification purposes.
- **Diagnose diseases**. PCR can be used to diagnose diseases by detecting the DNA of a pathogen in a clinical sample.
- **Develop new drugs and therapies**. PCR can be used to study the genes that are involved in diseases, which can help to develop new drugs and therapies.

The PCR process is typically divided into three stages:

1. **Denaturation**: The DNA is heated to a high temperature (95–98 °C) to separate the two strands of DNA.
2. **Annealing**: The temperature is specifically reduced to facilitate the primers to bind to their complementary sequence in the target DNA.
3. **Extension**: The temperature is raised to a moderate temperature (72–75 °C) to allow the DNA polymerase to synthesize new strands of DNA complementary to the primers.

There are many variants of PCR techniques. Some of the important types of PCR are,

(i) **qPCR**: qPCR stands for Quantitative PCR. It is also called real-time PCR or RT-PCR. In qPCR, the technique is similar to the standard PCR process except that a fluorescent dye is used so that real-time quantification of the amplified DNA in the reaction can be done. The thermal cyclers used in qPCR include an in-built fluorometer to detect the fluorescence.

(ii) **Rapid amplified polymorphic DNA analysis (RAPD)**: Conventional PCR protocol requires the use of specifically designed premiers of amplification of a pre-determined DNA segment. RAPD on the other hand employs short (~10 bases long), nonspecific primers, and lower annealing temperatures (~40 °C). This method is used to amplify random regions of genomic DNA. The main application of RAPDs is in the domain of genetic identity, particularly in studies involving closely related species.

(iii) **in situ PCR**: in situ PCR is a technique to exploit PCR reaction inside a cell on a microscopic slide. It is a powerful technique to study disease progression by enabling the identification of cellular markers.

(iv) **Multiplex PCR**: Compared with standard PCR systems where only 2 primers are used, multiplex PCR employs multiple and different primer pairs in the same reaction for simultaneous amplification of multiple targets.

3.4 Loop-Mediated Isothermal Amplification (LAMP) Assay

To eliminate the complexity in variable temperature PCR-based assay, a number of isothermal amplification techniques were developed in the past decade, like, recombinase polymerase amplification, rolling circle amplification, and helicase-dependent amplification (Zanoli and Spoto 2012). In the year 2000, Notomi (2000) introduced a groundbreaking isothermal amplification method, enabling the rapid amplification of limited DNA quantities into millions of copies within just one hour. They termed the technique **Loop-mediated isothermal amplification** (LAMP).

Loop-mediated isothermal amplification (LAMP) assay is a simple, rapid, specific, and cost-effective nucleic acid amplification method. In contrast to the PCR technique which involves extraction of DNA from the sample, and thermal cycling of the enzyme Taq polymerase (*Thermus aquaticus* polymerase) for the detection and quantification of plant pathogens thereby largely limiting its in-field application, the LAMP technique employs a thermostable enzyme, Bst DNA Polymerase (originally derived from *Bacillus stearothermophilus* DNA Polymerase I) (Aliotta et al. 1996). It is a single-pot reaction method. Instead of two primers used in conventional PCR assay, in LAMP assay a set of four (or six) different primers are used. The original LAMP method reported by Notomi et al. utilized four primers. The primer set consists of two outer primers, and two inner primers (forward and backward). In the year 2002, the Notomi group introduced two additional primers, termed loop primers (loop forward and loop backward) to accelerate the LAMP reaction rate (Nagamine et al. 2002). The features of the primers are as follows,

1. **Forward Inner Primer (FIP)** consists of two distinct regions: a 3′ end F2 region complementary to the template sequence's F2c region and a 5′ end F1c region identical to the template's F1c region.
2. **Backward Inner Primer (BIP)** comprises of B2 region (at 3′ end) complementary to the B2c region of the template sequence, and a B1c region (at 5′ end) identical to the B1c region of the template sequence.
3. **Forward Outer Primer (F3 primer)** comprises a F3 region complementary to the F3c region of the template sequence.
4. **Backward Outer Primer (B3 Primer)** comprises a B3 region complementary to the B3c region of the template sequence.
5. **Loop primer (loop forward)** has a position in the region between F2 and F1 in the direction of F1 to F2.
6. **Loop primer (loop backward)** has a position in the region between B1 and B2 in the direction of B1 to B2.

As mentioned above, LAMP utilizes two inner primers: the forward inner primer (FIP) and the backward inner primer (BIP). Each FIP and BIP harbors two distinct sequences, one complementary to the sense strand and the other to the antisense strand of the target DNA. All four primers take part in the initial phase of the reaction. During the later step involving the cycling reaction only the inner primers take part. The primers bind to six (or eight) distinct regions on the target gene. The Bst DNA polymerase enzyme is used in this method and the reaction follows the strand displacement DNA synthesis mechanism, occurring under isothermal conditions. The illustration of the principle of LAMP assay is given in Fig. 1. Probably the best ingenuity in LAMP is that the products and byproducts can also be visualized by the eye. Observation of color changes by added indicators like SYBR Green I, calcein, hydroxynaphthol blue, or formation of a white precipitate of Magnesium pyrophosphate produced during the reaction can be used to signal a positive reaction.

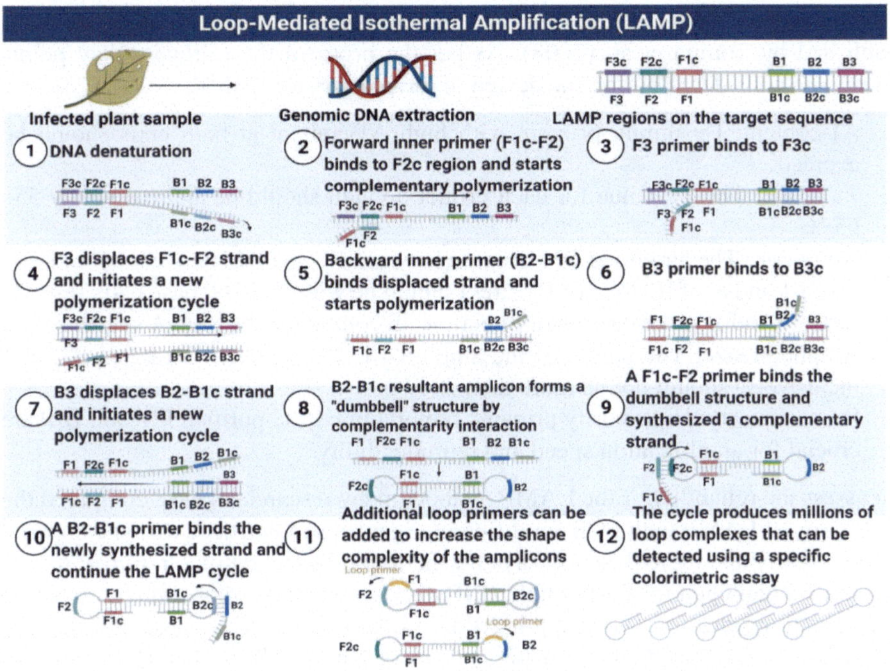

Fig. 1 Principle of LAMP assay (Patel et al. 2022, reproduced with permission)

Because of the continuous and isothermal nature of amplification, LAMP provides a very good tolerance against the interference of common PCR inhibitors. It does not require thermal cycler equipment and even can be carried out using a simple water bath (Feng et al. 2023). An important difference between traditional PCR and LAMP assay is summarised in Table 2.

Table 2 Important difference between traditional PCR and LAMP assay

Parameter	Polymerase chain reaction (PCR)	Loop-mediated isothermal amplification (LAMP)
Temperature	Variable	65 °C
Thermal cycler	Required	Not required
Time requirement	20–40 min	1.5–2 h
No. of primers	2 (recognizes 2 regions on the target DNA)	4–6 (recognizes 6 to 8 distinct regions on the target DNA)
Internal amplification control	Required	Not required
Protocol diversity	Different protocols for different organisms	Single protocol
Process controls	Multiple transfers	Two transfers
Amplicon	Discrete	Concatemeric

An elaborate experimental protocol for the LAMP assay technique has been published by Tomita et al. (2008). As per the protocol the following key points need to be considered during the design of the primers for LAMP,

1. AT-content: Designing primers with high AT-content at both ends should be avoided.
2. T_m value: The T_m value for each primer domain should be approximately 55–65 °C.
3. Amplicon: The amplified DNA region between F2 and B2 sites (e.g., 5′ of F2 to 5′ of F1 and 5′ of B2 to 5′ of B1 sites) should be within 40–60 base pairs (F is for forward and B is for backward directions in complementary DNA sequences)
4. Amplicon size: The length of the amplified DNA region (from F2 to B2 sites inclusively) should not be more than 200 bp.
5. Primers purity: High-purity primers, preferably HPLC-purified FIP and BIP are crucial for amplification speed and reproducibility.

To assess the reliability of the LAMP method, many researchers have compared the sensitivity of LAMP with PCR for different organisms (Khan et al. 2018) compared LAMP Assay and PCR-Based Assays for the Rapid Detection of *Alternaria solani*, a causative pathogen for Early blight (EB) disease which is a global threat to potato and tomato production (Song et al. 2011). In their study, Khan observed that the LAMP assay exhibited higher sensitivity than conventional PCR but had lower sensitivity compared to both RT-qPCR and nested PCR. In other reports, the LAMP assay showed sensitivity similar to nested PCR for *M. marinum* strains (Feng et al. 2023), and higher sensitivity compared to conventional PCR, nPCR, and qPCR for *Entamoeba histolytica* (Foo et al. 2020).

Tian and co-workers detected *Didymella bryoniae* in cucurbit seeds (having 0.05% or greater infestation) by LAMP assay with a detection threshold of 10 pg of genomic DNA. The LAMP primers were designed to target the RNA polymerase II RPB140 gene (*RPB2*) from *D. bryoniae* and Calcein was used as the visual indicator (Tian et al. 2017a).

Fusarium wilt in chickpeas is caused by the pathogen *Fusarium oxysporum* f. sp. *ciceris*. The detection of the organism cannot easily be done at the field level because of the similarity in symptoms caused by various other soil-borne pathogens. To address this issue, Ghosh et al. (2015) developed a LAMP assay targeting the elongation factor 1 alpha gene sequence for visual detection of *Fusarium oxysporum*. Hydroxynaphthol blue was added as a visual indicator.

LAMP assay has been successfully employed to differentiate between genetically very closely related pathogens. *Fusarium commune* and *Fusarium oxysporum* are two pathogens rhizome rot in lotus (*Nelumbo nucifera*) thereby severely reducing the production of edible rhizome from this perennial hydrophyte. Though the symptoms of attack by both species are similar, they differ significantly in disease severity. Comparing the mitochondrial genome of the two species, Deng et al. (2022) identified three specific DNA loci targets that are different in these two species, and developed primers for the individual detection of the two pathogenic species with LAMP detection limits of 10 pg/μl for *F. commune* and 1 pg/μl for *F. oxysporum*.

Recently, a development of LAMP method to detect and differentiate two Haplotypes of the Potato zebra chip pathogen *Candidatus liberibacte*r solanacearum Haplotypes A and B (Jiang et al. 2023a, b, c). On the genomic DNA for Haplotype A, they recorded a detection threshold of 2–20 ng/μL on haplotype A strains, and that for Haplotype B was 0.02–2 ng/μL.

For disease risk forecasting of late blight disease due to *Phytophthora infestans*, and early blight disease due to *Alternaria solani*, a real-time loop-mediated isothermal amplification (RT-LAMP) assay was developed by Lees and co-workers. Their method was based on the in-field detection of airborne inoculum of these pathogens. The airborne spores were collected with in-field air samplers and using the collected spores the pathogens were detected with detection sensitivity equivalent to ~1 sporangia per reaction (5 pg/μL of DNA for *P. infestans* and 4.4 pg/μL of DNA for *A. solani*) (Lees et al. 2019).

The sensitivity of the LAMP assay is much higher than conventional PCR and similar to that of RT-qPCR and nested PCR for most of the organisms. However, LAMP outperforms all PCR assays in terms of amplification time, ease of operation, and outdoor portability. Because of these practical benefits, the LAMP assay rapidly caught the attention of researchers. Different research groups designed and reported primers for a wide variety of plant pathogens (Table 3). SYBR Green I appears to be one of the most useful visual indicators.

3.5 Fluorescence In-Situ Hybridization (FISH)

Fluorescence in situ hybridization (FISH) is a powerful and sensitive technique for detecting target organisms directly within their environmental context. This method utilizes sequence-specific, fluorescently labeled oligonucleotide probes, which bind to their complementary targets, allowing for visualization and identification (Salgado-Salazar et al. 2018). This technique is used for the detection and location of a specific sequence of DNA. In the FISH technique, probes are prepared by attaching fluorescent dye to a small piece of DNA. Alternatively, the probes may be indirectly labeled with a hapten. The fluorescently tagged probe and the target DNA are mixed and set to denature. The denaturation converts both the fluorescent-tagged probe and the target DNA to single-stranded DNA. After denaturation, in the hybridization step where the single-stranded DNA combines with double-strand ones, the fluorescently tagged probe specifically hybridizes with complementary DNA sequences in the target DNA, and the signals can then be evaluated by fluorescence microscopy. If instead of a fluorescently tagged probe, a non-fluorescent haptan-tagged probe is used, then the visualization is done with an enzymatic or immunological detection system (Shakoori 2017).

Waite et al (2016) developed the FISH method for rapid detection of *Xanthomonas citri* subspecies citri. The pathogen can spread with wind and rain, and attacks different citrus plants, like orange, grapefruit, lemon, etc. Cardinale et al. (2018) studied the colonization behavior of the *Xylella fastidiosa* strain CoDiRO, the

Table 3 Detection of plant pathogens by LAMP technique

Organism	Host	Disease	Target	Visual Indicator	Genomic DNA Detection threshold	References
Didymella bryoniae	Cucurbit seeds	Gummy stem blight	DNA-dependent RNA polymerase II RPB140 gene (*RPB2*)	Calcein	10 pg	Tian et al. (2017a)
Fusarium oxysporum f. sp. *ciceris*	Chickpea	Fusarium wilt	Foc elongation factor 1 alpha (EF-1alpha) gene (FJ538243)	Hydroxynaphthol blue	10 fg	Ghosh et al. (2015)
Xanthomonas oryzae pv. oryzae	Rice (*Oryza sativa*)	Bacterial blight disease	*X. oryzae* pv. oryzae PXO_00080	Quant-IT Pico green reagent	1 pg–10 fg	Smit et al. (2012)
Xanthomonas oryzae pv. oryzicola	Rice (*Oryza sativa*)	Bacterial leaf streak disease	*X. oryzae* pv. oryzicola Xoryp_010100019045	Quant-IT Pico green reagent	1 pg–10 fg	Smit et al. (2012)
Bean pod mottle virus	Soybean seeds	Bean pod mottle	Coat protein (CP) gene	SYBR GREEN I	0.1 fg	Wei et al. (2012)
Colletotrichum truncatum	Soybean	Soybean anthracnose	*Rpb1*	SYBR Green I	100 pg	Tian et al. (2017b)
Botryosphaeria dothidea	Kiwifruit (*Actinidia chinensis*)	Rot diseases	β-tubulin gene and EF1α gene	SYBR Green I	10^{-5} ng	Wang et al. (2021)
Alternaria alternata (Fr.) Keissl	Apple	Alternaria Blotch Disease	Aapg-1 gene	SYBR Green I	1 fg	Liu et al. (2022)
Ralstonia solanacearum	Tomato	Bacterial wilt	*hrpB* gene	FQ-reporter	2×10^0 copies	Fan et al. (2023)
Ralstonia solanacearum phylotype I	Mulberry	Mulberry wilt disease	MG67 gene	SYBR Green I	160 fg	Huang et al. (2017)

(continued)

Table 3 (continued)

Organism	Host	Disease	Target	Visual Indicator	Genomic DNA Detection threshold	References
Ralstonia solanacearum	Potato	Bacterial wilt	*IpxC* gene	Gel Red (Fluorescence under UV)	2.5 pg	Okiro et al. (2019)
Phytophthora spinosum	Lettuce	Phytophthora wilt	Ribosomal DNA internal transcribed spacer region (rDNA–ITS)		10 fg	Feng et al. (2019)
Phytophthora uncinulatum	Lettuce	Phytophthora wilt	rDNA–ITS		100 fg	Feng et al. (2019)
Phytophthora pseudolactucae	Lettuce	Phytophthora wilt	*cox*1		100 fg	Feng et al. (2019)
Pythium irregulare	Diverse host	Seed, stem and root rot	rDNA–ITS	Hydroxynaphthol blue	100 fg	Feng et al. (2015)
Dickeya dianthicola	Potato	Blackleg of potato	*Alcohol dehydrogenase gene*	SYBR Green I	1 pg	Ocenar et al. (2019)
Peach latent mosaic viroid (PLMVd)	Peach, apricot, plum, etc	Peach latent mosaic (PLM) disease	tRNA	Loopamp Fluorescent Detection Reagent		Boubourakas et al. (2009)
Tomato leaf curl New Delhi virus	Crops belonging to *Cucurbitaceae* and *Solanaceae* families	Leaf curl disease	AV1 gene	6-carboxyfluorescein	50×10^{-5} ng	Caruso et al. (2023)
Leptosphaeria biglobosa 'brassicae'	Oilseed rape	Blackleg	Highly conserved 615-bp-long DNA sequence		132 fg	Du et al. (2021)

(continued)

Table 3 (continued)

Organism	Host	Disease	Target	Visual Indicator	Genomic DNA Detection threshold	References
Exobasidium vexans	Tea	Blister blight disease	rDNA-ITS	SYBR Green I		Chaliha et al. (2023)
Curtobacterium flaccumfaciens pv. *flaccumfaciens*	Leguminous crops	Bacterial wilt	Highly conserved 306 pb DNA fragment	Ethidium bromide	4 fg	Tegli et al. (2020)
Pyricularia oryzae	Wheat	Wheat blast	Triticum-specific polymorphism		5 pg	Thierry et al. (2020)
Valsa mali	Apple	Apple Valsa canker	Species-specific region of the elongation factor-1α sequence	SYBR Green I	1 ng	Xu et al. (2020)
Spongospora subterranea	Potato	Powdery scab of potato			2 pg	Jiang al. (2023a, b, c)
Wheat dwarf virus	Wheat	Wheat dwarf	Coat protein	SYBR Green I	500 fg	Trzmiel and Hasiów-Jaroszewska (2020)
Fusarium graminearum	Wheat	Fusarium head blight	218-bp region from a partial sequence of *F. graminearum* chromosome 1	Hydroxynaphthol Blue	100 fg	Gupta et al. (2020)
Puccinia triticina	Wheat	Leaf rust	$PtRA_{68}$ specific marker	Hydroxynaphthol Blue	100 fg	Manjunatha et al. (2018)

(continued)

Table 3 (continued)

Organism	Host	Disease	Target	Visual Indicator	Genomic DNA Detection threshold	References
Dickeya fangzhongdai	Onion	Soft rot and bleeding canker diseases	Major facilitator superfamily (MFS) transporter gene	SYBR Green I	100 fg	DeLude et al. (2022)
Colletotrichum gloeosporioides	Guava	Anthracnose of guava	β-tubulin 2 (*TUB2*) gene		10 fg	Lan et al. (2020)

pathogen responsible for causing olive quick decline syndrome using the FISH protocol.

4 Multiplexed Detection

Multiplexed detection of pathogens involves the identification of multiple pathogenic threats. This approach provides an opportunity for simultaneous detection of multiple biomarkers for different pathogens in a single assay (Huang et al. 2020). Multiplexed detection usually employs multiple transducers having different biorecognition probes (Cesewski and Johnson 2020).

Multiplexed mode can be applied to almost all types of classical detection techniques, e.g., PCR, LAMP, biosensors, etc. For instance, Sahu et al. developed a multiplex PCR technique for the detection of Salmonella contamination in seafood processing. Instead of using species-specific primers, they selected one pair of primers targeting the *invA* gene. The selected gene segment was common for a wide range of Salmonella serotypes, excluding a wide range of non-target species (Sahu et al. 2019).

5 Emerging Diagnostics Tools

There are a number of emerging diagnostics tools for plant pathogens that are having comparatively lesser real-life use yet but are under active development by researchers. Some of the notable tools and techniques are elaborated.

5.1 "Chemical-Nose"—Sensor-Based Detection of Plant Pathogens

Chemical sensors for the detection of specific analytes have been in development and use for quite a long time. Though most of the advancement in this sector targets diagnostics related to human diseases (Geng et al. 2019), recently the development of sensors for detection of plant pathogens has got a remarkable surge. A "chemical nose" sensor, in essence, is an array-based system that employs synthetic molecules to mimic the mammalian olfactory system's ability to detect and differentiate various odors. (Guerrini et al. 2017).

As defined by the IUPAC (Hulanicki et al. 1991) a chemical sensor is "*a device that converts chemical data, ranging from the concentration of a single sample component to complete composition analysis, into an analytically usable signal*". Chemical sensors have been designed for diverse class of analytes—from small molecules/ions (Zhang et al. 2023c; Cao et al. 2023; Jin et al. 2023; Murali et al. 2020; Ashraf et al. 2021; Butera et al. 2021), to biological macromolecules (Hu et al. 2022; Li et al. 2019; Kara et al. 2023; Ho et al. 2014), to whole cells (May et al. 2004) and organisms, like bacteria (Yao et al. 2021; Ouyang et al. 2021; Bhardwaj et al. 2019; Du et al. 2022; Fu et al. 2022; Zhang et al. 2023b; Liu and Su 2017; Kaur et al. 2015) and fungi (Thet et al. 2020).

The main advantages of this class of array sensors are simple protocol, fast response, and ease of handling. This type of sensor relies on specific interaction between the analyte and the specially designed organic receptors/probes. The organic receptors/probes are designed to show different affinity for the target analyte, and the outcome of such interaction is reported by characteristic color change (optical sensor) and/or by exhibition of fluorescence. Such sensors can be qualitative (presence or absence of the target analyte) or quantitative (amount of the target analyte) in nature.

Generally, array sensors are designed to "react" with the target analyte leading to the formation of covalent bonds between the analyte and the receptor, thereby forming a new chemical compound with a specific color or fluorescence property. However, non-covalent interactions, such as hydrogen bonds, $\pi-\pi$ staking (Harding et al. 2019; Walters et al. 2020), halogen bonding (Patrick et al. 2021), and electrostatic interaction (Gao et al. 2013), have been exploited to fine-tune the probe-analyte interactions leading to higher selectivity.

5.2 *Electrochemical Sensors for Plant Pathogens*

Electrochemical (EC) sensors are a subclass of chemical sensors (Sharma and Palatty 2023), in which the electrode is the transducer element (Simões and Xavier 2017). Electrochemical sensors are rapidly gaining prominence as powerful analytical tools due to their superior sensitivity and selectivity, quick response period, simplicity in operation, low production cost, miniaturization and portability, and requirement of an extremely small volume of the analyte (Simões and Xavier 2017; Hussain and Keçili 2020).

Similar to a chemical sensor, electrochemical sensors also rely on chemical interaction between the probe (here, the electrode) and the analyte, but unlike optical reporting (color change, fluorescence ON/OFF) of the outcome of the analysis as in the case of chemical sensors, the electrochemical sensors convert the probe-analyte reaction information into an electrical signal for the intended qualitative or quantitative analysis. Historically, the journey of electrochemical sensors application began with the monitoring of industrial oxygen in the 1950s (Simões and Xavier 2017). Since then, electrochemical sensors have been found in widespread commercial applications. Commercial glucose sensors can be cited as a classic real-life application of electrochemical sensors. Modified electrochemical sensors employing nanoparticle-based electrodes offer high surface area, high SNR, and better electron transfer rates (Gupte and Pradeep 2022; Kaya et al. 2022). A typical electrochemical sensor configuration consists of three electrodes: a working electrode, a reference electrode, and a counter electrode.

5.2.1 Important Types of Electrochemical Sensors

Important types of electrochemical sensors are:

Potentiometric sensors: Potentiometric sensors measure the potential difference between the working electrode compared to the reference electrode. The working principle of potentiometric sensors is based on the Nernst equation as shown in Eq. 1. This equation predicts a linear dependence of the sensor response, *E*, on the logarithm of a function of the activity of the ion in solution:

$$E = E^{\circ} + \frac{RT}{nF}\ \text{In}[a] \tag{1}$$

where *E* is the potential difference (in Volt) between the working and the reference electrodes, which is the sensor response for a particular analyte at a certain concentration. *R* is the gas constant (8.314 J/K), *F* is the Faraday constant (96,500 C/mol), *n* is the number of electrons (transferred in the redox reaction), and *a* is the activity of the principal ion (reflecting the effective concentration of the ion in solution). E° is the standard potential.

Amperometric sensors: This class of electrochemical sensors measures the current response at the working electrode surface when a constant voltage is applied between the working and the reference electrode. The resulting current at the sensor electrode is proportional to the analyte's concentration in accordance with the Cottrell equation (Eq. 2).

$$i = \frac{nFAc_j^0\sqrt{D_j}}{\sqrt{\pi t}} \tag{2}$$

where *i* is the Current (in ampere), *n* is the number of electrons, A is the area of the electrode (cm^2), $c^0{}_j$ is the initial concentration of the analyte, D_j is the diffusion coefficient, and t is the time in seconds (Baranwal et al., 2022).

Conductometric/Impedimetric sensors: Conductometric sensors measure the conductance (reciprocal of resistance) of the electrochemical cell comprising the analyte. The response of this class of sensors is governed by the generalized form of Ohm's law. Conductometric sensors work with dc. If ac signal is used in the measurement, the conductance becomes frequency-dependent conductance and the resistance *R* becomes impedance, hence the name Impedimetric sensors (Jaffrezic-Renault and Dzyadevych 2008).

Voltammetric sensor: In voltammetry, a controlled voltage sweep is applied on the working electrode relative to the reference electrode. The resultant current through the cell is then measured. Voltammetric sensors are further classified as linear sweeping voltammetry, cyclic voltammetry, pulsed potential voltammetry (or square wave voltammetry) (Barhoum et al. 2023).

5.2.2 General Design of Electrochemical Sensors

As mentioned earlier an electrochemical sensor typically includes a reference, a counter, and a working electrode. Among various reference electrodes, the Silver/silver-chloride electrode is the most frequently utilized reference electrode because of its simple construction, stability, and well-defined potential. Apart from that, Calomel Electrode, Copper–copper sulfate, iridium/iridium oxide (pseudo reference electrode), etc., have also been used. The counter electrode generally is a piece of platinum wire (or meshes) or graphite rod with a larger surface area and high chemical stability. The application of the sensor almost entirely depends on the design of the working electrode. Basic (non-specific) working electrodes can be constructed of inert materials like platinum, gold, silver, glassy carbon, etc. For practical purposes, the working electrode is modified with analyte-specific probe material so that the electrical behavior of the working electrode (and hence the whole electrochemical cell) changes significantly.

Schematic presentation of electrochemical signal transduction used in different types of electrochemical sensors and commercially available screen-printed electrodes (SPEs) are presented in Fig. 2.

5.2.3 Nanoparticle-Based Electrochemical Sensors

Nanomaterials are defined as "*materials where at least one of their dimensions is in the nanoscale, i.e. smaller than* 100 nm" (Joudeh and Linke 2022). Nanomaterials possess a higher surface area-to-volume ratio compared to their bulk counterpart. Due to their very small size, nanoparticles have a higher fraction of atoms at the surface. The surface atoms in nanomaterials have fewer direct neighbors, as a result of which binding energy per atom is also lower. Consequently, the physical and chemical properties of a material become significantly different in nanodomain than the same material in bulk.

To design nanoparticle-based electrochemical sensors, the surface of specific nanomaterials is functionalized with the receptor moieties (antibody, enzyme, etc.). The receptor-functionalized nanoparticles are then immobilized on the working electrode surface. As nanoparticles possess a higher surface area to volume ratio and different optical, electrical, and chemical properties compared to bulk material, the selectivity and sensitivity of the sensor can be remarkably enhanced. Further, this approach provides an additional tool for the design of electrochemical sensors where the receptor materials directly do not bind properly with the electrode surface. Graphene, carbon quantum dots, noble metal nanoparticles (Pt, Au, Ag), and metal chalcogenide nanoparticles (like, ZnO, TiO_2, ZnS, Ag_2O, etc.)

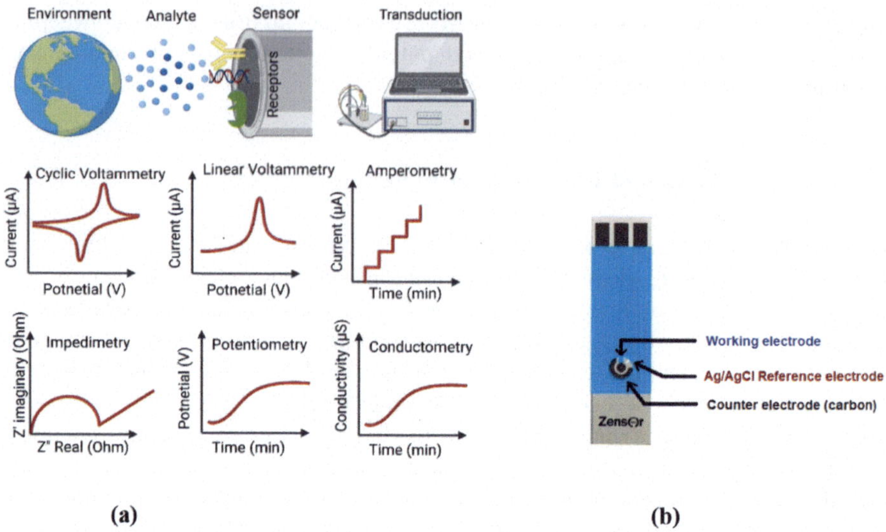

Fig. 2 **a** Schematic diagram of electrochemical signal transduction used in different types of electrochemical sensors: voltammetry, amperometry, impedimetry, potentiometry, and conductometry (adopted from Barhoum et al. 2023; Reproduced with permission from Elsevier), **b** a commercial Screen printed electrode (SPE)

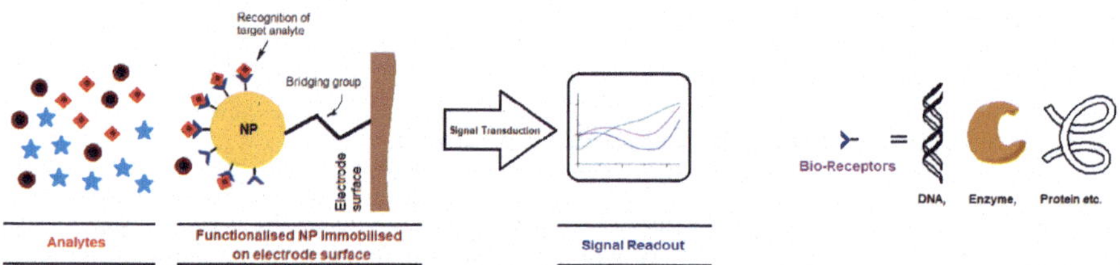

Fig. 3 Schematic presentation of nanoparticle-based electrochemical sensor

have been studied extensively for sensing applications. The schematic presentation of nanoparticle-based electrochemical sensors is presented in Fig. 3.

5.2.4 Detection of Plant Pathogens Using Electrochemical Sensors

A number of electrochemical biosensors have been developed for the detection of a varied array of plant pathogens. An electrochemical biosensor primarily has two core components: a molecular recognition layer and an electrochemical transducer (Dyussembayev et al. 2021). Different groups of researchers have documented the use of different types of receptors for the recognition layer (Table 4).

Table 4 Some electrochemical platforms developed for detection of plant pathogens

Type of biosensor	Description	Recognition probe	Pathogen	LOD	References
Impedimetric	Au/4-ATP/GA/anti-Psl/BSA	Antibody	*Pseudomonas syringae*	337 CFU/mL	Cebula et al. (2019)
Impedimetric	Screen-printed carbon electrode	Ssdna	Citrus tristeza virus	100 nM	Khater et al. (2019a)
Impedimetric	Isothermal recombinase polymerase amplification (RPA) based electrochemical sensor	Thiolated forward primer for P20 gene (387-bp)	Citrus tristeza virus	1000 fg μL^{-1}	Khater et al. (2019b)
Amperometric	Bi-enzyme modified electrode	Alcohol oxidase and horseradish peroxidase enzymes	Non specific	0.98 μM	Fang et al. (2016)
Cyclic Voltammetry	Bi-enzyme modified electrode	Alcohol oxidase and horseradish peroxidase enzymes	Non specific	22.95 μM	Fang et al. (2016)
Cyclic Voltammetry	Detected p-nitrophenol (enzymatically formed)	Biotinylated aptamer	Soybean Rust Fungi	100–200 collected spores per cm^2 of electrode	Krivitsky et al. (2021)
Cyclic Voltammetry	Streptavidin-coated screen-printed carbon electrode, using 1-ethyl-3-(3-dimethylaminopropyl)carbodiimide-*N*-hydroxysuccinimide linker	Streptavidin	miR393a	100 fM	Nehra et al. (2022)

Cassedy and the group highlighted antibodies as the most desirable for the recognition of target pathogens. Antibodies can recognize target antigens at very low concentrations and thus offer high sensitivity and specificity to the biosensor (Cassedy et al. 2020). Cebula and coworkers, (Cebula et al. 2019) reported an impedimetric immunosensor for the detection of *Pseudomonas syringae* pv. *lachrymans* (*Psl*). They modified gold disc electrodes of 3-mm diameter (the working electrode) with 4-aminothiophenol (4-ATP), glutaraldehyde (GA), and anti-Psl antibodies. A silver/silver chloride electrode served as the reference electrode, while a platinum sheet functioned as the counter electrode. The pathogenic antigen was detected by this Au/4-ATP/GA/anti-Psl/BSA electrode using Electrochemical impedance spectroscopy (EIS) with a detection limit (LOD) of 337 CFU/mL. Briefly, the working electrode fabrication process involved the following steps,

i. Incubation of clean gold electrode in ethanolic solution of 4-aminothiophenol. The thiol group of 4-aminothiophenol attaches to the gold surface.
ii. The 4-aminothiophenol attached gold electrode was treated with glutaraldehyde. Glutaraldehyde attaches with the free amino group of the already anchored 4-aminothiophenol. The linker for the attachment of the anti-Psl antibody was ready.
iii. The linker (4-ATP- GA) moiety-connected gold electrode was incubated with anti-Psl antibody at 4 °C.
iv. The antibody-modified electrode was treated with Bovin Serum albumin protein to cover any free gold surface

Lau et al. (2017) reported a nanoparticle-based electrochemical biosensor designed for rapid and sensitive detection of plant pathogen DNA on screen-printed electrodes. This sensor utilizes disposable screen-printed carbon electrodes, offering a cost-effective and portable approach.

Though most of the electrochemical biosensors have been developed targeting the detection of pathogenic bacteria, some work has been done for the detection of plant viruses. Khater et al. (2019a) reported an electrochemical sensor for citrus tristeza virus (CTV) using gold nanoparticle-modified electrodes. CTV is a threat to citrus production. The name 'tristeza' means "sadness" in Spanish and Portuguese. Citrus is the largest fruit crop in the world, and over the last 70 years, CVT has caused damage to millions of citrus trees (Bar-Joseph and Dawson 2008). In the work of Khater et al. (2019a), a screen-printed carbon electrode (SPCE) was electrodeposited with gold nanoparticles (AuNP). The reason for using gold nanoparticles over the carbon electrode was to immobilize thiolated ssDNA probes as the thiol group attaches efficiently to the gold surface. The deposition of AuNPs on SPCEs was done by dipping the SPCE in a $HAuCl_4$ solution maintaining pH 1 followed by reduction of gold (III) to gold (0) by applying a constant potential of −0.4 V. The thiolated ssDNA probes were then anchored over the gold surface by incubating the AuNP-modified SPCE in thiolated ssDNA probes solution for 2 h, followed by rinsing with Phosphate buffered saline. Later on, the Khater group reported an isothermal recombinase polymerase amplification (RPA) based electrochemical sensor for the citrus tristeza virus. The optimized RPA condition was transferred to the thiolated forward primer grafted AuNP-modified electrode surface to obtain the desired sensor with a nucleic acid detection limit of 1000 fg μL^{-1} (Khater et al. 2019b).

Methyl salicylate (MeSA) is a volatile long-distance signal transduction chemical released by plants in response to abiotic stress and pathogen attacks (Liu et al. 2018). Considering emission of MeSA by plants as an indicator of pathogen attacks, Fang et al. fabricated an amperometric sensor for Methyl salicylate (MeSA). The sensor is based on the bi-enzyme electrode, where two enzymes, viz, alcohol oxidase and horseradish peroxidase were grafted onto a carbon nanotube deposited electrode (Fang et al. 2016). The sensing mechanism follows the following reaction sequence:

(i) The MeSA hydrolyses to methanol,
(ii) Methanol is oxidized to formaldehyde by alcohol oxidase enzyme. At the same time, oxygen is reduced to hydrogen peroxide.
(iii) The enzyme horseradish peroxidase facilitates the conversion of hydrogen peroxide into water. This redox reaction (involving the transfer of electrons), generates an electrical current that is measurable as an amperometric signal.

5.3 *Electronic Nose*

The electronic nose (e-nose) primarily consists of a special electronic gas sensor array. As plants under stress or pathogenic attack emit certain Volatile organic compounds (VOCs), an E-nose for plant pathogen detection is designed to detect those VOCs and respond at the earliest. Such an e-nose contains one sensor array or multiple types of sensors. At the core of the detectors in e-noses, usually, there is a metal–oxide–semiconductor (MOS), or complementary metal-oxide semiconductor (CMOS) gas sensor array. These sensors detect the concentration of different VOCs by recording the change in electrical resistance of the metal oxide that occurred due to the adsorption of gases. E-nose sensors show lower specificity compared to molecular methods of pathogen detection (Cellini et al. 2017). Some of the commercially available e-nose detectors are, PEN3 (Airsense Analytics, Germany), EOS507C (Sacmi Scrl, Italy), Cyranose 320 (Smiths Detection Inc., USA), etc.

5.4 *Gas Chromatography-Mass Spectrometry (GC–MS)*

Gas chromatography–mass spectrometry (GC–MS) is a hyphenated technique for the analysis and identification of chemical species. It comprises two equipment—a gas chromatograph and a mass spectrometer. In GC–MS different components present in the test sample are separated by the GC unit. The GC is equipped with a capillary column (the stationary phase) through which the compounds of the sample travel in the vapor state. Helium is the commonly used eluent gas (the mobile phase) for GC, though inert gases like nitrogen or argon can also be used. As the sample mixture travels through the capillary column, its components separate based on their differing

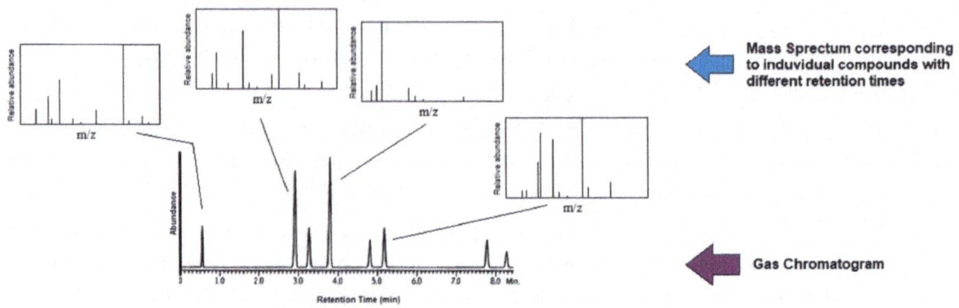

Fig. 4 Schematic presentation of GC–MS data readout

interactions (affinity) with the stationary phase. This differential interaction leads to each component eluting from the column at a specific time, known as the retention time of that component of the mixture.

Each of the individual chemical compounds separated by GC is fed to the Mass spectrometer for the study of their fragmentation pattern and molecular mass. Each of the mass spectra obtained can be analyzed for an unknown compound or can be compared with the available MS library to detect a known compound that already has an entry in the MS database. The most important benefit of GC–MS is its sensitivity, which requires tiny amounts of a substance for analysis. A representation of data outcome from a GC–MS analysis is shown in Fig. 4.

GC–MS based detection of pathogenic attach on plants mainly relies on the detection and identification of volatile organic compounds (VOCs) emitted by plants under stress. Karami et al. (2018) extracted VOCs using solid phase microextraction (SPME) fiber by obtaining bacterial headspace samples. The microextraction fiber was impregnated with divinyl benzene/carboxen/poly dimethyl siloxane for absorption of VOCs. The collected sample was then subjected to GC–MS. After ignoring the signature of the common VOCs, they successfully detected some bacterial species-specific VOCs. In another work, Wang et al. (2019) used GC–MS to detect biogenic volatile compounds produced by *Verticillium dahlia.* This soilborne fungal pathogen causes verticillium wilt in many plants as a result of which the leaves of the plant curl and discolor. They collected the fungal biogenic volatile compounds using solid-phase microextraction and analyzed the compounds through GC–MS.

Apart from the identification of pathogens, GC–MS has been used for metabolomics analysis of the plant-pathogen *Xanthomonas citri* subsp. citri to investigate the role of *N*-acetylcysteine as an antioxidant and anti-adhesive agent (Picchi et al. 2021).

Loulier et al. (2020) used a combination of e-nose and GC–MS to detect eight fungi, viz. *Armillaria gallica, Armillaria ostoyae, Fusarium avenaceum, Fusarium culmorum, Fusarium oxysporum, Fusarium poae, Rhizoctonia solani, Trichoderma asperellum*, and four oomycetes viz. *Phytophthora cactorum, P. cinnamomi, P. plurivora, P. ramorum.* The e-nose device to survey VOC emission, collected those by solid phase microextraction, and selectively identified the emitted molecules by BC-MS.

6 Plant Wearable Devices

Emerging wearable technology offers a promising avenue for plant health monitoring. These miniaturized sensors, worn by plants, can collect real-time physiological data in a non-invasive or minimally invasive manner. While still under development, plant wearables hold immense potential as powerful tools for early detection of plant pathogens, enabling timely intervention and improved crop health management.

Wearable sensors contain integrated analytical devices operating mostly in autonomous mode. Wearable monitoring devices developed for humans have been there for quite a long time, starting with the Holter monitor for cardiac diagnostics in 1962 (Ates et al. 2022). In the present era of connected sensor network systems for the Internet of Things (IoT), wearable disease diagnostic devices (Fig. 5) are becoming important for real-time monitoring with effective notification and warning (Wu et al. 2019) and hence the development of the wearable device is an emerging field with great potential.

Fig. 5 Photographs of wearable sensor patch attachment on the tomato leaf. **a** top view, **b** side view, **c** interface between the sensor and the leaf (zoomed-in). (Reproduced from Ref. Li et al. 2021, with permission from Elsevier)

Wearable devices are primarily deployed to detect early signs of disease in plants. These devices are usually flexible and can be directly attached to roots, stems, leaves, and other parts of host plants. Depending on requirement such devices contain one or more types of sensors with specific diagnostic purpose(s). For instance, the multi-modal plant healthcare leaf wearable sensor fabricated by Lu et al. (2020) using stacked $ZnIn_2S_4$(ZIS) nanosheets was capable of continuously monitoring light intensity, humidity, and temperature with a response time of $\sim$4 ms. Although the development of wearable sensors for plants is still in its infantile stage at present, a good number of works have been documented in this area (Mohammad-Razdari et al. 2022).

Plants emit different Volatile Organic Compounds (VOCs) when under stress. VOCs are released not only to respond to biotic/abiotic stress and attack by herbivores but also to protect them from various pathogens (Brilli et al. 2019). A substantial part of communication between plants regarding pathogen infection or herbivory occurs via airborne VOCs (Banerji et al. 2023). Therefore, continuous monitoring of VOC flux from plants has been regarded as a significant non-invasive technique for the detection of plant diseases. Li et al. (2021) designed a leaf-attachable chemiresistive wearable sensor for the detection of the pathogen *Phytophthora infestans* in tomato plants (Fig. 5). In their work, (*E*)-2-hexenal, 1-hexenal, methyl jasmonate, and 2-phenylethanol, were selected as VOC diagnostic markers for tomato plants infected by the pathogen *Phytophthora infestans* from literature reports. The emission level of (*E*)-2-hexenal has been reported to rise 4- to sixfold for *P. infestans*-infected tomato leaves compared to healthy plants (Raghava et al. 2010). Two different sensors were multiplexed in a single device. Reduced graphene oxide (rGO) was at the base of both sensor designs. In one sensor, rGO was modified with gold nanoparticles functionalized with chemical ligands; in the case of the other sensor the chemical ligands were directly grafted on the rGO nanosheet (without AuNP). The chemical ligands were selected to function as target recognition probes by forming Hydrogen-bond or halogen bonds with target VOCs, thereby increasing the electrical resistance of the sensor. The sensor containing the recognition probe ligand, 1,3-dis(3,5-bis(trifluoromethyl)phenyl)thiourea attached directly to the rGO nanosheet to work best with an LOD of 0.13–1.4 ppm, enabling the detection of infection within 4 days of inoculation.

7 Information and Communication Technology (ICT) for Phytopathological Management

Information and communication technology (ICT) in agriculture aims at the improvement of agricultural development by using information and communication processes. ICT in agriculture employs various data collection devices (sensors) that communicate with the host through private or public networks. Various plant health surveillance devices (optical images, IR sensors, plant wearable devices, electrochemical sensors, e-nose, etc.) constitute the sources of real-time (preferably continuous) data feed. The data can be received at a central facility in case of a large farming field or can be monitored through smartphones and other mobile devices.

8 Conclusion and Looking Forward

The development of new diagnostics tools for plant pathogens is essential for the effective management of plant diseases. These tools can help farmers quickly and accurately identify plant pathogens so that they can take action to prevent the spread of disease and protect their crops. Though a number of pathogen detection and identification techniques have been developed, still there are voids in the area of early detection and endemic forecast technologies. The rapid advancements in electronics, chemical technologies, and molecular biology, will promote the development of newer devices and techniques for the detection, monitoring, and management of plant pathogens. The coming era is going to witness the benefit of networked plant health monitoring and diagnostic devices effectively at every level of plant growers—from armature gardeners to large professional farmers.

Acknowledgements CORE Research Grant to SC from SERB, Govt of India, and financial support in the form of INSPIRE fellowship to SP and KD from DST, Govt of India, Research fellowship to RB from UGC, Govt of India, are thankfully acknowledged.

References

Agrios GN (2005) Introduction. In: Plant pathology. Elsevier, pp 3–75. https://doi.org/10.1016/B978-0-08-047378-9.50007-5

Aliotta JM, Pelletier JJ, Ware JL, Moran LS, Benner JS, Kong H (1996) Thermostable Bst DNA polymerase I lacks a 3'–>5' proofreading exonuclease activity. Genet Anal 12:185–195

Ashraf A, Islam M, Khalid M, Davis AP, Ahsan MT, Yaqub M, Syed A, Elgorban AM, Bahkali AH, Shafiq Z (2021) Naphthyridine derived colorimetric and fluorescent turn off sensors for Ni^{2+} in aqueous media. Sci Rep 11:19242. https://doi.org/10.1038/s41598-021-98400-2

Ates HC, Nguyen PQ, Gonzalez-Macia L, Morales-Narváez E, Güder F, Collins JJ, Dincer C (2022) End-to-end design of wearable sensors. Nat Rev Mater 7:887–907. https://doi.org/10.1038/s41578-022-00460-x

Banerji S, Hõrak H, Torop J, Huynh T (2023) Unravelling the secrets of plants: emerging wearable sensors for plants signals and physiology. Adv Sens Res. https://doi.org/10.1002/adsr.202300023

Baranwal J, Barse B, Gatto G, Broncova G, Kumar A (2022) Electrochemical sensors and their applications: a review. Chemosensors 10:363. https://doi.org/10.3390/chemosensors10090363

Barhoum A, Hamimed S, Slimi H, Othmani A, Abdel-Haleem FM, Bechelany M (2023) Modern designs of electrochemical sensor platforms for environmental analyses: principles, nanofabrication opportunities, and challenges. Trends Environ Anal Chem 38:e00199. https://doi.org/10.1016/j.teac.2023.e00199

Bar-Joseph M, Dawson WO (2008) Citrus tristeza virus. In: Encyclopedia of virology. Elsevier, pp 520–525. https://doi.org/10.1016/B978-012374410-4.00639-7

Behiry SI, Mohamed AA, Younes HA, Salem MZM, Salem AZM (2018) Antigenic and pathogenicity activities of *Ralstonia solanacearum* race 3 biovar 2 molecularly identified and detected by indirect ELISA using polyclonal antibodies generated in rabbits. Microb Pathog 115:216–221. https://doi.org/10.1016/j.micpath.2017.12.060

Bhardwaj N, Bhardwaj SK, Bhatt D, Lim DK, Kim K-H, Deep A (2019) Optical detection of waterborne pathogens using nanomaterials. TrAC, Trends Anal Chem 113:280–300. https://doi.org/10.1016/j.trac.2019.02.019

Boubourakas IN, Fukuta S, Kyriakopoulou PE (2009) Sensitive and rapid detection of peach latent mosaic viroid by the reverse transcription loop-mediated isothermal amplification. J Virol Methods 160:63–68. https://doi.org/10.1016/j.jviromet.2009.04.021

Brilli F, Loreto F, Baccelli I (2019) Exploiting plant volatile organic compounds (VOCs) in agriculture to improve sustainable defense strategies and productivity of crops. Front Plant Sci 10:264. https://doi.org/10.3389/fpls.2019.00264

Buja I, Sabella E, Monteduro AG, Chiriacò MS, De Bellis L, Luvisi A, Maruccio G (2021) Advances in plant disease detection and monitoring: from traditional assays to in-field diagnostics. Sensors 21:2129. https://doi.org/10.3390/s21062129

Butera E, Zammataro A, Pappalardo A, Trusso Sfrazzetto G (2021) Supramolecular sensing of chemical warfare agents. ChemPlusChem 86:681–695. https://doi.org/10.1002/cplu.202100071

Byzova N, Vinogradova S, Porotikova E, Terekhova U, Zherdev A, Dzantiev B (2018) Lateral flow immunoassay for rapid detection of grapevine leafroll-associated virus. Biosensors 8:111. https://doi.org/10.3390/bios8040111

Cao Y, Shi H, Zheng Y, Tan Z, Xie Z, Zhang C, Chen Z (2023) Polyaniline/Prussian blue nanolayer enhanced electrochemical sensing of H2O2 in EBC using an integrated condensation facemask. Sens Actuators, B Chem 393:134189. https://doi.org/10.1016/j.snb.2023.134189

Cardinale M, Luvisi A, Meyer JB, Sabella E, De Bellis L, Cruz AC, Ampatzidis Y, Cherubini P (2018) Specific fluorescence in situ hybridization (FISH) test to highlight colonization of *Xylem vessels* by *Xylella fastidiosa* in naturally infected olive trees (*Olea europaea* L.). Front Plant Sci 9:431. https://doi.org/10.3389/fpls.2018.00431

Caruso AG, Ragona A, Bertacca S, Montoya MAM, Panno S, Davino S (2023) Development of an in-field real-time LAMP assay for rapid detection of tomato leaf curl New Delhi virus. Plants 12:1487. https://doi.org/10.3390/plants12071487

Cassedy A, Mullins E, O'Kennedy R (2020) Sowing seeds for the future: the need for on-site plant diagnostics. Biotechnol Adv 39:107358. https://doi.org/10.1016/j.biotechadv.2019.02.014

Cebula Z, Żołędowska S, Dziąbowska K, Skwarecka M, Malinowska N, Białobrzeska W, Czaczyk E, Siuzdak K, Sawczak M, Bogdanowicz R, Nidzworski D (2019) Detection of the plant pathogen *Pseudomonas syringae* pv. lachrymans on antibody-modified gold electrodes by electrochemical impedance spectroscopy. Sensors 19:5411. https://doi.org/10.3390/s19245411

Cellini A, Blasioli S, Biondi E, Bertaccini A, Braschi I, Spinelli F (2017) Potential applications and limitations of electronic nose devices for plant disease diagnosis. Sensors 17:2596. https://doi.org/10.3390/s17112596

Cesewski E, Johnson BN (2020) Electrochemical biosensors for pathogen detection. Biosens Bioelectron 159:112214. https://doi.org/10.1016/j.bios.2020.112214

Chaliha C, Srivastava R, Kalita E, Sahoo L, Verma PK (2023) Rapid and precise detection of cryptic tea pathogen Exobasidium vexans: RealAmp validation of LAMP approach. World J Microbiol Biotechnol 39:52. https://doi.org/10.1007/s11274-022-03506-y

DeLude A, Wells R, Boomla S, Chuang S-C, Urena F, Shipman A, Rubas N, Kuehu DL, Bickerton B, Peterson T, Dobhal S, Arizala D, Klair D, Ochoa-Corona F, Ali ME, Odani J, Bingham J-P, Jenkins DM, Fletcher J, Stack JP, Alvarez AM, Arif M (2022) Loop-mediated isothermal amplification (LAMP) assay for specific and rapid detection of *Dickeya fangzhongdai* targeting a unique genomic region. Sci Rep 12:19193. https://doi.org/10.1038/s41598-022-22023-4

Deng S, Ma X, Chen Y, Feng H, Zhou D, Wang X, Zhang Y, Zhao M, Zhang J, Daly P, Wei L (2022) LAMP assay for distinguishing *Fusarium oxysporum* and *Fusarium commune* in lotus (*Nelumbo nucifera*) rhizomes. Plant Dis 106:231–246. https://doi.org/10.1094/PDIS-06-21-1223-RE

Du R, Huang Y, Zhang J, Yang L, Wu M, Li G (2021) LAMP detection and identification of the blackleg pathogen *Leptosphaeria biglobosa* 'brassicae.' Plant Dis 105:3192–3200. https://doi.org/10.1094/PDIS-08-20-1819-RE

Du J, Chen X, Liu K, Zhao D, Bai Y (2022) Dual recognition and highly sensitive detection of Listeria monocytogenes in food by fluorescence enhancement effect based on Fe_3O_4@ZIF-8-aptamer. Sens Actuators, B Chem 360:131654. https://doi.org/10.1016/j.snb.2022.131654

Dyussembayev K, Sambasivam P, Bar I, Brownlie JC, Shiddiky MJA, Ford R (2021) Biosensor technologies for early detection and quantification of plant pathogens. Front Chem 9:636245. https://doi.org/10.3389/fchem.2021.636245

Fan Z, Mei Y, Xing J, Chen T, Hu D, Liu H, Li Y, Liu D, Liu Z, Liang Y (2023) Loop-mediated isothermal amplification (LAMP)/Cas12a assay for detection of *Ralstonia solanacearum* in tomato. Front Bioeng Biotechnol 11:1188176. https://doi.org/10.3389/fbioe.2023.1188176

Fang Y, Umasankar Y, Ramasamy RP (2016) A novel bi-enzyme electrochemical biosensor for selective and sensitive determination of methyl salicylate. Biosens Bioelectron 81:39–45. https://doi.org/10.1016/j.bios.2016.01.095

FAO Website available at https://www.fao.org/news/story/en/item/1402920/icode/. Last accessed on 31 Aug 2023

Feng W, Hieno A, Kusunoki M, Suga H, Kageyama K (2019) LAMP detection of four plant-pathogenic oomycetes and its application in lettuce fields. Plant Dis 103:298–307. https://doi.org/10.1094/PDIS-05-18-0858-RE

Feng Y, Wang M, Jiang H, Shi Y, Zhang W, Wang H (2023) Comparative evaluation of LAMP and nested PCR for the rapid diagnosis of mycobacterium marinum infection. IDR 16:1601–1609. https://doi.org/10.2147/IDR.S404929

Feng W, Ishiguro Y, Hotta K, Watanabe H, Suga H, Kageyama K (2015) Simple detection of *Pythium irregulare* using loop-mediated isothermal amplification assay. FEMS Microbiol Lett 362:fnv174. https://doi.org/10.1093/femsle/fnv174

Foo PC, Nurul Najian AB, Muhamad NA, Ahamad M, Mohamed M, Yean Yean C, Lim BH (2020) Loop-mediated isothermal amplification (LAMP) reaction as viable PCR substitute for diagnostic applications: a comparative analysis study of LAMP, conventional PCR, nested PCR (nPCR) and real-time PCR (qPCR) based on Entamoeba histolytica DNA derived from faecal sample. BMC Biotechnol 20:34. https://doi.org/10.1186/s12896-020-00629-8

Fu X, Xie R, Wang J, Chen X, Wang X, Sun W, Meng J, Lai D, Zhou L, Wang B (2017) Development of colloidal gold-based lateral flow immunoassay for rapid qualitative and semiquantitative analysis of ustiloxins A and B in rice samples. Toxins 9:79. https://doi.org/10.3390/toxins9030079

Fu L, Chen Q, Jia L (2022) Carbon dots and gold nanoclusters assisted construction of a ratiometric fluorescent biosensor for detection of Gram-negative bacteria. Food Chem 374:131750. https://doi.org/10.1016/j.foodchem.2021.131750

Gao J, Lai Y, Wu C, Zhao Y (2013) Exploring and exploiting the synergy of non-covalent interactions on the surface of gold nanoparticles for fluorescent turn-on sensing of bacterial lipopolysaccharide. Nanoscale 5:8242. https://doi.org/10.1039/c3nr02490c

Geng Y, Peveler WJ, Rotello VM (2019) Array-based "Chemical Nose" sensing in diagnostics and drug discovery. Angew Chem Int Ed 58:5190–5200. https://doi.org/10.1002/anie.201809607

Ghosh D, Kokane S, Savita BK, Kumar P, Sharma AK, Ozcan A, Kokane A, Santra S (2022) Huanglongbing pandemic: current challenges and emerging management strategies. Plants 12:160. https://doi.org/10.3390/plants12010160

Ghosh R, Nagavardhini A, Sengupta A, Sharma M (2015) Development of loop-mediated isothermal amplification (LAMP) assay for rapid detection of *Fusarium oxysporum* f. sp. Ciceris—wilt pathogen of chickpea. BMC Res Notes 8:40. https://doi.org/10.1186/s13104-015-0997-z

Gregory PJ, Johnson SN, Newton AC, Ingram JSI (2009) Integrating pests and pathogens into the climate change/food security debate. J Exp Bot 60:2827–2838. https://doi.org/10.1093/jxb/erp080

Guerrini L, Garcia-Rico E, Pazos-Perez N, Alvarez-Puebla RA (2017) Smelling, seeing, tasting—old senses for new sensing. ACS Nano 11:5217–5222. https://doi.org/10.1021/acsnano.7b03176

Guo M, Qi D, Dong J, Dong S, Yang X, Qian Y, Zhou X, Wu J (2023) Development of dot-ELISA and colloidal gold immunochromatographic strip for rapid and super-sensitive detection of plum pox virus in apricot trees. Viruses 15:169. https://doi.org/10.3390/v15010169

Gupta S, Saharan MS, Gurjar MS, Singh J, Bashyal BM, Aggarwal R (2020) Molecular detection of Fusarium graminearum causing head blight of wheat by loop mediated isothermal amplification (LAMP) assay. Indian Phytopathol 73:667–672. https://doi.org/10.1007/s42360-020-00261-8

Gupte T, Pradeep T (2022) Nanosensors for water quality monitoring. In: Separation science and technology. Elsevier, pp 37–53. https://doi.org/10.1016/B978-0-323-90763-7.00010-X

Harding DP, Bootsma AN, Wheeler SE (2019) Better sensing through stacking: the role of non-covalent interactions in guanine-binding sensors. J Phys Chem B 123:487–495. https://doi.org/10.1021/acs.jpcb.8b12158

He S, Creasey Krainer KM (2020) Pandemics of people and plants: which is the greater threat to food security? Mol Plant 13:933–934. https://doi.org/10.1016/j.molp.2020.06.007

Henry Sum MS, Yee SF, Eng L, Poili E, Lamdin J (2017) Development of an indirect ELISA and dot-blot assay for serological detection of rice tungro disease. Biomed Res Int 2017:1–7. https://doi.org/10.1155/2017/3608042

Ho C-H, Frommer WB (2014) Fluorescent sensors for activity and regulation of the nitrate transceptor CHL1/NRT1.1 and oligopeptide transporters. eLife 3:e01917. https://doi.org/10.7554/eLife.01917

Hu S, Yang C, Li Y, Luo Q, Luo H (2022) Nanozyme sensor array based on manganese dioxide for the distinction between multiple amyloid β peptides and their dynamic aggregation process. Biosens Bioelectron 199:113881. https://doi.org/10.1016/j.bios.2021.113881

Huang W, Zhang H, Xu J, Wang S, Kong X, Ding W, Xu J, Feng J (2017) Loop-mediated isothermal amplification method for the rapid detection of *Ralstonia solanacearum* phylotype I mulberry strains in China. Front Plant Sci. https://doi.org/10.3389/fpls.2017.00076

Huang L, Tian S, Zhao W, Liu K, Ma X, Guo J (2020) Multiplexed detection of biomarkers in lateral-flow immunoassays. Analyst 145:2828–2840. https://doi.org/10.1039/C9AN02485A

Hulanicki A, Glab S, Ingman F (1991) Chemical sensors: definitions and classification. Pure Appl Chem 63:1247–1250. https://doi.org/10.1351/pac199163091247

Hussain CM, Keçili R (2020) Electrochemical techniques for environmental analysis. In: Modern environmental analysis techniques for pollutants. Elsevier, pp 199–222. https://doi.org/10.1016/B978-0-12-816934-6.00008-4

Jaffrezic-Renault N, Dzyadevych S (2008) Conductometric microbiosensors for environmental monitoring. Sensors 8:2569–2588. https://doi.org/10.3390/s8042569

Jiang J, Feindel W, Harding M, Feindel D, Bajema S, Feng J (2023a) Development and evaluation of a loop-mediated isothermal amplification (LAMP) method for detection of the potato powdery scab pathogen *Spongospora subterranea*. Plant Dis 107:136–141. https://doi.org/10.1094/PDIS-05-22-1000-SR

Jiang J, Feindel W, Swisher Grimm K, Harding M, Feindel D, Bajema S, Feng J (2023b) Development of a loop-mediated isothermal amplification (LAMP) method to detect the potato zebra chip pathogen ' *Candidatus* Liberibacter solanacearum' (Lso) and differentiate haplotypes A and B. Plant Dis 107:1697–1702. https://doi.org/10.1094/PDIS-09-22-2258-SR

Jiang L, Ren MX, Niu G, Shi J, Cao X, Duan Y, Wu H, Xie Z, Quan Y, Zhao L, Jiang Z, Gong Y, Ren W, Zhao G (2023c) Label-free, rapid and highly accurate identification and categorization of leukemia cells via Raman spectroscopy. Sens Actuators, B Chem 395:134497. https://doi.org/10.1016/j.snb.2023.134497

Jin K, Moon D, Kim M, Park J (2023) Tailoring the extrinsic porosity of a vapochromic metal–organic polyhedron for rapid VOC detection. Sens Actuators, B Chem 393:134205. https://doi.org/10.1016/j.snb.2023.134205

Joudeh N, Linke D (2022) Nanoparticle classification, physicochemical properties, characterization, and applications: a comprehensive review for biologists. J Nanobiotechnol 20:262. https://doi.org/10.1186/s12951-022-01477-8

Kara N, Ayoub N, Ilgu H, Fotiadis D, Ilgu M (2023) Aptamers targeting membrane proteins for sensor and diagnostic applications. Molecules 28:3728. https://doi.org/10.3390/molecules28093728

Karami N, Mirzajani F, Rezadoost H, Karimi A, Fallah F, Ghassempour A, Aliahmadi A (2018) Initial study of three different pathogenic microorganisms by gas chromatography-mass spectrometry. F1000Res 6:1415. https://doi.org/10.12688/f1000research.12003.3

Kaur G, Raj T, Kaur N, Singh N (2015) Pyrimidine-based functional fluorescent organic nanoparticle probe for detection of *Pseudomonas aeruginosa*. Org Biomol Chem 13:4673–4679. https://doi.org/10.1039/C5OB00206K

Kaya HK, Çağlayan T, Kuralay F (2022) Functionalized nanomaterial- based electrochemical sensors for point-of-care devices. In: Functionalized nanomaterial-based electrochemical sensors. Elsevier, pp 309–335. https://doi.org/10.1016/B978-0-12-823788-5.00021-1

Khan M, Wang R, Li B, Liu P, Weng Q, Chen Q (2018) Comparative evaluation of the LAMP assay and PCR-based assays for the rapid detection of *Alternaria solani*. Front Microbiol 9:2089. https://doi.org/10.3389/fmicb.2018.02089

Khater M, De La Escosura-Muñiz A, Quesada-González D, Merkoçi A (2019a) Electrochemical detection of plant virus using gold nanoparticle-modified electrodes. Anal Chim Acta 1046:123–131. https://doi.org/10.1016/j.aca.2018.09.031

Khater M, Escosura-Muñiz ADL, Altet L, Merkoçi A (2019b) In situ plant virus nucleic acid isothermal amplification detection on gold nanoparticle-modified electrodes. Anal Chem 91:4790–4796. https://doi.org/10.1021/acs.analchem.9b00340

Koczula KM, Gallotta A (2016) Lateral flow assays. Essays Biochem 60:111–120. https://doi.org/10.1042/EBC20150012

Krivitsky V, Granot E, Avidor Y, Borberg E, Voegele RT, Patolsky F (2021) Rapid collection and aptamer-based sensitive electrochemical detection of soybean rust fungi airborne urediniospores. ACS Sens 6:1187–1198. https://doi.org/10.1021/acssensors.0c02452

Lan C, Yao J, Yang X, Ruan H, Yu D, Jiang J (2020) Specific and sensitive detection of the guava fruit anthracnose pathogen (*Colletotrichum gloeosporioides*) by loop-mediated isothermal amplification (LAMP) assay. Can J Microbiol 66:17–24. https://doi.org/10.1139/cjm-2019-0099

Lau HY, Wu H, Wee EJH, Trau M, Wang Y, Botella JR (2017) Specific and sensitive isothermal electrochemical biosensor for plant pathogen DNA detection with colloidal gold nanoparticles as probes. Sci Rep 7:38896. https://doi.org/10.1038/srep38896

Lees AK, Roberts DM, Lynott J, Sullivan L, Brierley JL (2019) Real-time PCR and LAMP assays for the detection of spores of *Alternaria solani* and sporangia of *Phytophthora infestans* to inform disease risk forecasting. Plant Dis 103:3172–3180. https://doi.org/10.1094/PDIS-04-19-0765-RE

Li C, Ma X, Guan Y, Tang J, Zhang B (2019) Microcantilever array biosensor for simultaneous detection of carcinoembryonic antigens and α-fetoprotein based on real-time monitoring of the profile of cantilever. ACS Sens 4:3034–3041. https://doi.org/10.1021/acssensors.9b01604

Li Z, Liu Y, Hossain O, Paul R, Yao S, Wu S, Ristaino JB, Zhu Y, Wei Q (2021) Real-time monitoring of plant stresses via chemiresistive profiling of leaf volatiles by a wearable sensor. Matter 4:2553–2570. https://doi.org/10.1016/j.matt.2021.06.009

Liu Z, Su X (2017) A novel fluorescent DNA sensor for ultrasensitive detection of *Helicobacter pylori*. Biosens Bioelectron 87:66–72. https://doi.org/10.1016/j.bios.2016.07.061

Liu B, Kaurilind E, Jiang Y, Niinemets Ü (2018) Methyl salicylate differently affects benzenoid and terpenoid volatile emissions in Betula pendula. Tree Physiol 38:1513–1525. https://doi.org/10.1093/treephys/tpy050

Liu B, Li Z, Du J, Zhang W, Che X, Zhang Z, Chen P, Wang Y, Li Y, Wang S, Ding X (2022) Loop-mediated isothermal amplification (LAMP) for the rapid and sensitive detection of *Alternaria alternata* (Fr.) Keissl in apple alternaria blotch disease with Aapg-1 encoding the endopolygalacturonase. Pathogens 11:1221. https://doi.org/10.3390/pathogens11111221

López-Soriano P, Noguera P, Gorris MT, Puchades R, Maquieira Á, Marco-Noales E, López MM (2017) Lateral flow immunoassay for on-site detection of *Xanthomonas arboricola* pv. pruni in symptomatic field samples. PLoS ONE 12:e0176201. https://doi.org/10.1371/journal.pone.0176201

Loulier J, Lefort F, Stocki M, Asztemborska M, Szmigielski R, Siwek K, Grzywacz T, Hsiang T, Ślusarski S, Oszako T, Klisz M, Tarakowski R, Nowakowska JA (2020) Detection of fungi and oomycetes by volatiles using E-nose and SPME-GC/MS platforms. Molecules 25:5749. https://doi.org/10.3390/molecules25235749

Lu Y, Xu K, Zhang L, Deguchi M, Shishido H, Arie T, Pan R, Hayashi A, Shen L, Akita S, Takei K (2020) Multimodal plant healthcare flexible sensor system. ACS Nano 14:10966–10975. https://doi.org/10.1021/acsnano.0c03757

Manjunatha C, Sharma S, Kulshreshtha D, Gupta S, Singh K, Bhardwaj SC, Aggarwal R (2018) Rapid detection of *Puccinia triticina* causing leaf rust of wheat by PCR and loop mediated isothermal amplification. PLoS ONE 13:e0196409. https://doi.org/10.1371/journal.pone.0196409

May KML, Wang Y, Bachas LG, Anderson KW (2004) Development of a whole-cell-based biosensor for detecting histamine as a model Toxin. Anal Chem 76:4156. https://doi.org/10.1021/ac049810+

Mohammad-Razdari A, Rousseau D, Bakhshipour A, Taylor S, Poveda J, Kiani H (2022) Recent advances in E-monitoring of plant diseases. Biosens Bioelectron 201:113953. https://doi.org/10.1016/j.bios.2021.113953

Murali G, Reddeppa M, Seshendra Reddy C, Park S, Chandrakalavathi T, Kim M-D, In I (2020) Enhancing the charge carrier separation and transport via nitrogen-doped graphene quantum dot-TiO_2 nanoplate hybrid structure for an efficient NO gas sensor. ACS Appl Mater Interfaces 12:13428–13436. https://doi.org/10.1021/acsami.9b19896

Nagamine K, Hase T, Notomi T (2002) Accelerated reaction by loop-mediated isothermal amplification using loop primers. Mol Cell Probes 16:223–229. https://doi.org/10.1006/mcpr.2002.0415

Nehra A, Kumar A, Ahlawat S, Kumar V, Singh KP (2022) Substrate-free untagged detection of miR393a using an ultrasensitive electrochemical biosensor. ACS Omega 7:5176–5189. https://doi.org/10.1021/acsomega.1c06098

Notomi T (2000) Loop-mediated isothermal amplification of DNA. Nucleic Acids Res 28:63e–663. https://doi.org/10.1093/nar/28.12.e63

Ocenar J, Arizala D, Boluk G, Dhakal U, Gunarathne S, Paudel S, Dobhal S, Arif M (2019) Development of a robust, field-deployable loop-mediated isothermal amplification (LAMP) assay for specific detection of potato pathogen *Dickeya dianthicola* targeting a unique genomic region. PLoS ONE 14:e0218868. https://doi.org/10.1371/journal.pone.0218868

Oerke E-C (2006) Crop losses to pests. J Agric Sci 144:31–43. https://doi.org/10.1017/S0021859605005708

Okiro LA, Tancos MA, Nyanjom SG, Smart CD, Parker ML (2019) Comparative evaluation of LAMP, qPCR, conventional PCR, and ELISA to detect *Ralstonia solanacearum* in Kenyan potato fields. Plant Dis 103:959–965. https://doi.org/10.1094/PDIS-03-18-0489-RE

Ouyang Q, Yang Y, Ali S, Wang L, Li H, Chen Q (2021) Upconversion nanoparticles-based FRET system for sensitive detection of *Staphylococcus aureus*. Spectrochim Acta Part A Mol Biomol Spectrosc 255:119734. https://doi.org/10.1016/j.saa.2021.119734

Panferov VG, Safenkova IV, Varitsev YA, Drenova NV, Kornev KP, Zherdev AV, Dzantiev BB (2016) Development of the sensitive lateral flow immunoassay with silver enhancement for the detection of *Ralstonia solanacearum* in potato tubers. Talanta 152:521–530. https://doi.org/10.1016/j.talanta.2016.02.050

Patel R, Mitra B, Vinchurkar M, Adami A, Patkar R, Giacomozzi F, Lorenzelli L, Baghini MS (2022) A review of recent advances in plant-pathogen detection systems. Heliyon 8:e11855. https://doi.org/10.1016/j.heliyon.2022.e11855

Patrick SC, Hein R, Docker A, Beer PD, Davis JJ (2021) Solvent effects in halogen and hydrogen bonding mediated electrochemical anion sensing in aqueous solution and at interfaces. Chem A Eur J 27:10201–10209. https://doi.org/10.1002/chem.202101102

Paula T, Marc F (2013) Transgenic crops resistant to fungal, bacterial and viral pathogens. In: Christou P, Savin R, Costa-Pierce BA, Misztal I, Whitelaw CBA (eds) Sustainable food production. Springer New York, New York, NY, pp 1594–1612. https://doi.org/10.1007/978-1-4614-5797-8_168

Picchi SC, De Souza E Silva M, Saldanha LL, Ferreira H, Takita MA, Caldana C, De Souza AA (2021) GC-TOF/MS-based metabolomics analysis to investigate the changes driven by *N*-Acetylcysteine in the plant-pathogen *Xanthomonas citri* subsp. citri. Sci Rep 11:15558. https://doi.org/10.1038/s41598-021-95113-4

Prasanna BM, Carvajal-Yepes M, Kumar PL, Kawarazuka N, Liu Y, Mulema AA, McCutcheon S, Ibabao X (2022) Sustainable management of transboundary pests requires holistic and inclusive solutions. Food Sec 14:1449–1457. https://doi.org/10.1007/s12571-022-01301-z

Raghava T, Ravikumar P, Hegde R, Kush A (2010) Spatial and temporal volatile organic compound response of select tomato cultivars to herbivory and mechanical injury. Plant Sci 179:520–526. https://doi.org/10.1016/j.plantsci.2010.07.020

Razo SC, Panferov VG, Safenkova IV, Varitsev YA, Zherdev AV, Dzantiev BB (2018) Double-enhanced lateral flow immunoassay for potato virus X based on a combination of magnetic and gold nanoparticles. Anal Chim Acta 1007:50–60. https://doi.org/10.1016/j.aca.2017.12.023

Razo SC, Safenkova IV, Drenova NV, Kharchenko AA, Tsymbal YS, Varitsev YA, Zherdev AV, Pakina EN, Dzantiev BB (2021) New lateral flow immunoassay for on-site detection of *Erwinia amylovora* and its application on various organs of infected plants. Physiol Mol Plant Pathol 114:101637. https://doi.org/10.1016/j.pmpp.2021.101637

Ristaino JB, Anderson PK, Bebber DP, Brauman KA, Cunniffe NJ, Fedoroff NV, Finegold C, Garrett KA, Gilligan CA, Jones CM, Martin MD, MacDonald GK, Neenan P, Records A, Schmale DG, Tateosian L, Wei Q (2021) The persistent threat of emerging plant disease pandemics to global food security. Proc Natl Acad Sci USA 118:e2022239118. https://doi.org/10.1073/pnas.2022239118

Safenkova IV, Zaitsev IA, Varitsev YA, Byzova NA, Drenova NV, Zherdev AV, Dzantiev BB (2017) Development of a lateral flow immunoassay for rapid diagnosis of potato blackleg caused by Dickeya species. Anal Bioanal Chem 409:1915–1927. https://doi.org/10.1007/s00216-016-0140-6

Sahu B, Singh SD, Behera BK, Panda SK, Das A, Parida PK (2019) Rapid detection of Salmonella contamination in seafoods using multiplex PCR. Braz J Microbiol 50:807–816. https://doi.org/10.1007/s42770-019-00072-8

Salgado-Salazar C, Bauchan GR, Wallace EC, Crouch JA (2018) Visualization of the impatiens downy mildew pathogen using fluorescence in situ hybridization (FISH). Plant Methods 14:92. https://doi.org/10.1186/s13007-018-0362-z

Savary S, Willocquet L, Pethybridge SJ, Esker P, McRoberts N, Nelson A (2019) The global burden of pathogens and pests on major food crops. Nat Ecol Evol 3:430–439. https://doi.org/10.1038/s41559-018-0793-y

Selvarajan R, Kanichelvam PS, Balasubramanian V, Sethurama Subramanian S (2020) A rapid and sensitive lateral flow immunoassay (LFIA) test for the on-site detection of banana bract mosaic virus in banana plants. J Virol Methods 284:113929. https://doi.org/10.1016/j.jviromet.2020.113929

Shakoori AR (2017) Fluorescence in situ hybridization (FISH) and its applications. In: Bhat TA, Wani AA (eds) Chromosome structure and aberrations. Springer India, New Delhi, pp 343–367. https://doi.org/10.1007/978-81-322-3673-3_16

Sharma SS, Palatty S (2023) Advances in functionalized polyaniline nanocomposites for electrochemical sensing and energy storage applications. In: Applications of multifunctional nanomaterials. Elsevier, pp 177–196. https://doi.org/10.1016/B978-0-12-820557-0.00004-7

Simões FR, Xavier MG (2017) Electrochemical sensors. In: Nanoscience and its applications. Elsevier, pp 155–178. https://doi.org/10.1016/B978-0-323-49780-0.00006-5

Smit A-M, Strabala TJ, Peng L, Rawson P, Lloyd-Jones G, Jordan TW (2012) Proteomic phenotyping of novosphingobium nitrogenifigens reveals a robust capacity for simultaneous nitrogen fixation, polyhydroxyalkanoate production, and resistance to reactive oxygen species. Appl Environ Microbiol 78:4802–4815. https://doi.org/10.1128/AEM.00274-12

Song W, Ma X, Tan H, Zhou J (2011) Abscisic acid enhances resistance to *Alternaria solani* in tomato seedlings. Plant Physiol Biochem 49:693–700. https://doi.org/10.1016/j.plaphy.2011.03.018

Tegli S, Biancalani C, Ignatov AN, Osdaghi E (2020) A powerful LAMP weapon against the threat of the quarantine plant pathogen *Curtobacterium flaccumfaciens* pv. *flaccumfaciens*. Microorganisms 8:1705. https://doi.org/10.3390/microorganisms8111705

Thangavelu RM, Kadirvel N, Balasubramaniam P, Viswanathan R (2022) Ultrasensitive nano-gold labelled, duplex lateral flow immunochromatographic assay for early detection of sugarcane mosaic viruses. Sci Rep 12:4144. https://doi.org/10.1038/s41598-022-07950-6

Thet NT, Mercer-Chalmers J, Greenwood RJ, Young AER, Coy K, Booth S, Sack A, Jenkins ATA (2020) *SPaCE* Swab: point-of-care sensor for simple and rapid detection of acute wound infection. ACS Sens 5:2652–2657. https://doi.org/10.1021/acssensors.0c01265

Thierry M, Chatet A, Fournier E, Tharreau D, Ioos R (2020) A PCR, qPCR, and LAMP toolkit for the detection of the wheat blast pathogen in seeds. Plants 9:277. https://doi.org/10.3390/plants9020277

Thoraneenitiyan N, Choopara I, Nuanualsuwan S, Kokpol S, Somboonna N (2022) Rapid visual *Candidatus* Liberibacter asiaticus detection (citrus greening disease) using simple alkaline heat DNA lysis followed by loop-mediated isothermal amplification coupled hydroxynaphthol blue (AL-LAMP-HNB) for potential local use. PLoS ONE 17:e0276740. https://doi.org/10.1371/journal.pone.0276740

Tian Y, Liu D, Zhao Y, Wu J, Hu B, Walcott RR (2017a) Visual detection of *Didymella bryoniae* in cucurbit seeds using a loop-mediated isothermal amplification assay. Eur J Plant Pathol 147:255–263. https://doi.org/10.1007/s10658-016-0996-5

Tian Q, Lu C, Wang S, Xiong Q, Zhang H, Wang Y, Zheng X (2017b) Rapid diagnosis of soybean anthracnose caused by *Colletotrichum truncatum* using a loop-mediated isothermal amplification (LAMP) assay. Eur J Plant Pathol 148:785–793. https://doi.org/10.1007/s10658-016-1132-2

Tomita N, Mori Y, Kanda H, Notomi T (2008) Loop-mediated isothermal amplification (LAMP) of gene sequences and simple visual detection of products. Nat Protoc 3:877–882. https://doi.org/10.1038/nprot.2008.57

Tomlinson JA, Dickinson MJ, Boonham N (2010) Rapid detection of *Phytophthora ramorum* and *P. kernoviae* by two-minute DNA extraction followed by isothermal amplification and amplicon detection by generic lateral flow device. Phytopathology® 100:143–149. https://doi.org/10.1094/PHYTO-100-2-0143

Trzmiel K, Hasiów-Jaroszewska B (2020) Development of loop-mediated isothermal amplification assay for rapid detection of genetically different wheat dwarf virus isolates. Mol Biol Rep 47:8325–8329. https://doi.org/10.1007/s11033-020-05846-0

Waite DW, Griffin R, Taylor R, George S (2016) Rapid and accurate identification of *Xanthomonas citri* subspecies *citri* by fluorescence *in situ* hybridization. Lett Appl Microbiol 63:315–321. https://doi.org/10.1111/lam.12624

Walters F, Ali MM, Burwell G, Rozhko S, Tehrani Z, Daghigh Ahmadi E, Evans JE, Abbasi HY, Bigham R, Mitchell JJ, Kazakova O, Devadoss A, Guy OJ (2020) A facile method for the non-covalent amine functionalization of carbon-based surfaces for use in biosensor development. Nanomaterials 10:1808. https://doi.org/10.3390/nano10091808

Wang L, Hou H, Zhou Z, Tu H, Yuan H (2021) Identification and detection of botryosphaeria dothidea from kiwifruit (*Actinidia chinensis*) in China. Plants 10:401. https://doi.org/10.3390/plants10020401

Wang W, Li N, Liu X, Kang S (2019) I plate-based assay for studying how fungal volatile compounds (VCs) affect plant growth and development and the identification of VCs via SPME-GC-MS. Bio-Protocol 9. https://doi.org/10.21769/BioProtoc.3166

Wei Q-W, Yu C, Zhang S-Y, Yang C-Y, Miriam K, Zhang W-N, Dou D-L, Tao X-R (2012) One-step detection of Bean pod mottle virusin soybean seeds by the reverse-transcription loop-mediated isothermal amplification. Virol J 9:187. https://doi.org/10.1186/1743-422X-9-187

West JS, Canning GGM, Perryman SA, King K (2017) Novel technologies for the detection of Fusarium head blight disease and airborne inoculum. Trop Plant Pathol 42:203–209. https://doi.org/10.1007/s40858-017-0138-4

Wu F, Wu T, Yuce MR (2019) Design and implementation of a wearable sensor network system for iot-connected safety and health applications. In: 2019 IEEE 5th world forum on internet of things (WF-IoT). Presented at the 2019 IEEE 5th world forum on internet of things (WF-IoT' 19), IEEE, Limerick, Ireland, pp 87–90. https://doi.org/10.1109/WF-IoT.2019.8767280

Xu L, Wang Y, Zhu S, Li J, Chang Y, Huang L (2020) Development and application of a LAMP assay for the detection of the latent apple tree pathogen Valsa mali. Plant Dis 105:1065. https://doi.org/10.1094/PDIS-07-20-1449-RE

Yao S, Zhao C, Shang M, Li J, Wang J (2021) Enzyme-free and label-free detection of *Staphylococcus aureus* based on target-inhibited fluorescence signal recovery. Food Chem Toxicol 150:112071. https://doi.org/10.1016/j.fct.2021.112071

Zanoli L, Spoto G (2012) Isothermal amplification methods for the detection of nucleic acids in microfluidic devices. Biosensors 3:18–43. https://doi.org/10.3390/bios3010018

Zhang C, Guo M, Dong J, Liu L, Zhou X, Wu J (2023a) Visual and super-sensitive detection of maize chlorotic mottle virus by Dot-ELISA and Au nanoparticle-based immunochromatographic test strip. Viruses 15:1607. https://doi.org/10.3390/v15071607

Zhang J, Zhou M, Li X, Fan Y, Li J, Lu K, Wen H, Ren J (2023b) Recent advances of fluorescent sensors for bacteria detection-a review. Talanta 254:124133. https://doi.org/10.1016/j.talanta.2022.124133

Zhang Q, Zeng S-M-Z, Yuan L-C, Wu C-J, Wu S-Y, Ren S-Z, Zhao M-D, Wang X-M, Zhu H-L, Wang Z-C (2023c) Real-time diagnosis of glioblastoma by sensing peroxynitrite with a highly selective and rapid-responding fluorescent probe. Sens Actuators, B Chem 395:134502. https://doi.org/10.1016/j.snb.2023.134502

Molecular Diagnostics of Plant Viruses, Viroids, and Phytoplasma: An Updated Overview

Mohd. Aslam, Anjali Rani, Bhaskara Nand Pant, Prashant Singh, and Garima Pandey

Abstract Plant viruses, viroids, and phytoplasma have been related to high crop production costs and pose a severe threat to agricultural output globally. The great genetic diversity of plant viruses, viroids, and phytoplasma makes it challenging to create markers that can identify every type of pathogen. The identification and detection of these infections have been transformed by molecular diagnostics, allowing for quick and precise diagnosis and the facilitation of efficient disease management techniques. Various molecular diagnostic techniques, consisting of polymerase chain reaction (PCR), reverse transcription PCR (RT-PCR), loop-mediated isothermal amplification (LAMP), and quantitative real-time PCR, are used to analyze plant viruses, viroids, and phytoplasma and treat plant illnesses. The information provided in this chapter will be helpful to researchers, plant pathologists, and others interested in plant disease management.

Keywords Plants molecular diagnostics · Viruses · Viroid's · Phytoplasma

Abbreviations

LAMP	Loop-mediated isothermal amplification
PCR	Polymerase chain reaction
RT-PCR	Reverse transcription PCR
FISH	Fluorescence in situ hybridization
ELISA	Enzyme-linked immunosorbent assay
PMeV	Papaya meleira virus

Mohd. Aslam · A. Rani · G. Pandey (✉)
Department of Chemistry, SRM-Institute of Science and Technology, Delhi-NCR Campus, Modinagar, Ghaziabad, UP 201204, India
e-mail: garimapandey.pandey8@gmail.com

Mohd. Aslam · A. Rani · B. N. Pant · P. Singh
Department of Chemistry, Atma Ram Sanatan Dharma College, University of Delhi, New Delhi 110021, India

J. Chen et al. (eds.), *Molecular and Biotechnological Tools for Plant Disease Management*, Interdisciplinary Biotechnological Advances,
https://doi.org/10.1007/978-981-97-7510-1_7

VCV-M	Vicia cryptic virus M
PPV	Plum pox virus
CMV	Cucumber mosaic virus
CSVd	Chrysanthemum stunt viroid
PSTVd	Potato spindle tuber viroid
NGS	Next-generation sequencing
ToBRFV	Tomato Brown Rugose Fruit Virus
PAGE	Polyacrylamide gel electrophoresis
HSVd	Hop stunt viroid
CEVd	Citrus exocortis viroid

1 Introduction

Crops are the primary generators in the food chain and are largely used to feed the world's population. Intense research is aimed at the management of agricultural losses to assure global food security, which has been a significant worldwide issue recently (Ingram 2011). Food production per individual has grown, but demand is still rising significantly year after year. Crop losses are frequently caused by pests, and diseases make up about 40% of the main factors influencing crop yield (Savary et al. 2012). Crop output is disrupted because a variety of plant pathogens, including nematodes, viruses, fungi, bacteria, insects, mollicutes, and more, causing serious plant illnesses (Tewari and Sharma 2019). There has been a huge increase in the economic effect caused by plant viruses as a result of a growing need for agricultural products necessary to feed the rapidly expanding global population. For instance, the tomato-screened wilt virus, which affects different varieties of plants like groundnuts, tomatoes, tobacco, etc., causes damages of about $1 billion worldwide; the cocoa-screened fire virus damages cocoa beans worth $28 million in Africa; and the rice tungro virus results in losses of $1.5 billion among rice in Southeast Asia (Haňcinský et al. 2020). The geographic distribution of plant infections and pests has drastically shifted during the past few decades as a result of climate change. Many pathogens are now found in previously unanticipated geographic locations. The rapid growth in global trading and transportation in recent years has also contributed to the quick global transmission of both recently discovered and re-emerging plant diseases (Paul et al. 2021). Phytopathogenic microbial species such as viruses, phytoplasma, fungi, viroids, nematodes, and bacteria are the principal causes of illness in plants (Khater et al. 2017).

For the effective control of disease spread and related economic losses, plant disease identification is a crucial prerequisite (Strange and Scott 2005). Traditional methods of diagnosing plant diseases frequently involve physical or microscopic examination of morphological signs. A few frequently used methods for identifying plant diseases include measuring changes in the surface temperature of plant leaves using thermographic analysis, measuring changes in the fluorescent properties of leaves using fluorescence imaging techniques, and using hyperspectral imaging methods with a spectral range between 350 and 2500 nm (Chun et al. 2022). In addition to this, the crop plants' volatile organic molecules are also utilized as markers to detect plant diseases. Even though those techniques indirectly detect crop illnesses, they are far too subjective (Martinelli et al. 2015). The best method for protecting agricultural plants from various infections is to identify the infected crops and remove them from the field. The sample region and timing may produce false positives or false negatives, making it difficult to identify the virus-infected plants. Furthermore, plants may exhibit viral symptoms in response to severe weather, nutritional imbalances, other infections, pest damage, abiotic environments, etc. (Kalimuthu et al. 2022). The art of understanding these two "Ds," detection and diagnosis, as well as how they relate to one another, is complex. In most cases, detection refers to determining the presence of the desired microorganisms in the surrounding sample, such as soil, water, plant surfaces, etc. while diagnosis refers to determining the type of issue and the issue's root cause that led to the development of the disease (Louws et al. 1999). Early plant pathogen detection lays the groundwork for effective disease management and control in greenhouse and outdoor settings. The bulk of detection methods depends on the physical assessment of conditions and subsequent lab testing that are frequently labor and time-intensive processes. The shortcomings of traditional approaches have been addressed in the past few years by the advancement of methods using molecular technologies for plant disease identification, which have made it possible to quickly and accurately diagnose plant disease (Khaled et al. 2018). Plant identification and diagnosis of diseases is a costly procedure that demands a lot of work to produce a safer crop. Therefore, in this circumstance, accurate methodologies and low-cost strategies for diagnosing these plant infections are required. Hence, in this chapter, the authors focus on different molecular diagnosis methods for viruses, viroids, and phytoplasma.

2 Biotic Stress Impact on the Growth of Plants

Both abiotic and biotic stresses have an impact on plants when the environment is harsh. Heavy metals, temperature, salt, drought, UV-B radiation, light, flooding, and other temporary environmental factors are the principal abiotic stresses. Biotic stress in crops is worsened by pests, diseases, and herbivores. These environmental conditions immediately impact sessile plants (Kumar and Ohri 2023). Instead of infections by single pathogens, infections by bacteria, fungi, and viruses collectively are frequent which results in serious disease outcomes (Pandey et al. 2017). The interaction between the pathogen and the plant results in a variety of morphological, biochemical, and molecular responses, which indicate that it activates several signaling pathways. All the primary crops experience decreased photosynthesis due to different kinds of infections that impact leaves, which sharply affects productivity (Gimenez et al. 2018). The plant's primary effect and stress response is the formation of reactive species. Plant's various organs, including chloroplasts, mitochondria, peroxisomes, and apoplasts, all produce reactive species, which interfere with normal operation when under stress (Choudhury et al. 2017). Plant disease-related agricultural losses also seriously jeopardize the world's ability to feed its population and produce enough food (Chakraborty and Newton 2011).

3 Conventional Techniques for Identifying and Diagnosing Plant Diseases

Accurate diagnosis of the frequency, severity, and detrimental effects of plant sickness on significant farming and horticultural plants is urgently required. Visual inspection is the primary method used to identify plant diseases in conventional methods. The visual signs frequently appear after the crop has already suffered potentially significant harm. First is a visual examination of plant diseases which includes symptoms, stunted shoot and root growth, short internodes, distorted leaf structures, chlorosis, and fewer flowers and fruits, which are indicators of the presence of illness in plants (Broadley et al. 2011). These are significant and obvious signs that aid in identifying the pathogen responsible for the disease. Some examples of these symptoms include marks on leaves, culms (stem), vegetables, fruits, decaying roots, and other parts of the plant that affect its process of growth. The development of mosaic shapes on leaves and variations within the color of a plant's vegetative (stalks, leaves, etc.) and fertile (flower and seed) components are examples that show how the usual appearance has been transformed (Prokopy and Owens 1983).

Pathogen identification using microscopic analysis includes cultivating fungi from symptomatic plant tissues; the direct agar plating approach typically uses acidic potato dextrose agar media or the extract of malt. The pathogen that is invading is also easier to recognize when microscopic examinations of the spores and fungal mycelium are made (Mancini et al. 2016).

4 Molecular Diagnostics

Laboratory techniques that analyze molecules in a tissue or fluid sample, such as DNA, RNA, and proteins, help identify an illness or the risk of acquiring a disease, such as cancer. Additionally, molecular diagnostics can be used to discover new cases of a disease, monitor its progression, and plan treatment (Donoso and Valenzuela 2018). Molecular diagnostics is a field of diagnostics that uses molecular biology techniques to identify and analyze genetic and molecular markers that are linked with disease. It involves the identification and quantification of DNA, RNA, proteins, and other molecules to diagnose illnesses, gauge the likelihood of developing particular ailments, forecast the effectiveness of treatments, and track the development of diseases (Patrinos et al. 2017). Molecular diagnostics is important in many areas of healthcare, including infectious diseases, cancer, genetic disorders, pharmacogenomics, and personalized medicine. By detecting specific genetic changes, identifying infectious agents at the molecular level, and analyzing biomarkers linked with various diseases, it enables more precise and accurate diagnosis (Debnath et al. 2010). Numerous molecular and genetic assays, such as the polymerase chain reaction (PCR), nucleic acid sequencing, microarrays, fluorescence in situ hybridization (FISH), and others, are frequently employed in molecular diagnostics. These methods make it possible to find genetic polymorphisms, levels of gene expression, and other molecular alterations that can help with disease diagnosis and treatment (Lévesque 2001). Recent years have seen significant improvements in molecular diagnostics, and these advancements have encouraged the development of personalized medicine and targeted medicines. Understanding the molecular aspects of diseases will help us choose more effective treatments (Srivastava et al. 2021). In recent years molecular diagnostics have played a crucial role in the area of disease diagnosis in plants. The use of molecular diagnostics is extremely beneficial in the detection and identification of infectious agents in plants, such as bacteria, viruses, fungi, and parasites. Its high specificity and sensitivity allow for extremely accurate and precise detection of target molecules as well as their measurement and give rapid results. This technique is cost-effective and gives precise results and also helps to detect plant diseases (Patrinos and Ansorge 2009). Table 1 and Fig. 1 summarize some applications, detection, and techniques for plants based on molecular diagnostics.

5 Types of Plant Molecular Diagnostics

Precision and promptness in plant pathogen identification are essential for effective disease management in agriculture, which reduces output losses (Doddaraju et al. 2019). The traditional detection method focuses on morphological identification of the plant disease-causing organism, which is sometimes a difficult and time-consuming phenomenon. An in-depth understanding and expertise in recognizing certain bacterial and fungal structural traits are also required. High sensitivity and specificity, repeatability, and speed are key qualities that a successful diagnostic tool must have. It is more suited for diagnostic purposes if a procedure can multiplex and is affordable. To identify the plant pathogen, all diagnostic methods use either particular nucleic acid or protein locations (Fig. 2). The ability to detect these compounds is crucial, and several technologies have been created throughout time to help (Miller and Martin 1988).

5.1 DNA-Based Approaches

The Polymerase Chain Reaction (PCR) is a common method which is used for identifying different plant infections. PCR is required to make several copies of a

Table 1 Some detection techniques and their applications

S. no	Technique	Detection	Application
1	Polymerase chain reaction (PCR), real-time PCR, and DNA sequencing	Pathogens (bacteria, fungi, viruses, and microorganisms)	Plant pathogen detection
2	DNA sequencing and genotyping	Genetic traits (disease resistance, increased productivity, superior nutrition, or stress tolerance)	Genetic Trait Analysis
3	DNA fingerprinting or DNA barcoding	Plant varieties or cultivars for rights to intellectual property, plant breeding, and certification schemes	Plant Variety Identification
4	PCR	Presence of transgenes or genetic modifications (enforce laws, guarantee labeling adherence, and evaluate the environmental effect)	Genetically Modified Organism
5	Gene marker	Analyzing genetic markers	Plant Disease Resistance Testing
6	DNA barcoding	Plant genome (Identification of plant species, research on biodiversity, conservation initiatives, and surveillance of invasive species)	Plant DNA Barcoding
7	DNA sequencing	Quality control and authentication of plant-derived products	Quality control in plant products

certain DNA region in the shortest period (Kreitmann et al. 2023). The particular oligonucleotide sequences known as primers must bind to the template DNA for PCR to be successful. Specific primers for pathogen nucleic acids are capable of amplifying in PCR and show the existence of the infectious agent for sickness diagnosis. Strong reliance on the specificity of the primers and cross-contamination in the material might cause erroneous findings even though PCR is a quick and inexpensive procedure (Lau and Botella 2017). Several PCR techniques may be used to identify the agent causing plant diseases, including nested PCR (Green and Sambrook 2019), cooperative PCR (Drogue et al. 2014), multiplex PCR (Elnifro et al. 2000), and DNA fingerprinting (Nybom et al. 2014). To overcome the limitations of PCR methods, other amplification methods have been created, but they too have their advantages and disadvantages. Some of these include recombinase polymerase amplification (Babu et al. 2018), rolling circle amplification (Jeske 2018), loop-mediated isothermal amplification (Focke et al. 2013), and helicase-dependent amplification (Serhan et al. 2019). In addition to PCR techniques, a good approach for detecting numerous pathogens that are infecting the same plant concurrently and causing complicated disease signs is through DNA microarray. In a microarray, fluorescent dye-labeled DNA fragments unique to the pathogen are hybridized with probes to identify the infection (Serhan et al. 2019). Pathogen detection has improved with the development of bioinformatic evaluation along with DNA sequencing methods.

Fig. 1 Techniques used in molecular diagnostics

5.2 RNA-Based Approaches

RNA serves as the genetic material for many viruses that infect crops and cause illnesses; using PCR-based methods, identification of these viruses by transforming RNA into DNA can be possible. Complementary DNA (cDNA) is produced using RT-PCR. To determine whether fungi are still alive, RT-PCR can be useful (Aslam et al. 2017). In the natural world, dying cells quickly break down mRNA, and it cannot be transformed into cDNA. Currently, real-time PCR (qRT-PCR) is the accepted gold standard method for identifying plant infections. By utilizing a fluorescent signal that intensifies in proportion to the number of amplicons generated and the number of pathogens found in the specimen being studied, this technology enables the screening of the response and the amplification process to be done simultaneously (Tibebu and Nuh 2019). Pathogen load at a certain period may be identified and measured using qRT-PCR. This method has been used to accurately identify several phytopathogenic fungi in the sample (Black 2009). A technique called serial analysis

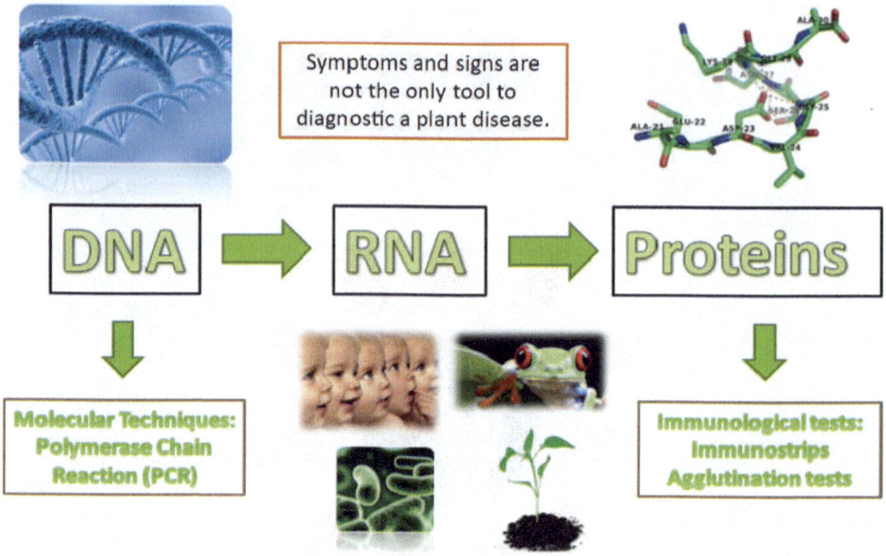

Fig. 2 Approaches for plant disease identification

of gene expression uses a particular gene sequence to quantitatively profile the gene expression of host–pathogen interactions and allows for the immediate detection of many transcripts (Fregene et al. 2004). This technique was first applied to detect the fungus Blumeriagraminis on the leaves of barley (Wang and Peng 2014).

5.3 *Protein-Based Approaches*

Protein-level disease diagnosis is primarily dependent on the specific, sensitive, and precise antigen and antibody response. However, because of how subtle this reaction is, fluorescent dyes must be used for labeling. In the enzyme-linked immunosorbent assay (ELISA), an enzyme associated with the secondary antibody converts the substrate into a colored component as a result of the antigen–antibody interaction (Konstantinou 2017). A particular primary antibody against a protein unique to the pathogen is required for ELISA to work properly in the identification of any infection (Edwards and Cooper 1985). The sensitivity of ELISA is affected by several variables, including the pathogen, the time since the sample was taken, and the pathogen load. As of now, fast detection kits are commercially available, and ELISA processes can employ mono and polyclonal antibodies that were produced for a variety of phytopathogenic fungi and bacteria (Martinelli et al. 2015).

6 Molecular Diagnosis of Plant Virus

One of the main causes of plant illnesses worldwide is plant viruses. Each year, viral infections cause billions of dollars in crop production damages, reducing the quality of plant products, and quantity (Thresh 2006). Amplification and hybridization-based molecular detection techniques for the majority of pathogenic viruses have been steadily produced and regularly reinforced. Diagnostic tools are being used because of the usage of recombinant DNA technologies in plant virology. Recombinant DNA technology has been successfully applied as a diagnostic tool that relies on viral genomic or protein sequences thanks to the development of molecular biology, and numerous molecular detection approaches have also been explored in parallel (Ulubaş Serçe and Ayyaz 2020). Efficient crop disease control requires precise virus disease identification and diagnosis. A vital component of preventing the spread of plant viruses is the creation of quick and accurate diagnostic techniques to confirm that plant material is free of viruses before dissemination.

Nucleic acid-based plant pathogen detection utilizes polymerase chain reaction (PCR) amplification of DNA sections that are particular to a certain pathogen. The bulk of molecular assays currently used rely on the amplification of a particular genome, utilizing the PCR technique (Kalimuthu et al. 2022). Due to its effectiveness in the areas of specificity and sensitivity, PCR has a significant impact on the field of diagnostics. By amplifying a genomic region, a small number of DNA or RNA fragments can be increased to a level that can be detected by the devices (Harshitha and Arunraj 2021). Silva et al. used a reverse transcription-recombinase polymerase amplification (RT-exoRPA)-based method for selective YMV detection which is reported in this paper. It has been demonstrated that this test is reliable and capable of detecting as small as 14 pg/l of pure RNA derived using plants with the YMV virus infection (Silva et al. 2015). Abreu et al. outline two techniques for Papaya meleira virus (PMeV) molecular diagnosis utilizing traditional and real-time PCR. These techniques were demonstrated to be more effective than existing viral detection techniques, which involve the collection of PMeV RNA and observation of numerous signs all through the region (Abreu et al. 2012). Zhang et al. to detect the Vicia cryptic virus M (VCV-M) genome in field samples developed a digoxigenin (DIG)-labeled DNA probe-based Northern blot test. The outcomes demonstrated that VCV-M in V. faba could be precisely and sensitively detected using both serological and nucleic acid assays (Zhang et al. 2022). They also developed a quick isothermal detection of Plum pox virus (PPV) employing reverse transcription-recombinase polymerase amplification, two AmplifyRP tests have been reported. The main enhancement of this assay is its ability to be carried out quickly in lab settings as well as in the field, taking only 20 min from sample preparation to result. The outcomes of this study showed that the AmplifyRP approach could identify all nine PPV strains (Zhang et al. 2014). Little Cherry Virus 2 (LChV2) is the primary contributory factor of little cherry disease (LCD) inside sweet cherry (Prunusavium) across North America along with other parts of the globe, which belongs to the genus Ampelovirus. A high-affinity lateral flow strip was used to detect terminally labeled amplicons. It has been demonstrated that the RT-RPA is easy, quick, and precise. In contrast, it is a more economical process even though RT-PCR's sensitivity is retained (Mekuria et al. 2014). The current diagnostic methods for Rose rosette virus (RRV), such as real-time PCR (RT-qPCR) and end-point reverse transcription-polymerase chain reaction (RT-PCR), are time-consuming as well as extremely sensitive, but they are only available in diagnostic labs with the necessary tools and knowledge. The isothermal reverse transcription-recombinase polymerase amplification (RT-RPA) assay that utilizes several gene domains for the targeted identification of RRV was created to get around this restriction. The assay's high degree of specificity prohibited a reaction with pathogens from the same species (Babu et al. 2017). RT-RPA and real-time RT-exo-RPA assays were used to test banana plants for Cucumber mosaic virus (CMV) infections employing crude extracts of leaves as templates which showed simple sample preparation and a 10 to 20-min run time. The outcomes were then compared with those obtained from an RT-PCR test using extracted total plant RNAs as layouts (Srivastava et al. 2019). For the identification of the Southern rice black-streaked dwarf virus (SRBSDV), an RT-RPA-based approach was created. A set of RPA primers was created to identify the conserved SP10 gene sequences on the genomic RNA S10 of the SRBSDV. The experiment, which took place in just one tube at 39 °C for a duration of 20 min, proved how RPA assay provides a reliable substitute for the quick detection of SRBSDV (Ma et al. 2022).

The Tomato Brown Rugose Fruit Virus (ToBRFV) was shown to produce yellow spots, deformation, and brown rugose necrotic lesions on fruits as well as minor to severe mosaic and blistering on necrosis on sepals, leaves, and necrosis. To distinguish ToBRFV from other economically significant tomato-affecting viruses, a quadruplex RT-PCR method was created. The results of this research improve our knowledge of the biology and molecular properties of ToBRFV and offer a quick and easy way to identify multiple infections (Yan et al. 2021). ToBRFV (tomato and Pepper) were identified by their nucleotide sequences by Fidan et al., and it was discovered that they resembled previously characterized strains of the virus that originated from Israel and Germany. They created an RT-PCR method for virus detection using the nucleotide sequences of the two strains. As part of a plan to keep the virus out of commercial crops, this assay might be employed for the detection of the virus in plant tissues (Fidan et al. 2021). Reverse transcriptase can be added to the LAMP reaction to change it to detect RNA targets. Many other commercially important crops, such as papaya, tomato, sugarcane, potato, banana, tobacco, yam, rice, cassava, and maize, have been infected by RNA and DNA viruses. To identify these viruses, the RT-LAMP and LAMP have been created (Lau and Botella 2017; Hema and Konakalla 2020; Tomlinson and Boonham 2008; Salar et al. 2013). Recently, Mehetre et al. used a nucleic acid base approach for plant virus detection. They show that the field of diagnostics requires new portable tools and technology that can deliver real-time assessments in a short amount of time (Fig. 3) (Mehetre et al. 2021).

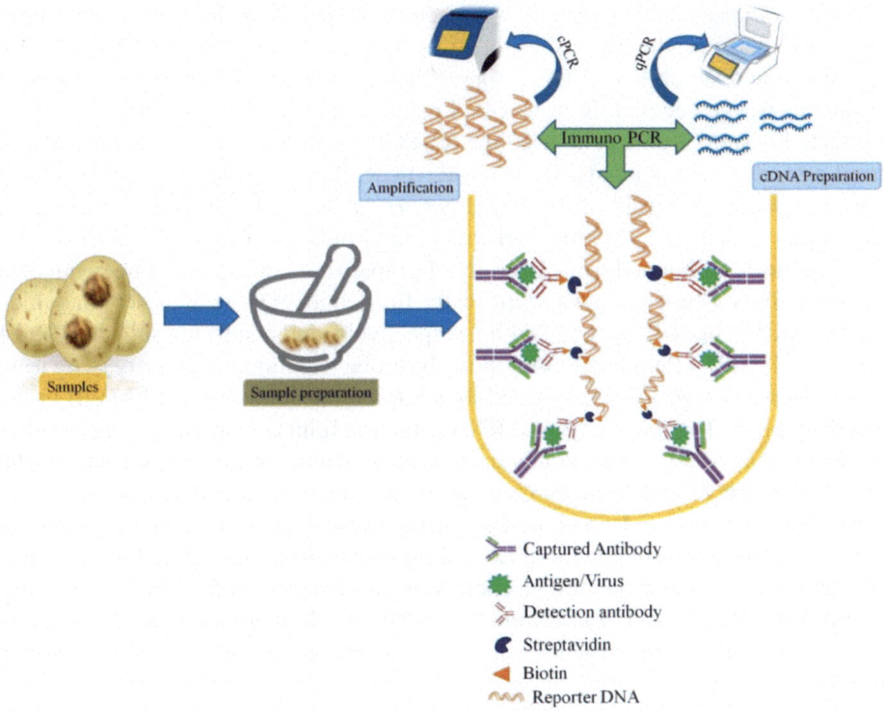

Fig. 3 Schematic representation of Immunocapture PCR (IC-PCR) used in plant viral diagnosis (Mehetre et al. 2021)

7 Molecular Diagnosis of Plant Viroid

Viroids are significant plant diseases, including the *Potato spindle tuber viroid* (PSTVd) and *Chrysanthemum stunt viroid* (CSVd). However, viroids have historically been challenging to identify due to their distinctive biological characteristics (Mumford et al. 2000). The first molecular technique that was employed in the fast (2–3 days) detection of plants with viroid infections was polyacrylamide gel electrophoresis (PAGE). Since PAGE under denaturing conditions is the sole diagnostic procedure that is sequence-independent, it continues to be essential for the detection of newly found viroids (Morris and Smith 1977). Dot-blot hybridization started to take the role of PAGE for the routine detection of PSTVd along with additional viroids at the beginning of the 1980s (Owens and Diener 1981). About 10 years later, the RT-PCR-based diagnostic techniques were developed. The search for group-specific primer pairs and the streamlining of the sample setup approach has received a lot of attention since then (Clementi et al. 1993). After the development of next-generation sequencing (NGS), the discovery of novel viruses, viroids, and hosts has grown quickly. While requiring no prior knowledge of the targeted pathogens,

NGS technologies enable a general approach (non-specific technique) to viral identification that can produce a species- or strain-specific result (Wang and Zhou 2016). The choice and suitability of these approaches for various viroid rely on the need for diagnosis, the accessibility of skilled persons, and the level of virosome titers in infected tissues. Kunta et al. used the RT-PCR technique for the identification of viroid present in citrus plants. By using RT-PCR, they managed to identify, clone, as well as sequence viroid of the *Hop stunt viroid* (HSVd), *Citrus exocortis viroid* (CEVd), *Citrus viroid-IV* (Citrus bark cracking viroid), and *Citrus viroid-III* (Citrus dwarfing viroid), obtained from a variety of grapefruit plants that had been infected with the viroids. Sequencing was done on the full-length cDNAs of these four, which had 370 bp, 303 bp, 295 bp, and 284 bp, respectively, and results are shown in Fig. 4 (Kunta et al. 2007). Peiro et al. used the poly-probe technique to identify two viroid namely *Hop stunt viroid* (HSVd) and *Peach latent mosaic viroid* (PLMVd) in the stone fruit plant. The range of viroid RNA detection limits recorded in all riboprobes was between 12.8 and 2.5 picograms and the standard temperature was set between 60 to 68 °C (Peiró et al. 2012). Ioanna et al. used one-step multiplex RT-PCR techniques for the detection of viroids in pome fruit plants. Results prove that the proposed test has the potential to be utilized for quick and extensive viroid screening. Using nucleic acid standards, the multiplex test had detection efficiency of 99.5%, 98.7%, and 98% for PBCVd, ASSVd, and "Ca. *P. mali*," respectively (Malandraki et al. 2015). Here Table 2 summarizes some plant viroid, their target, and the plant used for viroid's detection.

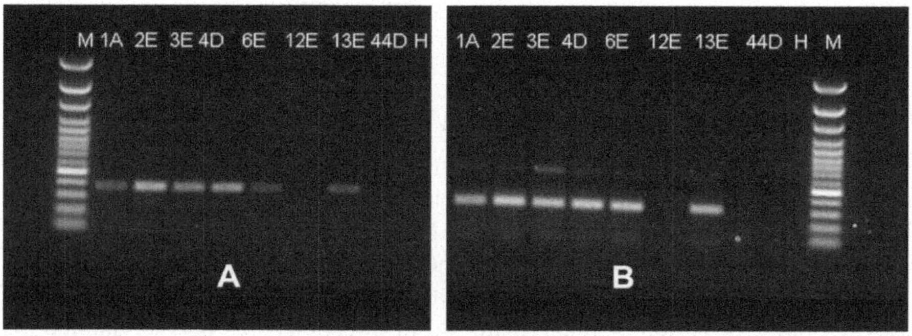

Fig. 4 Agarose gel showing reverse transcriptase-polymerase chain reaction products amplified with specific primers for **a** CEVd and **b** HSVd from viroid-infected grapefruit samples. 44D = noninoculated control; H = healthy (−); M = 100 bp DNA ladder (Kunta et al. 2007)

Table 2 Techniques used for plant viroid's detection and no. of targets

Plant type	Techniques	Viroid type	No. of targets	References
Citrus	Poly-probe, Dot-blot analyzation	*Citrus exocortis, bent leaf, hop stunt viroids, CVd-III*	Three	Jiang et al. (2007)
Stone fruit	Poly-probe dot-blot hybridization	HSVd and PLMVd	Two	Peiró et al. (2012)
Pome and stone fruit	Dot-blot Poly-probe, RT-PCR	*Apple Dimple Fruit Viroid (ADFVd), HSVd, Apple Scar Skin Viroid (ASSVd), Pear Blister Canker Viroid (PBCVd), Apple Fruit Crinkle Viroid (AFCVd), and PLMVd*	Six	Jiang et al. (2007)
Grapevine	Dot-blot hybridization (poly-probe)	*Grapevine yellow speckle viroid-1 and 2* (GYSV-1, 2), *Australian grapevine viroid* (AGVd), and HSVd	Four	Zhang et al. (2012)
Ornamental and vegetables	Dot-blot and northern blot hybridization	*Columnea latent, Mexican papita, Tomato apical stunt, Pepper chat fruit, Potato spindle tuber, Citrus exocortis, Chrysanthemum stunt, and Tomato planta macho viroid*	Eight	Torchetti et al. (2012)
Tomato	Micro-assay	*Tomato spotted wilt viroid* (TSWV)	One	Tiberini and Barba (2012)
Vitis vinifera	PCR	*Grapevine yellow speckle viroid 1*	One	Ward et al. (2011)
Prunus persica	SYM, IND, PCR	*Peach latent mosaic viroid* (Peach calico)	One	Guy (2013)
Apple	RT-PCR	*Apple scar skin viroid (ASSVd)*	One	Jiao et al. (2021)
Eggplant	S-page, northern blot hybridization	*Eggplant latent viroid*	One	Fadda et al. (2003)
Chrysanthemum	LAMP	*chlorotic mottle viroid*	One	Park et al. (2013)
Pear	Dot-blot hybridization	*Pear blister cankerviroid*	One	Joyce et al. (2006)
African oil palm	RT-LAMP	*Cocadviroid Coconut cadang-cadang viroid*	One	Thanarajoo et al. (2014)

(continued)

Table 2 (continued)

Plant type	Techniques	Viroid type	No. of targets	References
Grapefruit	S-Page RT-PCR	*Citrus barkcracking viroid*	One	Puchta et al. (1991)
Coleus (Plectranthusscutellarioides)	RT-qPCR	*Coleviroid Coleus blumeiviroids*	One	Botermans et al. (2020)
Columneaerythrophae	PAGE	*Columnea latent viroid*	One	Hammond et al. (1989)

8 Molecular Diagnosis of Plant Phytoplasma

Phytoplasmas in plants live in the cytoplasm of phloem sieve cells and spread throughout the plant by passing through the pores of the sieve plates that separate the sieve cells. Phytoplasmas seem to build up mostly in sink tissues, such as developing leaves and flowers, roots, and leaves. Numerous symptoms, including witches' brooms, phyllody, virescence, and altered volatile production, are displayed by infected plants (Sugio and Hogenhout 2012). Phytoplasmas are obligatory parasite bacteria that globally affect both low- and high-value annual and perennial crops, resulting in economically significant yield losses. Despite their importance, research into phytoplasma-host interactions and the development of efficient control methods have been slow since it is difficult to isolate these infections on culture media (Lee et al. 2000; Studiorum 2009). Bulgari et al. did co-localization of phytoplasmas and endophytic bacteria was detected using a quick and effective fluorescence in situ hybridization assay in nonembedded tissues of the model plant Catharanthus roseus, providing fresh insights into the relationship between these microorganisms (Bulgari et al. 2011). Buoso et al. analyzed modifications in macro- and micronutrient concentration in tomato plants after "CandidatusPhytoplasmasolani" infection. The mineral elements were independently studied in the leaf midrib, leaf lamina, and root to look into the potential impacts of infection on mineral element allocation (Buoso et al. 2022).

Cassava Witches' Broom Disease (CWB) is a phytoplasma infection that results in a 10 and 30% decrease in harvestable tuber yield and starch content, respectively. Researchers created a LAMP-based technique that provides precise phytoplasma detection from samples that were collected in the field. The assay quickly distinguished between CWB-infected and healthy cassava (Vu et al. 2016). Witches' broom disease of acid lime (WBDL), caused by phytoplasma, is detectable using PCR-based molecular techniques. In the current investigation, a phytoplasma-primer set was utilized for the initial round of PCR, and the amplified products were each subjected to three distinct nested primer pairs. All possible specimens were found to contain phytoplasma, and one reference amplicon was sequenced and added to

GenBank as a Candidatus Phytoplasma species (Ghosh et al. 2013). The Yellow Leaf Disease, which affects areca nut palms and has a phytoplasma etiology, is a devastating illness that significantly reduces yield. Using effective real-time PCR primers, a real-time PCR method utilized in the detection of phytoplasma that causes areca nut yellow leaf disease was developed to address the shortcomings of conventional PCR. The sequences identified 99% of their nucleotides with phytoplasma sequences known to cause Indian areca nut yellow leaf disease (Nair et al. 2014). The earlier identified phytoplasma strains of "Candidatus *P. asteris* related to Catharanthus leaf yellows" were analyzed using an entirely novel set of assays for the gyrA and gyrB genes within the phytoplasma gene and were effectively identified and taxonomically categorized. As promising genetic detection techniques for phytoplasma strains in plants and insects, these genes can be further investigated and used for the discovery of new phytoplasma groups infecting significant crops (Madhupriya et al. 2019). By the identification of the 16S rRNA sequence of phytoplasmas, diagnostic LAMP-based techniques for the quick detection of specific categories of phytoplasmas within 30 min were created. Using a real-time fluorometer with direct visualization of the amplification plots and annealing curves, the LAMP experiments were conducted. In comparison to traditional methods, assays based on LAMP for the in-field identification of sesame phytoplasmas that cause phyllody showed great upper hand and prospective use (Quoc et al. 2021). Real-time LAMP-based diagnostics for areca nut yellow leaf disease and phytoplasma 16S rDNA targeted LAMP were created by Nair et al. The samples that tested positive for phytoplasma formed the expected ladder-like bands on the agarose gel, displayed a color shift from violet to blue using hydroxy naphthol blue, and exhibited a prominent annealing peak at 850.5 °C in the live detection method (Nair et al. 2016). An effective and efficient nested PCR-based technique was created for the detection of related phytoplasmas to micropropagate phytoplasma-free Withaniasomnifera plants. In direct/nested PCR, universal primers created using phytoplasma 16S rDNA sequences were used. As opposed to the plants that looked healthy, phytoplasma was found in all the sickly plants. Additionally, it was quick enough to amplify phytoplasma DNA isolated from pure DNA of naturally infected plants diluted up to 2500 times, as well as from different growth stages of diseased plants produced in vitro (Khan et al. 2006). The development of a LAMP-based phytoplasma detection test focused on the groEL gene and 16S rDNA. It was demonstrated that a range of phytoplasma strains could use the groEL genome as a gene of interest for LAMP assays by the primer set's successful amplification of the groEL gene in multiple Candidatus Phytoplasma asteris strains (Sugawara et al. 2012).

9 Conclusion

Plant viruses, viroids, and phytoplasma molecular diagnostics are essential for maintaining natural ecosystems, defending global agriculture, and assuring food security. This chapter has emphasized the importance of these cutting-edge diagnostic methods in the rapid and precise identification of infectious agents that enable the prevention of disease epidemics. Our comprehension of viral, viroidal, and phytoplasma diversity, epidemiology, and evolution is further improved by the combination of molecular diagnostics with developments in bioinformatics and data analysis. This information is crucial for creating efficient management plans and creating resistant crop varieties through breeding initiatives. We can create a robust and fruitful future for global plant health by adopting cutting-edge technologies and encouraging teamwork.

References

Abreu PMV, Piccin JG, Rodrigues SP, Buss DS, Ventura JA, Fernandes PMB (2012) Molecular diagnosis of *Papaya meleira* virus (PMeV) from leaf samples of *Carica papaya* L. using conventional and real-time RT-PCR. J Virol Methods 180:11–17. https://doi.org/10.1016/j.jviromet.2011.12.003

Aslam S, Tahir A, Aslam MF, Alam MW, Shedayi AA, Sadia S (2017) Recent advances in molecular techniques for the identification of phytopathogenic fungi—a mini review. J Plant Interact 12:493–504. https://doi.org/10.1080/17429145.2017.1397205

Babu B, Washburn BK, Miller SH, Poduch K, Sarigul T, Knox GW, Ochoa-Corona FM, Paret ML (2017) A rapid assay for detection of Rose rosette virus using reverse transcription-recombinase polymerase amplification using multiple gene targets. J Virol Methods 240:78–84. https://doi.org/10.1016/j.jviromet.2016.11.014

Babu B, Ochoa-Corona FM, Paret ML (2018) Recombinase polymerase amplification applied to plant virus detection and potential implications. Anal Biochem 546:72–77. https://doi.org/10.1016/j.ab.2018.01.021

Black J (2009) Quantitative real-time polymerase chain reaction (qPCR) of filamentous fungi in carpet. https://doi.org/10.3768/rtipress.2009.mr.0011.0909

Botermans M, Roenhorst JW, Hooftman M, Verhoeven JTJ, Metz E, van Veen EJ, Geraats BPJ, Kemper M, Beugelsdijk DCM, Koenraadt H (2020) Development and validation of a real-time RT-PCR test for screening pepper and tomato seed lots for the presence of pospiviroids. PLoS ONE 15:e0232502

Broadley M, Brown P, Cakmak I, Rengel Z, Zhao F (2011) Function of nutrients: micronutrients. https://doi.org/10.1016/B978-0-12-384905-2.00007-8

Bulgari D, Casati P, Faoro F (2011) Fluorescence in situ hybridization for phytoplasma and endophytic bacteria localization in plant tissues. J Microbiol Methods 87:220–223. https://doi.org/10.1016/j.mimet.2011.08.001

Buoso S, Musetti R, Marroni F, Calderan A, Schmidt W, Santi S (2022) Infection by phloem-limited phytoplasma affects mineral nutrient homeostasis in tomato leaf tissues. J Plant Physiol 271:153659. https://doi.org/10.1016/j.jplph.2022.153659

Chakraborty S, Newton AC (2011) Climate change, plant diseases and food security: an overview. Plant Pathol 60:2–14. https://doi.org/10.1111/j.1365-3059.2010.02411.x

Choudhury FK, Rivero RM, Blumwald E, Mittler R (2017) Reactive oxygen species, abiotic stress and stress combination. Plant J 90:856–867. https://doi.org/10.1111/tpj.13299

Chun S, Gopal J, Muthu M (2022) A consolidative synopsis of the MALDI-TOF MS accomplishments for the rapid diagnosis of microbial plant disease pathogens. TrAC—Trends Anal Chem 156:116713. https://doi.org/10.1016/j.trac.2022.116713

Clementi M, Menzo S, Bagnarelli P, Manzin A, Valenza A, Varaldo PE (1993) Quantitative PCR and RT-PCR in virology. Genome Res 2:191–196. https://doi.org/10.1101/gr.2.3.191

Debnath M, Prasad GBKS, Bisen PS (2010) Molecular diagnostics: promises and possibilities. Springer Science & Business Media

Doddaraju P, Kumar P, Gunnaiah R, Gowda AA, Lokesh V, Pujer P, Manjunatha G (2019) Reliable and early diagnosis of bacterial blight in pomegranate caused by *Xanthomonas axonopodis* pv. *punicae* using sensitive PCR techniques. Sci Rep 9:1–9. https://doi.org/10.1038/s41598-019-46588-9

Donoso A, Valenzuela S (2018) In-field molecular diagnosis of plant pathogens: recent trends and future perspectives. Plant Pathol 67:1451–1461

Drogue B, Sanguin H, Chamam A, Mozar M, Llauro C, Panaud O, Prigent-Combaret C, Picault N, Wisniewski-Dyé F (2014) Plant root transcriptome profiling reveals a strain-dependent response during Azospirillum-rice cooperation, Front. Plant Sci 5:1–14. https://doi.org/10.3389/fpls.2014.00607

Edwards ML, Cooper JI (1985) Plant virus detection using a new form of indirect ELISA. J Virol Methods 11:309–319

Elnifro EM, Ashshi AM, Cooper RJ, Klapper PE (2000) Multiplex PCR: optimization and application in diagnostic virology. Clin Microbiol Rev 13:559–570. https://doi.org/10.1128/CMR.13.4.559-570.2000

Fadda Z, Daros JA, Fagoaga C, Flores R, Durán-Vila N (2003) Eggplant latent viroid, the candidate type species for a new genus within the family Avsunviroidae (hammerhead viroids). J Virol 77:6528–6532

Fidan H, Sarikaya P, Yildiz K, Topkaya B, Erkis G, Calis O (2021) Robust molecular detection of the new Tomato brown rugose fruit virus in infected tomato and pepper plants from Turkey. J Integr Agric 20:2170–2179. https://doi.org/10.1016/S2095-3119(20)63335-4

Focke F, Haase I, Fischer M (2013) Loop-mediated isothermal amplification (LAMP): methods for plant species identification in food

Fregene M, Matsumura H, Akano A, Dixon A, Terauchi R (2004) Serial analysis of gene expression (SAGE) of host-plant resistance to the cassava mosaic disease (CMD). Plant Mol Biol 56:563–571. https://doi.org/10.1007/s11103-004-3477-8

Ghosh D, Bhose S, Manimekalai R, Gowda S (2013) Molecular detection of *Candidatus* Phytoplasma spp. causing witches' broom disease of acid lime (*Citrus aurantifolia*) in India. J Plant Biochem Biotechnol 22:343–347. https://doi.org/10.1007/s13562-012-0160-z

Gimenez E, Salinas M, Manzano-Agugliaro F (2018) Worldwide research on plant defense against biotic stresses as improvement for sustainable agriculture. Sustainability 10:1–19. https://doi.org/10.3390/su10020391

Green MR, Sambrook J (2019) Nested polymerase chain reaction (PCR). Cold Spring Harb Protoc

Guy PL (2013) Ancient RNA? RT-PCR of 50-year-old RNA identifies peach latent mosaic viroid. Arch Virol 158:691–694

Hammond R, Smith DR, Diener TO (1989) Nucleotide sequence and proposed secondary structure of Columnea latent viroid: a natural mosaic of viroid sequences. Nucl Acids Res 17:10083–10094

Hančinský R, Mihálik D, Mrkvová M, Candresse T, Glasa M (2020) Plant viruses infecting solanaceae family members in the cultivated and wild environments: a review. Plants (2020). https://doi.org/10.3390/plants9050667

Harshitha R, Arunraj DR (2021) Real-time quantitative PCR: a tool for absolute and relative quantification. Biochem Mol Biol Educ 49:800–812. https://doi.org/10.1002/bmb.21552

Hema M, Konakalla NC (2020) Recent developments in detection and diagnosis of plant viruses. Elsevier Inc. https://doi.org/10.1016/B978-0-12-821406-0.00016-3

Ingram J (2011) A food systems approach to researching food security and its interactions with global environmental change. Food Secur 3:417–431. https://doi.org/10.1007/s12571-011-0149-9

Jeske H (2018) Barcoding of plant viruses with circular single-stranded DNA based on rolling circle amplification. Viruses 10. https://doi.org/10.3390/v10090469

Jiang W, Jiang P, Li Y, Wang X, Du Y (2007) Analysis of immunogenicity of minor envelope protein GP3 of porcine reproductive and respiratory syndrome virus in mice. Virus Genes 287–292. https://doi.org/10.1007/s11262-006-0067-7

Jiao J, Kong K, Han J, Song S, Bai T, Song C, Wang M, Yan Z, Zhang H, Zhang R, Feng J, Zheng X (2021) Field detection of multiple RNA viruses/viroids in apple using a CRISPR/Cas12a-based visual assay. Plant Biotechnol J 19:394–405. https://doi.org/10.1111/pbi.13474

Joyce PA, Constable FE, Crosslin J, Eastwell K, Howell WE, Rodoni BC (2006) Characterisation of Pear blister canker viroid isolates from Australian pome fruit orchards, Australas. Plant Pathol 35:465–471

Kalimuthu K, Arivalagan J, Mohan M, Samuel Selvan Christyraj JR, Arockiaraj J, Muthusamy R, Ju HJ (2022) Point of care diagnosis of plant virus: current trends and prospects. Mol Cell Probes 61:101779. https://doi.org/10.1016/j.mcp.2021.101779

Khaled AY, Abd Aziz S, Bejo SK, Nawi NM, Seman IA, Onwude DI (2018) Early detection of diseases in plant tissue using spectroscopy–applications and limitations. Appl Spectrosc Rev 53:36–64. https://doi.org/10.1080/05704928.2017.1352510

Khan JA, Srivastava P, Singh SK (2006) Sensitive detection of a phytoplasma associated with little leaf symptoms in *Withania somnifera*. Eur J Plant Pathol 115:401–408. https://doi.org/10.1007/s10658-006-9029-0

Khater M, de la Escosura-Muñiz A, Merkoçi A (2017) Biosensors for plant pathogen detection. Biosens Bioelectron 93:72–86. https://doi.org/10.1016/j.bios.2016.09.091

Konstantinou GN (2017) Enzyme-linked immunosorbent assay (ELISA). Food Allergens Methods Protoc 79–94

Kreitmann L, Miglietta L, Xu K, Malpartida-Cardenas K, D'Souza G, Kaforou M, Brengel-Pesce K, Drazek L, Holmes A, Rodriguez-Manzano J (2023) Next-generation molecular diagnostics: Leveraging digital technologies to enhance multiplexing in real-time PCR. TrAC Trends Anal Chem 160. https://doi.org/10.1016/j.trac.2023.116963

Kumar D, Ohri P (2023) Say "NO" to plant stresses: Unravelling the role of nitric oxide under abiotic and biotic stress. Nitric Oxide Biol Chem 130:36–57. https://doi.org/10.1016/j.niox.2022.11.004

Kunta M, Da Graça JV, Skaria M (2007) Molecular detection and prevalence of citrus viroids in Texas. Hort Sci 42:600–604. https://doi.org/10.21273/hortsci.42.3.600

Lau HY, Botella JR (2017) Advanced DNA-based point-of-care diagnostic methods for plant diseases detection, Front. Plant Sci 8:1–14. https://doi.org/10.3389/fpls.2017.02016

Lee I, Davis RE, Dawn E (2000) Hytoplasma: phytopathogenic mollicutes

Lévesque CA (2001) Molecular methods for detection of plant pathogens—what is the future? Can J Plant Pathol 23:333–336. https://doi.org/10.1080/07060660109506953

Louws FJ, Rademaker JL, De Bruijn FJ (1999) The three Ds of PCR-based genomic analysis of phytobacteria: diversity, detection, and disease diagnosis. Annu Rev Phytopathol 1:81–125

Ma W, Duan X, Tian Y, Jiao Z, Sun X, Zhao Z, Zhou T, Fan Z (2022) Recombinase polymerase amplification assay for sensitive and rapid detection of southern rice black-streaked dwarf virus in plants. J Virol Methods 301:114467. https://doi.org/10.1016/j.jviromet.2022.114467

Madhupriya AK, Rao GP, Khurana SMP (2019) Development and utilization of gyrA and gyrB gene-based diagnostics for the phytoplasma classified under 16Sr I group in plants and insects. Biotech 9:1–4. https://doi.org/10.1007/s13205-019-1706-8

Malandraki I, Varveri C, Olmos A, Vassilakos N (2015) One-step multiplex quantitative RT-PCR for the simultaneous detection of viroids and phytoplasmas of pome fruit trees. J Virol Methods 213:12–17. https://doi.org/10.1016/j.jviromet.2014.11.010

Mancini V, Murolo S, Romanazzi G (2016) Diagnostic methods for detecting fungal pathogens on vegetable seeds. Plant Pathol 65:691–703. https://doi.org/10.1111/ppa.12515

Martinelli F, Scalenghe R, Davino S, Panno S, Scuderi G, Ruisi P, Villa P, Stroppiana D, Boschetti M, Goulart LR, Davis CE, Dandekar AM (2015) Advanced methods of plant disease detection. A review. Agron Sustain Dev 35:1–25. https://doi.org/10.1007/s13593-014-0246-1

Mehetre GT, Leo VV, Singh G, Sorokan A, Maksimov I, Yadav MK, Upadhyaya K, Hashem A, Alsaleh AN, Dawoud TM, Almaary KS, Singh BP (2021) Current developments and challenges in plant viral diagnostics: a systematic review. Viruses 13:1–31. https://doi.org/10.3390/v13030412

Mekuria TA, Zhang S, Eastwell KC (2014) Rapid and sensitive detection of Little cherry virus 2 using isothermal reverse transcription-recombinase polymerase amplification. J Virol Methods 205:24–30. https://doi.org/10.1016/j.jviromet.2014.04.015

Miller SA, Martin RR (1988) Molecular diagnosis of plant disease. Annu Rev Phytopathol 26:409–432

Morris TJ, Smith EM (1977) Potato spindle tuber disease: procedures for the detection of viroid RNA and certification of disease-free potato tubers. Phytopathology 67:145–150

Mumford RA, Walsh K, Boonham N (2000) A comparison of molecular methods for the routine detection of viroids. EPPO Bull 30:431–435

Nair S, Roshna OM, Soumya VP, Hegde V, Suresh Kumar M, Manimekalai R, Thomas GV (2014) Real-time PCR technique for detection of arecanut yellow leaf disease phytoplasma, Australas. Plant Pathol 43:527–529. https://doi.org/10.1007/s13313-014-0278-7

Nair S, Manimekalai R, Ganga Raj P, Hegde V (2016) Loop mediated isothermal amplification (LAMP) assay for detection of coconut root wilt disease and arecanut yellow leaf disease phytoplasma. World J Microbiol Biotechnol 32:1–7. https://doi.org/10.1007/s11274-016-2078-4

Nybom H, Weising K, Rotter B (2014) DNA fingerprinting in botany: past, present, future. Investig Genet 5:1–35. https://doi.org/10.1186/2041-2223-5-1

Owens RA, Diener TO (1981) Sensitive and rapid diagnosis of potato spindle tuber viroid disease by nucleic acid hybridization. Science 213:670–672

Pandey P, Irulappan V, Bagavathiannan MV, Senthil-Kumar M (2017) Impact of combined abiotic and biotic stresses on plant growth and avenues for crop improvement by exploiting physio-morphological traits. Front Plant Sci 8:1–15. https://doi.org/10.3389/fpls.2017.00537

Park J, Jung Y, Kil E-J, Kim J, Tran DT, Choi S-K, Yoon J-Y, Cho WK, Lee S (2013) Loop-mediated isothermal amplification for the rapid detection of Chrysanthemum chlorotic mottle viroid (CChMVd). J Virol Methods 193:232–237

Patrinos GP, Ansorge W (2009) Molecular diagnostics. Academic Press

Patrinos GP, Danielson PB, Ansorge WJ (2017) Molecular diagnostics: past, present, and future. Elsevier Ltd. https://doi.org/10.1016/B978-0-12-802971-8.00001-8

Paul R, Ostermann E, Chen Y, Saville AC, Yang Y, Gu Z, Whitfield AE, Ristaino JB, Wei Q (2021) Integrated microneedle-smartphone nucleic acid amplification platform for in-field diagnosis of plant diseases. Biosens Bioelectron 187:113312. https://doi.org/10.1016/j.bios.2021.113312

Peiró A, Pallás V, Sánchez-Navarro JÁ (2012) Simultaneous detection of eight viruses and two viroids affecting stone fruit trees by using a unique polyprobe. Eur J Plant Pathol 132:469–475. https://doi.org/10.1007/s10658-011-9893-0

Prokopy RJ, Owens ED (1983) Visual detection of plants by herbivorous insects. Annu Rev Entomol 28:337–364. https://doi.org/10.1146/annurev.en.28.010183.002005

Puchta H, Ramm K, Luckinger R, Hadas R, Bar-Joseph M, Snger HL (1991) Primary and secondary structure of citrus viroid IV (CVd IV), a new chimeric viroid present in dwarfed grapefruit in Israel. Nucl Acids Res 19:6640

Quoc NB, Xuan NTT, Nghiep NM, Phuong NDN, Linh TB, Chau NNB, Chuong NDX, Nien NC, Dickinson M (2021) Loop-mediated isothermal amplification (LAMP) assay for detection of sesame phyllody phytoplasmas in Vietnam. Folia Microbiol (Praha) 66:273–283. https://doi.org/10.1007/s12223-020-00842-0

Salar RK, Siwach P, Gahlawat SK, Duhan JS (2013) Biotechnology: prospects and applications. Biotechnol Prospect Appl 1–315. https://doi.org/10.1007/978-81-322-1683-4

Savary S, Ficke A, Aubertot JN, Hollier C (2012) Crop losses due to diseases and their implications for global food production losses and food security. Food Secur 4:519–537. https://doi.org/10.1007/s12571-012-0200-5

Serhan M, Sprowls M, Jackemeyer D, Long M, Perez ID, Maret W, Tao N, Forzani E (2019) Total iron measurement in human serum with a smartphone. In: AIChE annual meeting, conference proceedings. https://doi.org/10.1039/x0xx00000x

Silva G, Bömer M, Nkere C, Lava Kumar P, Seal SE (2015) Rapid and specific detection of *Yam mosaic* virus by reverse-transcription recombinase polymerase amplification. J Virol Methods 222:138–144. https://doi.org/10.1016/j.jviromet.2015.06.011

Srivastava N, Kapoor R, Kumar R, Kumar S, Kumar S, Baranwal VK (2019) Rapid diagnosis of Cucumber mosaic virus in banana plants using a fluorescence-based real-time isothermal reverse transcription-recombinase polymerase amplification assay. J Virol Methods 270:52–58. https://doi.org/10.1016/j.jviromet.2019.04.024

Srivastava A, Srivastava S, Gupta T, Rishi N (2021) Next-generation molecular diagnostics for plant viruses. Virus Dis Ornam Plants 397–406. https://doi.org/10.1007/978-981-16-3919-7_22

Strange RN, Scott PR (2005) Plant disease: a threat to global food security. Annu Rev Phytopathol 43:83–116. https://doi.org/10.1146/annurev.phyto.43.113004.133839

Studiorum AM (2009) BERTACCINI and DUDUK 2009:355–378

Sugawara K, Himeno M, Keima T, Kitazawa Y, Maejima K, Oshima K, Namba S (2012) Rapid and reliable detection of phytoplasma by loop-mediated isothermal amplification targeting a housekeeping gene. J Gen Plant Pathol 78:389–397. https://doi.org/10.1007/s10327-012-0403-9

Sugio A, Hogenhout SA (2012) The genome biology of phytoplasma: modulators of plants and insects. Curr Opin Microbiol 15:247–254. https://doi.org/10.1016/j.mib.2012.04.002

Tewari S, Sharma S (2019) Molecular techniques for diagnosis of bacterial plant pathogens. Elsevier Inc. https://doi.org/10.1016/B978-0-12-814849-5.00027-7

Thanarajoo SS, Kong LL, Kadir J, Lau WH, Vadamalai G (2014) Detection of Coconut cadang-cadang viroid (CCCVd) in oil palm by reverse transcription loop-mediated isothermal amplification (RT-LAMP). J Virol Methods 202:19–23

Thresh JM (2006) Crop viruses and virus diseases. 9–32

Tibebu B, Nuh B (2019) Biotechnological tools for detection, identification and management of plant diseases. Afr J Biotechnol 18:797–807. https://doi.org/10.5897/ajb2018.16591

Tiberini A, Barba M (2012) Optimization and improvement of oligonucleotide microarray-based detection of tomato viruses and pospiviroids. J Virol Methods 185:43–51. https://doi.org/10.1016/j.jviromet.2012.05.028

Tomlinson J, Boonham N (2008) Potential of LAMP for detection of plant pathogens. PCAB Rev Perspect Agric Vet Sci Nutr Nat Resour 3:1–7. https://doi.org/10.1079/PAVSNNR20083066

Torchetti EM, Navarro B, Di Serio F (2012) A single polyprobe for detecting simultaneously eight pospiviroids infecting ornamentals and vegetables. J Virol Methods 186:141–146. https://doi.org/10.1016/j.jviromet.2012.08.005

Ulubaş Serçe Ç, Ayyaz M (2020) Diagnosis of the casual viruses of crop plants. Appl Plant Virol Adv Detect Antivir Strateg 55–67. https://doi.org/10.1016/B978-0-12-818654-1.00004-9

Vu NT, Pardo JM, Alvarez E, Le HH, Wyckhuys K, Nguyen KL, Le DT (2016) Establishment of a loop-mediated isothermal amplification (LAMP) assay for the detection of phytoplasma-associated cassava witches' broom disease. Appl Biol Chem 59:151–156. https://doi.org/10.1007/s13765-015-0134-7

Wang D, Peng Y (2014) Assessing the impact of dominant sequencing-based gene expression profiling techniques (SGEPTs) on phytopathogenic fungi. Chiang Mai J Sci 41:922–944

Wang A, Zhou X (2016). Curr Res Top Plant Virol. https://doi.org/10.1007/978-3-319-32919-2

Ward LI, Burnip GM, Liefting LW, Harper SJ, Clover GRG (2011) First report of grapevine yellow speckle viroid 1 and hop stunt viroid in grapevine (*Vitis vinifera*) in New Zealand. Plant Dis 95:617

Yan Z, Zhao M, Ma H, Liu L, Yang G, Geng C, Tian Y, Li X (2021) Biological and molecular characterization of tomato brown rugose fruit virus and development of quadruplex RT-PCR detection. J Integr Agric 20:1871–1879. https://doi.org/10.1016/S2095-3119(20)63275-0

Zhang Z, Peng S, Jiang D, Pan S, Wang H, Li S (2012) Development of a polyprobe for the simultaneous detection of four grapevine viroids in grapevine plants. Eur J Plant Pathol 132:9–16. https://doi.org/10.1007/s10658-011-9856-5

Zhang S, Ravelonandro M, Russell P, McOwen N, Briard P, Bohannon S, Vrient A (2014) Rapid diagnostic detection of plum pox virus in Prunus plants by isothermal AmplifyRP® using reverse transcription-recombinase polymerase amplification. J Virol Methods 207:114–120. https://doi.org/10.1016/j.jviromet.2014.06.026

Zhang K, Zhuang X, Xu H, Gan H, He Z, Chen J (2022) Development of polyclonal antibodies-based serological methods and a DIG-labelled DNA probe-based molecular method for detection of the Vicia cryptic virus-M in field plants. J Virol Methods 299:114331. https://doi.org/10.1016/j.jviromet.2021.114331

Detection and Identification of Plant Viruses, Viroids, and Phytoplasma Based on High-Throughput Molecular Approaches

Lellapalli Rithesh, Amitha Paul, Malavika Ram Amanthra Keloth, Shimi Jose, and Abhishek Kumar

Abstract Plants in field conditions are subjected to a range of biological and environmental stresses, with plant diseases induced by phytopathogens posing a significant risk to global food security. Consequently, the development of effective diagnostic technologies for early plant disease detection is imperative to ensure agricultural sustainability and safeguard the world's food supply. The advent of molecular diagnostics has revolutionized the diagnosis and profiling of plant viruses, viroids, and phytoplasma, offering swift and precise methods for infectious agent identification, quantification, and characterization. Therefore, this chapter has been aimed to give an in-depth look at a multitude of molecular tools employed in the detection and identification of plant viruses, viroids, and phytoplasma. Molecular techniques such as polymerase chain reaction (PCR), quantitative PCR (qPCR), recombinase polymerase amplification (RPA), loop-mediated isothermal amplification (LAMP), DNA microarray, etc. enable the precise and highly sensitive detection of these phytopathogens, even in asymptomatic plants. In addition, high-throughput sequencing (HTS) and metagenomics provide an extensive view of the pathogen diversity in a given specimen. The use of molecular diagnostics in plant virus, viroid, and phytoplasma management can aid in the early identification and swift reaction to outbreaks, ultimately leading to better disease management and crop protection.

Keywords Food security · PCR · qPCR · High-throughput sequencing (HTS) · NGS

L. Rithesh (✉) · A. Paul · M. R. Amanthra Keloth · S. Jose
Department of Plant Pathology, Kerala Agricultural University, Thiruvananthapuram, Kerala 695522, India
e-mail: rithesh132@gmail.com

A. Kumar
Faculty of Agriculture, Maharishi Markandeshwar (Deemed to Be University), Mullana, Ambala, Haryana 133 207, India

J. Chen et al. (eds.), *Molecular and Biotechnological Tools for Plant Disease Management*, Interdisciplinary Biotechnological Advances,
https://doi.org/10.1007/978-981-97-7510-1_8

1 Introduction

In the intricate network of agricultural ecosystems, the health of plants stands as a cornerstone, influencing not only food security but also the delicate balance of ecological harmony. Amidst the myriad threats that challenge the vitality of crops, the invisible realm of pathogens poses a formidable challenge. Among these silent invaders, viruses, viroids, and phytoplasmas emerge as elusive adversaries capable of wreaking havoc on agricultural landscapes with subtlety and stealth. In particular, the *Tomato Yellow Leaf Curl Virus* (TYLCV) has threatened food security and farmers' livelihoods by causing considerable production losses in tomatoes globally (Jones 2020). Similar to this, citrus greening has spread over the globe due to the bacteria *Candidatus* Liberibacter spp. destroying citrus orchards and necessitating massive eradication efforts in addition to financial losses (Li et al. 2021b).

The escalating global challenges brought about by newly emerging plant diseases, climate change, and the intensification of agricultural practices highlight the need for precise and effective diagnostic tools. Global trade and transportation frequently enable emerging plant diseases, such as the recent outbreak of Panama disease in banana plantations, allowing pathogens to quickly spread across continents (Ristiano et al. 2021). Climate change even aggravates the situation by influencing temperature and precipitation patterns, creating more favorable conditions for the proliferation of certain plant pathogens. Additionally, the intensification of agricultural practices, including monoculture and the use of agrochemicals, can weaken plant defenses and increase the susceptibility of crops to pathogen attacks (Fig. 1).

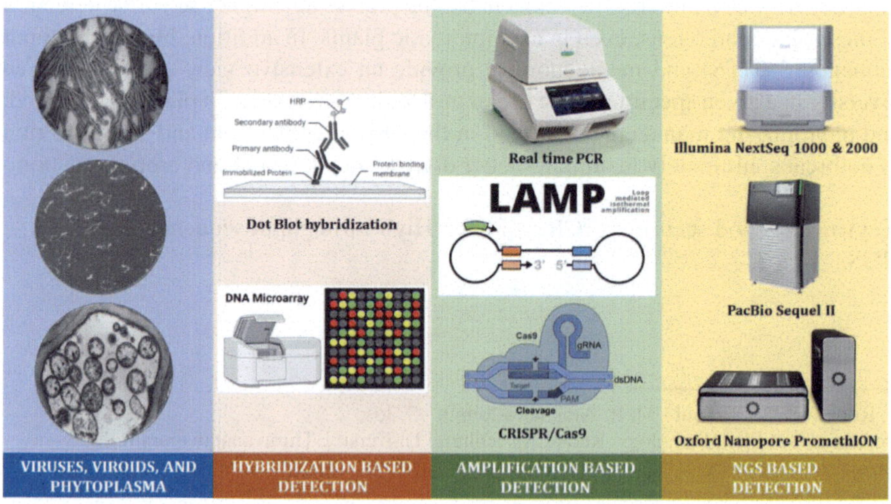

Fig. 1 Molecular diagnostics of viruses, viroids and phytoplasma

The advent of molecular diagnostics has revolutionized our ability to unravel the mysteries of plant-pathogen interactions, offering a powerful lens into the microscopic world that orchestrates both the success and failure of crops. This chapter embarks on a journey into the realm of molecular diagnostics, focusing specifically on the recognition and assessment of plant viruses, viroids, and phytoplasma. It delves into the intricate details of nucleic acid-based techniques, which aim to unveil the secrets hidden within the genetic codes of these minuscule foes.

Many serological and nucleic acid-based techniques are being used for the recognition of plant pathogens. For instance, DNA fingerprinting is a technique that analyzes specific regions of the pathogen's DNA to identify unique patterns, while reverse transcriptase-PCR (RT-PCR) is a method used to amplify and detect viral RNA. Nucleic acid sequence-based amplification (NASBA) is a technique that amplifies RNA sequences without the need for reverse transcription. Other methods, such as recombinase polymerase amplification (RPA) for isothermal amplification, loop-mediated isothermal amplification (LAMP) for rapid DNA amplification, microarray for analyzing multiple targets simultaneously, and macroarray (Dot Blot/Slot Blot) for high-throughput screening, along with next-generation sequencing (NGS), allow for the rapid sequencing of entire pathogen genomes (Martinelli et al. 2015; Yadav and Khurana 2016).

The exploration into molecular diagnostics serves not only as a scientific pursuit but as a crucial endeavor to fortify our defenses against the threats that jeopardize the very foundation of our agricultural systems. By illuminating cutting-edge advancements and the evolving landscape of diagnostic technologies, it aspires to empower researchers, practitioners, and policymakers alike with the knowledge needed to safeguard crops and ensure the resilience of our food systems in the face of ever-evolving biological threats.

2 Amplification Techniques

2.1 Loop-Mediated Isothermal Amplification (LAMP)

LAMP, a recently enhanced gene amplification method, offers a blend of rapidity, simplicity, and increased specificity. The LAMP reaction converts two distinct types of elongation reactions facilitated by the loop regions into a product, which results in the amplification of genes. The elongation reactions start with the template extending from the stem-loop structure that forms at the 3′ end, and then new primers attach to the loop region and extend. Then, a pair of inner as well as outer primers are employed. The sequences carried by each inner primer are identical to the inner region of the same chain at the 5′-terminal and complement one of the chains in the amplification region at the 3′-terminal. The strand-displacement synthesis mediated by DNA polymerase sequentially repeats the elongation reactions, utilizing the stem-loop segments as a stage. The basic idea principle of this method is to make a lot of precise DNA amplification pools that have a complementary sequence in a loop that repeats itself over and over (Notomi et al. 2015).

2.2 *Rolling Circle Amplification (RCA)*

In RCA, a circular DNA molecule is the template for exponentially amplifying the gene of interest through a robust enzymatic process. In RCA, random primers are used, which bind to the circularized DNA template, to initiate copying through DNA polymerase. This results in the production of long tandem repetitions of the original template sequence through a sequence of strand displacement events. The circular format of the template transforms it into an infinite linear template, a key characteristic that amplifies its efficacy. RCA, combined with selective enhancement techniques, has proven to be highly effective in improving the sensitivity and selectivity of current DNA typing systems. The circularization of DNA through RCA potentially enables the amplification of low-copy-number and fragmented DNA (Bhat et al. 2020).

The components of RCA reactions include a round DNA template, Φ29 DNA polymerase, short degenerated primers, and deoxynucleotide triphosphates (dNTPs), which are the distinct units of DNA required to synthesize a complementary sequence of the template. The enzyme Φ29 DNA polymerase is extensively used because of its high processivity. It extends the primer and adds nucleotides to the synthesized strand without interrupting or detaching from the template/synthesized strand combination. Interestingly, Φ29 DNA polymerase also aids in the displacement of downstream DNA strands during synthesis. In RCA, enrichment happens slowly after a single primer hybridization; during the reaction, several of these events happen at the same time, making long molecules that serve as models for an amplification hyperbranching cascade (Ali et al. 2014). The application of RCA contributes to increased sensitivity in detection assays. This is primarily attributed to the substantial increase in target availability for analysis, a particularly advantageous feature when dealing with fragmented or low-copy-number samples (Dean et al. 2001; Maciejewska et al. 2013).

2.3 *Helicase-Dependent Amplification (HDA)*

HDA is an isothermal process for amplification that uses DNA helicases to break the hydrogen bonds in the dsDNA matrix instead of heat, as in the case of thermal PCR. Here, helicase II, an enzyme obtained from *Escherichia coli*, uvrD helicase, is employed to separate the DNA. The uvrD helicase enzymes, along with another protein, the methyl-directed mismatch repair (MutL) protein, stimulate the helicase action and are used for the DNA mismatch repair mechanisms in bacterial systems. The mesophilic HDA (mHDA) technique utilizes this helicase to isothermally unwind the DNA duplex and two accessory proteins, MutL and single-stranded DNA-binding proteins (SSBs), to stabilize the separated strands. The oligonucleotide primers will pair with the 5′ and 3′ ends of complementary strands, and DNA polymerase working at 37 °C will extend the annealed primers with the addition of dNTPs. Later on, another modification of the HDA was developed in combination with a thermostable uvrD helicase obtained from *Thermoanaerobacter tengcongenesis* and the DNA polymerase from *Bacillus stearothermophilus* that functions at higher temperatures (60–65 °C), called thermophilic HDA (tHDA). HDA is particularly useful for situations where maintaining a constant temperature is easier than cycling through different temperatures, making it a suitable tool for field diagnostics (Wu et al. 2016; Barreda-García et al. 2018).

2.4 Recombinase Polymerase Amplification (RPA)

RPA is an isothermal amplification method that works at 37–42 °C. It is a selective and sensitive technique that includes minimal procedure for specimen preparation. It generates 1–10 copies of the DNA target in less than 20 min. RPA is a versatile technique as it can generate multiple targets, such as dsDNA, ssDNA, RNA and miRNA, sourced from a diverse array of organisms and samples. The expanding body of literature detailing RPA applications reflects its increasing popularity, with amplification occurring in the solid state, solution state and bridge amplification formats. Moreover, this technique seamlessly integrates multiple methods of detection, like real-time fluorescent detection and end-point lateral flow strips. The RPA process involves the initiation of the reaction by binding the recombinase protein to the primers with the help of a crowding agent (polyethylene glycol) and adenosine triphosphate (ATP). In order to facilitate invasion of the template strand at the cognate site by the primer, this creates a recombinase-primer complex that actively searches for dsDNA with a homologous sequence. The single-stranded binding proteins take a role in holding the stabilized displaced DNA strand, thus preventing the ejection of the inserted primer during branch migration. The recombinase breaks down at some point, and a DNA-polymerase enzyme that causes strand displacement attaches to the 3′ end of the primer and adds dNTPs to make it longer. The recurrence of this cyclic mechanism leads to exponential amplification (Lobato and O'Sullivan, 2018).

2.5 Polymerase Chain Reaction (PCR)

PCR is the most widely used nucleic acid-based diagnostic method used for the detection of phytoplasmas, plant viruses, and viroids (Table 1). It is a primer-based reaction, which enables the amplification of specific DNA sequences that fall between two known regions of the sequences. A standard PCR includes three basic steps, initially the denaturation step, where the double-stranded DNA unwinds at a high temperature into separated complementary single strands (90–95 °C) followed by annealing of 20–25 bp oligonucleotide primers (50–60 °C) and extension of these primers by a thermostable DNA polymerase at a temperature of 72 °C. These steps are carried out for 30 to 40 cycles. After the termination of each cycle, the copy number of DNA sequences doubles, yielding an exponential amplification of the DNA sequences. Several PCR variations have been developed over time to raise the sensitivity and specificity of conventional PCR. Plant viruses with genetic makeup of both single or double-stranded DNA, RNA and phytoplasmas with double-stranded DNA sequences can be used as a template for the amplification of either partial or complete genome sequences (Varma and Singh 2020).

2.5.1 Reverse Transcriptase (RT-PCR)

Most of the plant viruses have single-stranded RNA. Since conventional PCR amplifies double-stranded DNA, the RNA must first be converted to complementary DNA (cDNA) prior to being used in a PCR. In these circumstances, the reverse transcriptase enzyme, which serves as the PCR template, converts the RNA into cDNA (Mehetre et al. 2021a, b).

2.5.2 Nested PCR

Nested PCR requires two sequential amplification reactions using two sets of primers. An aliquot obtained from the initial PCR amplification is utilized as the template for the upcoming PCR amplification. The template is primed with primer pairs that are specifically designed to anneal internally to the first primer pair. Nested PCR assays are generally employed to reduce nonspecific binding products by the use of degenerate primers. It has been widely used to detect phytoplasma (Nair and Manimekalai 2021).

2.5.3 Multiplex PCR

Multiplex PCR is a technique that enables the sequential amplification of several pathogens in a single reaction. Multiple primer sets specific for each pathogen are present in the identical reaction mixture and conditions are optimized in such a way that the multiple primers and DNA polymerases function properly to amplify the specific target pathogens. In addition to that special care should be taken while designing the primers so that they shouldn't have complementarity among the primer and produce amplicons of different sizes (Lamilla et al 2023).

Table 1 List of different PCR techniques used in detection of plant pathogens

Pathogen/disease	Host	Technique	Principle	Sensitivity	Inferences	References
Tomato spotted wilt virus	Tomato	HDA	Detection	4 pg	Potential tool for early and sensitive detection	Wu et al. (2016)
Sugarcane streak mosaic virus	Sugarcane	IC-PCR	Detection	10^{-10} dilution	More sensitive than conventional PCR	Subekti et al. (2020)
Potato spindle tuber viroid (PSTVd)	Potato, tomato	Dot-blot hybridization	Detection	10^{-15} dilution	Appropriate for large scale sample diagnosis	Kekstidou et al. (2022)
Dendrobium viroid	Dendrobium	RT PCR	Detection	10^{-3} dilution	Minimises false positives and cross contamination between different samples	Li et al. (2021a)
Areca palm yellow leaf phytoplasma (AYL)	Areca palm	dd-PCR	Detection	0.07 copies/μl	A ddPCR detection technique built around the study's phytoplasma's *tuf* gene would be useful for both qualitative and quantitative phytoplasma detections of AYL	Yu et al. (2022)

(continued)

Table 1 (continued)

Pathogen/disease	Host	Technique	Principle	Sensitivity	Inferences	References
Tobacco mosaic virus (TMV), *Tomato mosaic virus* (ToMV), *Tomato mottle mosaic virus* (ToMMV), *Tomato brown rugose fruit virus* (ToBRFV)	Tomato	Quadruplex RT-PCR	Detection	0.02 pg	Quadruplex RT-PCR allowed for the instantaneous differentiation of ToBRFV from TMV, ToMV, and TSWV	Yan et al. (2021)
Tobacco streak virus (TSV)	Soyabean	RT-qPCR	Detection	10 fg/μl	Rapid and accurate detection	Zambrana-Echevarría et al. (2021)
Sisal purple leafroll	Sisal	Nested PCR	Detection	10 fg/μl	Reliable, efficient, and specific in identifying the 16SrI group phytoplasma	Wang et al. (2022)
Rice orange leaf phytoplasma	Rice	RT-qPCR	Detection	10 pg/μl	Reduces the chance of false positive in nested PCR	Wang et al. (2020)
Candidatus Liberibacter asiaticus (CLas), citrus tristeza virus (CTV), citrus tatter leaf virus (CTLV), citrus exocortis viroid (CEVd)	Citrus	Multiplex RT-PCR	Detection	CLas, CTV, CTLV, and CEVd were all positive at the maximum dilution of 10^{-3}, 10^{-2}, 10^{-2}, and 10^{-3}, respectively	Developed multiplex RT-PCR test with great sensitivity for the targeted identification of four citrus systemic diseases that are commercially significant	Yao et al. (2023)

(continued)

Table 1 (continued)

Pathogen/disease	Host	Technique	Principle	Sensitivity	Inferences	References
Indian citrus ringspot virus (ICRSV)	Kinnow mandarin	RT-LAMP	Detection	100 ag	Highly rapid and sensitive than traditional RT-PCR	Kokane et al. (2021)
Tomato chlorotic spot virus (TCSV)	Tomato	Reverse transcription (RT)-LAMP	Detection	0.9 pg μl^{-1}	In-field detection in less than 35 min	Yilmaz et al. (2023)
Tomato leaf curl New Delhi virus (ToLCNDV)	Zucchini squash, tomato, and pepper	Real-time LAMP	Detection	50×10^{-8} ng/μL	In-field detection within 45 min	Caruso et al. (2023)
Citrus exocortis viroid (CEVd)	Citrus	Reverse transcription (RT)-LAMP	Detection	236 pg	First RT-LAMP detection technique in any citrus-infecting viroid	Chambers et al. (2023)
Chrysanthemum stunt viroid (CSVd)	Chrysanthemum	Colorimetric LAMP	Detection	1 fg μl^{-1}	Highly precise; able to identify healthy plantlets from meristem tip culture	Supakitthanakorn et al. (2022)
Chrysanthemum chlorotic mottle viroid (CChMVd)	Chrysanthemum	Colorimetric LAMP	Detection	1 ng μl^{-1}	Extremely specific; able to detect viroid free plantlets	Supakitthanakorn et al. (2022)
Candidatus Phytoplasma asteris	Grapevine	LAMP	Detection	104 target dsDNA molecules/ml	Highly specific, good sensitivity and reproducibility	Alič et al. (2022)
16SrVI group phytoplasma strain (sweet cherry witches' broom disease)	Sweet cherry	LAMP	Detection	10 pg μl^{-1}	Much more sensitive and dependable than the traditional nested PCR assay	Shreenath et al. (2022)

(continued)

Table 1 (continued)

Pathogen/disease	Host	Technique	Principle	Sensitivity	Inferences	References
Cucurbit leaf crumple virus (CuLCrV), *Cucurbit yellow stunting disorder virus* (CYSDV) and *Cucurbit chlorotic yellows virus* (CCYV)	Cucurbits	Multiplex RT-RPA	Detection	Single copy	Virus infection with both DNA and RNA concurrently detected using the first multiplex RPA assay	Jailani and Paret (2023)
Potato leafroll virus (PLRV)	Potato	RT-RPA	Detection	3.27 × 106 copies μl^{-1}	First, a one-step, highly specific RT-RPA based RNA isolation from cellular disc paper	Kumar et al. (2023)
Tomato leaf curl New Delhi virus-potato (ToLCNDV [potato])	Potato	RPA	Detection	Dilution of 10^{-6}	First report of RPA based detection for ToLCNDV [potato]	Raigond et al. (2021)
Tobacco mosaic virus, *Tobacco etch virus*, and *Potato virus X*	*Nicotiana benthamiana*	CRISPR-Cas12a and CRISPR-Cas13a/d	Detection	10^{-10} dilutions	Diagnosis within 30 min of leaf harvest without pre-amplification	Marqués et al. (2022)
Candidatus Phytoplasma trifolii (potato purple top (PPT) and potato witches' broom (PWB) disease)	Potato	CRISPR Cas12	Detection	1aM	DNA endonuclease targeted CRISPR trans reporter (DETECTR) assay	Wheatley et al. (2022)

2.5.4 Immunocapture PCR (IC-PCR)

IC-PCR integrates the principle of antigen–antibody specificity and PCR. This technique uses an antibody to capture viral particles and eliminates the previous steps for extracting viral RNA. Additionally, the first step's involvement with antibodies boosts the selectivity of virus identification. The advantages of IC-PCR include its high sensitivity (it is capable of detecting target nucleic acid at picogram levels), automation, speed (it takes 2 h or less), and usefulness in situations where ELISA is ineffective (for example, geminiviruses) (Hajeri and Yokomi 2022).

2.5.5 Real-Time PCR (RT-QPCR)

RT-qPCR is designed to detect a few molecules of template among a complex background of nucleic acids. Here the PCR products are tracked in "real-time" as they amplify at each PCR reaction cycle. Along with that, this tool also allows for the quantification of the template DNA, which can be translated into an estimation of the titer of pathogens that may be present in the samples. The amplicons produced during each cycle are monitored through fluorescence via an amplicon non-specific (as SYBR Green) or amplicon-specific probes (TaqMan) binding (Rubio et al. 2020).

2.5.6 Droplet Digital PCR (Dd-PCR)

The dd-PCR is an advanced version of digital PCR, which utilizes the water in oil emersion or the partitioning of the DNA or RNA into tiny reactions. These individual droplets act as a separate PCR reaction and allow the target sequences with high precision. The dd-PCR can be used for quantifying viral loads in plant samples (Yu et al. 2022).

3 Clustered Regularly Interspaced Short Palindromic Repeats (CRISPR)

CRISPR is a groundbreaking genome editing technology derived from the natural immune systems of certain archaea and bacteria. This system acts as a host line of defense to combat foreign DNA, like viruses and plasmids. When these microorganisms encounter viral threats, they can acquire resistance by incorporating segments of the viral genome into specific regions known as CRISPR loci, along with the expression of cas genes. The most extensively studied CRISPR/Cas9 system originates from *Streptococcus pyogenes* and comprises two essential components, viz., a guide RNA (gRNA) that recognizes foreign elements and a Cas9 protein that causes its cleavage. The Cas9 protein consists of two nuclease domains, enabling it to snip both of the target strands of DNA, resulting in double-strand breaks (DSBs) and site mutations. It also specifically recognizes a trinucleotide sequence located in the target DNA that is called the protospacer adjacent motif (PAM). The gRNA helps in directing the Cas9 protein to a precise location in the DNA for editing. It consists of two distinct RNA molecules called CRISPR RNA (crRNA), containing sequences similar to the target DNA and trans-activating crRNA, or tracrRNA, that enable the processing of precursor crRNA molecules into mature crRNA. The two components of gRNA were combined to form a single guide RNA, or sgRNA, which became a significant advancement in CRISPR technology. This streamlined approach simplified the targeting process and propelled CRISPR-Cas9 as the most extensively adopted genome editing technology (Romay and Bragard, 2017). CRISPR/Cas9 technology continues to evolve, expanding its applications beyond DNA editing to include RNA and protein molecules. Its versatility, precision, and user-friendly nature have positioned it as a dominant tool in genetic engineering and biotechnology, holding the potential to revolutionize various research fields and applications.

3.1 Combined Application of Different Techniques

A visual test based on CRISPR/Cas12a has been designed to quickly and accurately identify the most prevalent RNA viruses and viroid infections that damage apple trees. This innovative diagnostic method can detect a range of pathogens, including *Apple chlorotic leaf spot virus* (ACLSV), *Apple necrotic mosaic virus* (ApNMV), *Apple stem grooving virus* (ASGV), *Apple stem pitting virus* (ASPV), and *Apple scar skin viroid* (ASSVd). The distinguishing feature of this approach is its simplicity and efficiency. Positive test results are discernible with the human eye due to the use of oligonucleotide-conjugated gold nanoparticles. Furthermore, this CRISPR/Cas12a-based diagnostic platform significantly reduces the time required for diagnosis, delivering results in an hour or less from the time of leaf sample collection. Additionally, the CRISPR-Cas12a-RT-RPA platform has exhibited enhanced sensitivity in comparison to the RT-qPCR method. This improved sensitivity ensures the accurate and early detection of apple tree pathogens, thereby enhancing disease management (Jiao et al. 2021).

A diagnostic approach utilizing the LAMP method combined with CRISPR-Cas12a has been devised to assess the presence of DNA viruses in plants. This method successfully identifies two significant geminiviruses, namely the *Tomato yellow leaf curl virus* (TYLCV) and the *Tomato leaf curl New Delhi virus* (ToLCNDV). It stands out for its efficiency, providing results within a span of just one hour. The visual readouts are straightforward and can be easily discerned using a cost-effective fluorescence visualizer, making this method not only accurate but also accessible and cost-efficient. This innovative diagnostic technique demonstrates unparalleled sensitivity and peculiarity, making it a reliable tool for virus detection (Mahas et al. 2021).

An innovative approach has been created for the in-field diagnosis of plant RNA viruses using a combination of two cutting-edge technologies, viz., RT-RPA and CRISPR/Cas12a. This powerful method, referred to as an in vitro specific CRISPR-based assay for nucleic acid detection-one-pot (iSCAN-OP), represents a quick and effective means of detecting three significant plant viruses, viz., *Tobacco mosaic virus* (TMV), *Potato virus Y* (PVY) and *Potato virus X* (PVX). This diagnostic method includes the utilization of LbCas12a protein sourced from Lachnospiraceae bacterium, which enhances the specificity of the assay. Additionally, the system employs a cost-effective commercially available P51 fluorescence visualizer that detects Hex-labeled single-stranded DNA reporter genes. This combination allows for real-time visualization of the diagnostic results. Remarkably, the entire diagnostic process can be completed at a single temperature and within a remarkably short timeframe of less than 30 min. This speed and simplicity make it a practical and valuable tool for in-field diagnosis, facilitating swift and accurate identification of plant-specific RNA viruses (Aman et al. 2020).

4 Molecular Hybridization Techniques

4.1 Dot-Blot Hybridization

The dot-blot hybridization technique involves molecular hybridization, where single-stranded sequences from the test samples bind to complementary single-stranded sequences of the probe. The probe, containing a reporter molecule, hybridizes with the plant sample's DNA or RNA, depending on the virus being tested. Detection of this hybrid molecule can be done through radiography or chemiluminescence, depending on the reporter molecule. DNA probes are labeled using methods like nick translation, random primed labeling, or polymerase chain reaction, while RNA probes are labeled through in vitro transcription (Bhat and Rao 2020).

4.2 DNA Microarray

Microarray techniques are extensively utilized for the identification of various infectious agents in a single assay. The plant viruses will be amplified using pathogen-specific primers through a polymerase chain reaction and labeled with appropriate molecules for further detection. These generated biological molecules are allowed to hybridize with virus-specific oligonucleotide sequences immobilized on a membrane or glass slide as probes in an array (Aslam et al. 2022).

5 Sequencing Techniques

5.1 NGS (Next Generation Sequencing)

Following the unveiling of the structure of DNA by James D. Watson and Francis Crick in 1953, various techniques emerged to examine genomic DNA sequences. These methods aided fundamental research on DNA's structure and functions, unraveling life's mysteries. DNA sequencing technologies have versatile applications for biologists, encompassing molecular cloning, breeding, pinpointing pathogenic genes, and conducting comparative and evolutionary studies. Ideally, DNA sequencing technologies should be swift, precise, user-friendly, and cost-effective.

Since 2009, NGS has been pivotal in plant virology (Mushtaq et al. 2021). NGS facilitates efficient, rapid, and cost-effective high-throughput sequencing of plant virus, viroid, and phytoplasma genomes, as well as specific small RNAs produced during infection (Table 2). Its applications cover diverse studies, encompassing characterization, profiling, distribution, accumulation, biogenesis, expansion of host range, virus strain differentiation, systemic movement, virus or viroid-host interactions, viroid evolution, pathogenesis, mutation, mRNA targeting, and more. The detection and characterization of phytoplasmas utilized NGS methods such as Anchored Hybrid Enrichment (AHE) (Trivellone et al. 2022). Before the era of next-generation sequencing, Sanger sequencing dominated biological research. However, the advent of NGS has marked significant advantages over Sanger sequencing as NGS allows the user to screen many samples cost-effectively in a stipulated time.

5.2 Short-Read Sequencing (Second Generation) Technology

After Sanger sequencing/first-generation sequencing, short-read technologies represent the subsequent advancement. Characterized by the massive sequencing of short DNA molecules (250–800 bp), these technologies involve parallel sequencing of clonally amplified DNA molecules (Stalder et al. 2023). The workflow includes library preparation, sequencing, and data analysis. Commercially, widely-used platforms encompass 454 Life Sciences/Roche's GS FLX, Illumina HiSeq, MiSeq, and NextSeq, ABI's SOLiD, as well as Ion Torrent PGM, Proton, S5, and BGISeq-500 (Mehetre et al. 2021a, b).

During viral infection in plants, the host generates short interfering RNAs (siRNA), specific RNA molecules. RNA silencing (RNAi), a cytoplasmic surveillance system, recognizes and targets homologous single and double-stranded RNA, guided by small interfering RNAs. These methods offer ample opportunities to detect plant viruses or viroids, even in symptomless infections with low titers, revealing previously unknown viral entities (Table 2). NGS can identify a multitude of siRNA sequences in virus-infected plant samples.

Table 2 NGS based detection of virus/viroid

Host plant	Disease	Virus/ Viroid identified	References
Papaya	Papaya ring spot disease	*Moroccan watermelon mosaic virus* (MWMV), *Cowpea mild mottle virus* (CpMMV) *Papaya mottle-associated virus* (PaMV)	Mumo et al. (2020)
Tomato	Begomovirus disease	Distinct ssDNA virus/ subviral agents	de Nazaré et al. (2020)
Grapevine	Leafroll disease	*Grapevine rupestris stem pitting virus* (GRVFV) *Grapevine leafroll-associated virus* (GLRaV)	Xiao et al. (2019)
Common bean	Virus symptomatic bean plants	*Pelargonium vein banding virus* (PVBV) *Bean common mosaic necrosis virus* (BCMNVA), *Aphid lethal paralysis virus* (ALPV)	Wainaina et al. (2019)
Tomato, lettuce	Phytovirome	Virion-associated nucleic acids (VANA)	Ma et al. (2019)
Cucumber	Virus	*Aphid lethal paralysis virus* (ALPV)	Maina et al. (2017)
Malus, Prunus, Pyrus	Viruses and viriods	*Apple stem pitting virus* (ASPV)	Rott et al. (2017)
Cowpea	Virion-associated nucleic acids (VANA)	*Bean common mosaic virus* (BCMV) *Cowpea mottle virus* (CMV)	Palanga et al. (2016)
Sweet Cherry	Little cherry viral infection	*Little cherry virus* 1 (LChV1)	Katsiani et al. (2018)
Vanilla	Viral diseases	Potexvirus, *Vanilla virus* X (VVX), *Vanilla latent virus* (VLV)	Grisoni et al. (2017)
Citrus	Viroid	Citrus dwarfing viroid	Dang et al. (2021)
Potato	Potato tuber necrotic ringspot disease (PTNRD)	*Potato virus Y* (PVY)	Della Bartola et al. (2020)
Potato	Chlorosis and mottling	*Arabis mosaic virus* (ArMV)	Monger et al. (2020)

(continued)

Table 2 (continued)

Host plant	Disease	Virus/ Viroid identified	References
Tomato weed	Rugose	*Tomato severe rugose virus* (ToSRV)	Duarte et al. (2020)
Cowpea	Yellow mosaic and mottling	*Cowpea bright yellow mosaic virus* (CABMV)	Naito et al. (2019)
Wheat	Streak, mottled mosaic	*Wheat streak mosaic virus* (WSMV)	Fellers et al. (2019)
Cassava	Mosaic	*Cassava mosaic virus* (CMV)	Boykin et al. (2018)
Watermelon	Chlorotic stunt	*Watermelon chlorotic stunt virus* (WmCSV)	Chalupowicz et al. (2019)
Prunus	Sharka disease	*Plum pox virus* (PPV)	Bronzato et al. (2018)
Grapes	Yellow speckle	*Grapevine yellow speckle viroid* 1 (GYSVd 1), *Grapevine yellow speckle viroid* 2 (GYSVd 2), *Japanese grapevine viroid* (JGVd), *Grapevine latent viroid* (GLVd)	Morgan et al. (2023)

NGS methods prove beneficial in exploring plant diseases with unknown causes. In vines suffering from syrah decline, a new marafivirus, *Grapevine Syrah virus-*1, was detected, alongside other RNA viruses affecting grapevines (Barba et al. 2014). Additionally, contemporary NGS techniques enable the concurrent detection of multiple viruses, potentially unveiling previously undiscovered ones. For example, siRNA sequencing unveiled *common bean viruses* in wild Tanzanian plants (Mwaipopo et al. 2021).

5.3 Long-Read Sequencing (Third Generation) Technology

Long-read sequencing, classified as third-generation sequencing technologies, is capable of directly generating sequences exceeding 10 kb from native DNA (Table 2). This technology is highly valuable for phytoplasma genotyping as it enables complete phasing of alleles and resolves typing ambiguities (Hamim et al. 2022). Pacific Biosciences (PacBio) and Oxford Nanopore Technology (ONT) are the main long-read technologies.

5.3.1 Pacific Biosciences (PacBio)

PacBio sequencing uses a single-stranded circular DNA (SMRTbell template) formed by ligating hairpin adaptors. In the "SMRT Cell" chip with pores (zero-mode waveguides), the SMRTbell library undergoes sequencing. Polymerase, bound to the adapter, initiates replication with fluorescently labeled nucleotides. Nucleotide-binding generates light pulses interpreted as 'Continuous Long Read' (CLR). The circular template allows sequencing through the adapter, generating sub-reads. Each pass produces a sub-read, with the polymerase's limited lifetime determining the number of passes (Mohammed et al. 2022).

5.3.2 Oxford Nanopore Technology (ONT)

ONT generates lengthy reads through single-stranded nucleic acid translocation via a protein pore (Liefting et al. 2021). It is based on the generation of electrical signals (Deamer et al. 2016) and is a sensitive platform for identifying single-strand DNA or RNA. Utilizing a protein channel (nanopore) on a membrane, biomolecules pass through nanopores under an electrophoretic force, temporarily altering current strength. The recorded signal undergoes base-calling, converting electrical signals to base sequences. This final step employs deep learning to enhance sequencing accuracy, necessitating efficient algorithms and large datasets for computational training (Lin et al. 2021).

5.3.3 MinION

The MinION device stands out as the sole portable, palm-sized, real-time tool for RNA and DNA sequencing. Its operation relies on nanopore sequencing, where the passage of RNA and DNA molecules through a nanopore disrupts electric current, enabling the determination of base sequences. The device includes a disposable flow cell into which the sample is introduced, and sensors detect signals as molecules are analyzed. Each MiniION flow cell features 512 nanopore channels available for simultaneous sequencing. Its versatility allows for utilization in both in vitro and in vivo conditions. The applications of MinION sequencing are vast, as demonstrated in its use to investigate the population of Bean common mosaic virus (BCMV), *Cucumber mosaic virus* (CMV), and *Bean yellow mosaic virus* (BYMV) (Tarquini et al. 2022).

6 Conclusion

In conclusion, the intricate associations between plants and pathogens within agricultural ecosystems necessitate advanced diagnostic tools to safeguard global food security. For the detection of plant viruses, viroids, and phytoplasmas in agricultural environments, molecular diagnostics is an essential tool. It provides tools for researchers to fight newly discovered plant diseases. The urgency for accurate and efficient diagnostics is heightened by climate change, intensified agricultural practices, and the emergence of novel pathogens. Techniques like LAMP, RCA, RPA, and CRISPR-based assays demonstrate their versatility and efficiency in detecting plant pathogens. The combination of different techniques, such as CRISPR/Cas12a-based visual assays and the innovative iSCAN-OP method, enhances sensitivity and specificity while providing rapid, point-of-care solutions for on-the-spot pathogen identification. Traditional methods like Dot-Blot Hybridization, DNA Microarray, and PCR variants remain integral to diagnostic protocols. NGS enables high-throughput sequencing, offering insights into the genomics of plant viruses, viroids, and phytoplasmas and facilitating comprehensive studies on their evolution, distribution, and host interactions. Molecular diagnostics offer hope for addressing agricultural challenges by uncovering the genetic makeup of plant pathogens, strengthening defenses, and ensuring global food system resilience against pathogens.

References

Ali MM, Li F, Zhang Z, Zhang K, Kang DK, Ankrum JA, Le XC, Zhao W (2014) Rolling circle amplification: a versatile tool for chemical biology, materials science and medicine. Chem Soc Rev 43(10):3324–3341

Alič Š, Dermastia M, Burger J, Dickinson M, Pietersen G, Pietersen G, Dreo T (2022) Genome-informed design of a LAMP assay for the specific detection of the strain of '*Candidatus* Phytoplasma asteris' phytoplasma occurring in grapevines in South Africa. Plant Dis 106(11):2927–2939

Aman R, Mahas A, Marsic T, Hassan N, Mahfouz MM (2020) Efficient, rapid, and sensitive detection of plant RNA viruses with one-pot RT-RPA–CRISPR/Cas12a assay. Front Microbiol 11:610872

Aslam HMU, Riaz H, Killiny N, Zhou XG, Thomashow LS, Peters NT, Chanda AK (2022) Microarray technology for the detection of plant diseases. Trends Plant Disease Assess 203–223

Barba M, Czosnek H, Hadidi A (2014) Historical perspective, development and applications of next-generation sequencing in plant virology. Viruses 6:106–136

Barreda-García S, Miranda-Castro R, de-Los-Santos-Álvarez N, Miranda-Ordieres AJ, Lobo-Castañón MJ (2018) Helicase-dependent isothermal amplification: a novel tool in the development of molecular-based analytical systems for rapid pathogen detection. Anal Bioanal Chem 410(3):679–693. https://doi.org/10.1007/s00216-017-0620-3

Bhat AI, Rao GP (2020) Dot-blot hybridization technique. In: Characterization of plant viruses springer protocols handbooks. Humana, New York, NY. https://doi.org/10.1007/978-1-0716-0334-5_34

Bhat AI, Rao GP, Bhat AI, Rao GP (2020) Rolling circle amplification (RCA)

Boykin L, Ghalab A, De Marchi BR, Savill A, Wainaina JM, Kinene T, Lamb S, Rodrigues M, Kehoe M, Ndunguru J, Tairo F (2018) Real-time portable genome sequencing for global food security [version 1; referees: 1 approved, 1 approved with reservations]

Bronzato A, Sherman D, Stone A, Gopakumar A, Wilson V, Schneider W, King J (2018) Nanopore sequencing as a surveillance tool for plant pathogens in plant and insect tissues. Plant Dis 102(8):1648–1652

Caruso AG, Ragona A, Bertacca S, Montoya MAM, Panno S, Davino S (2023) Development of an in-field real-time LAMP assay for rapid detection of tomato leaf curl New Delhi virus. Plants 12(7):1487

Chalupowicz L, Dombrovsky A, Gaba V, Luria N, Reuven M, Beerman A, Lachman O, Dror O, Nissan G, Manulis-Sasson S (2019) Diagnosis of plant diseases using the Nanopore sequencing platform. Plant Pathol 68(2):229–238

Chambers GA, Geering AD, Holford P, Kehoe MA, Vidalakis G, Donovan NJ (2023) A reverse transcription loop-mediated isothermal amplification assay for the detection of citrus exocortis viroid in Australian citrus. Australas Plant Pathol 52(2):121–132

Dang T, Lavagi-Craddock I, Bodaghi S, Vidalakis G (2021) Next-generation sequencing identification and characterization of MicroRNAs in dwarfed citrus trees infected with citrus dwarfing viroid in high-density plantings. Front Microbiol 12:646273

de Nazaré Almeida dos Reis L, Fonseca MEDN, Ribeiro SG, Naito FYB, Boiteux LS, Pereira-Carvalho RDC (2020) Metagenomics of neotropical single-stranded DNA viruses in tomato cultivars with and without the Ty-1 gene. Viruses 12(8):819

Deamer D, Akeson M, Branton D (2016) Three decades of nanopore sequencing. Nat Biotechnol 34(5):518–524

Dean FB, Nelson JR, Giesler TL, Lasken RS (2001) Rapid amplification of plasmid and phage DNA using phi29 DNA polymerase and multiply-primed rolling circle amplification. Genome Res 11(6):1095–1099

Della Bartola M, Byrne S, Mullins E (2020) Characterization of potato virus Y isolates and assessment of nanopore sequencing to detect and genotype potato viruses. Viruses 12(4):478

Duarte MF, Fonseca MEN, Boiteux LS, Costa H, Ribeiro BM, Melo FL, Pereira-Carvalho RC (2020) Identification of *Physalis angulata* (Solanaceae) as a natural alternative weed host of tomato severe rugose virus in Brazil. Plant Dis 104(2):600–600

Fellers JP, Webb C, Fellers MC, Shoup Rupp J, De Wolf E (2019) Wheat virus identification within infected tissue using nanopore sequencing technology. Plant Dis 103(9):2199–2203

Grisoni M, Marais A, Filloux D, Saison A, Faure C, Julian C, Theil S, Contreras S, Teycheney PY, Roumagnac P, Candresse T (2017) Two novel Alphaflexiviridae members revealed by deep sequencing of the *Vanilla* (*Orchidaceae*) virome. Adv Virol 162(12):3855–3861

Hajeri S, Yokomi R (2022) Immunocapture-reverse transcriptase loop-mediated isothermal amplification assay for detection of plant RNA viruses. Plant Virol: Methods Protocols 245–252

Hamim I, Sekine KT, Komatsu K (2022) How do emerging long-read sequencing technologies function in transforming the plant pathology research landscape? Plant Mol Biol 110(6):469–484

Jailani AAK, Paret ML (2023) Development of a multiplex RT-RPA assay for simultaneous detection of three viruses in cucurbits. Mol Plant Pathol 24(11):1443–1450

Jiao J, Kong K, Han J, Song S, Bai T, Song C, Wang M, Yan Z, Zhang H, Zhang R, Feng J (2021) Field detection of multiple RNA viruses/viroids in apple using a CRISPR/Cas12a-based visual assay. Plant Biotechnol J 19(2):394–405

Jones RA (2020) Disease pandemics and major epidemics arising from new encounters between indigenous viruses and introduced crops. Viruses 12(12):1388

Katsiani A, Maliogka VI, Katis N, Svanella-Dumas L, Olmos A, Ruiz-García AB, Marais A, Faure C, Theil S, Lotos L, Candresse T (2018) High-throughput sequencing reveals further diversity of little cherry virus 1 with implications for diagnostics. Viruses 10(7):385

Kekstidou O, Varveri C, Vassilakos N (2022) A simplified dot-blot hybridization protocol for Potato spindle tuber viroid detection in Solanaceae. Viroids: Methods Protocols 89–96

Kokane AD, Kokane SB, Warghane AJ, Gubyad MG, Sharma AK, Reddy MK, Ghosh DK (2021) A rapid and sensitive reverse transcription–loop-mediated isothermal amplification (RT-LAMP) assay for the detection of Indian Citrus Ringspot Virus. Plant Dis 105(05):1346–1355

Kumar R, Kaundal P, Tiwari RK, Lal MK, Kumari H, Kumar R, Naga KC, Kumar A, Singh B, Sagar V, Sharma S (2023) Development of reverse transcription recombinase polymerase amplification (RT-RPA): a methodology for quick diagnosis of potato leafroll viral disease in potato. Int J Mol Sci 24(3):2511

Lamilla J, Galvez A, Franco-Lara L (2023) Simultaneous detection and quantification by multiplex qPCR of' *Candidatus Phytoplasma asteris* and *Candidatus Phytoplasma fraxini* in a plant host and insect vectors. Tropical Plant Pathol 48(5):564–574

Li X, Ruan H, Zhou C, Meng X, Chen W (2021a) Controlling citrus Huanglongbing: green sustainable development route is the future. Front Plant Sci 12:760481

Li SY, Yang F, Xing HY, Che XR, Cao ZX, Zhang YW, Khoo CY, Zhou SF, Li A (2021b) Rapid sap-direct reverse transcription-polymerase chain reaction method for detection of dendrobium viroid in Dendrobium plants. Lett Appl Microbiol 73(1):26–30

Liefting LW, Waite DW, Thompson JR (2021) Application of oxford nanopore technology to plant virus detection. Viruses 13(8):1424

Lin B, Hui J, Mao H (2021) Nanopore technology and its applications in gene sequencing. Biosensors 11(7):214

Lobato IM, O'Sullivan CK (2018) Recombinase polymerase amplification: basics, applications and recent advances. TrAC, Trends Anal Chem 98:19–35

Ma Y, Marais A, Lefebvre M, Theil S, Svanella-Dumas L, Faure C, Candresse T (2019) Phytovirome analysis of wild plant populations: comparison of double-stranded RNA and virion-associated nucleic acid metagenomic approaches. J Virol 94(1):10–1128

Maciejewska A, Jakubowska J, Pawłowski R (2013) Whole genome amplification of degraded and nondegraded DNA for forensic purposes. Int J Legal Med 127:309–319

Mahas A, Hassan N, Aman R, Marsic T, Wang Q, Ali Z, Mahfouz MM (2021) LAMP-coupled CRISPR–Cas12a module for rapid and sensitive detection of plant DNA viruses. Viruses 13(3):466

Maina S, Edwards OR, de Almeida L, Ximenes A, Jones RAC (2017) Metagenomic analysis of cucumber RNA from East Timor reveals an Aphid lethal paralysis virus genome. Genome Announc 5:e01445-e1516

Marqués MC, Sánchez-Vicente J, Ruiz R, Montagud-Martínez R, Márquez-Costa R, Gómez G, Carbonell A, Daròs JA, Rodrigo G (2022) Diagnostics of infections produced by the plant viruses TMV, TEV, and PVX with CRISPR-Cas12 and CRISPR-Cas13. ACS Synth Biol 11(7):2384–2393

Martinelli F, Scalenghe R, Davino S, Panno S, Scuderi G, Ruisi P, Villa P, Stroppiana D, Boschetti M, Goulart LR, Davis CE (2015) Advanced methods of plant disease detection. A review. Agron Sustain Dev 35:1–25

Mehetre GT, Leo VV, Singh G, Sorokan A, Maksimov I, Yadav MK, Singh BP (2021a) Current developments and challenges in plant viral diagnostics: a systematic review. Viruses 13(3):412

Mehetre GT, Leo VV, Singh G, Sorokan A, Maksimov I, Yadav MK, Upadhyaya K, Hashem A, Alsaleh AN, Dawoud TM, Almaary KS (2021b) Current developments and challenges in plant viral diagnostics: a systematic review. Viruses 13(3):412

Mohammed AA, Senbeta B, Worku T, Mohammed AA, Senbeta B, Worku T (2022) Pacific bioscience sequence technology: review. Int J Vet Sci Res 8:27–33

Monger WA, Goodfellow HA, Back EJ (2020) First report of Arabis mosaic virus in potato (*Solanum tuberosum*), identified by nanopore sequencing. New Dis Rep 41(29):2044–588

Morgan S, Burger JT, Pietersen G (2023) Next generation RNA sequencing analysis of nuclear grapevines to identify viruses and viroids

Mumo NN, Mamati GE, Ateka EM, Rimberia FK, Asudi GO, Boykin LM, Machuka EM, Njuguna JN, Pelle R, Stomeo F (2020) Metagenomic analysis of plant viruses associated with papaya ringspot disease in *Carica papaya* L. in Kenya. Front Microbiol 11:205

Mushtaq M, Dar AA, Basu U, Bhat BA, Mir RA, Vats S, Dar MS, Tyagi A, Ali S, Bansal M, Rai GK (2021) Integrating CRISPR-Cas and next generation sequencing in plant virology. Front Genet 12:735489
Mwaipopo B, Rajamäki ML, Ngowi N, Nchimbi-Msolla S, Njau PJ, Valkonen JP, Mbanzibwa DR (2021) Next-generation sequencing-based detection of common bean viruses in wild plants from Tanzania and their mechanical transmission to common bean plants. Plant Dis 105(9):2541–2550
Nair S, Manimekalai R (2021) Phytoplasma diseases of plants: molecular diagnostics and way forward. World J Microbiol Biotechnol 37(6):102
Naito FY, Melo FL, Fonseca MEN, Santos CA, Chanes CR, Ribeiro BM, Gilbertson RL, Boiteux LS, de Cássia Pereira-Carvalho R (2019) Nanopore sequencing of a novel bipartite New World begomovirus infecting cowpea. Adv Virol 164:1907–1910
Notomi T, Mori Y, Tomita N, Kanda H (2015) Loop-mediated isothermal amplification (LAMP): principle, features, and future prospects. J Microbiol 53(1):1–5
Palanga E, Filloux D, Martin DP, Fernandez E, Gargani D, Ferdinand R, Zabré J, Bouda Z, Neya JB, Sawadogo M, Traore O (2016) Metagenomic-based screening and molecular characterization of cowpea-infecting viruses in Burkina Faso. PLoS ONE 11(10):e0165188
Raigond B, Pathania S, Verma A, Verma G, Kochhar T, Chakrabarti SK (2021) Recombinase polymerase amplification assay for rapid detection of a geminivirus associated with potato apical leaf curl disease. J Plant Dis Prot 128(4):1061–1071
Ristaino JB, Anderson PK, Bebber DP, Brauman KA, Cunniffe NJ, Fedoroff NV, Finegold C, Garrett KA, Gilligan CA, Jones CM, Martin MD (2021) The persistent threat of emerging plant disease pandemics to global food security. Proc Natl Acad Sci 118(23):e2022239118
Romay G, Bragard C (2017) Antiviral defenses in plants through genome editing. Front Microbiol 8:47
Rott M, Xiang Y, Boyes I, Belton M, Saeed H, Kesanakurti P, Hayes S, Lawrence T, Birch C, Bhagwat B, Rast H (2017) Application of next generation sequencing for diagnostic testing of tree fruit viruses and viroids. Plant Dis 101(8):1489–1499
Rubio L, Galipienso L, Ferriol I (2020) Detection of plant viruses and disease management: relevance of genetic diversity and evolution. Front Plant Sci 11:1092
Shreenath YS, Sunani SK, Rao GP, Dickinson M (2022) Development of loop-mediated isothermal amplification (LAMP) assay for the detection of clover proliferation phytoplasma-associated with sweet cherry witches' broom disease. Australas Plant Pathol 51(4):467–474
Stalder L, Oggenfuss U, Mohd-Assaad N, Croll D (2023) The population genetics of adaptation through copy number variation in a fungal plant pathogen. Mol Ecol 32(10):2443–2460
Subekti D, Hidayat SH, Damayanti TA (2020) Immunocapture reverse transcription polymerase chain reaction for detection of sugarcane streak mosaic virus. In: IOP conference series: earth and environmental science, vol 457, No 1. IOP Publishing, p 012067
Supakitthanakorn S, Vichittragoontavorn K, Kunasakdakul K, Ruangwong OU (2022) Development of the colorimetric loop-mediated isothermal amplification technique for rapid and sensitive detection of chrysanthemum stunt viroid in chrysanthemum. J Plant Protect Res
Tarquini G, Martini M, Maestri S, Firrao G, Ermacora P (2022) The virome of 'Lamon Bean': application of MinION sequencing to investigate the virus population associated with symptomatic beans in the Lamon Area, Italy. Plants 11(6):779
Trivellone V, Cao Y, Dietrich CH (2022) Comparison of traditional and next-generation approaches for uncovering phytoplasma diversity, with discovery of new groups, subgroups and potential vectors. Biology 11(7):977
Varma A, Singh MK (2020) Diagnosis of plant virus diseases. Appl Plant Virol 79–92
Wainaina JM, Ateka E, Makori T, Kehoe MA, Boykin LM (2019) A metagenomic study of DNA viruses from samples of local varieties of common bean in Kenya. PeerJ 7:e6465
Wang Z, Zhu Y, Li Z, Yang X, Zhang T, Zhou G (2020) Development of a specific polymerase chain reaction system for the detection of rice orange leaf phytoplasma. Plant Dis 104(2):521–526

Wang G, Wu W, Tan S, Liang Y, He C, Chen H, Yi K (2022) Development of a specific nested PCR assay for the detection of 16SrI group phytoplasmas associated with sisal purple leafroll disease in sisal plants and mealybugs. Plants 11(21):2817

Wheatley MS, Wang Q, Wei W, Bottner-Parker KD, Zhao Y, Yang Y (2022) Cas12a-based diagnostics for potato purple top disease complex associated with infection by '*Candidatus* Phytoplasma trifolii'-related strains. Plant Dis 106(8):2039–2045

Wu X, Chen C, Xiao X, Deng MJ (2016) Development of reverse transcription thermostable helicase-dependent DNA amplification for the detection of tomato spotted wilt virus. J AOAC Int 99(6):1596–1599

Xiao H, Li C, Al Rwahnih M, Dolja V, Meng B (2019) Metagenomic analysis of Riesling grapevine reveals a complex virome including two new and divergent variants of grapevine leafroll-associated virus 3. Plant Dis 103(6):1275–1285

Yadav N, Khurana SP (2016) Plant virus detection and diagnosis: progress and challenges. Front Discov Innov Interdiscip Microbiol 97–132

Yan ZY, Zhao MS, Liu LZ, Yang GL, Chao GENG, Yanping TIAN, Li XD (2021) Biological and molecular characterization of tomato brown rugose fruit virus and development of quadruplex RT-PCR detection. J Integr Agric 20(7):1871–1879

Yao SM, Wu ML, Hung TH (2023) Development of multiplex RT-PCR assay for the simultaneous detection of four systemic diseases infecting citrus. Agriculture 13(6):1227

Yilmaz S, Adkins S, Batuman O (2023) Field-portable, rapid, and low-cost RT-LAMP assay for the detection of tomato chlorotic spot Virus. Phytopathology® 113(3):567–576

Yu SS, Zhang XC, Song WW, Qin WQ (2022) Accurate and sensitive detection of areca palm yellow leaf phytoplasma in China by droplet digital PCR targeting tuf gene sequence. Ann Appl Biol 181(2):152–159

Zambrana-Echevarría C, Roth MG, Dasgupta R, German TL, Groves CL, Smith DL (2021) Sensitive and specific qPCR and nested RT-PCR assays for the detection of tobacco streak virus in soybean. PhytoFrontiers™ 1(4):291–300

Advances in Contemporary Tools for Detecting and Diagnosing Plant Pathogens

Anuj Chaudhary, Parul Chaudhary, Shivani Singh, Geeta Bhandari, and Vishal Tripathi

Abstract Plant infections critically limit yields and significantly reduce crop production, therefore posing substantial threats to food security and remaining a paramount concern in agriculture. While chemical measures are still the primary strategy to mitigate the spread of plant diseases, their excessive application can lead to reduced pathogen susceptibility. Moreover, excessive spraying can have adverse effects on soil microorganisms and ecosystem contamination. Thus, the development of effective diagnostic approaches for the early detection of phytopathogens is imperative to ensure agricultural sustainability and food security. Purposely, numerous molecular techniques have been adopted for rapid phytopathogen detection. This chapter explores various direct and indirect methods for such an aim including nanotechnology, biosensors, microfluids, among others. The utilization of nanotechnology in managing agricultural diseases is a significant research area. Nanotechnology can be considered a promising and potential tool to enhance sensitivity, precision, and speed in identifying plant pathogens while also facilitating high-throughput analysis.

Keywords Plant disease · Detection techniques · Nanoparticles · Pathogen identification · Microbial pathogens

---

A. Chaudhary · V. Tripathi
Department of Biotechnology, Graphic Era Deemed to Be University, Dehradun, Uttarakhand, India

P. Chaudhary (✉)
School of Agriculture, Graphic Era Hill University, Dehradun, Uttarakhand, India
e-mail: parulchaudhary1423@gmail.com

S. Singh
Govind Ballabh Pant University of Agriculture and Technology, Pantnagar, Uttarakhand, India

G. Bhandari
Department of Biosciences, Swami Rama Himalayan, University, Dehradun, Uttarakhand, India

J. Chen et al. (eds.), *Molecular and Biotechnological Tools for Plant Disease Management*, Interdisciplinary Biotechnological Advances,
https://doi.org/10.1007/978-981-97-7510-1_9

1 Introduction

Plant diseases stand as a major driver behind diminished agricultural output on a global scale. These diseases encompass both longstanding and emerging ailments that can be transmitted between plants. The speed at which they spread, their frequency, and the intensity of their impact collectively pose a real threat to the sustainable supply of global food resources (Ristaino et al. 2021). By manifesting through an array of symptoms on leaves, fruits, and stems, these diseases prove high discernibility through visual inspection—a crucial initial step in the diagnosis process. However, this method falters in identifying pathogen presence during the initial asymptomatic stage of plant infection. Timely recognition of plant diseases assumes paramount importance in safeguarding plant health. This recognition not only facilitates the management of infections across field and greenhouse settings throughout various disease progression phases but also curbs infection propagation. Furthermore, it serves to forestall novel plant diseases from infiltrating national borders. In light of the world's growing population and the need for increased crop yield, early detection of plant diseases stands as an indispensable tool for improving agricultural production to meet global demands.

Conventional methods for identifying plant diseases are plagued by their time-intensive and laborious nature, often necessitating 3–5 days to gauge disease incidence accurately. The evolving landscape of plant diseases demands innovative approaches that not only curtail the propagation and losses attributed to burgeoning and resurgent ailments but also accommodate global climate shifts (Patel et al. 2022). Within the context of this book chapter, we delve into recent strides in detection methodologies, particularly focusing on miniature devices. The realm of analytical techniques for identifying plant pathogens is bifurcated into direct and indirect recognition tools. Direct methods entail intricate laboratory protocols, while indirect methods hinge on the identification of signs of plant stress as indicators of disease presence (Patel et al. 2023a).

Numerous methods have been harnessed in conjunction with on-site visual inspection of plant symptoms to ascertain the incidence of plant diseases. Among these, DNA-based techniques enabled the detection of pathogens components i.e., proteins and nucleic acids within the plant host material (Khater et al. 2017). In parallel, alternative indirect approaches rely on the analysis of volatile organic compounds, which are emitted by plants as a defensive mechanism against pathogenic incursions. Research was mainly focused on the identification of plant pathogens and diagnosis of plant diseases, leveraging practices based on nucleic acids, notably polymerase chain reaction and DNA hybridization detection. These methods delve into the genetic constitution of pathogens, unraveling crucial information. Additionally, the immunoassay technique, harnessing the specificity of monoclonal antibodies, has proven to be adept at pinpointing plant viruses with high precision. This characteristic renders it particularly valuable for the assessment of substantial plant sample volumes and rapid on-site detection of plant infections (Gautam and Kumar 2020).

Pathogen sensing techniques encompass a diverse array of sensors including antibodies, DNA probes, phages, and other biological agents that contribute to pathogen recognition. Among these, antibody-based biosensors emerge as a label-free avenue, offering sensitive, and rapid quantitative and qualitative analyses of infections. Phase-based DNA was applied to identify plant diseases. Detection of *Pseudomonas cannabina*, which afflicts cruciferous vegetables, was achieved through bioluminescent-phage insights (Fang and Ramasamy 2015). Furthermore, the realm of DNA-based biosensors witnessed the creation of a non-faradaic electrochemical impedance spectroscopy-based biosensor for detecting the plant pathogenic bacterium *Ralstonia solanacearum* (Patel et al. 2023b).

While the exploration of nanoparticle-mediated plant disease resistance is still in its nascent stage, it presents a promising avenue for the advancement of sustainable agriculture. Nanotechnology demonstrates significant potential in regulating secondary metabolite levels, activating disease resistance genes, facilitating hormone delivery, and generating transgenic plants resistant to pathogenic threats (Dong et al. 2023).

2 Progress in Plant Pathogen Detection Methods

Nowadays, the agricultural sector confronts substantial threats from plant diseases. The impact of plant diseases and pests on commercially significant crops can result in up to 40% yield losses annually (Savary et al. 2019). These losses, with an estimated annual toll of $220 billion, exert a considerable financial burden (FAO 2019). In light of projections indicating a global population of around 9.7 billion by 2050, accompanied by heightened global food consumption, it is imperative to consider the socio-ecological repercussions of plant pathogens (FAO 2017).

Historically, addressing the escalating food demand involved expanding agricultural land, often at the expense of other vital spatial configurations, including residential, industrial, and notably, forested areas (McDonald and Stukenbrock 2016). With agriculture now claiming 50% of habitable land, the strategy of agricultural intensification—aiming to elevate yields—seems a prudent choice. However, the proliferation of host-specific viruses is expedited by this intensification, especially through extensive monoculture practices (Baldi and La Porta, 2020). International trade expansion further compounds the impact of infections by accelerating their introduction, leading to substantial crop damage and production reduction. These factors can exert devastating repercussions on the economy and society, particularly within developing countries (De Boer and Lopez 2012).

Mitigating losses stemming from plant diseases is of paramount importance, but even more vital is the pursuit of sustainable long-term solutions. In 2019, the European Commission introduced the Green Deal to address climate change and foster sustainable business and agricultural practices (EU 2019). To support this commitment's objectives, the implementation of integrated pest management (IPM) would be a viable approach. IPM centers on meticulous evaluation of all available plant protection techniques while minimizing health and environmental risks (stemming from pesticide use) below standard safe limits.

2.1 Non-Invasive Optical and Spectral Detection Methods

Visual identification of disease indications stands out as a straightforward technique for the perception of plant pathogens. However, this approach falters in identifying latent infections, where no visible signs have yet emerged. Moreover, the accurate identification of causal agents relies heavily on the observer's adeptness being susceptible to bias. To escape such limitations, more impartial methodologies have been devised. The advent of digitization has ushered in a growing reliance on imaging and optical methods for plant disease diagnosis. Notably, stressed or infected plants manifest distinct spectral signatures compared to healthy counterparts (Zubler and Yoon 2020). For instance, plants respond to biotic stressors by altering their chlorophyll concentration, thermal radiation, and even minute plant movements—changes that occur before visible signs like wilting or leaf lesions surface. Contemporary spectral techniques permit the measurement of these electromagnetic radiation fluctuations emitted by plants. This spectrum analysis can be implemented across various scales, spanning from capturing high-resolution leaf images to utilizing drones for assessing the spectral composition of expansive fields (Singh et al. 2021).

Further, as the size and cost of sensors are prone to continuous reduction, their agricultural integration may proliferate, promising substantial automation potential for routine surveillance. For instance, drone deployment facilitates regular monitoring of vast cultivation areas, pinpointing regions with pronounced biotic stress in plants, often referred to as "hotspots." By scaling this concept, satellite imaging, driven by high-resolution images, has showcased its efficacy. According to Raza et al. (2020), recent studies have demonstrated the ability of satellite imaging to detect *Fusarium virguliforme*, the culprit responsible for sudden death syndrome in soybean. Notably, this investigation accurately predicted the syndrome with over 75% accuracy, even preceding visible symptoms (Figs. 1, 2, 3).

The utilization of optical and spectral methods offers several advantages over alternative approaches: (i) continuous crop monitoring enables real-time detection, (ii) biotic stress can be identified, and (iii) such techniques are non-invasive and obviate the need for sample manipulation. However, interpreting data for plant biotic stress identification may be intricate, often necessitating the formulation of special algorithms, frequently relying on machine learning or neural networks. This complexity remains true despite the increasing simplicity of data collection using optical sensors. Furthermore, although imaging techniques can identify biotic stress before visible symptoms manifest, they still lack the resolution to pinpoint individual pathogens accurately. Thus, these methods should complement more accurate techniques in order to identify the responsible pathogen before devising effective disease management strategies. Yet, by identifying stress-displaying plant locations, more targeted sampling strategies can be implemented (Mahlein 2016).

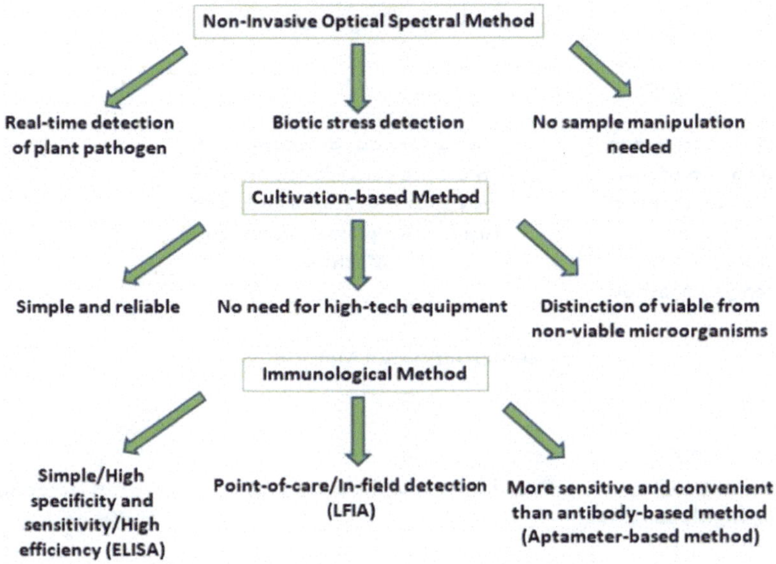

Fig. 1 Non-invasive optical spectral, cultivation-based and immunological methods for plant pathogen detection and their advantages

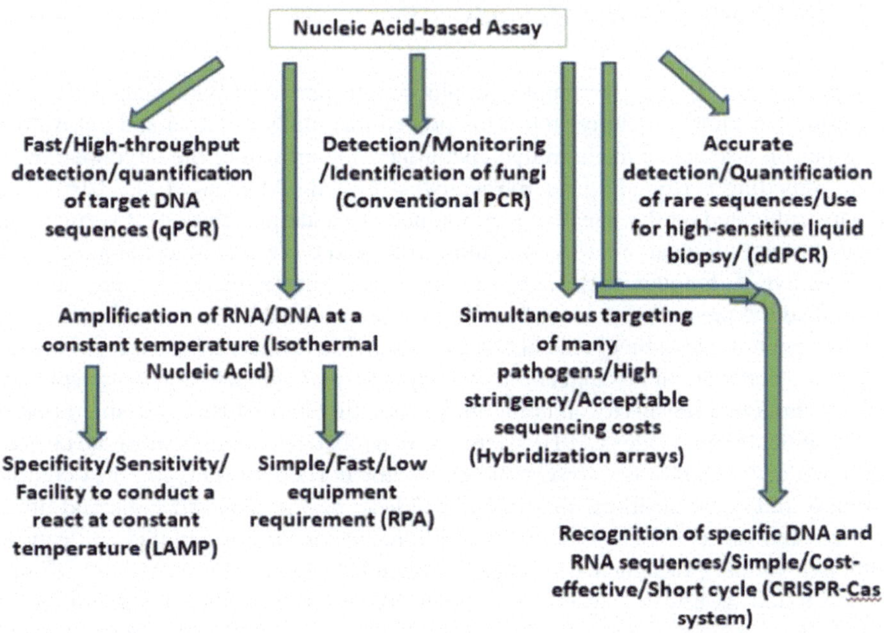

Fig. 2 Nucleic acid-based assay for plant pathogen detection and their advantages

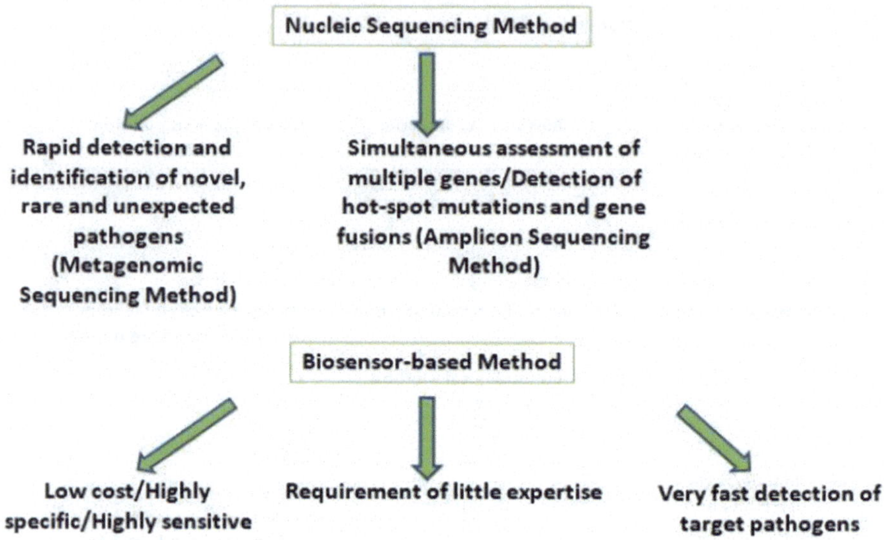

Fig. 3 Nucleic sequencing and biosensor-based methods for plant pathogen detection and their advantages

2.2 Cultivation-Based Methods

Among the most effective techniques for pinpointing and identifying plant pathogens are those that hinge on cultivation. This procedure entails isolating and cultivating microbes on a selective medium that encourages the growth of the target pathogen while impeding surrounding microorganisms' growth (Mancini et al. 2016). The isolates cultivated on the selective medium need to undergo subsequent verification through morphological, microscopic, molecular, and biochemical techniques.

Objective techniques such as biochemical and phenotypic assays are utilized to validate the presumptive identification of selective medium-cultivated clones. A range of phenotypic or biochemical assays is available, which can be executed manually or via commercial kits and automated equipment. Commercial testing typically offers heightened sensitivity and reliability. The utilization of Biolog™ microplates and analytical profile index (API) systems, which rely on the strain's substrate utilization tendencies, stands as a corresponding testament (Ferone et al. 2020). A combination of fatty acid profiling, matrix-assisted laser desorption/ionization, and time-of-flight analysis (MALDI-TOF) is an additional avenue for bacterial identification. Although initially designed for general biomarker monitoring, MALDI-TOF has demonstrated its effectiveness in microbial taxonomic identification (Chun et al. 2022). Each organism boasts a unique lipid fingerprint due to distinct fatty acid compositions, rendering fatty acid analysis as a rapid and cost-effective approach to identify plant pathogens down to the species level (Lacey et al. 2021). Furthermore, genetic marker sequencing, often referred to as DNA barcoding, is a widely adopted technique for taxonomic classification, involving a comparison of the genetic markers' DNA sequence with previously reported sequences of established species. Bacterial taxonomic characterization is made possible through the harnessing of 16S rRNA and rpoB genes, while the identification of fungi and oomycetes is driven by a common marker i.e., internal transcribed spacer (ITS) domain (Antil et al. 2023).

Cultivation-based approaches offer distinct advantages, including user-friendliness, reliability, and independence from sophisticated machinery. Moreover, they facilitate the differentiation between living and deceased organisms (Li et al. 2020). Additionally, these approaches enable the quantification of the target pathogen via a potent selective medium. The suite of phenotypic and biochemical tests conducted influences specificity. To ascertain the identity of the test strain, its biochemical and phenotypic traits are juxtaposed with those of reference strains (Castro-Escarpulli et al. 2015). The sensitivity range for detecting pathogens using cultivation-based approaches spans between 10 and 104 CFU/mL, contingent upon the target microbe and the matrix structure of the separation. Although enumeration might not be feasible, sensitivity in identification can be heightened by incorporating an enrichment step before plating.

However, cultivation-based techniques owe their limitations. Not all bacteria can be successfully cultured, it is known as the "great plate count anomaly." This method's notable drawback is the time required for microbial growth and subsequent multi-test confirmation of its identity, rendering the process time-consuming (varying from days to weeks) (Law et al. 2015; Li et al. 2020). Additionally, the host-dependence of viral plant diseases renders the use of (semi)-selective media unsuitable for their detection. Instead, viral infections can be identified by observing outward signs on the indicator or target plants, a method that can also be time-consuming depending on the tested plant (Legrand 2015; Mehetre et al. 2021).

Cultivation-based techniques continue to be employed extensively, primarily due to their simplicity. Numerous standardized cultivation-based methods for the identification of main plant diseases such as *Fusarium* and *Xanthomonas* spp. were described by regulatory organizations. To verify pathogen identity, European and Mediterranean Plant Protection Organization (EPPO) (2022) protocols often recommend DNA-based tests, e.g., traditional or real-time PCR experiments.

2.3 *Immunological Detection Methods*

Plant-–pathogen detection employs specialized antibodies in immunological or serological investigations. Microorganisms produce diverse antigenic compounds that can serve as targets for detection. These assays involve the use of antibodies designed to bind to specific epitopes on these antigens (Fang and Ramasamy 2015). By employing antibodies that have been conjugated with enzymes, fluorophores, or nanoparticles, it becomes possible to detect the interaction between antibodies and antigens. This approach facilitates the indirect detection of infections.

Many antibodies target the antigens of various pathogenic bacteria. Antibody-antigen interactions are notably specific. Two categories of antibodies are typically utilized: (i) polyclonal antibodies, which are a mixture of antibodies possessing varying affinities for the target antigen(s), or (ii) monoclonal antibodies, which are characterized by a singular epitope of specificity (Martinelli et al. 2015). Polyclonal antisera, due to their inclusion of antibodies targeting distinct epitopes, can more potentially cross-react with different antigens, resulting in unreal positive outcomes. In this context, the development of monoclonal antibodies using hybridoma and phage display technologies has revolutionized immunological testing. Monoclonal antibodies offer heightened specificity and reproducibility compared to polyclonal antibodies, thanks to their lack of inter-batch variability. However, they tend to be less sensitive and generally more expensive than their polyclonal counterparts (Ascoli and Aggeler 2018).

Immunological assays for pathogen detection encompass a wide array and many of them find application in clinical settings. These include fluorescence immunoassays, chemiluminescent immunoassays, latex agglutination assays, and western blotting, among others. The subsequent section delves into the most prevalent immunological methods adopted for the detection of plant pathogens. For a comprehensive list of additional immunological tests, please refer to Haselbeck et al. (2022).

2.3.1 Enzyme-Linked Immunosorbent Assay (ELISA)

ELISA stands as the foremost and widely adopted immunological method. Since the 1970s, ELISA has evolved into a dependable technique for global microbial infection detection, acclaimed for its speed, automation capabilities, and suitability for high-throughput testing (Martinelli et al. 2015). With its various formats i.e., sandwich, competitive, direct, and indirect, ELISA operates via the same fundamental principle. The latter consists of the utilization of distinct antibodies coupled with enzymes, owing to the capability to convert a colorless substrate into a colored one (Alhajj and Farhana 2023). The extent of the target antigen found in the sample is directly proportional to the color formation (Atmar 2014).

Assay type (competitive, direct, indirect, or sandwich ELISA), nature of antibodies used (monoclonal or polyclonal), linked enzyme, and corresponding substrate are all factors that affect the specificity and sensitivity of ELISA assays. An issue known as "matrix interference" can also arise from the sample matrix, exerting a substantial influence on ELISA's specificity and sensitivity. This can lead to increased unreal positive or negative outcomes due to decreased specificity.

Indirect ELISA tends to be more sensitive when signal amplification occurs resulting from the binding of numerous enzyme-labeled secondary antibodies to the primary antigen-attached antibody (Katikireddy and O'Sullivan 2011). Various strategies i.e., the application of avidin–biotin complexes for the facilitation of diverse avidin complexes binding (Atmar 2014), and peroxidase-antiperoxidase method (PAP) which couples a secondary antibody to a PAP complex containing multiple HRPs (Katikireddy and O'Sullivan 2011) were found to improve sensitivity.

Performing ELISA assay without a prior enrichment step is a relatively straightforward process that takes from one to several hours to complete (Bari and Kawasaki 2014). The concurrent analysis of multiple samples is facilitated by employing multi-well plates in various configurations–a process that can be further streamlined through automation. However, certain drawbacks accompany ELISA usage. Antibodies demand specialized buffers and refrigeration for storage due to their low chemical and physical stability. Furthermore, producing new antibodies can be challenging and costly. Nonetheless, new ELISA tests for plant pathogen identification are still being developed (Gorris et al. 2021).

2.3.2 Lateral Flow Immunoassays (LFIA)

LFIA is a widely employed immunological technique for detecting plant pathogens. These tests tend to use nitrocellulose membrane strips housed within a plastic container (Lopez-Soriano et al. 2017). The analyzed sample is prior applied to a designated area on the device. Capillary forces propel the sample through the release pad containing (primary) antibody conjugates, where the antibodies bind to specific antigens. Then, the sample progresses toward the device's test area. Antibodies specific to the target antigen, recognizing a different epitope, are permanently immobilized on the nitrocellulose membrane. Afterward, the antibodies serving against the primary antibody as positive control are detected in the second test zone. Latex or colloidal gold nanoparticles correspond to the label of primary antibodies found in the release pad. Upon reaching the test zone, these labeled antibodies enable visual detection of (non-)availability of the target antigen. While LFIAs are commonly designed for single analyte detection, multiplex LFIAs have been developed for medical diagnostics applications. This can involve incorporating multiple test zones onto the device or using distinct labels or different colored labels to differentiate between different analytes, thus achieving multiplexing (Anfossi et al. 2019).

LFIA excels as a point-of-care qualitative diagnostic tool for target pathogen detection due to its user-friendliness, portability, cost-effectiveness, and ability to generate timely outcomes (within very few minutes) excluding sample preparation time (Lopez-Soriano et al. 2017). Whereas, semi-quantification is feasible with sensor inclusion. However, the sensitivity of the sample application area is limited by the small volume of sample that can be loaded onto it (Koczula and Gallotta 2016). It is worth noting that LFIA is solely applicable to liquid samples. When evaluating solid materials or complex matrices i.e., soil or plant materials, a sample pre-treatment of LFIA is needed to extract relevant target antigens.

2.3.3 Aptamer-Based Detection as an Alternative to Antibodies

In specific scenarios, aptamers can serve as a viable substitute for antibodies. Aptamers, short oligonucleotides (DNA or RNA), exhibit a distinct three-dimensional structure and possess a strong affinity for the target analyte or pathogen.

These molecules typically consist of 10 to 100 nucleotides (Zhao et al. 2018). Aptamers function analogously to antibodies that target specific antigens. With attributes like cost-effectiveness, easy manufacturing, resistance to degradation, compact size, and straightforward labeling, aptamers hold great potential for replacing antibodies. Consequently, aptamers can be integrated into diverse applications, such as lateral flow devices and enzyme-linked apta-sorbent assays (ELASAs) (Toh et al. 2015).

Despite their potential, several challenges still hinder the widespread adoption of aptasensors. For instance, there exists a notable interplay between selectivity and affinity, while parameters like sample viscosity, pH, and ionic strength can exert influences (Shahdordizadeh et al. 2017). Although not yet widely employed, a few studies have showcased the efficacy of aptamer-based selection in detecting plant pathogens, e.g., apple stem spitting virus and soybean rust fungus (Krivitsky et al. 2021).

2.4 Nucleic Acid-Based Assays for Detection

Nucleic acid sequences serve as ideal molecular targets for the identification and detection of pathogenic microorganisms, spanning bacterial, fungal, or viral species. Techniques like PCR and hybridization-based approaches are employed to pinpoint the presence of specific genetic sequences unique to the target pathogen. While the subsequent section delves into the most extensively employed approaches for identifying plant pathogens, it is worth noting that variations of these procedures abound. Alternatively, sequencing techniques can be adapted to detect distinctive DNA or RNA sequences (Paul et al. 2020).

Nucleic acid-based assays can encounter reduced performance as a result of phenolic compounds and humic acid presence. To address this, various DNA extraction methods have been developed to obtain pure DNA. These methods range from straightforward commercial kits to more intricate procedures designed for challenging matrices, often involving steps such as liquid nitrogen treatment and additional enzymatic processes. However, these diverse DNA extraction techniques are unsuitable for cautious applications due to their complexity outside of a laboratory setting. Consequently, efforts have been directed toward developing simplified DNA extraction techniques that require minimal equipment (Paul et al. 2020).

RNA, prone to swift degradation outside the cell, poses extraction challenges. Addressing this issue involves directly targeting RNA. Yet, efficient and powerful RNA extraction from sample matrices used for plant pathogen identification can be intricate. Moreover, RNA has demonstrated greater stability than anticipated in certain matrices (Schostag et al. 2020).

2.4.1 Conventional PCR and Variants

The amplification of specific DNA fragments can be possible through Polymerase Chain Reaction (PCR) using components like primers, dNTPs, a DNA polymerase enzyme, and a heat cycler. Specific DNA segments to the targeted pathogen can be amplified using well-designed primers. Identifying a PCR fragment helps in the research and detection of the possible availability of the target pathogen. Traditional PCR setup, known as conventional PCR (cPCR), often involves endpoint detection, e.g., agarose gel electrophoresis, to verify the obtained product. However, a notable drawback of cPCR is its time-consuming nature. Additionally, the process of accessing tubes for gel electrophoresis expands the threat of cross-contamination, potentially affecting the laboratory environment, reagents, or other samples (Maurer 2011).

PCR stands as a precise and highly sensitive method. The limit of detection relies on factors like sample type, DNA extraction efficiency, and amplification efficacy, which can be influenced by PCR setup and primer design. Theoretically, PCR can amplify a few copies of target nucleic acids (3 copies), as predicted by Poisson statistics (Kralik and Ricchi 2017). Compared to traditional culture-based techniques, PCR is notably faster and offers greater sensitivity and specificity, and results delivery within a few hours (Priyanka et al. 2016). However, in cases where multiple infections must be perceived in a single sample, distinct PCR methods must be employed, potentially escalating costs.

2.4.2 Quantitative PCR (qPCR)

qPCR or real-time PCR stands apart from conventional PCR by enabling real-time measurement of amplified DNA throughout the PCR reaction, rather than relying on endpoint detection as previously mentioned by Shen (2019). In qPCR, the intensity of fluorescence serves as a gauge for DNA amplification. This is achieved by incorporating sequence-specific probes, allowing the quantification of amplified DNA. While dsDNA-binding dyes are cost-effective, they pose the challenge of non-specific binding to amplification products. Post-qPCR can be employed to identify non-specific amplification.

Additionally, qPCR employs various sequence-specific probes, such as TaqMan and scorpion probes. These probes exhibit enhanced specificity as fluorescence is only emitted upon hybridization with the intended target sequence. This approach also permits limited multiplexing, accommodating 4–6 genetic targets per reaction when distinct fluorescently labeled probes are utilized (Rajagopal et al. 2019). The use of a standard curve involves CT values calculated for samples with known DNA template amounts, DNA concentration, and consequently the target pathogen within a sample (Shen 2019). A qPCR designed to detect *P. cactorum* in strawberry samples demonstrated exceptional specificity and sensitivity, successfully identifying up to 10 zoospores per gram of plant part (Verdecchia et al. 2021).

2.4.3 Droplet Digital PCR (ddPCR)

ddPCR offers a precise approach for nucleic acid quantification. Briefly, a sample's DNA is divided into around 20,000 minuscule water-in-oil droplets, each preferably encapsulating either a single copy of the template DNA or none at all (Chen et al. 2021). Every droplet serves as an isolated PCR reaction vessel for amplifying a DNA segment specific to the target pathogen, if present. Intercalating dyes are employed to ascertain whether a PCR reaction has happened. These subsequent dews are then individually analyzed using microfluidics technology to measure their fluorescence (Hindson et al. 2011).

The number of droplets containing an amplified product allows the calculation of the original sample's template DNA quantity through Poisson statistics. Compared to real-time PCR, ddPCR offers several advantages. Most notably, it directly quantifies the amount of DNA present without the need for a calibration curve. This reduces the quantification's reliance on amplification efficiency, which can vary between real samples and calibration curves, thus enhancing reliability (Hayden et al. 2013).

In comparison with qPCR, ddPCR demonstrates increased resistance to PCR inhibitors and enhanced sensitivity. Unlike qPCR, which relies on amplification efficiency, ddPCR's end-point detection makes it less susceptible to these inhibitors (Taylor et al. 2017). Detection limits of 36 CFUs/20L and 5 CFUs/20L were revealed from the examination of sensitivity and inhibitor resistance of qPCR and ddPCR for *Xanthomonas citri* sp. citri detection, respectively. Notably, ddPCR exhibited superior performance with increased citrus leaf extract concentrations, highlighting its robustness against inhibitors (Zhao et al. 2016).

Although ddPCR holds promising potential for complex sample matrices like soil, it remains 2.3-fold more expensive than qPCR (Maheshwari et al. 2021) and lasts 2–3 times more due to its more intricate process, the requirement for reactions to reach the endpoint and microfluidics fluorescence measurements (Morcia et al. 2020). Finally, ddPCR's quantification dynamic range is narrower than that of qPCR due to the limited number of droplets that can lead to saturation with the target DNA, posing challenges for accurate quantification in high-concentration scenarios (Ricchi et al. 2017).

2.4.4 Isothermal Nucleic Acid Amplification Methods

While PCR-based techniques are commonly employed for detecting plant pathogens, their applicability is often limited by the necessity for a heat cycler and purified DNA. Alternatively, isothermal amplification methods can be used. These techniques utilize strand-displacing DNA polymerases as their central amplification mechanism, thus eliminating the need for thermal cycling equipment (Lau and Botella 2017). This enables assays to be conducted using cost-effective tools like heating blocks rather than requiring specialized thermal cyclers (Inanov et al. 2021). Isothermal amplification owes to the ability to work with less purified DNA. Loop-mediated isothermal amplification (LAMP) and recombinase polymerase amplification (RPA) are some examples of isothermal amplification methods (Becherer et al. 2020).

Loop-Mediated Isothermal Amplification (LAMP)

LAMP stands out as a highly cited and effective technique (Becherer et al. 2020). It revolves around the use of strand-displacing DNA polymerase and a set of inner and outer primers (2 primers each). The core principle involves the formation of single-stranded DNA with dumbbell-like structures at both ends due to the intramolecular similarity of the inner primers and the strand-displacing polymerase's action. The outer and inner primers target these dumbbell structures, and the polymerase generates new amplified DNA endlessly, consequential in elongated concatemers of a specific DNA sequence (Ivanov et al. 2021). Typically conducted at an average temperature, the results of LAMP are obtained in around 30 min excluding sample preparation (de Paz et al. 2020).

LAMP offers several advantages including the ability to perform isothermal amplification and measurement with cost-effective equipment, making it suitable for field use. The short analysis time and minimal sample preparation required (e.g., plant tissue grinding, lysis buffer addition, and sample boiling) improve its efficiency. Additionally, LAMP assays often tolerate PCR inhibitors, thus allowing the use of crude extracts. However, the sensitivity can be further enhanced with proper DNA extraction. Although traditionally considered as a qualitative technique, recent research suggests the feasibility of quantification, akin to qPCR, by linking the initial DNA or pathogen amount to the time taken to reach a signal threshold (Nguyen et al. 2020). While LAMP may not match qPCR's quantification accuracy, its utility is extended by the fact that signal quantification can be achieved using smartphone cameras.

However, a challenge associated with LAMP lies in its primer design process, which is more intricate than other methods, despite the availability of primer design software like "LAMP Designer" or "NEB LAMP". Additionally, LAMP is generally more suitable for qualitative assays, and while quantification is possible, it may be less accurate compared to qPCR (Jia et al. 2019).

Recombinase Polymerase Amplification (RPA)

RPA comprises several key components such as binding proteins, DNA recombinase, nucleotides, and primers. The process initiates with the formation of a complex between recombinase proteins and primers, which then seeks out the complementary sequence on the DNA template. Once identifying a match, the recombinase-primer composite invades the opposite sequence, displacing the complementary strand. To prevent rehybridization of the dislocated strand with the original template DNA, the complex is stabilized by ssDNA-binding proteins. The strand-displacement polymerase extends the hybridized primer using dNTPs. Isothermal strand invasion and elongation typically occur between 37 and 42 °C (Oliveira et al. 2021).

Unlike traditional PCR, RPA achieves rapid target DNA amplification without the need for temperature cycling. The continuous process often reaches its endpoint within 20 min (Ivanov et al. 2021). Detection is typically performed using a lateral flow device or gel electrophoresis at the endpoint. Point-of-care applications are facilitated using FAM and biotin-labeled forward and reverse primers in a lateral flow device, thus enabling quick visual identification of amplification products. The test line, consisting of immobilized streptavidin, interacts with nanoparticles containing conjugated anti-FAM antibodies on the test strip's conjugated release pad. The accumulation of the amplification product can be visually assessed at the test line. Real-time variations are also available and use fluorescent probes to monitor amplification (Li et al. 2018).

RPA's recent rapid growth can be attributed to its user-friendly operation, rapid study periods, and low working temperature (Oliveira et al. 2021). It can even function at body heat, albeit the slightly lower operating temperature might necessitate a more energy-efficient heating block (Li et al. 2018). Minimal sample preparation is required for RPA, significantly reducing preparation time. It exhibits enhanced resistance to PCR inhibitors, with some studies demonstrating its success with crude plant extracts (Li et al. 2018). This makes RPA particularly suitable for point-of-care applications. Being remarkably sensitive, RPA can detect 1 to 10 copies of master DNA per reaction. It can also be adapted to target RNA templates, such as to identify RNA viruses (Babu et al. 2018).

2.4.5 Hybridization Arrays

Hybridization arrays harness the remarkable ability of DNA strands to hybridize with their complementary counterparts, thus offering a valuable method for pathogen detection. This technique employs labeled sequence-specific probes that bind to specific genetic sequences of the target pathogen enabling its detection. While hybridization-based assays like FISH, and southern and northern blotting are often less effective for identifying plant diseases, hybridization allows for an immediate recognition of numerous diseases, overcoming the limited multiplexing capabilities of other methods discussed so far (Thies 2015).

Micro- and macro-arrays are two prominent types of hybridization arrays, sharing core functionalities while differing in probe density, quantity, and detection techniques. Micro-arrays consist of densely packed probe patches, with less than 200 μm in diameter, and immobilized on a glass slide. Typically, fluorescent probes label the target DNA, and laser-induced fluorescence microscopes are used for detection. On the other hand, macro-arrays feature less densely packed spots, larger than 300 μm, and halted on a membrane made up of nylon. Chemiluminescent markers are commonly employed for detection in this case (Urbez-Torres et al. 2015). Several hybridization arrays have been developed for the detection of bacterial, fungal, and viral plant diseases, showcasing their utility (Aslam et al. 2022).

2.4.6 CRISPR-Cas Detection System

The introduction of CRISPR-Cas molecular tools has been transformative in the field of molecular biology, enabling site-directed mutagenesis across various organisms (Doudna and Charpentier 2014). While these systems are prominently utilized for genome editing due to their specificity and adaptability, their potential in molecular diagnostics has been garnering increasing attention. Pathogen detection methods harnessing various Cas variants have been explored, often involving DNA extraction followed by Cas protein binding to a specific pathogen DNA motif (Huang et al. 2022). This requisite event generates a detectable signal that indicates the presence of pathogenic DNA, visible through fluorescence, electrochemistry, or on lateral flow devices (Wang et al. 2020).

CRISPR-Cas-based pathogen recognition offers several advantages. Notably, these methods are cost-effective, highly specific, sensitive, and adaptable to point-of-care diagnostics, and often not require advanced technology. The majority of assays can be completed within 2 h, providing timely results (Wang et al. 2020). While the inherent sensitivity of most CRISPR-Cas-based detection techniques lies in the picomolar range, coupling with target sequence preamplification can significantly enhance sensitivity. They frequently collaborate with isothermal amplification methods like LAMP, though amplification on the PCR-based is also a possibility, aligning with their potential as point-of-care tools (Kaminski et al. 2021).

Despite notable advancements, several limitations still restrict the broad application of CRISPR-Cas-based detection in real-world scenarios. Multiplexing capacity, though expanding, remains limited, with various ways to evaluate multiple pathogens in a single run. Sample preparation for CRISPR-Cas-based diagnostics often involves time-consuming steps, and pre-amplification stages are frequently necessary to enhance sensitivity and detect low titer infections, thus increasing cost and analysis time. Additionally, the single-nucleotide specificity, while being advantageous, can lead to false-negative results if a single mutation in the target gene is present, which becomes particularly concerning for viruses (Benzigar et al. 2021).

2.5 Nucleic Acid Sequencing Methods

The landscape of microbial identification has been transformed by the advent of DNA sequencing, Microorganism identification can be accurately determined through the sequencing of definite markers and comparison of the resultant sequences to reference databases (Barghouthi 2011). In contrast to traditional morphological and phenotypic tests, this method offers enhanced reliability and accuracy in microorganism identification (Tewari et al. 2011). In the last 15 years, the remarkable strides in sequencing technologies have propelled their utilization for both identification and detection purposes.

The first-generation sequencing machinery, chiefly the Sanger method, produce relatively lengthy reads of up to 1000 bp. In contrast, second-generation sequencing such as IonTorrent and Illumina yield shorter reads (100–300 bp) while boasting substantial throughput capabilities. Third-generation sequencers, e.g., Nanopore and PacBio, can use sequences for single molecules and produce ultra-extensive reads. PacBio sequencing, while requiring a lab setting and initial investment, offers heightened accuracy. Notably, the MinION platform of nanopore sequencing stands out due to its portability and affordability, thus holding promising plant pathogen identification (Loit et al. 2019). It is worth acknowledging that while nanopore sequencing presents portable and cost-effective benefits, its accuracy falls slightly short of the PacBio and older group sequencers (Hu et al. 2021).

2.5.1 Metagenomic Sequencing Method

Metagenomics involves the shotgun sequencing of every fragment of DNA present in a sample, providing a comprehensive view of its genetic diversity (Quince et al. 2017). In essence, DNA is extracted and fragmented into small parts, which are then sequenced on a massive scale (Sharpton 2014). Numerous sequencing reads are obtained and subsequently assembled to generate contiguous sequences, known as contigs, forming a metagenome. The subsequent analysis aims to extract valuable insights from the metagenome, either for microorganisms' identification or for information gleaning about their functional genes (Lapidus and Korobeynikov, 2021; Semenow 2021).

Assembly-based approaches, however, come with limitations when dealing with metagenome sequencing data. The assembly process demands substantial computational resources and intricate data analysis techniques. Challenges arise when distinguishing highly similar genomes of closely related species, hampering sequence overlap identification during assembly. Additionally, low-abundance species often lack sufficient sequencing coverage for generating substantial contigs (Quince et al. 2017). In such scenarios, assembly-free analytical techniques present a viable alternative. This method directly compares individual sequencing reads with a reference database (Lind and Pollard 2021), significantly reducing analysis time and obviating the need for complex assemblies. The acquired sequences are directly matched to reference genomes, mitigating issues related to low-abundance species detection. However, this strategy necessitates pertinent reference databases (Ayling et al. 2020) and can sometimes yield more unreal positive hits due to the potential extraction of sequences from globally conserved regions of the wrong microbe.

Metagenomic sequencing offers several advantages: (i) the wealth of information allows identification down to the strain level; (ii) it facilitates unbiased immediate detection of fungal and prokaryotic strains; (iii) it requires minimal prior heritable information of the specific phytopathogen (Aragona et al. 2022); and (iv) it enables the retrieval of genomes as of uncultivated microbes (Piombo et al. 2021).

This approach proves to be particularly valuable for detecting viral plant pathogens, given their lack of universally targeted genes in amplicon sequencing. Moreover, the metagenomic approach eliminates the need for prior knowledge, allowing the detection of yet-undiscovered plant pathogens (Adams and Fox 2016).

2.5.2 Amplicon Sequencing Method

Amplicon sequencing, a technique widely used for analyzing microbial community structures, involves sequencing a precise marker gene. This process depends on the amplification of a chosen marker, with its DNA sequences subsequently compared to a suitable reference database to determine organisms' identity (Bush et al. 2019). The marker gene must meet specific criteria for PCR amplification through designed universal primers: (i) it should be present across the entire spectrum of target microorganisms; (ii) it shall encompass conserved regions; and (iii) the latter must enclose variable regions acting as unique signatures for microorganism differentiation (Semenov 2021). Internal transcribed spacer within the ribosomal RNA of fungi is generally utilized as the marker gene for fungal identification, while the 16S ribosomal RNA gene is preferred for determining bacterial populations (Semenov 2021).

Amplicon sequencing offers several advantages. Firstly, it focuses solely on sequencing marker genes for identification, rendering it more cost-effective compared to metagenome sequencing which involves sequencing plant DNA or less informative sections of microbial genomes (Piombo et al. 2021). Secondly, user-friendly data analysis pipelines are widely available, thus simplifying the process (Vasar et al. 2021). Thirdly, numerous reference sequences can be employed for identification. Lastly, amplicon sequencing is more sensitive than metagenome sequencing, as PCR amplification enables the recognition of low-abundance species (Sekse et al. 2017).

However, this method does have limitations. It generally provides limited taxonomic resolution, usually identifying organisms only up to the genus or species levels. This constraint arises from the fact that only a short segment of a single marker gene is typically amplified (Abdelfattah et al. 2018). This becomes crucial for phytopathogen detection, as closely related species may exhibit infective or non-infective traits. Nevertheless, PacBio and nanopore platforms, enable the sequencing of larger marker genes compared to traditional methods like Illumina. This extended sequencing capability significantly enhances precise taxonomic classification, potentially enabling a better identification up to species or strain level (Graf et al. 2021).

2.6 Biosensor-Based Methods

Biosensors are innovative strategies composed of a biorecognition component and a physical–chemical transducer that are combined to produce a measurable signal.

These sensors are auspicious for accurate appliances due to their affordability, user-friendliness, and rapid results. The three prevalent types of transducers are electrochemical, mass-based, and optical. Electrochemical transducers detect binding events through variations in voltage and conductance. Mass-based transducers rely on alterations in mass to identify shifts in resonance frequency. Optical transducers detect binding events through variations in reflected incoming light (Hameed et al. 2018).

Numerous biorecognition essentials are employed in biosensors, mainly antibodies, enzymes, aptamers, and probes. Aptamers, due to their ease of labeling and conformational changes upon target binding, offer a robust choice for biosensors (Shahdordizadeh et al. 2017). Enzymes facilitate high-specificity and high-affinity conversion of specific target analytes. Although the enzyme-based approach is less effective for pathogen detection, it finds frequent use in detecting specific substrates. The selection of the optimal biorecognition component hinges on the transducer type and target molecule. Aptamers and antibodies are directed at specific antigens, while nucleic acid probes target distinct DNA sequences. Factors like labeling simplicity, immobilization feasibility, and sample type should be also considered as they impact the biosensor's cost, robustness, and sensitivity. The proper control of biorecognition components on the perception surface is vital for biosensor efficiency. It aims to secure the biorecognition element to the electrode, ensuring optimal packing and orientation of recognition elements (Cardoso et al. 2022).

Patel et al. (2023a, b) have advanced a DNA-based impedimetric sensor capable of detecting the agricultural disease *Ralstonia solanacearum* in soil samples. Whole-cell-based biosensors, such as genetically engineered *E. coli* strains to respond to volatile compounds during the infection process, are employed for *Penicillium digitatum* pathogen detection in oranges (Chalupowicz et al. 2020). Metal-based biosensors have proven to be effective in detecting the *Colletotrichum gloeosporioides* pathogen (Harpaz et al. 2022). These examples underscore the potential of biosensors across a range of applications within the field of plant-pathogen identification.

3 The Role of Nanotechnology in Pest Management

Nanotechnology, derived from the Greek word meaning "dwarf," refers to the manipulation of materials at a nanoscale, typically within the range of 1 to 100 nm (nm) (Bhattacharyya et al. 2010). This emerging field combines science and technology to address a spectrum of challenges across various sectors, including agriculture, food, medicine, and engineering, among others (Kumar et al. 2017; El Wakeil et al. 2017; Khan et al. 2023; Sarmah et al. 2023; Rani et al. 2023; Parveen et al. 2023). Nanotechnology offers innovative solutions to complex issues like resource scarcity, climate change, and pest management, making it particularly relevant in agricultural practices (Duhan et al. 2017; Adisa et al. 2019; Bahrulolum et al. 2021; Kumar et al. 2017) (Figs. 4, 5).

Fig. 4 Role of nanotechnology in plant disease management

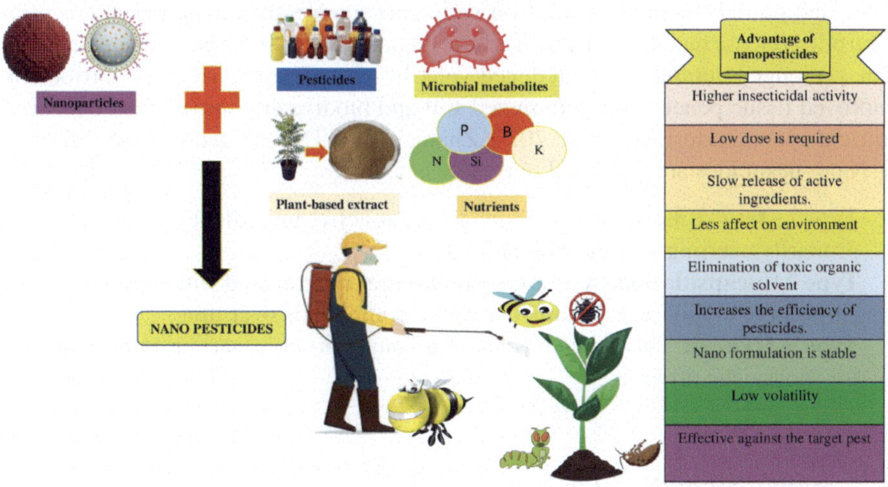

Fig. 5 Nanopesticides and their application

3.1 Applications in Pest Management

Traditional pest management methods, such as pesticide application and integrated pest management, often come with negative consequences like environmental contamination, non-target species impact, and soil degradation (Rai and Ingle 2012). For instance, Christopher et al. (2020) reported that a substantial portion of insecticides applied to crops ends up affecting non-target organisms and polluting air, water, and soil. More than 90% of sprayed pesticides can leach into the environment (Urkude 2019).

To address these issues, a transition toward green technologies, involving plant-derived materials and microbes for nanoparticle synthesis, is emerging (Bhandari et al. 2023). Nanotechnology employs nano-biosensors for early-stage pathogen detection and eradication (Prasad et al. 2014). These biosensors incorporate biological sensing elements (enzymes, cells, antibodies, nucleic acids, aptamers, among others) that detect signals from diseased plants (Dhaigude et al. 2023). Transducers, often composed of semiconducting nanomaterials, convert these signals into electrical ones for further analysis.

3.2 Nano-Pesticides

Nanotechnology's impact is evident in the creation of nano-pesticides, which involve encapsulating active pesticide ingredients on a nanoscale (Gahukar and Das 2020; de Albuquerque et al. 2021). These nano-formulations exhibit repellent, toxic, and growth-retarding effects on various pests. Notably, nano-pesticides offer controlled release mechanisms, high efficacy against target species, enhanced adsorption, prolonged durability, and reduced volatilization. These attributes contribute to improved tissue penetration, minimized soil and biodiversity impact, and enhanced nutrient cycling (Hofmann et al. 2020; Zhang et al. 2021; Srivastava et al. 2023).

Nano-pesticides are classified into three categories:

1. **Type I Metal-Based:** Inorganic metals act as active ingredients without carriers, primarily exerting antimicrobial effects.
2. **Type 2 Encapsulation:** Active ingredients are encapsulated within polymers like chitosan or cellulose, releasing the active compounds over time.
3. **Type 3 Hybrid:** This type consists of a combination of various nanostructures for synergistic effects.

3.3 Examples and Effects

The application of 125–250 ppm nano alumina dust, comprising aluminum oxide, exhibited a 100% mortality rate among rice weevils (*Sitophilus oryzae*) (Buteler et al. 2015). Green-synthesized silver nanoparticles from *Nerium oleander* demonstrated larvicidal activity against *Tribolium castaneum* (Jafor and Annon 2018). Additionally, diacyl hydrazine-based nano-formulations exhibited anti-pest activity against *Spodoptera litura* (Pandey et al. 2020) (Table 1).

Table 1 Nanotechnology for disease management

Nano-pesticides	Property	Target pest	References
pH-responsive λ-cyhalothrin nano pesticides	Release of active ingredients in a controlled and sustainable manner	*Harmonia axyridis* (Asian lady beetle)	Hou et al. (2023)
Silicon dioxide-based micro-electro-mechanical system	Early pest detection	Pheromones of *Helicoverpa armigera* (cotton ball worm) and *Bactrocera oleae* (olive fruit fly)	Srujana and Bhagat (2022)
Neem nano formulation with adjuvant	Insecticidal activity	Whitefly (*Bemisia tabaci* G.) in brinjal (*Solanum melongena*)	Iqbal et al. (2022)
Emamectin Benzoate functionalized polysuccinimide with glycine methylester nanoparticles	Insecticidal activity	*Plutella xylostella* (Diamond back moth)	Ye et al. (2022)
Green synthesized nanoparticles	Fumigant, acute toxicity, attractancy, repellency, reproductive inhibition	All stages of Mosquito (egg, larva, and adult)	Pathipati and Kanurparthi (2021)
Neem gum nano-formulations	Pupicidal, antifeedant, larvicidal activity	*Helicoverpa armigera* (Cotton ball worm) and *Spodoptera litura* (tobacco cutworm)	Kamaraj et al. (2018)
Silica nanoparticles	Direct abrasion of insect cuticles	*Spodoptera littoralis* (cotton leafworm)	Ayoub et al. (2017)
Nano calcium formulation	Protection efficacy against oviposition punctures	California red scale and *Bactrocera dorsalis*	Hua et al. (2015)
Imidacloprid-loaded sodium alginate NPs	Insecticidal	Sucking pest (Leaf hopper)	Pradhan et al. (2013)
Nano-silver	Antifungal effect	*Sphaerotheca pannosa* (causes rose powdery mildew)	

4 Conclusion

The timely and accurate diagnosis of plant diseases plays a pivotal role in mitigating losses and minimizing the impact of plant damage on a global scale. Traditional diagnostic methods, while being effective, often suffer from drawbacks such as time-intensive processes, the need for specialized equipment, and reliance on experienced personnel. However, the landscape of disease diagnosis is undergoing a transformative shift with the advent of nanotechnologies. Recent advancements in these fields have led to the emergence of biosensors, which offer innovative solutions to these challenges. Biosensors leverage the power of nanoscale technologies to revolutionize the way plant diseases are detected and managed. These cutting-edge tools have provided faster and more accurate identification of pathogens, enabling early intervention measures that can significantly curtail crop damage and economic losses. One of the notable advantages of biosensors lies in their potential for large-scale applications, such as monitoring infections in agricultural settings. By swiftly detecting the initial stages of disease, biosensors empower farmers and experts to implement control strategies at the earliest feasible phase. This proactive approach not only prevents the spread of infections but also promotes sustainable agricultural practices by minimizing the need for excessive pesticide use or other interventions. As the nanotechnology field continues to evolve, the potential for biosensors to transform plant disease diagnosis and management is only beginning to be realized. Through the integration of these innovative tools into various aspects of agriculture, from precision farming to disease monitoring, biosensors hold the promise of revolutionizing the way crops are safeguarded and food security is ensured in a rapidly changing world. The era of biosensor-enabled early disease diagnosis is ushering in a new chapter in the history of plant protection and crop management.

References

Abdelfattah A, Malacrinò A, Wisniewski M, Cacciola SO, Schena L (2018) Metabarcoding: a powerful tool to investigate microbial communities and shape future plant protection strategies. Biol Control 120:1–10. https://doi.org/10.1016/j.biocontrol.2017.07.009

Adams I, Fox A (2016) Diagnosis of plant viruses using next-generation sequencing and metagenomic analysis. In: Wang A, Zhou X (eds) Current research topics in plant virology. Springer International Publishing. https://doi.org/10.1007/978-3-319-32919-2_14

Adisa IO, Pullagurala VLR, Peralta-Videa JR, Dimkpa CO, Elmer WH, Gardea-Torresdey JL, White JC (2019) Recent advances in nano-enabled fertilizers and pesticides: a critical review of mechanisms of action. Environ Sci Nano 6(7):2002–2030. https://doi.org/10.1039/C9EN00265K

Alhajj M, Farhana A (2023) Enzyme linked immunosorbent assay. In: StatPearls (Treasure Island (FL). StatPearls Publishing. Available at: http://www.ncbi.nlm.nih.gov/books/NBK555922/

Anfossi L, Di Nardo F, Cavalera S, Giovannoli C, Baggiani C (2019) Multiplex lateral flow immunoassay: an overview of strategies towards high-throughput point-of-need testing. Biosensors 9(1). https://doi.org/10.3390/bios9010002

Antil S, Abraham JS, Sripoorna S, Maurya S, Dagar J, Makhija S (2023) DNA barcoding, an effective tool for species identification: a review. Mol Biol Rep 50(1):761–775. https://doi.org/10.1007/s11033-022-08015-7

Aragona M, Haegi A, Valente MT, Riccioni L, Orzali L, Vitale S (2022) New-generation sequencing technology in diagnosis of fungal plant pathogens: a dream comes true? J. Fungi 8(7):737. https://doi.org/10.3390/jof8070737

Ascoli CA, Aggeler B (2018) Overlooked benefits of using polyclonal antibodies. Biotechniques 65(3):127–136. https://doi.org/10.2144/btn-2018-0065

Aslam HMU, Riaz H, Killiny N, Zhou X-G, Thomashow L, Peters N (2022) Microarray technology for detection of plant diseases. In: Ul Haq I, Ijaz S (eds) Trends in plant disease assessment. Springer, Singapore, pp 203–223. https://doi.org/10.1007/978-981-19-5896-0_11

Atmar RL (2014) Immunological detection and characterization. In: Kaslow RA, Stanberry LR, Le Duc JW (eds) Viral infections of humans: epidemiology and control. Springer, Boston, MA, pp 47–62. https://doi.org/10.1007/978-1-4899-7448-8_3

Ayling M, Clark MD, Leggett RM (2020) New approaches for metagenome assembly with short reads. Briefings Bioinf 21(2):584–594. https://doi.org/10.1093/bib/bbz020

Ayoub HA, Khairy M, Rashwan FA, Abdel-Hafez HF (2017) Synthesis and characterization of silica nanostructures for cotton leaf worm control. J Nanostruct Chem 7:91–100. https://doi.org/10.1007/s40097-017-0229-2

Babu B, Ochoa-Corona FM, Paret ML (2018) Recombinase polymerase amplification applied to plant virus detection and potential implications. Anal Biochem 546:72–77. https://doi.org/10.1016/j.ab.2018.01.021

Bahrulolum H, Nooraei S, Javanshir N, Tarrahimofrad H, Mirbagheri VS, Easton AJ, Ahmadian G (2021) Green synthesis of metal nanoparticles using microorganisms and their application in the agrifood sector. J Nanobiotechnol 19(1):1–26. https://doi.org/10.1186/s12951-021-00834-3

Baldi P, La Porta N (2020) Molecular approaches for low-cost point-of-Care pathogen detection in agriculture and forestry. Front Plant Sci. https://doi.org/10.3389/fpls.2020.570862

Barghouthi SA (2011) A universal method for the identification of bacteria based on general PCR primers. Indian J Microbiol 51(4):430–444. https://doi.org/10.1007/s12088-011-0122-5

Bari ML, Kawasaki S (2014) Rapid methods for food hygiene inspection. In: Batt CA, Tortorello ML (eds) Encyclopedia of food microbiology. Academic Press, pp 269–279. https://doi.org/10.1016/B978-0-12-384730-0.00287-1

Becherer L, Borst N, Bakheit M, Frischmann S, Zengerle R, von Stetten F (2020) Loop-mediated isothermal amplification (LAMP)—review and classification of methods for sequence-specific detection. Anal Methods 12(6):717–746. https://doi.org/10.1039/C9AY02246E

Benzigar MR, Bhattacharjee R, Baharfar M, Liu G (2021) Current methods for diagnosis of human coronaviruses: pros and cons. Anal Bioanal Chem 413(9):2311–2330. https://doi.org/10.1007/s00216-020-03046-0

Bhandari G, Dhasmana A, Chaudhary P, Gupta S, Gangola S, Gupta A, Rustagi S, Shende SS, Rajput VD, Minkina T, Malik S, Slama P (2023) A perspective review on green nanotechnology in agro-ecosystems: opportunities for sustainable agricultural practices and environmental remediation. Agriculture 13:668. https://doi.org/10.3390/agriculture13030668

Bhattacharyya A, Bhaumik A, Rani PU, Mandal S, Epidi TT (2010) Nano-particles-a recent approach to insect pest control. Afr J Biotechnol 9(24):3489–3493

Bush A, Compson ZG, Monk WA, Porter TM, Steeves R, Emilson E (2019) Studying ecosystems with DNA metabarcoding: lessons from biomonitoring of aquatic macroinvertebrates. Front Ecol Evol. https://doi.org/10.3389/fevo.2019.00434

Buteler M, Sofie SW, Weaver DK, Driscoll D, Muretta J, Stadler T (2015) Development of nanoalumina dust as insecticide against *Sitophilus oryzae* and *Rhyzopertha dominica*. Int J Pest Manage 61(1):80–89. https://doi.org/10.1080/09670874.2014.1001008

Cardoso RM, Pereira TS, Facure MHM, dos Santos DM, Mercante LA, Mattoso LHC (2022) Current progress in plant pathogen detection enabled by nanomaterials-based (bio)sensors. Sens Actuators Rep 4:100068. https://doi.org/10.1016/j.snr.2021.100068

Castro-Escarpulli G, Alonso-Aguilar NM, Rivera G, Bocanegra-Garcia V, Guo X, Jurez-Enrquez SR (2015). Identification and typing methods for the study of bacterial infections: a brief review and mycobacterial as case of study. Arch Clin Microbiol 7(1)

Chalupowicz D, Veltman B, Droby S, Eltzov E (2020) Evaluating the use of biosensors for monitoring of *Penicillium digitatum* infection in citrus fruit. Sens Actuators, B Chem 311:127896

Chen B, Jiang Y, Cao X, Liu C, Zhang N, Shi D (2021) Droplet digital PCR as an emerging tool in detecting pathogens nucleic acids in infectious diseases. Clin Chim Acta 517:156–161. https://doi.org/10.1016/j.cca.2021.02.008

Christopher FC, Kumar PS, Christopher FJ, Joshiba GJ, Madhesh P (2020) Recent advancements in rapid analysis of pesticides using nano biosensors: a present and future perspective. J Clean Prod 269:122356. https://doi.org/10.1016/j.jclepro.2020.122356

Chun S, Gopal J, Muthu M (2022) Comprehensive synopsis of the maldi tof Ms accomplishments in rapid diagnosis of microbial. Plant Dis Pathog. https://doi.org/10.2139/ssrn.4100342

de Albuquerque FP, de Oliveira JL, dos Santos Machado L, Richardi VS, da Silva MAN, Pompêo MLM, Carlos VM (2021) Use of nontarget organism *Chironomus sancticaroli* to study the toxic effects of nanoatrazine. Ecotoxicology 30:733–750. https://doi.org/10.1007/s10646-021-02400-x

de Paz HD, Brotons P, Esteva C, Muñoz-Almagro C (2020) Validation of a loop-mediated isothermal amplification assay for rapid diagnosis of invasive pneumococcal disease. Front Cell Infect Microbiol. https://doi.org/10.3389/fcimb.2020.00115

De Boer SH, López MM (2012) New grower-friendly methods for plant pathogen monitoring. Annu Rev Phytopathol 50(1):197–218. https://doi.org/10.1146/annurev-phyto-081211-172942

Dhaigude V, Naliyapara HB, Chaudhary P, Fayssal SA, Nadda AK, Minkina T, Rajput VD (2023) Role of nanobiosensors in agriculture advancement. In: Chaudhary P, Chaudhary A, Nadda AK, Khati P (eds) Advances in nanotechnology for smart agriculture: techniques and applications, 1st ed. CRC Press. https://doi.org/10.1201/9781003345565-9

Dong B, Jiang R, Chen J-F, Xiao Y, Zong-you L, Chen W-S (2023) Strategic nanoparticle-mediated plant disease resistance. Crit Rev Biotechnol 43:22–37. https://doi.org/10.1080/07388551.2021.2007842

Doudna JA, Charpentier E (2014) The new frontier of genome engineering with CRISPR-Cas9. Science 346(6213):1258096. https://doi.org/10.1126/science.1258096

Duhan JS, Kumar R, Kumar N, Kaur P, Nehra K, Duhan S (2017) Nanotechnology: the new perspective in precision agriculture. Biotechnology Reports 15:11–23. https://doi.org/10.1016/j.btre.2017.03.002

El Wakeil N, Alkahtani S, Gaafar N (2017) Is nanotechnology a promising field for insect pest control in IPM programs? In: New pesticides and soil sensors. Academic Press, pp 273–309. https://doi.org/10.1016/B978-0-12-804299-1.00008-4

EPPO (2022) EPPO standards. Available at: https://www.eppo.int/RESOURCES/https%3A%2F%2Feppo_standards

EU (2019) A European green deal. In: European Commission—European commission. Available at: https://ec.europa.eu/info/strategy/priorities-2019-2024/european-green-deal_en

Fang Y, Ramasamy RP (2015) Current and prospective methods for plant disease detection. Biosensors 5(3):537–561. https://doi.org/10.3390/bios5030537

FAO (2017) The future of food and agriculture: trends and challenges. Food and Agriculture Organization of the United Nations, pp 11–16

FAO (2019) New standards to curb the global spread of plant pests and diseases. Available at: https://www.fao.org/news/story/en/item/1187738/icode/

Ferone M, Gowen A, Fanning S, Scannell AGM (2020) Microbial detection and identification methods: bench top assays to omics approaches. Compr Rev Food Sci Food Saf 19(6):3106–3129. https://doi.org/10.1111/1541-4337.12618

Gahukar RT, Das RK (2020) Plant-derived nanopesticides for agricultural pest control: challenges and prospects. Nanotechnol Environ Eng 5:1–9. https://doi.org/10.1007/s41204-020-0066-2

Gautam AK, Kumar S (2020) Techniques for the detection, identification, and diagnosis of agricultural pathogens and diseases. In: Natural remedies for pest, disease and weed control. Academic Press, pp 135–142

Gorris MT, Sanz A, Peñalver J, López MM, Colomer M, Marco-Noales E (2021) Detection and diagnosis of *Xylella fastidiosa* by specific monoclonal antibodies. Agronomy 11(1). https://doi.org/10.3390/agronomy11010048

Graf J, Ledala N, Caimano MJ, Jackson E, Gratalo D, Fasulo D (2021) High-resolution differentiation of enteric bacteria in premature infant fecal microbiomes using a novel rRNA amplicon. Mbio 12(1):e03656-e13620. https://doi.org/10.1128/mBio.03656-20

Hameed S, Xie L, Ying Y (2018) Conventional and emerging detection techniques for pathogenic bacteria in food science: a review. Trends Food Sci Technol 81:61–73. https://doi.org/10.1016/j.tifs.2018.05.020

Harpaz D, Duanis-Assaf D, Alkan N, Eltzov E (2022) Detection of postharvest pathogenic fungi by RNA markers in high-throughput platform on streptavidin plate. Postharvest Biol Technol 183:111728. https://doi.org/10.1016/j.postharvbio.2021.111728

Haselbeck AH, Im J, Prifti K, Marks F, Holm M, Zellweger RM (2022) Serology as a tool to assess infectious disease landscapes and guide public health policy. Pathogens 11(7). https://doi.org/10.3390/pathogens11070732

Hayden RT, Gu Z, Ingersoll J, Abdul-Ali D, Shi L, Pounds S (2013) Comparison of droplet digital PCR to real-time PCR for quantitative detection of cytomegalovirus. J Clin Microbiol 51(2):540–546. https://doi.org/10.1128/JCM.02620-12

Hindson BJ, Ness KD, Masquelier DA, Belgrader P, Heredia NJ, Makarewicz AJ (2011) High-throughput droplet digital PCR system for absolute quantitation of DNA copy number. Anal Chem 83(22):8604–8610. https://doi.org/10.1021/ac202028g

Hofmann T, Lowry GV, Ghoshal S, Tufenkji N, Brambilla D, Dutcher JR, Wilkinson KJ (2020) Technology readiness and overcoming barriers to sustainably implement nanotechnology-enabled plant agriculture. Nat Food. 1(7):416–425. https://doi.org/10.1038/s43016-020-0110-1

Hou R, Zhou J, Song Z, Zhang N, Huang S, Kaziem AE, Zhang Z (2023) PH-responsive λ-cyhalothrin nanopesticides for effective pest control and reduced toxicity to Harmonia axyridis. Carbohydrate Poly 302:120373. https://doi.org/10.1016/j.carbpol.2022.120373

Hu T, Chitnis N, Monos D, Dinh A (2021) Next-generation sequencing technologies: an overview. Hum Immunol 82(11):801–811. https://doi.org/10.1016/j.humimm.2021.02.012

Hua KH, Wang HC, Chung RS, Hsu JC (2015) Calcium carbonate nanoparticles can enhance plant nutrition and insect pest tolerance. J Pesticide Sci 40(4):208–213. https://doi.org/10.1584/jpestics.D15-025

Huang T, Zhang R, Li J (2022) CRISPR-cas-based techniques for pathogen detection: retrospect, recent advances, and future perspectives. J Adv Res. https://doi.org/10.1016/j.jare.2022.10.011

Iqbal N, Hazra DK, Purkait A, Agrawal A, Kumar J (2022) Bioengineering of neem nano-formulation with adjuvant for better adhesion over applied surface to give long term insect control. Colloids Surf B 209:112176. https://doi.org/10.1016/j.colsurfb.2021.112176

Ivanov AV, Safenkova IV, Zherdev AV, Dzantiev BB (2021) The potential use of isothermal amplification assays for in-field diagnostics of plant pathogens. Plants 10(11):2424. https://doi.org/10.3390/plants10112424

Jafer FS, Annon MR (2018) Larvicidal effect of pure and green-synthesized silver nanoparticles against *Tribolium castaneum* (herb.) and *Callosobruchus maculatus* (fab.). J Global Pharma Technol 10(3):448–454

Jia B, Li X, Liu W, Lu C, Lu X, Ma L (2019) GLAPD: whole genome-based LAMP primer design for a set of target genomes. Front Microbiol. https://doi.org/10.3389/fmicb.2019.02860

Kamaraj C, Gandhi PR, Elango G, Karthi S, Chung IM, Rajakumar G (2018) Novel and environmental friendly approach; Impact of Neem (*Azadirachta indica*) gum nano formulation (NGNF) on *Helicoverpa armigera* (Hub.) and *Spodoptera litura* (Fab.). Int J Biol Macromol 107:59–69. https://doi.org/10.1016/j.ijbiomac.2017.08.145

Kaminski MM, Abudayyeh OO, Gootenberg JS, Zhang F, Collins JJ (2021) CRISPR-based diagnostics. Nat Biomed Eng 5(7). https://doi.org/10.1038/s41551-021-00760-7

Katikireddy K, O'Sullivan F (2011) Immunohistochemical and immunofluorescence procedures for protein analysis. Methods Mol Biol 784:155–167. https://doi.org/10.1007/978-1-61779-289-2_11

Khan N, Sudhakar K, Mamat R (2023) Seaweed farming: a perspectives of genetic engineering and nano-technology application. Heliyon. https://doi.org/10.1016/j.heliyon.2023.e15168

Khater M, Escosura-Muñiz A, Merkoçi A (2017) Biosensors for plant pathogen detection. Biosens Bioelectron 93:72–86. https://doi.org/10.1016/j.bios.2016.09.091

Koczula KM, Gallotta A (2016) Lateral flow assays. Essays Biochem 60(1):111–120. https://doi.org/10.1042/EBC20150012

Kralik P, Ricchi M (2017) A basic guide to real time PCR in microbial diagnostics: definitions, parameters, and everything. Front Microbiol. https://doi.org/10.3389/fmicb.2017.00108

Krivitsky V, Granot E, Avidor Y, Borberg E, Voegele RT, Patolsky F (2021) Rapid collection and aptamer-based sensitive electrochemical detection of soybean rust fungi airborne urediniospores. ACS Sens 6(3):1187–1198. https://doi.org/10.1021/acssensors.0c02452

Kumar M, Shamsi TN, Parveen R, Fatima S (2017) Application of nanotechnology in enhancement of crop productivity and integrated pest management. Nanotechnol Agric Paradigm 361–371

Lacey RF, Sullivan-Hill BA, Deslippe JR, Keyzers RA, Gerth ML (2021) The fatty acid methyl ester (FAME) profile of phytophthora agathidicida and its potential use as diagnostic tool. FEMS Microbiol Lett 368(17):fnab113. https://doi.org/10.1093/femsle/fnab113

Lapidus AL, Korobeynikov AI (2021) Metagenomic data assembly—the way of decoding unknown microorganisms. Front Microbiol. https://doi.org/10.3389/fmicb.2021.613791

Lau HY, Botella JR (2017) Advanced DNA-based point-of-care diagnostic methods for plant diseases detection. Front Plant Sci. https://doi.org/10.3389/fpls.2017.02016

Law JW-F, Ab Mutalib N-S, Chan K-G, Lee L-H (2015) Rapid methods for the detection of food-borne bacterial pathogens: principles, applications, advantages and limitations. Front Microbiol. https://doi.org/10.3389/fmicb.2014.00770

Legrand P (2015) Biological assays for plant viruses and other graft-transmissible pathogens diagnoses: a review. EPPO Bull 45(2):240–251. https://doi.org/10.1111/epp.12222

Li J, Macdonald J, Von SF (2018) Review: a comprehensive summary of a decade development of the recombinase polymerase amplification. Analyst 144(1):31–67. https://doi.org/10.1039/C8AN01621F

Li J, Zhu Y, Wu X, Hoffmann MR (2020) Rapid detection methods for bacterial pathogens in ambient waters at the point of sample collection: a brief review. Clin Infect Dis 71(Supplement_2):S84–S90. https://doi.org/10.1093/cid/ciaa498

Lind AL, Pollard KS (2021) Accurate and sensitive detection of microbial eukaryotes from whole metagenome shotgun sequencing. Microbio 9(1):58. https://doi.org/10.1186/s40168-021-01015-y

Loit K, Adamson K, Bahram M, Puusepp R, Anslan S, Kiiker R (2019) Relative performance of MinION (Oxford nanopore technologies) versus sequel (Pacific biosciences) third-generation sequencing instruments in identification of agricultural and forest fungal pathogens. Appl Environ Microbiol 85(21):e01368-e11319. https://doi.org/10.1128/AEM.01368-19

Lopez-Soriano P, Noguera P, Gorris MT, Puchades R, Maquieira Á, Marco-Noales E (2017) Lateral flow immunoassay for on-site detection of *Xanthomonas arboricola* pv. pruni in symptomatic field samples. PloS One 12(4). https://doi.org/10.1371/journal.pone.0176201

Maheshwari Y, Selvaraj V, Godfrey K, Hajeri S, Yokomi R (2021) Multiplex detection of "*Candidatus liberibacter asiaticus*" and *Spiroplasma citri* by qPCR and droplet digital PCR. PLoS ONE 16(3):e0242392. https://doi.org/10.1371/journal.pone.0242392

Mahlein A-K (2016) Plant disease detection by imaging sensors – parallels and specific demands for precision agriculture and plant phenotyping. Plant Dis 100(2):241–251. https://doi.org/10.1094/PDIS-03-15-0340-FE

Mancini V, Murolo S, Romanazzi G (2016) Diagnostic methods for detecting fungal pathogens on vegetable seeds. Plant Pathol 65(5):691–703. https://doi.org/10.1111/ppa.12515

Martinelli F, Scalenghe R, Davino S, Panno S, Scuderi G, Ruisi P (2015) Advanced methods of plant disease detection: a review. Agron Sustain Dev 35(1):1–25. https://doi.org/10.1007/s13593-014-0246-1

Maurer JJ (2011) Rapid detection and limitations of molecular techniques. Annu Rev Food Sci Technol 2(1):259–279. https://doi.org/10.1146/annurev.food.080708.100730

McDonald BA, Stukenbrock EH (2016) Rapid emergence of pathogens in agro-ecosystems: Global threats to agricultural sustainability and food security. Philos Trans R Soc b: Biol Sci 371(1709):20160026. https://doi.org/10.1098/rstb.2016.0026

Mehetre GT, Leo VV, Singh G, Sorokan A, Maksimov I, Yadav MK (2021) Current developments and challenges in plant viral diagnostics: a systematic review. Viruses 13(3):412. https://doi.org/10.3390/v13030412

Morcia C, Ghizzoni R, Delogu C, Andreani L, Carnevali P, Terzi V (2020) Digital PCR: what relevance to plant studies? Biology 9(12):433. https://doi.org/10.3390/biology9120433

Nguyen HQ, Nguyen VD, Van Nguyen H, Seo TS (2020) Quantification of colorimetric isothermal amplification on the smartphone and its open-source app for point-of-care pathogen detection. Sci Rep 10(1). https://doi.org/10.1038/s41598-020-72095-3

Oliveira BB, Veigas B, Baptista PV (2021) Isothermal amplification of nucleic acids: the race for the next "Gold standard." Front Sens. https://doi.org/10.3389/fsens.2021.752600

Pandey A, Srivastava S, Aggarwal N, Srivastava C, Adholeya A, Kochar M (2020) Assessment of the pesticidal behaviour of diacyl hydrazine-based ready-to-use nanoformulations. Chem Biol Technol Agric 7:1–15. https://doi.org/10.1186/s40538-020-0177-9

Parveen H, Chaudhary P, Srivastava P, Priya S, Bhandari G, Chaudhary A (2023) Role of nanoparticles in agriculture. In: Chaudhary P, Chaudhary A, Nadda AK, Khati P (eds) Advances in nanotechnology for smart agriculture: techniques and applications, 1st edn. CRC Press. https://doi.org/10.1201/9781003345565-1

Patel R, Vinchurkar M, Shaikh AM, Patkar R, Adami A, Giacomozzi F, Ramesh R, Pramanick B, Lorenzelli L, Baghini MS (2023b) Part II: Impedance-based DNA biosensor for detection of isolated strains of phytopathogen *Ralstonia solanacearum*. Bioelectrochemistry 153:108500. https://doi.org/10.1016/j.bioelechem.2023.108500

Patel R, Mitra B, Vinchurkar M, Adami A, Patkar R, Giacomozzi F, Lorenzelli L, Baghini MS (2022) A review of recent advances in plant-pathogen detection systems. Heliyon 8:e11855. https://doi.org/10.1016/j.heliyon.2022.e11855

Patel R, Mitra B, Vinchurkar M, Adami A, Patkar R, Giacomozzi F, Lorenzelli L, Baghini MS (2023a) Plant pathogenicity and associated/related detection systems. A review. Talanta 251:123808. https://doi.org/10.1016/j.talanta.2022.123808

Pathipati UR, Kanuparthi PL (2021) Silver nanoparticles for insect control: bioassays and mechanisms. In: Silver nanomaterials for agri-food applications. Elsevier, pp 471–494. https://doi.org/10.1016/B978-0-12-823528-7.00007-X

Paul R, Ostermann E, Wei Q (2020) Advances in point-of-care nucleic acid extraction technologies for rapid diagnosis of human and plant diseases. Biosens Bioelectron 169:112592. https://doi.org/10.1016/j.bios.2020.112592

Piombo E, Abdelfattah A, Droby S, Wisniewski M, Spadaro D, Schena L (2021) Metagenomics approaches for the detection and surveillance of emerging and recurrent plant pathogens. Microorganisms 9(1). https://doi.org/10.3390/microorganisms9010188

Pradhan S, Roy I, Lodh G, Patra P, Choudhury SR, Samanta A, Goswami A (2013) Entomotoxicity and biosafety assessment of PEGylated acephate nanoparticles: a biologically safe alternative to neurotoxic pesticides. J Environ Sci Health Part B 48(7):559–569. https://doi.org/10.1080/03601234.2013.774891

Prasad R, Kumar V, Prasad KS (2014) Nanotechnology in sustainable agriculture: present concerns and future aspects. Afr J Biotech 13(6):705–713. https://doi.org/10.5897/AJBX2013.13554

Priyanka B, Patil RK, Dwarakanath S (2016) A review on detection methods used for foodborne pathogens. Indian J Med Res 144(3):327–338. https://doi.org/10.4103/0971-5916.198677

Quince C, Walker AW, Simpson JT, Loman NJ, Segata N (2017) Shotgun metagenomics, from sampling to analysis. Nat Biotechnol 35(9). https://doi.org/10.1038/nbt.3935

Rai M, Ingle A (2012) Role of nanotechnology in agriculture with special reference to management of insect pests. Appl Microbiol Biotechnol 94:287–293. https://doi.org/10.1007/s00253-012-3969-4

Rajagopal A, Yurk D, Shin C, Menge K, Jacky L, Fraser S (2019) Significant expansion of real-time PCR multiplexing with traditional chemistries using amplitude modulation. Sci Rep 9(1). https://doi.org/10.1038/s41598-018-37732-y

Rani A, Kumar T, Trivedi B, Chaudhary P (2023) Nanotechnology in climate smart agriculture. In: Chaudhary P, Chaudhary A, Nadda AK, Khati P (eds) Advances in nanotechnology for smart agriculture: techniques and applications, 1st edn. CRC Press. https://doi.org/10.1201/9781003345565-14

Raza MM, Harding C, Liebman M, Leandro LF (2020) Exploring the potential of high-resolution satellite imagery for the detection of soybean sudden death syndrome. Remote Sens 12(7). https://doi.org/10.3390/rs12071213

Ricchi M, Bertasio C, Boniotti MB, Vicari N, Russo S, Tilola M (2017) Comparison among the quantification of bacterial pathogens by qPCR, dPCR, and cultural methods. Front Microbiol. https://doi.org/10.3389/fmicb.2017.01174

Ristaino JB, Anderson PK, Bebber DP, Brauman KA, Cunniffe NJ, Fedoroff NV, Finegold C, Garrett KA, Gilligan CA, Jones CM, Martin MD, MacDonald GK, Neenan P, Records A, Schmale DG, Tateosian L, Wei Q (2021) The persistent threat of emerging plant disease pandemics to global food security. Proc Natl Acad Sci U S A 118(23):e2022239118. https://doi.org/10.1073/pnas.2022239118

Sarmah K, Borah R, Boruah S, Sarmah D (2023) Nanopesticides: revolutionizing pest management with nanotechnology

Savary S, Willocquet L, Pethybridge SJ, Esker P, McRoberts N, Nelson A (2019) The global burden of pathogens and pests on major food crops. Nat Ecol Evol 3(3):430–439. https://doi.org/10.1038/s41559-018-0793-y

Schostag MD, Albers CN, Jacobsen CS, Priemé A (2020) Low turnover of soil bacterial rRNA at low temperatures. Front Microbiol. https://doi.org/10.3389/fmicb.2020.00962

Sekse C, Holst-Jensen A, Dobrindt U, Johannessen GS, Li W, Spilsberg B (2017) High throughput sequencing for detection of foodborne pathogens. Front Microbiol. https://doi.org/10.3389/fmicb.2017.02029

Semenov MV (2021) Metabarcoding and metagenomics in soil ecology research: achievements, challenges, and prospects. Biol Bull Rev 11(1):40–53. https://doi.org/10.1134/S2079086421010084

Shahdordizadeh M, Taghdisi SM, Ansari N, Alebooye Langroodi F, Abnous K, Ramezani M (2017) Aptamer based biosensors for detection of *Staphylococcus aureus*. Sens Actuators b: Chem 241:619–635. https://doi.org/10.1016/j.snb.2016.10.088

Sharpton TJ (2014) An introduction to the analysis of shotgun metagenomic data. Front Plant Sci. https://doi.org/10.3389/fpls.2014.00209

Shen C-H (2019) Chapter 9–amplification of nucleic acids. In: Shen C-H (ed) Diagnostic molecular biology. Academic Press, pp 215–247. https://doi.org/10.1016/B978-0-12-802823-0.00009-2

Singh A, Jones S, Ganapathysubramanian B, Sarkar S, Mueller D, Sandhu K (2021) Challenges and opportunities in machine-augmented plant stress phenotyping. Trends Plant Sci 26(1):53–69. https://doi.org/10.1016/j.tplants.2020.07.010

Srivastava P, Chaudhary P, Parveen P (2023) Nanoparticles and their effect on plant health and productivity. In: Chaudhary P, Chaudhary A, Nadda AK, Khati P (eds) Advances in nanotechnology for smart agriculture: techniques and applications, 1st edn. CRC Press. https://doi.org/10.1201/9781003345565-6

Srujana S, Bhagat D (2022) Sensors for agricultural pest management using nanotechnology/nanomaterials. Nanotechnol Environ Eng 7(1):319–323. https://doi.org/10.1007/s41204-022-00241-5

Taylor SC, Laperriere G, Germain H (2017) Droplet digital PCR versus qPCR for gene expression analysis with low abundant targets: from variable nonsense to publication quality data. Sci Rep 7(1). https://doi.org/10.1038/s41598-017-02217-x

Tewari D, Cieply S, Livengood J (2011) Identification of bacteria recovered from animals using the 16S ribosomal RNA gene with pyrosequencing and Sanger sequencing. J Vet Diagn Invest 23(6):1104–1108. https://doi.org/10.1177/1040638711425583

Thies J (2015) Chapter 6—molecular approaches to studying the soil biota. In: Eldor Paul A (ed) Soil microbiology, ecology and biochemistry, 4th edn. Academic Press, pp 151–185. https://doi.org/10.1016/B978-0-12-415955-6.00006-2

Toh SY, Citartan M, Gopinath SCB, Tang T-H (2015) Aptamers as a replacement for antibodies in enzyme-linked immunosorbent assay. Biosens Bioelectron 64:392–403. https://doi.org/10.1016/j.bios.2014.09.026

Urbez-Torres JR, Haag P, Bowen P, Lowery T, O'Gorman DT (2015) Development of a DNA macroarray for the detection and identification of fungal pathogens causing decline of young grapevines. Phytopathol 105(10):1373–1388. https://doi.org/10.1094/PHYTO-03-15-0069-R

Urkude R (2019) Application of nanotechnology in insect pest management. Int Res J Sci Eng 7(6):151–156

Vasar M, Davison J, Neuenkamp L, Sepp S-K, Young JPW, Moora M (2021) User-friendly bioinformatics pipeline gDAT (graphical downstream analysis tool) for analysing rDNA sequences. Mol Ecol Resour 21(4):1380–1392. https://doi.org/10.1111/1755-0998.13340

Verdecchia E, Ceustermans A, Baets D, Ferreira J, Bonants P, Melis P (2021) Quantitative PCR for detection and quantification of *Phytophthora cactorum* in the cultivation of strawberry. Eur J Plant Pathol 160(4):867–882. https://doi.org/10.1007/s10658-021-02290-z

Wang X, Shang X, Huang X (2020) Next-generation pathogen diagnosis with CRISPR/Cas-based detection methods. Emerg Microbes Infect 9(1):1682–1691. https://doi.org/10.1080/22221751.2020.1793689

Ye X, Liu M, Zhao N, Xiao C, Xu H, Jia J (2022) Targeted delivery of emamectin benzoate by functionalized polysuccinimide nanoparticles for the flowering cabbage and controlling *Plutella xylostella*. Pest Management Sci 78(2):758–769. https://doi.org/10.1002/ps.6689

Zhang P, Guo Z, Ullah S, Melagraki G, Afantitis A, Lynch I (2021) Nanotechnology and artificial intelligence to enable sustainable and precision agriculture. Nat Plants 7(7):864–876. https://doi.org/10.1038/s41477-021-00946-6

Zhao S, Wang S, Zhang S, Liu J, Dong Y (2018) State of the art: lateral flow assay (LFA) biosensor for on-site rapid detection. Chin Chem Lett 29(11):1567–1577. https://doi.org/10.1016/j.cclet.2017.12.008

Zhao Y, Xia Q, Yin Y, Wang Z (2016) Comparison of droplet digital PCR and quantitative PCR assays for quantitative detection of *Xanthomonas citri* subsp. *citri*. PloS One 11(7):e0159004. https://doi.org/10.1371/journal.pone.0159004

Zubler AV, Yoon J-Y (2020) Proximal methods for plant stress detection using optical sensors and machine learning. Biosensors 10(12):193. https://doi.org/10.3390/bios10120193

Advanced Molecular Techniques in the Identification of Phytopathogenic Fungi

Sarita Yadav, Preeti Kumari, Shikha Sharma, Vijay Kalra, Minakshi Sharma, and Bhawna Batra

Abstract A great range of fungal pathogens infect plants, especially economically significant crops, and these infections lead to the most harmful illnesses and significant yield loss worldwide. There are estimated to be about 1.5 million fungi species. Among them, the number of described species is less than 5%. The global agricultural sector has significant economic challenges due to phytopathogenic fungal species, which resulted in a substantial drop in crop yield in both quality and quantity. These illnesses have a significant negative economic and production impact on the global agricultural sector. The most crucial phases in creating management plans for phytopathogenic fungi are detection and diagnosis. Because there are many different fungal pathogen species, races, biotypes, and strains within each species, and because crop cultivation practices vary from region to region, it is difficult to develop direct detection assays. Diagnostics for fungal plant diseases rely on a wide range of technologies, from conventional taxonomy to innovative molecular equipment. The potential of the organism to be cultured, the length of time involved, and the need for in-depth taxonomical knowledge are some of the major drawbacks of traditional approaches. Diseases must be diagnosed early and accurately to anticipate outbreaks and allow enough time to create mitigation techniques. Modern molecular techniques, such as DNA-based techniques, now offer crucial equipment for precise plant disease identification. Efficient amplification platforms, DNA barcoding, varied quantitative PCR, probe creation, and next-generation sequencing based on RNA sequence have revolutionized the study of fungus differentiation and diagnosis. Although molecular diagnosis techniques have made tremendous advancements over the past two decades, there are still more advancements that need to be developed and utilized by

S. Yadav · P. Kumari · S. Sharma · M. Sharma
Department of Zoology, Maharshi Dayanand University, Rohtak, HR 124001, India

B. Batra (✉)
Department of Biotechnology, DCR University of Science and Technology, Murthal, Sonipat, HR 131001, India
e-mail: bhanukalra22@gmail.com

V. Kalra
Department of Biochemistry, University of Health Science, Rohtak, HR 124001, India

J. Chen et al. (eds.), *Molecular and Biotechnological Tools for Plant Disease Management*, Interdisciplinary Biotechnological Advances,
https://doi.org/10.1007/978-981-97-7510-1_10

phyto-diagnosticians. Ultimately, molecular diagnosis methods employed in phyto-diagnostic clinics must be simple, reliable, effective, affordable, and user-friendly to compete with and more advanced than conventional methods.

Keywords Phytopathogenic fungi · Detection · Molecular technique

1 Introduction

1.1 A Brief Introduction

Fungal phytopathogens are the principal biotic agent for causing disastrous crop diseases. About 8,000 fungi species and oomycetes are responsible for causing various infections in plants from the step of seedling to seed maturation step in the natural climatic environment. These phytopathogens cause serious illness either alone or in combination with other phytopathogens. The most disastrous diseases by phytopathogenic fungi are canker, dieback, leaf spot, blight, damping off, gall, Root rot, wilt, rust, scab, and powdery mildew, etc. The phytopathogenic illness resulted in a significant dropping in quantity, quality, and yield of the agricultural system including economic aspects such as horticulture, agronomy, forest plant species, ornamental, and floriculture. Day-by-day increase in population also raises their demands for agricultural food security and safety that ultimately need highly organized plant disease management plan. A well-planned and productive framework is very important for early and rapid response to fight against plant pathogenic fungi. The early and rapid diagnostics of fungal phytopathogens are imperative for plant protection and also contribute to upgrading crop health and vigor (Zeilinger et al. 2016; Li et al. 2022).

1.2 Enzymes Degrading Plant Cell Wall

The enzymes that degrade cell walls are responsible for cuticle and cell wall degradation of host plants and also promote colonization, invasion, and pathogenic fungi expansion. The phyto-pathogenic fungi produce enzymes that degrade cell walls interact with plants and cause infection. The interaction between the two is most importantly studied via proteomics and molecular biology domains. The principal enzymes that degrade cell walls include chitinase, protease, pectinase, and cellulase produced by *Rhizoctoniasolani* cause serious degradation of maize radicle. This degradation capability is enhanced by an increase in the concentration of enzymes that degrade cell walls. The enzyme cellulase secreted by plant pathogens in tissues of living plants has the job of softening and degrading the materials of cell walls. The fungi *Fusarium graminearum* secretes xylanase, cellulose, and pectinase enzymes that cause infection and expansion in the host cell wall, resulting in cell wall relaxation and degradation of cell wall components that are very helpful in penetrating the host tissue and resulting in expansion of infection in the host tissue. A cell wall degrading enzyme named β-galactosidase results in lactose degradation in the cell wall that produces glucose and galactose and quickens the fruit softening process that shows β-galactosidase is plentiful in the stage of fruit softening and is responsible for the breakdown of galactosyl bonds in cell wall result in decline in the integrity of cell wall. Enzymes like protease, hemicellulase, phospholipase and amylase degrade protein, hemicellulose, lipid, and starch and are also involved in pathogenic processes (Fig. 1). Pathogenic fungi cause infections in plants and illness is not just by enzymes that degrade cell walls but via toxins, hormones, and various additional factors. Simultaneously pathogenic infection also results in the activation of the enzyme defense system and production of antifungal substances that, result in cell wall degrading enzyme inhibition and attain disease resistance. It shows that plant infection via pathogenic fungi is a sophisticated process (Hamel et al. 2012; Andika et al. 2017; Carneiro et al. 2008; Chatterjee et al. 2016; Guerriero et al. 2015).

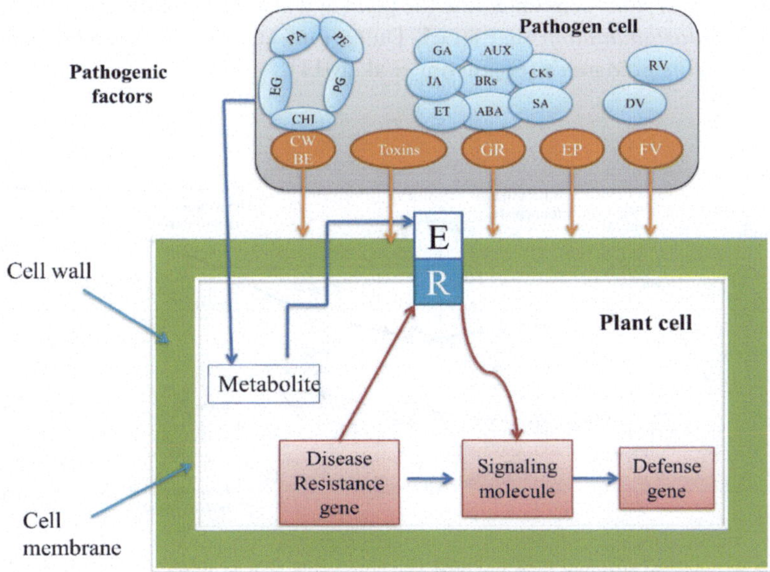

Fig. 1 The pathogenic factors of phytopathogenic fungi. CWBE: Cell wall-degrading enzymes, PE: pectin methylesterase, EG: Endo-1,4-β-D-glucanase, PA: protease, CHI: chitinase, GR: growth regulator, AUX: auxin, CKs: cytokinins, PG: pectin polygalacturonase, BRs: brassinolides, ABA: abscisic acid, GA: gibberellic acid, SA: salicylic acid, JA: jasmonic acid, ET: ethylene, EP: effector proteins, FV: fungal virus, DV: DNA virus, RV: RNA virus (Peng et al. 2021)

1.3 Interaction Representing Molecular Mechanism Between the Plants and Phytopathogenic Fungi

Plant infections via pathogenic fungi are the utmost pitfall for food security at the global level. An understanding of the complex interaction of pathogenic fungi and flora is crucial to control plant diseases (Li et al. 2022).

The interaction between host and pathogen gives rise to a continuous series of events known as the disease cycle responsible for the development and continuation of disease (Fig. 2). The disease cycle consists of two phases:

(a) Spread and contact: in this firstly spreading of fungi and then interaction with the appropriate host via environmental mechanisms like water, insects, wind, etc. (Travadon et al. 2012).
(b) Pre-penetration:—It involves spore germination, attachment of pathogen to the host structures, and then recognition events due to host signals and also by environment factors (Tucker and Talbot 2001).
(c) By natural openings like wounds and direct penetration (by structures like appressoria) pathogens enter into the plant and also by wounds caused by insect-like *Ophiostomataulmi* that attack Dutch elm and *Grosmanniaclavigera* that attack *lodge pole pines* (Diguistini et al. 2011).

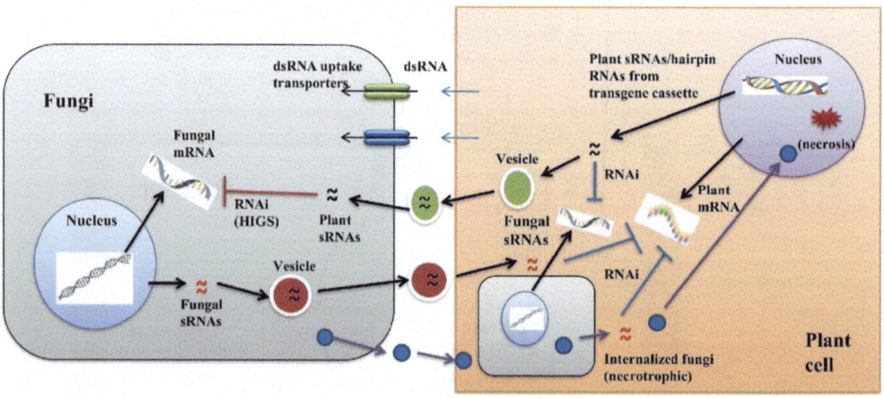

Fig. 2 In the context of plant host-induced gene silencing that is RNAi-mediated, the diagram represents a fungal pathogen and plant cell interaction. During a host–pathogen interaction, the success of plant resistance or fungal colonization depends on which pathogen, organism, or host plant can overcome the defense response of another. Plant sRNAs (small RNAs), which are created via a plant's natural defense mechanism or from hairpin RNAs present in the transgenic RNAi plants (that target a fungal gene), can pass between plants and fungi via vesicles or RNA uptake transporters. Following their entry into the fungus, the sRNAs are released through their vesicles and the host-induced silencing of the gene begins. Similar to bacterial sRNAs, via vesicles the fungal sRNAs can also enter plant cells and cause the genes to be silenced and responsive for host defense. In addition to sRNAs, fungal cells secreted effectors infiltrate pathogen and plant cells directly enter the host cell and suppress the genes responsible for host defense or activate host genes which are favorable for the growth of fungus, ultimately causing necrosis (Majumdar et al. 2017)

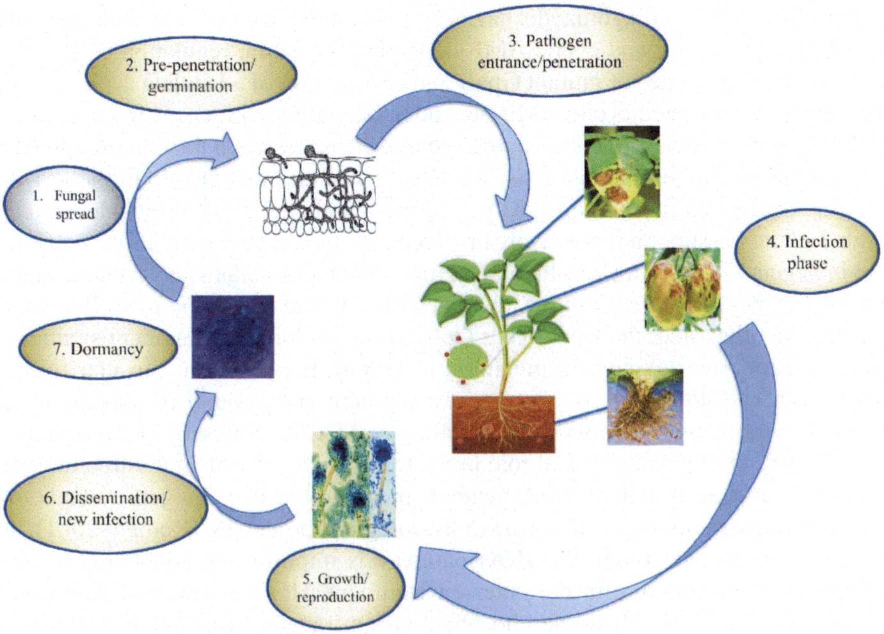

Fig. 3 Disease cycle (Rajam and Chauhan 2021)

(d) After infecting the host the pathogen starts invading and exhibiting contact with the host cells and infection starts spreading from cell to cell ultimately giving visible symptoms.
(e) The infected host tissues result in the production of numerous fungal spores.
(f) spores get dispersed from the reproduction site and come into contact with the surface of the new host plant.
(g) Pathogen undergoes a dormant stage in unfavorable conditions for survival (Brown and Ogle 1997) (Fig. 3).

1.4 Signal Recognition of the Pathogenic Fungi an Infection Process

The continuous run of enzymatic reactions that transmit the extracellular signals via the cellular membrane is known as the signaling pathway. The membrane receptors sense external signals. The cell membrane receptors include tyrosinase kinase receptors, GPCR (G-protein coupled receptors), ion channel receptors, and others that regulate gene expression. The G protein (heterotrimer) in fungi, is involved in vegetative growth, sporulation, pathogenicity, and infection structure differentiation regulation by regulating adenyl cyclase, phospholipase, and ion channels (Bölker 1998; Lengeler et al. 2000). The G protein-coupled receptors and heterotrimer G proteins, in case of Ustilagomaydis having role in mating as well as in pathogenicity via hormone response and cyclic AMP-dependent signaling regulation (Liebmann et al. 2003). Phenotypes of mutant GpaB and acA mutant of adenylate cyclase show that the GpaB-mediated cyclic AMP-dependent signaling pathway is involved in *A. fumigatus* pathogenesis. In the case of *Cryphonectriaparasitica* the Gα, protein Cpg 1 has a crucial role in pathogenesis. The Cpg-1 has an integral role in the growth and trophic mycelia sporulation and the process of pathogenesis is Gβγ independent. Cpgb-1 of the Gβ subunit is positively involved in pathogenicity regulation but has no role in vegetative growth (Kasahara and Nuss 1997). Cacl adenylate cyclase, in the case of *Cryptococcus neoformans,* regulates the capsule formation as well as toxicity. On the other side, the Gβ subunit Gpb1 is not important in pigment synthesis, pod formation, and toxicity. In mutant gpr4 GPCR, the defective capsular formation is seen and no change is observed for pigment and toxicity (Chaoyang et al. 2006). The transmission network of signaling of MAPK is present in the center of the cell (Fig. 4). It plays a crucial role in photosynthesis, growth and differentiation of cells, neurotransmitter release, metabolism, infection of pathogens, adaptation in adverse environments, and various physiological processes (Romeis 2001). In numerous pathogenic fungi, the HOG pathway is important in adaptation toward elevated osmotic pressure in the outer environment. HOG pathway of *Neurosporacrassa* having Hpt1 (Histidine phosphate group transfer proteins), Os1 (Osmotically responsive histidine kinase), RRG1/2, and the downstream coupling for the phosphoric acid system including Os4, Os2, and Os5 (Fujimura et al. 2003; Jones et al. 2007).

In numerous pathogenic fungi, the HOG pathway has a role in adaptation to the elevated osmotic pressure for the outer environment. The HOG pathway of *Neurosporacrassa* consists of Hpt1 (Histidine phosphate group transfer proteins), Os1 (Osmotically responsive histidine kinase), RRG1/2, and the downstream coupling system for phosphoric acid including Os4, Os2, and Os5 (Fujimura et al. 2003; Jones et al. 2007). It has been observed that the *N. crassa* HOG pathway not only belongs to environmental stress yet also involved in the regulation of bacterial pigment production, diimide fungicide resistance, and pathogenicity of bacteria. It is also observed that the *Botrytis cinerea* HOG pathway is sensitive to osmotic stress, DCFs, and pathogenicity for various hosts (Motoyama et al. 2008). In *Magnaportheoryzae* Osm1 that is homologous protein of Hog1 is a significant element for the HOG pathway having a role in osmotic stress and drug sensitivity but having no close association with pathogen pathogenicity (Motoyama et al. 2008). The Fus3/Kss1 pathway has a role in pathogenic fungi sexual reproduction and also in pathogenicity regulation. The Ste5 scaffold protein is missing in *Aspergillus nidulans* but yet forms complex: AnSte50-AnSte11-AnSte7-AnFus3. Upstream signals activate the complex and then AnFus3 makes an entry in the nucleus resulting in the activation of Ste12 (AnSteA) and AnVeA (homologous protein) activity. The activated protein AnSteA regulates the sexual reproduction and mycelial fusion in *A. nidulans* and the activated AnVeA regulates the *A. nidulans* secondary metabolism (Bayram et al. 2012). *M. oryzae,* homologous protein Fus3 mutant (pmk1) unable to produce appressorium and unable to penetrate the host surface (Bruno et al. 2004). The CWI pathway has a similarity

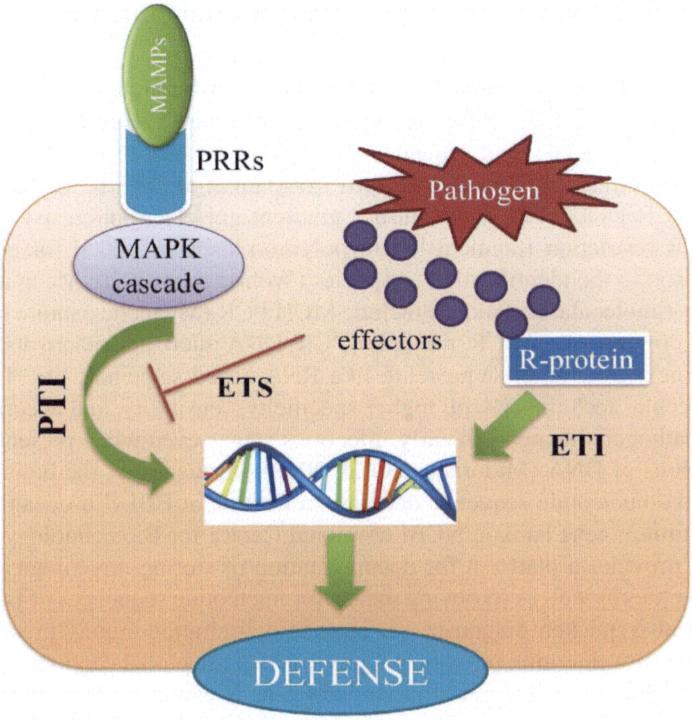

Fig. 4 Signaling represents the interaction between plant-fungal pathogens. Plants' first line of defense is comprised of protein receptors that are found in the plasma membrane. When PRRs (Pattern recognition receptors) detect the conserved structures of microbes (MAMPs (Microbial-associated molecular patterns) or PAMPs (pathogen-associated molecular patterns), calcium signaling and MAPK cascades are activated, activating PAMP-triggered immunity (PTI). A condition known as ETS (effector-triggered susceptibility) is caused due to pathogens interfering with PTI through effectors and preventing the response of PTI. On the opposite side, ETI (Effector-triggered immunity) is a response brought on by effector recognition through the plant R proteins (Zeilinger et al. 2016)

with the MAPK pathway that is studied in pathogenic fungi. In the CWI pathway kinase Slt2 is a central component having a role in the regulation of cell wall integrity of *M. oryzae*, *A. brassicicola,* and *A. nidulans* (Mey et al. 2002). It has been found that Slt2 is associated with pathogen pathogenicity, e.g. *Candida albicans* (Monge et al. 2006).

2 Molecular Methods to Detect Phytopathogenic Fungi

The modern strategies for fungal detection include high-throughput molecular detection techniques. It includes PCR, nested PCR, real-time PCR, rolling circle amplification (RCA), Nucleic acid sequence-based amplification (NASBA), and Loop-mediated isothermal amplification (LAMP) (Aslam et al. 2017b, a). The methods involving PCR-DGGE (PCR denaturing gradient gel electrophoresis) and PCR–RFLP (PCR restriction fragment length polymorphism) are good for genotyping in comparison to the identification of species (Wöhrle et al. 2020; Meng and Doyle 2002). Other molecular techniques include MCH PCR (Magnetic capture hybridization PCR), co-operational PCR, multiplex PCR, DNA micro and macro array, In-situ PCR, and next-generation sequencing like RNA-based sequence, etc. The DNA-based molecular techniques with higher specificity, sensitivity, and accuracy allow the phytopathogen diagnosis in early infectious stages even when present in lower concentrations of DNA (Midorikawa et al. 2018). Furthermore, the bioinformatics database like nucleotide sequence database collection at EBI (European Bioinformatics Institute), gene bank at NCBI (National Centre for Biotechnology Information), etc., provides a platform for documentation of storing, mycological nomenclature novelties as well as retrieving means for nucleotide sequences of fungi infect plants that also quicken diagnosis and species delimitation among present fungal species. The new emerging phytopathogens threaten the global economy that's why the identification and detection of phytopathogens is essential. The present review summarizes the different molecular techniques of fungal phytopathogen diagnosis including advantages as well as disadvantages. This also inspects the molecular tools utilized for diagnosis of present, emerging, and re-emerging fungal phytopathogen in a wide variety of crops (Hariharan and Prasannath 2021).

2.1 Identification of Phytopathogenic Fungi

There are limitations associated with the existing conventional method that's why molecular techniques came into existence that are helpful in classification and identification problems. This new approach to molecular diagnostics is highly valuable in fungal diagnostics (Fig. 5). It includes nucleic acid (NA) based probe technology, Polymerase chain reaction (PCR), and immunological methods. Earlier methods depend on phenotypic features and the later methods depend upon genotypic features resulting in increased specificity, effectiveness, and high accuracy in results.

By considering molecular approaches, it is important to perform early analytical activities including extraction of genomic DNA in which effective lysis of fungal cells and recovery of DNA then finally purification and then quantification of the extracted DNA. By employing various procedures fungal DNA can be isolated from

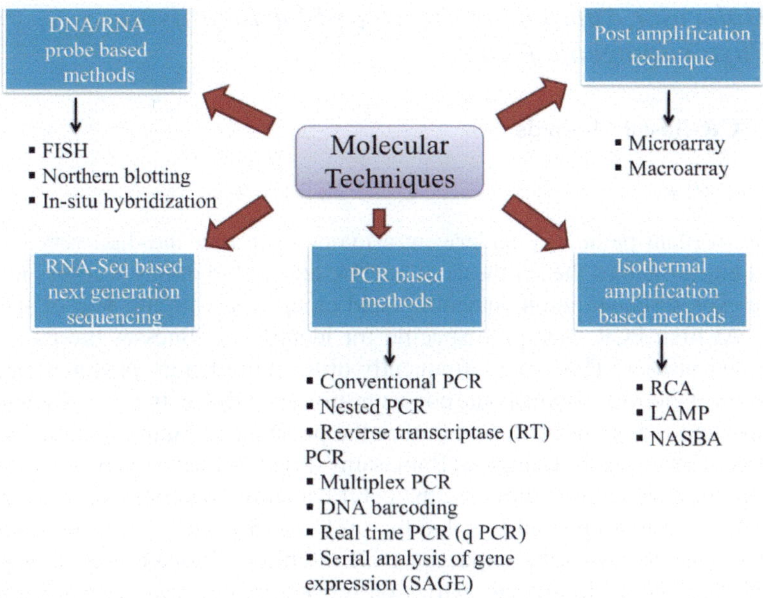

Fig. 5 An overview of molecular techniques (Gautam et al. 2022)

the infected plants. A recent approach using commercial DNA extraction kits to diagnose fungal plant infections (Abdullah et al. 2018). The set processes of the laboratory involve chitin cell wall disruption via grinding, lyophilizing mycelia, isolation of DNA using a chemical-containing buffer, protein removal by a mixture of phenol and chloroform, and precipitation by propanol. The purification of extracted fungal DNA via conventional techniques includes magnetic particle separation by silica coating, pelleting, and spin filtration across silica membranes. To get a satisfactory concentration of DNA the quantity of fungal DNA in samples is measured by UV spectrophotometer and finally diluted by ultrapure PCR grade water (Hariharan and Prasannath 2021). To control the spread of fungal disease a precise diagnosis of disease is required that is based on the causative agents' diagnosis. Out of 1.5 million species, <5% fungal species have been reported. Now the recording of the remaining 95% fungal species is comparatively easy with molecular techniques in comparison to the conventional methods that result in the addition of 1000 new species every year (Hawksworth 1991).

The chief target in the diagnostic field is the identification of disease-causing organisms in a rapid, reliable, and accurate manner. The fundamental methods for detection depend upon cultural, microscopic, and morphological ways that need extensive labor, time, and knowledge of classical taxonomy. These ways resulted in unreliable results because of identification problems. Ultimately highly expert and specialists in the field are required to have practice in identification resulting in effective outcomes (Chalupová et al. 2014).

2.2 *Molecular Approaches Developed for Identifying Phytopathogenic Fungi*

2.2.1 PCR-Based Methods

Traditional PCR

The field of plant pathology underwent a major shift after the discovery of PCR. This technique enables the synthesis of a DNA segment in multiple copies utilizing alternate cycles; denaturation, annealing, and elongation with the help of specified primers. At first, PCR was quite specific for identifying illnesses brought on via bacteria and viruses. This is now frequently utilized to identify phytopathogens as well. Deoxynucleoside triphosphate concentration and DNA extraction efficiency are both important factors in PCR. Sometimes the presence of inhibitors in the sample assay affects performance (Fang and Ramasamy 2015). With the inclusion of specific chemical and enzymatic treatments, the cetyl trimethyl ammonium bromide technique is frequently employed to solve his problem in plants. There are numerous additional ways to lessen the impact of PCR inhibitors. Instead of ethanol precipitation in the phenol–chloroform extraction procedure, purification of silica matrix can be employed for DNA recovery (Mancini et al. 2016). Additionally, primer designing is necessary for PCR technology to identify infections, and to start a series of DNA replications, that may restrict the usefulness of the technique for disease field sampling. To get around this restriction, a single pair of primers may not always produce findings that are precise and accurate. Nowadays, the most used tools are DNA probes and nested primers. PCR technology is used to identify *Mycosphaerellafijiensis* and *M. musicola*, which are responsible for the leaf spot disease that affects bananas and plantain (Henson and French Citation 1993). Even though this method produces incredibly precise results, it is extremely expensive and labor-intensive (Kuzdraliński et al. 2017).

End-point PCR

The PCR invention has revolutionized plant disease management by allowing for the precise identification of many plant diseases that also involve fungi (Ma & Michailides). The vitro method exponentially amplifies a fragment of DNA template (Lilit and Nidhi 2013) by employing particular dNTPs (deoxyribonucleotide triphosphates), primers, and Taq DNA polymerase via several denaturation, annealing, and extension cycles, final extension, and then final-hold reactions at various temperatures (Griffiths 1999; Capote et al. 2012). Reliable fungal plant disease detection is accomplished by synthesizing either specialized oligonucleotides target fungal species and universal primers for amplification of several infections and then sequencing. By comparing each pair of nucleotide sequences from fungal isolation to ex-type cultures found in the database GenBank, NCBI, Basic local alignment search tool (BLAST) analysis can establish the identity of each isolate. Targeted phytopathogenic fungi detected by AGE (Agarose gel electrophoresis) (Moher et al. 2015; Suzuki et al. 2017). Comparing end-point PCR methods with other available molecular diagnosis methods for plant disease caused by fungi, they are thought to be a more affordable option. However, designing primer sets for end-point PCR assays to distinguish between closely related fungal diseases can be challenging and time-consuming (Sikdar et al. 2014). They discovered that for quick diagnosis, the quantitative real-time PCR technique is highly sensitive in comparison to the end-point PCR (Hariharan and Prasannath 2021).

Nested PCR

To get a better level of sensitivity and specificity, nested PCR is accomplished. For instance, the sensitivity of nested PCR for fungus identification is a thousand times greater in comparison to the single PCR. In the nested PCR procedure, a pair of primers is utilized in the first round for the amplification of a substantial DNA portion, and the amplified sequence of DNA acts as the target sequence for the second round, in which two internal primers are used. The amplification of two cycles is completed in distinct tubes, there is a significant risk of contamination with this method of PCR. Therefore, the likelihood of false-positive findings from contamination and the amount of work required are the main downsides of this method (Rahman et al. 2013). To conquer the restrictions, laboratories should take very specific safety precautions, such as using separate instruments and locations for every cycle of PCR (Trtkova and Raclavsky 2006). Nested PCR was utilized to identify the orchid disease caused by *Phytophthora species.* By using nested PCR, the disease-causing *Phytophthora species. P. parasitica* and *P. palmivora* were identified (Tsui et al. 2011). Using nested PCR, *P. cactorum* was exclusively found in the strawberry-diseased plant (Bhat and Browne 2010; Capote et al. 2012; Green and Sambrook 2019; Suzuki et al. 2017).

Multiplex PCR

In the same reactions many pairs of primers are used that enable simultaneous detection of various target sequences of DNA, demonstrating the time and cost savings known as multiplex PCR. When plants become infected with many pathogens, the multiplex PCR technique is extremely important in plant pathology (Pallás et al. 2018). On agarose gel, several fragments that are unique for the targeted pathogenic fungi undergo simultaneous amplification and identification based on their molecular sizes. Amplicon size has a major impact on how accurately DNA is synthesized. Proper selection of primers must be done, including their annealing temperature and relative concentration, to avoid pitfalls (Dasmahapatra and Mallet 2006). These days, the multiplex approach for identifying pathogenic fungi uses padlock probes (PLPs). Fungal infections such as *B. cactivora, F. oxysporum, P. cactorum,* and *P. nicotinae* have been simultaneously detected using the multiplex PCR approach, and they severely afflict grafted cacti (Cho et al. 2016; Wang et al. 2019; Bangratz et al. 2020).

Quantitative PCR

Real-time quantification and detection of specific RNA or DNA sequences from plant pathogenic fungi in a reaction mixture of PCR is made possible by quantitative PCR (qPCR). Via extrapolating a value of Ct that is the cycle threshold from fungi samples utilizing sequence-specific primers, it is possible to determine the relative copy number for the target sequences of RNA/DNA (Balodi et al. 2017). SYBE Green I, Molecular Beacons, Eva Green, and sequence-specific reporter probes that are fluorescence-labeled like TaqMan probe are examples of fluorescent dyes (Badali and Nabili 2012) that are employed to keep an eye on the reaction as it is being amplified. The fundamental idea is that the fluorescent signal, which is created via an intercalating dye/dye-labeled reporter probe breaks down while amplification, is proportional to the quantity of amplicon generated in every cycle. Chestnut trees, *Castaneadentate,* and *Castaneasativa* are all affected by the hypervirulent developing fungal phytopathogen that is *Cryphonectriaparasitica*, which also causes deadly bark cankers, blight, wilting, and dieback (Murolo et al. 2018; Rahman et al. 2020). *C. parasitica* detection utilizing rDNA ITS sequences using molecular diagnosis by qPCR was possible having a sensitivity of 2 fg for genomic DNA, which is comparable to a single pathogen spore (Chandelier et al. 2019). An emerging pathogenic fungus *Ramulariacollo-cygni* exhibits distinctive symptoms of brown and small spots on leaves, awns, and sheaths making it challenging for correct diagnosis of disease utilizing conventional approaches (Havis et al. 2015). The first molecular detection report of *R. collo-cygni* of the barley seed was created and submission was done via qPCR (Havis et al. 2014). One instance of a novel fungus pathogen-identification via a quantitative PCR strategy is an aggressive as well as developing British -*Verticillium longisporum* (Depotter et al. 2017). The Fungi responsible for the sunflower's phomopsis stem canker are *Diaporthegulyae* and *D. helianthi*, which may be identified and quantified using qPCR. This assay was successful in screening the same genus of causal compounds (Elversion et al. 2020). It is difficult to diagnose the diseases using traditional methods since *Pyrenophoratritici-repentis* and *Parastagono sporanodorum* co-infect the wheat resulting in similar physiognomies. A duplex qPCR assay was performed using 2 dual-labeled probes using distinct fluorogenic rep TaqMan probes (Shuey et al. 2014). The results of a duplex qPCR assay using 2 dual-labeled probes using distinct fluorogenic-reporters, which allowed sequences of DNA from *P. triticirepentis* and *P. anodorum* amplification simultaneously yet independent of one another, accurate, good for multiple pathogens high throughput screening and appropriate for continuous differentiation. The target DNA high throughput quantification in the various biological areas (Schena et al. 2013) and trustworthy information on pathogen load may be obtained using this technology, which is quick and very sensitive (Griffiths 1999). Additionally, the probe (TaqMan) provides an additional specificity degree (Shuey et al. 2014). However quantitative PCR requires special equipment. The instrument and the probe price might be high (Abdullah et al. 2018; Sanzani et al. 2014; Lukianova et al. 2021; Chaves et al. 2023).

Reverse Transcriptase PCR (RT-PCR)

The modifications resulted in enhanced usefulness and diagnostic power of Polymerase chain reaction in numerous medical and biological disciplines (Tang et al. 1997). The inability of PCR types to distinguish between living and dead fungi is a significant drawback. The invention of Reverse transcriptase PCR eliminates this drawback. Since mRNA of dead cells gets degraded, mRNA detection via RT-PCR is utilized to determine the vitality of the cells (Capote et al. 2012). The following approach results in the conversion of RNA to cDNA firstly using random primers then Reverse transcriptase enzyme, amplify the result using any PCR-based technique. So, RNA-containing virus infections (such as those caused by retroviruses) are primarily detected and diagnosed by RT-PCR. When evaluating the efficacy of antimicrobial vaccines or treatments, the detection of viruses that contain RNA can be useful. The fungi named *Fusarium graminearum* cause FEB (Fusarium ear blight) illness in cereals like rye, maize, barley, wheat, and oats have been quantified using RT-PCR (Brown et al. 2011).

Real-time PCR (Q PCR)

PCR conventional techniques provide target DNA amplification, that is not sufficient for early disease diagnosis. A novel technology called RT-PCR has been made available in recent years. This technique is a traditional PCR improved version because it allows for both DNA quantification and amplification (Mackay 2004). The fluorescent dyes introduction like SYBR Green I dye by utilizing sequence-specific probes that are fluorescently labeled for instance Taq Man probe made it possible to examine reactions during the steps of amplification (Badali and Nabili 2012). There is the generation of a fluorescent signal when the intercalation of fluorescent dye with DNA. This signal gets stronger with every amplification cycle as the quantity of target DNA increases. The fluorescent dye has limitations because of its non-specific nature but it is of less cost to employ as the monitoring agent. The binding of the intercalating dye to the DNA sample can result in false positives such as primer dimers. The remarkable specificity of fluorogenic probes led to their adoption later (Schena et al. 2004).

Two different fluorescence dyes are used to connect the probes i.e. reporter dye which attaches to the 5′ end and quencher dye which attaches to the 3′ end. Fluorescence emission is prevented via reporter dye and the quenching dye nearby. The exonuclease activity of Taq polymerase results in the detachment of reporter dye from quenching dye and begins fluorescing (Dasmahapatra and Mallet 2006). Fungus responsible for diseases *Clado sporiumclado sporoiodes*, *Aspergillusversi color*, *Stachybotry schartarum*, and *Alternaria* alternative have been quantified and identified utilizing qPCR. Secondary metabolites of *Fusarium sp.* are harmful to agriculture products, animals, humans, and wildlife. Members of this genus are also responsible for human and plant diseases, making them extremely important (Yadav and Singh 2017). In the production of berries, fungi from the genera Botrytis, Colletotrichum, and Verticillium are a major issue. For this goal, a multiplex real-time PCR test was developed. Three multiplexable fluorogenic probes tailored to the selected fungi were used along with three universal fungal primers targeting the D2 region of the large subunit ribosomal DNA. The study's triplex approach to molecular detection of these fungi enables quick and accurate identification of key berry pathogens, which speeds up the implementation of preventative measures in plantations and significantly lowers losses brought on by fungal diseases (Malarczyk et al. 2020).

BIO-PCR

A pre-assay incubation phase is used in the BIO-PCR assay and this technique is modified end-point PCR, to enhance the biomass of the causative agent. Via cultivation in the growing medium, the target pathogen inhibits the growth of the non-targeted microorganisms for enhanced detection, this technology is primarily used to concentrate target pathogens (Shaad et al. 1995; Tsui et al. 2011) and successfully utilized to find seed-borne fungi (Kumar et al. 2020). The BIO-PCR technique was used to identify the lupin anthracnose disease, a seed and airborne condition brought on by *Colletotrichumlupini*. To enrich the *C. lupine* biomass, the incubation of seeds with modified Malt Broth of Yeast and a primer pair that is species-specific was created using the IGS sequence of rDNA. By using the BIO-PCR methodology, *C. lupine* in the Lupinus species could be found (Pecchia et al. 2019). The advantages of BIO-PCR over the endpoint PCR technique include increased sensitivity, removal of PCR inhibitors, and the living cells identification to prevent false positive results Fatmi et al. (2005). When selective media are utilized for the experiment, this approach has the drawbacks of being time-consuming and expensive (Mancini et al. 2016).

Magnetic capture hybridization PCR

Hybridization with magnetization DNA is purified using ssDNA is single-stranded DNA probe hybridization on the magnetic beads as part of the PCR (MCH-PCR) process, which is then used to amplify the target DNA. This PCR test was primarily developed to address PCR inhibitors that were present in extracts of plants during the isolation of DNA isolation. Biotinylated oligonucleotide having specificity for the region of DNA of the targeted fungus pathogen is coated on the magnetic beads (Walcott et al. 2004). Due to double-stranded DNA (dsDNA) hybridization with magnetic beads, the resulting complex may be separated from the inhibitors. Two cucurbit seed pathogens that produce gummy stem blight and bacterial fruit blotch, *Didymellabryoniae and Acidovoraxavenae* subspecies. *Citrulli* respectively, were tested using MCH and the real-time-based PCR tests. The technique made it possible to identify both examined pathogens simultaneously in cucurbit seed samples (Mancini et al. 2016). A highly conserved area of a fungal pathogen is used to design an oligonucleotide probe sequence for a capture probe used in MCH-PCR (Levesque 2001). Using BLAST, the chosen sequence specificity can be checked in silico. A probe is subsequently biotin-labeled at the 5′ end to enable attachment with a streptavidin-coated magnetic bead. MCH-PCR may shorten the overall time of detection, improve the sensitivity of PCR, and get rid of various inhibitors of amplification reaction as well as excess non-target DNA (Johnson et al. 2013).

Digital droplet PCR (DdPCR)

The sample partitioning method called ddPCR can only detect a single gene. Easy to use, reliable, and increased specificity in quantifying plant pathogens like fungus, fastidious bacteria, bacteria, viroid, and viruses are its benefits over standard PCR procedures (del Pilar Martinez-Diz et al. 2020). Another illustration is the creation of a sensitive, dependable, quick, and affordable isothermal amplification method that is recombinase polymerase amplification-based and the lateral-flow test to identify *Candidatus Liberi bacterasiaticus*. Gold nanoparticles (AuNPs) are also used in localized surface plasmon resonance (SPR) to create colorimetric nano-sensing devices for detection (Ghosh et al. 2018). A single molecule is multiplied in a highly diluted droplet using a contemporary technique termed droplet digital PCR (ddPCR). A probe that is fluorescently tagged is used to locate the target molecule. ddPCR has shown to be a reliable method for detection even in low concentrations because of its less sensitivity to inhibitors. In this method, the target material is divided into 20,000 droplets, and the template molecules are amplified by PCR in each droplet. The ddPCR method was able to identify *Tilletiacontroversateliospores* in the soil for the first time having a sensitive limit of detection (LOD) of 2.1 copies/L, outperforming PCR and real-time PCR (Liu et al. 2020). *Ilyonectria liriodendra*, which causes black foot disease that affects grapevine nurseries, was also identified using the ddPCR approach. According to the author, the method's accuracy, efficacy, and specificity were related to those RT PCR, and the degree of correlation between ddPCR and ddPCR was $R^2 = 0.95$. The research team created a format of ddPCR for *Acidovoraxcitrulli*, responsible for the serious threat of cucurbits. The limit of detection was significantly more sensitive in comparison to the qPCR, which detected 102 CFU/mL of bacteria and 2 fg/DNA. Comparing ddPCR to qPCR for quantifying *Plasmodio phorabrassicae*, a required soil-borne pathogen, is responsible for club root infections in brassica crops. By calculating results in several soil types, the proposed approach was confirmed. According to the findings of their study, ddPCR is a more adaptable method than qPCR for locating and measuring spore seeds of *P. brassicae*. Another study identified the PMTV (Potato mop-top virus), a danger to the global output of potatoes, using qRT PCR and RT ddPCR testing. While both assays were capable of PMTV detection in various types of soil, RT ddPCR displayed good linearity for the RNA isolated via samples of soil. A successful ddPCR was used in a different investigation to identify the pathogen *Erwiniaamylovora*, responsible for the fire blight that infects apples and pears. The QuantStudio 3D (QS3D) chip-based dPCR system was included in the suggested method for use in the field. The internal *Alternaria* spp. transcribed spacer and *B. cinerea* intergenic spacer both were used in the application. The threshold level of detection for each target was 0.1 pg of the DNA. The test was validated across an entire year to comprehend pathogen detection at various seasons of plant development (Patel et al. 2022).

2.2.2 Serial Analysis of Gene Expression (SAGE)

A method based on the analysis of sequence for profiling of quantitative expression of a gene is SAGE (Serial analysis of gene expression) helps in the simultaneous detection of many transcripts. In this method finding the corresponding expressed genes is based on the sequencing, 15 bp quantification or large oligonucleotides, and sequence similarity against the known genomic sequences (Velculescu et al. 1995). For SAGE, a tag is a nucleotide sequence that has a specific length and is located next to the restriction enzyme's 3′-most restriction site. A set of tags and their count values, which provide a digital representation of cellular gene expression, make up the SAGE data product. The original SAGE methodology has undergone several technical changes to increase its effectiveness, including decreasing the amount of input RNA, lengthening SAGE tags, and enabling subsequent application of SAGE data (Sharma et al. 2007). This technique amplifies two samples that have been ligated together and labeled with different primers. Primer removal results in sticky ends that help to create concatemers. They are sequenced after being cloned into a vector and subjected to a thorough computer analysis. SAGE has some shortcomings. It first requires a significant amount of mRNA. Second, with more complicated genomes, a 15 bp tag may not always be sufficient to precisely identify the gene of origin. *B. graminis* was originally detected on leaves of barley using SAGE (Wang and Peng 2014).

2.2.3 DNA Barcoding

A tiny section of DNA is employed in the molecular diagnostic procedure known as "DNA barcoding" to determine species that belong to eukaryotic life in various domains. The barcode region is present in the Targeted lineages. It shows enough variations in sequences of DNA to discriminate various species. Standardized sequences of 500–800 bps are utilized in DNA barcoding to identify species having markers showing a broad range for the taxonomic groups (Krishna and Francis 2012). This PCR-based method is useful for identifying species of microorganisms including fungi, algae, and bacteria as well as uncultured taxa from the natural environment. It may also be used to identify species of microbial organisms like fungi, bacteria, and algae. The amount of representative species contained in the database determines how well a sp. may be identified via barcode. Fungal detection during stages of life-cycle is not good for the identification of morphological features by DNA barcoding. The effectiveness of barcoding is predicated on the idea that genetic differences within a sp. are substantially smaller than across species. By using DNA barcoding, morphological identification, discovery and recognition at the molecular level for fungi is possible. Effective markers are highly helpful in this strategy to identify the poorly understood fungus (Roe et al. 2010). With Pythium which is the most prevalent genus found, DNA meta-barcoding revealed that Phytophthora and Pythium sp. present in an extensive manner in the globally transported rhizospheric soil. Golden-brown algae increased in the soil samples as a result of baiting, a standard enrichment technique for Phytophthora species, but possibly plant-harmful oomycetes did not increase in relative or absolute abundance. DNA sequences representing the oomycete isolates acquired after the enrichment were produced by metabarcoding of rhizospheric soil, and they were successfully recognized (Rossmann et al. 2021). A standardized tool for identification at the species level can be utilized for the assessment of biodiversity, diagnostics, ecological studies and the monitoring of fungal phytopathogens has the potential to be provided by recent advancements in

DNA barcoding technology (Kashyap et al. 2017). DNA barcoding has several scientific benefits, such as (i) making it easier to identify sp. At any life stage (ii) making it possible to discover new species based on NA sequences (Nucleic acid) phylogenetic (iii) offering the diversification of life (iv) encouraging tool development for sequencing of DNA, which is helpful in the study of biodiversity (Savolainen et al. 2010; Schoch et al. 2012, Prassan et al. 2021).

2.2.4 Methods Based on DNA/RNA Probe (Sequencing Independent Methods)

Using RNA/DNA probe techniques, it is possible to quickly and sensitively identify plant infections brought on by microorganisms like fungi. The majority of today's knowledge is said to be supported by this technology. In these procedures, the probe is utilized to analyze nucleic acid without amplifying it beforehand. Probes are single-stranded, small DNA sequences, that have been radiolabeled with isotopes like ^{33}P, ^{32}P, or ^{35}S or with a chemiluminescent reporter molecule. To locate the homologous sequence on target DNA, these are used. For the identification of fungus, DNA probes are typically employed in place of PCR in conventional procedures. However, in more recent techniques, PCR is frequently used (McCartney et al. 2003).

Northern Blotting

Northern blotting also referred to as RNA blotting, transfers the RNA onto a carrier to identify fungal pathogens. Except for using RNA material rather than DNA, the Northern blotting is identical to the Southern blotting. To check the expression of the target gene, the RNA from every tissue must purified first. Afterward, the RNA is placed onto an agarose gel. The gel is then exposed to an electric current, which causes RNA to migrate to the gel's bottom. Smaller RNAs move more quickly than larger RNAs (Kim et al. 2010). The RNA molecule keeps up its location about every other molecule by being blotted on a particular filter paper after being separated from other RNA pieces. The filter is subsequently subjected to radioactive probes to enable complementary sequence hybridization. The autoradiography filter is then applied to develop the film. Finally, if the probe has successfully hybridized to an RNA fragment on the filter, a band is seen on the autoradiograph. Northern blotting is useful for analyzing the gene expression. The size of RNA is first resolved by the band arrangement on the blot. The transcript coding capacity and the protein size can both be roughly estimated if the amount of the RNA is known. One of the biggest drawbacks of the blotting approach is the potential for mRNA degradation during the electrophoresis. Second, it is difficult to detect with several probes. In comparison to RT-PCR, the northern blotting approach has a lower level of sensitivity. Real-time PCR and Northern Blotting were used to identify *Magnaporthegrisea* in rice plants (Qi and Yang 2002).

In Situ *Hybridization (ISH)*

The purpose of the ISH technique is to find mRNAs in fixed samples. The fundamental phase of ISH is designing an antisense single-stranded RNA (ssRNA) probe intended to bind to the target mRNA. However, this is also possible to employ complementary DNA probes and synthesized oligonucleotide probes (Jensen 2014). The radioactive isotopes: ^{125}I, ^{35}S, and ^{32}P are frequently utilized for probe labeling due to their increased sensitivity as well as detection simplicity. The labeling of nonisotopic probes using tyramide, alkaline phosphatase, digoxigenin, biotin, or bromodeoxyuridine is used. The signal detection methods include liquid emulsion, autoradiography using X-ray film, photography, and microscopic methods (Corthell 2014). Using the ISH technique, rust pathogens *Uromyces transversalis* isolate CA-07, *Puccini-ahoriana* isolate PA-11, and *Phakopsorapachyrhizi* isolate Taiwan 72–1 that infect *Gladiolus hortulanus, Chrysanthemum morifolium* and the *Glycine max*, respectively, were identified. Although ISH maximizes the use of difficult-to-obtain tissue, its principal drawbacks include the high cost of radioactive probes, their danger, and the challenge of target identification with low RNA and DNA quantities (Ellison et al. 2016).

Fluorescent in situ hybridization (fish)

In terms of diagnosing plant diseases, FISH (Fluorescent in situ hybridization) acts as a new and advanced technique. This combines the sensitivity of fluorochrome-based detection techniques with the specificity of DNA sequences. FISH assays use probes of RNA or DNA that are either directly or directly fluorescently labeled to find sequences of RNA or DNA in cells/tissues. The ribosomal RNA (rRNA) of the microbial cells is hybridized by fluorescently mono-labeled oligonucleotide probes in conventional FISH procedures, and stained cells are seen using confocal laser scanning microscopy and wide-field epifluorescence. It is still debatable how the strain of *Xylella fastidiosa*, which causes olive rapid decline syndrome (OQDS), colonizes *Oleaeuropaea* L. xylem (Cardinale et al. 2018). A pathogen found in soil called *Sclerotium rolfsii* causes southern blight in tomato plants. By using the FISH and oligonucleotide probe labeled via cyanine dyes Cy5 and Cy3, the DNA isolation soil smears for *S. rolfsii* having 0.06 pg μ^{-1} were detected successfully. The best characteristics of FISH include sensitivity, reproducibility, accuracy, specificity, and speed. Additionally, FISH has the potential to give data providing morphology, resolution, and the recognition of the primary pathogens in mixed species specimens. Auto-fluorescence materials decrease the assay's specificity by giving false positive results (Hariharan and Prasannath 2021).

Next-Generation Sequencing (NGS)

A novel method used for diagnosing is NGS (Next-generation sequencing) or HTS (high throughput sequencing). Innovative detection methods and phytopathogen identification have been made possible by the advent of NGS technologies. The main processes in DNA-based NGS involve library preparation, bioinformatics analysis, DNA isolation, and DNA fragmentation, enormous parallel sequencing, variant/mutation annotation, and interpretation. Some frequently used advanced techniques of sequencing in HTS include pyrosequencing, colony sequencing, massively parallel signature sequencing, and sequencing via oligonucleotide ligation and detection (solid). Greater coverage along with resolution of the dynamic characteristics of the transcriptome are provided by RNA-Sequencing (RNA-Seq). The platform Illumina Hi-sequence has become the industry standard for NGS and is the widely usable NGS plan for RNA-sequencing. Mi-sequence, a desktop sequencer that was most recently published by the platform. NGS based on RNA-Seq can be used to quickly identify fungal plant infections causing novel illnesses. Sarcococca blight is caused by a novel fungus pathogen found in ornamental plants utilizing genome whole sequencing method utilizing Illumina MiSeq. The pathogen *Pseudo-Peronospora cubensis* responsible for cucurbit downy mildew was studied to improve molecular diagnostics. Seven distinct areas in *P. cubensis* were found using comparative genomics utilizing RNA-Sequence of the closely related *P. humuli* species that could be used to create diagnostic markers. Stone and pome fruits are damaged both after and before harvest by the fungus pathogen *Moniliniafructicola*, which causes brown rot disease. The *M. fructicola* Mfrc123 strain's genome was sequenced using a combination of the short-term read illumine NGS and long-term read PacBio (pacific biosciences) third-generation platforms for sequencing using a hybrid and the hierarchical denovo technique. When it comes to NGS methods, third-generation sequencing uses single molecular technology advancements, having an advantage over the second-generation sequencing approach. The main drawback of NGS is the length of time required for the assembly and processing of enormous volumes of sequence data. Low yield of RNA, RNA stability, and contaminants along with DNA, chemicals, or salts frequently impose limitations on next-generation applications. Even though data can be easily and quickly obtained, mycological and bioinformatics expertise are required for the NGS analysis, therefore understanding bioinformatics and fungal analyses is essential to prevent any kind of misinterpretation (Aragona et al. 2022).

2.2.5 Methods Based on Isothermal Amplification

Rolling circle amplification

An isothermal enzymatic assay called rolling circle amplification (RCA) uses RNA or DNA polymerase to produce ssDNA/RNA molecules. A homologous buffer, DNA polymerase, is a short primer for DNA/RNA, deoxynucleotide triphosphates (dNTPs), and a circular template are needed for an RCA experiment. In this test, it is crucial to prolong a single/number of primers annealing to the circular DNA template utilizing amplification of DNA via phi29 DNA polymerase, which has the activity of strand displacement. To release ssDNA, the strand displacement procedure moves the previously synthesized DNA molecule out of the way of the freshly created DNA template (Javaheri et al. 2014). RCA test has superiority in terms of simplicity, efficiency, and lack of temperature cycling devices. Additionally, this approach can be utilized to examine protein post-translational modification, mRNA splicing, SNP (Single nucleotide polymorphism), and gene expression (Aslam et al. 2017b, a).

Loop-mediated Amplification

For a decade, loop-mediated amplification which is LAMP technology is significantly advanced because it is a diagnostic technique for various plant diseases and has enormous potential for the management of plant diseases (Liu et al. 2017). The LAMP reaction is divided into two primary steps, starting with cyclic amplification and ending with a step of elongation (Panno et al. 2019). In the LAMP test, six distinct sequences on targeted NA (Nucleic acid) are recognized using 2 internal primer sets including FIP (Forward inner primer) and the BIP (Backward inner primer), one is B-Loop (Backward loop primer), and a second 2 outer primers set (B3 and F3). FIP and BIP contain two separate sequences that correspond to sense and the anti-sense strands of the DNA that is being targeted. The LAMP method may be sped up by adding two additional loop primers LF (Loop forward) and the LB (Loop backward) and also including DNA polymerase (Gomez-Gutierrez and Goodwin 2022). Using the LAMP assay, which focused on the DNA sequence of cytochrome b, the sugar beetroot rust-causing fungus *Uromycesbetae* was found in less than 30 min (Kaczmarek et al. 2019). The diagnosis of *Fusarium circinatum,* the quarantine fungal pathogen, which resulted in pitch canker disease in the pine and conifers, was made using conventional LAMP (Cl AMP) and a qLAMP (quantitative LAMP) assay. Assays of qLAMP demonstrated increased specificity in comparison to the cLAMP for *F. circinatum* detection, according to LAMP probes targeting TEF-1 (Stehlíková et al. 2020). Additionally, the initial qLAMP and LAMP detection techniques were created with MGG04322 that target *Magnaportheoryzae* for the swift and precise diagnosis of blast disease in rice (Li et al. 2019).

The panama disease in banana plants is caused by *Fusarium odoratissimum* TR4 (tropical race 4), which was correctly identified using a LAMP test employing TR4 markers sequencing from diversity arrays technology sequencing (DArTseq) technology (Ordóñez et al. 2019). The nucleic acid amplification using up to 6 primers gives the technique its high sensitivity and specificity (Becherer et al. 2020). LAMP technology is a strong candidate for quick and affordable alternative tests due to its isothermal and energy-efficient intensification requirements (Waliullah et al. 2020). As a result, LAMP has a solid reputation in several industries, such as those in the medical, agricultural, and food sectors (Panno et al. 2020).

Nucleic acid-based amplification (NASBA)

Under isothermal circumstances, nucleic acids are amplified using NASBA. Reverse transcriptase converts a ssRNA template into cDNA, which is used frequently for RNA amplification. The assay uses avian by myeloblastosis virus reverse transcriptase (RT), RNase H, and the T7 RNA polymerase around 41 °C temperature (Chang et al. 2012). It can produce more than 108 copies of the necessary NA sequences in less than 30 min, and the amplicons are found utilizing hybridization of the captured probe with molecular beacons (Heo et al. 2019). Additionally, NASBA is an ideal lab-on-a-chip system since it only calls for brief reactions has great sensitivity with strict control, is unaffected by inhibitors, and is ideal for such gadgets (Hønsvall and Robertson 2017). The application of phytofungal pathogen detection is extremely uncommon. It may be utilized to identify fungi-related disorders in the future (Aragona et al. 2022).

2.2.6 Post-Amplification Techniques

DNA Microarray

DNA microarrays (DNA chip/gene chip/biochip) is a technique where a collection of minute patches of DNA, is strategically connected to a solid surface, e.g. glass. To hybridize target cDNA, thousands of DNA probes are assembled into tiny patches. To determine the relative transcripts abundance of a sample, hybridized cDNA probe complexes can be found and quantified via silver, fluorophore, and chemiluminescence-labeled targets (Guigó 2013). The high throughput multiple detection of diverse pathogens of plants like viroids, viruses, fungi, and bacteria made it possible by DNA microarray advancements (Krawczyk et al. 2017). Using qPCR microarray technology, 48-well silicon microarrays were used to target fungal potato pathogens like *Rhizoctoniasolani* (Nikitin et al. 2018).

Detection of many pathogens of sugar beetroot root rot, including *Botrytis cinerea, Aphano-mycescochlioides,* and *Penicillium expansum*, was made possible using a novel microarray test (array-tube) combining genes marker: TEF-1, ITS, and 16S rDNA via the help of efficient probes (Olaya et al. 2016). It is quick, simple, uniform, and economical to generate DNA microarrays in batches using conventional microscope slides (Aslam et al. 2022). The fact that a microarray can detect sequences for that array intended to be recognized is a significant disadvantage.

DNA Microarray

DNA microarrays are created by constructing 15–30 base oligonucleotide species-specific probes, which are then arranged onto well plates and glued on a nitrocellulose/nylon membrane. The target DNA sequence can subsequently be identified by hybridization of a probe with amplified PCR and target DNA that is labeled (Zhang et al. 2008). Dioxigenin-labeled RNA probes were used in an oligo-DNA microarray test to identify the bacteria in apple plants' phyllospheres. For fungi identification, 40 bps of oligo-DNA sequence from the rDNA-ITS region of fungus were chosen, and several non-phytopathogenic fungi and bacteria, as well as the fungi that cause scab (Venturiainaequalis), *Alternaria blotch* (*Alternariamali*), and Marssonina blotch (*Diplocarponmali*) were found (He et al. 2012). A complicated disease of the grapevine called "young vine decline" (YVD) causes a significant amount of mortality in young vineyards. A DNA microarray was used to diagnose 61 species, including 34 YVD fungal infections (Úrbez-Torres et al. 2015). Even if the sample contains many infections, microarray provides a trustworthy and efficient way to identify the pathogens. Major drawbacks of microarray include the life of filters, identification of novel genes exhibiting low transcript abundance, and the huge filter volume required for hybridization (Gammill and Lee 2008). Other drawbacks of this array include its inability to quantify pathogens and its inability to determine whether DNA found belongs to a living organism (Úrbez-Torres et al. 2015).

2.2.7 RNA Interference

The most recent method for identifying and managing phytopathogenic fungi is RNA interference (RNAi). This is an effective method that may be applied in a highly tissue-specific manner to combat the mycotoxigenic fungus that infects crop plants and causes disease (Panwar et al. 2013). Execution of positive transgenic RNA interference based on various elements including (a) Vector designing in a way that resulted in dsRNAs that provide siRNA (Small interfering RNA) sp. For the ideal silencing of gene, (b) availability of the target siRNAs at the place of infection, (c) Proper uptake of siRNAs via fungus, (d) siRNA half-life, (e) silencing effects of amplification. RNAi combats the worrisome growth of fungicide-resistant plant diseases and removes the negative effects of existing disease management. RNAi uses a cellular defense mechanism to completely kill down the genes. Employing sequence homology necessary for dsRNA/hairpin RNA (hpRNA) destruction and silencing of fungal cells can identify the target transcripts (Nakayashiki et al. 2005). Cell-specific/dual RNA sequencing data was presented to determine the distinct fungal targets. Then, hpRNA/dsRNA can be altered to produce a specific transcript that will directly restrict the development of fungal disease. Short-interfering RNA molecules help the cell's inherent machinery, where RNAi technology takes advantage to properly knock down the target gene expression. The potential for off-target effects, which silence genes that tolerate partial complementarity to sense/anti-sense strands of the target gene, is a major drawback of RNA interference. Zea-mays has been subjected to RNAi to identify the fungus *Aspergillus flavus* (Masanga et al. 2015). In *Hordeumvulgare*, this is also used to pinpoint *Fusarium graminearum* (Chen et al. 2006).

3 Conclusion and Future Prospectus

ELISA, an immunological technique was used for fungi detection and quantification when the morphological characteristics were not evident. The following immunological techniques depend on the Ag-Ab (antigen–antibody) response in combination with readily accessible fluorescent dye/enzyme. The development of Abs has generally been successful against viruses, but it has been less useful against complex organisms like fungi. Therefore, making highly specific antibodies is expensive and complicated. These techniques only consistently function at the genus level, among other restrictions. In diagnostic applications, specificity and accuracy are crucial, which may limit the use of immunological techniques. When no obvious symptoms are present, NA-based techniques can be utilized to identify closely similar species and find the little pathogen. In-situ hybridization, southern hybridization, fluorescent in-situ hybridization, microarrays, and macroarrays are examples of DNA/

RNA probe technology used in molecular methods. RCA (Rolling circle amplification), LAMP (Loop-mediated isothermal amplification), and NASBA (nucleic acid sequence-based amplification) are examples of isothermal amplification technology. RT-PCR, nested PCR, multiplex PCR, real-time PCR, and DNA barcoding are all examples of PCR technology. Table 1 lists the benefits and restrictions of each approach. New, developing, previously known, and re-emerging fungal plant diseases are now more easily detected and diagnosed thanks to recent developments in molecular biology techniques. These molecular-based methods have been effective in locating and diagnosing symptomatic/asymptomatic diseases caused by culturable/non-culturable fungus pathogens in isolated cases as well as in co-infections of agricultural significant field, floricultural, forest plant species, ornamental and horticultural. Quantitative PCR is widely employed in the quantification and separation of causative agents when a small sample is loaded for detection using other PCR-centred methods. Currently, LAMP is demonstrating victory in the field of fungal disease diagnosis and made it possible to identify a variety of pathogens that cause devastating diseases in plants, including *Colletotrichum* sp., *Alternaria* sp., *Verticillium* sp., *Fusarium* sp., *Botrytis* spp., and *Puccinia* spp. NGS can be used to find new emerging pathogenic infections by sequencing fungal genomes on a variety of platforms without knowledge of pathogenic sequence beforehand. The molecular techniques discussed in the study are highly precise, efficient, lab-based, and used for high-end equipment to identify fungal plant diseases. To prevent any incorrect interpretation of the results of molecular analyses, however, mycology and bioinformatics skills are crucial. By integrating with other emerging technical breakthroughs, molecular methods should become a POCT (point of care testing) for fungal disease detection. The difficulties lie in the need for scientists to create workable methods for crop disease molecular diagnosis.

Table 1 Molecular diagnostic methods to identify phytopathogenic fungus and their merits and demerits (Aslam et al. 2017b, a; Mirmajlessi et al. 2015; Patel et al. 2022)

Molecular approaches	Merit	Demerits
Conventional PCR	Provide fast and accurate results. when the use of species-specific primer	Requires extensive labor and high-cost
Nested PCR	Need a dual set of primers resulted in increased yield and increased specificity for amplification of targeted DNA	Contamination risk due to two amplification cycles
Multiplex PCR	Act as a good money and time saver by utilizing several primer pairs in the same reaction	Probes and primer interference reduce the sensitivity
Reverse transcriptase PCR	Provide quantitative data related to pathogens, highly sensitive in comparison to the conventional PCR	Each assay formation is time-consuming and needs expensive reagents and equipment
Real-time PCR (q PCR)	Automated and no requirement for analysis of post-amplification	Cost and complexity because of simultaneous thermal cycling and detection via fluorescence
Digital droplet PCR	Extensive sensitivity, absolute quantification as no requirement of extrapolation	Instrument cost is high for initial purchase; Need for rigorous quality-control action
In situ hybridization	Extreme utilization of short-supply tissue	Difficulty in target identification having low RNA and DNA copies
FISH	Can be utilized for non-dividing cells	The probe-preparation approach is highly difficult because it is imperative to tailor probes to identify particular DNA sequences
Microarray	Simple to use due to no requirement of the huge scale DNA sequencing	mRNA large amount is required
LAMP	Sensitive, rapid, and very specific	Complex primer design; identify only a single specific pathogen; there is a risk of contamination of the sample
NASBA	No need for expensive equipment. Better in comparison to RT-PCR	Reaction specificity is dependent on the thermolabile enzymes. The temperature of the reaction cannot be exceeded more than 42Â°C with no compromise
RNA interference (RNAi)	Can simultaneously interrogate the thousands of genes	Variability and the incompleteness of knockdowns and the non-specificity of reagents potential
Northern Blotting	Detect size of RNA	Applied on small scale genes sample

(continued)

Table 1 (continued)

Molecular approaches	Merit	Demerits
SAGE	There is no need of prior knowledge of genome subject	Specificity for the tag sequence
RNA-Seq	Increased sensitivity and specificity	Need of costly equipment. Need of bioinformatics knowledge for analysis of data

References

Abdullah AS, Turo C, Moffat CS, Lopez-Ruiz FJ, Gibberd MR, Hamblin J, Zerihun A (2018) Real-time PCR for diagnosing and quantifying co-infection by two globally distributed fungal pathogens of wheat. Front Plant Sci 9:1086

Andika IB, Wei S, Cao C, Salaipeth L, Kondo H, Sun L (2017) Phytopathogenic fungus hosts a plant virus: a naturally occurring cross-kingdom viral infection. Proc Natl Acad Sci 114(46):12267–12272

Aragona M, Haegi A, Valente MT, Riccioni L, Orzali L, Vitale S, Infantino A (2022) New-generation sequencing technology in diagnosis of fungal plant pathogens: a dream comes true? J Fungi 8(7):737

Aslam S, Tahir A, Aslam MF, Alam MW, Shedayi AA, Sadia S (2017a) Recent advances in molecular techniques for the identification of phytopathogenic fungi–a mini review. J Plant Interact 12(1):493–504

Aslam S, Tahir A, Aslam MF, Alam MW, Shedayi AA, Sadia S (2017b) Recent advances in molecular techniques for the identification of phytopathogenic fungi—a mini review. J Plant Interact 12(1):493–504. https://doi.org/10.1080/17429145.2017.1397205

Aslam HMU, Riaz H, Killiny N, Zhou XG, Thomashow LS, Peters NT, Chanda AK (2022) Microarray technology for detection of plant diseases. Trends Plant Disease Assess 203–223

Badali H, Nabili M (2012) Molecular tools in medical mycology; where we are! Jundishapur J Microbiol 6(1):1–3

Balodi R, Bisht S, Ghatak A, Rao KH (2017) Plant disease diagnosis: technological advancements and challenges. Indian Phytopathol 70(3):275–281

Bangratz M, Wonni I, Kini K, Sondo M, Brugidou C, Béna G, Tollenaere C (2020) Design of a new multiplex PCR assay for rice pathogenic bacteria detection and its application to infer disease incidence and detect co-infection in rice fields in Burkina Faso. PLoS ONE 15(4):e0232115

Bayram Ö, Ahmed YL, Maruyama JI, Valerius O, Rizzoli SO, Ficner R, Irniger S, Braus GH (2012) The *Aspergillus nidulans* MAPK module AnSte11-Ste50-Ste7-Fus3 controls development and secondary metabolism. PLOS Genet 8:e1002816

Becherer L, Borst N, Bakheit M, Frischmann S, Zengerle R, von Stetten F (2020) Loop-mediated isothermal amplification (LAMP)–review and classification of methods for sequence-specific detection. Anal Methods 12(6):717–746

Bhat RG, Browne GT (2010) Specific detection of *Phytophthora cactorum* in diseased strawberry plants using nested polymerase chain reaction. Plant Pathol 59(1):121–129

Bölker M (1998) Sex and crime: heterotrimeric G proteins in fungal mating and pathogenesis. Fungal Genet Biol 25(3):143–156

Brown J, Ogle H (1997) Fungal diseases and their control. Plant pathogens and plant diseases. APPS, Australia, 1(4):443–467

Brown NA, Bass C, Baldwin TK, Chen H, Massot F, Carion PW, Hammond-Kosack KE (2011) Characterisation of the *Fusarium graminearum*-wheat floral interaction. J Pathogens

Bruno KS, Tenjo F, Li L, Hamer JE, Xu JR (2004) Cellular localization and role of kinase activity of PMK1 in *Magnaporthe grisea.* Eukaryot Cell 3(6):1525–1532

Capote N, Pastrana AM, Aguado A, Sánchez-Torres P (2012) Molecular tools for detection of plant pathogenic fungi and fungicide resistance. Plant Pathol 374

Cardinale M, Luvisi A, Meyer JB, Sabella E, De Bellis L, Cruz AC, Ampatzidis Y, Cherubini P (2018) Specific fluorescence *in Situ* hybridization (FISH) test to highlight colonization of xylem vessels by *Xylella fastidiosa* in naturally infected olive trees (*Olea europaea* L.). Front Plant Sci 9:431. https://doi.org/10.3389/fpls.2018.00431

Carneiro AA, Gomes EA, Guimarães CT, Fernandes FT, Carneiro NP, Cruz I (2008) Molecular characterization and pathogenicity of isolates of *Beauveria* spp. to fall armyworm. Pesq Agrop Brasileira 43:513–520

Chalupová J, Raus M, Sedlářová M, Šebela M (2014) Identification of fungal microorganisms by MALDI-TOF mass spectrometry. Biotechnol adv 32(1):230–241

Chandelier A, Massot M, Fabreguettes O, Gischer F, Teng F, Robin C (2019) Early detection of *Cryphonectria parasitica* by real-time PCR. Eur J Plant Pathol 153:29–46

Chang CC, Chen CC, Wei SC, Lu HH, Liang YH, Lin CW (2012) Diagnostic devices for isothermal nucleic acid amplification. Sensors 12(6):8319–8337

Chaoyang X, Yong-Sun B, Gary MC, Joseph H (2006) G protein-coupled receptor Gpr4 senses amino acids and activates the cAMP-PKA pathway in *Cryptococcus neoformans.* Mol Biol Cell 17(2):667–679

Chatterjee S, Kuang Y, Splivallo R, Chatterjee P, Karlovsky P (2016) Interactions among filamentous fungi *Aspergillus niger*, *Fusarium verticillioides* and *Clonostachys rosea*: fungal biomass, diversity of secreted metabolites and fumonisin production. BMC Microbiol 16:1–13

Chen S, Tao L, Zeng L, Vega-Sanchez ME, Umemura K, Wang GL (2006) A highly efficient transient protoplast system for analyzing defence gene expression and protein–protein interactions in rice. Mol Plant Pathol 7(5):417–427

Corthell JT (2014) In situ hybridization. Basic Mol Protocols Neurosci: Tips, tricks Pitfalls 105–111

Dasmahapatra KK, Mallet J (2006) DNA barcodes: recent successes and future prospects. Heredity 97(4):254–255

de Chaves MQG, Morán F, Barbé S et al (2023) A new and accurate qPCR protocol to detect plant pathogenic bacteria of the genus '*Candidatus* Liberibacter' in plants and insects. Sci Rep 13:3338. https://doi.org/10.1038/s41598-023-30345-0

del Pilar Martínez-Diz M, Andrés-Sodupe M, Berbegal M, Bujanda R, Díaz-Losada E, Gramaje D (2020) Droplet digital PCR technology for detection of *Ilyonectria liriodendri* from grapevine environmental samples. Plant Dis 104(4):1144–1150

Depotter JR, Rodriguez-Moreno L, Thomma BP, Wood TA (2017) The emerging British *Verticillium longisporum* population consists of aggressive Brassica pathogens. Phytopathology 107(11):1399–1405

DiGuistini S, Wang Y, Liao NY, Taylor G, Tanguay P, Feau N, Breuil C (2011) Genome and transcriptome analyses of the mountain pine beetle-fungal symbiont Grosmannia clavigera, a lodgepole pine pathogen. Pro Natl Acad Sci 108(6):2504–2509

Ellison MA, McMahon MB, Bonde MR, Palmer CL, Luster DG (2016) In situ hybridization for the detection of rust fungi in paraffin embedded plant tissue sections. Plant Methods 12:1–11

Elverson TR, Kontz BJ, Markell SG, Harveson RM, Mathew FM (2020) Quantitative PCR assays developed for *Diaporthe helianthi* and *Diaporthe gulyae* for phomopsis stem canker diagnosis and germplasm screening in sunflower (*Helianthus annuus*). Plant Dis 104(3):793–800

Fatmi M, Damsteegt VD, Schaad NW (2005) A combined agar-absorption and BIO-PCR assay for rapid, sensitive detection of *Xylella fastidiosa* in grape and citrus. Plant Pathol 54(1):1–7

Fang Y, Ramasamy RP (2015) Current and prospective methods for plant disease detection. Biosensors 5(3):537–561

Fujimura M, Ochiai N, Oshima M, Motoyama T, Ichiishi A, Usami R, Yamaguchi I (2003) Putative homologs of SSK22 MAPKK Kinase and PBS2 MAPK Kinase of Saccharomyces cerevisiae Encoded by os-4 and os-5 genes for osmotic sensitivity and fungicide. Biosci Biotechnol Biochem 67(1):186–191

Gammill LS, Lee VM (2008) Gene discovery: macroarrays and microarrays. Methods Cell Biol 87:297–312

Gautam AK, Verma RK, Avasthi S, Sushma, Bohra Y, Devadatha B, Suwannarach N (2022) Current insight into traditional and modern methods in fungal diversity estimates. J Fungi 8(3):226

Ghosh G, Panicker L, Kumar NN, Mallick V (2018) Surface plasmon resonance of counter ions coated charged silver nanoparticles and application in bio-interaction. Mater Res Express 5(5):055005

Gomez-Gutierrez SV, Goodwin SB (2022) Loop-mediated isothermal amplification for detection of plant pathogens in wheat (*Triticum aestivum*). Front Plant Sci 13:857673. https://doi.org/10.3389/fpls.2022.857673

Green MR, Sambrook J (2019) Nested polymerase chain reaction (PCR). Cold Spring Harbor Protocols 2019(2):pdb-prot095182

Griffiths MW (1999) Nucleic acid-based assaysl overview

Guerriero G, Hausman JF, Strauss J, Ertan H, Siddiqui KS (2015) Destructuring plant biomass: focus on fungal and extremophilic cell wall hydrolases. Plant Sci 234:180–193

Guigó R (2013) The coding and the non-coding transcriptome. In: Handbook of systems biology. Academic Press, pp 27–41

Hawksworth DL (1991) The fungal dimension of biodiversity: magnitude, significance, and conservation. Mycol Res 95(6):641–655

Hamel LP, Nicole MC, Duplessis S, Ellis BE (2012) Mitogen-activated protein kinase signaling in plant-interacting fungi: distinct messages from conserved messengers. Plant Cell 24(4):1327–1351

Hariharan G, Prasannath K (2021) Recent advances in molecular diagnostics of fungal plant pathogens: a mini review. Front Cell Infect Microbiol 10:600234. https://doi.org/10.3389/fcimb.2020.600234

Havis ND, Gorniak K, Carmona MA, Formento AN, Luque AG, Scandiani MM (2014) First molecular detection of Ramularia leaf spot (*Ramularia collo-cygni*) in seeds and leaves of barley in Argentina. Plant Dis 98(2):277–277

Havis ND, Brown JK, Clemente G, Frei P, Jedryczka M, Kaczmarek J, Hess M (2015) Ramularia collo-cygni—an emerging pathogen of barley crops. Phytopathology 105(7):895–904

Henson JM, French RC (1993) The polymerase chain reaction and plant disease diagnosis. Papers Plant Pathol 2

He YH, Isono S, Shibuya M, Tsuji M, Adkar Purushothama CR, Tanaka K, Sano T (2012) Oligo-DNA custom macroarray for monitoring major pathogenic and non-pathogenic fungi and bacteria in the phyllosphere of apple trees. PLoS ONE 7(3):e34249

Heo S, Kim HR, Lee HJ (2019) Development of a quantitative real-time nucleic acid sequence based amplification (NASBA) assay for early detection of apple scar skin viroid. Plant Pathol J 35(2):164

Hønsvall BK, Robertson LJ (2017) From research lab to standard environmental analysis tool: will NASBA make the leap? Water Res 109:389–397

Ito Y (2002) Cell culture engineering using intelligent biomaterials. In: Animal cell technology: challenges for the 21st century: proceedings of the joint international meeting of the Japanese association for animal cell technology (JAACT) and the European society for animal cell technology (ESACT) 1998, Kyoto, Japan. Springer Netherlands, pp 171–175

Javaheri Tehrani S, Aliabadian M, Fata A, Najafzadeh MJ (2014) Rolling Circle Amplification (RCA): an approach for quick detection and identification of fungal species. J Mycol Res 1(1):55–62

Jensen E (2014) Technical review: in situ hybridization. Anat Rec 297(8):1349–1353

Ji Cho H, Hong SW, Kim HJ, Kwak YS (2016) Development of a multiplex PCR method to detect fungal pathogens for quarantine on exported cacti. Plant Pathol J 32(1):53–57

Johnson KL, Zheng D, Kaewnum S, Reid CL, Burr T (2013) Development of a magnetic capture hybridization real-time PCR assay for detection of tumorigenic *Agrobacterium vitis* in grapevines. Phytopathology 103(6):633–640

Jones CA, Greer-Phillips SE, Borkovich KA (2007) The response regulator RRG-1 functions upstream of a mitogen-activated protein kinase pathway impacting asexual development, female fertility, osmotic stress, and fungicide resistance in *Neurospora crassa*. Mol Biol Cell 18(6):2123–2136

Kaczmarek AM, King KM, West JS, Stevens M, Sparkes D, Dickinson MJ (2019) A loop-mediated isothermal amplification (LAMP) assay for rapid and specific detection of airborne inoculum of Uromyces betae (sugar beet rust). Plant Dis 103(3):417–421

Kasahara S, Nuss DL (1997) Targeted disruption of a fungal G-protein β subunit gene results in increased vegetative growth but reduced virulence. Mol Plant Microbe Interact 10(8):984–993

Kashyap PL, Rai P, Kumar S, Chakdar H, Srivastava AK (2017) DNA barcoding for diagnosis and monitoring of fungal plant pathogens. In: Singh BP, Gupta VK (eds) Molecular markers in mycology. fungal biology. Springer, Cham. https://doi.org/10.1007/978-3-319-34106-4_5

Kim SW, Li Z, Moore PS, Monaghan AP, Chang Y, Nichols M et al (2010) A sensitive non-radioactive northern blot method to detect small RNAs. Nucl Acids Res 38:e98–e98

Krawczyk K, Uszczyńska-Ratajczak B, Majewska A, Borodynko-Filas N (2017) DNA microarray-based detection and identification of bacterial and viral pathogens of maize. J Plant Dis Prot 124:577–583

Krishna Krishnamurthy P, Francis RA (2012) A critical review on the utility of DNA barcoding in biodiversity conservation. Biodivers Conserv 21:1901–1919

Kumar R, Gupta A, Srivastava S, Devi G, Singh VK, Goswami SK, Aggarwal R (2020) Diagnosis and detection of seed-borne fungal phytopathogens. Seed-borne diseases of agricultural crops: detection, diagnosis and management, pp 107–142

Kuzdraliński A, Kot A, Szczerba H, Nowak M, Muszyńska M (2017) A review of conventional PCR assays for the detection of selected phytopathogens of wheat. Microbial Physiol 27(3):175–189

Lengeler KB, Davidson RC, D'souza C, Harashima T, Shen WC, Wang P, Heitman J (2000) Signal transduction cascades regulating fungal development and virulence. Microbiol Mol Biol Rev 64(4):746–785

Lévesque AC (2001) Molecular methods for detection of plant pathogens—what is the future? Can J Plant Path 23(4):333–336. https://doi.org/10.1080/07060660109506953

Li YG, Jiang WY, Li S, Jiang D, Wang RT, Ji P, Pan HY (2019) Occurrence of cotyledon spot on soybean caused by *Fusarium oxysporum* in China. Plant Dis 103(11):2967–2967

Li L, Zhu XM, Zhang YR, Cai YY, Wang JY, Liu MY, Lin FC (2022) Research on the molecular interaction mechanism between plants and pathogenic fungi. Int J Mol Sci 23(9):4658

Liebmann B, Gattung S, Jahn B, Brakhage AA (2003) CAMP signaling in *Aspergillus fumigatus* is involved in the regulation of the pathogenicity determinant-encoding gene pksP and the defense against killing by macrophages. Mol Gen Genomics 269:420–435

Lilit W, Nidhi A (2013) Polymerase chain reaction. J Investig Dermatol 133(6):392–395

Liu W, Huang S, Liu N, Dong D, Yang Z, Tang Y, Huang L (2017) Establishment of an accurate and fast detection method using molecular beacons in loop-mediated isothermal amplification assay. Sci Rep 7(1):40125

Liu J, Li C, Muhae-Ud-Din G, Liu T, Chen W, Zhang J, Gao L (2020) Development of the droplet digital PCR to detect the teliospores of *Tilletia controversa* Kühn in the soil with greatly enhanced sensitivity. Front Microbiol 11:4. https://doi.org/10.3389/fmicb.2020.00004

Lukianova AA, Evseev PV, Stakheev AA, Kotova IB, Zavriev SK, Ignatov AN, Miroshnikov KA (2021) Quantitative real-time PCR assay for the detection of *Pectobacterium parmentieri*, a causal agent of potato soft rot. Plants 10:1880. https://doi.org/10.3390/plants10091880

Ma Z, Michailides TJ (2007) Approaches for eliminating PCR inhibitors and designing PCR primers for the detection of phytopathogenic fungi. Crop Prot 26(2):145–161

Mackay IM (2004) Real-time PCR in the microbiology laboratory. Clin Microbiol Infect 10(3):190–212

Majumdar R, Rajase Karan K, Cary JW (2017) RNA interference (RNAi) as a potential tool for control of mycotoxin contamination in crop plants: concepts and considerations. Front Plant Sci 8:200

Malarczyk DG, Panek J, Frąc M (2020) Triplex real-time PCR approach for the detection of crucial fungal berry pathogens—*Botrytis* spp., *Colletotrichum* spp. and *Verticillium* spp. Int J Mol Sci 21(22):8469

Mancini V, Murolo S, Romanazzi G (2016) Diagnostic methods for detecting fungal pathogens on vegetable seeds. Plant Pathol 65(5):691–703

Masanga JO, Matheka JM, Omer RA, Ommeh SC, Monda EO, Alakonya AE (2015) Downregulation of transcription factor aflR in *Aspergillus flavus* confers reduction to aflatoxin accumulation in transgenic maize with alteration of host plant architecture. Plant Cell Rep 34:1379–1387

McCartney HA, Foster SJ, Fraaije BA, Ward E (2003) Molecular diagnostics for fungal plant pathogens. Pest Manage Sci: Formerly Pesticide Sci 59(2):129–142

Meng J, Doyle MP (2002) Introduction. Microbiological food safey. Microbes Infect 4(4):395–397

Mey G, Held K, Scheffer J, Tenberge KB, Tudzynski P (2002) CPMK2, an SLT2-homologous mitogen-activated protein (MAP) kinase, is essential for pathogenesis of *Claviceps purpurea* on rye: evidence for a second conserved pathogenesis-related MAP kinase cascade in phytopathogenic fungi. Mol Microbiol 46(2):305–318

Midorikawa GE, Miller RN, Bittencourt DMDC (2018) Molecular identification and detection of foodborne and feedborne mycotoxigenic fungi. Mol Tech Food Biol: Saf Biotechnol Authenticity Traceability 385–407

Mirmajlessi SM, Destefanis M, Gottsberger RA, Mänd M, Loit E (2015) PCR-based specific techniques used for detecting the most important pathogens on strawberry: a systematic review. Syst Rev 4:1–11

Moher D, Shamseer L, Clarke M, Ghersi D, Liberati A, Petticrew M, Stewart LA (2015) Preferred reporting items for systematic review and meta-analysis protocols (PRISMA-P) 2015 statement. Syst Rev 4(1):1–9

Monge RA, Roman E, Nombela CPLA, Pla J (2006) The MAP kinase signal transduction network in *Candida albicans*. Microbiology 152(4):905–912

Motoyama T, Ochiai N, Morita M, Iida Y, Usami R, Kudo T (2008) Involvement of putative response regulator genes of the rice blast fungus *Magnaporthe oryzae* in osmotic stress response, fungicide action, and pathogenicity. Curr Genet 54:185–195

Murolo S, De Miccolis Angelini RM, Faretra F, Romanazzi G (2018) Phenotypic and molecular investigations on hypovirulent *Cryphonectria parasitica* in Italy. Plant Dis 102(3):540–545

Nakayashiki T, Kurtzman CP, Edskes HK, Wickner RB (2005) Yeast prions [URE3] and [PSI+] are diseases. Proc Natl Acad Sci 102(30):10575–10580

Nikitin M, Deych K, Grevtseva I, Girsova N, Kuznetsova M, Pridannikov M, Golikov A (2018) Preserved microarrays for simultaneous detection and identification of six fungal potato pathogens with the use of real-time PCR in matrix format. Biosensors 8(4):129

Olaya G, Linley R, Edlebeck K, Kousik S, Kuhn P (2016) Oxathiapiprolin baseline sensitivity distribution of *Phytophthora capsici* isolates. In: Phytopathology, vol 106, no 12. Amer Phytopathological Soc., USA, pp 76–76

Ordóñez N, Salacinas M, Mendes O, Seidl MF, Meijer HJG, Schoen CD, Kema GHJ (2019) A loop-mediated isothermal amplification (LAMP) assay based on unique markers derived from genotyping by sequencing data for rapid in planta diagnosis of Panama disease caused by Tropical Race 4 in banana. Plant Pathol 68(9):1682–1693

Pallás V, Sánchez-Navarro JA, James D (2018) Recent advances on the multiplex molecular detection of plant viruses and viroids. Front Microbiol 9:2087

Panno S, Matić S, Tiberini A, Caruso AG, Bella P, Torta L, Davino S (2020) Loop mediated isothermal amplification: principles and applications in plant virology. Plants 9(4):461

Panno S, Ruiz-Ruiz S, Caruso AG, Alfaro-Fernandez A, Font San Ambrosio MI, Davino S (2019) Real-time reverse transcription polymerase chain reaction development for rapid detection of Tomato brown rugose fruit virus and comparison with other techniques. Peer J 7:e7928

Panwar V, McCallum B, Bakkeren G (2013) Endogenous silencing of *Puccinia triticina* pathogenicity genes through in planta-expressed sequences leads to the suppression of rust diseases on wheat. Plant J 73(3):521–532

Patel R, Mitra B, Vinchurkar M, Adami A, Patkar R, Giacomozzi F, Lorenzelli L, Baghini MS (2022) A review of recent advances in plant-pathogen detection systems. Heliyon 8(12):e11855. https://doi.org/10.1016/j.heliyon.2022.e11855

Pecchia S, Caggiano B, Da Lio D, Cafà G, Le Floch G, Baroncelli R (2019) Molecular detection of the seed-borne pathogen *Colletotrichum lupini* targeting the hyper-variable IGS region of the ribosomal cluster. Plants 8(7):222

Peng Y, Li SJ, Yan J, Tang Y, Cheng JP, Gao AJ, Xu BL (2021) Research progress on phytopathogenic fungi and their role as biocontrol agents. Front Microbiol 12:670135

Prassan C, Narayan SB, Hillol C, Saxena AK (2021) DNA barcoding of phytopathogens for disease diagnostics and bio-surveillance. World J Microbiol Biotechnol 37(3)

Qi M, Yang Y (2002) Quantification of *Magnaporthe grisea* during infection of rice plants using real-time polymerase chain reaction and northern blot/phosphoimaging analyses. Phytopathology 92(8):870–876

Rahman MT, Uddin MS, Sultana R, Moue A, Setu M (2013) Polymerase chain reaction (PCR): a short review. Anwer Khan Modern Med College J 4(1):30–36

Rahman CR, Arko PS, Ali ME, Khan MAI, Apon SH, Nowrin F, Wasif A (2020) Identification and recognition of rice diseases and pests using convolutional neural networks. Biosys Eng 194:112–120

Rajam MV, Chauhan S (2021) Host-induced gene silencing (HIGS): An emerging strategy for the control of fungal plant diseases. Genome Eng Crop Improvement 97–116

Roe AD, Rice AV, Bromilow SE, Cooke JEK, Sperling FAH (2010) Multilocus species identification and fungal DNA barcoding: insights from blue stain fungal symbionts of the mountain pine beetle. Mol Ecol Resourc 10:946–959

Romeis T (2001) Protein kinases in the plant defence response. Curr Opin Plant Biol 4(5):407–414

Rossmann S, Lysøe E, Skogen M, Talgø V, Brurberg MB (2021) DNA metabarcoding reveals broad presence of plant pathogenic oomycetes in soil from internationally traded plants. Front Microbiol 12:637068. https://doi.org/10.3389/fmicb.2021.637068

Sanzani SM, Li Destri Nicosia MG, Faedda R, Cacciola SO, Schena L (2014) Use of quantitative PCR detection methods to study biocontrol agents and phytopathogenic fungi and oomycetes in environmental samples. J Phytopathol 162(1):1–13

Savolainen V, Cowan RS, Vogler AP, Roderick GK, Lane R (2010) Towards writing the encyclopaedia of life: an introduction to

Schena L, Nigro F, Ippolito A, Gallitelli D (2004) Real-time quantitative PCR: a new technology to detect and study phytopathogenic and antagonistic fungi. Eur J Plant Pathol 110:893–908

Schena L, Li Destri Nicosia MG, Sanzani SM, Faedda R, Ippolito A, Cacciola SO (2013) Development of quantitative PCR detection methods for phytopathogenic fungi and oomycetes. J Plant Pathol 7–24

Schoch CL, Seifert KA, Huhndorf S, Robert V, Spouge JL, Levesque CA, Chen W, Fungal Barcoding Consortium (2012) Nuclear ribosomal internal transcribed spacer (ITS) region as a universal DNA barcode marker for Fungi. In: Proceedings of the national academy of sciences of the United States of America, vol 109, pp 6241–6246

Shaad NW, Cheong SS, Tamaki S, Hatziloukas E, Panopoulos NJ (1995) A combined biological and enzymatic amplification technique to detect *Pseudomonas syringae*. Phytopathology 85:243–248

Sharma PC, Matsumura H, Terauchi R (2007) Use of serial analysis of gene expression (sage) for transcript profiling in plants. Genomics-assisted crop improvement: vol. 1: genomics approaches and platforms, pp 227–244

Shuey MM, Drees KP, Lindner DL, Keim P, Foster JT (2014) Highly sensitive quantitative PCR for the detection and differentiation of *Pseudogymnoascus destructans* and other *Pseudogymnoascus species*. Appl Environ Microbiol 80(5):1726–1731

Sikdar P, Okubara P, Mazzola M, Xiao CL (2014) Development of PCR assays for diagnosis and detection of the pathogens *Phacidiopycnis washingtonensis* and *Sphaeropsis pyriputrescens* in apple fruit. Plant Dis 98(2):241–246

Stehlíková D, Luchi N, Aglietti C, Pepori AL, Diez JJ, Santini A (2020) Real-time loop-mediated isothermal amplification assay for rapid detection of *Fusarium circinatum*. Biotechniques 69(1):11–17

Suzuki JY, Amore T, Horrell JR, Imamura JS, Matsumoto T (2017) Isolation of anthurium nuclei for molecular applications. Acta Hort 52(9S):S372

Tang YW, Procop GW, Persing DH (1997) Molecular diagnostics of infectious diseases. Clin Chem 43(11):2021–2038

Trtkova J, Raclavsky V (2006) Molecular-genetic approaches to identification and typing of pathogenic Candida yeasts. Biomed Papers-Palacky Univ Olomouc 150(1):51

Travadon R, Baumgartner K, Rolshausen PE, Gubler WD, Sosnowski MR, Lecomte P, Péros JP (2012) Genetic structure of the fungal grapevine pathogen Eutypa lata from four continents. Plant Pathol 61(1):85–95

Tsui CKM, Woodhall J, Chen W, Lévesque CA, Lau A, Schoen CD, Najafzadeh MJ, Sybren de Hoog G (2011) Molecular techniques for pathogen identification and fungus detections in the environment. Ima Fungus 2:177–189

Tucker SL, Talbot NJ (2001) Surface attachment and pre-penetration stage development by plant pathogenic fungi. Annual Rev phytopathology 39(1):385–417

Úrbez-Torres JR, Haag P, Bowen P, Lowery T, O'Gorman DT (2015) Development of a DNA macroarray for the detection and identification of fungal pathogens causing decline of young grapevines. Phytopathology 105(10):1373–1388

Velculescu VE, Zhang L, Vogelstein B, Kinzler KW (1995) Serial analysis of gene expression. Science 270(5235):484–487

Walcott RR, Gitaitis RD, Langston DB (2004) Detection of *Botrytis aclada* in onion seed using magnetic capture hybridization and the polymerase chain reaction. Seed Sci Technol 32(2):425–438

Waliullah S, Ling KS, Cieniewicz EJ, Oliver JE, Ji P, Ali ME (2020) Development of loop-mediated isothermal amplification assay for rapid detection of Cucurbit leaf crumple virus. Int J Mol Sci 21(5):1756

Wang D, Peng Y (2014) Assessing the impact of dominant sequencing-based gene expression profiling techniques (SGEPTs) on phytopathogenic fungi. Chiang Mai J Sci 41(4):922–944

Wang Z, Zuo J, Gong J et al (2019) Development of a multiplex PCR assay for the simultaneous and rapid detection of six pathogenic bacteria in poultry. AMB Expr 9:185. https://doi.org/10.1186/s13568-019-0908-0

Wöhrle J, Krämer SD, Meyer PA, Rath C, Hügle M, Urban GA, Roth G (2020) Digital DNA microarray generation on glass substrates. Sci Rep 10(1):1–9

Yadav MK, Singh BP (2017) Real-time polymerase chain reaction (PCR) based identification and detection of fungi belongs to genus *Fusarium*. In: Singh BP, Gupta VK (eds) Molecular markers in mycology. Fungal biology. Springer, Cham. https://doi.org/10.1007/978-3-319-34106-4_4

Zeilinger S, Gupta VK, Dahms TE, Silva RN, Singh HB, Upadhyay RS, Nayak SC (2016) Friends or foes? Emerging insights from fungal interactions with plants. FEMS Microbiol Rev 40(2):182–207

Zhang N, McCarthy ML, Smart CD (2008) A macroarray system for the detection of fungal and oomycete pathogens of solanaceous crops. Plant Dis 92(6):953–960

Plant MicroRNAs: Identification and Their Application in Disease Management

Sandeep Kour, Deepak Kumar, Mohd. Ali, Parkirti, Roohi Sharma, Vikram, Rajesh Kumari Manhas, Harish Changotra, and Puja Ohri

Abstract Plants are exposed to a plethora of disease-causing agents like bacteria, fungi, viruses, and nematodes in their natural habitat. To counteract the harmful effects of these phytopathogens, plants have developed certain mechanisms where they stimulate or suppress a diverse array of genes. Among these, are the plant microRNAs (miRNAs), which are small regulatory RNAs, responsible for gene silencing in plants. These miRNAs play vital roles in developing resistance against phytopathogens by suppressing various genes at the post-transcriptional levels. The alterations in the expression levels, both of genes and miRNAs, mediate defense responses in plants and are also involved in managing resistances in plants in response to different stresses. Lately, high-throughput sequencing has led researchers to identify and characterize several miRNAs in plants and their involvement during pathogen invasion. Hence, the current book chapter highlights the recent advances and mechanisms that are involved in the defense responses of plant miRNAs against various phytopathogenic organisms.

Keywords Phytopathogens · miRNAs · Biotic stressors · Bacteria · Nematodes

S. Kour · D. Kumar · Mohd. Ali · Parkirti · R. Sharma · Vikram · P. Ohri (✉)
Department of Zoology, Guru Nanak Dev University, Amritsar, Punjab 143005, India
e-mail: ohri_puja_11@rediffmail.com

R. K. Manhas
Department of Microbiology, Guru Nanak Dev University, Amritsar, Punjab 143005, India

H. Changotra
Department of Molecular Biology and Biochemistry, Guru Nanak Dev University, Amritsar, Punjab 143005, India

J. Chen et al. (eds.), *Molecular and Biotechnological Tools for Plant Disease Management*, Interdisciplinary Biotechnological Advances,
https://doi.org/10.1007/978-981-97-7510-1_11

1 Introduction

Phytopathogens, including bacteria, fungi, nematodes, viruses, and various other parasites compromise crop productivity and challenge food security (Liu et al. 2017). These pathogens pose hinderances in the way of achieving sustainable agricultural productivity. To guard themselves against these pathogenic microorganisms, plants have developed small RNAs (sRNAs) that provide robust immunity toward any pathogen attack (Islam et al. 2017, 2018a). These sRNAs can be categorized into micro RNAs (miRNAs) and small interference RNAs (siRNAs) (Islam et al. 2018b). Among these two types of sRNAs, miRNAs are considered highly diverse and dynamic and are being focused upon by researchers of the world to improve plant defense against pathogens (Harris et al. 2013; Borges and Martienssen 2015; Stepien et al. 2017; Noman and Aqeel 2017). They were first detected in *Caenorhabditis elegans* (Lee et al. 1993) and since then, extensive research has been done to describe their functions in the life cycle of plants along with the understanding of their regulatory roles in plant-pathogen interactions (Djami-Tchatchou et al. 2017; Noman et al. 2017).

miRNAs are small, 21–24 nucleotides long, single-stranded RNAs encoded by microRNAs (MIR) genes (Pandita 2022). They are transcribed by RNA polymerase-II and converted into primary miRNAs (pri-miRNAs), and further recognized by Dicer-like 1 (DCL1) in plants (Yu et al. 2017). Pri-miRNAs are in turn converted into precursor miRNAs (pre-miRNAs) and eventually to miRNA duplexes (Auyeung et al. 2013; Alarcón et al. 2015). The biogenesis of the plant mi RNAs has been documented in several studies. The functioning of plant miRNAs has been expanded through advancements in genetics, bioinformatics, molecular and biochemical strategies as well as next-generation sequencing techniques (Goodwin et al. 2016; Noman et al. 2017). The flexibility of the functional activities of miRNAs is ensured by various processes such as RNA editing, complementarity, mutations in genes, availability of targets, etc. (Islam et al. 2018b). miRNAs have been reported to play crucial roles in enhancing plant growth and development by providing resistance against various types of stresses (Sunkar et al. 2012; Yuan et al. 2023). Most of the studies conducted on plant miRNAs have focused on their role against various types of abiotic stressors such as drought, cold, salinity, heat, heavy metal, etc., rather than biotic stressors (Zhang et al. 2015a; Samad et al. 2017). However, in the present book chapter, we have summarized the biogenesis and mode of action of plant miRNAs and the significance of diversified miRNAs in protecting plants by providing defense against bacterial, fungal, viral, and nematode attacks.

2 Plant MicroRNA Biosynthesis

microRNAs (miRNAs) are short, endogenous, noncoding RNAs that control the expression of a target gene associated with a variety of plant activities (Rawal et al. 2023). Recent years have seen significant advancements in our understanding of the miRNA production process in plants (Ledda et al. 2020; Sun et al. 2022). A canonical (Fig. 1) and a non-canonical miRNA biogenesis pathway have therefore been proposed (Komatsu et al. 2023; Shang et al. 2023). The steps involved in the multistep process that results in miRNA formation include RNA polymerase II transcription of *MIR* genes (miRNA encoding genes), primary miRNA (pri-miRNA) processing and RISC (RNA-induced silencing complex) assembly (Al-Noshokaty et al. 2023). The majority of the *MIR* genes in the plant genome are transcribed into pri-miRNAs rather than being translated into proteins. Transcription factors that are influenced by the various miRNA forms start the transcription of MIR genes. The transcriptional control of MIR genes is greatly influenced by both cis and trans-acting regulatory elements. The effectiveness of pri-miRNA processing is additionally influenced by nucleotide alteration, pri-miRNA structure, splicing and alternate splicing (Bofill-De Ros et al. 2022; Liang et al. 2022). Additionally, the GCN5 and PRC2 complex effectively controls the expression of MIR genes by altering histones (Jeena et al. 2022). Further, for improving pri-miRNA stability, a 5′ cap (7-methylguanosine) and a 3′ poly-A tail is added after the transcription of *MIR* genes into pri-miRNAs. The precursor miRNA (pre-miRNA), which is produced by the pri-miRNA, is cleaved into a duplex that contains the guide and passenger strands (miRNA-5/-3p) (Zhang et al. 2015b). A type III RNase named DICER LIKE-1 (DCL1), which is primarily facilitated by a dsRNA-binding protein called HYPONASTIC LEAVES 1 (HYL1) and the zinc finger protein SERRATE (SE), catalyzes the processing of pri-miRNA in plants (Bielewicz et al. 2023; Jozwiak et al. 2023). These collectively create a dicing complex, which is represented by D bodies, in the nucleus. Both the efficiency and fidelity of DCL1 processing depend heavily on HYL1. The pri-miRNA processing pathway is made more effective by the SE and HYL1 proteins, respectively (Hoang et al. 2020). While the zinc finger protein SE is necessary for the action of DCL1, the HYL1 protein dimer interacts with the miRNA-5/-3p duplex (Li and Yu 2021). The nuclear HUA Enhancer-1 protein 2′-O-methylates the 3′ nucleotides of the original miRNA/miRNA duplex. This change inhibits non-templated 3′-polymerization, which speeds up miRNA turnover (Hajieghrari and Niazi 2023). According to Wu et al. (2010), pri-miRNAs can be processed by DCL3 in rice in addition to DCL1, creating a class of 24 nucleotide-long miRNAs that control DNA methylation within the nucleus. Additionally, the structure of the plant DCL3 complex with pre-siRNA was examined. In this complex, DCL3 uses a positively charged region and an aromatic cap to precisely recognize the complementary strand and the guide strand (Jeena et al. 2022). Modified miRNA-5/-3p duplex is transported to the cytoplasm by the HASTY protein in plants, an animal EXPORTIN 5 homolog (Waititu et al. 2020).

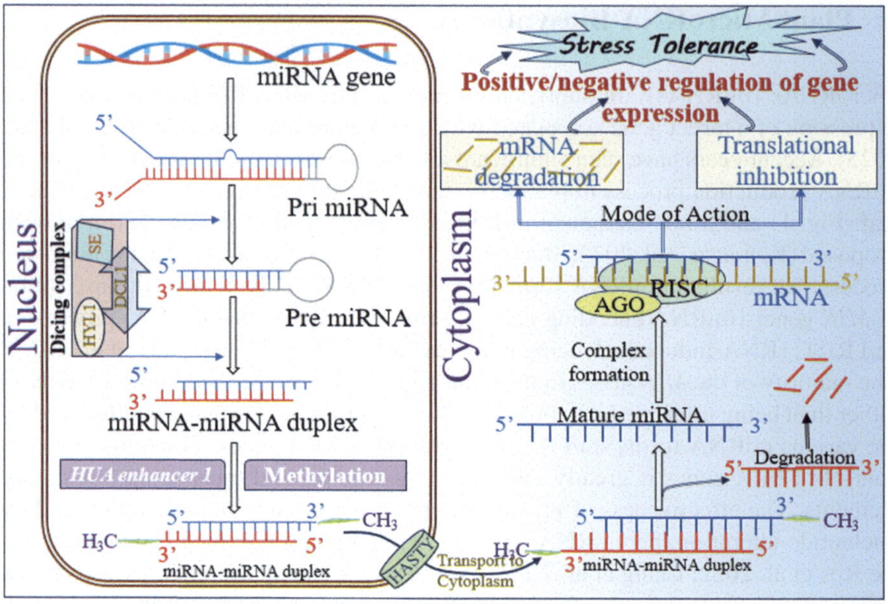

Fig. 1 Plant miRNA biosynthesis and mode of action. Following transcription, the pri-miRNA is converted into miRNA/miRNA duplex by the Dicer activity of the DICER LIKE-1 (DCL1)/ HYPONASTIC LEAVES 1 (HYL1) complex. The zinc finger protein SERRATE's (SE) assistance is also necessary for this phase. The methylase activity of HUA Enhancer-1 further methylates the miRNA duplex. HASTY is necessary for the export of methylated miRNA duplex from the nucleus. The mature miRNA is added to a specific RISC (RNA-induced silencing complex) that contains AGO1 (ARGONAUTE1), while the other miRNA is degraded. Through its "slicer" function or by inhibiting translation activity, the miRISC assembly causes target mRNA cleavage and translational repression

A non-canonical pathway of miRNA biosynthesis has been reported to be derived from rRNA. In the mammalian system, the mirtron pathway was found to be the first non-canonical miRNA biogenesis pathway. The miRNAs produced by this system go through specific maturation processes and are processed by spliceosomes and debranching enzymes in the nucleus (Wani et al. 2020). The first potential mirtron, osaMIR1429, was discovered in rice, although several mirtron genes have been discovered in animal systems (Zhu et al. 2008). Recently, osa-MIR1429 expression was only found in the shoot tissues of rice plants that were exposed to high temperatures (Mangrauthia et al. 2018), pointing to the gene's critical function in stress responses. In eukaryotes, the pre-rRNA gene *RN45s* transcript has two internal transcribed spacers (ITS1 and 2) and 18S, 5.8S and 28S rRNA regions. During the ensuing rRNA splicing process, the sequence of rRNA domains is totally preserved and conserved. More species-specific sequences of nucleotide that serve as phylogenetic markers can be found in the ITSs (Schultz and Wolf 2009). In the murine ITS2 region, where the mature body was changed without the usual dicer-dependent ways, the first pre-rRNA-derived miRNA (miR-712) was discovered (Son et al. 2013).

Lonicera japonica and *Papaver somniferum* both contain rRNA-derived miRNAs. A lengthy polycistronic pre-rRNA region had a clustered *P. somniferum* miRNA site. With miRNA sequences that were highly comparable to those of 13 plant species, five of these clustering miRNAs were species constant (Davoodi Mastakani et al. 2018). According to this finding, the clustered *P. somniferum* miRNA site might not be found in the ITS regions. It's interesting to note that plants rarely have such clustered miRNAs with a comparable conserved pattern. Reports have demonstrated the functional significance of pre-rRNA-generating miRNA in plants, but the mechanism by which they are silenced is yet unknown (Qiu et al. 2009; Sun et al. 2022). Further, used a high-throughput sequencing data- and structure-based approach to identify five *Arabidopsis* and eighteen rice noncanonical miRNAs in another attempt to identify the miRNAs produced from the mirtron pathway. By using degradome sequencing, the target genes of these potential mirtrons were predicted and verified. The non-canonical mechanism of miRNA synthesis in plants and its importance in stress response would be further supported by a thorough analysis of these mirtrons and their biological implications.

3 Mode of Action

Plant miRNAs control target genes at the post-transcriptional phase by means of two different mechanisms (Raza et al. 2023): translational repression of target genes and target gene mRNA degradation (Fig. 1). Mature miRNAs must enter an RISC to operate during the degradation of target gene mRNA. The miRNA-ribonucleoprotein complex (miRNP), which includes mature miRNAs and proteins, is another name for this complex (Iwakawa and Tomari 2022; Kasprzyk et al. 2023). The most significant protein in this RISC complex is the ARGONAUTE (AGO) protein. When ATP attaches to a dimer of the Heat Shock Protein 90 (HSP90) in the cytoplasm, the protein AGO1 forms a complex with the HSP90 protein and the miRNA strand is integrated into the AGO1 protein (Yu et al. 2017). The nucleotides of plant miRNAs are complementary to those of the target mRNAs because they are typically obtained from the reverse copy of the target genes. The AGO will break down the target mRNA at the 10th and 11th nucleotides of the miRNA binding region once the plant miRNA recognizes and binds to the target mRNA (Jiang et al. 2020). In *Arabidopsis*, ten AGOs have been discovered; AGO1 has four structural domains, PAZ, Mid, PIWI and the N-terminal domain (Xu et al. 2023). While the PIWI domain has endonucleolytic activity and is known to cleave the target mRNAs, the MID and PAZ domains interact with the 5′ phosphate and 3′ ends of the short RNA, respectively (Svoboda 2020; Xiao and MacRae 2022).

Numerous investigations have shown that for AGO1 to perform its shear function, complete complementary pairings of bases near the plant miRNA shear region with the target mRNA are required. According to Franco-Zorrilla et al. (2007), the IPS1 (induced by phosphate starvation 1, a non-coding gene) transcript was detected bound by miR399, but it was unable to mate with miR399 at the miR399 splice site, preventing AGO1 from splicing it. Based on this idea, scientists have created a number of transcripts that can bind to target miRNAs but are resistant to being cut by AGO1 (Franco-Zorrilla et al. 2007; Yan et al. 2012; Liu et al. 2014). According to Eamens and colleagues, the DRB3 and DRB5 proteins are necessary for the cleavage of miRNA target mRNAs, while the DRB1 and DRB2 proteins are crucial for the correct processing of miRNA precursors (Eamens et al. 2009). Only wild-type plants showed the photo-bleached phenotype after receiving the artificial miRNA PRI-MIR159B, which targets the Phytoene Desaturase (PDS) gene, was introduced into the drb1, drb2, drb3, drb5 and triple mutant drb235. However, only the drb1, drb2 and drb235 transgenic lines had poor pri-miRNA processing, indicating that the DRB3 and DRB5 proteins are crucial for miRNA-mediated cleavage (Eamens et al. 2012). Repression of target genes during translation is another way that plant miRNAs work. According to Aukerman and Sakai (2003), overexpression of miR172 dramatically decreased the quantities of proteins encoded by target mRNAs but did not affect the abundance of the target mRNAs' expression. They therefore postulated that plant miRNAs can also suppress the translation of target genes and that the same miRNA may even regulate target genes in a way that both shears them and inhibits their translation. This is one of the causes of the partial complementarity between the expression of plant miRNAs and their target genes (Ó'Maoiléidigh et al. 2021; Ma and Hu 2023).

4 Plant MicroRNAs Against Phytopathogens

Plants are significant for the survival of all organisms but they are continuously attacked by a number of pathogens. These phytopathogens can be bacteria, fungi, nematodes, or other parasites that utilize host machinery for their survival, cause diseases and ultimately destroy the whole plant. However, plants have developed several natural mechanisms to defend themselves against all pathogens. These mechanisms involve chemical defense, structural defense, systemic acquired resistance and hypersensitive response. One such defense mechanism in plants involves the role of miRNA in the functioning of the immune system (Zehra et al. 2021). To detect pathogens and activate the first line of defense, recognition of pathogen-associated molecular patterns (PAMPs) is required. However, pathogens tend to suppress the first line of defense of plants by disturbing the signaling pathway of PAMP-triggered immunity (PTI). So, to counter the pathogenic action, plants activate their second line of defense called effector-triggered immunity (ETI) (Yang et al. 2021). During this process, miRNA are either activated or repressed to regulate gene expression of target genes by altering signaling pathways that may involve jasmonic acid (JA), abscisic acid (ABA), and auxin as well (Šečić et al. 2021). For example, miR482/2118 in eudicots regulates *NUCLEOTIDE BINDING SITE-LEUCINE-RICH REPEAT* (*NBS-LRR*) genes that play a significant role in plant disease resistance. In monocots, it mainly targets long noncoding RNAs (lncRNAs) that play a pivotal role in plant reproduction (Zhang et al. 2021). Thus, miRNA actively participates in the defense system of plants against phytopathogens.

4.1 Bacteria

The response of plant miRNA in defense against bacteria has been observed during plant-bacterial infection (Fig. 2). Various studies have investigated that the miRNAs play an essential role in plant protection against disease-causing bacteria (Table 1). For example, a study on *Solanum melongena* plants infected with *Ralstonia solanacearum*, causing bacterial wilt in plants, revealed the presence of 375 miRNA of which miR414 was highly conserved in the whole plant system and miR5021 and miR5658 were corresponding to *Arabidopsis thaliana* (Kapadia et al. 2023). Similarly, in another study, the miRNA profiles of bacterial blight-resistant rice and the susceptible cultivar Huang-Hua-Zhan (HHZ) exposed to one of the destructive bacterial diseases caused by *Xanthomonas oryzae*, i.e., bacterial blight were studied and, in both varieties, 538 known and 312 novel miRNAs were found, of which only 26 and 17 were responsive to infection, respectively. The target genes of their miRNAs were associated with a regulatory pathway of reactive oxygen species, flavonoid synthesis, defense response, growth and development, and plant hormone signal transduction (Lu et al. 2021). Bioinformatic analysis revealed three kinds of immunity-related clusters of differentially expressed miRNAs in rice. The functions of different immunity clusters are defined by overexpressing osa-miR167d, osa-miR159b, and osa-miR164a. Also, during infection, 80 regulatory units participated consisting of 38 cleaved targets and 29 known miRNAs. However, the disruption of osa-miR159b/*OsMYBGA, OsMPK20-4, OsLRR-RLK2* and osa-miR164a/*OsNAC60,* osa-miR167d-5p/*OsWD40-174* regulatory units adjusted the defense response of rice to *Xoo* infestation (Jia et al. 2020). In *A. thaliana*, miR167 overexpression overcame the disease caused by the bacterial pathogen, *Pseudomonas syringae* by regulating the target genes ARF8 and ARF6, while the other functions of miR167 were reported as its response toward various processes such as root development, osmotic stress and flower development (Caruana et al. 2020). Table 1 represents the effect of miRNAs in various plant species against bacterial pathogens.

4.2 Fungi

Worldwide, fungal infections in plants have affected agricultural production and economy for a long time. Several crop plants such as wheat, rice, cotton, barley, etc. are frequently targeted by fungal diseases like fusarium wilt, fire blight, powdery mildew, leaf spot, etc. that affect the plant growth and crop yield. The main cause behind these diseases is fungal pathogens such as *Drechslera oryzae, Alternaria solani, Penicillium expansum, Rhizoctonia solani, Macrophomina phaseolina*, etc.

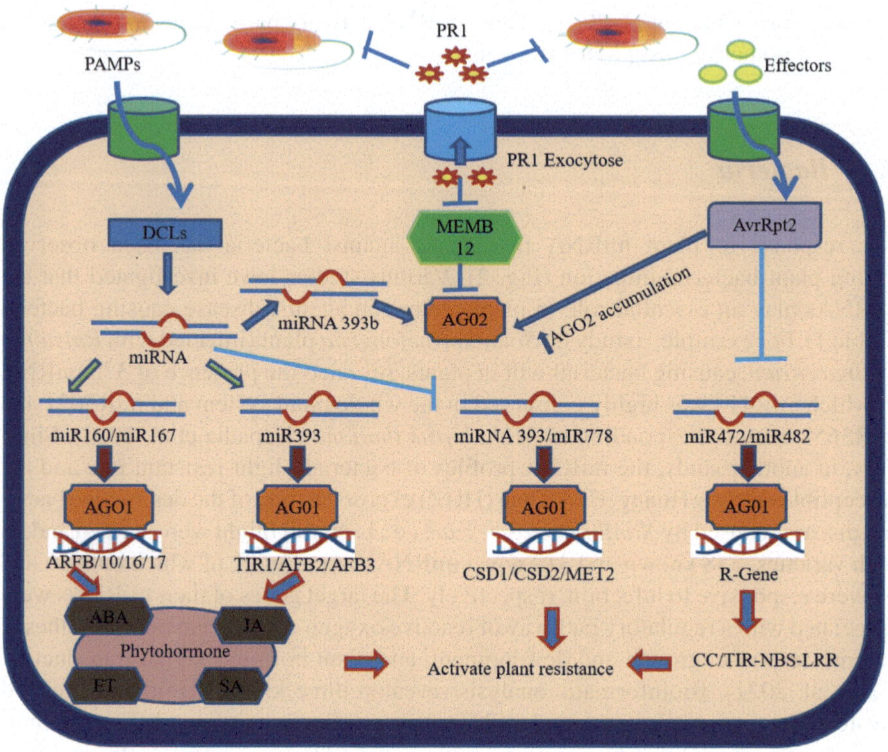

Fig. 2 Plant miRNAs actively promote defense against bacterial attack by modulating disease resistance through their ability to fine-tune several plant hormones. After infection, plants identify PAMPs and control the accumulation of miRNAs. Certain miRNAs (miR160, miR167, and miR393) control the activation of R protein, while other miRNAs (miR482/miR472) control the fine-tuning of plant hormone networks to control disease resistance. The partnering strand of miR393, miR393b increases plant immunity by encouraging the exocytosis of antimicrobials

These pathogens have a wide host range and high potential to parasitize over 400 or more species of plants. Plants use various strategies to combat these disease-causing pathogenic fungi, one of which includes miRNAs, which also play a significant role in maintaining homeostasis in plants. Usually, only specific miRNAs function as regulators of some specific target gene expressions, by either translational repression or sequence-specific cleavage (Yadav and Ramakrishna 2023). When plants get infected with fungi, their miRNAs either up or down-regulate in order to suppress diseases (Fig. 3). Plant miRNAs control the expression of their target transcripts to modulate the physiological balance of phytohormones. To date, a number of miRNAs have been found to provide resistance during the fungal attack (Kulshrestha et al. 2020). For example, in an experiment on sorghum plants, out of 75 miRNAs, 36 novel miRNAs were found active in response to fungal infection in both resistant and susceptible genotypes and 50 of them were significantly different between susceptible and resistant genotypes. Moreover, for 35 differentially expressed miRNAs, 149 target genes

Table 1 Role of miRNAs against bacterial pathogens

Pathogen	miRNA	Plant species	Target genes	Effect	References
Candidatus liberibacter	miR164, MiR482	*Citrus aurantifolia*	TIR-NBS-LRR	Regulated response of plants towards disease	Bojórquez-Orozco et al. (2023)
Magnaporthe oryzae	Chr12_41899, Chr12_40110 and Chr8_26996	*Oryza sativa*	Zinc finger A20, serine carboxypeptidase and Chr8_26996	Improved resistance to blast disease	Javed et al. (2023)
X. oryzae pv. *oryzae (Xoo)*	osa-miR1846-5p,	*O. sativa*	ACO4	Enhanced disease resistance in plants, *cis*- and *trans*- networking was formed along with IncRNAs and mRNAs	Li et al. (2023)
Xanthomonas albilineans (*Xa*) and *Acidovorax avenae* sub sp. *avenae* (*Aaa*)	sof-miR167	*Saccharum spontaneum*	*SpapARFs* and *ShARFs*	Regulated bacterial infection and enhanced plant survival under salicylic acid stress	Lin et al. (2023)
HLB bacteria	miR171b	*Citrus plants*	Orange1.1t00200.1, Orange1.1t00200.2, Cs5g08980.1 and Orange1.1t00199.1	Improved resistance to HLB	Lv et al. (2023)
Ralstonia solanacearum	Sly-miR482e-3p	*Solanum lycopersicum*	eTM482e-3p-1	Decreased sensitivity of plants to bacterial infection	Si et al. (2023)
Lonsdalea populi	pto-miR159a	*Populus ciliata*	MYB33, GAMYB	Acquired resistance towards bacteria, altered crosstalk between downstream hub gene	Yang et al. (2023)

(continued)

Table 1 (continued)

Pathogen	miRNA	Plant species	Target genes	Effect	References
Verticillium dahliae	ghr-miR393	*Gossypium hirsutum*	GhTIR1	Enhanced susceptibility to bacterial stress	Shi et al. (2022)
P. syringae	DmiR	*Arabidopsis*	DRN1	Immunity against bacterial stress, fungal growth inhibition, survival rate enhanced under salt stress	Dhar et al. (2020)
P. syringae	ath-miR164c	*A. thaliana*	AtP5CS1	Resisted to bacterial stress, modulated metabolism of proline mitigated drought stress	Gupta et al. (2020)
X. oryzae pv. *oryzae (Xoo)*	Osa-miR1876	*O. sativa*	NBS8R	Activated immune signaling pathway by triggering PTI in rice, Developed resistance to *Xoo*	Jiang et al. (2020)
P. syringae pv. *Tomato* and *Erwinia amylovora*	MdmiR285N	*Malus x domestica*	GUS	Enhanced plant immunity to pathogen, regulated seed development, reproductive development	Pompili et al. (2020)

(continued)

Table 1 (continued)

Pathogen	miRNA	Plant species	Target genes	Effect	References
R. solanacearum	gma-miR4415, osa-miR169, gra-miR482, mtr-miR5261, aly-miR398, ath-miR1886, mtr-miR164, ath-miR167 and ppt-miR1223	*Zingiber officinale* and *Curcuma amada*	ABCG11 and DNAJ1, AAP4, WRKY19, PRT6, RSP2, 4CLL1, ABCC5 and FRS6	Increased resistance to bacterial wilt disease, regulated multiple biological processes like cell differentiation, signal transduction etc	Snigdha and Prasath (2020)
Phytophthora infestans	miR482/2118	*S. lycopersicum*	NLR genes	Enhanced resistance to infection from bacteria and oomycetes	Canto-Pastor et al. (2019)
X. campestris pv. *campestris*	miR390, miR169, miR156 and miR167	*Brassica oleracea*	ARFs	Resistance against pathogen	Santos et al. (2019)

were spotted (Fu et al. 2020). *Verticillium dahliae* is a soil-born fungus that causes verticillium wilt which is a highly disastrous disease and was reported to infect cotton plants. In a study, it was observed that during posttranscriptional changes, miR477 of cotton cleaved the miRNA of its target gene *GhCBP60A*. The knocking down of this gene, however, increased the plant's resistance, while its central role was to determine salicylic acid level by regulating *GhICS1* expression (Hu et al. 2020). In another study, *V. dahliae* infection-induced secondary miRNAs such as miR166 production in *Arabidopsis* plant, that entered the fungal hyphae via extracellular vesicles and silenced specific target genes to suppress pathogenicity (Cai et al. 2018). Recently, in maize plants affected by *F. verticillioides*, fungal miRNAs such as FvTPP or their transcription factors such as ATG 13, ATG 17 and many others were found involved in causing infection to the host but, plant miRNAs such as zma-miR156i-3p, zma-miR171k-5p and zma-miR528b-5p and others enhanced the plant immunity by either downregulating or upregulating their target genes in the fungus (Qu et al. 2023).

Similarly, miRNA396 alters the innate immunity of plants without affecting growth thus, acting as a potent means of protection against fungal infection. This was observed in the case of *Arabidopsis* where a reduction in the activity of miR396 conferred a wider range of resistance to hemibiotrophic and necrotrophic fungal pathogens (Soto-Suárez et al. 2017). In rice plants also, 34 families of miRNAs were identified, of which most were involved in resistance toward plant fungal disease

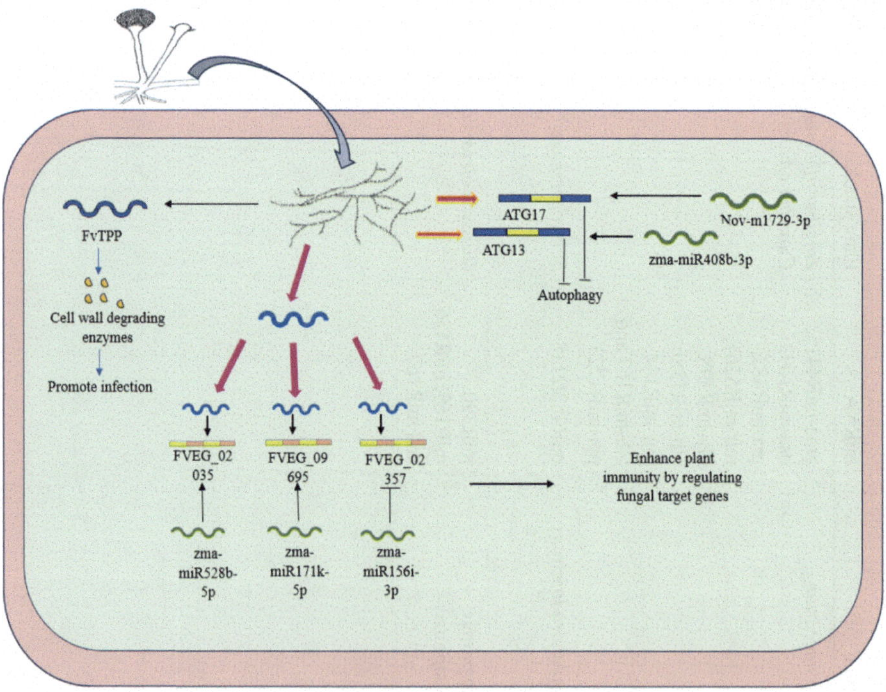

Fig. 3 Plant miRNAs regulating fungal infection

caused by *R. solani,* and expression of some miRNAs such as miR531a, mir1861i, miR444b.2, novel_miR135 and novel_miR1956 were responsible for controlling rice fungal disease by enhancing the immunity of plants (Wenlei et al. 2020). Some other studies regarding the effect of miRNAs on their target genes in different plants are listed in Table 2.

4.3 Viruses

Plants have developed a defense mechanism by naturally producing their own miRNAs in response to viral or viroid attacks (Ramesh et al. 2021). The initial findings indicated that *Arabidopsis* mutant DCL1 displayed the lowest vulnerability to RCNMV infection (Qasim et al. 2018). Given that DCL1 plays an essential role in the early stages of pre-miRNA processing, these observations suggest a hypothesis regarding the intrinsic utilization of miRNAs to inhibit and alter host gene expression. Various miRNAs such as miRNA168, miRNA169, miRNA398 and miRNA482 have been identified as key regulators of plant immunity against various viruses in different plant species (Rego et al. 2023). These miRNAs influence the expression of viral genes and pathways involved in defense mechanisms, offering valuable insights into potential strategies for inducing and enhancing plant resistance against viral infections.

Plant growth and morphology can be significantly impacted by viral infections and miRNAs can play a critical role in regulating plant immunity against these infections. Several studies have delved into their involvement in this process. One such investigation focused on the Tombus virus, which was found to induce the expression of miRNA168. This specific miRNA acts as a regulator of ARGONAUTE1 (AGO1) mRNA. Consequently, the enhanced expression of miRNA168 serves to suppress AGO1 mRNA levels thereby, influencing plant immunity (Kar and Raichaudhuri 2021). Another study examined the response of tobacco plants to infection by the tobacco mosaic virus (TMV) strain TMV-43A. This viral infection led to the differential expression of various miRNAs (Guo et al. 2017). These differentially expressed miRNAs are potential candidates for regulating plant resistance mechanisms post-TMV infection, suggesting their importance in enhancing immunity. In the case of soybeans, researchers investigated both resistant and susceptible cultivars in response to the mungbean yellow mosaic India virus (MMYIV) (Ramesh et al. 2017). Their findings revealed significant upregulation of miRNA169 and miRNA398 in soybean-resistant cultivars. Conversely, the susceptible cultivar exhibited considerably lower expressions of miRNA169 and miRNA398. This contrast highlights the crucial role these miRNAs play in bolstering plant resistance against MMYIV. Similarly, the study of common beans involved infecting virus-resistant cultivars with MMYIV to identify potential target miRNAs (Patwa et al. 2019). Among the candidate miRNAs discovered, miRNA482 stood out. miRNA482 exhibited a down-regulated expression pattern post-MMYIV infection and was found to target NB-LRR-resistant genes (Raza et al. 2023). This downregulation of miRNA482 subsequently led to an increase

Table 2 Role of miRNAs against fungal pathogens

Pathogen	Plant species	miRNA	Target gene	Effect	References
Verticillium dahliae	Cotton	ghr-miR530, miR171a	GhSAP6, GhSCL6	Regulated SA-relative gene expression that finally regulated plant defence, and increased plant deficiency	Hu et al. (2023a, b)
Puccinia triticina	Wheat	tae-miR9677b, miR9653b and tae-miR9773	Lr67, Lr34	Downregulated genes developed disease resistance	Spychala et al. (2023)
Peronophythora litchii	Litchi	LcmiR167, *LcmiR160*	LcARF8, LcARF18	Increased disease resistance	Yin et al. (2023)
V. dahliae	Cotton	miR398b	Gh_D05G3257 (5NB) and Gh_A04G0380 (4NB)	Directed the cleavage of mRNAs involved in homeostasis and ROS accumulation	Miao et al. (2022)
Alternaria longipes	Tobacco	nbe-miR482b and nbe-miRN2031*	R1B-23, R1B-16, R1B-23, TMV resistance protein N gene and NBS-coding resistance gene analogue	Targeted expression of resistance genes of plants and decreased injury levels	Wang et al. (2022)
Fusarium verticillioides	Maize	zma-unmiR4	ZmGA2ox4	Perturbed accumulation of bioactive gibberellin	Xu et al. (2022)
Magnaporthe oryzae	Rice	miR1871	LOC_Os06g22850	Encoded microfibrillar associated protein and enhanced immunity	Li and Yu (2021)

(continued)

Table 2 (continued)

Pathogen	Plant species	miRNA	Target gene	Effect	References
A. alternata	Apple	mdm-miR390a	MdRPK2, *MdLRR8*	Downregulated target gene enhanced resistance in plant	Qin et al. (2021)
F. oxysporum	Cucumber	*miR6300, miR319a*	DAHP1, PERK2, BEE1 and JRL3	Helpful in breeding program and regulated defence against disease	Xu et al. (2021)
Botrytis cinerea	Tomato	miR1001	Bcin10g01400.1 and Bcin03g02170.1	Inhibited germination of conidiospore	Meng et al. (2020)
Puccinia triticina	Wheat	miR164	Lr46 gene	Helped in resistance of leaf rust disease	Tomkowiak et al. (2020)
M. oryzae	Rice	miR156	WRKY45 and SPL14	Improved resistance towards rice blast disease	Zhang et al. (2020)
F. graminearum	Wheat	MiR1023	FGSG_03101	Silenced target gene and supressed hydrolase accumulation	Jiao and Peng (2018)

in the mRNA levels of NB-LRR encoding genes ultimately, enhancing the common bean's resistance against MMYIV.

In a separate study, researchers investigated the response of tomato plants against tomato leaf curl virus (ToLCV) and its interaction with three specific miRNAs (miRNA156, miRNA159, and miRNA403). In tomato plants resistant to ToLCV, there were no observable changes in the expression levels of miRNA159 and miRNA403 while miRNA156 showed a slight up-regulation. In contrast, susceptible tomato plants exhibited a significant increase in the expression of all three miRNAs (miRNA156, miRNA159, and miRNA403) upon infection with ToLCV. (Tousi et al. 2017). Furthermore, a study involving black gram (*Vigna munga* L.) identified several miRNAs that were induced following infection with the Mungbean Yellow Mosaic India Virus (MMYIV) (Kundu et al. 2017). These induced miRNAs included miR156, miRNA159, miRNA160, miRNA166, miRNA398, miRNA1511, miRNA1514, miRNA2118 as well as novel miRNAs named vmu-miRNAn7, vmu-miRNAn8, vmu-miRNAn13 and vmu-miRNAn14. These miRNAs have been identified as crucial players in regulating several important transcription factors (TFs), such as NB-LRR, NAC, MYB, Zinc finger, CCAAT-box TF, fructose 2–6 bisphosphate, and HD-ZIP. These TFs are well-recognized as major regulators of essential processes within plants, including immune system modulation, hormonal regulation, and antioxidant defense mechanisms (Wani et al. 2021; Yang et al. 2015). There is mounting evidence linking miRNA to viral illnesses and virus-induced gene silencing. Plant viruses have been shown to include more than 30 RNA silencing suppressors, also known as pathogenic variables such as p19, p21, p25, and p69 (Varallyay et al. 2014). Additionally, solid proof for using viral gene silencing to strengthen plant immunity is growing every day. Since siRNAs and miRNAs have several characteristics in common, hence it is speculated that miRNAs may also act as silencing invaders. For instance, miR171 in *Arabidopsis* demonstrated cleavage of mRNAs expressing scarecrow-like transcription factors, which guided an RNAi-like process (Kamitani et al. 2016).

While the aforementioned research findings have demonstrated the significance of miRNAs in these processes, the precise working of these regulatory networks remains a subject of intrigue and requires further investigation. Understanding the intricate interactions between miRNAs and their target, TFs is imperative for unraveling the precise mechanism behind the defensive strategies of plants against viral stressors (Begum 2022). Such insights could ultimately pave the way for innovative strategies to enhance plant resilience and crop yield in the face of viral challenges, representing a promising avenue for future research in plant biology and biotechnology. Examples of some plant miRNAs and their target genes against viral infections are listed in Table 3.

4.4 Nematodes

The most devastating plant-parasitic nematodes (PPNs) that cause substantial crop yield losses worldwide are sedentary endoparasitic nematodes like root-knot nematodes (RKNs) of the genus *Meloidogyne*, cyst nematodes (CNs) of the genus *Heterodera* and *Globodera*, root-lesion nematodes etc. (Blok et al. 2008; Jones et al. 2013). Plant immunity relies heavily on miRNA-mediated gene silencing. Resistance to RKNs is achieved by silencing the genes that are involved in housekeeping or parasitism in afflicted plants (Vos et al. 2013). For instance, RNAi was used to establish resistance in *Arabidopsis thaliana* by targeting four parasitism-related genes in sugar beetroot cyst nematode (*H. schachtii*) (Sindhu et al. 2008). *M. incognita*-induced gall development in soybean roots was successfully reduced by inhibiting many tyrosine phosphatase (TP) and mitochondrial stress-70 protein precursor (MSP) genes (Ibrahim et al. 2011). According to various reports, several miRNAs are involved in plant-nematode interactions. Down-regulation of miR161, miR164, miR167a, miR172c, miR396c, miR396a, b, and miR398a was reported in

Table 3 Target genes of plant miRNAs in some viral pathogens

Pathogen	Plant species	miRNAs	Target gene	Effect	References
Tomato yellow leaf curl virus (TYLCV)	*Solanum lycopersicum*	Rep-amiRNA1 and Rep-amiRNA2	AC1/Rep gene	A decrease in viral multiplication was observed led to disease resistance	
Bean common mosaic virus (BCMV)	*Populus alba*	miR156	squamosa promoter-binding-like protein	Increased chlorophyll content, the net photosynthesis rate, and stomatal conductance	Wang et al. (2023)
Cucumber mosaic virus (CMV)	*Capsicum annuum*	miR171, miR172, miR156, miR159	Conserved sequences	Increased the immune response against viral pathogenicity	Luo et al. (2023)
Potato virus X (PVX), Potato virus Y (PVY) and Potato Spindle Tuber Viroid (PSTVd)	*S. tuberosum*	amiR-159P25, amiR-159HCPro and amiR-159Brp1	*P25*, *HC-Pro* and *Brp1*	Provided resistance against the infection of PVY, PVX and PSTVd	Jiang et al. (2023)
Maize Iranian mosaic virus (MIMV)	*Zea mays*	miR166, miR167, miR169, miR395, miR399, miR408, miR482	Hub genes	Diagnosed the disease and provided aid in disease resistance	Ghorbani et al. (2023)
Sugarcane mosaic disease (SMD)	*Saccharum* spp.	miR395b	*ATPs*-3 and low-affinity sulfate transporter 3-like (*LAST-3*) genes	Provided resistance against pathogen	
RSV	*Oryza sativa*	osa-MIR528	RSV MP gene	Transgenic plants were resistant to pathogenicity	Zhou et al. (2022)

(continued)

Table 3 (continued)

Pathogen	Plant species	miRNAs	Target gene	Effect	References
CGMMV	*Nicotiana benthamiana* and *Cucumis sativus*	amiR-GUS	GUS gene	Prevented the occurrence and accumulation of disease	Miao et al. (2021)
Citrus psorosis virus (CPsV)	*Citrus sinensis*	miR169, miR172, miR393	CCAAT-binding TF	Reduced symptom severity of disease	Marmisolle et al. (2020)
Barley yellow dwarf virus (BYDV)	*Hordeum vulgare*	hvu-miR169 and hvu-miR6182,	MLOC_37399.1 gene and LRR receptor-like serine/threonine-protein kinase gene	Provided resistance and novel strategies against the virus	Jarošová et al. (2020)
Sugarcane Bacilliform Guadeloupe A virus (SCBGAV)	*Saccharum officinarum*	sof-miR396	vital ORF3 Region	Silenced the gene against pathogen	Ashraf et al. (2020)
Cowpea severe mosaic virus (CPSMV)	*Vigna unguiculata*	miR1533, miR1530, miR1528	Kat-p80, DEAD-Box, GST and SPB9	Expressed the target genes for providing resistance strategies	Martins et al. (2020)
Cucumber green mottle mosaic virus (CGMMV)	*N. benthamiana*	miR156, miR164, miR171	amiR1-CP, amiR4-MP and amiR6-Rep	Provided disease resistance to the host plant	Liang et al. (2019)
MYMIV	*Phaseolus vulgaris*	pvu-miR160j, miR482	Auxin response factor-16, 10 and NB-LRR	Targeted and regulated the viral gene for pathogenicity	Patwa et al. (2019)

(continued)

Table 3 (continued)

Pathogen	Plant species	miRNAs	Target gene	Effect	References
Soybean mosaic virus (SMV)	*Glycine max*	miR1507a, miR1507c, miR482a, miRNAs, miR168a, miR403a, miR162b, miR1515a	NBS-LRR, AGO1, AGO2, DCL1 and DCL2	Coded the disease resistance proteins and downregulated the genes for viral symptoms	Bao et al. (2018)
Mungbean yellow mosaic virus (MYMIV)	*V. mungo*	miR156/miR157	NB-LRR, NAC, MYB, Zinc finger, CCAAT-box TF	Lead to stress regulation of MYMIV	Kundu et al. (2017)
Grapevine leafroll-associated virus 3 (GLRaV-3)	*Vitis vinifera*	vvi-miR159c and vvi-miR167a	GSVIVT01018057001 sequence	Identification of target genes lead to better host–pathogen interactions	Bester et al. (2017)
Tobacco mosaic virus 43-A (TMV 43-A)	*N. benthamiana*	Nbe-MIR1a, Nbe-MIR1b	Leucine rich protein domain	Poly (A) mutant lead to host resistance against the gene	Guo et al. (2017)
Wheat dwarf virus (WDV)	*N. benthamiana*	pre-miR171	Conserved region	Provided highly efficient mutants for the WDV resistance	Kis et al. (2016)
Rice stripe virus (RSV), rice black-streaked dwarf virus (RBSDV)	*O. sativa*	pre-miR528	Middle segment, 30 end	Lead to degradation of viral genes	Sun et al. (2022)

(continued)

Table 3 (continued)

Pathogen	Plant species	miRNAs	Target gene	Effect	References
Cassava brown streak virus (CBSV), Ugandan cassava brown streak virus (UCBSV)	*N. benthamiana*	pre-miR159a	P1, P3, CI, Nib and CP	Resistance against the pathogenicity of the disease was observed	Wagaba et al. (2016)
Tomato spotted wilt virus (TSWV)	*N. benthamiana*	pre-miR159a	N, NSs	Symptoms against the virus were eliminated for better growth	Mitter et al. (2016)

Arabidopsis following *H. schachtii* infection (Kammerhofer et al. 2015). Recently, 60 miRNAs from 25 families were reported as potentially active in reducing soybean cyst nematode infection (Tian et al. 2017). The miRNAs involved in gall development induced by RKNs were studied in *Arabidopsis* roots at 3, 7 and 14 dpi (days post inoculation) and it was revealed that 62 miRNAs were differentially expressed (DE) in *M. javanica*-induced galls at 3 dpi and 24 miRNAs were differentially expressed in *M. incognita*-induced galls at 7 and/or 14 dpi (Cabrera et al. 2015). Only two DE miRNAs i.e., miR390 and miR319, with the same expression profiles, were found in all three stages of gall formation. Using RT-qPCR, 17 miRNAs were identified as DE in tomato galls at one or more of the five developmental phases studied (Kaur et al. 2017). Plants overexpressing a miR319-resistant TCP4 had fewer galls and higher levels of endogenous JA whereas, lines overexpressing Ath-MIR319 had the opposite effect. These findings implied that the miR319/TCP4 module was important in tomato galls because it modulated JA biosynthesis produced by RKN invasion (Zhao et al. 2015). The cleavage of TAS3 transcripts by miR390 in *Arabidopsis* creates secondary siRNAs (tasiRNAs), which promote post-transcriptional suppression of the auxin-responsive transcription factors ARF2, ARF3 and ARF4 (Marín et al. 2010). In *Arabidopsis* root galls, a role for the regulatory gene module formed of miR172 and the two transcription factors TOE1 (target of early activation tagged 1) and FT (flowering locus T) has been revealed during the production of gigantic cells (Díaz-Manzano et al. 2017). At 3 and 7 dpi, the expression of eight tomato miRNAs controlling defense-related proteins was particularly analyzed by qPCR, and an inverse association was identified between the expression of these miRNAs and their targets in response to cyst nematode infection (Święcicka et al. 2017). In *Arabidopsis*, miR396 was repressed at the start of syncytium formation and increased subsequently, whereas its target transcription factors, the growth-regulating factors (GRF) i.e., GRF1, GRF3 and GRF8, showed the reverse pattern (Hewezi et al. 2012). An inverse correlation of transcript levels was identified between miR858 and its target MYB83 at 7, 10, and 14 dpi, depicting the role of miR858/MYB83 module in *A. syncytia* produced by *H. schachtii* (Piya et al. 2017).

5 Conclusion

Throughout the world, crop yield is threatened by a plethora of phytopathogens. Recently, significant development has been made in elaborating miRNA cascades occurring in response to attacks by plant pathogens. In spite of the discovery of different components involved in the biosynthesis, functions, and degradation of miRNAs, much is still not known about the subcellular sites where these processes occur. Extensive research needs to be done to investigate miRNAs-based processes involved in plant-pathogen interactions in order to devise novel tactics for the management of diseases and improvement in crop productivity. Also, understanding the molecular mechanisms involved in the interactions taking place between plants and plant stressors, with specific emphasis on miRNA-mediated processes can have immense benefits in the field of agriculture.

References

Alarcón CR, Lee H, Goodarzi H, Halberg N, Tavazoie SF (2015) N 6-methyladenosine marks primary microRNAs for processing. Nature 519(7544):482–485

Al-Noshokaty TM, Elballal MS, Helal GK, Abulsoud AI, Elshaer SS, El-Husseiny AA, Doghish AS (2023) miRNAs driving diagnosis, prognosis and progression in Merkel cell carcinoma. Pathol Res Pract 154763

Ashraf F, Ashraf MA, Hu X, Zhang S (2020) A novel computational approach to the silencing of Sugarcane *Bacilliform Guadeloupe* A Virus determines potential host-derived miRNAs in sugarcane (*Saccharum officinarum* L.). Peer J 8:e8359

Aukerman MJ, Sakai H (2003) Regulation of flowering time and floral organ identity by a microRNA and its APETALA2-like target genes. Plant Cell 15(11):2730–2741

Auyeung VC, Ulitsky I, McGeary SE, Bartel DP (2013) Beyond secondary structure: primary-sequence determinants license pri-miRNA hairpins for processing. Cell 152(4):844–858

Bao D, Ganbaatar O, Cui X, Yu R, Bao W, Falk BW, Wuriyanghan H (2018) Down-regulation of genes coding for core RNAi components and disease resistance proteins via corresponding microRNAs might be correlated with successful Soybean mosaic virus infection in soybean. Mol Plant Pathol 19(4):948–960

Begum Y (2022) Regulatory role of microRNAs (miRNAs) in the recent development of abiotic stress tolerance of plants. Gene 821:146283

Bester R, Burger JT, Maree HJ (2017) Differential expression of miRNAs and associated gene targets in grapevine leafroll-associated virus 3-infected plants. Arch Virol 162:987–996

Bielewicz D, Dolata J, Bajczyk M, Szewc L, Gulanicz T, Bhat SS, Szweykowska-Kulinska Z (2023) Hyponastic leaves 1 interacts with RNA Pol II to ensure proper transcription of microRNA genes. Plant Cell Physiol 64(6):571–582

Blok VC, Jones JT, Phillips MS, Trudgill DL (2008) Parasitism genes and host range disparities in biotrophic nematodes: the conundrum of polyphagy versus specialisation. BioEssays 30:249–259

Bofill-De Ros X, Hong Z, Birkenfeld B, Alamo-Ortiz S, Yang A, Dai L, Gu S (2022) Flexible pri-miRNA structures enable tunable production of 5'isomiRs. RNA Biol 19(1):279–289

Bojórquez-Orozco AM, Arce-Leal ÁP, Montes RAC, Santos-Cervantes ME, Cruz-Mendívil A, Méndez-Lozano J, Castillo AG, Rodrigues-Negrete EA, Leyva-López NE (2023) Differential expression of miRNAs involved in response to *Candidatus Liberibacter asiaticus* infection in mexican lime at early and late stages of Huanglongbing disease. Plants 12(5):1039

Borges F, Martienssen RA (2015) The expanding world of small RNAs in plants. Nat Rev Mol Cell Biol 16:727–741

Cabrera J, Barcala M, García A et al (2015) Differentially expressed small RNAs in Arabidopsis galls formed by *Meloidogyne javanica*: a functional role for miR390 and its TAS3-derived tasiRNAs. New Phytol 209:1625–1640

Cai Q, Qiao L, Wang M, He B, Lin FM, Palmquist J, Huang SD, Jin H (2018) Plants send small RNAs in extracellular vesicles to fungal pathogen to silence virulence genes. Science 60(6393):1126–1129

Canto-Pastor A, Santos BA, Valli AA, Summers W, Schornack S, Baulcombe DC (2019) Enhanced resistance to bacterial and oomycete pathogens by short tandem target mimic RNAs in tomato. Proc Natl Acad Sci 116(7):2755–2760

Caruana JC, Dhar N, Raina R (2020) Overexpression of Arabidopsis microRNA167 induces salicylic acid-dependent defense against *Pseudomonas syringae* through the regulation of its targets ARF6 and ARF8. Plant Direct 4(9):e00270

Davoodi Mastakani F, Pagheh G, Rashidi Monfared S, Shams-Bakhsh M (2018) Identification and expression analysis of a microRNA cluster derived from pre-ribosomal RNA in *Papaver somniferum* L. and *Papaver bracteatum* L. PLoS One 13(8):e0199673

Dhar N, Caruana J, Erdem I, Raina R (2020) An Arabidopsis disease related nonspecific lipid transfer protein 1 is required for resistance against various phytopathogens and tolerance to salt stress. Gene 753:144802

Díaz-Manzano FE, Cabrera J, Ripoll JJ et al (2017) A role for the gene regulatory module *microRNA172/target of early activation tagged 1/flowering locus T* (*miRNA172/TOE1/FT*) in the feeding sites induced by *Meloidogyne javanica* in *Arabidopsis thaliana*. New Phytol 217:813–827

Djami-Tchatchou AT, Sanan-Mishra N, Ntushelo K, Dubery IA (2017) Functional roles of microRNAs in agronomically important plants—potential as targets for crop improvement and protection. Front Plant Sci 8:378

Eamens AL, Smith NA, CurtinSJ WMB, Waterhouse PM (2009) The *Arabidopsis thaliana* double-stranded RNA binding protein DRB1 directs guide strand selection from microRNA duplexes. RNA 15(12):2219–2235

Eamens AL, Wook Kim K, Waterhouse PM (2012) DRB2, DRB3 and DRB5 function in a non-canonical microRNA pathway in *Arabidopsis thaliana*. Plant Signal Behav 7(10):1224–1229

Franco-Zorrilla JM, Valli A, Todesco M, Mateos I, Puga MI, Rubio-Somoza I, Paz-Ares J (2007) Target mimicry provides a new mechanism for regulation of microRNA activity. Nat Genet 39(8):1033–1037

Fu F, Girma G, Mengiste T (2020) Global mRNA and microRNA expression dynamics in response to anthracnose infection in sorghum. BMC Genom 21:1–16

Ghorbani A, Rostami M, Izadpanah K (2023) Gene network modeling and pathway analysis of maize transcriptomes in response to Maize Iranian mosaic virus. Genomics 115(3):110618

Goodwin S, McPherson JD, McCombie WR (2016) Coming of age: ten years of nextgeneration sequencing technologies. Nat Rev Genet 17:333–351

Guo S, Hsueh YC, Tucker-Kellogg G, Wong SM (2017) Differential expression of novel microRNAs in response to the infection of a TMV mutant with an internal poly (A) tract in *N. benthamiana*. Virus Res 239:143–171

Gupta A, Patil M, Qamar A, Senthil-Kumar M (2020) Ath-miR164c influences plant responses to the combined stress of drought and bacterial infection by regulating proline metabolism. Environ Exp Bot 172:103998

Hajieghrari B, Niazi A (2023) Phylogenetic and evolutionary analysis of plant small RNA 2′-O-methyltransferase (HEN1) protein family. J Mol Evol 1–17

Harris JF, Micheva-Viteva S, Li N, Hong-Geller E (2013) Small RNA-mediated regulation of host–pathogen interactions. Virulence 4(8):785–795

Hewezi T, Maier T, Nettleton D, Baum TJ (2012) The Arabidopsis MicroRNA396-GRF1/GRF3 regulatory module acts as a developmental regulator in the reprogramming of root cells during cyst nematode infection. Plant Physiol 159:321–335

Hoang NT, Tóth K, Stacey G (2020) The role of microRNAs in the legume–*Rhizobium* nitrogen-fixing symbiosis. J Exp Bot 71(5):1668–1680

Hu G, Hao M, Wang L, Liu J, Zhang Z, Tang Y, Peng Q, Yang Z, Wu J (2020) The cotton miR477-CBP60A module participates in plant defense against *Verticillium dahlia*. Mol Plant Microbe Interact 33(4):624–636

Hu G, Wang B, Jia P, Wu P, Lu C, Xu Y, Shi L, Zhang F, Zhong N, Chen A, Wu J (2023b) The cotton miR530-SAP6 module activated by systemic acquired resistance mediates plant defense against *Verticillium dahliae*. Plant Sci 330:111647

Hu G, Ge X, Wang P, Chen A, Li F, Wu J (2023a) The cotton miR171a-SCL6 module mediates plant resistance through regulating GhPR1 expression. Plant Physiol Biochem 107995

Ibrahim HMM, Alkharouf NW, Meyer SD et al (2011) Post-transcriptional gene silencing of root-knot nematode in transformed soybean roots. Exp Parasitol 127:90–99
Islam W, Islam SU, Qasim M, Wang L (2017) Host-Pathogen interactions modulated by small RNAs. RNA Biol 4(7):891–904
Islam W, Noman A, Qasim M, Wang L (2018a) Plant responses to pathogen attack: small RNAs in focus. Int J Mol Sci 19(2):515
Islam W, Qasim M, Noman A, Adnan M, Tayyab M, Farooq TH, Wei H, Wang L (2018b) Plant microRNAs: front line players against invading pathogens. Microb Pathog 118:9–17
Iwakawa HO, Tomari Y (2022) Life of RISC: Formation, action, and degradation of RNA-induced silencing complex. Mol Cell 82(1):30–43
Jarošová J, Singh K, Chrpová J, Kundu JK (2020) Analysis of small RNAs of barley genotypes associated with resistance to Barley Yellow Dwarf Virus. Plants 9(1):60
Javed M, Reddy B, Sheoran N, Ganesan P, Kumar A (2023) Unraveling the transcriptional network regulated by miRNAs in blast-resistant and blast-susceptible rice genotypes during *Magnaporthe oryzae* interaction. Gene 886:147718
Jeena GS, Singh N, Shikha Shukla RK (2022) An insight into microRNA biogenesis and its regulatory role in plant secondary metabolism. Plant Cell Rep 41(8):1651–1671
Jia Y, Li C, Li Q, Liu P, Liu D, Liu Z, Wang Y, Jiang G, Zhai W (2020) Characteristic dissection of *Xanthomonas oryzae* pv. *oryzae* responsive microRNAs in rice. Int J Mol Sci 21(3):785
Jiang P, Lian B, Liu C, Fu Z, Shen Y, Cheng Z, Qi Y (2020) 21-nt phasiRNAs direct target mRNA cleavage in rice male germ cells. Nat Commun 11(1):5191
Jiang L, Mu R, Wang Z, Liu S, Lu D (2023) Silencing P25, HC-Pro and Brp1 of potato virus (viroid) using artificial microRNA confers resistance to PVX, PVY and PSTVd in transgenic potato. Potato Res 66(1):231–244
Jiao J, Peng D (2018) Wheat microRNA1023 suppresses invasion of *Fusarium graminearum* via targeting and silencing FGSG_03101. J Plant Interact 13(1):514–521
Jones JT, Haegeman A, Danchin EGJ et al (2013) Top 10 plant-parasitic nematodes in molecular plant pathology. Mol Plant Pathol 14:946–961
Jozwiak M, Bielewicz D, Szweykowska-Kulinska Z, Jarmolowski A, Bajczyk M (2023) SERRATE: a key factor in coordinated RNA processing in plants. Trends Plant Sci 28(7):P841-853
Kamitani M, Nagano AJ, Honjo MN, Kudoh H (2016) RNA-Seq reveals virus–virus and virus–plant interactions in nature. FEMS Microbiol Ecol 92(11):176
Kammerhofer N, Radakovic ZS, Regis JMA et al (2015) Role of stress-related hormones in plant defence during early infection of the cyst nematode *Heterodera schachtii* in Arabidopsis. New Phytol 207:778–789
Kapadia C, Datta R, Mahammad SM, Tomar RS, Kheni J K and Ercisli S (2023) Genome-wide identification, quantification, and validation of differentially expressed miRNAs in eggplant (*Solanum melongena* L.) based on their response to *Ralstonia solanacearum*infection. ACS omega 8(2):2648–2657
Kar MM, Raichaudhuri A (2021) Role of microRNAs in mediating biotic and abiotic stress in plants. Plant Gene 26:100277
Kasprzyk ME, Kazimierska M, Sura W, Dzikiewicz-Krawczyk A, Podralska M (2023) Non-coding RNAs: mechanisms of action. In: Sztuba-Solinska J (ed) Navigating non-coding RNA. Academic Press, pp 89–138
Kaur P, Shukla N, Joshi G et al (2017) Genome-wide identification and characterization of miRNAome from tomato (*Solanum lycopersicum*) roots and root-knot nematode (*Meloidogyne incognita*) during susceptible interaction. PLoS ONE 12:e0175178
Kis A, Tholt G, Ivanics M, Várallyay É, Jenes B, Havelda Z (2016) Polycistronic artificial miRNA-mediated resistance to W heat dwarf virus in barley is highly efficient at low temperature. Mol Plant Pathol 17(3):427–437
Komatsu S, Kitai H, Suzuki HI (2023) Network regulation of microRNA biogenesis and target interaction. Cells 12(2):306

Kulshrestha C, Pathak H, Kumar D, Dave S, Sudan J (2020) Elucidating micro RNAs role in different plant–pathogen interactions. Mol Biol Rep 47:8219–8227

Kundu A, Paul S, Dey A, Pal A (2017) High throughput sequencing reveals modulation of microRNAs in *Vigna mungo* upon Mungbean Yellow Mosaic India Virus inoculation highlighting stress regulation. Plant Sci 257:96–105

Ledda B, Ottaggio L, Izzotti A, Sukkar SG, Miele M (2020) Small RNAs in eucaryotes: new clues for amplifying microRNA benefits. Cell Biosci 10(1):1–13

Lee RC, Feinbaum RL, Ambros V (1993) The *C. elegans* heterochronic gene lin-4 encodes small RNAs with antisense complementarity to lin-14. Cell 75:843–854

Li M, Yu B (2021) Recent advances in the regulation of plant miRNA biogenesis. RNA Biol 18(12):2087–2096

Li M, Li W, Zhao M, Li Z, Wang GL, Liu W, Liang C (2023) Transcriptome analysis reveals a lncRNA-miRNA-mRNA regulatory network in OsRpp30-mediated disease resistance in rice. BMC Genom 24(1):643

Liang C, Hao J, Li J, Baker B, Luo L (2019) Artificial microRNA-mediated resistance to cucumber green mottle mosaic virus in *Nicotiana benthamiana*. Planta 250:1591–1601

Liang C, Cai Q, Wang F, Li S, You C, Xu C, Chen X (2022) *Arabidopsis* RBV is a conserved WD40 repeat protein that promotes microRNA biogenesis and ARGONAUTE1 loading. Nat Commun 13(1):1217

Lin JX, Ali A, Chu N, Fu HY, Huang MT, Mbuya SN, Gao SJ, Zhang HL (2023) Identification of ARF transcription factor gene family and its defense responses to bacterial infection and salicylic acid treatment in sugarcane. Front Microbiol 14

Liu Q, Wang F, Axtell MJ (2014) Analysis of complementarity requirements for plant microRNA targeting using a *Nicotiana benthamiana* quantitative transient assay. Plant Cell 26(2):741–753

Liu Z, Shi L, Yang S, Lin Y, Weng Y, Li X, Hussain A, Noman A, He S (2017) Functional and promoter analysis of ChiIV3, a Chitinase of pepper plant, in response to *Phytophthora capsici* infection. Int J Mol Sci 18(8):1661

Lu J, Wang C, Zhang F, Zeng D, Zhou Y (2021) Comparative microRNA profiling reveals microRNAs involved in rice resistant response to bacterial blight. Crop J 9(4):834–842

Luo Y, Qin C, Qiu H, Zhang X, Tang X, Luo X, Chen X (2023) Novel microRNAs associated with the immune response to cucumber mosaic virus in hot pepper (*Capsicum annuum* L.). Physiol Mol Plant Pathol 124:101963

Lv Y, Zhong Y, Jiang B, Yan H, Ren S, Cheng C (2023) MicroRNA miR171b positively regulates resistance to Huanglongbing of citrus. Int J Mol Sci 24(6):5737

Ma Z, Hu L (2023) MicroRNA: a dynamic player from signalling to abiotic tolerance in plants. Int J Mol Sci 24(14):11364

Mangrauthia SK, Sailaja B, Pusuluri M, Jena B, Prasanth VV, Agarwal S, Voleti SR (2018) Deep sequencing of small RNAs reveals ribosomal origin of microRNAs in *Oryza sativa* and their regulatory role in high temperature. Gene Rep 11:270–278

Marín E, Jouannet V, Herz A et al (2010) MiR390, *Arabidopsis TAS3*tasiRNAs, and their *AUXIN RESPONSE FACTOR* targets define an autoregulatory network quantitatively regulating lateral root growth. Plant Cell 22:1104–1117

Marmisolle FE, Arizmendi A, Ribone A, Rivarola M, García ML, Reyes CA (2020) Up-regulation of microRNA targets correlates with symptom severity in *Citrus sinensis* plants infected with two different isolates of citrus psorosis virus. Planta 251:1–1

Martins TF, Souza PF, Alves MS, Silva FD, Arantes MR, Vasconcelos IM, Oliveira JT (2020) Identification, characterization, and expression analysis of cowpea (*Vigna unguiculata* [L.] Walp.) miRNAs in response to cowpea severe mosaic virus (CPSMV) challenge. Plant Cell Rep 39:1061–1078

Meng X, Jin W, Wu F (2020) Novel tomato miRNA miR1001 initiates cross-species regulation to suppress the conidiospore germination and infection virulence of *Botrytis cinerea* in vitro. Gene 759:145002

Miao S, Liang C, Li J, Baker B, Luo L (2021) Polycistronic artificial microRNA-mediated resistance to cucumber green mottle mosaic virus in cucumber. Int J Mol Sci 22(22):12237

Miao Y, Chen K, Deng J, Zhang L, Wang W, Kong J, Klosterman SJ, Zhang X, Aierxi A, Zhu L (2022) MiR398b negatively regulates cotton immune responses to *Verticillium dahliae* via multiple targets. Crop J 10(4):1026–1036

Mitter N, Zhai Y, Bai AX, Chua K, Eid S, Constantin M, Mitchell R, Pappu HR (2016) Evaluation and identification of candidate genes for artificial microRNA-mediated resistance to tomato spotted wilt virus. Virus Res 211:151–158

Noman A, Aqeel M (2017) MiRNA-based heavy metal homeostasis and plant growth. Environ Sci Pollut Res 24:10068–10082

Noman A, Fahad S, Aqeel M, Ali U, Amanullah AS, Baloch SK, Zainab M (2017) MiRNAs: major modulators for crop growth and development under abiotic stresses. Biotechnol Lett 39:685–700

Ó'Maoiléidigh DS, van Driel AD, Singh A, Sang Q, Le Bec N, Vincent C, Coupland G (2021) Systematic analyses of the MIR172 family members of *Arabidopsis* define their distinct roles in regulation of APETALA2 during floral transition. PLoS Biol 19(2):e3001043

Pandita D (2022) Role of miRNA technology and miRNAs in abiotic and biotic stress resilience. In: Aftab T, Roychoudhury A (eds) Plant perspectives to global climate changes. Academic Press, pp 303–330

Patwa N, Nithin C, Bahadur RP, Basak J (2019) Identification and characterization of differentially expressed *Phaseolus vulgaris* miRNAs and their targets during mungbean yellow mosaic India virus infection reveals new insight into Phaseolus-MYMIV interaction. Genomics 111(6):1333–1342

Piya S, Kihm C, Rice JH et al (2017) Cooperative regulatory functions of miR858 and MYB83 during cyst nematode parasitism. Plant Physiol 174:1897–1912

Pompili V, Piazza S, Li M, Varotto C, Malnoy M (2020). Transcriptional regulation of MdmiR285N microRNA in apple (*Malus x domestica*) and the heterologous plant system *Arabidopsis thaliana*. Hortic res 7

Qasim M, Husain D, Islam SU, Ali H, Islam W, Hussain M, Wang F, Wang L (2018) Effectiveness of *Trichogramma chilonis* Ishii against spiny bollworm in Okra and susceptibility to insecticides. J Entomol Zool Stud 6(1):1576–1581

Qin L, Zhao L, Wu C, Qu S, Wang S (2021) Identification of microRNA transcriptome in apple response to *Alternaria alternata* infection and evidence that miR390 is negative regulator of defense response. Sci Hortic 289:110435

Qiu D, Pan X, Wilson IW, Li F, Liu M, Teng W, Zhang B (2009) High throughput sequencing technology reveals that the taxoid elicitor methyl jasmonate regulates microRNA expression in Chinese yew (*Taxus chinensis*). Gene 436(1–2):37–44

Qu Q, Liu N, Su Q, Liu X, Jia H, Liu Y, Sun M, Cao Z, Dong J (2023) MicroRNAs involved in the trans-kingdom gene regulation in the interaction of maize kernels and *Fusarium verticillioides*. Int J Biol Macromol 242:125046

Ramesh SV, Chouhan BS, Kumar G, Praveen S, Chand S (2017) Expression dynamics of *Glycine max* (L.) Merrill microRNAs (miRNAs) and their targets during Mungbean yellow mosaic India virus (MYMIV) infection. Physiol Mol Plant Pathol 100:13–22

Ramesh SV, Yogindran S, Gnanasekaran P, Chakraborty S, Winter S, Pappu HR (2021) Virus and viroid-derived small RNAs as modulators of host gene expression: molecular insights into pathogenesis. Front Microbiol 11:614231

Rawal HC, Ali S, Mondal TK (2023) Role of non-coding RNAs against salinity stress in Oryza species: strategies and challenges in analyzing miRNAs, tRFs and circRNAs. Int J Biol Macromol 125172

Raza A, Charagh S, Karikari B, Sharif R, Yadav V, Mubarik MS, Habib M, Zhuang Y, Zhang C, Chen H, Varshney RK (2023) MiRNAs for crop improvement. Plant Physiol Biochem 24:107857

Rego EC, Pinheiro TD, Fonseca FC, Gomes TG, Costa ED, Bastos LS, Alves GS, Cotta MG, Amorim EP, Ferreira CF, Togawa RC (2023) Characterization of microRNAs and target genes in *Musa acuminata* subsp. burmannicoides, var. Calcutta 4 during interaction with *Pseudocercospora musae*. Plants 12(7):1473

Samad AFA, Sajad M, Nazaruddin N, Fauzi IA, Murad AMA, Zainal Z, Ismail I (2017) MicroRNA and transcription factor: key players in plant regulatory network. Front Plant Sci 8:565. https://doi.org/10.3389/fpls.2017.00565

Santos LS, Maximiano MR, Megias E, Pappas M, Ribeiro SG and Mehta A (2019) Quantitative expression of microRNAs in *Brassica oleracea* infected with *Xanthomonas campestris* pv. *campestris*. Mol BiolRep 46(3):3523–3529

Schultz J, Wolf M (2009) ITS2 sequence–structure analysis in phylogenetics: a how-to manual for molecular systematics. Mol Phylogenet Evol 52(2):520–523

Šečić E, Kogel KH, Ladera-Carmona MJ (2021) Biotic stress-associated microRNA families in plants. J Plant Physiol 263:153451

Shang R, Lee S, Senavirathne G, Lai EC (2023) microRNAs in action: biogenesis, function and regulation. Nat Rev Genet 1–18

Shi G, Wang S, Wang P, Zhan J, Tang Y, Zhao G, Li F, Ge X, Wu J (2022) Cotton miR393-TIR1 module regulates plant defense against *Verticillium dahliae* via auxin perception and signaling. Front Plant Sci 13:888703

Si X, Liu H, Cheng X, Xu C, Han Z, Dai Z, Wang R, Pan C, Lu G (2023) Integrative transcriptomic analysis unveils lncRNA-miRNA-mRNA interplay in tomato plants responding to *Ralstonia solanacearum*. Int J Biol Macromol 253:126891

Sindhu A, Maier T, Mitchum MG et al (2008) Effective and specific in planta RNAi in cyst nematodes: expression interference of four parasitism genes reduces parasitic success. J Exp Bot 60:315–324

Snigdha M, Prasath D (2020) Unravelling the differential expression of potential microRNAs in bacterial wilt-resistant and susceptible ginger species. PREPRINT (Version 1) available at Research Square. https://doi.org/10.21203/rs.3.rs-115945/v1

Son DJ, Kumar S, Takabe W, Woo Kim C, Ni CW, Alberts-Grill N, Jo H (2013) The atypical mechanosensitive microRNA-712 derived from pre-ribosomal RNA induces endothelial inflammation and atherosclerosis. Nat Commun 4(1):3000

Soto-Suárez M, Baldrich P, Weigel D, Rubio-Somoza I, San Segundo B (2017) The Arabidopsis miR396 mediates pathogen-associated molecular pattern-triggered immune responses against fungal pathogens. Sci Rep 7(1):44898

Spychała J, Tomkowiak A, Noweiska A, Bobrowska R, Bocianowski J, Książkiewicz M, Sobiecg A and Kwiatek MT (2023) Expression profiling of the slow rusting resistance genes Lr34/Yr18 and Lr67/Yr46 in common wheat (*Triticum aestivum* L.) and associated miRNAs patterns. Genes 14(7):1376

Stepien A, Knop K, Dolata J, Taube M, Bajczyk M, Barciszewska-Pacak M, Pacak A, Jarmolowski A, Szweykowska-Kulinska Z (2017) Posttranscriptional coordination of splicing and miRNA biogenesis in plants. Wiley Interdiscip Rev RNA 8(3):e1403

Sun M, Xu S, Mei Y, Li J, Gu Y, Zhang W, Wang J (2022) MicroRNAs in medicinal plants. Int J Mol Sci 23(18):10477

Sunkar R, Li YF, Jagadeeswaran G (2012) Functions of microRNAs in plant stress responses. Trends Plant Sci 17(4):196–203

Svoboda P (2020) Key mechanistic principles and considerations concerning RNA interference. Front Plant Sci 11:1237

Swięcicka M, Skowron W, Cieszyński P et al (2017) The suppression of tomato defence response genes upon potato cyst nematode infection indicates a key regulatory role of miRNAs. Plant Physiol Biochem 113:51–55

Tian B, Wang S, Todd TC et al (2017) Genome-wide identification of soybean microRNA responsive to soybean cyst nematodes infection by deep sequencing. BMC Genom 18:527

Tomkowiak A, Jędrzejewski T, Spychała J, Kuczyński J, Kwiatek MT, Tyczewska A, Skoweonska R and Twardowski T (2020). Analysis of miRNA expression associated with the Lr46 gene responsible for APR resistance in wheat (*Triticum aestivum* L.). J Appl Genet 61(4):503–511

Tousi N, Eini O, Ahmadvand R, Carra A, Miozzi L, Noris E, Accotto GP (2017) In silico prediction of miRNAs targeting ToLCV and their regulation in susceptible and resistant tomato plants. Australas Plant Pathol 46:379–386

Varallyay E, Oláh E, Havelda Z (2014) Independent parallel functions of p19 plant viral suppressor of RNA silencing required for effective suppressor activity. Nucl Acids Res Spec Publ 42(1):599–608

Vos C, Schouteden N, Van Tuinen D et al (2013) Mycorrhiza-induced resistance against the root-knot nematode *Meloidogyne incognita* involves priming of defense gene responses in tomato. Soil Biol Biochem 60:45–54

Wagaba H, Patil BL, Mukasa S, Alicai T, Fauquet CM, Taylor NJ (2016) Artificial microRNA-derived resistance to Cassava brown streak disease. J Virol Methods 231:38–43

Waititu JK, Zhang C, Liu J, Wang H (2020) Plant non-coding RNAs: origin, biogenesis, mode of action and their roles in abiotic stress. Int J Mol Sci 21(21):8401

Wang L, Zhang W, Shen W, Li M, Fu Y, Li Z, Li J, Liu H, Su X, Zhang B, Zhao J (2023) Integrated transcriptome and microRNA sequencing analyses reveal gene responses in poplar leaves infected by the novel pathogen bean common mosaic virus (BCMV). Front Plant Sci 14:1163232

Wang X, Zhan S, Li M, Chen A, Lu Y, Wang P, Zhang Z and Gu L (2022) Identification and functional analysis of miRNAs in *N. benthamiana* in response to *A. longipes* infection. Physiol Mol Plant Pathol 117:101762

Wani SH, Kumar V, Khare T, Tripathi P, Shah T, Ramakrishna C, Mangrauthia SK (2020) MiRNA applications for engineering abiotic stress tolerance in plants. Biol 75:1063–1081

Wani SH, Anand S, Singh B, Bohra A, Joshi R (2021) WRKY transcription factors and plant defense responses: latest discoveries and future prospects. Plant Cell Rep 40:1071–1085

Wenlei C, Xinxin C, Jianhua Z, Zhaoyang Z, Zhiming F, Shouqiang O, Shimin Z (2020) Comprehensive characteristics of microRNA expression profile conferring to *Rhizoctonia solani* in Rice. Rice Sci 27(2):101–112

Wu L, Zhou H, Zhang Q, Zhang J, Ni F, Liu C, Qi Y (2010) DNA methylation mediated by a microRNA pathway. Mol Cell 38(3):465–475

Xiao Y, MacRae IJ (2022) The molecular mechanism of microRNA duplex selectivity of Arabidopsis ARGONAUTE10. Nucleic Acids Res 50(17):10041–10052

Xu J, Xian Q, Zhang N, Wang K, Zhou X, Li Y, Dong J, Chen X (2021) Identification of miRNA-target gene pairs responsive to fusarium wilt of cucumber via an integrated analysis of miRNA and transcriptome profiles. Biomolecules 11(11):1620

Xu Y, Wang R, Ma P, Cao J, Cao Y, Zhou Z, Li T, Wu J, Zhang H (2022) A novel maize microRNA negatively regulates resistance to *Fusarium verticillioides*. Mol Plant Pathol 23(10):1446–1460

Xu Y, Zhang Y, Li Z, Soloria AK, Potter S, Chen X (2023) The N-terminal extension of Arabidopsis ARGONAUTE 1 is essential for microRNA activities. PLoS Genet 19(3):e1010450

Yadav R, Ramakrishna W (2023) MicroRNAs involved in nutritional regulation during plant–microbe symbiotic and pathogenic interactions with rice as a model. Mol Biotechnol 1–18

Yan J, Gu Y, Jia X, Kang W, Pan S, Tang X, Tang G (2012) Effective small RNA destruction by the expression of a short tandem target mimic in *Arabidopsis*. Plant Cell 24(2)L:415–427

Yang X, Zhang L, Yang Y, Schmid M, Wang Y (2021) MiRNA mediated regulation and interaction between plants and pathogens. Int J Mol Sci 22(6):2913

Yang X, Fu T, Yu R, Zhang L, Yang Y, Xiao D, Wang YY, Wang Y, Wang Y (2023) MiR159a modulates poplar resistance against different fungi and bacteria. Plant Physiol Biochem 201:107899

Yang YX, Ahammed GJ, Wu C, Fan SY, Zhou YH (2015) Crosstalk among jasmonate, salicylate and ethylene signaling pathways in plant disease and immune responses. Curr Protein Pept Sci 16(5):450–461

Yin C, Xie L, Wu Y, Qu H, Yang B, Gong L, Jiang Y, Li T (2023) Involvement of miRNAs-mediated senescence and salicylic acid defense in postharvest litchi downy blight. Food Chem 404:134662

Yu Y, Jia T, Chen X (2017) The 'how' and 'where' of plant microRNAs. New Phytol 216(4):1002–1017

Yuan J, Wang X, Qu S, Shen T, Li M, Zhu L (2023) The roles of miR156 in abiotic and biotic stresses in plants. Plant Physiol Biochem 204:108150
Zehra A, Raytekar NA, Meena M, Swapnil P (2021) Efficiency of microbial bio-agents as elicitors in plant defense mechanism under biotic stress: a review. Curr Res Microbial Sci 2:100054
Zhang S, Liu Y, Yu B (2015a) New insights into pri-miRNA processing and accumulation in plants. Wiley Interdiscip REv: RNA 6(5):533–545
Zhang K, Shi X, Zhao X, Ding D, Tang J, Niu J (2015b) Investigation of miR396 and growth-regulating factor regulatory network in maize grain filling. Acta Physiol Plant 37:28. https://doi.org/10.1007/s11738-014-1767-6
Zhang T, Chang H, Zhang B, Liu S, Zhao T, Zhao E, Zhao H, Zhang H (2020) Transboundary pathogenic microRNA analysis framework for crop fungi driven by biological big data and artificial intelligence model. Comput Biol Chem 89:107401
Zhang Y, Waseem M, Zeng Z, Xu J, Chen C, Liu Y, Zhai J, Xia R (2021) MicroRNA482/2118, a miRNA superfamily essential for both disease resistance and plant development. New Phytol 233(5):2047–2057
Zhao W, Li Z, Fan J et al (2015) Identification of jasmonic acid-associated microRNAs and characterization of the regulatory roles of the miR319/TCP4 module under root-knot nematode stress in tomato. J Exp Bot 66:4653–4667
Zhou L, Yuan Q, Ai X, Chen J, Lu Y, Yan F (2022) Transgenic rice plants expressing artificial miRNA targeting the rice stripe virus MP gene are highly resistant to the virus. Biology 11(2):332
Zhu QH, Spriggs A, Matthew L, Fan L, Kennedy G, Gubler F, Helliwell C (2008) A diverse set of microRNAs and microRNA-like small RNAs in developing rice grains. Genome Res 18(9):1456–1465

RNAi as a Potential Tool for Control and Management of Plant Disease: An Updated Overview

Sheuli Kangsa Banik, Rajat Nath, Sharmistha Sharma, Sibashish Kirtaniya, Deepa Nath, Jayanta Kumar Patra, Manabendra Dutta Choudhury, and Anupam Das Talukdar

Abstract A diverse spectrum of pathogens and pests has been identified to infest various plant species and cause havoc on horticultural, agricultural, and forestry crops. Several improved crop varieties that are resistant to a variety of plant diseases brought on by biotic (pests, pathogens) and/or abiotic (climate change) factors are the main focus of plant biotechnologists across the world. Although numerous conventional methods might potentially minimize agricultural production loss; improved, cost-effective, ecologically sound, and efficacious solutions are required to fulfill the needs of a rising population. Thus, RNA interference (RNAi) finds paramount importance in resolving many hazardous plant diseases. RNAi is a method of gene silencing that uses double-stranded RNA of both plants and animals as triggers and targets homologous RNA to degrade or inhibit gene expression at the transcriptional, translational level in a sequence-specific manner or by chromatin remodeling. siRNA and miRNA are the two types of small single-stranded RNAs that play a significant role in the RNAi pathways. Recently several transgenic and non-transgenic plant-based RNAi techniques have been proven to have great potential in treating plant diseases when no other alternatives were available. Therefore RNAi is considered to serve as a novel gene therapy technique for plant disease control and management. In this chapter, an attempt has been made to elucidate the development of RNA interference, the molecular mechanism underlying this technique, and various strategies and applications to effectively control plant diseases through a pest management approach.

S. K. Banik · R. Nath · S. Sharma · S. Kirtaniya · M. D. Choudhury · A. Das Talukdar (✉)
Department of Life Science and Bioinformatics, Assam University, Silchar, Cachar, Assam 788011, India
e-mail: anupam@bioinfoaus.ac.in

D. Nath
Department of Botany, Guru Charan College, Silchar, Cachar, Assam 788004, India

J. K. Patra
Research Institute Integrative Life Sciences, Dongguk University-Seoul, Goyangsi 10326, Republic of Korea

J. Chen et al. (eds.), *Molecular and Biotechnological Tools for Plant Disease Management*, Interdisciplinary Biotechnological Advances,
https://doi.org/10.1007/978-981-97-7510-1_12

Keywords RNAi · Plant disease · Biotic stress · Abiotic stress

1 Introduction

Globally, biotic (pests, pathogens) and abiotic (climate change) stress are persistently threatening horticultural, agricultural, and forestry crops. Almost every plant grown by humans is subjected to being infested by plant pests and pathogens (Bocos-Asenjo et al. 2022; Kuo and Falk 2020). These plant pests and pathogens have reduced worldwide food crop yields by up to 40% annually which poses a direct impact on food scarcity, the economy, and the agro-ecosystem (He et al. 2016; Kuo and Falk 2020). To prevent such crop losses various interventions based on plant species, the pathogen, the environment, and geographic location have been employed since the onset of the green revolution (He et al. 2016). Conventional, traditional approaches that are in use to manage and control plant diseases include five principle strategies: (a) pathogen **avoidance or evasion** which is the growing of plants where the occurrence of the pathogen is null, (b) **exclusion** that prevents the introduction of pathogens in a new area, (c) **eradication** that is the elimination of pathogens from cropland, (d) **direct protection**, it includes pathogen-targeting chemicals to hinder the pathogens' capacity to infect plants and cause illness, and (e) **resistance** development in host plant against the pathogen (Fig. 1) (Ul Haq and Ijaz 2020).

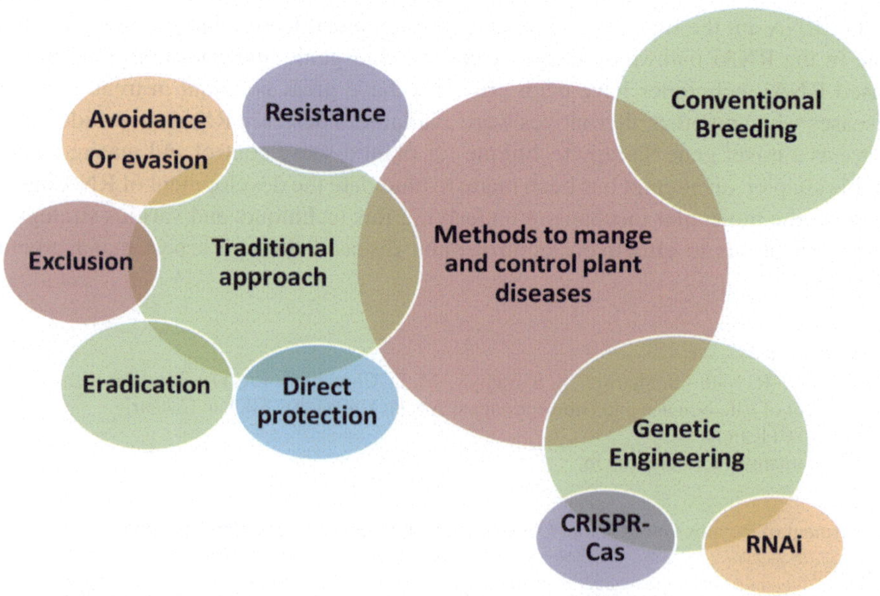

Fig. 1 Strategies of plant disease control and management

Generally plant breeding technique is used extensively to produce host plant resistance; however certain drawbacks have limited the scope of plant disease control and management by this technique. This could be attributed to a variety of factors: such as the appearance of novel pathogens for which resistance is unknown or a lack of sufficient genetic resources of resistance for use in plant breeding programs (Wani et al. 2010). Also, plants that are resistant to recognized pathogens may not be genetically suitable with plants that require protection, it can also be anticipated that evolution may allow pathogens that cause diseases to be selected and take benefit of new niches and circumvent the resistance, and if virulent pathogens enter huge anthropogenic homozygous cropland where it finds favorable environmental conditions to flourish the pathogen can spread quickly and can cause severe diseases (Kuo and Falk 2020). Therefore to minimize agricultural production loss; cost-effective, ecologically sound, and efficacious solutions are required to fulfill the needs of a rising population.

In recent times newer approaches in plant breeding techniques like genetic engineering and gene transfer technologies have aided in plant disease control and management strategies. Genetic engineering involves altering the genetic material of the organism to develop resistance against diseases. Recently, CRISPR-Cas and RNA interference techniques have gained rapid popularity (Fig. 1) (Rajput et al. 2021; Wani et al. 2010). Both the CRISPR-Cas and RNAi approaches are equally potent, however, due to the lack of on-target efficiency of CRISPR-Cas, the RNAi approach has gained surplus benefits in mitigating some of the hazardous plant diseases (Halder et al. 2022). In eukaryotes, RNAi is a natural gene control mechanism. The process initiates by recognizing target mRNA sequences and thereby silencing the specific gene expressions at transcriptional and translational levels or complete degradation of the target sequences (Kuo and Falk 2020). When a pathogenic DNA or RNA enters a host system, RNAi instantly triggers an immune response against it (Halder et al. 2022). The RNAi approach is extensively used in protecting plants from severe diseases, also attempts have been made to exploit the RNAi technique to battle insects, fungi, and nematodes that attack the host plants (Matranga and Zamore 2007).

2 Inception of RNA Interference Concept

Since the discovery of short non-coding RNAs, which are crucial to a process known as RNA silencing, our knowledge of RNA-mediated functions has greatly increased (Wani et al. 2010). Two concurrent research fields have been established in response to the identification of these short non-coding RNAs; one is to understand the systemic findings of the effects of short non-coding RNAs and the other is the application of small non-coding RNAs to alter cellular defenses and processes (Rosa et al. 2018). Also, the short RNA strands (siRNA or miRNA) can be used to either degrade the target mRNA or impede the protein translation processes in eukaryotes (De Backer et al. 2002). By inhibiting the genes responsible for diverse stressors and enhancing unique features in plants, such as disease resistance, the RNAi approach has opened up new avenues in the development of several molecular tools that can be used in crop improvement. Therefore, it has been established as a potent therapeutic tool to fight against pathogens invading plants (Sharma et al. 2013).

Early studies on virus-infected plants found that under some circumstances the plants might recuperate from symptoms and acquire resistance to reinfection. A plethora of studies on *Tobacco ringspot virus* (TRSV) revealed that there is a resistance/immunity development in the host plant against the present and subsequent infections, whereas the plants were subjected to infection upon the attack of unrelated viruses (Ratcliff et al. 1999; Rosa et al. 2018; Wingard 1928), this type of occurrence was described using terminology such as 'acquired immunity', 'superinfection exclusion', and 'cross-protection' (Fulton 1986; Johnson 1937). Further investigations revealed that RNA molecules functioned during viral and even viroid cross-protection in a sequence-specific manner thereby generating the idea that the resistant factor was brought about by the genetic material of the pathogen. Therefore, to produce resistant plant varieties, the functional genetic constituents could be engineered rather than using the whole infectious viral strain (Palukaitis and Zaitlin 1984). Subsequent experiments on *Caenorhabditis elegans par-1* gene revealed that along with the antisense strand, the sense strand could also exhibit the similar effect of gene silencing (Fire et al. 1998; Guo et al. 1995). Later Fire et al. coined the term '**RNA interference**' upon demonstrating that foreign sense and antisense RNA may obstruct the endogenous gene's activity. The dsRNA acts as a powerful tool in the RNAi-based approach and even exhibits better results than ssRNA. For this crucial finding, Andrew Fire and Craig Mello were awarded the Nobel Prize for Medicine in 2006 (Fire et al. 1998). Later, RNAi has been reported in petunia plants, various insects, fungi (*Neurospora crassa)*, and mammals (Elbashir et al. 2001; Hammond et al. 2000; Napoli et al. 1990; Romano and Macino 1992).

3 Mechanism of RNAi Action

RNAi is the inherent mechanism in eukaryotes for defending them against viruses and transposable elements, which can compromise their genomic DNA integrity (Obbard et al. 2009). RNAi operating pathways involve the production of non-coding RNA (siRNA and miRNA) essential for the downregulation of gene expression (Kuo and Falk 2020). Certain criteria determine the generation of either siRNA or miRNA: siRNA is derived from (a) dsRNA obtained from viral replication or transcripts of DNA and RNA virus, (b) dsRNA obtained from plant's endogenous RNA-dependent RNA polymerase (RdRp), (c) dsRNAs obtained naturally from sense and antisense mRNA strands as well as single-stranded RNAs (ssRNA). Whereas miRNA is derived from endogenously derived noncoding RNA transcripts called primary miRNA (pri-miRNA) (Rosa et al. 2018). Therefore, the distinguishing feature is that siRNAs are believed to be predominantly exogenous in origin, procured from transposons or viruses, while miRNAs were assumed to be endogenous and purposely produced through the organism's genome (Tomari and Zamore 2005). However, there exist similarities too between them regarding their size similarities and sequence-specific inhibitory functions suggesting their relatedness in the mode of their action like a dependence on dicer enzyme and the Argonaute family of proteins (Ago) (Meister and Tuschl 2004).

Thus, the fundamental principle of target gene silencing is based on the formation of small RNA sequences (approx 20–30 nucleotides long) by the dicer complex, its incorporation or loading into the argonaute proteins and thereby developing the RNA Induced Silencing Complex (RISC). This complex will then help to silence target gene expressions (Svoboda 2020).

In the siRNA pathway, the dsRNAs obtained from different origins are cleaved by varied types of plant-specific dicer enzymes, a ribonuclease enzyme type-III (DCL) to generate siRNA duplex having 21–24 nucleotides. Thereafter, the dsRNA unwinds to create one passenger strand that is eventually destroyed and one guide strand that is intended to target the corresponding endogenous mRNA. Later, the RNA-induced silencing complex (RISC) is established to carry out the gene knockout process, which leads to the suppression or inhibition of protein expression by directing the guide RNAs to bind complementarily to the target mRNA (Mroz and Rocco 2006). It is also possible for certain siRNAs, often 24-nucleotides siRNAs, to penetrate the nucleus and target transcripts produced by plant DNA-directed RNA polymerase to cause RNA-directed DNA methylation (RdDM). Then, siRNA can interact with DNA methyltransferase-like domains rearranged methyltransferase 2 (DRM2) to induce suppression of RdDM to gene expressions (Fig. 2a) (Halder et al. 2022; Kuo and Falk 2020; Rosa et al. 2018).

While in the miRNA pathway, single-stranded RNA transcripts (MIR genes) serve as the endogenous source of the hairpin-folded pri-miRNA RNA transcript. At this point, MIR genes are acted upon by RNA pol II to create Pri-miRNA, which is subsequently treated inside the nucleus by DCL cleavage to create pre-miRNA. Once more, DCL breaks down the freshly synthesized pre-miRNA to create duplex micro-RNA, which is then carried to the cytoplasm where the guide strand joins with Ago proteins to create RISC (RNA-induced Silencing Complex). Additionally, the guide RNAs attach to the target RNA after Watson and Crick's complementary base pairing, causing either the degradation of the target sequences or interference with translation (Fig. 2b) (Halder et al. 2022; Kuo and Falk 2020; Rosa et al. 2018).

Both transcriptional and post-transcriptional repression could be possible through the process of RNAi. Transcriptional regulation is quite rare in animals, while plants are much more familiar with it. Transcriptional regulations involve repression in simple processes of de novo DNA methylation or histone modifications. On its other extreme, post-transcriptional modifications involve either endonucleolytic cleavage of cognate RNA or mRNA destabilization (Svoboda 2020).

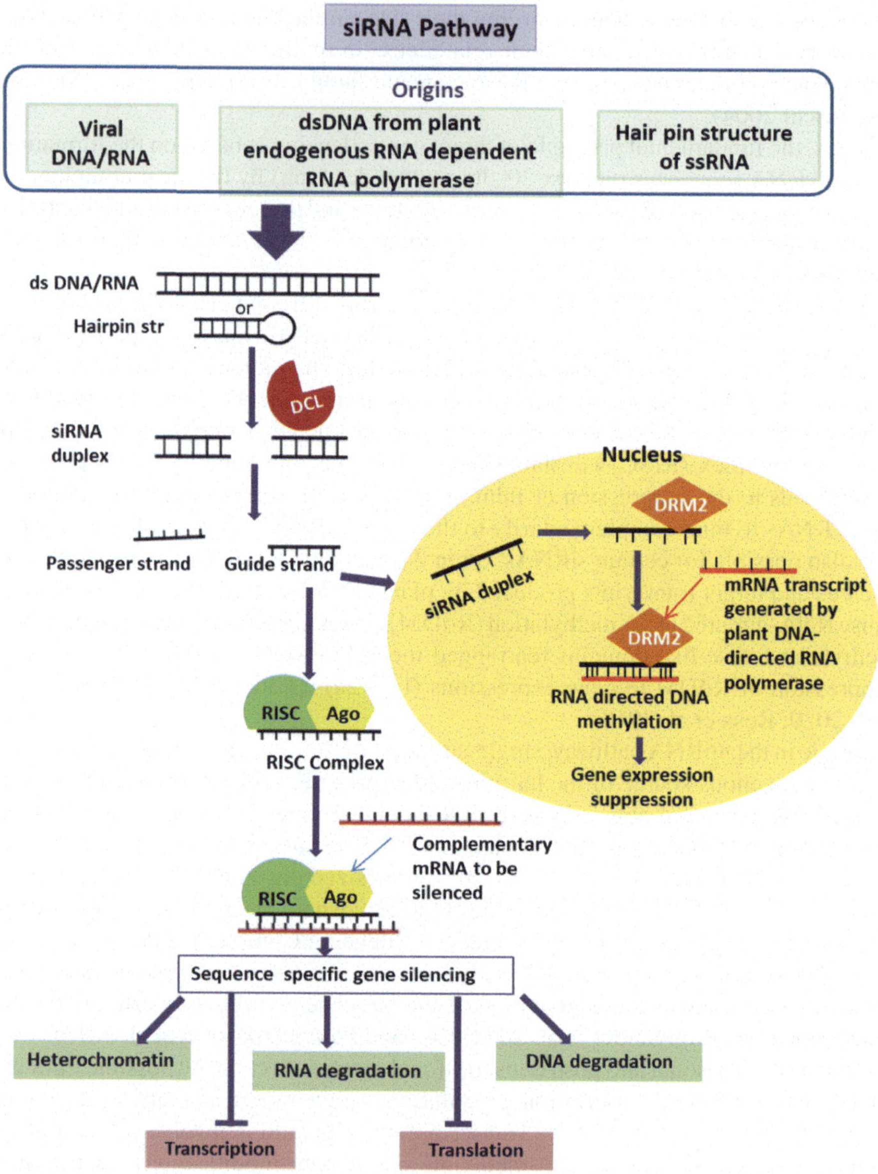

a *siRNA mediated gene silencing*

Fig. 2 A siRNA mediated gene silencing. **b** miRNA mediated gene silencing

Fig. 2 (continued)

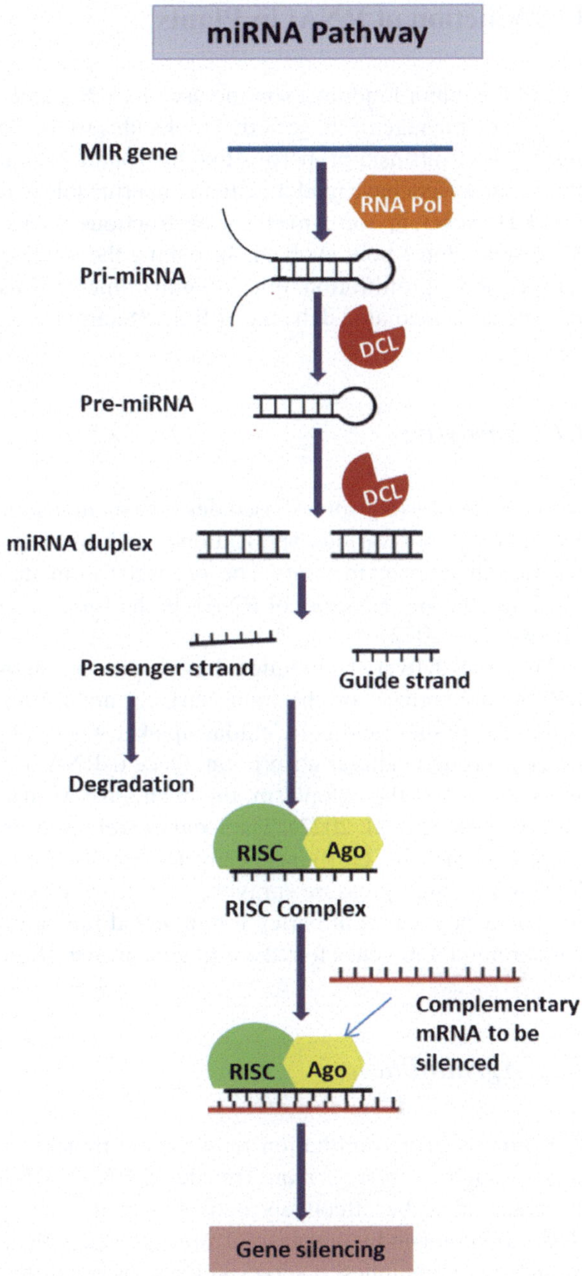

b *miRNA mediated gene silencing*

4 Induction of RNAi in Plants

One of the major hindrances in the use of siRNA and miRNA as therapeutic agents is the poor carriage of these active molecules to the appropriate cell type, tissue, or organ. Their intrinsic properties like hydrophilic nature, negative charge, and high molecular weight have rendered them impermeable to the cell membranes (Lam et al. 2015). However, to overcome these obstructions and facilitate their effective delivery, the systems have now evolved. In plants, the siRNAs and miRNAs are delivered via spraying, agrofiltration, micro-bombardment, virus-induced gene silencing, and nanomaterial-mediated delivery of RNA (Sharma et al. 2013; Zhang et al. 2021).

4.1 Spraying

The concept of using spray-based delivery methods for RNAi induction is an attractive approach for various applications, such as crop protection, pest control, and disease management in plants. The idea is to formulate and deliver siRNA molecules, which are the key effectors of RNAi, in the form of a spray to the target organisms (Hoang et al. 2022).

The spray delivery technique is based upon two steps: (a) Foliar absorption, where dsRNAs are sprayed on the foliar surfaces and eventually penetrate to the interior of the leaf tissues, and (b) Cellular uptake, where plant cells take up the dsRNAs directly through cellular absorption. Once dsRNA has passed through the cell wall pores and enters the cytoplasm, the plant's RNAi machinery may transform it into siRNAs (Hoang et al. 2022). The local as well as systemic suppression of the green fluorescent protein (GFP) has been successfully mitigated using exogenous siRNA delivered by high-pressure spraying, also when dsRNAs and siRNAs were sprayed upon detached leaves of barley it minimized fungal infections by suppressing three cytochrome P450 genes to reduce fungus growth (Das and Sherif 2020).

4.2 Agroinfiltation

The process of agroinfiltation requires the transfer of DNA/RNA to plants naturally through *Agrobacterium.* Transfer DNA (T DNA) of the *Agrobacterium has* the potential to be effectively delivered in the host plant. When genes of the Ti plasmid (Tumor-inducing plasmid) are replaced with desired hairpin RNA by genetic engineering techniques, it serves as a potent vector for plants (Chen et al. 2013).

The introduction of *Agrobacterium* into the intercellular spaces of tissues in plants involves the use of a syringe and vacuum infiltration. Firstly, the posterior side of the leaf epidermis is nicked with the aid of a needle without piercing both sides. The Agrobacterium is then introduced into the nick. After the entry, the leaf color changes from light green to dark color and confirms the successful infiltration of the host plant (Vaghchhipawala et al. 2011; Wroblewski et al. 2005). Syringe infiltration has been the method of choice for various applications such as research on plant-pathogen interactions, abiotic stresses, protein localization, and protein–protein interaction, along with plant gene functional analysis and transient silencing assays (Vaghchhipawala et al. 2011;). However, for plant species where syringe infiltration becomes difficult an alternative has been developed, vacuum infiltration. In this approach, the plant leaves are initially immersed in an infiltration mix containing *Agrobacterium* strain containing the target recombinant DNA (rDNA). Now, inside a vacuum chamber, negative atmospheric pressure is created, where the submerged plants are housed. The vacuum sucks air from the submerged plant leaf's interstitial spaces. When the vacuum is released, again Agrobacterium-containing medium fills the area that was previously filled with air (Leuzinger et al. 2013; Chen et al. 2013). In an experiment to protect soybean plants from a mite *Tetranychusurticae*, agrofiltration was found to be very useful. *Tetranychusurticae* is known for causing serious damage to crop plants by gaining resistance toward the conventional disease control method. As an alternative measure RNAi technique was utilized in soybean plants where agroinfiltration was used to rapidly produce the mite's hairpin structured RNAs for COPA and AQ9 genes, following the feeding of those soybean plants to *T. urticae*high mortality of the pest was exhibited (Dubey et al. 2017).

4.3 Micro-Bombardment

This approach is well known for transferring both linear and circular DNA templates into the chloroplast and nuclear genomes. The process is equally potent for both transient and stable gene silencing. This procedure involves covering gold or tungsten particles with dsRNA, and siRNA that encode hairpin structures and they are then fired into plant cells, protoplasts, meristems, embryos, or calluses. Thus, the process is also well known as the 'particle or microprojectile bombardment approach, particle gun transformation, and biolistic method' (Klahre et al. 2002; Davies et al. 2013). Due to the quick processing time, cheap manufacturing costs, and ease of use for introducing numerous genes or chimeric DNA, the most well-known direct method for changing monocotyledons' genetic makeup is biolistics. Additionally, the method neither depends on the cell's electro-physiological properties nor requires a vector of a certain sequence (Rivera et al. 2012). In a study, it was found that when synthetic siRNAs were delivered into plants by biolistic pressure, it initiated a silencing process of GFP (green fluorescent protein) expression that can spread both locally and throughout the entire plant system. When the leaves are systematically silenced through siRNAs, the levels of target RNA sequences get reduced and this confirms that RNA can trigger a phenomenon called systemic PTGS (Post-Transcriptional Gene Silencing). Our findings suggest that siRNAs or intermediates created by siRNAs may include silencing signals, and these signals eventually cause the self-synthesis of siRNAs (Klahre et al. 2002).

4.4 Virus-Induced Gene Silencing (VIGS)

VIGS is an effective technique used in plant molecular biology to explore gene function and examine the part that certain genes play in plant virus defense. This technique takes advantage of the natural defense mechanism of RNAi to silence specific sequences of genes and understand their impact on plant-virus interactions (Lu et al. 2003; Voinnet 2001). Researchers engineered viral vectors that express short oligonucleotide stretches of the target gene's RNA sequence. These viral vectors are typically derived from plant viruses, like Tobacco mosaic virus (TMV), potato virus X (PVX), Cucumber mosaic virus (CMV), Tomato golden mosaic virus (TGMV), Zucchini yellow mosaic virus (ZYMV), and Tobacco rattle virus (TRV), which are modified to carry the gene fragments of interest instead of their genetic material. When these viral vectors infect the host organism, they deliver the engineered RNA fragments, which trigger the RNAi machinery in the cells (Dinesh-Kumar et al. 2003).

TMV is the first modified virus used to successfully apply the VIGS method to *Nicotiana benthamiana* plant for pds gene silencing (Unver and Budak 2009). VIGS vector should have the potential to infect the plant and disseminate swiftly and evenly, including the plant's meristematic areas. Furthermore, it should not create any powerful silence suppressors. TRV has the most of these characteristics among all viruses. TRV-based vectors have been used to silence genes in *Nicotiana benthamiana*, tomato, and potato with success. It infects plants without exhibiting chlorotic or necrotic signs, making the VIGS-induced phenotype easier to identify. TRV infects every cell of the plant, causing a consistent silencing phenotype (Dinesh-Kumar et al. 2003). Proliferating cell nuclear antigen (PCNA), a meristematic gene, was silenced by tomato golden mosaic virus in *N. benthamiana* (Peele et al. 2001).

VIGS has become an extensively used tool in plant research, mainly due to its simplicity, versatility, and high-throughput capabilities. The process is quite rapid, does not require extensive plant transformation protocols, and is easy to formulate on a plethora of plant species. This approach shows promise in developing new strategies for crop protection, as it allows for targeted and environmentally friendly control of plant viruses (Burch-Smith et al. 2004; Unver and Budak 2009). Figure 3 shows the induction of RNAi in plants via various methods.

4.5 Nanotechnology-Mediated Delivery of RNA

Nanotechnology-mediated delivery involves the use of either nanoparticles themselves as protectants or as carriers for prevailing pesticides and dsRNA for RNAi-mediated protection. As carriers, nanoparticles offer a wide range of advantages, including (i) longer durability, (ii) better pesticide solubility in water, (iii) lower toxicity, and (iv) more site-specific absorption into the target pest. Another potential advantage of nanocarriers is that they give better environmental stress regulation

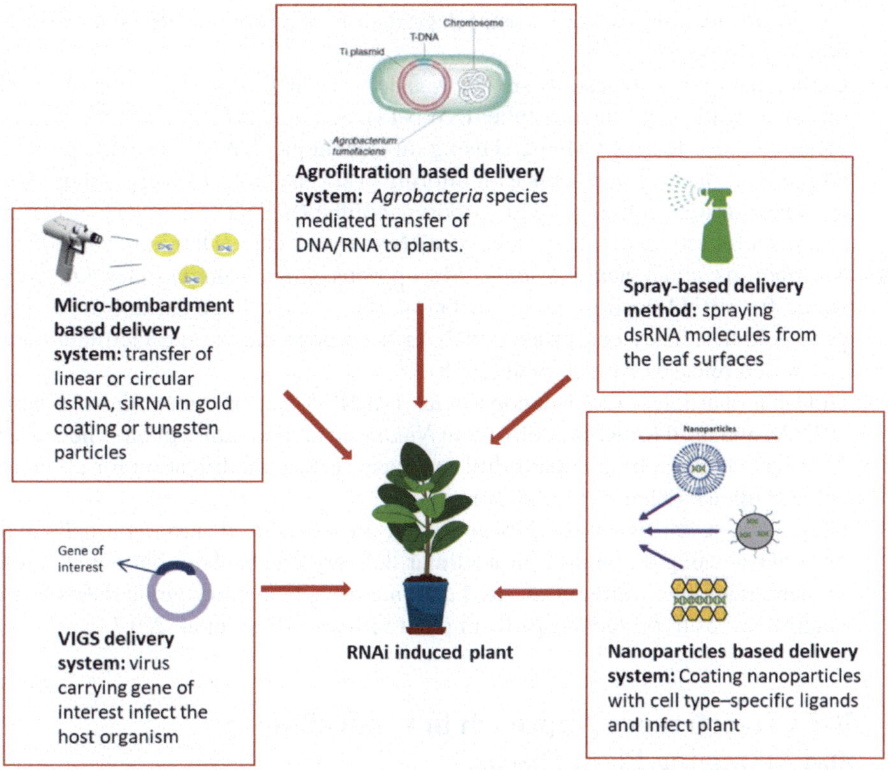

Fig. 3 Induction of RNAi in plants via various methods

opportunities to the nanopesticides and would rather result in a reduction in the number of applications, very cost-effective and less toxic (Worrall et al. 2018; Zhang et al. 2021).

Nanomaterials offer several advantages for RNAi delivery in plants. They can protect RNA molecules from degradation, facilitate cellular uptake, and enable controlled release of siRNAs at the target site (Pugsley et al. 2021). Some commonly used nanomaterials for RNAi delivery in plants are:

(a) Lipid-based nanoparticles: Liposomes are surfactants consisting of different phospholipids that can function as vesicles to enter plant cell walls. Lipid nanoparticles can encapsulate and protect siRNAs. They can be engineered to facilitate efficient cellular uptake and release of siRNAs into plant cells. Nutrients have been successfully delivered into tomato leaves using a 100 nm PEGylated liposome. Several studies reveal the effectiveness of liposome particles in delivering the dsRNAs into the cells (Mat Jalaluddin et al. 2023).
(b) Polymer-based nanoparticles: Polymers such as chitosan, polyethyleneimine (PEI), or poly (lactic-co-glycolic acid) (PLGA) can be used to form nanoparticles for RNAi delivery. These nanoparticles hold the siRNAs very well and

provide protection from any sort of degradation, and promote effective cellular delivery (Singha et al. 2011)

(c) Carbon nanotubes: In a study siRNA was delivery method was developed based on single-walled carbon nanotubes (SWNTs). In plants, *Nicotiana benthamiana* genes can be effectively silenced using this technique. SWNTs provide passive administration and fluorescence monitoring of siRNA molecules in plant tissues, as demonstrated. Single-walled carbon nanotube (SWNTs) is a very effective approach for the successful uptake of siRNAs by the cells (Demirer et al. 2020).
(d) Mesoporous silica nanoparticles: Mesoporous silica nanoparticles are well suited for siRNA loading and distribution due to their huge surface areas and pore volumes. They can protect siRNAs from degradation and facilitate their controlled release (Worrall et al. 2018).
(e) Gold nanoparticles: Gold nanoparticles (AuNPs) can be functionalized with siRNAs and used for RNAi delivery in *Nicotiana benthamiana* plant. They offer excellent stability, biocompatibility, and easy surface modification for efficient cellular uptake (Zhang et al. 2021).
(f) Clay nanosheets: Plasmid DNA and RNA are introduced into plant cells and protoplasts using a nonviral intracellular delivery mechanism that is mediated by nanomaterials. Mitter et al. used clay nanosheets to carry guide dsRNAs to healthy leaves to protect crops from plant viruses (Mitter et al. 2017).

5 RNA Interference Approach in Controlling and Managing Plant Diseases

It is a powerful genetic mechanism that can be used to control and manage diseases caused by both biotic and abiotic stresses. By targeting genes involved in stress responses, RNAi can enhance plant tolerance and survival under adverse conditions (Bharathi et al. 2022; Wani et al. 2010).

5.1 RNAi-Based Approaches to Control Plant Diseases Against Biotic Factors

5.1.1 RNAi Against Plant Pathogenic Fungi

Fungal pathogens are responsible for a significant portion of crop losses globally. RNAi has been utilized to target crucial genes involved in fungal pathogenesis, such as those encoding cell wall-degrading enzymes or virulence factors. By silencing these genes, fungal growth and infection can be suppressed (Halder et al. 2022). Some experimental studies highlighting the evidence of RNAi-mediated control of fungal diseases in plants are elucidated herein (Table 1).

Table 1 Use of RNAi technology in various plants against fungi

Target organism	Target gene	Disease	Plant	Result	References
Fusarium graminearum	Chitin synthase gene (CHS3)	Fusarium head blight	Wheat	Decreased severity of disease and mycotoxin production	Koch et al. (2013)
Alternaria cyamopsidis	AcCHS7 gene	Blight or leaf spots	Guar	Slowed disease development by destroying the hyphae and limiting mycelium growth and conidiation	Thakur and Prasad (2020)
Botrytis cinerea	TOR gene	Grey mould disease	Potato	Enhanced resistance to *B. cinerea*	Xiong et al. (2019)
Verticillium dahlia	VdH1 gene	Verticillium wilt	Cotton	Reduced leaf wilting, defoliation, and complete drying of whole seedlings	Zhang et al. (2016)
Magnaporthe grisea	PKS/NRPS gene	Rice blast	Rice	A drastic reduction in fungal virulence and disease severity in rice plants	Böhnert et al. (2004)
Cucumber mosaic Virus, Zucchini mosaic Virus, Watermelon mosaic virus	CMV, ZYMV and WMV coding sequences	Mosaic diseases	Different types of squashes	Decreased mottling and dwarfism	Tricoll et al. (1995)

Koch et al. (2013) demonstrated the successful use of RNAi in wheat to control Fusarium head blight. By silencing the chitin synthase gene (CHS3) in the fungal pathogen *Fusarium graminearum*, they observed a significant reduction in disease severity and mycotoxin production. This study proved that RNAi targeting critical genes involved in fungal pathogenesis can effectively suppress fungal growth and disease symptoms.

Due to Guar's limited natural resistance to Alternaria cyamopsidis blight, improved varieties with a high level of resistance are required. Therefore in an investigation, it was found that pSilentCHS7-2 and pARTCHS7 constructs effectively silenced the AcCHS7 gene in guar plants of the HG 2–20 variety and *Alternaria*, respectively. Silencing of AcCHS7 in *Alternaria* destroys the hyphae and reduces the progression of the disease (Thakur and Prasad 2020).

The fungus *Verticillium dahlia*is known to cause Verticillium wilt to cotton plants. To protect the plant from this disease RNAi was used as a tool by Zhang et al. They targeted *V. dahliae* hygrophobins1 (VdH1) gene and created a knockout mutant, VdaDvdh1. It was observed that, in contrast to the wild-type strain, the VdaDvdh1 mutant exhibited significantly better results. Several symptoms such as enhanced hyphae growth and limited formation of melanized microsclerotia, could be delayed or reduced through twenty days post-inoculation (dpi), defoliation, and complete drying of whole seedlings (Zhang et al. 2016).

In potato plants, *Botrytis cinerea*is are known to cause grey mold disease. The target of rapamycin (TOR) signaling was studied by researchers to understand whether TOR may be used to control plant fungal infections. Xiong et al. highlighted the efficacy of RNAi in HIGS for crop protection. By targeting BcTOR, transgenic potato lines exhibited enhanced resistance to *B. cinerea*, demonstrating the potential of RNAi as a valuable tool in combatting plant pathogens (Xiong et al. 2019). This is well illustrated in the Fig. 4 (Sharma et al. 2019) (https://www.potatogrower.com/2017/06/shades-of-gray#grey%20mold%20fig%202.jpg).

The study by Heidi U. Böhnert investigated the role of a putative PKS/NRPS gene in signaling pathogen attack to resistant rice plants. Through RNAi-mediated gene silencing, they demonstrate that the PKS/NRPS gene was crucial for the effective infection of the fungus *Magnaporthe grisea*. By introducing siRNA molecules that specifically targeted the PKS/NRPS gene in *Magnaporthe grisea*. They observed that the silenced strains of *Magnaporthe grisea* were unable to infect resistant rice plants effectively.

5.2 RNAi Against Plant Pathogenic Virus

RNAi can be used to combat plant pathogenic viruses by silencing specific viral genes, thereby reducing viral replication, transmission, and symptom development. Here are a few key studies that demonstrate the effectiveness of RNAi against plant pathogenic viruses (Table 2).

Researchers utilized artificial microRNAs (amiRNAs) to target the pre-coated (AV2) and coated proteins (AV1) of ToLCNDV (Tomato leaf curl New Delhi virus). The genes encoded by the protein cause Tomato leaf curl disease. These amiRNAs were incorporated into the tomato plants through genetic transformation. The artificially synthesized miRNAs especially function to silence the specific target gene expression and thereby degrade viral RNA, thus suppressing viral replication and spread through out the plant body. The recombinant gene showed better results

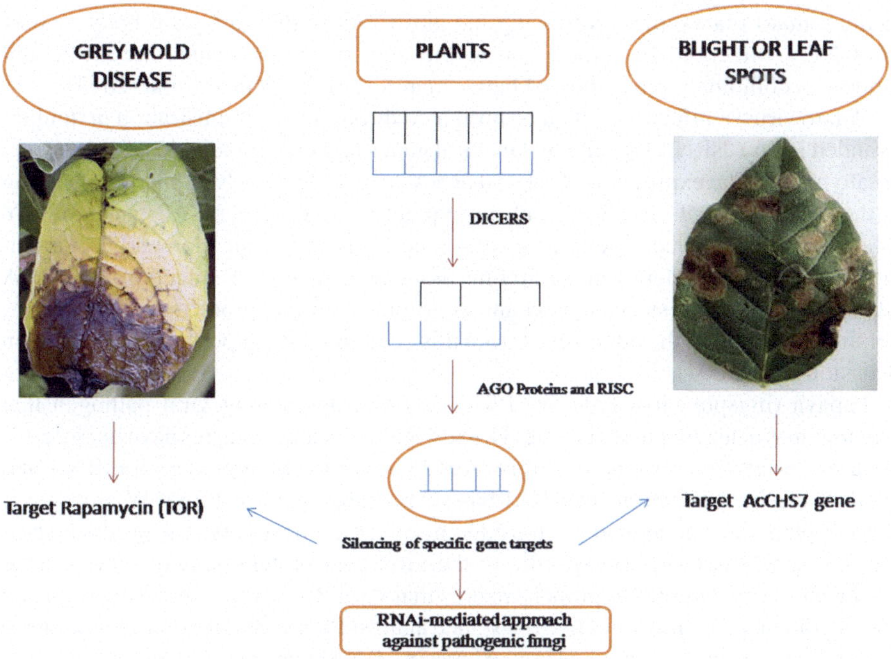

Fig. 4 RNAi-mediated pathogenic fungi management in plants

Table 2 Use of RNAi technology in various plants against virus

Target organism	Target Gene	Disease	Plant	Result	References
ToLCNDV	AV2 and AV1	Tomato leaf curl disease	Tomato	Reduced viral DNA accumulation and mild symptoms in the transgenic species	Van Vu et al. (2013)
Potato virus Y	Coat protein	Potato virus Y disease	Potato	Reduced viral replication and accumulation, resulting in milder or no symptoms associated with PVY infection	Missiou et al. (2004)
Papaya ringspot virus type W	Coat protein	Papaya ringspot disease	Papaya	Reduced viral symptoms, such as mosaic patterns on leaves, stunted growth, and fruit deformities	Jia et al. (2017)
Rice tungro bacilliform virus	Viral RNA	Rice tungro	Rice	Reduced symptoms associated with RTBV infection, including stunted growth, yellowing of leaves, and reduced yield losses	Tyagi et al. (2008)
Ipomovirus	Coat protein	Cassava brown streak disease	Cassava	Reduced viral replication and accumulation, resulting in milder or no symptoms associated with CBSD infection	Patil et al. (2011)

in the tomato plants, thus expressing the amiRNAs in the host plant body against ToLCNDV infection. In comparison to the non-transgenic plants, the artificially induced recombinant plant showed better manifestations (Van Vu et al. 2013).

Potato virus Y infections could be mitigated through the administration of double-stranded RNA (dsRNA) specific to key viral genes (e.g coat protein) in the transgenic potato plants. The expression of these dsRNAs triggered the RNAi pathway, leading to the degradation of viral RNA and subsequently a high level of resistance to PVY infection was observed. They demonstrate reduced viral replication and accumulation, resulting in milder or no symptoms associated with PVY infection. The RNA silencing-mediated resistance mechanism provides durable protection against PVY, helping to safeguard the potato crop from this economically important viral pathogen (Missiou et al. 2004).

Papaya ringspot virus type W (PRSV-W) is a devastating viral pathogen that infects papaya, leading to severe yield losses. RNAi-based strategies have been developed to confer resistance against this virus by specifically targeting viral RNA and suppressing its replication. In RNAi-based resistance against PRSV-W, transgenic papaya plants are generated by introducing genetic constructs that produce short interfering RNAs (siRNAs) specific to the viral coat protein gene or other crucial viral genes. This targeted approach ensures that viral RNA is specifically recognized and degraded, providing effective resistance against PRSV-W infection. Studies have shown that transgenic papaya plants expressing RNAi constructs exhibit high levels of resistance to PRSV-W. These plants demonstrate reduced viral replication and accumulation, resulting in milder or no disease symptoms (Jia et al. 2017).

By employing RNAi technology, researchers have tried to combat the Rice tungro bacilliform virus (RTBV) in rice plants through target gene silencing in RTBV, thereby mitigating its detrimental effects on rice crops. They have developed genetic constructs introduced into rice plants, which produce small RNA molecules specifically targeting the viral RNA of RTBV. The expression of RNAi constructs had led to a decreased accumulation of the virus in inoculated rice plants. As a result, these transgenic rice plants exhibited reduced symptoms associated with RTBV infection, including stunted growth, yellowing of leaves, and reduced yield losses (Tyagi et al. 2008).

The study conducted by Patil et al. (2011) relied on RNAi technology, where resistance against Cassava brown streak disease (CBSD) was provided through target gene silencing of the two virus species in the genus *Ipomovirus*. In their study, genetic constructs were developed and introduced into cassava plants, targeting the coat protein of the viral genomes. These constructs triggered the RNAi pathway, leading to the production of small RNA molecules that could recognize and degrade the viral RNA, thereby inhibiting viral replication. The study demonstrated that RNAi-mediated resistance was effective against diverse isolates belonging to the two virus species responsible for CBSD. The transgenic cassava plants expressing the RNAi constructs exhibited reduced viral replication and accumulation, resulting in milder or no symptoms associated with CBSD infection (Patil et al. 2011).

5.3 RNAi Against Plant Pathogenic Nematode

RNAi has shown promising results in combating plant parasitic nematodes, which are microscopic worms that parasitize plant roots and cause significant agricultural losses and several studies have demonstrated its effectiveness. Here are a few examples of RNAi-based approaches targeting plant nematodes (Table 3).

Walnut cultivation faces a serious threat due to Crown gall disease and root lesions caused by the pathogenic bacterium *Agrobacterium tumefaciens* and nematode *Pratylenchus vulnus* respectively. Researchers have used RNAi silencing to address both problems simultaneously. The researchers employed artificially synthesized small interfering RNA sequences targeting the *iaaM* and *ipt* genes of *Agrobacterium*, and the Pv010 gene of *P. vulnus*. Due to this, there was an increase in crop yield and a reduction in chemical dependency (Walawage et al. 2013).

Root-knot nematodes belonging to the genus *Meloidogyne*, a parasitic species that infect thousands of plants (e.g *Arabidopsis.* It affects plants' defense response by inhibiting K + channels mediated by oesophageal gland effector genes like Mi-msp2. When two separate RNAi lines of *Arabidopsis* were produced, Mi-msp2 dsRNA showed reduced results in the number of females, egg masses, and gall formations. Additionally, Mi-msp2 mRNA abundance was significantly affected in the female

Table 3 Applications of RNAi approach on various plants against nematode

Target organism	Target Sequences	Disease	Plant	Result	References
Pratylenchusvulnus	Pv010	Root lesion	Walnut	Increased crop yield and reduction in chemical dependency	Walawage et al. (2013)
Meloidogyne spp.	Mi-msp2	Root-knot disease	*Arabidopsis*	Reductions in the number of galls, females, and egg masses	Joshi et al. (2019)
Heteroderaavenae	Life cycle/ housekeeping genes	–	Wheat	Reduction in yield losses	Blyuss et al. (2019)
Heteroderaglycines	Hg-rps23, Hg-snb1, and Hg-cpn1	Soybean cyst	Soyabean	Mitigated crop losses	Zhang et al. (2022)
Heteroderaschachtii	3B05, 4G06, 8H07, and 10A06	Root-knot disease	Arabidopsis	Reduced number of mature female	Sindhu et al. (2009)

nematodes feeding on artificially synthesized plants producing dsRNA. These findings implied that the Mi-msp2 mRNA might be a successful target for Root-knot nematodes, controlling significant economic losses (Joshi et al. 2019).

Blyuss et al. have investigated the use of biostimulants derived from soil streptomycetes in wheat (*Triticum aestivum* L.) against the nematode *Heterodera avenae* through RNAi approach. In the study, numerous streptomycete-derived biostimulants were used to determine their efficacy in promoting plant growth, development, and resistance to the cereal cyst nematode. These biostimulants resulted in increased synthesis of si/miRNA that are complementary to plant and nematode mRNA in plant cells which are particularly efficient at suppressing the translation of nematode mRNA essential for nematode life cycle, thereby lowering nematode infestation and increasing plant resistance to worms (Blyuss et al. 2019).

Similarly, Zhang et al. (2022), focused on developing genetically modified soybean plants that provided resistance to the nematode (*Heterodera glycines*) in soybeans. It invades the roots of soybean plants and causes significant yield losses. By designing and introducing specific siRNA molecules (Hg-rps23, Hg-snb1, and Hg-cpn1) that target vital genes in the soybean cyst nematode, the researchers aimed to disrupt the nematode's essential biological processes and hinder its ability to cause damage to soybean plants. The transformation of soybean plants exhibited enhanced resistance to nematode infestation, as the silenced genes in the nematode disrupted its normal development, reproduction, or other vital functions (Zhang et al. 2022).

In an investigation, four parasitism genes (3B05, 4G06, 8H07, and 10A06) were targeted through a host-induced RNAi-mediated gene silencing approach in a biologically engineered *Arabidopsis thaliana* plant. *A. thaliana* acts as the host for the sugar beetroot nematode *Heterodera schachtii.* It was revealed worms that feed on plants expressing the appropriate RNAi, constructs specifically lowered the mRNA abundances of the targeted nematode genes. The number of mature female worms decreased as a result of this host-induced RNAi approach in all four nematode parasitizing genes (Sindhu et al. 2009).

5.4 RNAi Against Plant Insects

Insects cause damage to approximately 20–40% of crops. In this regard, RNAi has been used as a powerful tool in developing transgenic plants that can defend themselves against herbivorous insects by producing dsRNA (Cagliari et al. 2019). Here are a few examples of RNAi-based approaches to control insect pests (Table 4).

RNAi approach was utilized to produce mortality in western corn rootworms (*Diabrotica virgifera virgifera* LeConte) either via feeding them an artificial diet or transgenic corn plant expressing dsRNA. (V-ATPase subunits, along with α-tubulin) and when coleopteran pests feed on these plants, the ingested dsRNA triggers RNAi, silencing vital genes in their digestive system and causing their death (Baum et al. 2007).

Table 4 Applications of RNAi approach on various plants against insect

Target organism	Target gene	Disease	Plant	Result	References
Diabroticavirgiferavirgifera	V-ATPase subunits, α-tubulin	Rootworm disease	Corn	Increased crop yield	Baum et al. (2007)
Helicoverpaarmigera	CYP6AE14	Bollworm disease	Cotton	Larvae exhibited reduced weight gain, increased mortality, and delayed development	Mao et al. (2007)
Diabroticaspp	Cry3Bb1, Cry34Ab1/ Cry35Ab1 and DvSnf7	Rootworm disease	Maize	Reduced adult emergence of pests and reduced root damage	Head et al. (2017)
Dalbulusmaidis	*V-ATPase B and V-ATPase D*	–	Maize	Significant mortality and wing deformation of insects	Jones et al. (2021)
Dalbulusmaidis	AChE1 and AChE2	–	Cabbage	Reduced larval survival, growth inhibition, and deformities in treated larvae	Gong et al. (2013)

Cytochrome P450 gene (CYP6AE14) of cotton bollworm (*Helicoverpaarmigera*) allows herbivores to withstand the cotton metabolite, gossypol. When gossypol gets introduced into the diet, CYP6AE14 is strongly expressed in the gut and leads to larval growth. However, when plant-producing CYP6AE14-specific dsRNA is fed to the larva, RNAi aids in lowering the expression of CYP6AE14 in the midgut of the larva. As a result, the larvae exhibit reduced weight gain, increased mortality, and delayed development (Mao et al. 2007).

Monsanto and Dow AgroSciences performed separate field studies to assess the efficacy of corn rootworm (CRW) protected maize hybrids expressing the coleopteran-active (*Diabrotica spp.* complex) characteristics and developed two GMO varieties SmartStax and SmartStax PRO. SmartStax expressed Cry3Bb1 and Cry34Ab1/Cry35Ab1 whereas SmartStax PRO expressed an additional gene DvSnf7. Eventually, it was observed that SmartStax PRO was more efficient in reducing adult emergence and in reducing root damage and was the first commercially approved GMO variety (Head et al. 2017).

Dalbulusmaidis is a phloem-feeding insect of maize crops. In a study, two subunits B & D of vacuolar ATP synthase (V-ATPase B and V-ATPase D) were silenced using the RNAi technique. These caused significant mortality and wing deformation of insects (Jones et al. 2021).

In a study, the efficacy of chemically synthesized small interfering RNA (siRNA) molecules targeting genes AChE1 and AChE2 of the insect against the *Plutellaxylostella*, a destructive pest of cruciferous crops was evaluated. Under laboratory and field conditions siRNA molecules effectively induced RNAi in *Plutellaxylostella*, leading to reduced larval survival, growth inhibition, and deformities in treated larvae (Gong et al. 2013).

5.5 RNAi Against Plant Pathogenic Bacteria

It has been discovered that RNA interference has significant potential in managing diseases and providing resistance to plants against bacterial infections. Remarkable cases of bacterial illness management in which RNAi demonstrated an extraordinary form of gene control have been recorded (Table 5).

In an investigation carried out by Escobar et al. (2001), the RNAi technique was successfully applied to provide resistance against crown gall disease. The disease was again caused by the pathogenic bacteria *Agrobacterium tumefaciens*. Firstly, they targeted the expressions of sequences *iaaM* and *ipt* to create transgenic *Arabidopsis thaliana and Lycopersicon esculentum* plants. Secondly, dsRNAs are synthesized within the plants to initiate the silencing of iaaM and ipt through the RNAi approach. Thus, these newly synthesized plants are extremely resistant to crown gall disease (Escobar et al. 2001).

In the rice plant (*Oryzae sativa*), the RNAi approach is used to silence OsSSI2 (OsSSI2-kd), which is mainly responsible for the dehydrogenation of fatty acids and hence increases the resistance to the leaf blight disease (Jiang et al. 2009).

Table 5 Applications of RNAi approach on various plants against bacteria

Target organism	Target Gene	Disease	Plant	Result	References
Agrobacterium tumefaciens	*iaaM* and *ipt*	Crown gall disease	*Arabidopsis thaliana Lycopersicon esculentum*	Resistance of the plants from crown gall disease	Escobar et al. (2001)
Xanthomonas oryzae	OsSSI2	Leaf blight	Rice plant	Increased resistance to the leaf blight disease	Jiang et al. (2009)

6 RNAi-Based Approaches to Control Plant Diseases Against Abiotic Factors

It has been estimated that abiotic stress is directly responsible for roughly 70% of agricultural production decline. Climate serves as a major factor in exacerbating the intensity of various abiotic stressors, namely drought and salt stress, leading to considerable production losses of staple crops including wheat, maize, and barley. Abiotic stresses can also weaken plants' defense systems and make them more susceptible to various diseases (Sewelam et al. 2021; Younis et al. 2014).

6.1 Drought Stress Tolerance

Drought causes highly significant ecological impacts on agricultural output across the world. Huge efforts have been devised by various plant scientists to boost crop yield to deal with decreasing water availability. About drought RNAi approach can be applied to treat various diseases in plants (Younis et al. 2014).

Drought-resistant transgenic rice varieties were introduced into the market, by employing the RNAi approach to restrict the expression level of the RACK1 gene in those rice varieties. RACK1 gene is known to control plant growth and development, against drought stress in rice (Li et al. 2009).

The fungi, *Fusarium* are soil-borne and can cause wilting in many plant species. Drought conditions can weaken the plant's defenses, making it more susceptible to *Fusarium* infections. In tomato plants when the cloned RNA construct of the fungal ornithine decarboxylase (ODC) gene was inserted, the transgenic RNAi lines exhibited a decline in fungal growth in the plants (Bouqellah 2023; Singh et al. 2020).

6.2 Salt Stress Tolerance

Soil salinity is an important factor in maintaining agricultural productivity. RNAi technique has been used in many plants to restore productivity yield.

In another study on rice, meristem-specific rice cytokinin oxidase (OsCKX2) causing decreased yield under salt-stressed conditions was studied. Using an RNAi-based technique, it was shown that knocking down OsCKX2 expression results in increased cellular cytokinin levels, which leads to a greater number of spikelets than wild-type plants under control circumstances (Joshi et al. 2018).

A transcription factor, OsERF922 of rice has been found to negatively regulate both resistance to fungi *Magnaporthe oryzae*, which causes rice blast disease, as well as provides salt tolerance to the rice plants. Through RNAi-induced gene knock-outs, the downregulation of OsERF922 expression has been shown to enhance salt tolerance by improving plant growth, physiological characteristics, and ion homeostasis under salt stress conditions whereas the enhanced expression levels of the PR gene, PAL gene, and other genes associated with enzymes that regulate phytoalexin biosynthesis was related with enhanced disease resistance in RNAi plants (Liu et al. 2012).

6.3 Commercialization of RNAi-Based Products

Researches in RNAi-based technique for crop protection and development from biotic stresses have resulted in a few commercial RNAi-based products. In 2017, the very first commercial product "SmartStax Pro" maize approved by US regulators was introduced. The "SmartStax Pro" maize involves RNAi-mediated silencing of *Snf7* gene of *Diabrotica virgifera virgifera* that leads to reduced root damage of maize andprovides improved target pest control. The product was made available to farmers of the US and Canada in the years 2022 and 2023 respectively (De Schutter et al. 2022). Thus, only improved quality crop products were authorized for commercialization. Monsanto Company developed Vistive Gold, which are high-oleic soybeans and comprises of genes that are involved in fatty acid biosynthesis. When these genes were targeted through the RNAi approach, this resulted in reduced desaturation of oleic acid. The product has been thus approved as food, as well as for cultivation purposes across different countries between the years 2011 and 2015 (ISAAA.Inc.). Soon "Super-High Oleic" (SHO) safflower will be released, in which the developing seeds would have the essential requirements for the conversion of oleic acid. This will act as a surplus to the biofuel, chemical, and lubricant industries. "HarvXtra" alfalfa from Forrage genetics made a more digestible crop for cattle by reducing lignin content by RNAi-mediated silencing. Recently, Syngenta and Greenlight Biosciences have completed separate field studies demonstrating improved efficacy of dsRNA-treated plants in controlling the Colorado potato beetle (Bramlett et al. 2020; Rodrigues et al. 2021). Therefore for a sustainable and biotic stress-resistant agricultural sector, the ultimate goal should be to emphasize more efficient research and develop evidence-based products that are approved by the authorized committee.

7 Conclusion

In the current era, the continuous stress on plants due to biotic and abiotic factors has made them vulnerable to various pests and pathogens. These pests and pathogens exhibit a direct effect on food scarcity, the economy, and the agro-ecosystem. To mitigate this problem, RNAi has been employed as a tool for plant disease control and management. It is a natural mechanism present in plants to regulate gene expression. An elaborate understanding of RNAi pathways and their components has enabled several scientists to specifically silence those genes that are responsible for disease susceptibility in plants. By selectively targeting and degrading the messenger RNA (mRNA) molecules, which are essential for protein synthesis, RNAi disrupts the production of specific proteins involved in disease progression. This targeted gene silencing approach has proven effective against various plant pathogens, making RNAi a promising strategy for plant disease control.

RNAi has been successfully implemented to control a wide array of plant diseases caused by bacteria, nematodes, fungi, or viruses. In the case of viral diseases, small interfering RNAs (siRNAs) are designed to target viral RNA sequences, preventing viral replication and spread within the plant. RNAi-mediated resistance against bacterial pathogens involves the silencing of genes essential for the bacteria's survival or virulence. Fungal diseases can be managed by targeting specific genes involved in fungal pathogenesis, including those responsible for toxin production or cell wall degradation. Similarly, RNAi has been used to suppress nematode infestations by interfering with the essential genes required for their development or feeding. RNAi-based technique has also been successfully used to generate commercialized products like SmartStax Pro and Vistive Gold for crop protection and improvement. With further research, discoveries and knowledge will create new prospects for creating unique and broader techniques to plant disease control and enhance agricultural outputs.

References

Baum JA, Bogaert T, Clinton W, Heck GR, Feldmann P, Ilagan O, Pleau M (2007) Control of coleopteran insect pests through RNA interference. Nat Biotechnol 25(11):1322–1326

Bharathi JK, Anandan R, Benjamin LK, Muneer S, Prakash MAS (2022) Recent trends and advances of RNA interference (RNAi) to improve agricultural crops and enhance their resilience to biotic and abiotic stresses. Plant Physiol Biochem

Blyuss KB, Fatehi F, Tsygankova VA, Biliavska LO, Iutynska GO, Yemets AI, Blume YB (2019) RNAi-based biocontrol of wheat nematodes using natural poly-component biostimulants. Front Plant Sci 10:483

Bocos-Asenjo IT, Niño-Sánchez J, Ginésy M, Diez JJ (2022) New insights on the integrated management of plant diseases by RNA strategies: mycoviruses and RNA interference. Int J Mol Sci 23(16):9236

Böhnert HU, Fudal I, Dioh W, Tharreau D, Notteghem J-L, Lebrun M-H (2004) A putative polyketide synthase/peptide synthetase from Magnaporthe grisea signals pathogen attack to resistant rice. Plant Cell 16(9):2499–2513

Bouqellah NA (2023) In silico and in vitro investigation of the antifungal activity of trimetallic Cu–Zn-magnetic nanoparticles against *Fusarium oxysporum* with stimulation of the tomato plant's drought stress tolerance response. Microb Pathog 178:106060

Bramlett M, Plaetinck G, Maienfisch P (2020) RNA-based biocontrols—a new paradigm in crop protection. Engineering 6(5):522–527

Burch-Smith TM, Anderson JC, Martin GB, Dinesh-Kumar SP (2004) Applications and advantages of virus-induced gene silencing for gene function studies in plants. Plant J 39(5):734–746

Cagliari D, Dias NP, Galdeano DM, Dos Santos EÁ, Smagghe G, Zotti MJ (2019) Management of pest insects and plant diseases by non-transformative RNAi. Front Plant Sci 10:1319

Chen Q, Lai H, Hurtado J, Stahnke J, Leuzinger K, Dent M (2013) Agroinfiltration as an effective and scalable strategy of gene delivery for production of pharmaceutical proteins. Adv Tech Biol Med 1(1)

Das PR, Sherif SM (2020) Application of exogenous dsRNAs-induced RNAi in agriculture: challenges and triumphs. Front Plant Sci 11:946

Davies KM, Deroles SC, Boase MR, Hunter DA, Schwinn KE (2013) Biolistics-based gene silencing in plants using a modified particle inflow gun. Biolistic DNA Deliv: Methods and Protocols 63–74

De Backer MD, Raponi M, Arndt GM (2002) RNA-mediated gene silencing in non-pathogenic and pathogenic fungi. Curr Opin Microbiol 5(3):323–329

Demirer GS, Zhang H, Goh NS, Pinals RL, Chang R, Landry MP (2020) Carbon nanocarriers deliver siRNA to intact plant cells for efficient gene knockdown. Sci Adv 6(26):eaaz0495

Dinesh-Kumar S, Anandalakshmi R, Marathe R, Schiff M, Liu Y (2003) Virus-induced gene silencing. Plant Funct Genom 287–293

Dubey VK, Lee UG, Kwon DH, Lee SH (2017) Agroinfiltration-based expression of hairpin RNA in soybean plants for RNA interference against *Tetranychus urticae*. Pestic Biochem Physiol 142:53–58

Elbashir SM, Harborth J, Lendeckel W, Yalcin A, Weber K, Tuschl T (2001) Duplexes of 21-nucleotide RNAs mediate RNA interference in cultured mammalian cells. Nature 411(6836):494–498

Escobar MA, Civerolo EL, Summerfelt KR, Dandekar AM (2001) RNAi-mediated oncogene silencing confers resistance to crown gall tumorigenesis. Proc Natl Acad Sci 98(23):13437–13442

Fire A, Xu S, Montgomery MK, Kostas SA, Samuel ED, Craig CM (1998) Potent and specific genetic interference by double-stranded RNA in caenorhabditis elegans. Nature 391(6669):806–811

Fulton RW (1986) Practices and precautions in the use of cross protection for plant virus disease control. Annu Rev Phytopathol 24(1):67–81

Gong L, Chen Y, Hu Z, Hu M (2013) Testing insecticidal activity of novel chemically synthesized siRNA against *Plutella xylostella* under laboratory and field conditions. PLoS ONE 8(5):e62990

Guo L, Degenstein L, Dowling J, Yu Q-C, Wollmann R, Perman B, Fuchs E (1995) Gene targeting of BPAG1: abnormalities in mechanical strength and cell migration in stratified epithelia and neurologic degeneration. Cell 81(2):233–243

Halder K, Chaudhuri A, Abdin MZ, Majee M, Datta A (2022) RNA interference for improving disease resistance in plants and its relevance in this clustered regularly interspaced short palindromic repeats-dominated era in terms of dsRNA-based biopesticides. Front Plant Sci 13

Hammond SM, Bernstein E, Beach D, Hannon GJ (2000) An RNA-directed nuclease mediates post-transcriptional gene silencing in Drosophila cells. Nature 404(6775):293–296

He D-C, Zhan J-S, Xie L-H (2016) Problems, challenges and future of plant disease management: from an ecological point of view. J Integr Agric 15(4):705–715

Head GP, Carroll MW, Evans SP, Rule DM, Willse AR, Clark TL, Meinke LJ (2017) Evaluation of SmartStax and SmartStax PRO maize against western corn rootworm and northern corn rootworm: efficacy and resistance management. Pest Manag Sci 73(9):1883–1899

Hoang BTL, Fletcher SJ, Brosnan CA, Ghodke AB, Manzie N, Mitter N (2022) RNAi as a foliar spray: efficiency and challenges to field applications. Int J Mol Sci 23(12):6639

https://www.potatogrower.com/2017/06/shades-of-gray#grey%20mold%20fig%202.jpg. Accessed on 23 July 2023

https://www.isaaa.org/gmapprovaldatabase/event/default.asp?EventID=177. Accessed on 11 July 2023

Jia R, Zhao H, Huang J, Kong H, Zhang Y, Guo J, Zuo J (2017) Use of RNAi technology to develop a PRSV-resistant transgenic papaya. Sci Rep 7(1):12636

Jiang C-J, Shimono M, Maeda S, Inoue H, Mori M, Hasegawa M, Takatsuji H (2009) Suppression of the rice fatty-acid desaturase gene OsSSI2 enhances resistance to blast and leaf blight diseases in rice. Mol Plant Microbe Interact 22(7):820–829

Johnson J (1937) An acquired partial immunity to the tobacco streak disease. Trans Wisconsin Acad Sci Arts Lett 30:27

Jones T-KL, Bernal JS, Medina RF (2021) Assessing the Functionality of RNA interference (RNAi) in the phloem-feeding maize pest *Dalbulus maidis*. bioRxiv 2021.2009. 2029.462424

Joshi R, Sahoo KK, Tripathi AK, Kumar R, Gupta BK, Pareek A, Singla-Pareek SL (2018) Knockdown of an inflorescence meristem-specific cytokinin oxidase–OsCKX2 in rice reduces yield penalty under salinity stress condition. Plant, Cell Environ 41(5):936–946

Joshi I, Kumar A, Singh AK, Kohli D, Raman K, Sirohi A, Jain PK (2019) Development of nematode resistance in Arabidopsis by HD-RNAi-mediated silencing of the effector gene Mi-msp2. Sci Rep 9(1):17404

Klahre U, Crété P, Leuenberger SA, Iglesias VA, Meins F Jr (2002) High molecular weight RNAs and small interfering RNAs induce systemic posttranscriptional gene silencing in plants. Proc Natl Acad Sci 99(18):11981–11986

Koch A, Kumar N, Weber L, Keller H, Imani J, Kogel K-H (2013) Host-induced gene silencing of cytochrome P450 lanosterol C14α-demethylase–encoding genes confers strong resistance to Fusarium species. Proc Natl Acad Sci 110(48):19324–19329

Kuo Y-W, Falk BW (2020) RNA interference approaches for plant disease control. Biotechniques 69(6):469–477

Lam JK, Chow MY, Zhang Y, Leung SW (2015) SiRNA versus miRNA as therapeutics for gene silencing. Mol Therapy-Nucl Acids 4:e252

Leuzinger K, Dent M, Hurtado J, Stahnke J, Lai H, Zhou X, Chen Q (2013) Efficient agroinfiltration of plants for high-level transient expression of recombinant proteins. J Visualized Exp JoVE (77)

Li D-H, Hui L, Yang YL, Zhen P, Liang J (2009) Down-regulated expression of RACK1 gene by RNA interference enhances drought tolerance in rice. Rice Sci 16(1):14–20

Liu D, Chen X, Liu J, Ye J, Guo Z (2012) The rice ERF transcription factor OsERF922 negatively regulates resistance to *Magnaporthe oryzae* and salt tolerance. J Exp Bot 63(10):3899–3911

Lu R, Martin-Hernandez AM, Peart JR, Malcuit I, Baulcombe DC (2003) Virus-induced gene silencing in plants. Methods 30(4):296–303

Mao Y-B, Cai W-J, Wang J-W, Hong G-J, Tao X-Y, Wang L-J, Chen X-Y (2007) Silencing a cotton bollworm P450 monooxygenase gene by plant-mediated RNAi impairs larval tolerance of gossypol. Nat Biotechnol 25(11):1307–1313

Mat Jalaluddin NS, Asem M, Harikrishna JA, Ahmad Fuaad AAH (2023) Recent progress on nanocarriers for topical-mediated RNAi strategies for crop protection—a review. Molecules 28(6):2700

Matranga C, Zamore PD (2007) Small silencing RNAs. Curr Biol 17(18):R789–R793

Meister G, Tuschl T (2004) Mechanisms of gene silencing by double-stranded RNA. Nature 431(7006):343–349

Missiou A, Kalantidis K, Boutla A, Tzortzakaki S, Tabler M, Tsagris M (2004) Generation of transgenic potato plants highly resistant to potato virus Y (PVY) through RNA silencing. Mol Breeding 14:185–197

Mitter N, Worrall EA, Robinson KE, Li P, Jain RG, Taochy C, Xu ZP (2017) Clay nanosheets for topical delivery of RNAi for sustained protection against plant viruses. Nat Plants 3(2):1–10

Mroz EA, Rocco JW (2006) RNA interference: natural, experimental, and clinical roles in cancer biology. Head Neck 28(12):1132–1141

Napoli C, Lemieux C, Jorgensen R (1990) Introduction of a chimeric chalcone synthase gene into petunia results in reversible co-suppression of homologous genes in trans. Plant Cell 2(4):279–289

Obbard DJ, Gordon KH, Buck AH, Jiggins FM (2009) The evolution of RNAi as a defence against viruses and transposable elements. Philos Trans Roy Soc B: Biol Sci 364(1513):99–115

Palukaitis P, Zaitlin M (1984) A model to explain the" cross-protection" phenomenon shown by plant viruses and viroids. Plant-Microbe Interact: Mol Genet Perspect (USA)

Patil BL, Ogwok E, Wagaba H, Mohammed IU, Yadav JS, Bagewadi B, Alicai T (2011) RNAi-mediated resistance to diverse isolates belonging to two virus species involved in Cassava brown streak disease. Mol Plant Pathol 12(1):31–41

Peele C, Jordan CV, Muangsan N, Turnage M, Egelkrout E, Eagle P, Robertson D (2001) Silencing of a meristematic gene using geminivirus-derived vectors. Plant J 27(4):357–366

Pugsley CE, Isaac R, Warren NJ, Cayre OJ (2021) Recent advances in engineered nanoparticles for RNAi-mediated crop protection against insect pests. Front Agron 3:652981

Rajput M, Choudhary K, Kumar M, Vivekanand V, Chawade A, Ortiz R, Pareek N (2021) RNA interference and CRISPR/Cas gene editing for crop improvement: paradigm shift towards sustainable agriculture. Plants 10(9):1914

Ratcliff FG, MacFarlane SA, Baulcombe DC (1999) Gene silencing without DNA: RNA-mediated cross-protection between viruses. Plant Cell 11(7):1207–1215

Rivera AL, Gómez-Lim M, Fernández F, Loske AM (2012) Physical methods for genetic plant transformation. Phys Life Rev 9(3):308–345

Rodrigues T, Sridharan K, Manley B, Cunningham D, Narva K (2021) Development of dsRNA as a sustainable bioinsecticide: from laboratory to field crop protection products for sustainable agriculture. ACS Publications, pp 65–82

Romano N, Macino G (1992) Quelling: transient inactivation of gene expression in *Neurospora crassa* by transformation with homologous sequences. Mol Microbiol 6(22):3343–3353

Rosa C, Kuo Y-W, Wuriyanghan H, Falk BW (2018) RNA interference mechanisms and applications in plant pathology. Annu Rev Phytopathol 56:581–610

De Schutter K, Taning CNT, Van Daele L, Van Damme EJ, Dubruel P, Smagghe G (2022) RNAi-based biocontrol products: market status, regulatory aspects, and risk assessment. Front Insect Sci 1

Sewelam N, El-Shetehy M, Mauch F, Maurino VG (2021) Combined abiotic stresses repress defense and cell wall metabolic genes and render plants more susceptible to pathogen infection. Plants 10(9):1946

Sharma V, Sanghera GS, Kashyap PL, Sharma BB, Chandel C (2013) RNA interference: a novel tool for plant disease management. Afr J Biotechnol 12(18)

Sharma S, Singh R, Kumar A (2019) Screening of cluster bean genotypes against Alternaria blight (*Alternaria cucumerina* var. cyamopsidis) in field condition. J Pharmacognosy Phytochem 8(3):596–599

Sindhu AS, Maier TR, Mitchum MG, Hussey RS, Davis EL, Baum TJ (2009) Effective and specific in planta RNAi in cyst nematodes: expression interference of four parasitism genes reduces parasitic success. J Exp Bot 60(1):315–324

Singh N, Mukherjee SK, Rajam MV (2020) Silencing of the ornithine decarboxylase gene of *Fusarium oxysporum* f. sp. *lycopersici* by host-induced RNAi confers resistance to *Fusarium wilt* in tomato. Plant Mol Biol Report 38:419–429

Singha K, Namgung R, Kim WJ (2011) Polymers in small-interfering RNA delivery. Nucleic Acid Ther 21(3):133–147

Svoboda P (2020) Key mechanistic principles and considerations concerning RNA interference. Front Plant Sci 11:1237

Thakur O, Prasad R (2020) Engineering resistance to *Alternaria cyamopsidis* by RNAi mediated gene silencing of chitin synthase export chaperone CHS7 in guar. Physiol Mol Plant Pathol 112:101541

Tomari Y, Zamore PD (2005) Perspective: machines for RNAi. Genes Dev 19(5):517–529

Tricoll DM, Carney KJ, Russell PF, McMaster JR, Groff DW, Hadden KC, Quemada HD (1995) Field evaluation of transgenic squash containing single or multiple virus coat protein gene constructs for resistance to cucumber mosaic virus, watermelon mosaic virus 2, and zucchini yellow mosaic virus. Bio/technology 13(12):1458–1465

Tyagi H, Rajasubramaniam S, Rajam MV, Dasgupta I (2008) RNA-interference in rice against Rice tungro bacilliform virus results in its decreased accumulation in inoculated rice plants. Transgenic Res 17:897–904

Ul Haq I, Ijaz S (2020) History and recent trends in plant disease control: an overview. Plant disease management strategies for sustainable agriculture through traditional and modern approaches, pp 1–13

Unver T, Budak H (2009) Virus-induced gene silencing, a post transcriptional gene silencing method. Int J Plant Genom

Vaghchhipawala Z, Rojas CM, Senthil-Kumar M, Mysore KS (2011) Agroinoculation and agroinfiltration: simple tools for complex gene function analyses. Plant Reverse Genet Methods Protocols 65–76

Van Vu T, Choudhury NR, Mukherjee SK (2013) Transgenic tomato plants expressing artificial microRNAs for silencing the pre-coat and coat proteins of a begomovirus, Tomato leaf curl New Delhi virus, show tolerance to virus infection. Virus Res 172(1–2):35–45

Voinnet O (2001) RNA silencing as a plant immune system against viruses. Trends Genet 17(8):449–459

Walawage SL, Britton MT, Leslie CA, Uratsu SL, Li Y, Dandekar AM (2013) Stacking resistance to crown gall and nematodes in walnut rootstocks. BMC Genomics 14(1):1–13

Wani SH, Sanghera GS, Singh NB (2010) Biotechnology and plant disease control-role of RNA interference. Am J Plant Sci 1(02):55

Wingard SA (1928) Hosts and symptoms of ring spot, a virus disease of plants

Worrall EA, Hamid A, Mody KT, Mitter N, Pappu HR (2018) Nanotechnology for plant disease management. Agronomy 8(12):285

Wroblewski T, Tomczak A, Michelmore R (2005) Optimization of Agrobacterium-mediated transient assays of gene expression in lettuce, tomato and Arabidopsis. Plant Biotechnol J 3(2):259–273

Xiong F, Liu M, Zhuo F, Yin H, Deng K, Feng S, Zhang S (2019) Host-induced gene silencing of BcTOR in *Botrytis cinerea* enhances plant resistance to grey mould. Mol Plant Pathol 20(12):1722–1739

Younis A, Siddique MI, Kim C-K, Lim K-B (2014) RNA interference (RNAi) induced gene silencing: a promising approach of hi-tech plant breeding. Int J Biol Sci 10(10):1150

Zhang T, Jin Y, Zhao J-H, Gao F, Zhou B-J, Fang Y-Y, Guo H-S (2016) Host-induced gene silencing of the target gene in fungal cells confers effective resistance to the cotton wilt disease pathogen *Verticillium dahliae*. Mol Plant 9(6):939–942

Zhang H, Cao Y, Xu D, Goh NS, Demirer GS, Cestellos-Blanco S, Yang P (2021) Gold-nanocluster-mediated delivery of siRNA to intact plant cells for efficient gene knockdown. Nano Lett 21(13):5859–5866

Zhang Y, Zhao Q, Zhang J, Niu L, Yang J, Liu X, Yang X (2022) Enhanced resistance to soybean cyst nematode in transgenic soybean via host-induced silencing of vital *Heterodera glycines* genes. Transgenic Res 31(2):239–248

RNA Interference for Plant Disease Management: Updated Methods, Current Applications and Future Directions

Nivedita Singh, Tarushi Attri, Rajina, Renu Thakur, and Monica Sharma

Abstract Plant diseases cause significant losses in crops and deteriorate the quantity and quality of the crop produced. Although there are traditional methods such as synthetic pesticides, cultural management, resistance varieties, etc. for managing plant diseases, these methods have their limitations. In contrast to synthetic pesticides, which have a detrimental effect on human and environmental health and promote pathogen resistance, RNA-based approaches to plant disease control provide an alluring alternative. Double-stranded RNA (dsRNA) activates RNA interference (RNAi), a conserved process that silences certain genes. Small RNAs with a sequence of 21–30 nucleotides are generated during RNAi, and they influence the expression of genes in a sequence-specific way. The natural occurrence of cross-kingdom RNAi provides a basis for host-induced gene silencing. RNAi protects the host plants against diseases by silencing the pathogen's target genes. This is an effective disease management technique. The use of RNA interference against numerous plant diseases is updated in the current book chapter.

Keywords Plant diseases · dsRNA · Genes · RNAi · Disease management

N. Singh · T. Attri · Rajina · R. Thakur · M. Sharma (✉)
Department of Plant Pathology, College of Horticulture and Forestry, Dr Y S Parmar University of Horticulture and Forestry (Solan), Neri, Hamirpur, HP 177001, India
e-mail: dmonicasharma@gmail.com

N. Singh
e-mail: nivedita716@gmail.com

T. Attri
e-mail: attritarushi947@gmail.com

Rajina
e-mail: rajinakhan1694@gmail.com

R. Thakur
e-mail: renu6660@gmail.com

J. Chen et al. (eds.), *Molecular and Biotechnological Tools for Plant Disease Management*, Interdisciplinary Biotechnological Advances,
https://doi.org/10.1007/978-981-97-7510-1_13

The ever-increasing population of the earth is expected to reach 9.8 billion by 2050. To meet the projected net demand for one billion tons of grain by 2050, there needs to increase of 60–70% in the global food supply. Constantly changing climatic conditions make this goal even harder to achieve owing to the various abiotic and biotic stresses. These stresses have varying effects on plant growth, development, and production, making it challenging to stop yield losses. Pathogenic fungi, bacteria, nematodes, and viruses are the worst enemies among biotic stresses, as they are capable of destroying the whole crop. Researchers have been looking into ways to protect crops from these organisms, but in today's intensive agricultural systems, synthetic pesticides are a must to ensure the world's food supply. They shield crops from diseases that would otherwise result in yield losses of over 30% (Gebremichael et al. 2021; Zewdu et al. 2023). Utilizing synthetic pesticides has been the conventional strategy for most parts of disease management. Considering their harmful effects on the environment, beneficial microorganisms, and human health, there is a growing desire for competent, cost-effective, and environmentally friendly alternative methods to control pests and diseases (Nicolopoulou et al. 2016). Genetic resistance in plants is ideally the most desirable and eco-friendly approach, but too often the ability to use it for plant protection is lacking. Even in the presence of valuable genetic resistance sources, the conventional process of plant breeding is time-consuming. Additionally, plants that are resistant to known pathogens may not be genetically compatible with those in need of protection. Despite the implementation of successful genetic resistance, the evolutionary process might enable pathogens to adapt and exploit the new environment, overcoming the resistance. Given the widespread practice of cultivating genetically uniform crops in large populations, the rapid spread of virulent pathogens and severe disease outbreaks can occur under favorable environmental conditions. Consequently, the continuing struggle to control plant diseases necessitates the development of novel, efficient, and environmentally friendly approaches (Kuo and Falk 2020).

The discovery of molecular biology gave scientists a new edge in research and in developing novel ideas to replace conventional methods of crop improvement. One of these technologies is RNA interference, a process with great potential for crop improvement. This technique is an alternative to conventional methods of breeding to meet the food requirements of today's population, which is increasing at an alarming rate. It is also called pathogen-derived resistance, as the resistance results from the pathogen itself (Ratcliff et al. 1999). RNA interference (RNAi) is an ideal approach as it acts at the required genetic level and has the prospective to work well with the current challenges in plant security (Hernández and Chacón 2021). In violet petunias, the introduction of a transgene-induced over-expression of chalcone synthase gene (CHS). This was the first report of RNA interference. The CHS gene is required for flavonoid biosynthesis which encodes Chalcone synthase enzyme and its over-expression resulted in an unexpected phenotype of white petunia. Further studies showed that exogenously introduced CHS genes simultaneously repressed endogenous CHS genes, reducing the expression of both endogenous and exogenous CHS genes (Napoli et al. 1990). A comparable occurrence, termed quelling or silencing in fungi, was observed in the filamentous fungus *Neurospora crassa*, which involved the introduction of the 'albino-1' transgene into the fungus, leading to the suppression of expression in the endogenous gene. In *Caenorhabditis elegans*, silencing of the unc-22 gene because of the injection of dsRNAs was reported (Fire et al. 1998). All through the RNA interference, dsRNAs are transformed into shorter small interfering RNAs (siRNAs) and bind to their homologous messenger RNAs (mRNAs) (Zamore et al. 2000; Hammond et al. 2001). Moreover, siRNAs also guide and instruct the multi-component RNA-induced silencing complexes (RISCs) to cut the specific targeted RNAs (Hammond et al. 2000). The RNase III family enzymes are involved in the cleavage of dsRNAs and produce putative guide RNAs (Bernstein et al. 2001). In viral infections, it was observed that after infection, the plants were resistant to any secondary infection from that virus or any other highly homologous strain.

The management of plant diseases through the RNAi approach involves the utilization of small double-stranded RNA (dsRNA) molecules. These molecules are employed to induce the silencing of key pathogenicity genes in causal organisms which consequently restricts or inhibits the growth of pathogens, contributing to disease control (Muhammad et al. 2019). RNA interference or post-transcription gene silencing, is a conserved process in eukaryotic organisms where complementary small RNA (sRNA) can target a gene and suppress its expression through sequence-specific degradation. The initial step in the RNAi pathway is the formation of double-stranded RNA (dsRNA). Double-stranded RNAs are recognized and cleaved by Dicer ribonucleases along with siRNA duplexes formation (Bernstein et al. 2001; Fire et al. 1998). In the process, the first one of the strands of double-stranded RNA is loaded into an ARGONAUTE (AGO) protein, which leads to the formation of RISC (RNA-induced silencing complex), which, after recognition, silences the complementary target RNA by a variety of mechanisms (Hammond et al. 2000, 2001). RNAi is found to occur naturally in plants as a cellular defense mechanism, and different pathogens also induce RNAi in plants. It is interesting to note that despite possessing such a robust defense mechanism, the plants are regularly infected by pathogens. The reason lies in the process of evolution, as it plays a key role in plant-pathogen interaction and works both ways. Plant pathogens overcome this defense by encoding for proteins known as 'silencing suppressors' (Navarro et al. 2008; Qiao et al. 2013). These suppressors are the proteins that interfere with the RNAi at different steps of the pathway by making available small-interfering RNA molecules and thus safely preventing the degradation of double-stranded RNAs (dsRNA) or preventing their direct interaction with the RNA interference enzymes of the host.

Since its discovery, RNA interference (RNAi) has been harnessed across diverse methods to manage plant diseases effectively. The focus of these strategies was predominantly on amplifying the expression of double-stranded RNA (dsRNA), utilizing transgenes equipped with robust constitutive promoters to achieve extensive gene silencing and generate abundant populations of transgene-derived small RNAs (sRNAs). In spite of being widely used, these methods need specificity, as there is a common occurrence of inadvertently targeting cellular transcripts that share high-sequence complementarity with certain small RNAs derived from transgenes (López et al. 2020). To overcome the limited specificity, 'second-generation RNAi' strategies based on artificial sRNAs (art-sRNAs) have come to light as a promising technique for plant protection against various diseases (Carbonell and Daròs 2017). Furthermore, traditional RNAi has focused more on the development of resistant transgenic plants, which has garnered significant attention and also raised public concern over the genetic transformation of crops. The issue of the transformation of crops has been resolved with the introduction of techniques like virus-induced gene silencing (VIGS) and spray-induced gene silencing (SIGS). All these techniques, including host-induced gene silencing (HIGS), are discussed in this chapter. It summarizes the current RNAi techniques, their applications, and lastly, the challenges and future perspectives surrounding the production, delivery systems, and risk assessment of RNAi for crop protection.

1 Mechanism of RNA Interference

Post-transcriptional gene silencing is a biological mechanism in eukaryotes where short, noncoding RNAs are used for gene expression control and defense (Hernández and Chacón 2021). A dogmatic mechanism that identifies mRNA and suppresses its gene expression is known as Post-transcriptional gene silencing (PTGS). This process is initiated by long double-stranded RNA (dsRNA) or hairpin-structured RNAs in a sequence-specific manner. Dicer is an RNase III-like enzyme that breaks the dsRNA, generating approximately twenty-one nucleotide RNAs known as small interfering RNAs (siRNAs) (Fig. 1). These siRNAs are double-stranded and undergo unwinding to produce two ssRNAs. The guiding strand then integrates into the RNA-induced silencing complex (RISC). Subsequently, the complementary strand, known as the antisense strand, aligns with the mRNA transcript, guiding the RNA-induced silencing complex (RISC) to cleave the target mRNA. This cleavage is facilitated by the endonucleolytic activity of the Argonaute protein (AGO), leading to the disintegration of the target mRNA (Senthil and Mysore 2010; Nunes and Dean 2012).

Non-coding RNAs, typically ranging from 20 to 500 nucleotides in length, are commonly referred to as small regulatory RNAs as they play a crucial role in regulating transcription and translation processes in organisms (Watkins and Arya 2019). These originate from long dsRNA precursors or the RNA UTR region in the cell. This gene regulation is carried out by restricting specific genes at the transcriptional or translational level (Soltani et al. 2021). sRNAs are associated with growth, differentiation, death, and defense of cells in eukaryotes (Barathi et al. 2023) and the various stress responses in prokaryotes and eukaryotes (Li et al. 2020; Lu et al. 2021; Tang et al. 2021). Various classes of small RNA (sRNAs) molecules include short interfering RNAs (siRNAs), microRNAs (miRNAs), tRNA-derived RNA fragments (tRFs), and Piwi-interacting small RNAs (piRNAs). At present, small interference RNAs and micro RNAs are the most thoroughly understood and frequently utilized for translational applications and will be discussed further.

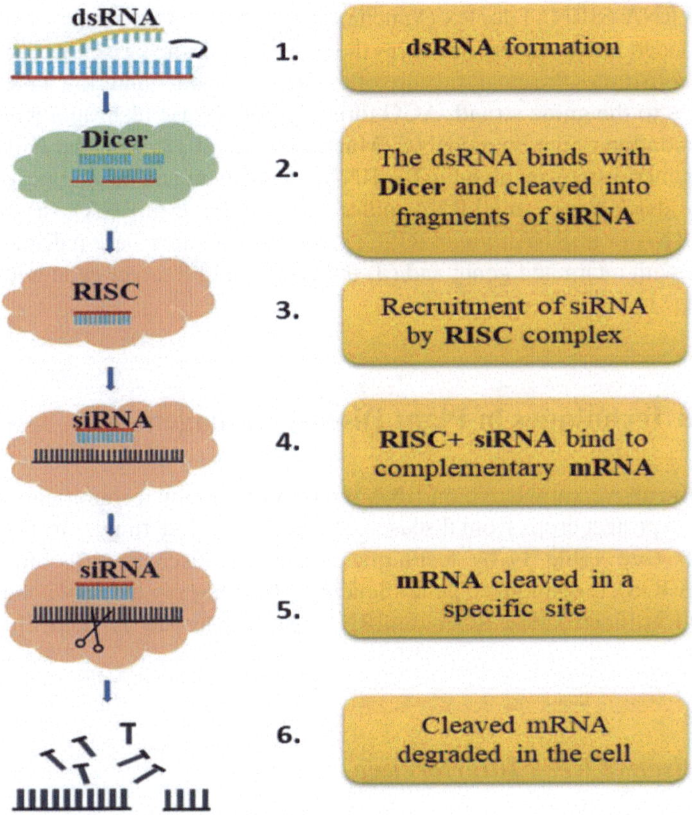

Fig. 1 Major steps in the RNAi mechanism

Micro RNAs are of small (20–22 nucleotide sequences), generated from miRNA genes (MIRs); here, the ssRNA molecule folds back on itself to create a double-stranded region with a loop called the RNA hairpin (hpRNAs). They mostly function in the regulation of gene expression. Whereas siRNAs are 20 to 24 nt sequences that are generated from dsRNAs synthesized by host endogenous RNA-dependent RNA polymerase (RdRP), RNA virus replication, and ss RNAs that form hairpin-loop secondary structures. Plants also have an RNA surveillance system that recognizes and degrades specific RNAs, resulting in the generation of siRNAs (Kuo and Falk 2020). The host's DICER binds with the long dsRNA or hpRNA and cleaves them into 21–25 nt long siRNA (Borges and Martienssen 2015; D'Ario et al. 2017). A DICER enzyme has an N-terminus and a C-terminus. The N-terminus consists of (i) a helicase domain, (ii) a Piwi/Argonaute/Zwille (PAZ) motif, (iii) a dsRNA binding domain, and (iv)the C-terminus has two RNase III motifs. These newly formed siRNAs are then incorporated with the RISC. This multicomponent protein complex contains the Argonaute protein (AGO) with a small RNA-binding domain and endo-nuclease activity for RNA cleavage, caused by the ATP-dependent unwinding of the small interfering RNA (siRNA) duplex (Vaucheret et al. 2004). One strand of siRNA (the guide) is loaded into the RISC, whereas the other strand (the passenger) is degraded. The AGO constitutes the catalytic core of the RISC. On encountering an RNA that is homologous to the guide strand, AGO directs ribonuclease cleavage, which results in the degradation of the bound RNA (Machado et al. 2018; Kuo and Falk 2020).

So the general steps involved in the RNAi mechanism can be summarized as: (a) cleavage of dsRNA into small RNA molecules (siRNA or miRNA); (b) the loading of these siRNA or miRNA into an RISC; (c) complementary base pairing of siRNA or miRNA with a targeted gene; and (d) the cleavage of a target gene with an AGO protein (Fig. 1).

2 RNAi Techniques in Plant Disease Management

Based on the above knowledge on RNA interference, many technologies have been developed to protect crops from disease-causing entities like fungi, viruses, bacteria, and viroids (see Table 1) by techniques such as antisense, virus-induced gene silencing (VIGS), host-induced gene silencing (HIGS), spray induced gene silencing (SIGS) and artificial small RNA (amiRNA and syn-tasi-RNA) and illustrated in Fig. 2.

2.1 Antisense RNA Interference

Antisense RNA interference (RNAi) is a molecular biology technique used to hinder the expression of specific genes by introducing complementary RNA molecules, known as antisense RNA or antisense oligonucleotides, to target the mRNA. The antisense RNA binds to the target messenger RNA (mRNA) through base pairing, preventing the mRNA from being translated into protein or promoting its degradation. This interference with gene expression offers a way to study gene function, and in therapeutic applications, it can be used to selectively inhibit the expression of disease-related genes (Krishna and Mishra 2017).

The following steps are required for the working of Antisense RNAi:

1. Design of Antisense Molecules: Antisense RNA or antisense oligonucleotides are designed to be complementary to a specific region of the target mRNA. The goal is to create a double-stranded RNA structure by base pairing with the target mRNA.
2. Delivery of Antisense Molecules: Antisense RNA molecules can be introduced into cells or organisms through various methods, such as direct injection, transfection, or viral vectors. In therapeutic applications, delivery methods need to be carefully considered for effectiveness and safety.

Table 1 A list of RNAi techniques for plant disease management in fungi, viruses, bacteria, and viroids

S. no	Method	Host	Pathogen	Target/coordinates	Effect/result	Reference
Fungi						
1	VIGS	*Phaseolus vulgaris*	*Uromyces appendiculatus*	Four candidate effector mRNAs	Resistant plants	Cooper and Campbell (2017)
2	VIGS	*Solanum lycopersici*	*Stemphylium lycopersici*	*NBS-LRR-SLNLC1*	Verify gene function	Cui et al. (2018)
3	VIGS	*Triticum aestivum*	*P. striiformis* f. sp. *tritici*	*Pst GSRE1*	Reduction in sporulation and the fungi biomass	Qi et al. (2019)
4	SIGS	*Glycine max*	*Phakopsora pachyrhizi*	CHS Involved in chitin synthesis	Reduction in fungal growth and the number of urediniospore	Saito et al. (2022)
5	SIGS	*Malus domestica*	*Venturia inaequalis*	*SsGAP1*, *SsRAS1* and *SsRAS2*	Reduction of growth and conidial spores formation	Bhagta et al. (2023)
6	SIGS	*Lactuca sativa*	*Botrytis cinerea*	*BcBmp1* and *BcPls1*	BcBmp1 exhibited a delay in the germination of conidia, BcPls1 showed noticeable stunted growth, and there was an evident reduction in the necrotic area	Spada et al. (2023)
7	SIGS	*Pinus radiata*	*Dothistroma septosporum*	DsAflR	Reduced virulence	Mosan et al. (2023)

(continued)

Table 1 (continued)

S. no	Method	Host	Pathogen	Target/coordinates	Effect/result	Reference
8	HIGS	*Allium cepa*	*Fusarium oxysporum* f. sp. *cepae (Foc)*	*Foc*-sRNAs (Fx-sR3, Fx-sR4 and Fx-sR9)	Increased resistance	Mahanty et al. (2023)
9	HIGS	*Zea mays*	*Aspergillus flavus*	Polygalacturonase (p2c)	Reduced aflatoxin production	Raruang et al. (2023)
10	VIGS	*Triticum aestivum*	*Puccinia striiformis* f. sp. *tritici* (Pst)	*TaPR1-7* (host gene)	Resistance gene	Liu et al. (2023)
11	SIGS	*Arabidopsis thaliana*	*B. cinerea, S. sclerotioru, R. solani, Aspergillus niger, Verticillium dahliae*	–	Less infection when dsRNA was taken up by fungal species	Qiao et al. (2021)
12	SIGS	*Hevea brasiliensis*	*Erysiphe quercicola*	β-tubulin, CYP51 and CHS	Reduced expression and disease severity	Cao et al. (2024)
13	HIGS	*Nicotiana benthamiana* & *Arabidopsis thaliana*	*Sclerotinia sclerotiorum*	sGAP1, SsRAS1 and SsRAS2	Reduced virulence	Xu et al. (2024)
14	HIGS & VIGS	*Triticum aestivum*	*F. culmorum*	Secreted lipase (Fg. 11), MAP kinase (Fmk1), Beta 1,3-Glucan synthase (Gls1)	Decreased virulence	Chen et al. (2016)
15	HIGS	*Oryza sativa*	*Magnaporthe oryzae*	CRZ1,MAGB, CYP51A, CYP51B	Resistance in transgenic rice	Wang and Dean (2022)

(continued)

Table 1 (continued)

S. no	Method	Host	Pathogen	Target/coordinates	Effect/result	Reference
Viruses						
1	Art-siRNA	*Nicotiana benthamiana*	CBSV, UCBSV	*P1* and *NIb* CBSV & *P1* and *CP* UCBSV	Resistant plants	Wagaba et al. (2016)
2	Art-siRNA	*N. benthamiana*	CGMMV	*1; AthMIR156*	Reduction in replication and disease	Liang et al. (2019)
3	SIGS (NPs)	*N. tabacum*	TMV	CP & RdRP	Reduction in infected seed and incidence	Xu et al. (2023)
4	HIGS-SIGS	*S. lyopersicum*	GBNV	miRNA	—	Nivedha et al. (2023)
5	Art-siRNA	*Vitis vinifera*	GFLV	CP	Resistant Plants	Qu et al. (2007)
6	Art-siRNA	*Arabidopsis thaliana*	TuMV	CP + HC-Pro	High protection	Lafforgue et al. (2013)
7	Art-siRNA	*N. benthamiana*	TBSV	50 terminus TBSV(+) RNA	80% of plants were symptom-free	Gago et al. (2019)
8	SIGS	*N. benthamiana*	TMV	Replication gene	Virus resistance	Niehl et al. (2018)
9	Art-siRNA	*S. lyopersicum, N. benthamiana*	TYLCV	AV1	Reduced incidence and severity	Liu et al. (2020)
10	HIGS	*Citrus sinensis*	CPsV	ihpCP	Resistant transgenic plants	Francesco et al. (2020)
11	HIGS	*Prunus* spp.	PPV	ihpCP	Resistant transgenic plants	Montes et al. (2014)
12	SIGS	*N. benthamiana, Vigna unguiculata*	BCMV	CP & NIb	Protection from aphid-mediated virus transmission	Worrall et al. (2019)

(continued)

Table 1 (continued)

S. no	Method	Host	Pathogen	Target/coordinates	Effect/result	Reference
Viroid						
1	HIGS	*N. benthamiana*	*PSTVd*	Virulence gene	Resistant Plants	Adkar-Purushothama et al. (2015)
2	SIGS	*S. lycopersicum*	*PSTVd*	Virulence Specific gene	Resistant Plants	Carbonell et al. (2008)
3	Art-siRNA	*S. lycopersicum*	PSTVd-Nb	1–359 (+), all	Reduced accumulation	Schwind et al. (2009)
4	Art-siRNA	*N. benthamiana*	PSTVd-RG1	130–150 (+), V	Accumulation delay	Carbonell and Daròs (2017)
5	SIGS	*S. lycopersicum*	CChMVd	Viroid specific gene	Reduced virulence	Carbonell et al. (2008)
6	SIGS	*Citrus* spp.	CEVd	Viroid specific gene	Reduced infection	Carbonell et al. (2008)
Bacteria						
1	HIGS	*Citrus* spp.	*Xanthomona citri* subsp. *citri*	*PDS* and *CalS1*	Resistant Plants	Enrique et al. (2011)
2	HIGS	*A. thaliana*	*Agrobacterium tumefaciens*	*ipt* and *iaaM*	Resistant Plants	Dunoyer et al. (2006)
3	HIGS	*Juglans regia*	*A. tumefaciens*	*ipt* and *iaaM*	Resistant Plants	Walawage et al. (2013)

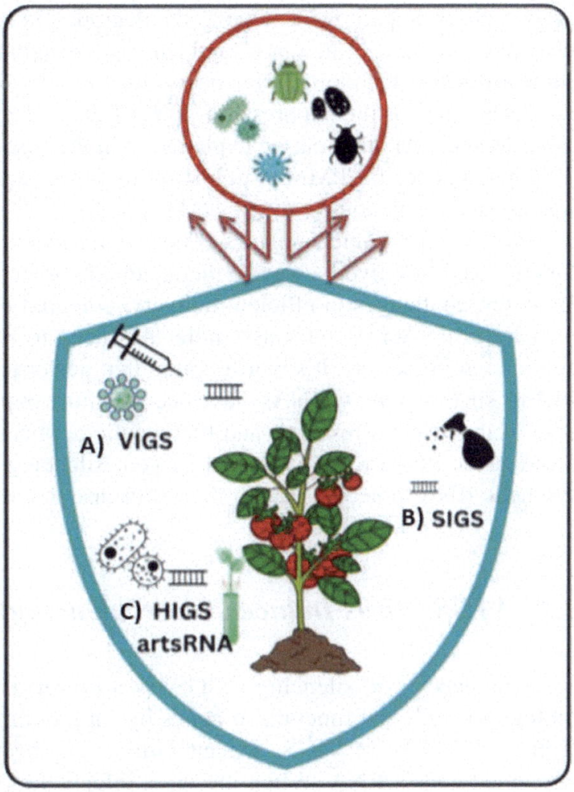

Fig. 2 RNAi techniques for plant protection

3. Binding to Target mRNA: The introduced antisense RNA binds to the complementary region of the target mRNA, forming a double-stranded RNA structure. This binding can occur in the nucleus or the cytoplasm, depending on the cellular location of the target mRNA.
4. Inhibition of Gene Expression: The binding of antisense RNA to the target mRNA interferes with the normal cellular processes involved in gene expression. This interference can lead to the degradation of the mRNA or the inhibition of translation, preventing the synthesis of the corresponding protein.
5. Downstream Effects: The reduced expression of the target gene can result in specific phenotypic changes in the cells or organisms. In a research context, this allows researchers to examine the function of the targeted gene. In therapeutic applications, it provides a means to selectively inhibit the expression of disease-associated genes.

Applications: This technique was first used by **Calgene Company** to develop tomato having a qualitative trait which is long shelf life. This trait was achieved by silencing the effect of a gene that codes for polygalacturonase. This is a type of pectinase that degrades the pectin cell wall which provides firmness and structural defence against different pathogens. Due to the degradation of pectin, the tomatoes become susceptible to fungal infections and cause its softening. A beautiful example of the antisense RNAi technology has been illustrated by Wang et al. (2020). They used this technique for the suppression of *TaPUB1* and rendered the wheat plants with salt tolerance. Another classy explanation in the case of viruses for the suppression of targeted genes for PMMoV replicase in pepper against Pepper mild mottle virus was given by Dalakouras and coworkers in 2020.

Antisense RNA interference has been extensively used in research to study gene function and has given a way to therapeutic approach for treating various diseases. However, challenges in efficient delivery, potential off-target effects, and the need for careful design of antisense molecules need to be addressed for its successful application in therapy. It's worth noting that newer RNA interference technologies, such as siRNAs) and shRNAs, have become more popular due to their efficiency and ease of use in both research and therapeutic applications. These technologies also leverage the RNAi pathway to achieve gene silencing but involve the introduction of synthetic RNA molecules rather than antisense RNA.

2.2 VIGS (Virus Induced Gene Silencing)

Virus-induced gene silencing (VIGS) is a powerful tool used in plant molecular biology to study the function of genes by suppressing their expression temporarily (Liu et al. 2023). In VIGS, a plant virus is engineered to transmit a fragment of the aimed gene's RNA. When the virus infects the plant, this RNA fragment triggers the natural defense mechanisms of the plant, resulting in degradation of the introduced viral RNA and the endogenous mRNA corresponding to the target gene. Consequently, the expression of the target gene is suppressed, allowing researchers to note the effects of gene silencing on the plant. Here are the key steps involved in virus-induced gene silencing:

1. Selection of Target Gene: Researchers choose a gene of interest to study and understand its function or silence it.
2. Construction of VIGS Vector: A viral vector, typically a plant virus, is tailored to transmit a fragmented part of the target gene's RNA. The viral vector is usually engineered to be non-pathogenic and to infect the plant efficiently.
3. Introduction of VIGS Vector into Plants: The modified viral vector is introduced into the plant by various methods, such as agroinfiltration or biolistic delivery.
4. Virus Infection and Gene Silencing: Once inside the plant, the virus infects plant cells and releases the modified RNA, which triggers the RNA interference (RNAi) pathway. The plant's RNAi machinery recognizes the introduced RNA as foreign and initiates a sequence of events leading to the degradation of both the viral RNA and the endogenous mRNA of the target gene.
5. Observation of Phenotypic Changes: As the expression of the target gene is suppressed, researchers can observe the resulting phenotypic changes in the plant.

These changes may include alterations in morphology, development, or responses to various stimuli.

Advantages of VIGS in plant functional genomics research:

1. **Rapid Results**: VIGS allows for relatively quick and transient gene silencing, making it faster than generating stable transgenic plants.
2. **Reverse Genetics**: It provides a reverse genetics approach, enabling researchers to study gene function without relying on mutants.
3. **High Throughput**: VIGS can be applied to go through the function of multiple genes in parallel, making it suitable for high-throughput analyses.
4. **Temporally Controlled**: The effects of gene silencing are often temporary, allowing researchers to study the reversible aspects of gene function.

However, it's important to note that VIGS has some limitations, such as potential off-target effects, variable silencing efficiency, and the transient nature of the silencing effect. Despite these challenges, VIGS remains a valuable tool in plant research for functional genomics studies and understanding gene function.

Applications:

VIGS is increasingly used to determine the function of new genes in various plants (Table 1). Many of these plants are important crops that provide nutrients and fibres or produce products for medicinal uses. Based on a recent article, VIGS emerges as a high-throughput tool capable of inducing heritable epigenetic alterations in plants by transiently suppressing targeted gene expression via the viral genome. The VIGS technology involves the development of vectors derived from viruses keeping in view the selection of the ideal orientation and size of the gene to be targeted, and the implementation of the most appropriate inoculation method (Zulfiqar et al. 2023; Jagram and Dasgupta 2023).

2.3 HIGS (Host Induced Gene Silencing)

Host-induced gene silencing (HIGS) is a biotechnological strategy used to control plant diseases by employing the RNA interference (RNAi) mechanism. In this, we talk about the **cross-kingdom RNAi** as our target is the silencing or protection of plants from entities that cause economic losses and degrade food quality. The entities belong to different kingdoms such as Insects (Animalia), Fungi, and Protista. Hence, the name is cross-kingdom RNAi. This technique is achieved via two approaches, i.e., HIGS and SIGS. Now, we elaborate on the use of HIGS with the help of important pieces of work done by renowned scientists (Schaefer et al. 2020). In RNA interference gene expression in the eukaryotic cells is regulated by degradation of mRNA or inhibition of translation of an mRNA molecule. In the context of plant diseases, HIGS involves the transfer of genetic material, mostly small RNA molecules, from the host plant to the pathogen to suppress the expression of essential genes in the pathogen, ultimately leading to its reduced virulence or death.

The general steps involved in HIGS are as follows:

1. Identification of Target Genes: Researchers identify specific genes in the pathogen that are essential for its survival, growth, or virulence.
2. Design of RNA Molecules: Small RNA molecules viz., small interfering RNAs (siRNAs) or microRNAs (miRNAs), are designed to specifically target the identified genes. These RNA molecules are usually designed to be complementary to the messenger RNA (mRNA) of the target gene.
3. Transformation of Host Plants: Transformation by various methods like agrobacterium is done to introduce the desired RNA molecules into the host plant.
4. RNAi Machinery Activation: Once inside the host plant cells, the introduced RNA molecules trigger the plant's RNA interference machinery. The RNAi machinery processes the introduced RNAs into small RNA molecules that are then used to guide the degradation or suppression of the target gene's mRNA.
5. Transfer of RNA Molecules to Pathogen: The small RNA molecules move from the host plant cells to the invading pathogen, where they recognize and bind to the complementary mRNA of the target gene.
6. Silencing of Target Genes in Pathogen: The binding of small RNA molecules to the target gene's mRNA in the pathogen triggers its degradation or inhibits its translation, leading to a reduction in the expression of the target gene.
7. Impact on Pathogen: Reduced expression of essential genes in the pathogen can result in compromised virulence, impaired growth, or even death of the pathogen.

Applications:

For the first time in 2010, Nowara and coworkers reported this technique in barley (*Hordeum vulgare*) against the infection of the obligate pathogen *Blumeria graminis* which causes powdery mildew. Avra10 (effector gene) was silenced which led to reduced fungal development in the host plant (Nowara et al. 2010). The transgenic plants were resistant to *B. gramins* and similar results were obtained in oomycete pathogens as well as against *F. oxysporum* and *R. solani* (Govindarajulu et al. 2015; Hu et al. 2020; Sanju et al. 2015; Zhou et al. 2016).

The recent application of HIGS has been elucidated by Rosa et al. (2018). They have used this technique in *Zea mays* crop by preparing a transgenic named Smartstax Pro. The target of this transgenic was corn rootworm (*Diabrotica vwirgifera virgifera*). After the successful experiment, Smartstax Pro was approved for commercial use by the U.S. Environmental Protection Agency, the U.S. Food and Drug Administration, and the U.S. Department of Agriculture.

HIGS is a potential tool for controlling plant diseases caused by fungi such as *Fusarium* species. It is based on the silencing of pathogenicity genes of pathogens by using host-derived RNA molecules. Recently, the successful silencing of important genes encoding cytochrome P450, chitin synthase, protein kinases, and others in *F. graminearum, F. oxysporum, F. culmorum*, and *F. verticillioides*, led to reduced mycotoxin production has been done. HIGS has also been applied to reduce pathogen aggressiveness in other crops such as wheat, tall fescue, potato, lettuce, and cotton (Ding et al. 2021).

HIGS renders the plant community a sustainable and environmentally friendly approach to controlling plant diseases. It offers an alternative to traditional methods such as chemical pesticides and can be designed to specifically target pathogens, minimizing the impact on non-target organisms. However, challenges remain, including the delivery of RNA molecules to the target organisms and potential concerns related to off-target effects. Ongoing research aims to address these challenges and further optimize the application of HIGS in crop protection.

RNAi applications in plant protection against biotic stressors

(A) RNAi is induced by using virus vectors by infecting the plant with weak viral pathogens or inoculating the plant with viral vector in VIGS. (B) dsRNA is exogenously applied in SIGS, which can be taken up by the pathogen or absorbed by the plant and transported to the pathogen (C) dsRNA and art-sRNA constructs are incorporated into the plant with a vector (*Agrobacterium*) and transgenic resistant plants are raised.

2.4 Spray Induced Gene Silencing (SIGS)

It is a novel approach that involves the application of double-stranded RNA (dsRNA) molecules exogenously into plants.

The general steps involved in SIGS are as follows:

1. The dsRNA is sprayed onto plant surfaces and it is proposed to work in two ways to target pathogen genes critical for causing disease. Either direct uptake of dsRNAs takes place to induce the RNA silencing or the host plant absorbs the dsRNAs.
2. Either way, the dsRNA undergoes processing into smaller siRNAs by the plant or pathogen DCL (Dicer-like) proteins.
3. Both the siRNAs and unprocessed dsRNAs are transferred into the pathogen cells in the former; where the pathogen's DCL proteins further process the dsRNAs into siRNAs (Sang and Kim 2020).
4. RNAi machinery proceeds further to degrade the targeted complementary site on the pathogen genome.

Thus, this technique silences the pathogen's gene without transforming the plant host and avoids any heritable modifications to the genome of the plant (Koch et al. 2016; Wang and Jin 2017). SIGS was first demonstrated against Tobamovirus, Potyvirus, and Alfalfa mosaic virus groups (Tenllado and Diaz 2001) and has effectively targeted a variety of plant pathogens, including viruses (Dubrovina and Kiselev 2019; Das and Sherif 2020), viroids (Carbonell et al. 2008), fungal pathogens (Koch et al. 2016; McLoughlin et al. 2018) and even insect vectors of plant pathogens till date (Dalakouras et al. 2020). It is a sustainable and excellent technique for pathogen control at both pre- and post-harvesting stages without generating a large number of off-target effects (Vetukuri et al. 2021). It differs from host-induced gene silencing

(HIGS) in that no transformation of the agricultural plant is involved. The utilization of HIGS is limited as in many crop plants, including fruit trees, vegetables, and flowers, there is a lack of efficient transformation techniques. Furthermore, many consumers still do not think highly of genetically modified (GMO) products, and GMOs are also prohibited in some countries. Therefore, SIGS comes out to be advantageous over HIGS in cases where non-transgenic plant protection strategies are more appropriate (Qiao et al. 2021). For this technique to be fully effective a crucial step is the efficient uptake of RNA from the environment by pathogens, which may vary according to the species.

Applications

In viruses like Alfalfa mosaic virus (AMV), Pepper mild mottle virus (PMMoV) and Tobacco etch virus (TEV), this topical application of dsRNAs has been observed to be effective in protecting plants against them (Mitter et al. 2017; Dalakouras et al. 2020; Das and Sherif 2020). In a recent study, SIGS effectively controlled the sheath blight of rice (*Rhizoctonia solani*). Here, SUPPRESSOR OF ACTIN 1 (SAC1) and DYNACTIN 1 (DCN1) genes for vesicle trafficking pathway, and or a virulence factor gene for *R. solani* (polygalacturonase), were targeted, which considerably reduced the disease symptoms in rice plants (Qiao et al. 2021). In another study, a potential control for *Phytophthora infestans* causing late blight disease was demonstrated by spraying dsRNAs onto the leaves, which targeted the pathogen genes coding for guanine-nucleotide binding (G) protein β-subunit (PiGPB1), haustorial membrane protein (PiHMP1), cutinase (PiCUT3), or endo-1,3(4)-βglucanase (PiENDO3) in *P. infestans* (Kalyandurg et al. 2021).

Advantages of SIGS in plant functional genomics:

- It is a quick method suitable for both pre and post-harvest plant protection.
- It is a more acceptable technique as transgenic plants are not involved.
- Less costly and time-consuming than transformations requiring techniques.

Regardless of the potential that this technique has shown, the short-term environmental stability of dsRNA is one of the major challenges to its application in the field. This shortcoming can be overcome by developing carriers for dsRNA within which it can be transported to the target without any degradation. The materials used must be environmentally safe and favor the slow release of RNAi. Some of the tested alternatives are formulations with nanoparticles (Wang et al. 2023), artificial vesicles (Qiao et al. 2023), and bioclay (Nino et al. 2023). In a recent study conducted by Mukherjee et al. (2024), they loaded dsRNA with magnesium/ferrous-layered double hydroxides nanosheets which proved to be more efficient in reducing the white mould symptoms of *S. sclerotiorum.* In addition, mechanical inoculation through the spreading of dsRNA with soft sterile brushes and gentle rub inoculation can also aid in dsRNA delivery to induce RNAi (Lau et al. 2015; Dubrovina and Kiselev 2019).

2.5 *Artificial Small RNA (Art-SRNA)*

Art-sRNAs are artificially designed, typically twenty-one nucleotide sRNAs from endogenous sRNA precursors, to recognize and silence complementary RNA targets (Carbonell 2017). These constructs also require a medium for the transfer into the plant. The high specificity and effectiveness of these constructs give them an edge over other techniques and these traits are attributed to no off-targets, in vivo accumulation to a high level, and highly sequenced complement with the target RNA (Cisneros and Carbonell 2020). This results in serendipitous targeting of complementary cellular transcripts. There are two main classes of artificial sRNAs: artificial microRNAs (amiRNAs) and artificial/synthetic trans-acting small interfering RNAs (syn-tasiRNAs) (Carbonell and Daros 2017). Both sRNAs have different biogenesis pathways while performing similarly by associating with AGO1 for specific silencing of target RNAs by hydrolyzing the nucleotide bond or repression of protein synthesis.

Ami-RNAs:

For the production of AmiRNAs, amiRNA transgenes, and a functional plant MIRNA precursor are used. These lab-generated amiRNA guide/amiRNA* (artificial) sequences substitute the endogenous miRNA guide/miRNA* sequences (Carbonell and Daros 2017). The transgene undergoes transcription into the amiRNA primary transcript (pri-amiRNA), which is subsequently processed by the DCL1 protein to produce the pre-amiRNA. Following this, the amiRNA duplex is formed, and the amiRNA guide strand is incorporated into AGO1 to target complementary pathogenic RNA.

Applications:

This technique was first used against RNA viruses like TYMV) and TuMV, in the transgenic plant *Arabidopsis thaliana* having amiRNA transgenes (Niu et al. 2006). *A. thaliana* miR159 precursor was modified to express amiRNAs targeting viral mRNA sequences encoding two gene silencing suppressors which were P69 of TYMV and HC-Pro of TuMV. Another study done by Roshdi et al. (2023) resulted in the making of TYLCV-OM-resistant tomato plants. They designed amiRNA-based constructs named Rep-amiRNA1 and Rep-amiRNA2a which targeted the AC1/Rep gene of TYLCV-OM for the production of TYLCV-OM-resistant tomato plants. The study revealed that these transgenic tomato plants exhibited increased expression of amiRNA, resulting in the effective silencing of the AC1/Rep transcripts of TYLCV-OM.

Syn-Tasi RNA:

Zhang (2014) worked on syn-tasiRNAs that are produced *in-situ* in plants by expressing a functional *TAS* precursor (endogenous tasiRNA sequence(s)). Synthesis of syn-tasiRNA occurs via endogenous tasiRNA-generating pathways, initiated by the cleavage of the syn-tasiRNA precursor by a miRNA/AGO complex. The conversion of these breakdown products into double-stranded RNA (dsRNA) is facilitated by RNA-dependent RNA polymerase 6, and subsequent processing is carried out by DCL4 to form syn-tasiRNA duplexes that align with the trigger miRNA target site. The guide strand of the syn-tasiRNA duplex is then loaded onto Argonaute protein, which mediates the silencing of targeted complementary RNAs. (Cisneros et al. 2021).

Applications:

Chen and coworkers (2016) induced multiple virus resistance by using syn-tasiRNAs in *Arabidopsis* where co-expression of three syn-tasiRNAs against CMV and three syn-tasiRNAs against TuMV from a single TAS3a precursor was observed. Another study conducted on Tomato leaf curl virus, New Delhi strain (ToLCNDV) in *Arabidopsis* utilized this technique targeting the ATP/GTP binding domain of AC1 gene of ToLCNDV. AC1 gene of (ToLCNDV), is responsible for replication and pathogenicity of this virus. Therefore, silencing of this gene conferred resistance ranging from 40 to 80% (Sharma and Prasad 2020).

Advantages of art-siRNA in plant functional genomics:

- Efficient and reliable as host-derived miRNA constructs are used.
- Fewer off-target effects-the ability to degrade just the target genes without affecting the expression of non-target genes.
- High accuracy and RNA promoter compatibility.
- More constancy within a living system.
- Environmentally safe.
- Heritability of phenotypes.

However, few problems are discovered in this approach. For instance, the amiRNAs are not durable enough if the non-conserved regions of plant viruses are targeted, transgenic plants expressing only a single amiRNA may be challenged by strong virus pressure under field conditions and the resistance against viruses may not be sustained. To solve this, the artificial multiplexing of amiRNAs can be done using different precursor configurations or syn-tasiRNA precursors can be used, which allow the simultaneous expression of more than one art-sRNAs. It minimizes the possibility of simultaneous mutation of all separate target sites of the virus to completely prevent degradation by escaping each art-sRNA, and thus increase the antiviral resistance (Lafforgue et al. 2013; Cisneros and Carbonell 2020). In addition, the efficient use of polycistronic amiRNA-mediated resistance to WSMV was useful in *Triticum aestivum* and *Hordeum vulgare*, respectively (Fahim et al. 2012; Kis et al. 2016). There are some web designing tools developed for art-sRNA such as 'Web MicroRNA Designer 3' (WMD3) (Ossowski et al. 2008) and the 'Plant Small RNA Maker Suite' (P-SAMS) (Fahlgren et al. 2016). The system comprises modules for target specificity that leverage annotated plant transcriptomes. These modules can analyze the potential base-pairing interactions between the candidate sRNA and the entire set of cellular transcripts of a specific species.

3 Conclusion

Finally, this chapter concludes RNA interference (RNAi) stands as a transformative and promising technology with far-reaching implications across various fields, particularly in the realm of molecular biology and crop protection. The ability to selectively silence the expression of genes at the post-transcriptional level has opened new doors for a better understanding of various gene functions, developing therapeutic techniques, and controlling plant diseases. Here, are some key points to consider in summarizing the significance and potential of RNA interference:

- RNAi enables researchers to achieve a high level of precision in gene regulation by selectively suppressing the expression of target genes.
- In the medical field, RNAi has shown promise for therapeutic applications, allowing for the development of novel treatments for various genetic disorders and diseases.
- RNAi has come up as a powerful tool in plant pathology for developing crop varieties with enhanced resistance to pathogens. The ability to target specific genes in plant pests and pathogens offers a sustainable and environmentally friendly approach to disease control in agriculture.
- RNAi has had a profound impact on genetics and agriculture. Its versatility and adaptability make it a valuable tool for studying and manipulating gene expression in diverse biological systems.
- Despite its potential, there are some challenges and considerations such as off-target effects, delivery methods, and regulatory considerations need to be addressed. Ongoing research seeks to refine RNAi technologies, enhance delivery systems, and establish guidelines for responsible use.

4 Future Perspectives of Rna Interference in Plant Pathology

RNA interference (RNAi) in plant pathology has shown great promise as a powerful tool for controlling plant diseases. However, keep in mind that developments in the field may have occurred since then. Here are some future perspectives on the employment of RNA interference in plant pathology:

Specific Targets: Researchers are likely to focus on improving the specificity and efficiency of RNAi in targeting pathogens. This may involve the identification of unique and essential genes in pathogens to develop more precise RNAi constructs.

RNAi for Different Pathogens: Tailoring RNAi approaches for specific plant-pathogen interactions may become more common. Customized RNAi molecules could be designed to target specific genes or pathways in different pathogens, providing a more effective and tailored approach to disease control.

Delivery Systems: Developing efficient delivery systems for RNAi molecules remains a challenge. Future research may focus on innovative delivery methods, such as nanoparticles, viral vectors, or other nanocarriers, to improve the stability and targeted delivery of RNAi molecules to plant cells.

Integration with Breeding Programmes: Integrating RNAi technology with traditional breeding programs could lead to the development of disease-resistant crop varieties. By incorporating RNAi-mediated resistance traits into crop plants, it may be possible to enhance their resilience against a broad spectrum of pathogens.

Understanding the off-targets: Efforts have to be done, to better understand and minimize off-target effects of RNAi. Advances in bioinformatics and gene expression analysis could contribute to the design of RNAi molecules with reduced unintended consequences.

Combination with other Control Strategies: Combining RNAi with other disease control strategies, such as fungicides or cultural practices, may provide a holistic approach to plant disease management. Synergistic effects could enhance the overall effectiveness of disease control measures. Also combining RNAi with gene editing tools like CRISPR is an interesting avenue to explore for disease resistance.

Public Acceptance: As RNAi technologies advance, addressing regulatory and public acceptance issues will be crucial. Clear guidelines and regulations regarding the use of RNAi in agriculture need to be established to ensure its responsible and safe deployment.

Global Adoption and Commercialization

The adoption of RNAi-based products in agriculture may increase globally as research progresses and commercial applications become available. Industry collaboration and investment in RNAi technologies for plant pathology could drive commercialization.

Field Trials and Real-World Applications:

Conducting extensive field trials to evaluate the long-term efficacy, environmental impact, and practicality of RNAi-based solutions will be essential for the widespread adoption of this technology in real-world agricultural settings.

Continued research, technological advancements, and collaboration between academia, industry, and regulatory bodies will play crucial roles in shaping the future of RNA interference in plant pathology.

References

Adkar PCR, Kasai A, Sugawara K, Yamamoto H, Yamazaki Y, He YH, Takada N, Goto H, Shindo S, Harada T, Sano T (2015) RNAi mediated inhibition of viroid infection in transgenic plants expressing viroid-specific small RNAs derived from various functional domains. Sci Rep 5(1):17949

Barathi S, Sabapathi N, Aruljothi KN, Lee JH, Shim JJ, Lee J (2023) Regulatory small RNAs for a sustained eco-agriculture. Int J Mol Sci 24:1041

Bernstein E, Caudy AA, Hammond SM, Hannon GJ (2001) Role for bidentate ribnuclease in the initiation site of RNA interference. Nature 409:363–366

Bhagta S, Bhardwaj V, Kant A (2023) Exogenous dsRNA trigger RNAi in *Venturia inaequalis* resulting in down regulation of target genes and growth reduction. Mol Biol Rep 50:8421–8429

Borges F, Martienssen RA (2015) The expanding world of small RNAs in plants. Nat Rev Mol Cell Biol 16:727–741

Cao X, Han Q, West JS (2024) Spray-induced gene silencing as a potential tool to control rubber tree powdery mildew disease. Physiol Mol Plant Pathol 129:102182

Carbonell A (2017) Artificial small RNA-based strategies for effective and specific gene silencing in plants. In: Dalmay T (ed) Plant gene silencing: mechanisms and applications. CABI Publishing, Wallingford, Oxfordshire, pp 110–127

Carbonell A, Daròs JA (2017) Artificial microRNAs and synthetic trans-acting small interfering RNAs interfere with viroid infection. Mol Plant Pathol 18(5):746–753

Carbonell A, de Alba ÁE, Flores R, Gago S (2008) Double-stranded RNA interferes in a sequence-specific manner with the infection of representative members of the two viroid families. Virology 371(1):44–53

Chen W, Kastner C, Nowara D, Oliveira-Garcia E, Rutten T, Zhao Y, Deising HB, Kumlehn J, Schweizer P (2016) Host-induced silencing of *Fusarium culmorum* genes protects wheat from infection. J Exp Bot 67(17):4979–4991

Cisneros AE, Carbonell A (2020) Artificial small RNA-based silencing tools for antiviral resistance in plants. Plants 9(6):669

Cisneros AE, De la TMA, Carbonell A (2021) Artificial small RNAs for functional genomics in plants. RNA-based technologies for functional genomics in plants, concepts and strategies in plant sciences. Springer Nature, Berlin, pp 1–29

Cooper B, Campbell KB (2017) Protection against common bean rust conferred by a gene-silencing method. Phytopathology 107:920–927

Cui Y, Jiang J, Yang H, Zhao T, Xu X, Li J (2018) Virus-induced gene silencing (VIGS) of the NBS-LRR gene SLNLC1 compromises Sm-mediated disease resistance to Stemphylium lycopersici in tomato. Biochem Biophys Res Commun 503(3):1524–1529

D'Ario M, Griffiths JS, Kim M (2017) Small RNAs: big impact on plant development. Trends Plant Sci 22:1056–1068

Dalakouras A, Wassenegger M, Dadami E, Ganopoulos I, Pappas ML, Papadopoulou K (2020) Genetically modified organism-free RNA interference: exogenous application of RNA molecules in plants. Plant Physiol 182(1):38–50

Das PR, Sherif SM (2020) Application of exogenous dsRNAs-induced RNAi in agriculture: challenges and triumphs. Front Plant Sci 25(11):946

Ding Y, Chen Y, Yan B, Liao H, Dong M, Meng X, Wan H, Qian W (2021) Host-induced gene silencing of a multifunction gene sscnd1 enhances plant resistance against *Sclerotinia sclerotiorum*. Front Microbiol 12:693334

Dubrovina AS, Kiselev KV (2019) Exogenous RNAs for gene regulation and plant resistance. Int J Mol Sci 20(9):2282

Dunoyer P, Himber C, Voinnet O (2006) Induction, suppression and requirement of RNA silencing pathways in virulent *Agrobacterium tumefaciens* infections. Nat Genet 38(2):258–263

Enrique R, Siciliano F, Favaro MA, Gerhardt N, Roeschlin R, Rigano L, Sendin L, Castagnaro A, Vojnov A, Marano MR (2011) Novel demonstration of RNAi in citrus reveals importance of citrus callose synthase in defence against *Xanthomonas citri* subsp. *citri*. Plant Biotechnol J 9(3):394–407

Fahim M, Millar AA, Wood CC, Larkin PJ (2012) Resistance to Wheat streak mosaic virus generated by expression of an artificial polycistronic microRNA in wheat. Plant Biotechnol J 10(2):150–163

Fahlgren N, Hill ST, Carrington JC, Carbonell A (2016) P-SAMS: a web site for plant artificial microRNA and synthetic trans-acting small interfering RNA design. Bioinfo 32:157–215

Fire A, Xu S, Montgomery MK, Kostas SA, Driver SE, Mello CC (1998) Potent and specific genetic interference by double stranded RNA in *Caenorhabditis elegans*. Nature 391:806–811

Francesco DA, Simeone M, Gómez C, Costa N, Garcia ML (2020) Transgenic sweet orange expressing hairpin CP-mRNA in the interstock confers tolerance to citrus psorosis virus in the nontransgenic scion. Transgenic Res 29:215

Gago ZS, Schuck J, Weinholdt C, Knoblich M, Pantaleo V, Grosse I, Gursinsky T, Behrens SE (2019) Highly efficacious antiviral protection of plants by small interfering RNAs identified in vitro. Nucleic Acids Res 47:9343–9357

Gebremichael DE, Haile ZM, Negrini F, Sabbadini S, Capriotti L, Mezzetti B, Baraldi E (2021) RNA interference strategies for future management of plant pathogenic fungi: prospects and challenges. Plants 10(4):650

Govindarajulu M, Epstein L, Wroblewski T, Michelmore RW (2015) Host-induced gene silencing inhibits the biotrophic pathogen causing downy mildew of lettuce. Plant Biotechnol J 13:875–883. https://doi.org/10.1111/pbi.12307

Hammond SM, Bernstein E, Beach D, Hannon GJ (2000) An RNA-directed nuclease mediates post-transcriptional gene silencing in *Drosophila* cells. Nature 404:293–296

Hammond SM, Boettcher S, Caudy AA, Kobayashi R, Hannon GJ (2001) Argonaute2, a link between genetic and biochemical analyses of RNAi. Science 293:1146–1150

Hernández SA, Chacón CR (2021) RNAi crop protection advances. Int J Mol Sci 22(22):12148

Hu D, Chen ZY, Zhang C, Ganiger M (2020) Reduction of *Phakopsora pachyrhizi* infection on soybean through host-and spray-induced gene silencing. Mol Plant Pathol 21:794–807. https://doi.org/10.1111/mpp.12931

Jagram N, Dasgupta I (2023) Principles and practice of virus induced gene silencing for functional genomics in plants. Virus Genes 59:173–187

Kalyandurg PB, Sundararajan P, Dubey M, Ghadamgahi F, Zahid MA, Whisson SC, Vetukuri RR (2021) Spray-induced gene silencing as a potential tool to control potato late blight disease. Phytopathology 111(12):2168–2175

Kis A, Tholt G, Ivanics M, Varallyay E, Jenes B, Havelda Z (2016) Polycistronic artificial miRNA-mediated resistance to wheat dwarf virus in barley is highly efficient at low temperature. Mol Plant Pathol 17(3):427–437

Koch A, Biedenkopf D, Furch A, Weber L, Rossbach O, Abdellatef E, Linicus L, Johannsmeier J, Jelonek L, Goesmann A, Cardoza V, McMillan J, Mentzel T, Kogel H (2016) An RNAi-based control of *Fusarium graminearum* infection through spraying of long dsRNAs involves a plant passage and is controlled by the fungal silencing machinery. PLOS Pathog 12:e1005901

Krishna R, Mishra CB (2017) Application of antisense RNA technology for crop improvement. A review. Int J Appl Res 3(7):1319–1321

Kuo YW, Falk BW (2020) RNA interference approaches for plant disease control. Bio Techniques 69(6):469–477

Lafforgue G, Martinez F, Niu QW, Chua NH, Daros JA, Elena SF (2013) Improving the effectiveness of artificial microRNA (amiR)-mediated resistance against Turnip mosaic virus by combining two amiRs or by targeting highly conserved viral genomic regions. J Virol 87:8254–8256

Lau SE, Schwarzacher T, Othman RY, Harikrishna JA (2015) DsRNA silencing of an R2R3-MYB transcription factor affects flower cell shape in a dendrobium hybrid. BMC Plant Biol 15:194. https://doi.org/10.1186/s12870-015-0577-3

Li XP, Ma XC, Wang H, Zhu Y, Liu XX, Li TT, Zheng YP, Zhao JQ, Zhang JW, Huang YY, Pu M, Feng H, Fan J, Li Y, Wang WM (2020) Osa-miR162a finetunes rice resistance to *Magnaporthe oryzae* and yield. Rice (n Y) 13:38

Liang C, Hao J, Li J, Baker B, Luo L (2019) Artificial microRNA-mediated resistance to cucumber green mottle mosaic virus in Nicotiana benthamiana. Planta 250:1591–1601

Liu Q, Li Y, Xu K, Li D, Hu H, Zhou F (2020) Clay nanosheet-mediated delivery of recombinant plasmids expressing artificial miRNAs via leaf spray to prevent infection by plant DNA viruses. Hortic Res 7:179

Liu R, Lu J, Xing J, Xue L, Wu Y, Zhang L (2023) Characterization and functional analyses of wheat TaPR1 genes in response to stripe rust fungal infection. Sci Rep 13(1):3362

López DL, Spada M, Daròs JA, Carbonell A (2020) Fine-tune control of targeted RNAi efficacy by plant artificial small RNAs. Nucl Acids Res 48(11):6234–6250

Lu J, Yang X, Chen J, Li T, Li Y (2021) Osa-miR439 negatively regulates rice immunity against *Magnaporthe oryzae*. Ric Sci 28:156–165

Machado AK, Brown NA, Urban M, Kanyuka K, Hammond-Kosack KE (2018) RNAi as an emerging approach to control Fusarium head blight disease and mycotoxin contamination in cereals. Pest Manag Sci 74(4):790–799

Mahanty B, Mishra R, Joshi RK (2023) *Fusarium oxysporum* f. sp. *cepae* small RNAs (Foc-sRNAs) promote disease susceptibility in onion (*Allium cepa* L.) through cross kingdom RNA interference. Physiol Mol Plant Pathol 125:102018

McLoughlin AG, Wytinck N, Walker PL, Girard IJ, Rashid KY, de Kievit T, Fernando WD, Whyard S, Belmonte MF (2018) Identification and application of exogenous dsRNA confers plant protection against *Sclerotinia sclerotiorum* and *Botrytis cinerea*. Sci Rep 8(1):7320

Mitter N, Worrall EA, Robinson KE, Xu ZP, Carroll BJ (2017) Induction of virus resistance by exogenous application of double-stranded RNA. Curr Opin Virol 26:49–55

Montes C, Castro A, Barba P, Rubio J, Sánchez E, Carvajal D, Aguirre C, Tapia E, Dell Orto P, Decroocq V, Prieto H (2014) Differential RNAi responses of Nicotiana benthamiana individuals transformed with a hairpin-inducing construct during Plum pox virus challenge. Virus Genes 49:325–338

Mosen AM, Guo Y, Hassing B, Mesarich CH, Bradshaw RE (2023) An RNA interference (RNAi) target with potential to control Dothistroma needle blight. N Z Plant Prot 4(76):35–53

Muhammad T, Zhang F, Zhang Y, Liang Y (2019) RNA interference: a natural immune system of plants to counteract biotic stressors. Cells 8(1):38

Mukherjee S, Beligala G, Feng C, Marzano SY (2024) Double-stranded RNA targeting white mold *Sclerotinia sclerotiorum* argonaute 2 for disease control via spray-induced gene silencing. Phytopathology

Napoli C, Lemieux C, Jorgensen R (1990) Introduction of a chimeric chalcone synthase gene into petunia results in reversible co-suppression of homologous genes in trans. Plant Cell 2:279–289

Navarro L, Jay F, Nomura K, He SY, Voinnet O (2008) Suppression of the microRNA pathway by bacterial effector proteins. Sci 321:964–967

Nicolopoulou SP, Maipas S, Kotampasi C, Stamatis P, Hens L (2016) Chemical pesticides and human health: the urgent need for a new concept in agriculture. Front Public Health 4:148

Niehl A, Soininen M, Poranen MM, Heinlein M (2018) Synthetic biology approach for plant protection using dsRNA. Plant Biotechnol J 16:1679–1687

Niño SJ, Sambasivam PT, Sawyer A, Hamby R, Chen A, Czislowski E, Li P, Manzie N, Gardiner DM, Ford R, Xu ZP (2022) BioClay™ prolongs RNA interference-mediated crop protection against *Botrytis cinerea*. J Integr Plant Biol 64(11):2187–2198

Niu QW, Lin SS, Reyes JL, Chen KC, Wu HW, Yeh SD, Chua NH (2006) Expression of artificial microRNAs in transgenic *Arabidopsis thaliana* confers virus resistance. Nat Biotechnol 24:1420–1428

Nivedha M, Harish S, Angappan K, Karthikeyan G, Kumar KK, Murugan M, Jayakanthan M (2023) In silico identification and validation of microRNAs from the genome of Solanum lycopersicum targeting Groundnut bud necrosis orthotospovirus. Physiol Mol Plant Pathol 127:102086

Nowara D, Gay A, Lacomme C, Shaw J, Ridout C, Douchkov D, Hensel G, Kumlehn J, Schweizer P (2010) HIGS: host-induced gene silencing in the obligate biotrophic fungal pathogen *Blumeria graminis*. Plant Cell 22:3130–3141

Nunes CC, Dean RA (2012) Host-induced gene silencing: a tool for understanding fungal host interaction and for developing novel disease control strategies. Mol Plant Pathol 13:519–5295

Ossowski S, Schwab R, Weigel D (2008) Gene silencing in plants using artificial microRNAs and other small RNAs. Plant J 53:674–690

Qi T, Guo J, Liu P, He F, Wan C, Islam MA, Tyler BM, Kang Z, Guo J (2019) Stripe rust effector PstGSRE1 disrupts nuclear localization of ROS-promoting transcription factor TaLOL2 to defeat ROS-induced defense in wheat. Mol Plant 12(12):1624–1638

Qiao Y, Liu L, Xiong Q, Flores C, Wong J, Shi J, Wang X, Liu X, Xiang Q, Jiang S, Zhang F (2013) Oomycete pathogens encode RNA silencing suppressors. Nat Genet 45(3):330–333

Qiao L, Lan C, Capriotti L, Ah-Fong A, Nino Sanchez J, Hamby R, Heller J, Zhao H, Glass NL, Judelson HS, Mezzetti B (2021) Spray-induced gene silencing for disease control is dependent on the efficiency of pathogen RNA uptake. Plant Biotechnol J 19(9):1756–1768

Qiao L, Niño SJ, Hamby R, Capriotti L, Chen A, Mezzetti B, Jin H (2023) Artificial nanovesicles for dsRNA delivery in spray-induced gene silencing for crop protection. Plant Biotech J 21(4):854–865

Qu J, Ye J, Fang R (2007) Artificial microRNA-mediated virus resistance in plants. J Virol 81:6690–6699

Raruang Y, Omolehin O, Hu D, Wei Q, Promyou S, Parekattil LJ, Rajasekaran K, Cary JW, Wang K, Chen ZY (2023) Targeting the *Aspergillus flavus* p2c gene through host-induced gene silencing reduces *A. flavus* infection and aflatoxin contamination in transgenic maize. Front Plant Sci 14:1150086

Ratcliff FG, MacFarlane SA, Baulcombe DC (1999) Gene silencing without DNA. RNA-mediated crossprotection between viruses. Plant Cell 11:1207–1216

Rosa C, Kuo YW, Wuriyanghan H, Falk BW (2018) RNA interference mechanisms and applications in plant pathology. Annu Rev Phytopathol 56:581–610

Roshdi AMR, Ammara U, Khan J, Al-Sadi AM, Shahid MS (2023) Artificial microRNA-mediated resistance against Oman strain of tomato yellow leaf curl virus. Front Plant Sci 14:1164921

Saito H, Sakata N, Ishiga T, Ishiga Y (2022) Efficacy of RNA-spray-induced silencing of *Phakopsora pachyrhizi* chitin synthase genes to control soybean rust. J Gen Plant Pathol 88:203–206

Sang H, Kim JI (2020) Advanced strategies to control plant pathogenic fungi by host-induced gene silencing (HIGS) and spray-induced gene silencing (SIGS). Plant Biotechnol Rep 14(1):1–8

Sanju S, Siddappa S, Thakur A, Shukla PK, Srivastava N, Pattanayak D (2015) Host-mediated gene silencing of a single effector gene from the potato pathogen Phytophthora infestans imparts partial resistance to late blight disease. Funct Integr Genomics 15:697–706. https://doi.org/10.1007/s10142-015-0446-z

Schaefer LK, Parlange F, Buchmann G, Jung E, Wehrli A, Herren G, Müller MC, Stehlin J, Schmid R, Wicker T, Keller B, Bourras S (2020) Cross-kingdom RNAi of pathogen effectors leads to quantitative adult plant resistance in wheat. Front Plant Sci 11:253. https://doi.org/10.3389/fpls.2020.00253

Schwind N, Zwiebel M, Itaya A, Ding B, Wang MB, Krczal G, Wassenegger M (2009) RNAi-mediated resistance to Potato spindle tuber viroid in transgenic tomato expressing a viroid hairpin RNA construct. Mol Plant Pathol 10(4):459–469

Senthil-Kumar M, Mysore KS (2010) RNAi in plants: recent developments and applications in agriculture. Gene silencing: theory techniques and applications. Nova Science Publishers, New York, pp 183–199

Sharma N, Prasad M (2020) Silencing *AC1* of Tomato leaf curl virus using artificial microRNA confers resistance to leaf curl disease in transgenic tomato. Plant Cell Rep 39:1565–1579

Soltani FE, Taghvimi S, Abedi KZ, Weber C, Shabaninejad Z, Taheri AM, Hossein KS, Mousavi P, Movahedpour A, Natarelli L (2021) Insights into the function of regulatory RNAs in bacteria and Archaea. J Transl Med 1:403–423

Spada M, Pugliesi C, Fambrini M, Palpacelli D, Pecchia S (2023) Knockdown of *Bmp1* and *Pls1* virulence genes by exogenous application of RNAi-inducing dsRNA in *Botrytis cinerea*. Int J Mol Sci 24(5):4869

Tang J, Gu X, Liu J, He Z (2021) Roles of small RNAs in crop disease resistance. Stress Biol 1(1):6

Tenllado F, Diaz RJR (2001) Double-stranded RNA-mediated interference with plant virus infection. J Virol 75(24):12288–12297

Vaucheret H, Vazquez F, Crété P, Bartel DP (2004) The action of ARGONAUTE1 in the miRNA pathway and its regulation by the miRNA pathway are crucial for plant development. Genes Dev 18:1187–1197

Vetukuri RR, Dubey M, Kalyandurg PB, Carlsson AS, Whisson SC, Ortiz R (2021) Spray-induced gene silencing: an innovative strategy for plant trait improvement and disease control. Crop Breeding Appl Biotech 21: 387921S11

Wagaba H, Patil BL, Mukasa S, Alicai T, Fauquet CM, Taylor NJ (2016) Artificial microRNA-derived resistance to Cassava brown streak disease. J Virol Methods 231:38–43

Walawage SL, Britton MT, Leslie CA, Uratsu SL, Li Y, Dandekar AM (2013) Stacking resistance to crown gall and nematodes in walnut rootstocks. BMC Genomics 14(1):1–3

Wang M, Jin H (2017) Spray-induced gene silencing: a powerful innovative strategy for crop protection. Trends Microbiol 25(1):4–6

Wang M, Dean RA (2022) Host induced gene silencing of *Magnaporthe oryzae* by targeting pathogenicity and development genes to control rice blast disease. Front Plant Sci 13:959641

Wang W, Wang W, Wu Y, Li Q, Zhang G, Shi R, Yang J, Wang Y, Wang W (2020) The involvement of wheat U-box E3 ubiquitin ligase TaPUB1 in salt stress tolerance. J Integr Plant Biol 62(5):631–651. https://doi.org/10.1111/jipb.12842

Wang Y, Yan Q, Lan C, Tang T, Wang K, Shen J, Niu D (2023) Nanoparticle carriers enhance RNA stability and uptake efficiency and prolong the protection against *Rhizoctonia solani*. Phytopath Res 5(1):2

Watkins D, Arya DP (2019) Regulatory roles of small RNAs in prokaryotes: parallels and contrast with eukaryotic miRNA. Non-Coding RNA. Investig 3:28

Worrall E, Bravo CA, Nilon AT, Fletcher SJ, Robinson KE, Carr J, Mitter N (2019) Exogenous application of RNAi-Inducing double-stranded RNA inhibits aphid-mediated transmission of a plant virus. Front Plant Sci 10:265

Xu X, Jiao Y, Shen L, Li Y, Mei Y, Yang W, Li C, Cao Y, Chen F, Li B, Yang J (2023) Nanoparticle-dsRNA treatment of pollen and root systems of diseased plants effectively reduces the rate of tobacco mosaic virus in contemporary seeds. ACS Appl Mater Interfaces 15(24):29052–29063

Xu Y, Tan J, Lu J, Zhang Y, Li X (2024) RAS signalling genes can be used as host-induced gene silencing targets to control fungal diseases caused by *Sclerotinia sclerotiorum* and *Botrytis cinerea*. Plant Biotechnol J 22(1):262–277

Zamore PD, Tuschl T, Sharp PA, Bartel DP (2000) RNAi: double-stranded RNA directs the ATP-dependent cleavage of mRNA at 21 to 23 nucleotide intervals. Cell 101:25–33

Zewdu GA, Tamene W, Assefa Z, Tamiru T, Bankseni L, Borja M, Hasen T (2023) The role of modern plant breeding to control plant disease: a review. Agric J 17:19–27

Zhang ZJ (2014) Artificial trans-acting small interfering RNA: a tool for plant biology study and crop improvements. Planta 239:1139–1146

Zhou B, Bailey A, Niblett C, Qu R (2016) Control of brown patch (*Rhizoctonia solani*) in tall fescue (*Festuca arundinacea* Schreb.) by host induced gene silencing. Plant Cell Rep 35:791–802. https://doi.org/10.1007/s00299-015-1921-7

Zulfiqar S, Farooq MA, Zhao T, Wang P, Tabusam J, Wang Y, Xuan S, Zhao J, Chen X, Shen S, Gu A (2023) Virus-induced gene silencing (VIGS): a powerful tool for crop improvement and its advancement towards epigenetics. Int J Mol Sci 24(6):5608

Emerging Molecular Tools and Breeding Strategies for Plant Bacterial Disease Management

Rikky Rai and Maruti Nandan Rai

Abstract Plant bacterial diseases pose an enormous challenge to the global food supply chain. Traditionally, the management of plant diseases has relied on applying chemical control measures, antibiotics, biocontrol agents, and the use of resistant varieties. Many of these measures impact human health and the environment adversely due to either being toxic and non-biodegradable or risks of evolving antibiotic resistance in bacterial pathogens. The molecular tools and breeding strategies have the potential to be a pivotal frontier in mitigating the impact of plant-bacterial diseases. The molecular tools encompassing genomics, transcriptomics, and proteomics unravel the complexities of host–pathogen interactions at the molecular depth. Likewise, advanced breeding strategies, including marker-assisted selection and genome editing techniques such as CRISPR-Cas9 are emerging as the methods of choice to develop disease-resistant plant varieties more sustainably. This chapter digs into applications of these tools and studies illustrating successful implementations and challenges in deploying these methodologies. The chapter envisions the trajectory of future innovations, anticipating the role of artificial intelligence and synthetic biology in reshaping the landscape of plant bacterial disease management.

Keywords Bacterial pathogens · CRISPR/Cas9 · Crop breeding · Biotic stress · Omics

---

R. Rai (✉)
Department of Botany, University of Allahabad, Uttar Pradesh, Prayagraj, India
e-mail: rikkyrai@allduniv.ac.in

M. N. Rai
College of Agricultural, Consumer, and Environmental Sciences (ACES), University of Illinois, Urbana-Champaign, IL, USA

J. Chen et al. (eds.), *Molecular and Biotechnological Tools for Plant Disease Management*, Interdisciplinary Biotechnological Advances,
https://doi.org/10.1007/978-981-97-7510-1_14

1 Introduction

The ever-increasing global population needs a continuous and stable food supply to meet the food and nutritional demands worldwide. Plant bacterial diseases put a major constraint on food security and are estimated to cause ~10% loss in global crop production annually. Effective management strategies are paramount to mitigate the impact of plant bacterial diseases on sustainable agricultural ecosystems. Contrary to managing human or animal diseases where every individual is attended to, the plants are generally treated as populations, and applied measures are mostly preventive rather than curative. Therefore, a clear understanding of plant bacterial diseases and their epidemiology, disease symptoms, and pathogenesis mechanisms is important in devising effective management strategies. Disease management practices vary substantially from one disease to another depending on several aspects such as the pathogen, the host, and biotic and abiotic factors associated with the disease. Traditional approaches for mitigating plant bacterial diseases have mostly relied on cultural practices, applying chemical control measures, antibiotics, biocontrol agents, and the use of resistant crop varieties. Due to the intrinsic limitations of many of these methods, either they are not very effective or pose other hazardous challenges. For example, one cultural practice, mulching the soil with organic matter, helps reduce disease by increasing the number of beneficiary microorganisms that can thrive in the soil, however, it may also increase the disease severity by disseminating already existing pathogens. Chemical control measures, though effective, impact human health and the environment adversely due to their toxic and non-biodegradable nature. Similarly, antibiotic spraying presents the risks of the evolution of antibiotic-resistant pathovars because of higher antibiotic exposure.

Sustainable disease management strategies to mitigate the impact of plant bacterial pathogens on global agriculture require an integrated and multi-faceted approach with larger roles for precision agricultural practices based on the latest molecular tools and breeding strategies. Early detection and accurate diagnosis are fundamental in implementing timely and targeted management measures. Advances in molecular diagnostics and remote sensing technologies play a crucial role in enhancing our ability to identify bacterial pathogens swiftly. Emerging molecular tools, including genomics, transcriptomics, and genome editing tools offer unprecedented insights into the molecular basis of plant-pathogen interactions. Coupled with precision breeding strategies, these tools have the potential to accelerate the development of resistant plant varieties more efficiently. Sustainable agriculture hinges on the development of strategies that are not only effective in disease control but also environmentally friendly and economically viable for farmers. While these technological advancements hold great promise, they also bring ethical concerns related to genetic modifications and their impacts on our society. Striking a balance between innovation and its responsible deployment is therefore critical for sustainable and socially acceptable disease management practices.

This chapter explores an in-depth amalgamation of emerging molecular tools and advanced breeding strategies for the effective management of plant bacterial diseases. At the molecular level, we delve into the details of host–pathogen interactions, unraveling adaptive mechanisms and evolutionary dynamics through the lens of cutting-edge multi-omics approaches. We also discuss advanced breeding strategies like marker-assisted selection and the transformative landscape of genomic editing. Through an array of relevant studies, we explore practical applications, successes, and challenges faced in implementing emerging molecular tools and breeding strategies at a larger scale. Overall, this chapter transcends conventional paradigms offering a profound scientific discourse on the intricacies of plant bacterial disease management.

2 Molecular Tools and Breeding Techniques: Definition and Scope

Molecular tools refer to a diverse set of techniques used to study and manipulate biological molecules such as DNA, RNA, and protein in vitro, or in cells and organisms at the molecular level. These tools are pivotal in studying and analyzing complex biological questions across multiple fields including molecular biology, biochemistry, genetics, genomics, epidemiology, and disease management. Molecular tools are wide-ranging and include simple as well as advanced methods suited for different objectives (Table 1). For example, simple tools like PCR and ELISA are used for the detection of a particular DNA sequence (PCR) or peptide (ELISA) and are useful in molecular diagnosis of human and plant diseases. Advanced tools like CRISPR and RNAi are used to edit the genome of an organism for the desired phenotype. Generally, these molecular tools are employed in various combinations for different experimental purposes.

Breeding techniques are methods and practices used in the controlled reproduction of plants, animals, or microorganisms to achieve desired biological traits or characteristics. The goal of breeding is to enhance specific traits of interest, usually of commercial values, such as yield, resistance to diseases, adaptability to environmental conditions, and certain aesthetic features. Selective breeding has been in practice for thousands of years to boost the value of crops as well as farm animals. In the modern post-genomic era, the availability of large multi-omics datasets has led to the onset of more advanced and less time-consuming breeding approaches compared to classical selective breeding methods. An analysis of yearly research articles published on 'plant breeding' and genomics-derived phrases indicates a correlated and exponential increase in the number of scholarly articles related to these two topics post-year 2000 indicating that access to large genomic data has indeed revolutionized the field of plant breeding (Fig. 1).

To tackle the challenge of bacterial plant disease management effectively and sustainably, the combination of emerging molecular tools and breeding techniques presents a unique frontier. Molecular tools like genome sequencing help identify and characterize genes associated with resistance to specific bacterial pathogens.

Table 1 Frequently used keywords in this chapter, associated with molecular tools, breeding, and bacterial disease management, and their description

Key-word	Description
PCR	Polymerase chain reaction (PCR) is a laboratory technique that rapidly amplifies copies of a specific segment of DNA. Frequently used as a molecular diagnostic tool for various applications
ELISA	Enzyme-Linked ImmunoSorbent Assay is an immunological technique. It is used as a diagnostic test to detect the presence of a particular antibody/antigen
CRISPR-Cas9	A genetic engineering technique, based on a bacterial CRISPR-Cas9 antiviral defense system, by which the genomes of a living organism may be modified
Genetic markers	A genetic marker is a DNA sequence with a known location on a chromosome, usually associated with a particular trait. It is used as a molecular identifier in genetic breeding
Marker-assisted selection	Marker-assisted selection (MAS) is an indirect selection process in breeding where a trait of interest is selected indirectly based on a genetic marker linked to a phenotypic trait of interest
QTL	Quantitative Trait Locus (QTL) analysis is a method that links phenotype data with genotypic data to explain the genetic basis of variation in complex traits. QTL analysis is used in breeding to identify specific regions of chromosomes associated with desired phenotypes
Genomics	Genomics is the study of the structure, function, evolution, variations, and genetic makeup of living organisms
Metagenomics	Metagenomics is the study of the structure and function of entire nucleotide sequences obtained from all the organisms present in a microbial community
Transcriptomics	Transcriptomics is the study of the complete gene expression profile of an organism under a certain environmental condition
R genes	R genes (Resistance genes) are genes in plant genomes that confer plant disease resistance against pathogens by producing R proteins
S genes	S genes (Susceptibility genes) are present in plants and make them susceptible to pathogenic bacteria. Bacterial pathogens use effector proteins to target S genes, resulting in effector-triggered susceptibility
GWAS	Genome-Wide Association Study is a tool to analyze the genome-wide set of genetic variants and their association with a particular phenotypic trait
SNP	A single nucleotide polymorphism (SNP) is a genomic variant at a single base position in the DNA. SNPs in a genome may influence phenotypic traits
VOCs	Volatile organic compounds (VOCs) are produced by soil bacteria and possess strong antimicrobial activity. They are used as biocontrol agents in plant disease management
PGPB	Plant Growth-Promoting Bacteria (PGPB) are found in the rhizosphere and promote plant growth by various means

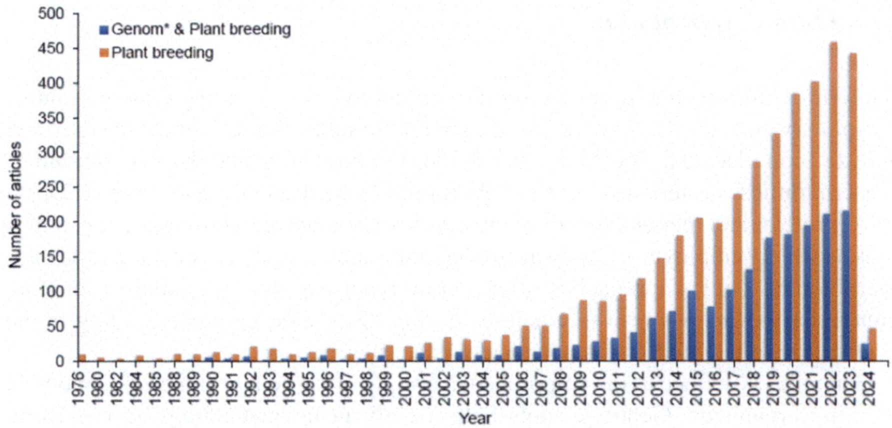

Fig. 1 Yearly number of publications with "Plant breeding" and genomics-derived phrases as keywords in their title/abstract retrieved from the MEDLINE database using the PubMed search engine. The Y-axis indicates the number of publications retrieved in the last ~75 years (till 24 Feb 2024) against the indicated year on the X-axis for a search string—"Plant breeding" AND Genom* [Title/Abstract]

This genetic information, the knowledge of DNA sequences associated with certain beneficial traits often referred to as genetic markers, is then integrated into breeding programs using techniques like marker-assisted selection (MAS), enabling the rapid and accurate selection of plant varieties with enhanced resistance traits. Genetic engineering tools offer the ability to introduce or enhance specific resistance genes in plants leading to large-effect qualitative disease resistance. In addition, molecular insights about the disease-resistance quantitative trait loci (QTL), and the regions of DNA associated with a disease-resistance phenotype contribute to polygenic and durable resistance. Molecular assays (Table 1) aid in the swift detection of bacterial pathogens, informing breeders about the prevalent and evolving pathogen populations. The molecular tools improve our understanding of plant-pathogen interactions at the molecular level and thus guide breeding efforts to optimize the selection of plants with robust defense mechanisms. The incorporation of genomic selection further accelerates the breeding process by predicting resistance levels early in the cycle. Together, these integrated approaches offer a comprehensive and targeted means to develop crops with improved and resilient resistance to bacterial diseases. Case studies on emerging molecular tools and breeding strategies for plant disease management as discussed later in this chapter, demonstrate the successful implementation of these techniques, paving the way for more resilient and disease-resistant crop varieties to ensure global food security.

2.1 Omics Approaches

The term 'Omics technologies' commonly refers to a set of rapidly evolving multidisciplinary and cutting-edge tools ending with the suffix 'omics' aimed to study and analyze biomolecules. The four most common omics technologies are genomics, transcriptomics, proteomics, and metabolomics, which are used to analyze DNA, RNA, protein, and metabolites, respectively. Multi-omics, combining various omics tools, is widely employed to study the molecular interactions between plants and their microbiomes. Omics approaches offer a comprehensive understanding of intricate plant–microbe-environment interactions during disease development, aiding in the formulation of targeted disease management strategies.

Genomics is defined as the comprehensive study of the genetic material (genomes) of living organisms. Genomic tools help greatly in understanding the molecular mechanisms of how plants and pathogens respond to each other at the genetic level and thus play a crucial role in developing strategies for disease management in plants. Comparative genomics, the analysis of conserved genomic regions, sheds light on host–pathogen co-evolution and identifies molecular markers associated with resistance genes thereby facilitating marker-assisted selection in breeding. The evolution and affordability of next-generation sequencing (NGS) has enormously facilitated genome sequencing of bacterial pathogens thereby enabling the identification of virulence and genetic factors associated with pathogenesis. For example, the recent completion of the first reference-quality genome of *Xanthomonas cucurbitae* sheds light on virulence factors crucial for its pathogenicity in cucurbits (Rai et al. 2021). Similarly, genome sequencing and comparative genomics studies of *X. campestris* pv. *campestris*, *X. arboricola* pv. *juglandis*, and *Pseudomonas syringae* pv. *syringae* provided significant insights into the virulence of these pathogens and their host specificities (Assis et al. 2021; Chen et al. 2022; He et al. 2007; Qian et al. 2005; Ruinelli et al. 2022). Likewise, genome sequencing of the host of bacterial phytopathogens and their comparative genomic analysis to previous well-annotated genomes helps in identifying resistance genes against bacterial diseases and proves crucial for developing resistant crop varieties through breeding or genome editing. Another useful molecular tool dual RNA-seq is used to get information on mutual transcriptomic responses of the host plants and their pathogen simultaneously during plant-pathogen interactions. Dual RNA-seq unveils key genes involved in pathogenesis and defense pathways offering significant insights into disease progression and resistance mechanisms.

Integrating genomics with transcriptomics, proteomics, metabolomics, and phenomics, provides a holistic understanding of plant-pathogen interactions. Proteomic and metabolomic profiles during infection uncover the molecular players, proteins, and metabolites crucial for pathogen recognition and defense responses. Analysis of secreted protein effectors from bacterial pathogens, as demonstrated in *Xyllela fastidiosa*, *Xanthomonas campestris* pv. *campestris*, *Pantoea stewartii*, and *Ralstonia solanacearum*, enhance our understanding of virulence factors and their significance in pathogenesis (Alexander Watt et al. 2005; Bustamante Smolka et al. 2003; Jacobs et al. 2012; Wu et al. 2007). Similarly, examining the plant microbiome in both diseased and healthy states elucidates the influence of microbial communities on plant health and disease progression. This information helps to get a better understanding of the dynamics of microbial communities and their interactions with the host plant required for manipulating the plant microbiome for enhanced defense against bacterial pathogens. Integrating these multi-omics approaches with CRISPR/Cas9-based gene editing and high-throughput phenotypic screenings identifies specific genes and proteins critical for resistance or susceptibility. To summarize, the analysis of multi-omics data provides pivotal insights for developing targeted and effective strategies for the early detection and deterrence of diseases in economically important crops. Implementing these strategies holds promise for sustainable and environmentally friendly approaches to managing bacterial diseases in crops.

2.1.1 Genome Sequencing

Genome sequencing is a powerful tool that reveals the complete DNA sequence of an organism. Sequencing the genomes of bacterial pathogens can pinpoint the genes responsible for pathogenicity, paving the way for targeted disease management strategies. Knowing these key genes governing host–pathogen interaction is instrumental in disease diagnostics, population-level studies, and epidemiological investigations allowing the rapid identification of bacterial pathogens, tracking disease spread, and predicting potential threats. Monitoring genomic diversity aids in preparing for emerging diseases and devising proactive disease management strategies. There are examples of several successful genome sequencing projects, that have significantly advanced our understanding of plant-bacterial interactions. The genome sequencing of *Xanthomonas oryzae* pv. *oryzae*, *Pseudomonas syringae* pv. *tomato*, *Erwinia amylovora*, *Ralstonia solanacearum*, *Clavibacter michiganensis* subsp. *michiganensis*, *Xylella fastidiosa*, and *Candidatus Liberibacter asiaticus* have offered valuable insights into virulence factors, effector molecules, and potential resistance targets of these pathogens (Almeida et al. 2009; Duan et al. 2009; Gartemann et al. 2008; Lee et al. 2005; Sebaihia et al. 2010; Simpson 2000).

Bioinformatic analysis of the sequenced genome helps narrow down suitable candidate genes for disease resistance (R genes) and thereby accelerates the timeline of developing disease-resistance varieties (Wulft and Moscou 2014). Similarly, 'mapping by sequencing' involves correlating genetic variations and phenotypic traits facilitating the identification of genes responsible for specific functions and therefore stands out as a powerful tool in plant disease management (Austin et al. 2011; Mascher et al. 2014). Another powerful tool used in enhancing plant breeding efforts is Genome-Wide Association Studies (GWAS). GWAS is a method to study the association between a genome-wide set of single-nucleotide polymorphisms (SNPs) and desired phenotypes. GWAS plays a crucial role in identifying genetic markers associated with disease resistance. For example, GWAS studies on large rice germplasm collection have identified several R genes that are promising targets for developing disease-resistance varieties (Andras et al. 2020; Demirjian et al. 2023; Liu et al.

2020; Lu et al. 2021). To summarize, genome sequencing not only enhances our understanding of the underlying mechanisms of plant-bacterial interactions but also provides practical applications for developing strategies for disease management and crop improvement. Ongoing and future projects in genomics hold great promise to shape the landscape of agriculture by providing sustainable solutions for managing bacterial diseases in crops.

2.1.2 Comparative Genomics

Comparative genomics is defined as the comparison of genomic features of different organisms, particularly closely related species. These features include DNA sequences, genes, and their physical locations, gene order (synteny), and other regulatory elements. Comparative genomics offers a comprehensive understanding of the molecular intricacies that shape host–pathogen relationships and unravels the core genes shared among different bacterial strains, elucidating essential housekeeping functions critical for bacterial survival. Moreover, the construction of phylogenetic trees based on genomic data provides evolutionary insights, tracing the historical relationships among bacterial strains and the origins of plant diseases. This approach also helps in the detection of pathogenicity islands, genomic regions housing virulence factors, and effector proteins that manipulate host defenses. Similarly, comparative genomic analysis of host plants identifies their counterparts like susceptibility or resistance genes in the plant host. It also leads to the detection of horizontally transferred genes, shedding light on the acquisition of new virulence factors or antibiotic-resistance genes.

Phylogenomic analysis enhances the resolution for distinguishing closely related strains, a crucial aspect for epidemiological studies and precision in disease management. Mobile genetic elements, such as plasmids and transposons, carry genes related to virulence and antibiotic resistance, influencing the evolution and adaptability of bacterial pathogens. Genomic changes in both bacterial pathogens and host plants indicate an evolutionary arms race and the emergence of novel resistance mechanisms. Insights into highly variable genomic regions signify genome plasticity, providing a view into how bacterial pathogens adapt to changing environments. These findings through comparative genomics help in the development of molecular diagnostic tools enabling accurate and rapid identification of bacterial strains in agricultural settings. For example, the population-wide genomic analysis was conducted on 65 isolates of the cassava bacterial blight pathogen *Xanthomonas axonopodis* pv. *manihotis*, spanning 70 years and three distinct countries, identified conserved effectors (Bart et al. 2012). These effectors serve as promising targets for developing durable plant resistance against bacterial pathogens. This milestone work underscores the genetic basis of host–pathogen interactions and provides a foundation for developing strategies to enhance cassava resistance against bacterial blight over the long term. Another example is a population genomic analysis of *Xanthomonas*, which infected both pepper and tomato across various locations, years, and hosts.

This analysis uncovered the diversity within effector repertoires and identified determinants of host specificity for *Xanthomonas* field strains (Schwartz et al. 2015). The study suggests that there is potential to enhance disease control strategies through extensive sampling of field strains and gaining a profound understanding of effector content. Moreover, this research highlights the importance of comprehending the genetic variations in pathogen populations to develop effective measures for disease management in crops like pepper and tomato.

In phytopathogen *Erwinia tracheiphila*, a high proportion of recent horizontal gene acquisitions, mobile element proliferation, and genome decay was discovered suggesting that *E. tracheiphila* has recently emerged as a host-restricted pathogen (Shapiro et al. 2018). The observed genomic characteristics, including horizontal gene transfers and mobile element activities, imply dynamic evolutionary processes that contribute to the adaptation and specialization of *E. tracheiphila* as a pathogen with a limited host range. A series of other similar studies employing population and comparative genomics have offered valuable insights into the scales of variation and changes observed in plant pathogens over time, within and among geographic regions, and across phylogenetic trees. Timilsina and colleagues have reviewed the comparative studies on diverse bacterial genera, encompassing Gram-negative bacteria composed of facultative anaerobes such as *Dickeya, Pantoea*, and *Pectobacterium*, as well as obligate aerobes including *Pseudomonas, Xanthomonas, and Xylella* (Timilsina et al. 2022). Overall, comparative genomics offers a holistic understanding of the molecular basis of plant-bacterial pathogen interactions, the coevolution of host–pathogen armory, and the evolutionary responses of plant pathogens to various selection pressures, contributing to our knowledge of their adaptation and guiding strategies for disease management and crop protection.

2.1.3 Metagenomics and Microbiome Studies

Metagenomics is defined as the study of genomes of whole microbial communities in their natural habitats. Microbial communities are considered gold mines for the discovery of new genes, proteins, enzymes, and microbial metabolites. Metagenomic analysis of these microbial systems helps identify such microbes and provides a deeper understanding of their interactions and dynamics, which is crucial for managing plant diseases. With its ability to unravel the complex microbial communities in the plant environment, metagenomics emerges as a promising field for the management of plant bacterial diseases.

The plant microbiome consists of a diverse community of microorganisms, including bacteria, fungi, archaea, and viruses, residing in and around plant tissues. Microbial communities can vary among plant species, tissues, and environmental conditions. Rhizosphere microbes in the soil around roots and endophytes residing within the plant tissues harbor beneficial microbes that contribute to nutrient recycling, growth promotion, stress tolerance, and overall plant health. A balanced and diverse microbiome correlates with improved plant health and resilience, fostering positive interactions through nutrient exchange and disease suppression. In terms of disease suppression, some microbes in the plant microbiome act as biocontrol agents, suppressing the growth and virulence of pathogenic organisms. Beneficial microbes can induce systemic resistance in plants, enhancing their ability to defend against a broad range of pathogens. Notably, microbes produce secondary metabolites and prime the plant immune system for a more rapid and effective response against pathogens. To enhance plant defense against bacterial pathogens, manipulating the microbiome provides promising solutions. Selective microbial inoculation involves introducing specific strains of beneficial microbes known for their biocontrol abilities against bacterial pathogens. The targeted selection and introduction of microbes enhance disease resistance without negatively impacting plant health. Disease resistance can also be engineered by modifying strains to boost antimicrobial compound production and designing synthetic communities with desired characteristics. Microbes with diverse mechanisms of action can ably provide comprehensive protection against bacterial pathogens.

Microbial Volatile Organic Compounds (VOCs) play a significant role, acting as signaling molecules to activate plant defense responses. VOCs prime the plant immune system, preparing for a rapid and effective response to bacterial pathogens. Rhizosphere modification through soil amendments and enhanced microbial diversity creates conditions favorable for beneficial microbes. Stress-adapted microbes, selected for stress tolerance or drought resistance, contribute to plant resilience in challenging environments. By identifying and introducing microbes adapted to stress conditions, plant resilience against bacterial pathogens is enhanced. Interestingly, several studies have reported that when a plant is infected with a pathogen such as harmful fungi or bacteria, the plant growth-promoting bacteria (PGPB) play a role in inhibiting the growth or activity of these pathogens, thereby reducing the damage caused to the plant (Compant et al. 2005; Galambos et al. 2021; Santoyo et al. 2016). Engineering this beneficial PGPB can have a significant positive impact on agricultural production (Mitter et al. 2017). While some research has been conducted on manipulating microbiomes for disease control, understanding the long-term effects and their role in biocontrol remains limited. Bioengineering of plant microbiomes, though in its early stages, is seen as a promising strategy for enhancing plant biological capabilities, with potential significant agricultural importance. To conclude, understanding and manipulating the plant microbiome offers potential innovative strategies to strengthen plant defense mechanisms.

2.2 Molecular Breeding

Molecular breeding, also known as marker-assisted breeding, is a powerful approach in managing plant bacterial diseases. Molecular breeding combines cutting-edge molecular techniques with traditional plant breeding methods. It involves the use of molecular markers linked to desirable traits, such as disease resistance, to accelerate the breeding process. DNA markers, identified through techniques like quantitative trait loci (QTL) mapping, facilitate the identification and mapping of genes associated with resistance to bacterial diseases. In a review by Li and colleagues, broad-spectrum resistance (BSR) is mentioned as an important trait to provide resistance against multiple pathogen species as well as diverse strains of the same pathogen (Li et al. 2020). BSR candidates in plants include pattern recognition receptors, nucleotide binding, and leucine-rich repeat receptors, as well as defense signaling and pathogenesis-related proteins. Utilizing molecular markers linked to these resistance genes allows for the rapid and accurate identification of plants carrying desirable traits, and accelerates the selection process, reducing the time required for traditional breeding methods. Molecular breeding enables the simultaneous incorporation of multiple resistance genes into a single plant through pyramiding and enhances the durability and broad-spectrum nature of resistance, reducing the risk of pathogen adaptation. Notably, the process facilitates the introgression of resistance genes from wild or exotic germplasm into elite cultivars, overcoming challenges associated with linkage drag by using molecular markers to select the desired traits during the breeding process. Wild species and landraces with their rich genetic diversity serve as valuable sources for new resistance genes, contributing to the development of resilient crops through molecular breeding approaches. Many widely used resistance genes such as Xa21, Yr36, Fhb7, Xa23, and Xa27 originated from wild species (Fu et al. 2009; Gu et al. 2005; Song et al. 1995; Wang et al. 2015, 2020).

Molecular breeding approaches can also be used to (i) monitor and update molecular markers based on evolving pathogens, (ii) enhance the overall efficiency of breeding programs by integrating seamlessly with other methods, (iii) integrate with other 'omics' technologies such as transcriptomics and proteomics, providing a comprehensive understanding of the molecular mechanisms underlying resistance to bacterial diseases. Subsequently, genome editing technologies, such as CRISPR-Cas9, precisely modify target genes associated with disease resistance which allows for the development of crops with specific and tailored resistance traits. Singh and colleagues (2012) presented a groundbreaking achievement in rice breeding, marking the first successful transfer of resistance genes for bacterial blight (Xa13, Xa21), blast (Pi54), and sheath blight (qSBR11-1) through marker-assisted backcross breeding (Singh et al. 2012). This development establishes these lines as valuable donors of resistance genes for Basmati rice breeding programs targeting multiple diseases. In another study, Moya and colleagues (2023) conducted genetic mapping to identify tolerance to bacterial stem blight caused by *Pseudomonas syringae* pv. *syringae* in Alfalfa (*Medicago Sativa* L.) (Moya et al. 2023). The study yields valuable insights into host responses and identifies markers that can be utilized in alfalfa breeding programs for the effective management of bacterial stem blight. Engaging farmers and local communities in the development and selection of resistant varieties is a proactive approach to addressing specific bacterial disease challenges in their regions. This participatory method empowers local communities to foster disease management strategies. Furthermore, involving farmers in the process ensures continuous monitoring and allows for the timely update of molecular markers based on evolving pathogen populations. The integration of molecular tools with traditional breeding methods enhances the speed, accuracy, and effectiveness of breeding programs, ultimately contributing to more resilient and disease-resistant crops.

2.2.1 Genome Editing

Genome editing through RNA interference (RNAi) and CRISPR-Cas9 are two distinct technologies with different mechanisms for modifying genes. While both RNAi and CRISPR-Cas9 can be used for genome editing, CRISPR-Cas9 generally offers higher precision and versatility for making targeted changes to the DNA sequence. Several Susceptibility (S) genes have been successfully edited to develop rice and wheat plants with broad-spectrum resistance. Key studies by Oliva et al. 2019; Wang et al. 2015; Zaidi et al. 2018 have demonstrated successful applications of CRISPR-Cas9 in enhancing resistance in these crops.

RNA Interference (RNAi)

RNA interference (RNAi) is a powerful molecular tool that has shown promise in the management of plant bacterial diseases. RNAi is a natural mechanism that regulates gene expression by silencing or inhibiting the activity of specific genes through the introduction of small RNA molecules, such as small interfering RNA (siRNA) or microRNA (miRNA). In the context of plant bacterial diseases, RNAi can be used to silence key genes in either the plant host or the bacterial pathogen, thereby modulating host–pathogen interactions. The key aspects of RNAi in the management of plant bacterial diseases are: (i) Host-Induced Gene Silencing (HIGS), (ii) Silencing Plant Genes for Immune Responses, and (iii) RNAi-Mediated Silencing of Virulence Factors.

Host-Induced Gene Silencing (HIGS) represents a notable approach, involving the introduction of RNA molecules targeting essential bacterial genes into the plant host. This triggers the plant to express small interfering RNAs (siRNAs) that selectively degrade bacterial RNA, thereby suppressing crucial pathogenic genes. The outcome is a reduction in bacterial virulence, ultimately enhancing the plant's resistance. Another facet of RNAi application involves silencing specific plant genes associated with susceptibility or immune responses to bacterial pathogens. Through RNAi-mediated downregulation, genes facilitating infection or compromising the plant's defense mechanisms can be targeted. This results in plants exhibiting improved resistance to bacterial infections due to the modulation of host gene expression. RNAi also finds utility in directly targeting bacterial virulence factors. By delivering RNA molecules that specifically silence these factors, the pathogenicity and virulence of the bacterial pathogen are reduced. This approach proves effective in decreasing disease severity and mitigating the impact of bacterial infections.

Efficient delivery methods for RNAi molecules are crucial for successful implementation. Techniques such as transgenic plants, viral vectors, or nanoparticles enable targeted delivery of RNA molecules to plant cells or bacterial pathogens.

This enhances the uptake of RNAi molecules, ensuring effective gene silencing and, consequently, robust disease management. Furthermore, the construction of RNAi vectors plays a pivotal role in designing constructs that target specific genes in both plants and bacterial pathogens. These constructs typically express short hairpin RNAs (shRNAs) or small interfering RNAs (siRNAs), leading to the production of transgenic plants capable of sustained and targeted gene silencing. RNA interference (RNAi) has been employed successfully in the management of bacterial diseases in plants. For example, transgenic citrus plants engineered to express hairpin RNA molecules targeting essential genes in Xcc exhibited reduced bacterial growth and enhanced resistance to citrus canker (Enrique et al. 2011). Similarly, transgenic plants with RNAi constructs targeting virulence genes in *Ralstonia solanacearum*, showed reduced pathogenicity of the bacteria (Li et al. 2011). Although RNAi methods have produced more resistant plants against viruses, fungi, and nematodes (Halder et al. 2022), these examples showcase the adaptability and efficacy of RNAi-mediated strategies in conferring resistance to diverse bacterial pathogens too.

RNAi-mediated strategies provide a targeted and versatile means for managing plant bacterial diseases. However, RNAi techniques face challenges such as off-target effects, incomplete gene silencing, and transient effects. Researchers continue to address these limitations and explore alternative gene-editing technologies such as CRISPR-Cas9 to enhance precision and efficiency in genetic manipulation. As our understanding of host–pathogen interactions and RNAi mechanisms advances, these strategies hold significant promise for developing resilient crops and mitigating the impact of bacterial diseases in agriculture.

CRISPR-Cas9 and Its Applications in Plant Breeding

CRISPR-Cas9 (Clustered Regularly Interspaced Short Palindromic Repeats and CRISPR-associated protein 9) is a revolutionary genome-editing tool that has significant applications in plant breeding for managing bacterial plant diseases. CRISPR-Cas9 uses a guide RNA to direct the Cas9 enzyme to specific DNA locations, enabling precise modifications. It has widespread applications in genetic engineering, offering the ability to edit genes for therapeutic, agricultural, and research purposes. This method allows the editing of targeted regions of desired genes associated with disease susceptibility or resistance pathways and therefore accelerates the development of resistant plant varieties. The precision of CRISPR-Cas9 allows for (i) targeted disruption of host susceptibility factors, (ii) introduction of resistance genes into susceptible plant varieties, (iii) enhancement of plant immune responses, (iv) targeted disabling of bacterial virulence genes, (v) enabling the simultaneous editing of multiple genes (multiplexing) in a single plant and, (vi) facilitating large-scale genomics studies for a comprehensive understanding of plant-pathogen interactions. Additionally, its high specificity minimizes off-target effects, and multiplexed editing enhances the robustness and durability of resistance. Importantly, CRISPR-Cas9 reduces dependency on antibiotics/selectable markers, addressing associated concerns with the use of antibiotics/markers in the development of genetically modified crops.

Several studies have successfully demonstrated the precision and efficiency of genome editing, particularly CRISPR-Cas9, in managing plant bacterial diseases. In rice, targeted editing of the susceptibility gene, OsSWEET14, resulted in edited plants exhibiting resistance to *Xanthomonas oryzae* pv. *oryzae* (Xoo), the causative agent of bacterial blight (Zafar et al. 2020; Zeng et al. 2020). CRISPR-Cas9 editing of susceptibility factors DIPM-1, DIPM-2, and DIPM-4 in apples, led to elevated resistance against fire blight disease caused by the pathogen *Erwinia amylovora* (Malnoy et al. 2016). Likewise, many more studies have successfully utilized CRISPR-Cas9 editing to develop citrus varieties resistant to *Xanthomonas citri* subsp. *citri* (Xcc), the bacterial pathogen responsible for citrus canker (Huang et al. 2022; Jia et al. 2017; Peng et al. 2017; Su et al. 2023). The susceptibility gene *CsLOB1* was precisely targeted and modified using CRISPR-Cas9, resulting in citrus plants with enhanced resistance to Xcc. These examples collectively underscore the potential of CRISPR-Cas9 in engineering resistance against diverse bacterial pathogens in economically important crops. Overall, CRISPR-Cas9 is a transformative tool in plant breeding for managing bacterial plant diseases. Its precision, efficiency, and versatility contribute to developing disease-resistant crop varieties with the potential to address global challenges in agriculture (Fig. 2).

2.3 Synthetic Biology

Synthetic biology involves the application of genetic engineering principles to design and construct new biological components, systems, and organisms for specific desired purposes. Synthetic biology offers multiple innovative approaches for plant bacterial disease management. Crop resistance against pathogens can be increased by synthetic resistance genes with enhanced specificity and durability against pathogens, thus creating crops with tailored and effective resistance. Similarly, synthetic guide RNAs (sgRNAs) designed to target bacterial virulence genes or the plant's defense pathways result in providing better protection against the pathogen. Programmable CRISPR-based technologies enable precise control of gene expression in both plants and bacterial pathogens, thereby fine-tuning the resistance of crops against specific bacterial pathogens. Another interesting approach is to use synthetic microbial consortia as probiotics, which enhances plant health by suppressing pathogens and promoting plant growth, competing with pathogens, and inducing systemic resistance through engineered beneficial microorganisms. RNA-based interference networks offer adjustable gene expression regulation for improved resistance. Cellular immunity enhancement involves engineering robust immune responses in plant cells against bacterial infections.

Synthetic biosensors allow early pathogen detection in plants, facilitating timely intervention. For instance, engineering plants to produce fluorescent or colorimetric signals in response to the presence of specific bacterial pathogens allows early and accurate detection of bacterial infections, enabling effective disease management.

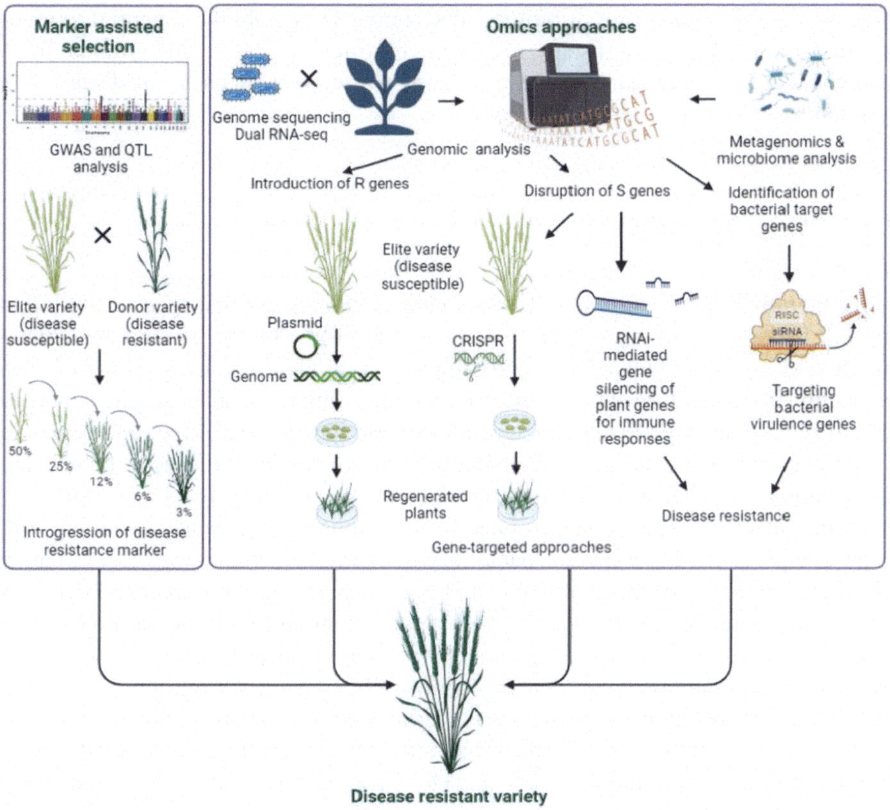

Fig. 2 Emerging molecular tools and breeding strategies in managing plant bacterial diseases. A summarized graphical representation of the emerging molecular tools and breeding strategies discussed in this chapter (Figure generated in Biorender)

Phage-based therapies design synthetic bacteriophages for targeted control of bacterial pathogens by engineering phages with increased specificity and efficacy against bacterial strains causing plant diseases. Metabolic engineering enhances plant resistance by producing antibacterial compounds, synthesizing antimicrobial secondary metabolites, or modifying existing pathways to boost their production.

Synthetic biology approaches can be extended toward developing synthetic plant–microbe communication which establishes symbiotic relationships that contribute to disease resistance by designing synthetic signaling systems. These systems facilitate communication between plants and beneficial microorganisms, promoting colonization and activation of plant defense responses. Modular synthetic constructs offer rapid deployment for diverse bacterial pathogens, streamlining disease management efforts. These synthetic biology approaches represent a transformative shift, providing tailored and sustainable solutions for plant disease control by harnessing molecular-level engineering of biological systems. They hold the potential to revolutionize the field of plant disease management, offering precise and efficient tools for enhancing plant resistance and mitigating the impact of bacterial diseases on agriculture.

2.4 Phage Therapy

Phage therapy involves the use of bacteriophages, the viruses that specifically infect and kill bacteria, to manage bacterial diseases in crops. Phage therapy has gained attention as a potential alternative to traditional antibiotics and chemical treatments, especially in agriculture where concerns about antibiotic resistance and environmental impact are significant. Bacteriophages are viruses that infect and replicate within bacterial cells, ultimately causing the lysis of the bacterial host. Phages are often highly specific to particular bacterial strains, reducing the impact on non-target organisms. Phages can be isolated from various environments, including soil, water, and plants, to target specific bacterial pathogens. Understanding the host range, specificity, and effectiveness of isolated phages against target bacteria is crucial for successful phage therapy. Phages can be formulated into different preparations, such as liquid suspensions, powders, or gels, suitable for application to crops. Formulations are designed to maintain phage stability and viability during storage and application. Phages exhibit specificity to particular bacterial species or strains, allowing targeted control of specific pathogens. Unlike broad-spectrum antibiotics, phages tend to have minimal impact on non-target organisms. However, bacterial pathogens can develop resistance to phages over time. Ongoing coevolutionary dynamics between phages and bacteria may influence the efficacy of phage therapy. Phage formulations are applied to plants by the following methods: (i) Spray application: phage suspensions can be applied directly to crops using conventional spraying equipment, (ii) Seed treatment: seeds are treated with phage formulations before planting to prevent bacterial infections from the early stages of plant development, (iii) Greenhouse application: phage therapy is particularly feasible in controlled environments like greenhouses, where the spread of phages can be managed more effectively.

Phage therapy can be integrated with Integrated Pest Management (IPM) strategies, combining biological, cultural, and chemical control measures in agriculture. However, challenges and considerations must be carefully addressed including the impact of environmental factors, such as UV light, temperature, and pH on phage persistence and efficacy. The establishment of regulatory frameworks for the approval and registration of phage-based products in agriculture is crucial. The effectiveness of phage therapy has been demonstrated in managing fire blight disease in apple and pear caused by *Erwinia amylovora* (Park et al. 2022). Additionally, phage therapy has shown promising results against *Xanthomonas campestris* pv. *campestris* (Holtappels et al. 2022). Stefani and colleagues have reviewed the bacteriophage-based biocontrol strategies on various *Xanthomonas* groups of phytopathogens comprehensively (Stefani et al. 2021).

Despite its potential, phage therapy comes with several limitations. The narrow host range of phages may limit their efficacy against diverse strains of a bacterial pathogen. Moreover, the timing of application is crucial to align with bacterial infection events for effective control. To address these challenges, future research and development efforts are essential. The continued exploration and discovery of novel phages with a broader host range is warranted for the development of more stable and effective phage formulations for practical application.

2.5 Precision Agriculture Technologies

Precision agriculture employs data-driven techniques for effective bacterial disease management in crops, incorporating advanced technologies such as remote sensing, Geographic Information Systems (GIS), Global Positioning Systems (GPS), data analytics, machine learning, and weather forecasting. Remote sensing technologies play a crucial role in disease outbreak inspection, utilizing high-resolution images captured by multispectral cameras and thermal sensors. Sensors monitoring environmental conditions, including temperature, humidity, wind, and moisture content, contribute to understanding disease development. Pathogen-based sensors enable rapid bacterial pathogen detection for swift disease diagnosis. GIS and GPS technologies integrate diverse data sources, creating precise disease maps based on location. Data analysis and machine learning algorithms provide models for disease pattern analysis, enabling early disease outbreak warnings. Examples of such applications include the use of hyperspectral imaging based on convolutional neural network (CNN) models for detecting black spot disease in roses and early detection of peanut bacterial wilt disease through hyperspectral remote sensing utilizing aerial images acquired by Unmanned Aerial Vehicles (UAVs) with built-in sensors to successfully identify citrus greening disease for HLB detection in plants, showcasing the potential of these tools (Chen et al. 2020; Garcia-Ruiz et al. 2013; Ma et al. 2020). Precision agriculture tools not only assist in disease detection but also aid in optimizing the timing for applying fertilizers, pesticides, or biocontrol agents to combat bacterial pathogens. The precision irrigation system adjusts water application based on soil moisture content. Integrated platforms and traceability systems provide recommendations to farmers and trace the origin of produce, preventing distribution during disease outbreaks. In conclusion, the effective utilization of precision agriculture techniques empowers farmers to make informed decisions and implement targeted management strategies for bacterial diseases.

2.6 Big Data and Machine Learning

The integration of big data and machine learning has transformed the landscape of bacterial plant disease management, offering sophisticated tools for prediction, detection, and control. Diverse data sources can be analyzed in combination for disease outbreak prediction. As mentioned earlier in the chapter, genomic data encompassing the sequence of bacterial pathogens involved in plant diseases and identification of virulence genes and factors associated with disease severity, combined with environmental data, such as climate and geographical information provide a comprehensive understanding of disease dynamics. Epidemiological Data, historical Outbreaks, and real-time disease incidence contribute valuable information on the prevalence and severity of bacterial infections. Additionally, crop and soil health data obtained through monitoring overall crop health and assessing soil characteristics that influence pathogen survival and plant susceptibility further enrich the dataset. Furthermore, deploying sensors facilitates real-time monitoring of environmental conditions and plant health. The integration of machine learning for data fusion becomes instrumental, where algorithms seamlessly combine diverse datasets through feature engineering. Predictive modeling, employing machine learning algorithms like Random Forest, Support Vector Machines, or Neural Networks, aids in forecasting disease outbreaks. Early warning systems, incorporating thresholds, alarms, and automated alerts help increase preparedness for disease management. Decision support systems simulate different scenarios and optimize intervention strategies, leveraging machine learning algorithms and optimization techniques. Model validation and calibration ensure the robustness and accuracy of predictive models.

Machine learning algorithms further optimize disease management strategies against bacterial diseases of plants. For example -regression models predict disease severity based on environmental and genomic data while clustering algorithms and deep learning aid in pathogen identification. Reinforcement learning and game theory analyze interactions and suggest optimal intervention strategies. Genomic Prediction and Genome-Wide Association Studies (GWAS) guide resistance breeding. Dynamic modeling, employing agent-based and epidemic models, simulates disease spread. Precision agriculture practices can be driven by machine learning interpretation.

Machine learning models come into play for predicting Cas9 specificity for precision genome editing with CRISPR-Cas9, ensuring accurate outcomes, and conducting off-target analysis to minimize unintended effects. To monitor and control resistance development, evolutionary algorithms are employed, modeling the dynamic bacterial populations to predict the emergence of resistance, while adaptive management strategies adjust control approaches in response to changing pathogen populations. Additionally, for phage therapy optimization, machine learning contributes by developing models predicting the effectiveness of specific phages against bacterial pathogens and optimizing delivery strategies for maximal impact. Integrated Pest Management (IPM) strategies benefit from machine learning-driven multi-modal optimization, combining chemical, biological, and cultural control methods, along with risk-based decision-making tools to predict and manage

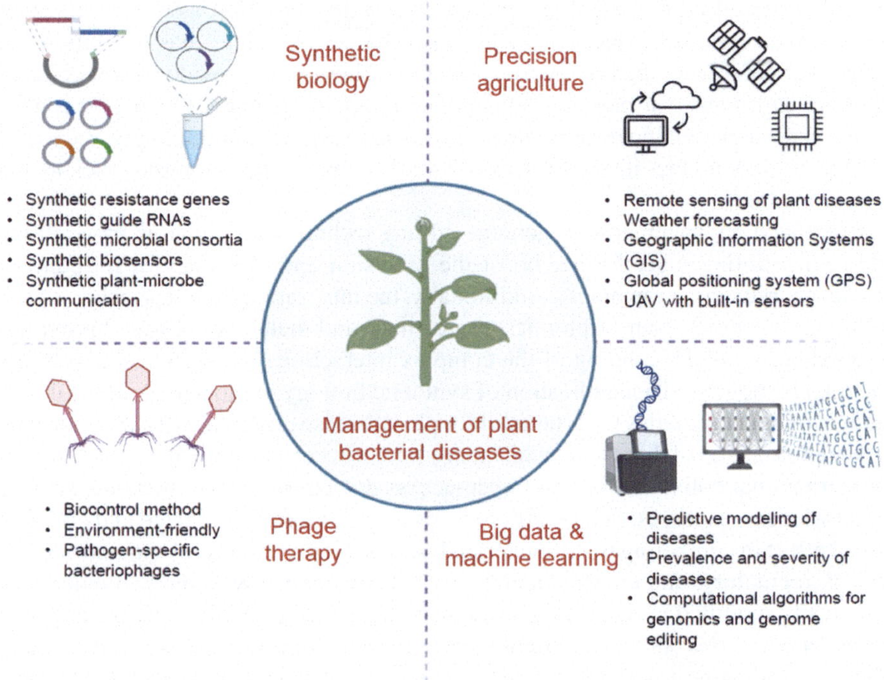

Fig. 3 Data-driven cutting-edge approaches for plant bacterial disease management. A pictorial representation of the summarized view of data-driven cutting-edge approaches used in bacterial disease management (Figure generated in Biorender)

disease risks. Adaptive management strategies, facilitated by learning models and feedback loop integration, allow continuous adjustments to disease management approaches based on real-time data, contributing to more efficient and sustainable agriculture, and reducing reliance on broad-spectrum treatments (Fig. 3).

3 Conclusion and Prospects

The successful management of bacterial diseases in plants demands continuous analysis and updates of the strategies outlined in this chapter. Exploration of emerging molecular tools and breeding practices for plant disease management represents a crucial advancement in agriculture. The integration of novel approaches holds great promise for improving crop yield, ensuring food security, and mitigating economic losses associated with bacterial diseases in plants. The use of molecular tools allows for a more precise understanding of plant-pathogen interactions, enabling the development of targeted and effective management strategies. The introduction of advanced and cutting-edge technologies such as CRISPR-Cas9, RNA interference, next-generation sequencing, and sensor-based techniques has revolutionized the field of plant breeding for disease resistance. These tools provide researchers and breeders with powerful means to manipulate the plant genome, identify resistance genes, and accelerate the development of resistant varieties. Additionally, the use of molecular markers facilitates the selection of desirable traits in a more efficient and precise manner.

Continued advancements in genome editing technologies, such as base editing and prime editing, may further boost the precision and efficiency of introducing resistance traits into crop plants. Additionally, the integration of omics technologies, including genomics, transcriptomics, proteomics, and metabolomics, will provide a comprehensive understanding of the complex interactions between host plants and bacterial pathogens. The application of synthetic biology approaches and the development of bioinformatics tools tailored for plant disease research will also contribute to the optimization of disease management strategies. Furthermore, the establishment of international collaborations and open-access databases will facilitate the sharing of knowledge, resources, and germplasm, fostering a more coordinated and effective global effort in combating plant-bacterial diseases. However, it is essential to address ethical, regulatory, and environmental concerns associated with the deployment of genetically modified crops. Public awareness and engagement in discussions about the benefits and risks of these technologies are crucial for informed decision-making. Moreover, adapting to effective disease management must consider global climate changes affecting the distribution of bacterial pathogens. Addressing these challenges requires a multidisciplinary approach encompassing research, education, technology, and policy development. Collaboration among scientists, farmers, policymakers, and industry stakeholders is crucial for devising and implementing effective disease control strategies in agriculture.

References

Alexander Watt S, Wilke A, Patschkowski T, Niehaus K (2005) Comprehensive analysis of the extracellular proteins from *Xanthomonas campestris* pv. *campestris* B100. Proteomics 5. https://doi.org/10.1002/pmic.200400905

Almeida NF, Yan S, Lindeberg M, Studholme DJ, Schneider DJ, Condon B, Liu H, Viana CJ, Warren A, Evans C, Kemen E, MacLean D, Angot A, Martin GB, Jones JD, Collmer A, Setubal JC, Vinatzer BA (2009) A draft genome sequence of *Pseudomonas syringae pv. tomato T1* reveals a type III effector repertoire significantly divergent from that of *Pseudomonas syringae pv. tomato DC3000*. Mol Plant-Microbe Interact 22. https://doi.org/10.1094/MPMI-22-1-0052

Andras JP, Fields PD, Pasquier LD, Fredericksen M, Ebert D (2020) Genome-wide association analysis identifies a genetic basis of infectivity in a model bacterial pathogen. Mol Biol Evol 37. https://doi.org/10.1093/molbev/msaa173

Assis RAB, Varani AM, Sagawa CHD, Patané JSL, Setubal JC, Uceda-Campos G, da Silva AM, Zaini PA, Almeida NF, Moreira LM, Dandekar AM (2021) A comparative genomic analysis of *Xanthomonas arboricola* pv. *juglandis* strains reveal hallmarks of mobile genetic elements in the adaptation and accelerated evolution of virulence. Genomics 113. https://doi.org/10.1016/j.ygeno.2021.06.003

Austin RS, Vidaurre D, Stamatiou G, Breit R, Provart NJ, Bonetta D, Zhang J, Fung P, Gong Y, Wang PW, McCourt P, Guttman DS (2011) Next-generation mapping of Arabidopsis genes. Plant J 67. https://doi.org/10.1111/j.1365-313X.2011.04619.x

Bart R, Cohn M, Kassen A, McCallum EJ, Shybut M, Petriello A, Krasileva K, Dahlbeck D, Medina C, Alicai T, Kumar L, Moreira LM, Neto JR, Verdier V, Santana MA, Kositcharoenkul N, Vanderschuren H, Gruissem W, Bernal A, Staskawicz BJ (2012) High-throughput genomic sequencing of cassava bacterial blight strains identifies conserved effectors to target for durable resistance. Proc Natl Acad Sci U S A. https://doi.org/10.1073/pnas.1208003109

Bustamante Smolka M, Martins D, Winck FV, Santoro CE, Castellari RR, Ferrari F, Brum IJ, Galembeck E, Coletta Filho HD, Machado MA, Marangoni S, Novello JC (2003) Proteome analysis of the plant pathogen *Xylella fastidiosa* reveals major cellular and extracellular proteins and a peculiar codon bias distribution. Proteomics 3. https://doi.org/10.1002/pmic.200390031

Chen T, Yang W, Zhang H, Zhu B, Zeng R, Wang X, Wang S, Wang L, Qi H, Lan Y, Zhang L (2020) Early detection of bacterial wilt in peanut plants through leaf-level hyperspectral and unmanned aerial vehicle data. Comput Electron Agric 177. https://doi.org/10.1016/j.compag.2020.105708

Chen D, Zhong X, Cui J, Li H, Han R, Yue X, Xie J, Kang J (2022) Comparative genomic analysis of *Xanthomonas campestris* pv. *campestris* isolates BJSJQ20200612 and GSXT20191014 provides novel insights into their genetic variability and virulence. Front Microbiol 13. https://doi.org/10.3389/fmicb.2022.833318

Compant S, Duffy B, Nowak J, Clément C, Barka EA (2005) Use of plant growth-promoting bacteria for biocontrol of plant diseases: Principles, mechanisms of action, and future prospects. Appl Environ Microbiol. https://doi.org/10.1128/AEM.71.9.4951-4959.2005

Demirjian C, Vailleau F, Berthomé R, Roux F (2023) Genome-wide association studies in plant pathosystems: success or failure? Trends Plant Sci. https://doi.org/10.1016/j.tplants.2022.11.006

Duan Y, Zhou L, Hall DG, Li W, Doddapaneni H, Lin H, Liu L, Vahling CM, Gabriel DW, Williams KP, Dickerman A, Sun Y, Gottwald T (2009) Complete genome sequence of citrus huanglongbing bacterium, "*Candidatus liberibacter asiaticus*" obtained through metagenomics. Mol Plant Microbe Interact. https://doi.org/10.1094/MPMI-22-8-1011

Enrique R, Siciliano F, Favaro MA, Gerhardt N, Roeschlin R, Rigano L, Sendin L, Castagnaro A, Vojnov A, Marano MR (2011) Novel demonstration of RNAi in citrus reveals importance of citrus callose synthase in defence against *Xanthomonas citri* subsp. *citri*. Plant Biotechnol J 9. https://doi.org/10.1111/j.1467-7652.2010.00555.x

Fu D, Uauy C, Distelfeld A, Blechl A, Epstein L, Chen X, Sela H, Fahima T, Dubcovsky J (2009) A kinase-START gene confers temperature-dependent resistance to wheat stripe rust. Science 1979:323. https://doi.org/10.1126/science.1166289

Galambos N, Compant S, Wäckers F, Sessitsch A, Anfora G, Mazzoni V, Pertot I, Perazzolli M (2021) Beneficial insects deliver plant growth-promoting bacterial endophytes between tomato plants. Microorganisms. https://doi.org/10.3390/microorganisms9061294

Garcia-Ruiz F, Sankaran S, Maja JM, Lee WS, Rasmussen J, Ehsani R (2013) Comparison of two aerial imaging platforms for identification of Huanglongbing-infected citrus trees. Comput Electron Agric. https://doi.org/10.1016/j.compag.2012.12.002

Gartemann KH, Abt B, Bekel T, Burger A, Engemann J, Flügel M, Gaigalat L, Goesmann A, Gräfen I, Kalinowski J, Kaup O, Kirchner O, Krause L, Linke B, McHardy A, Meyer F, Pohle S, Rückert C, Sehneiker S, Zeilermann EM, Pühler A, Eichenlaub R, Kaiser O, Bartels D (2008). The genome sequence of the tomato-pathogenic actinomycete *Chvibacter michiganensis* subsp. *michiganensis* NCPPB382 reveals a large island involved in pathogenicity. J Bacteriol 190. https://doi.org/10.1128/JB.01595-07

Gu K, Yang B, Tian D, Wu L, Wang D, Sreekala C, Yang F, Chu Z, Wang GL, White FF, Yin Z (2005) R gene expression induced by a type-III effector triggers disease resistance in rice. Nature. https://doi.org/10.1038/nature03630

Halder K, Chaudhuri A, Abdin MZ, Majee M, Datta A (2022) RNA interference for improving disease resistance in plants and its relevance in this clustered regularly interspaced short palindromic repeats-dominated era in terms of dsRNA-based biopesticides. Front Plant Sci. https://doi.org/10.3389/fpls.2022.885128

He YQ, Zhang L, Jiang BL, Zhang ZC, Xu RQ, Tang DJ, Qin J, Jiang W, Zhang X, Liao J, Cao JR, Zhang SS, Wei ML, Liang XX, Lu GT, Feng JX, Chen B, Cheng J, Tang JL (2007) Comparative and functional genomics reveals genetic diversity and determinants of host specificity among reference strains and a large collection of Chinese isolates of the phytopathogen *Xanthomonas campestris* pv. *campestris*. Genome Biol 8. https://doi.org/10.1186/gb-2007-8-10-r218

Holtappels D, Fortuna KJ, Moons L, Broeckaert N, Bäcker LE, Venneman S, Rombouts S, Lippens L, Baeyen S, Pollet S, Noben JP, Oechslin F, Vallino M, Aertsen A, Maes M, Van Vaerenbergh J, Lavigne R, Wagemans J (2022) The potential of bacteriophages to control *Xanthomonas campestris* pv. *campestris* at different stages of disease development. Microb Biotechnol 15. https://doi.org/10.1111/1751-7915.14004

Huang X, Wang Y, Wang N (2022) Highly efficient generation of canker-resistant sweet orange enabled by an improved CRISPR/Cas9 system. Front Plant Sci. https://doi.org/10.3389/fpls.2021.769907

Jacobs JM, Babujee L, Meng F, Milling A, Allen C (2012) The in planta transcriptome of *Ralstonia solanacearum*: conserved physiological and virulence strategies during bacterial wilt of tomato. mBio 3. https://doi.org/10.1128/mBio.00114-12

Jia H, Zhang Y, Orbović V, Xu J, White FF, Jones JB, Wang N (2017) Genome editing of the disease susceptibility gene CsLOB1 in citrus confers resistance to citrus canker. Plant Biotechnol J. https://doi.org/10.1111/pbi.12677

Lee B-M, Park Y-J, Park D-S, Kang H-W, Kim J-G, Song E-S, Park I-C, Yoon U-H, Hahn J-H, Koo B-S, Lee G-B, Kim H, Park H-S, Yoon K-O, Kim J-H, Jung C, Koh N-H, Seo J-S, Go S-J (2005) The genome sequence of *Xanthomonas oryzae pathovar oryzae* KACC10331, the bacterial blight pathogen of rice. Nucl Acids Res 33:577–586. https://doi.org/10.1093/nar/gki206

Li W, Deng Y, Ning Y, He Z, Wang GL (2020) Exploiting broad-spectrum disease resistance in crops: from molecular dissection to breeding. Annu Rev Plant Biol. https://doi.org/10.1146/annurev-arplant-010720-022215

Li CW, Su RC, Cheng CP, Sanjaya YSJ, Hsieh TH, Chao TC, Chan MT (2011) Tomato RAV transcription factor is a pivotal modulator involved in the AP2/EREBP-mediated defense pathway. Plant Physiol 156. https://doi.org/10.1104/pp.111.174268

Liu MH, Kang H, Xu Y, Peng Y, Wang D, Gao L, Wang X, Ning Y, Wu J, Liu W, Li C, Liu B, Wang GL (2020) Genome-wide association study identifies an NLR gene that confers partial resistance to *Magnaporthe oryzae* in rice. Plant Biotechnol J. https://doi.org/10.1111/pbi.13300

Lu J, Wang C, Zeng D, Li J, Shi X, Shi Y, Zhou Y (2021) Genome-wide association study dissects resistance loci against bacterial blight in a diverse rice panel from the 3000 rice genomes project. Rice. https://doi.org/10.1186/s12284-021-00462-3

Ma J, Pang L, Yan L, Xiao J (2020) Detection of black spot of rose based on hyperspectral imaging and convolutional neural network. Agri Eng. https://doi.org/10.3390/agriengineering2040037

Malnoy M, Viola R, Jung MH, Koo OJ, Kim S, Kim JS, Velasco R, Kanchiswamy CN (2016) DNA-free genetically edited grapevine and apple protoplast using CRISPR/Cas9 ribonucleoproteins. Front Plant Sci. https://doi.org/10.3389/fpls.2016.01904

Mascher M, Jost M, Kuon JE, Himmelbach A, Abfalg A, Beier S, Scholz U, Graner A, Stein N (2014) Mapping-by-sequencing accelerates forward genetics in barley. Genome Biol. https://doi.org/10.1186/gb-2014-15-6-r78

Mitter B, Pfaffenbichler N, Flavell R, Compant S, Antonielli L, Petric A, Berninger T, Naveed M, Sheibani-Tezerji R, von Maltzahn G, Sessitsch A (2017) A new approach to modify plant microbiomes and traits by introducing beneficial bacteria at flowering into progeny seeds. Front Microbiol. https://doi.org/10.3389/fmicb.2017.00011

Moya YS, Medina C, Herrera B, Chamba F, Yu L-X, Xu Z, Deborah AS (2023) Genetic mapping of tolerance to bacterial stem blight caused by *Pseudomonas syringae* pv. *syringae* in Alfalfa (*Medicago sativa* L.). Plants. https://www.mdpi.com/2223-7747/13/1/110

Oliva R, Ji C, Atienza-Grande G, Huguet-Tapia JC, Perez-Quintero A, Li T, Eom JS, Li C, Nguyen H, Liu B, Auguy F, Sciallano C, Luu VT, Dossa GS, Cunnac S, Schmidt SM, Slamet-Loedin IH, Vera Cruz C, Szurek B, Frommer WB, White FF, Yang B (2019) Broad-spectrum resistance to bacterial blight in rice using genome editing. Nat Biotechnol. https://doi.org/10.1038/s41587-019-0267-z

Park J, Kim B, Song S, Lee YW, Roh E (2022) Isolation of nine bacteriophages shown effective against *Erwinia amylovora* in Korea. Plant Pathol J (Faisalabad). https://doi.org/10.5423/PPJ.NT.11.2021.0172

Peng A, Chen S, Lei T, Xu L, He Y, Wu L, Yao L, Zou X (2017) Engineering canker-resistant plants through CRISPR/Cas9-targeted editing of the susceptibility gene CsLOB1 promoter in citrus. Plant Biotechnol J. https://doi.org/10.1111/pbi.12733

Qian W, Jia Y, Ren S-X, He Y-Q, Feng J-X, Lu L-F, Sun Q, Ying G, Tang D-J, Tang H, Wu W, Hao P, Wang L, Jiang B-L, Zeng S, Gu W-Y, Lu G, Rong L, Tian Y, Yao Z, Fu G, Chen B, Fang R, Qiang B, Chen Z, Zhao G-P, Tang J-L, He C (2005) Comparative and functional genomic analyses of the pathogenicity of phytopathogen *Xanthomonas campestris* pv. *campestris*. Genome Res 15:757–767. https://doi.org/10.1101/gr.3378705

Rai R, Pasion J, Majumdar T, Green CE, Hind SR (2021) Genome sequencing and functional characterization of *Xanthomonas cucurbitae*, the causal agent of bacterial spot disease of cucurbits. Phytopathology. https://doi.org/10.1094/PHYTO-06-20-0228-R

Ruinelli M, Blom J, Smits THM, Pothier JF (2022) Comparative genomics of prunus-associated members of the *Pseudomonas syringae* species complex reveals traits supporting co-evolution and host adaptation. Front Microbiol. https://doi.org/10.3389/fmicb.2022.804681

Santoyo G, Moreno-Hagelsieb G, del Carmen Orozco-Mosqueda M, Glick BR (2016) Plant growth-promoting bacterial endophytes. Microbiol Res. https://doi.org/10.1016/j.micres.2015.11.008

Schwartz AR, Potnis N, Timilsina S, Wilson M, Patané J, Martins J, Minsavage GV, Dahlbeck D, Akhunova A, Almeida N, Vallad GE, Barak JD, White FF, Miller SA, Ritchie D, Goss E, Bart RS, Setubal JC, Jones JB, Staskawicz BJ (2015) Phylogenomics of *Xanthomonas* field strains infecting pepper and tomato reveals diversity in effector repertoires and identifies determinants of host specificity. Front Microbiol. https://doi.org/10.3389/fmicb.2015.00535

Sebaihia M, Bocsanczy AM, Biehl BS, Quail MA, Perna NT, Glasner JD, DeClerck GA, Cartinhour S, Schneider DJ, Bentley SD, Parkhill J, Beer SV (2010) Complete genome sequence of the plant pathogen *Erwinia amylovora* strain ATCC 49946. J Bacteriol. https://doi.org/10.1128/JB.00022-10

Shapiro LR, Paulson JN, Arnold BJ, Scully ED, Zhaxybayeva O, Pierce NE, Rocha J, Klepac-Ceraj V, Holton K, Kolter R (2018) An introduced crop plant is driving diversification of the virulent bacterial pathogen *Erwinia tracheiphila*. mBio 9. https://doi.org/10.1128/mBio.01307-18

Simpson AJG (2000) The complete genome sequence of the plant pathogen *Xylella fastidiosa*. Biochem Soc Trans. https://doi.org/10.1042/bst028a102b

Singh A, Singh VK, Singh SP, Pandian RTP, Ellur RK, Singh D, Bhowmick PK, Gopala Krishnan S, Nagarajan M, Vinod KK, Singh UD, Prabhu KV, Sharma TR, Mohapatra T, Singh AK (2012) Molecular breeding for the development of multiple disease resistance in Basmati rice. AoB Plants. https://doi.org/10.1093/aobpla/pls029

Song WY, Wang GL, Chen LL, Kim HS, Pi LY, Holsten T, Gardner J, Wang B, Zhai WX, Zhu LH, Fauquet C, Ronald P (1995) A receptor kinase-like protein encoded by the rice disease resistance gene, Xa21. Science 270:1804–1806

Stefani E, Obradović A, Gašić K, Altin I, Nagy IK, Kovács T (2021) Bacteriophage-mediated control of phytopathogenic xanthomonads: a promising green solution for the future. Microorganisms. https://doi.org/10.3390/microorganisms9051056

Su H, Wang Y, Xu J, Omar AA, Grosser JW, Calovic M, Zhang L, Feng Y, Vakulskas CA, Wang N (2023) Generation of the transgene-free canker-resistant *Citrus sinensis* using Cas12a/crRNA

ribonucleoprotein in the T0 generation. Nat Commun. https://doi.org/10.1038/s41467-023-39714-9
Timilsina S, Goss EM, Koebnik R, Potnis N, Jones JB (2022) Editorial: population and comparative genomics of plant pathogenic bacteria. Front Microbiol. https://doi.org/10.3389/fmicb.2022.1012034
Wang C, Zhang X, Fan Y, Gao Y, Zhu Q, Zheng C, Qin T, Li Y, Che J, Zhang M, Yang B, Liu Y, Zhao K (2015) XA23 Is an executor r protein and confers broad-spectrum disease resistance in rice. Mol Plant. https://doi.org/10.1016/j.molp.2014.10.010
Wang S, Liu W, Lu D, Lu Z, Wang X, Xue J, He X (2020) Distribution of bacterial blight resistance genes in the main cultivars and application of Xa23 in rice breeding. Front Plant Sci. https://doi.org/10.3389/fpls.2020.555228
Wu Q, Jiang Z, Liao J, Chen Z, Li H, Mei M, Zhang LH (2007) Identification of genetic markers to distinguish the virulent and avirulent subspecies of *Pantoea stewartii* by comparative proteomics and genetic analysis. Appl Microbiol Biotechnol. https://doi.org/10.1007/s00253-006-0656-3
Wulft BBH, Moscou MJ (2014) Strategies for transferring resistance into wheat: from wide crosses to GM cassettes. Front Plant Sci. https://doi.org/10.3389/fpls.2014.00692
Zafar K, Khan MZ, Amin I, Mukhtar Z, Yasmin S, Arif M, Ejaz K, Mansoor S (2020) Precise CRISPR-Cas9 mediated genome editing in super basmati rice for resistance against bacterial blight by targeting the major susceptibility gene. Front Plant Sci. https://doi.org/10.3389/fpls.2020.00575
Zaidi SS, Mukhtar MS, Mansoor S (2018) Genome editing: targeting susceptibility genes for plant disease resistance. Trends Biotechnol. https://doi.org/10.1016/j.tibtech.2018.04.005
Zeng X, Luo Y, Vu NTQ, Shen S, Xia K, Zhang M (2020) CRISPR/Cas9-mediated mutation of OsSWEET14 in rice cv. Zhonghua11 confers resistance to *Xanthomonas oryzae* pv. *oryzae* without yield penalty. BMC Plant Biol 20. https://doi.org/10.1186/s12870-020-02524-y

Disease-Resistant Genes and Signal Transduction Pathways and Their Applications in Disease Management

Srinivasan Kameswaran, Gopikrishna Pitchika, Bellamkonda Ramesh, Manjunatha Bangeppagari, Battana Swapna, Mandala Ramakrishna, and P. Sudhakar Reddy

Abstract The intricate defense mechanisms that plants evolved to fend off various infections and pests. When infections can get past infection-preventing mechanical challenges, plant receptors start signaling pathways that cause the production of genes involved in the defense response. The ability of plant immune mechanisms to recognize pathogen chemicals, carry out signal transduction, and mount a defensive response through pathways involving multiple genes and their byproducts is essential to their ability to operate. Pathogens choose a decentralized, multicomponent immune system in a deliberate effort to dodge and disrupt response pathways. The study of plant immunity has significantly increased because of recent developments in molecular approaches, partly because of their potential applications to farming systems. There, we go through the key elements of the plant defense system, the current state of knowledge, and the route that plant-pathogen association research will take going forward. In this overview, we will go over how the decentralization of crop defense mechanisms has enhanced the potential for pathogen resistance evolution as well as given pathogens more ways to counteract such defensive responses. We conclude that the rapid growth of information being sparked by the fast advances in the fields of molecular biology along bioinformatics shall boost agricultural output and illuminate how complex molecular linkages emerge.

S. Kameswaran (✉) · B. Swapna · M. Ramakrishna
Department of Botany, Vikarama Simhapuri University College, Kavali, Andhra Pradesh 524201, India
e-mail: kambharath@gmail.com

G. Pitchika · P. S. Reddy
Department of Zoology, Vikarama Simhapuri University College, Kavali, Andhra Pradesh 524201, India

B. Ramesh
Department of Internal Medicine, University of Nebraska Medical Center, Omaha, NE 68198, USA

M. Bangeppagari
Department of Cell Biology and Molecular Genetics, Sri Devaraj Urs Academy of Higher Education and Research (Deemed to Be University), Tamaka, Kolar, Karnataka 563103, India

J. Chen et al. (eds.), *Molecular and Biotechnological Tools for Plant Disease Management*, Interdisciplinary Biotechnological Advances,
https://doi.org/10.1007/978-981-97-7510-1_15

Keywords Disease resistance · Defense response · Plant immunity · Coevolution · Plant-pest interaction · Resistance gene · Pathogenesis · Phytopathology

1 Introduction

1.1 The Emergence of a Complex, Multicomponent System for Plant Disease Resistance

A revolution in agriculture occurred about 10,000 years ago, deepening human interaction with plants. Early agriculturalists were likely to choose plants based on a variety of criteria, notably their capacity to withstand illness since life depended heavily on output. "Theophrastus, an Aristotelian" pupil, wrote on phytopathology in his writings, which date to the "third or fourth centuries B.C." Numerous plant illnesses described in antiquity were frequently attributed to supernatural sources (Whetzel 1918). The scientific categorization of pathogenic organisms, which were now visible using compound microscopes, was made possible by microbiology through the Age of Enlightenment. Early nineteenth-century experiments employing chemicals like the "Bordeaux copper sulfate and calcium oxide combination" helped demonstrate the efficiency of fungicides (Moutinho-Pereira et al. 2001). Heinrich Anton de Bary also established harmful microorganisms as the root causes of plant illnesses in the nineteenth century (Kutschera and Hossfeld 2012). Resistance genes (R genes), which are responsible for heritable resistance, were discovered by researchers thanks to an early twentieth-century understanding of heredity and genetics (Rhoades 1935; Bushnell 2012). The "Harold Henry Flor created a revolutionary gene-for-gene model" (Flor 1942) went on to characterize R genes, which are linked to "avirulence genes" that can be found in the pathogenic organism and result in an unfavorable response. Later in the twentieth century, discoveries in chemistry and molecular biology led to the elucidation of mechanisms for resistance via R genes (Hammond-Kosack and Jones 1996, 1997). Several classes of genes may be encoding controllers of disease resistance, including dependence, have been discovered because of advances in "molecular techniques and genomics" at the beginning of the "twenty-first century" (Antolín-Llovera et al. 2012). It was discovered that the R genes were only one component of a larger network of interconnected variables. The modern molecular investigation has significantly improved upon the straightforward framework of "The gene-for-gene concept proposed by Harold Henry Flor", demonstrating "that plant resistance" depends upon a complicated mechanism of regulation that controls plant immune mechanisms. Plant parts of the immune system are involved in pathogen detection, signal transduction, and defense response. According to the zig-zag paradigm, selection forces drive the evolution of viruses with sophisticated evasion mechanisms and plants with intricate detection mechanisms (Jones and Dangl 2006). These elements are the topic of this review, along with how they affect immunity. Because entire assessments are devoted to one category of a single factor, we are not going to address every feature of each element in detail. However, we intend to illustrate the theoretical underpinnings of phytopathogen resistance, backed up by instances. We will first quickly examine the elements of the plant's defense mechanism and the way pathogens evolved to get around defenses.

1.2 Detection and Reaction Are Key Elements in Plant Defence

The study of the molecular mechanisms of disease resistance led to the identification of a family of "cellular receptors" that carry out direct sensing of toxic substances. In a cell's membranes, "structure identification receptors (PRRs)" may detect both "pathogen-related molecular structures" and "cell wall-associated kinases enzymes". Recognizing "damage-related patterns of molecules" caused by infection-related cell damage (Zipfel 2014; Decreux and Messiaen 2005). The effectors of pathogens are recognized by receptors with "nucleotide-irrevocable sites along with leucine-more repeat (NLRs)" (Dangl et al. 2013). One of the numerous signaling cascades that have not yet been fully understood is started by WAKs, PRRs, and NLRs. G proteins, "mitogen-expressed kinase proteins", calcium, "ubiquitin, transcription elements", hormones, and epigenetic changes all influence "diseases-associated (PR) gene communication" (Meng and Zhang 2013; Marino et al. 2012; Zhang et al. 2012; Robert-Seilaniantz et al. 2011; Lecourieux et al. 2006). To stop the spread of the infection, "reactive O_2 molecules" are created, cell walls are altered, stomata close, and anti-pest proteins and chemicals (such as protease inhibitors, chitinases, phytoalexins, and defensins) are released (Juge 2006; van Loon et al. 2006). As is currently understood through molecular methods, pathogen tolerance within plants involves a range of organelles and categories, including proteins and "non-protein" molecules, all of which are important in regulating the immune mechanism. Each of these jobs has factors that impact development and the abiotic stress response, among other signaling pathways. We must completely explain these molecular relationships that happen while an appropriate pathogen engages with plant tissue if we are to have a better knowledge of plant-pathogen interactions. But first, let us quickly go through the pathogens that cause these reactions as well as how their evolution produced plants' sophisticated immune systems.

1.3 Virulence of the Pathogen Depends on Attack and Evasion

Plant communities have evolved an extremely diverse collection of immunological receptors and mechanisms of response in response to various pathogen species and races (Dangl and Jones 2001; Gururani et al. 2012; Hammond-Kosack and Jones

1997). Pathogens react by developing strategies to counteract host defenses or elude observation by the host (Jones and Dangl 2006). PAMP-triggered immunity (PTI) is frequently defeated by pathogens using effects, which may be produced in a variety of ways. The microorganism "*Xanthomonas translucens*", for instance, produces between forty and twenty receptors that are transported to the cytoplasm of wheat by an instance of the III secretion channel that resembles a tube (Boch and Bonas 2010). A type VI secretion system is used by other organisms of the *Xanthomonas* genus to transport the effectors extracellularly (Gardiner et al. 2014). By "shifting activator genes" between one organism to another one, a process known as "horizontal gene transfer," infections may additionally improve their pathogenicity. According to a theory put up by Friesen et al. (2006), this event took place when the "necrosis influence ToxA mutation" was transmitted from "*Pyrenophoratritici-repentis*" to "*Parastagonosporanodorum*", resulting in the illness that causes "tan spot in wheat." These effectors are recognized by plant effector-triggered immunity (ETI), which then starts a defense reaction. Such responses must vary since infections vary in the way they receive nutrients from bacteria that feed on healthy tissue as well as "necrotrophs that feed on dead tissue". HR that is meant to prevent biotrophs from growing only promotes necrotroph infection (Glazebrook 2005). By converting from biotrophy to necrotrophy, hemi-biotrophs deceive the plant (Vleeshouwers and Oliver 2014), generating effectors in waves that vary depending on the tissue and stage of infection (Toruño et al. 2016). Pathogens have evolved effectors that undermine many of the different signaling processes, which have caused the complex mechanism that leads to pathogen resistance to be subverted at several levels. As we will go over in more detail later, the initial stage of pathogen detection focuses on the identification of molecules that are specific to pathogens or produced when a pathogen is present (Fig. 1).

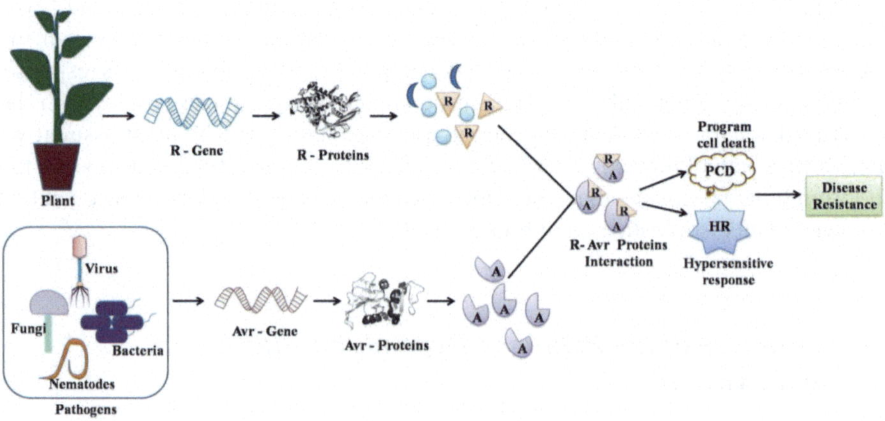

Fig. 1 Development of disease resistance and plant-pathogen interaction

2 Detection of Pathogens

2.1 WAKs and PRRs Track Down PAMPs and DAMPs

A variety of microorganisms can be recognized by PRRs, including proteins from bacteria, fungal polysaccharides, and nucleic acids from viruses (Niehl et al. 2016; Sun et al. 2013; de Jonge et al. 2010). These kinds of receptors often have "leucine-rich repeats" and can attach to external ligands, cytoplasmic kinase domains for phosphorylation-based "signal transduction," and transmembrane regions required to allow insertion in the cell membrane (Zipfel 2014). Due to its ability to bind to a variety of elicitors, LRRs are extremely varied. The regulation of various PRRs depends on the "protein brassinosteroid insensitive 1-associated receptor kinase 1" and "somatic embryogenesis trigger-like kinases" (Monaghan and Zipfel 2012; Prince et al. 2014). The release of "kinase motifs can go within the nucleus" and the beginning of "transcription reprogramming" by activated PRRs may stop the continuous induction of extensive signaling (Park and Ronald 2012). Molecules recognized via diverse PRRs include: Peptidoglycan, flagellin, and EF-Tu are components of bacteria (Zipfel et al. 2004; Kunze et al. 2004; Gust et al. 2007), "fungal (chitin, xylanase) (Kaku et al. 2006; Ron and Avni 2004), oomycete (β-glucan and elicitins) (Fliegmann et al. 2004; Du et al. 2015), viral (double-stranded RNA) (Niehl et al. 2016), and insect (aphid-derived elicitors) (Prince et al. 2014)". Identification of fungi reactive oxygen species connected to the wheat protein recognition receptors TaRLP1.1, TaRLK-R1-3, and TaLRK10 to susceptibility to rusty of the genus *Puccinia* Sp., because numerous studies of this nature were conducted to comprehend specific molecular links in "*Arabidopsis*" (Feuillet et al. 1997; Zhou et al. 2007; Jiang et al. 2013).

Additional receptors detect damage by identifying components of cells that are currently damaged by pathogenic enzymes, in contrast to PRRs can sense non-self-pathogen molecules upon infection. These were previously demonstrated in Arabidopsis with the sensing of extracellular ATP by "DORN1/LecRK-I.9 or oligogalacturonides" by WAK1 (Brutus et al. 2010; Choi et al. 2014). Like PRRs, WAKs include cytoplasmic kinase domains and an exogenous "galacturonan-binding domain" at their "N-terminus" that engages in pectin molecules within the cell walls. Oligogalacturonic acid, which is produced when fungal enzymes break down the pectin in plant cell walls, is detected by WAK1 and WAK2 (Brutus et al. 2010). Plant lectins can distinguish between carbohydrates that come straight from pathogens or from harm caused by infection (Lannoo and Van Damme 2014). Protein recognition receptors (Lannoo and Van Damme 2014). Plants can identify many extracellular molecules that serve as indicators of pathogen infection, including exogenous ATP, NAD(P), and DNA (Gust et al. 2017). To prevent PAMPs from being detected and to decrease PTI's effectiveness, pathogens have developed. Chitin-binding proteins, such as Avr4 and Slp1, are produced by *Cladosporium fulvum* and *Magnaportheo-ryzae*, respectively, to block plant perception (Lo Presti et al. 2015; Mentlak et al. 2012; van den Burg et al. 2006). According to the zig-zag hypothesis (Jones and

Dangl 2006), vegetation evolved ways to avoid the effectors that infections release to thwart several aspects of plant defense. They use a distinct, greater variety of proteins to recognize these pathogen effectors that encourage infection in plants.

2.2 NLRs Find Effectors of Pathogens

The two gene groups that have evolved the quickest are "NLRs" and R genes. For the beginning of forward signaling, the shape of the molecules makes switches from an open, ATP-binding phase having accessible the N-terminus motifs to a compacted, ADP-binding phase (Cui et al. 2015; Michelmore et al. 2013; Takken and Goverse 2012). A naturally preserved field containing a site of Nucleotides binding that appears in "R Proteins, *Caenorhabditis elegans* Death-4 Protein (NB-ARC), and Apoptotic Protease-Activating Factor" comes before a naturally preserved domain with a "Nucleotide-Binding site" (Takken and Goverse 2012). While the toll-interleukin receptor (TIR) domain at the N-terminal is only present in dicotyledon "N-terminal coiled-coil (CC)" domain, nucleotide-binding site leucine-rich repeat (NBS-LRR) proteins are present in both monocots and dicots (Jacob et al. 2013). Because the LRR's variability is the same as that of PRRs, these receptor types may recognize a wide variety of effector forms. Unlike PRRs, which are typically found outside the cytoplasm, LRRs are situated at terminated by a free carboxyl group (-COOH) of NLRs. In the "NB-ARC, there are several maintained motifs, including the phosphate-binding loop of P-loop NTPases (the Walker A motif), resistant nucleotide-binding site, and leucine-rich repeat (NLR, also known as NB-LRR), RNBS-C, RNBS-B, GLPL, Kinase-2/Walker-B, MHDV", and RNBS-D which are designated according to the preserved amino-acid sequences they contain. Since none of the domains are necessarily over-functioning, the lack of a "P-loop in the rice Pb1 CNL protein" proves this. Nucleotide required for cell death (NRCs), helper NLRs, and sensors were used to demonstrate how NLRs function in networks (Wu et al. 2017, 2016). "Through non-self and modified self" identification, NLRs have developed into sophisticated receptors that can identify a wide range of alterations (Baggs et al. 2017). NLR interactions are not always beneficial since in hybrids, mismatched NLRs can cause autoimmunity (Tran et al. 2017). According to the genomes of *Streptomyces coelicolor* (Horinouchi et al. 1990), nematodes (Hengartner and Horvitz 1994), and humans (Zou et al. 1997), the latter two of which have been linked to the programmed death of cells, nucleotide-binding domain-ARCs resembling the ones detected in Nucleotide oligomerization domain (NOD)-like receptors are present in many different organisms. Extra-pericentromeric clusters (Andersen et al. 2016), where chromosomal recombination is common, are where many NLR genes are found. These genes can change rapidly because of chromosomal rearrangements, duplications, and uneven cross-over (Marone et al. 2013). In the development of sequences of regulation like promoters, transposable elements are also involved (Le et al. 2014; Seidl and Thomma 2017). Increased chances of functional diversity result from the translocation of NLR genes to unrelated sites (Wu et al. 2017). Numerous interactions between effectors and NLRs are being identified in *Arabidopsis*, like PRR interactions (Meyers et al. 2003). NLRs can be directly triggered by pathogen effectors (Dodds et al. 2006), or they may be stimulated by identifying changes to the effector's targeted protein; (Van Der Biezen and Jones 1998; Dangl and Jones 2001), a decoy protein that mimics the effector's target(van der Hoorn and Kamoun 2008), or the NLR itself (Wu et al. 2015; Kroj et al. 2016; Cesari et al. 2014). The receptors Resistance to *Pseudomonas syringae*1 and Resistance to *Pseudomonas syringae* 2, which detect the targeted expression of the resistance negative repressor of interacting protein4 via *Pseudomonas syringae* effectors AvrRpt2 and AvrRpm1, respectively," are some of the least thoroughly studied "NLR-effector interactions" (Marathe and Dinesh-Kumar 2003). While the guardian protein, "RIN4", also attaches to the plasma membrane, Resistance to *Pseudomonas syringae*1 and Resistance to *Pseudomonas syringae*2 dwell inside the cell's membranes in contrast to other NLRs (Axtell and Staskawicz 2003).RPM1 connects to the negative regulation of transcription by competitive promoter binding (AtTIP49a) after activation, which strengthens resistance mechanisms (Holt et al. 2002).

As targets for effectors, which include the "transcription factor sequences," the triggers often the goal, NLRs may contain integrated decoy domains with a variety of structural variations. The WRKY TF domain enables *Arabidopsis thaliana* immunity with "*Ralstonia solanacearum*1" to bind directly to promoter sequences (Noutoshi et al. 2005). The WRKY domain is altered by the *Ralstonia solanacearum* effector PopP2, resulting in the stimulation of the NLR (Sarris et al. 2015; Le Roux et al. 2015). NLRs can dimerize as well. For instance, the "*Magnaporthe oryzae*" triggers "Avr-Pia" and "Avr1-Co39" can be directly bound by the "CNL proteins RGA5 and RGA4" from rice. There is still more research to be conducted to completely understand the way "NLRs self-associate or dimerize", particularly economically relevant genes like wheat stem rust resistance genes (Casey et al. 2016). NLRs utilize additional proteins to convey signals, like how PRRs depend on brassinosteroid-insensitive 1-associated receptor kinase 1 (BAK1). EDS1 regulator protein (Enhanced Disease Susceptibility 1) and homolog of NDR1 (Non-Race Specific Disease Resistance 1) proteins are linked to CNLs and TNLs, respectively (Day et al. 2006; Knepper et al. 2011; Wiermer et al. 2005). NLRs also can localize to specific cell parts, including the nucleus as well as endosomes. Because of the Phytophthora infestans activator Avr3aKI, potato-specific Late Blight Resistance Protein R3a migrates toward the endosomes and then interacts with additional effectors (Engelhardt et al. 2012). Polymorphic barley CNL and mildew A (MLA) proteins accumulate within the cell nucleus to prevent WRKY TFs from downregulating immunity (Shen et al. 2007). Exocytosis may be utilized by plants to produce antimicrobial chemicals and transfer immunity sensors across membranes in cells. Certain pathogen effectors obstruct endocytosis (Chaparro-Garcia et al. 2015), protease secretion (Bozkurt et al. 2011), and vesicular trafficking to gain an advantage from the interruption of antibacterial chemical synthesis. Effectors from "*Pyrenophoranodorum*" and "*Pyrenophoratritici-repentis*", "SnTox1" and

"PtrToxA", exploit the vulnerability "genes Snn1" and "Tsn1", and these encode the NLR and WAK proteins (Faris et al. 2010; Shi et al. 2016; Liu et al. 2017; Ciuffetti et al. 2010; Liu et al. 2016), respectively. Due to this immune system hijacking, necrotrophic diseases can cause HR through the development of "reactive oxygen species" (Manning et al. 2009). To start defense reactions, PRRs, WAKs, and NLRs depend on intricate signaling systems. Significant contributions to this signal transduction are played by MAPKs, hormones, TFs, and other elements.

3 Transduction of Signals

3.1 *Numerous Signaling Pathways Mediate Resistance.*

Numerous cellular activities involve the signaling pathways that receptors activate, including ubiquitin, G-proteins, calcium fluctuations, and MAPKs. The usual Mitogen-activated protein kinases signaling paradigm states that Ras proteins, which are membrane-attached, aid in the conversion of guanosine triphosphate to guanosine diphosphate by "phosphorylating Raf proteins are MAPKKKs and thus phosphorylate and activate a MAPKK (MEK) proteins," "phosphorylating MAPK (ERK) proteins," and so on (Meng and Zhang 2013). The finding of "Mitogen-activated protein kinases genes in *Arabidopsis*" that contain 20 MAPKs, 10 MAPKKs, and 60 MAPKKKs (Ichimura et al. 2002) was made possible by the significance of Mitogen-activated protein kinases in numerous cell processes. PRRs/Coreceptors flagellin-sensing 2 and EF-TU receptor merges with BCL2 antagonist/killer 1 and Mitogen-activated protein kinases signaling are triggered in response to interactions between bacterial flagellin and the extension factor (Zipfel et al. 2006; Chinchilla et al. 2007). A MAPK cascade is also started by the pathogen pectin degradation that wall-associated kinases 1(WAK1) and wall-associated kinases 2 (WAK2) detect (Brutus et al. 2010; Galletti et al. 2011; Kohorn et al. 2012). MAPK genes are also implicated in the signal transduction of NLR perception, according to studies in the tomato (Melech-Bonfil and Sessa 2010; del Pozo et al. 2004; Oh and Martin 2011). MAPK signaling may also suppress defense systems (Frye et al. 2001), and pathogens synthesis triggers may collaborate with Mitogen-activated protein kinases (MAPK) signaling to suppress defense mechanisms (Zhang et al. 2007). This is because it participates in many biological functions, "the large G protein allosterically activated by G-protein-coupled receptor (GPCR) system" receives a lot of attention. When external ligands bind to the membrane-bound GPCR of the α-subunit of the protein known as the G-protein multifaceted, the GDP in the subunit is changed to GTP, resulting in the subunit's separation from the complex with the β-γ-subunit and subsequent signaling (Temple and Jones 2007). The α-subunits then reassociate after the subunit hydrolyzes GTP. Although G-proteins play a part in HR and stomatal closure (Zhang et al. 2012), metazoan systems employ G-protein signaling more frequently (Trusov and Botella 2016; Temple et al. 2010; Jones and Assmann 2004).

Numerous signaling systems, particularly those involved in defense, are activated by ubiquitination and the subsequent proteasomal destruction of proteins. Components are adversely controlled by targeted degradation or favorably by the inhibition of their degradation (Marino et al. 2012). To interfere with this signaling and spread infection, infectious pathogens create effectors that interact with the proteasome-ubiquitin pathway (Marino et al. 2012). Plants additionally utilize small ubiquitin-like modifiers (SUMOs) to control response, and pathogens also interfere with this signaling (Park et al. 2011).

Receptors that change the number of the mineral calcium (Ca_2^+) serve as signaling networks that start responses to friendly or hazardous bacteria (Lecourieux et al. 2006; Baggs et al. 2017; Inoue et al. 2013). Many families of transcription factors, particularly CAMTAs (calmodulin-binding transcription activators), are activated by calcium when it is present. This is accomplished via calcium-binding messenger protein, calcium-dependent protein kinases (CDPKs) superfamily genes, and proteins like "calcineurin B" (Reddy et al. 2011; Poovaiah et al. 2013). According to Takahashi et al. (2011), CaM activates the MAPK cascade, which results in the production of ROS. The expressions of the NDR1 and EDS1 proteins, as well as hormone activation, are both regulated by calcium signaling (Poovaiah et al. 2013). CDPKs move to their nucleus and "phosphorylate WRKY TFs" associated with RPM1 ETI and RPS2 (Gao et al. 2013). The hormones, which play a variety of functions in stress and developmental responses, may deliver such a chemical signal (Meng and Zhang 2013). Hormonal changes regulate the different expressions of defense response genes like calcium signaling. Transient Mitogen-Activated Protein Kinase activity during Pamp-triggered immunity is significantly controlled by hormone signaling, but protracted MAPK expression during ETI indicates less dependence on hormone control (Tsuda et al. 2013) (Fig. 2).

3.2 Resistance is Systematically Initiated and Repressed by Plant Hormones

Auxin, jasmonate, cytokinin (CK), abscisin II, salicylic acid, nitrogen monoxide, gibberellic acid (GA), ethylene (ET), as well as steroid hormones (BR), are a few of the hormones that work after pathogen detection and provide an additional level of regulation. In addition to altering an essential number among different development as well as reaction processes, performs a crucial function throughout both regional and systemic immune defenses against "biotrophic" along hemibiotrophic disorders (Vlot et al. 2009). Upstream interactions between MAPK and SA pathways may result in certain pathways stimulating salicylic acid activities while others are triggered by SA (Vlot et al. 2009). Phytoalexin deficient 4 (PAD4) and EDS1 work together to transmit signals from receptors to SA in the case of TNL receptors and CNL receptors, respectively (Wiermer et al. 2005; Aarts et al. 1998; Century et al. 1997). When a pathogen is present via a complex web of interconnections, SA delivers

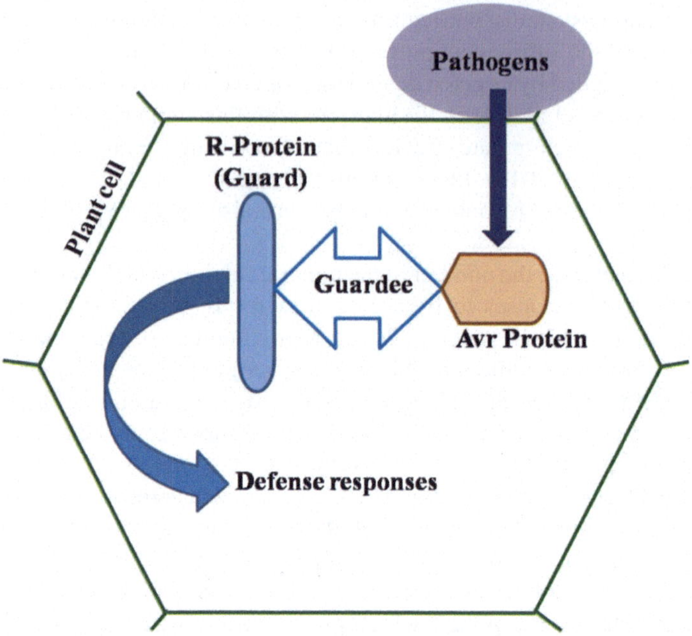

Fig. 2 Hypothesis of Guard. R genes are not directly influenced by the factor encoded by avirulence (Avr). Guardee protein, which initially interacts with Avr protein, then forms a complex with R protein and activates plant defense mechanisms, either directly or indirectly

the signal to activate transcription factors to boost the production of "defense-related genes". The "oligomer protein" NPR1's disulfide-bonded structures are broken down by thioredoxins after being activated by signaling, allowing its "monomers" to move across the cytoplasm toward the cell's nucleus, adhere with the transcription factor TGA ("binding site: TGACG"), and activate the genes involved in resistance (Robert-Seilaniantz et al. 2011; Tada et al. 2008). "The *Cochliobolusvictoriae* pathogenic effector, victorin", makes use of this mechanism to target the thioredoxin TRX-h5, which is necessary for the "monomerization of NPR1", and additionally to induce cells to die via controlling the LOV1 *Arabidopsis* susceptibility protein (Lorang et al. 2012).

The plant's reaction to herbivorous insects (McConn et al. 1997; Ruther and Kleier 2005) and necrotrophic diseases (Robert-Seilaniantz et al. 2011; Geraats et al. 2003; Vijayan et al. 1998; Thomma et al. 2001) is significantly influenced by JA and ET. When a caterpillar begins to eat plants, JA and ET increase the emission of volatile chemicals, which is triggered by the oral secretions of the caterpillar (Schmelz et al. 2003). Insect gut bacteria can weaken plants' JA-mediated defenses to sabotage plant responses (Chung et al. 2013). For the signaling process, the detection of "bacterial flagellin" causes a rise in ET production (Denoux et al. 2008). "Transcription factors ethylene insensitive 3 (EIN3)" does destroy through ubiquitin–proteasome system activity when ET" is not present. Raf-like protein kinase constitutive triple response1 and receptors are inactivated by ET, which prevents the inhibition of "EIN2 and EIN3" and promotes the overexpression of ethylene signaling, the production of "defense genes, and necrotroph resistance" (Robert-Seilaniantz et al. 2011; Tsuda and Somssich 2015). This is a target of pathogen intervention as well since the TF SIERF4 is Desumoylated by the "XopD effector of the tomato disease *Xanthomonas euvesicatoria*" to disrupt hormone signaling, specifically to inhibit ET synthesis and resistance (Kim et al. 2013). The essential functions of the regulators' nitric oxide, abscisic acid, auxin, gibberellic acid, cytokinin, and brassinosteroid in the immune while growing highlight the close ties between defense and growth, which are usually antagonistic (Pieterse et al. 2009). Two ways that ABA is connected to various plant stressors are the suppression and stimulation of resistance responses in an occurrence as well as a lack of environmental stress, respectively (Ton et al. 2009). ABA signaling must be present in essential stomata closing to regulate water loss, gas exchange, and pathogen access to tissue (Lind et al. 2015; Melotto et al. 2008). Deficits in GA and ascorbic acid (AA) increase resistance (Barth et al. 2004; Pavet et al. 2005). Peptides can act as plant hormones as hormones are frequently referred to as systemic regulators. The symbiotic herbivory response is linked to modifications in the expression of genes, especially in neighboring crops that have not been under biotic stresses, according to recent studies on the short peptide hormone system (Coppola et al. 2017). It suggests plant hormones may trigger interplant interaction to enhance defensive responses. We will now concentrate on those proteins that directly influence transcription in the genome after briefly discussing some of the key signaling components.

3.3 Transcription Factors Are Important for Transcriptional Reprogramming

At various levels of resistance, transcriptional reprogramming is facilitated by TF activity. These levels include (1) proteins associated with resistance that are normally expressed including kinases, receptors, and reaction-suppressing proteins. (2) triggered receptor proteins regulate Transcription factor action, and (3) Transcription factors that are triggered after receptor initiation (i.e., the MAPK cascade that phosphorylates TFs to activate them) (Qiu et al. 2008). In contrast to metazoan systems, plant TFs were varied. The final six of these include particularly significant resistance (Tsuda and Somssich 2015), and they are the MADS-box, homeodomain, AP2/ERF, C_2H_2 zinc finger, TGA/bZIP, bHLH, MYB, WRKY and NAC, families. Rice tolerance to *Chilo suppressalis* as well as *Arabidopsis* tolerance to *Botrytis cinerea* is positively regulated by APETALA2/ethylene response factor (AP2/ERF) and is connected to ET control (Lu et al. 2011; Zhao et al. 2012; Moffat et al. 2012).In this last instance, ET responsive factor 6 (ERF6) stimulates plants' defensins 1.1 as well as 1.2 after MPK3/MPK6 phosphorylation (Meng et al. 2013). Nicotine is a secondary metabolite derived from *Nicotiana attenuata* that inhibits herbivory, and

NaMYC2 is an MYC2 TF that controls its synthesis (Woldemariam et al. 2013). Rice bZIP TF OsTGAP1, which is activated by chitin, increases the production of antimicrobial substances (Okada et al. 2009). Abiotic stressors including cold, drought, or injury may also influence resistance signaling (Seo et al. 2010; Wu et al. 2009). To counteract the effects of ethylene response factor 1 (ERF1), the bHLH TF MYC2 upregulates wounding responses while downregulating resistance (Lorenzo et al. 2004). A modified TF that can be activated in addition to kinase phosphorylation is the no apical meristem (NAM)-Arabidopsis transcriptional stimulation factor, cup-shaped cotyledon genes (Seo et al. 2010). Infection-specific TFs may additionally target genes implicated in *Blumeriagraminis* penetration resistance in barley HvNAC6 upregulated genes, for example (Jensen et al. 2007). A lot of the extensively researched resistance responses also seem to include WRKYs in addition to these kinds of TFs.

"Zinc finger domains and conserved WRKYGQK amino acid residues" allow WRKY TFs to attach with "W-box regions (TTGACC/T)" (Rinerson et al. 2015). WRKY transcription factors both beneficially and detrimentally regulate the regulation of genes connected to defense, such as those that affect the viral response, in addition to playing a role in abiotic circumstances including salt and drought tolerance (Chen et al. 2013). WRKY TFs are split into three categories: I, IIa-e, and III (Rinerson et al. 2015) to identify potential evolutionary connections. These categories are based on the examination of ancestral species, such as algae. MPK3/MPK6 phosphorylates WRKY33, which then attaches the transcription regulators for genes required in the synthesis of ethylene gas (Li et al. 2012) and antibacterial compounds (through phytoalexin deficiency 3 Mao et al. 2011), which is a part of *Arabidopsis* immunity to *B. cinerea.* On the other hand, PTI is suppressed by barley HvWRKY1 and HvWRKY2, but when a bacterium *B. graminis* effector AVRA10 is present, "CNL protein mildew locus A10 (MLA10)" disturbs PTI (Shen et al. 2007). Interactions between the WRKY1 and MYB6 TFs and the barley CNL protein MLA1 increase *B. graminis* resistance (Chang et al. 2013). When triggered by the CNL protein Pb1, rice WRKY45 beneficially modulates blast resistance (Inoue et al. 2013). When "rice immune receptors regulators Xa21 is digested", the enzyme's motif is sent to the cell nucleus where it links against the inhibited nuclear transportation, demonstrating receptors transcription factors function against *Xanthomonas oryzae* (Park and Ronald 2012; Peng et al. 2008). The sequence variations of the allelic genes OsWRKY45-1 as well as OsWRKY45-2", respectively, have downregulation and upregulation protection against *Xanthomonas oryzae* even though each of them favorably modifies the resistance of *Magnaporthe grisea* (Tao et al. 2009).To control BR signaling, WRKY TFs collaborate with members of other TF families, such as the bHLH TFs AtBES1/BZR2 and AtBZR1 (Lozano-Durán et al. 2013). It has been shown that the wheat TFs TaWRKY49 and TaWRKY62 influence the activation of genes linked to jasmonic acid, ethylene, salicylic acid, and ROS, contributing to resistance to *Puccinia striiformis* (Wang et al. 2017a, b).

To inhibit the activity of crop transcription factors or to directly increase plants' gene regulation, infection-produced receptors may disrupt transcription transformation. The nuclear translocation of sweet potato transcription factors NAC is prevented by *Phytophthora infestans* effectors (McLellan et al. 2013). The nicotine mosaic virus appears as interfering with an *N*-acetyl cysteine TFs Arabidopsis activating factor2 to diminish resistance to force, just as the *P. syringae* effector HopD1 interacts with the membrane-bound TF NTL9 to reduce "effector-triggered immunity" (Block et al. 2014; Wang et al. 2009a, b). Effectors that activate as plant TFs are likewise produced by pathogens (Toruño et al. 2016). TALEs are produced by the bacteria *Xanthomonas* and *Ralstonia* and bind to the promoters of plant genes that confer susceptibility (Boch and Bonas 2010; Schornack et al. 2013). The transcription activator-like AvrXa10 and AvrBs3 of known Xoo bacterial strain implicated in defense to *Xanthomonas oryzae* along with *Xanthomonas campestris*, accordingly", have TALE-binding promoter sites in the rice gene Xa10 as well as pepper R gene Bs3, respectively (Tian et al. 2014; Römer et al. 2007). While transcriptional reprogramming is directly triggered by TFs, response-related genes are also regulated.

3.4 PDR and Epigenetic Regulation Offer Additional Regulatory Mechanisms.

In addition to being important targets for pathogen as well as plant degradation, nucleic acids also play a function in regulating plant immunity. By upregulating response proteins or downregulating response inhibitors, epigenetic modifications can promote pathogen resistance. The expression of several components of the plant defense response is impacted by chromatin alteration brought on by pathogen exposure (Jaskiewicz et al. 2011). Plant defense is influenced by recombination among homologous chromosomes, RNA interference (RNAi), DNA methylation, histone methylation/acetylation, and histone methylation/acetylation (Holeski et al. 2012). Pathogen resistance can be demonstrated to decrease when resistance-related genes are methylated (Le et al. 2014; Dowen et al. 2012; López Sánchez et al. 2016). In response to *Pseudomonas syringae*, the histone deacetylase HDA19 interacts with WRKY transcription factors to promote salicylic acid repression activity (Choi et al. 2012; Kim et al. 2008). According to Jaskiewicz et al. (2011) and Mosher et al. (2006), histone acetyltransferase, as well as methylation of gene regulators, are caused by the administration of salicylic acid analogs and *Pseudomonas syringae* infection in *Arabidopsis thaliana*, respectively. Epigenetic regulation enhances the way the host responds to probable dangers along with reducing the use of resources on uncertain threat defense. Due to their metabolic expense of resistance as well as the permanence in the pathogenic culture, plants that can inhibit and stimulate responses have a selective advantage. A further degree of control over a plant's resistance responses comes from epigenetic factors.

Diseased-derived resistance employs the usage of pathogenic components that provide immunity against infection, in contrast to the ligand-receptor types of resistance previously addressed. If the plant has been exposed to a less harmful strain of the pathogen, it may develop increased resistance to infection with a later, more virulent race or pathotype (Burketova et al. 2015). *Rhynchosporium commune* resistance was improved in the offspring of barley that had been chemically primed (Walters and Paterson 2012). It is fascinating to note that wheat primed *Fusarium graminearum* to produce more mycotoxin (Ameye et al. 2015) by increasing its tolerance to the fungus. PDR to viruses usually involves the activity of virus capsid proteins, replication, and interference RNAs (Beachy 1997; Abel et al. 1986). Plants that produce a virus' coat protein are resistant to different viral strains because they interfere with virus disassembly, which is necessary for viral replication (Namba et al. 1992). The plant makes microRNAs from viral sequences to stop virus transcription, and this is necessary for viral reproduction in the cell of the host. Pathogenic RNA is degraded by the antiviral process known as RNAi. According to Cruz et al. (2014), RNAi acts as the fundamental mechanism that created genetically modified wheat resistant against the wheat streak mosaic virus. According to some infections, the silencing of RNA is inhibited by effectors (Navarro et al. 2008; Qiao et al. 2013; Pumplin and Voinnet 2013). The establishment of soybean nematode resistance by RNAi has been additionally suggested (Tian et al. 2016).MicroRNAs, alternative splicing, and alternative polyadenylation are all used by plants to control defense-related components (Zhai et al. 2011; Li et al. 2012; Liu et al. 2017; Yang et al. 2014; Tsuchiya and Eulgem 2013; Sela et al. 2012; Saintenac et al. 2013; Halterman et al. 2001, 2003). According to Wang et al. (2017a, b), pathogens can now use small RNAs are effectors to infiltrate plant cells along with inhibit the host immune systems. "*Botrytis cinerea*" employs "RNA silencing of the host" (Weiberg et al. 2013) and *Phytophthora sojae* inhibits RNA silencing (Qiao et al. 2013; Qiao et al. 2015) to enable infection. Additionally, improved resistance to RNA viruses and antifungal activity have been linked to ribonucleases (RNases) in the apoplast (Sugawara et al. 2016; Galiana et al. 1997; Hugot et al. 2002; Filipenko et al. 2013). Strong RNase and antifungal activities are seen within the wheatwin1 protein (Caporale et al. 2004). In response to the numerous sensors and messengers that detect the presence of insects, plants have an extensive arsenal of immune systems that stop diseases and insects from spreading, growing, herbivory, and reproducing.

4 Defense Reaction

4.1 Pathogen Infection is Inhibited by Cell Wall, HR, and ROS Modification

In the area of a pathogen infection, the immune responses that are most often used cause the programmed death of cells. This establishes an incubation zone to stop a virus it is propagating; this approach is effective against diseases that depend on living tissue (biotrophs). When a pathogen infects a host, peroxidases are produced to produce ROS, which is employed in several elements of the resistance responses (Passardi et al. 2005; Almagro et al. 2008). Nicotinamide adenine dinucleotide phosphate (NADPH) oxidase must produce superoxide before peroxidases may use it to produce the chemical compound hydrogen peroxide. To produce ROS, a single nicotinamide adenine dinucleotide phosphate (NADPH) oxidase, respiratory burst oxidase protein D (RBOHD), interacts through the protein recognition receptors FLS2 and EFR is phosphorylated by BIK1 (Kadota et al. 2014). When the compound hydrogen peroxide spreads throughout cells that surround it to start the creation of chemicals to prevent oxidation injury, ROS leads to a programmed death of cells (Levine et al. 1994). It was discovered that transgenic plants with impaired detoxification of ROS molecules reacted more strongly to infections that cause HR (Mittler et al. 1999). The process of an oxidative burst, which is how ROS are utilized in HR, also makes conditions unfit for pathogen survival and reproduction (Lamb and Dixon 1997). As a result, ROS directly interferes with signal transduction and the defense response, preventing the germination of fungal spores (Lamb and Dixon 1997). Other enzymes, including amine and oxalate oxidases, along with peroxidases and NADPH oxidase, also result in the production of ROS (Kärkönen and Kuchitsu 2015). Oxalic acid is a byproduct created by the necrotrophic fungus *Sclerotinia sclerotiorum* that decreases plant oxidative burst during the initial phases of infection but boosts ROS generation once it has established itself (Cessna et al. 2000). Oxalate oxidase proteins sometimes referred to as germins, are produced by wheat and barley and increase their pathogen resistance (Dumas et al. 1995) by degrading oxalic acid (Dong et al. 2008; Donaldson et al. 2001; Liu et al. 2015). Genetically modified crops employing barley or wheat "oxalate oxidase genes" were shown to have increased defense against *S. sclerotiorum* as well as various insects (Liang et al. 2001; Ramputh et al. 2002; Wei et al. 1998; Zimmermann et al. 2006).

Glycoprotein crosslinking is mediated by ROS and strengthens the walls of cells (Lamb and Dixon 1997; Bradley et al. 1992). Even though bacteria, fungi, and nematodes must break through their cell membranes to reproduce, doing so hinders pathogenic mobility as well as restricts their ability to reach the nutrition required for growth. Since they are deficient in some essential degrading enzymes that numerous fungi require, bacterial pathogens use wounds and stomata to get plant nutrients. Stomatal guard cells close the stomata to prevent ingress by sensing bacterial PAMPs (like flg22) and lipopolysaccharides (Melotto et al. 2008, 2006). *P. syringae* responds by producing coronatine, which mimics phytohormones and interferes with the production of hormones to start the reopening of closed stomata (Melotto et al. 2006). Earlier, it was thought that stomata acted as an inactive conduit for pathogen penetration via plant cells; however, it has been demonstrated that control of stomata involves an innovative defense system along with its reaction to stress caused by abiotic factors. Plants and diseases both create several other chemicals that can impact a host's capacity to metabolize their proteins, carbohydrates, and lipids in addition to oxidase.

4.2 The Enzymes and Inhibitors of Enzymes Prevent Harmful Effects and Help in the diagnosis

Cellulose-like catalysts are used by fungi to break down the cell walls of plants. Plants produce inhibitors of enzymes by the discovery of such fungal proteins, and they also deposit callose and lignin to fortify the cell wall (Bellincampi et al. 2014). Pectinases and xylanases are also used by diseases like *Fusarium culmorum* and *F. graminearum* to break down plant polysaccharides along with cellulases (Kang and Buchenauer 2000; Wanjiru et al. 2002). As a kind of evolutionary retaliation, plants have evolved enzymes like "chitinases and -1–3-glucanases" and disintegrate polysaccharides contained in infections (Kong et al. 2005; Anguelova-Merhar et al. 2001). In addition to stopping the development of bacteria, this breakdown of the fungal cellular components also renders PAMPs, which are readily accessible to the plant's PRRs, undermining the insect's ability to flee. Chitinase activity in wheat stops the germination of fungus spores by impairing a vital component that makes up cell membranes (Collinge et al. 1993; Broekaert et al. 1988). It has been shown that transgenic wheat chitinases serve as an antifungal agent against a variety of species of fungi, and not just those that are damaging to wheat (Singh et al. 2007). Chitin is shielded from plants chitinases by the Avr4 binds to chitin of *Cladosporium fulvum* (van den Burg et al. 2006). Hevamine possesses cylindrical structural domains and lysozyme and chitinase activity, which are frequently related to cellular metabolism (Nagano et al. 2002). Hevamine also performs a variety of additional tasks. Plant 1,3-glucanases hydrolyze 1,3-glucan in fungal cell membranes similarly to the enzyme chitinases along with produce monomers that additionally activate plant defense mechanisms (Leubner-Metzger et al. 1999; Doxey et al. 2007). This comprehensive strategy attempts to increase plant defenses while decreasing the potency of pathogenic components. To propagate to the host organism, diseases must have a broad array of proteins, which complicates the structure of plant defense proteins and receptors. Cell walls are made up of several diverse components, including cellulose, hemicellulose, pectin, and lignin. Therefore, the struggle during evolution for the capacity to penetrate and strengthen the outer layer of cells may be viewed as an analogy for the full mutual evolution in the general plant-infection connection.

Its efficacy of enzyme molecules (such as chitinase from plants and cellulase from fungi) has been decreased because of the synthesis of proteases caused by infections and plants. Protease inhibitors are utilized as well by diseases and plants to stop the action of these proteases (Jashni et al. 2015). To demonstrate the way proteases, which function across multiple stages in plant and pathogen interactions, the authors address 10 important types of proteases, classifying them according to their position: cytonuclear, apoplastic, endomembrane, and vacuolar (Thomas and van der Hoorn 2018). Slavokhotova et al. (2014) claim that wheat produces hevein-like antimicrobial compounds that bind to fungalysin without cleaving it, stopping the degeneration of the enzyme chitinases which is essential for avoiding disease. Defense chitinases are broken down to produce *Fusarium verticillioides* Zn-metalloproteinase fungalysin (Naumann et al. 2011). Trypsin, chymotrypsin, and α-amylase from *Fusarium* are inhibited by barley protease inhibitors (Pekkarinen et al. 2007). Inhibitors of wheat and barley-amylase, which are utilized by insects and fungus to metabolize starches, prevent pests from digesting starch (Franco et al. 2000, 2002; Feng et al. 1996; Mundy et al. 1983; Petrucci et al. 1976). Both non-proteinaceous and proteinaceous types of α-amylase inhibitors are produced; the latter are naturally occurring substances that mimic α-amylase substrate (Franco et al. 2002). *N. attenuata* generates trypsin enzyme inhibitors in addition to nicotine to fight *Spodoptera exigua* (Steppuhn and Baldwin 2007). Through protease inhibitors, pathogens can also be found in plants. For example, "tomato RLP Cf-2 detection" of *C. fulvum* detects "Avr2" which was initially developed to prevent the Rcr3 and Pip1 are paralogous secreted papain-like proteases of tomato (Shabab et al. 2008; Rooney et al. 2005). The interference of tomato protease activity by the oomycete *P. infestans* and the worm *G. rostochiensis* shows a complicated network of protein–protein interactions (Song et al. 2009; Lozano-Torres et al. 2012).

Lipids take an active role in several biological processes, like what proteins do. A host plant's main defense against potential pathogens is formed by lipids. Once a pathogen detects cutin or other components of the waxy cuticle, the infection begins. After coming into touch with the wax on the surface of wheat (Reisige et al. 2006) and barley (Tsuba et al. 2002), respectively, *Puccinia graminis* and *Blumeriagraminis* start to develop appressoria. To get across the cuticle, fungi use cutinases to hydrolyze cutin into cutin monomers (Woloshuk and Kolattukudy 1986). Wheat (Gottwald et al. 2012) and *Arabidopsis* (Oh et al. 2005) both contain plant lipases that prevent fungus growth. The antibacterial properties of lipid-transfer proteins are also correlated with enhanced pathogen permeability to the membrane. Nevertheless, further information regarding how these kinds of proteins impact bacteria has yet to be clarified (Finkina et al. 2016). Effectors from *P. syringae*, for example, need lipid modification to enter the host cell (Nimchuk et al. 2000), and instead of the cytoplasm, the membrane of plasma has NLR receptors and can detect such effectors (Boyes et al. 1998; Leister et al. 1996). The interference with sphingolipid metabolism through the *Fusarium* toxin fumonisin (Abbas et al. 1994) or the defense mechanisms set off by bacterial lipopolysaccharide PAMPs (Desaki et al. 2006; Zeidler et al. 2004) are examples of how lipids can also serve as specific targets of pathogens or plant sensing. To produce defense-related components, phospholipids, and galactolipids are broken down into free fatty acids by phospholipases and lipoxygenases (LOXs), which can also play a role in stomatal closure (Montillet et al. 2013). Phospholipase activity, known to be connected to a variety of hormonal and stress-related reactions (Zhao 2015), can produce substances that may be specifically connected to the inflammatory response, including phosphatidic acid (Canonne et al. 2011; Rentel et al. 2004). Along with the general activity of enzymes related to the components of cells, plants have evolved specific proteins and defend themselves against infection by pathogenic organisms.

4.3 Defensins and Proteins that Resemble Thaumatins Actively Prevent Infections Caused by Pathogenic Bacteria

The classes of tiny plant-derived proteins known as defensins actively repel or prevent parasitic plants and encroaching microorganisms (de Zélicourt et al. 2007). Despite being initially identified as γ-thionins in both barley and wheat, it has been demonstrated that they have structural similarities with animal defensins (Colilla et al. 1990; Mendez et al. 1990). Like how many pharmaceutical antibiotics function, plant protein defensins disrupt the infections' ability to produce proteins and regulate their enzymes. The barley defensins ω-hordothionin and γ-hordothionin can penetrate the cytoplasm of pathogens (Van Der Weerden et al. 2008, 2010) and stop the protein translation (Mendez et al. 1990; Méndez et al. 1996). Tad1 ("*Triticum aestivum* defensin 1"), which has antipathogen characteristics, is expressed in the crown (Koike et al. 2002). Defensins disrupt ions-signaling (Spelbrink et al. 2004), inhibit the enzyme proteases (Wijaya et al. 2000), and cause infection ROS generation (Aerts et al. 2007). Numerous plant defensins are anti-fungal and particularly potent in seeds, in contrast to mammal defensins, which prevent the development of germs. According to Stotz et al. (2009), they account for 0.5% of the total amount of proteins in seeds as well as 30% of the protein molecules that are released from seed coverings. Numerous tissues contain defensins, which can also be produced as the seasons change (Koike et al. 2002; Garcia-Olmedo et al. 1998). The γ-core domains and C-terminal hydrophobic serve as vital membrane contact and anti-fungal properties, respectively (De Paula et al. 2011; Lacerda et al. 2014). With bridges formed by disulfide, the purothionin, knottin, and scorpion toxin-like domains of defensins create complex "cysteine-knot" structures. Similar protein domains are used to inhibit potassium channels by scorpion poisons and several plant defensins (Vriens et al. 2016; Almeida et al. 2002). The antibody such as defensins enhance the permeability of the pathogen membrane, which starts necrosis (Brogden 2005). Over 300 defensin-like proteins with high cysteine content have been found in *Arabidopsis* (Silverstein et al. 2005) and *M. truncatula* (Graham et al. 2004). Defensins have been divided into 18 classes according to species, structure, and function (Van Der Weerden and Anderson 2013). Vegetation's defense compound defensins have the potential to be used as artificial sweeteners, anticancer drugs, and medicinal antibiotics (Wang et al. 2009a, b; Ming and Hellekant 1994).

About the thaumatin peptide from *T. daniellii*, proteins involved in pathogenesis are referred to as thaumatin-like substances (Wel and Loeve 1972). The 1,3-D-glucans are bound by proteins that resemble barley thaumatin and are linked to an anti-fungal property (Hejgaard et al. 1991), resistance to *F. graminearum* and powdery mildew (Bryngelsson and Geen 1989), or both (Trümper et al. 2016). Zeamatin (maize), osmotin (tobacco), avematin (oat), trimatin (wheat), and hordomatin (barley) are antifungal permatin-like proteins that break the membrane gaps in fungal walls (Hejgaard et al. 1991; Vigers et al. 1991; Robers and Selitrennikoff 1990; Abad et al. 1996). Permatins build up in seeds similarly to defensins (Skadsen et al.

2000). According to studies conducted on barley, wheat, and comparable grains, the "thaumatin-like proteins PRHv-1 and PWIR2", which share sequences among oat (Lin et al. 1996), are expressed alongside additional genes during fungal resistance responses. "Thaumatin-like proteins" affect the grinding performance of barley and have the potential to be artificially sweetened (Singh et al. 2017). Thaumatin peptides generate a single in the seventeen classes of PR proteins: oxidase and oxidase, β-1,3-glucanases, chitinases, proteinase inhibitors, endoproteinases, lipid-transfer protein molecules, defensins, and thionins, thaumatin, and the lesser recognized anti-fungal as well as both functionally uncategorized (Pr-17) groups (van Loon et al. 2006; Van Loon and Van Strien 1999; Ebrahim et al. 2011; Sinha et al. 2014). As additional pest-repelling techniques, plants also use "non-protein" compounds as well as assistance via varied species.

4.4 Chemical and Biological Plant Weapons Include Phytoalexins and Beneficial Symbionts

In reacting to invading insects organic compounds known as "phytoalexins" are produced to hinder reproduction, development, and metabolism. It was once thought that "phytoalexins" protected potatoes with "*Phytophthora infestans*" as a form of defense. Numerous classes of plant chemicals have the potential to serve as pesticides; those which can do so have been termed "phytoanticipins". The plant *Arabidopsis* produces malalexin, the prototypical "phytoalexin", in reactions to a variety of microbe illnesses and pests (Kettles et al. 2013). WRKY TFs (Mao et al. 2011; Pandey et al. 2010; Ahuja et al. 2012) and MAPK cascades (Xu et al. 2008) control the activity of camalexin. Some resilient infections have the capacity to detoxify this compound (Pedras et al. 2011). The benzoxazinoids (BXs) that wheat generates, like "2,4-dihydroxy-7-methoxy-1,4-benzoxazin-3-one" (benzoxazinones), protect the plant from pathogens and pests and prevent the "barley yellow dwarf virus" from proliferating through aphid feeding (Weibull and Niemeyer 1995; Wouters et al. 2016; Givovich and Niemeyer 1991).The wheat-associated infections "*Gaeumannomycesgraminis* and *F. culmorum*" have the ability to detoxify BXs (Friebe et al. 1998).It is likely that wheat acquires the genes required to change "2,4-dihydroxy-1,4-benzoxazin-3-one (DIBOA)" to "DIMBOA" by means of the progenitors part of the "B genome", as *Triticum urartu* and *Aegilops tauschii* do not accumulate DIMBOA (Niemeyer and Jerez 1997). Additional phytoalexins reported in cereals include "avenanthramides" from oat as well as "diterpenoids" from rice (Ahuja et al. 2012). Avenacin A-1, a phytoanticipin belonging to the saponin group, is generated by the oat root epidermis and interacts with the sterols in fungal membranes to form holes (Osbourn et al. 1994; Morrissey and Osbourn 1999; González-Lamothe et al. 2009). Avenacin A-1 can be detoxified by *G. graminis* var. *avenae* (Osbourn et al. 1991). Numerous species are being reported to produce phytoalexins, some of which have been classed as having allelopathic effects (Wu et al. 2001), but the signaling pathways that cause their synthesis are still unknown. Additionally, phytoalexins could be advantageous in medical settings (Ahuja et al. 2012; Adhikari et al. 2015). Exogenous chemicals used on crops, including herbicides based on glyphosate, may make them more susceptible to disease (Martinez et al. 2018).

In addition to making insecticides, plants may create herbivore-induced plant volatiles as a kind of defense that might assist them in drawing in local "predators of herbivorous insects". The invasive wasps that survive by feasting the larval worms are attracted to them because of the terpenoid and indole compounds released by the caterpillars eating the maize leaves (Turlings et al. 1990, 1991, 1993). Volicitin from caterpillar oral secretions stimulates the activation of JA signaling (Schmelz et al. 2003). In addition to producing sticky chemicals like resin and latex that capture insects to ward off herbivory (McKay et al. 2003; Agrawal and Konno 2009; Becerra 1994), plants may also create more photosynthetic material (Thomson et al. 2003) to shield themselves from the predatory eating of their leaves. In addition to numerous cellular constituents including terpenoids, tannins, flavonoids, phenolics, and alkaloids, morphological traits like trichomes also inhibit insect herbivory (War et al. 2012). In addition to providing the host with access to nutrition, certain symbiotics help the recipient defend against illness. While wheat rhizobacteria generate antibiotic chemicals that are effective in fighting the soil-borne fungus "*G. graminis*", paddy "arbuscular mycorrhiza" induces increased defenses that safeguard their hosts (Cook et al. 1995). Corn mycorrhizae can increase DIMBOA production (Song et al. 2011). Symbionts are being demonstrated to influence resistance mechanisms by suppressing JA-mediated defense (Plett et al. 2014) or interfering with the formation of ROS and β-1,3-glucanase (Cárdenas et al. 2008; Mitra and Long 2004). These interactions show how a plant's defense mechanism has several layers.

5 Conclusions and Future Directions

Thanks to new molecular techniques and more potent computation, the twenty-first century will undoubtedly witness more advancement in the comprehension of "plant-pathogen interactions". When more information on plant-pathogen interactions becomes available, phytopathology, like other areas, will expand. The demand for long-lasting pathogen resistance in crops is pushed by several reasons, including disease pressures linked to contemporary agricultural practices and climate change (Zhan et al. 2015). Research to change crop genetics to create enhanced resistance will keep going along with improvements in our understanding of plant immunity. The most effective way to establish resistance is to keep modifying the receptors required to start defense responses. NLRs may develop into a significant biotechnological tool that is utilized to modify the CRISPR/Cas9 system's activity to build resistance to any disease. One new method makes use of *P. syringae* protease AvrPphB's cleavage of PBS1 to activate *Arabidopsis* NLR RPS5 (Shao et al. 2003; Qi et al. 2014). According to Kim et al. (2016), additional infections can trigger immune responses that cause the PBS1 cutting region to be replaced with a cutting region for different pathogens' proteases. Despite the current drawbacks of this approach, future research will change crops utilizing distinct "R-genes" that could not be derived from varied species. For a receptor to trigger an efficiently effective defense action, it is necessary to match receptors alongside an appropriate signal transmission pathway when engineering novel resistance routes to different diseases. By enabling vegetation to produce disease genes that promote resistance, "PDR" could potentially have uses within the field of biotechnology (Sarma et al. 2016). Future research is going to concentrate on understanding "qualitative resistance and gene pyramiding" due to the long-lasting resistance they possess (De Wit 1994), including the decades-long resistance seen in the barley cultivar NDB 112 (Zhou and Steffenson 2013; Wilcoxson et al. 1990; Haas et al. 2016; Valjavec-Gratian and Steffenson 1997). Future research will find more mechanisms of resistance control and response with implications for farming systems. Recognizing disease resistance along with resistance in plants can significantly boost agricultural output by reducing crop loss. Additionally, it will increase our understanding of the complex molecular relationships as well as evolutionary cooperation and underlie such a subject and the numerous ramifications for other biological systems that it has.

References

Aarts N, Metz M, Holub E, Staskawicz BJ, Daniels MJ, Parker JE (1998) Different requirements for EDS1 and NDR1 by disease resistance genes define at least two R gene-mediated signaling pathways in *Arabidopsis*. Proc Natl Acad Sci USA 95:10306–10311

Abad LR, D'Urzo MP, Liu D, Narasimhan ML, Reuveni M, Zhu JK, Niu X, Singh NK, Hasegawa PM, Bressan RA (1996) Antifungal activity of tobacco osmotin has specificity and involves plasma membrane permeabilization. Plant Sci 118:11–23

Abbas HK, Tanaka T, Duke SO, Porter JK, Wray EM, Hodges L, Sessions AE, Wang E, Merrill AH Jr, Riley RT (1994) Fumonisin-and AAL-toxin-induced disruption of sphingolipid metabolism with accumulation of free sphingoid bases. Plant Physiol 106:1085–1093

Abel PP, Nelson RS, De B, Hoffmann N, Rogers SG, Fraley RT, Beachy RN (1986) Delay of disease development in transgenic plants that express the tobacco mosaic virus coat protein gene. Science 232:738–744

Adhikari KB, Tanwir F, Gregersen PL, Steffensen SK, Jensen BM, Poulsen LK, Nielsen CH, Høyer S, Borre M, Fomsgaard IS (2015) Benzoxazinoids: cereal phytochemicals with putative therapeutic and health-protecting properties. Mol Nutr Food Res 59:1324–1338

Aerts AM, François IE, Meert EM, Li Q-T, Cammue BP, Thevissen K (2007) The antifungal activity of RsAFP2, a plant defensin from *Raphanus sativus*, involves the induction of reactive oxygen species in *Candida albicans*. J Mol Microbiol Biotechnol 13:243–247

Agrawal AA, Konno K (2009) Latex: A model for understanding mechanisms, ecology, and evolution of plant defense against herbivory. Annu Rev Ecol Evol Syst 40:311–331

Ahuja I, Kissen R, Bones AM (2012) Phytoalexins in defense against pathogens. Trends Plant Sci 17:73–90

Almagro L, Gómez Ros L, Belchi-Navarro S, Bru R, Ros Barceló A, Pedreno M (2008) Class III peroxidases in plant defense reactions. J Exp Bot 60:377–390

Almeida MS, Cabral KM, Kurtenbach E, Almeida FC, Valente AP (2002) Solution structure of *Pisum sativum* defensin 1 by high-resolution NMR: plant defensins, an identical backbone with different mechanisms of action. J Mol Biol 315:749–757

Ameye M, Audenaert K, De Zutter N, Steppe K, Van Meulebroek L, Vanhaecke L, De Vleesschauwer D, Haesaert G, Smagghe G (2015) Priming of wheat with the green leaf volatile Z-3-hexenyl acetate enhances defense against *Fusarium graminearum* but boosts deoxynivalenol production. Plant Physiol 167:1671–1684

Andersen EJ, Ali S, Reese RN, Yen Y, Neupane S, Nepal MP (2016) Diversity, and evolution of disease resistance genes in barley (*Hordeum vulgare* L.). Evol Bioinformat Online 12:99–108

Anguelova-Merhar V, Westhuizen A, Pretorius Z (2001) β-1,3-glucanase and chitinase activities and the resistance response of wheat to leaf rust. J Phytopathol 149:381–384

Antolín-Llovera M, Ried MK, Binder A, Parniske M (2012) Receptor kinase signaling pathways in plant-microbe interactions. Ann Rev Phytopathol 50:451–473

Axtell MJ, Staskawicz BJ (2003) Initiation of RPS2-specified disease resistance in *Arabidopsis* is coupled to the AvrRpt2-directed elimination of RIN4. Cell 112:369–377

Baggs E, Dagdas G, Krasileva K (2017) NLR diversity, helpers, and integrated domains: making sense of the NLR identity. Curr Opin Plant Biol 38:59–67

Barth C, Moeder W, Klessig DF, Conklin PL (2004) The timing of senescence and response to pathogens is altered in the ascorbate-deficient arabidopsis mutant vitamin C-1. Plant Physiol 134:1784–1792

Beachy RN (1997) Mechanisms and applications of pathogen-derived resistance in transgenic plants. Curr Opin Biotechnol 8:215–220

Becerra JX (1994) Squirt-gun defense in bursera and the chrysomelid counterploy. Ecology 75:1991–1996

Bellincampi D, Cervone F, Lionetti V (2014) Plant cell wall dynamics and wall-related susceptibility in plant-pathogen interactions. Front Plant Sci 5:228

Block A, Toruño TY, Elowsky CG, Zhang C, Steinbrenner J, Beynon J, Alfano JR (2014) The *Pseudomonas syringae* type III effector HopD1 suppresses effector-triggered immunity, localizes to the endoplasmic reticulum, and targets the *Arabidopsis* transcription factor NTL9. New Phytol 201:1358–1370

Boch J, Bonas U (2010) *Xanthomonas* AvrBs3 family-type III effectors: discovery and function. Ann Rev Phytopathol 48:419–436

Boyes DC, Nam J, Dangl JL (1998) The *Arabidopsis thaliana* RPM1 disease resistance gene product is a peripheral plasma membrane protein that is degraded coincident with the hypersensitive response. Proc Natl Acad Sci USA 95:15849–15854

Bozkurt TO, Schornack S, Win J, Shindo T, Ilyas M, Oliva R, Cano LM, Jones AM, Huitema E, van der Hoorn RA (2011) *Phytophthora infestans* effector AVRblb2 prevents secretion of a plant immune protease at the haustorial interface. Proc Natl Acad Sci USA 108:20832–20837

Bradley DJ, Kjellbom P, Lamb CJ (1992) Elicitor-and wound-induced oxidative cross-linking of a proline-rich plant cell wall protein: a novel, rapid defense response. Cell 70:21–30

Broekaert W, Van Parijs J, Allen A, Peumans W (1988) Comparison of some molecular, enzymatic, and antifungal properties of chitinases from thornapple, tobacco, and wheat. Physiol Mol Plant Pathol 33:319–331

Brogden KA (2005) Antimicrobial peptides: pore formers or metabolic inhibitors in bacteria? Nat Rev Microbiol 3:238–250

Bruix M, Jimenez M, Santoro J, Gonzalez C, Colilla F, Mendez E, Rico M (1993) Solution structure of γ1-H and γ1-P thionins from barley and wheat endosperm determined by ^{1}H-NMR: a structural motif common to toxic arthropod proteins. Biochemistry 32:715–724

Brutus A, Sicilia F, Macone A, Cervone F, De Lorenzo G (2010) A domain swap approach reveals a role of the plant wall-associated kinase 1 (WAK1) as a receptor of oligogalacturonides. Proc Natl Acad Sci USA 107:9452–9457

Bryngelsson T, Green B (1989) Characterization of a pathogenesis-related, thaumatin-like protein isolated from barley challenged with an incompatible race of mildew. Physiol Mol Plant Pathol 35:45–52

Burketova L, Trda L, Ott PG, Valentova O (2015) Bio-based resistance inducers for sustainable plant protection against pathogens. Biotechnol Adv 33:994–1004

Bushnell W (2012) The cereal rusts: origins, specificity, structure, and physiology. Elsevier, Amsterdam, The Netherlands

Canonne J, Froidure-Nicolas S, Rivas S (2011) Phospholipases in action during plant defense signaling. Plant Signal Behav 6:13–18

Caporale C, Di Berardino I, Leonardi L, Bertini L, Cascone A, Buonocore V, Caruso C (2004) Wheat pathogenesis-related proteins of class 4 have ribonuclease activity. FEBS Lett 575:71–76

Cárdenas L, Martínez A, Sánchez F, Quinto C (2008) Fast, transient and specific intracellular ROS changes in living root hair cells responding to Nod factors (NFs). Plant J 56:802–813

Casey LW, Lavrencic P, Bentham AR, Cesari S, Ericsson DJ, Croll T, Turk D, Anderson PA, Mark AE, Dodds PN (2016) The CC domain structure from the wheat stem rust resistance protein SR33 challenges paradigms for dimerization in plant NLR proteins. Proc Natl Acad Sci USA 113:12856–12861

Century KS, Shapiro AD, Repetti PP, Dahlbeck D, Holub E, Staskawicz BJ (1997) NDR1, a pathogen-induced componentrequired for arabidopsis disease resistance. Science 278:1963–1965

Césari S, Kanzaki H, Fujiwara T, Bernoux M, Chalvon V, Kawano Y, Shimamoto K, Dodds P, Terauchi R, Kroj T (2014) The NB-LRR proteins RGA4 and RGA5 interact functionally and physically to confer disease resistance. EMBO J 33:1941–1959

Cesari S, Bernoux M, Moncuquet P, Kroj T, Dodds PN (2014) A novel conserved mechanism for plant NLR protein pairs: The "integrated decoy" hypothesis. Front Plant Sci 5:606

Cessna SG, Sears VE, Dickman MB, Low PS (2000) Oxalic acid, a pathogenicity factor for *Sclerotinia sclerotiorum*, suppresses the oxidative burst of the host plant. Plant Cell 12:2191–2199

Chang C, Yu D, Jiao J, Jing S, Schulze-Lefert P, Shen Q-H (2013) Barley MLA immune receptors directly interfere with antagonistically acting transcription factors to initiate disease resistance signaling. Plant Cell 25:1158–1173

Chaparro-Garcia A, Schwizer S, Sklenar J, Yoshida K, Petre B, Bos JI, Schornack S, Jones AM, Bozkurt TO, Kamoun S (2015) *Phytophthora infestans* RXRL-WY effector AVR3a associates with dynamin-related protein 2 required for endocytosis of the plant pattern recognition receptor FLS2. PLoS ONE 10:e0137071

Chen L, Zhang L, Li D, Wang F, Yu D (2013) WRKY8 transcription factor functions in the TMV-cg defense response by mediating both abscisic acid and ethylene signaling in *Arabidopsis*. Proc Natl Acad Sci USA 110:E1963–E1971

Chinchilla D, Zipfel C, Robatzek S, Kemmerling B, Nürnberger T, Jones JD, Felix G, Boller T (2007) A flagellin-induced complex of the receptor FLS2 and BAK1 initiates plant defense. Nature 448:497–500

Choi SM, Song HR, Han SK, Han M, Kim CY, Park J, Lee YH, Jeon JS, Noh YS, Noh B (2012) HDA19 is required for the repression of salicylic acid biosynthesis and salicylic acid-mediated defense responses in *Arabidopsis*. Plant J 71:135–146

Choi J, Tanaka K, Cao Y, Qi Y, Qiu J, Liang Y, Lee SY, Stacey G (2014) Identification of a plant receptor for extracellular ATP. Science 343:290–294

Chung SH, Rosa C, Scully ED, Peiffer M, Tooker JF, Hoover K, Luthe DS, Felton GW (2013) Herbivore exploits orally secreted bacteria to suppress plant defenses. Proc Natl Acad Sci USA 110:15728–15733

Ciuffetti LM, Manning VA, Pandelova I, Betts MF, Martinez JP (2010) Host-selective toxins, PtrtoxA and PtrtoxB, as necrotrophic effectors in the *Pyrenophoratritici-repentis*wheat interaction. New Phytol 187:911–919

Colilla FJ, Rocher A, Mendez E (1990) γ-Purothionins: Amino acid sequence of two polypeptides of a new family of thionins from wheat endosperm. FEBS Lett 270:191–194

Collinge DB, Kragh KM, Mikkelsen JD, Nielsen KK, Rasmussen U, Vad K (1993) Plant chitinases. Plant J 3:31–40

Cook RJ, Thomashow LS, Weller DM, Fujimoto D, Mazzola M, Bangera G, Kim D-S (1995) Molecular mechanisms of defense by rhizobacteria against root disease. Proc Natl Acad Sci USA 92:4197–4201

Coppola M, Cascone P, Madonna V, Lelio I, Esposito F, Avitabile C, Romanelli A, Guerrieri E, Vitiello A, Pennacchio F (2017) Plant-to-plant communication triggered by systemin primes anti-herbivore resistance in tomato. Sci Rep 7:15522

Cruz LF, Rupp JLS, Trick HN, Fellers JP (2014) Stable resistance to Wheat streak mosaic virus in wheat mediated by RNAi. In vitro cell. Dev Biol Plant 50:665–672

Cui H, Tsuda K, Parker JE (2015) Effector-triggered immunity: from pathogen perception to robust defense. Ann Rev Plant Biol 66:487–511

Dangl JL, Horvath DM, Staskawicz BJ (2013) Pivoting the plant immune system from dissection to deployment. Science 341:746–751

Dangl JL, Jones JD (2001) Plant pathogens and integrated defense responses to infection. Nature 411:826–833

Day B, Dahlbeck D, Staskawicz BJ (2006) NDR1 interaction with RIN4 mediates the differential activation of multiple disease resistance pathways in *Arabidopsis*. Plant Cell 18:2782–2791

de Jonge R, van Esse HP, Kombrink A, Shinya T, Desaki Y, Bours R, van der Krol S, Shibuya N, Joosten MH, Thomma BP (2010) Conserved fungal LysM effector Ecp6 prevents chitin-triggered immunity in plants. Science 329:953–955

de Zélicourt A, Letousey P, Thoiron S, Campion C, Simoneau P, Elmorjani K, Marion D, Simier P, Delavault P (2007) Ha-def1, a sunflower defensin, induces cell death in Orobanche parasitic plants. Plainta 226:591–600

De Wit PJ (1994) Host resistance to a fungal tomato pathogen lost by a single base-pair change in an avirulence gene. Nature 367:384–386

De Paula VS, Razzera G, Barreto-Bergter E, Almeida FC, Valente AP (2011) Portrayal of complex dynamic properties of sugarcane defensin 5 by NMR: multiple motions associated with membrane interaction. Structure 19:26–36

Decreux A, Messiaen J (2005) Wall-associated kinase WAK1 interacts with cell wall pectins in a calcium-induced conformation. Plant Cell Physiol 46:268–278

del Pozo O, Pedley KF, Martin GB (2004) MAPKKKα is a positive regulator of cell death associated with both plant immunity and disease. EMBO J 23:3072–3082

Denoux C, Galletti R, Mammarella N, Gopalan S, Werck D, De Lorenzo G, Ferrari S, Ausubel FM, Dewdney J (2008) Activation of defense response pathways by OGs and Flg22 elicitors in *Arabidopsis* seedlings. Mol Plant 1:423–445

Desaki Y, Miya A, Venkatesh B, Tsuyumu S, Yamane H, Kaku H, Minami E, Shibuya N (2006) Bacterial lipopolysaccharides induce defense responses associated with programmed cell death in rice cells. Plant Cell Physiol 47:1530–1540

Dodds PN, Lawrence GJ, Catanzariti A-M, Teh T, Wang C-I, Ayliffe MA, Kobe B, Ellis JG (2006) Direct protein interaction underlies gene-for-gene specificity and coevolution of the flax resistance genes and flax rust avirulence genes. Proc Natl Acad Sci USA 103:8888–8893

Donaldson PA, Anderson T, Lane BG, Davidson AL, Simmonds DH (2001) Soybean plants expressing an active oligomeric oxalate oxidase from the wheat gf-2.8 (germin) gene are resistant to the oxalate-secreting pathogen *Sclerotinasclerotiorum*. Physiol Mol Plant Pathol 59:297–307

Dong X, Ji R, Guo X, Foster SJ, Chen H, Dong C, Liu Y, Hu Q, Liu S (2008) Expressing a gene encoding wheat oxalate oxidase enhances resistance to *Sclerotinia sclerotiorum* in oilseed rape (*Brassica napus*). Planta 228:331–340

Dowen RH, Pelizzola M, Schmitz RJ, Lister R, Dowen JM, Nery JR, Dixon JE, Ecker JR (2012) Widespread dynamic DNA methylation in response to biotic stress. Proc Natl Acad Sci USA 109:E2183–E2191

Doxey AC, Yaish MW, Moffatt BA, Griffith M, McConkey BJ (2007) Functional divergence in the *Arabidopsis* β-1,3-glucanase gene family inferred by phylogenetic reconstruction of expression states. Mol Biol Evol 24:1045–1055

Du J, Verzaux E, Chaparro-Garcia A, Bijsterbosch G, Keizer LP, Zhou J, Liebrand TW, Xie C, Govers F, Robatzek S (2015) Elicitin recognition confers enhanced resistance to phytophthora infestans in potato. Nat Plants 1:15034

Dumas B, Freyssinet G, Pallett KE (1995) Tissue-specific expression of germin-like oxalate oxidase during development and fungal infection of barley seedlings. Plant Physiol 107:1091–1096

Ebrahim S, Usha K, Singh B (2011) Pathogenesis-related (PR) proteins in plant defense mechanism. Sci Against Microb Pathog 2:1043–1054

Engelhardt S, Boevink PC, Armstrong MR, Ramos MB, Hein I, Birch PR (2012) Relocalization of late blight resistance protein R3A to endosomal compartments is associated with effector recognition and required for the immune response. Plant Cell 24:5142–5158

Faris JD, Zhang Z, Lu H, Lu S, Reddy L, Cloutier S, Fellers JP, Meinhardt SW, Rasmussen JB, Xu SS (2010) A unique wheat disease resistance-like gene governs effector-triggered susceptibility to necrotrophic pathogens. Proc Natl Acad Sci USA 107:13544–13549

Feng GH, Richardson M, Chen MS, Kramer KJ, Morgan TD, Reeck GR (1996) α-amylase inhibitors from wheat: amino acid sequences and patterns of inhibition of insect and human α-amylases. Insect Biochem Mol Biol 26:419–426

Feuillet C, Schachermayr G, Keller B (1997) Molecular cloning of a new receptor-like kinase gene encoded at the Lr10 disease resistance locus of wheat. Plant J 11:45–52

Filipenko E, Kochetov A, Kanayama Y, Malinovsky V, Shumny V (2013) PR-proteins with ribonuclease activity and plant resistance against pathogenic fungi. Russ J Genet Appl Res 3:474–480

Finkina E, Melnikova D, Bogdanov I (2016) Lipid transfer proteins as components of the plant innate immune system: structure, functions, and applications. Acta Nat 8:47–61

Fliegmann J, Mithöfer A, Wanner G, Ebel J (2004) An ancient enzyme domain hidden in the putative β-glucan elicitor receptor of soybean may play an active part in the perception of pathogen-associated molecular patterns during broad host resistance. J Biol Chem 279:1132–1140

Flor HH (1942) Inheritance of pathogenicity in melampsora lini. Phytopathology 32:653–669

Franco OL, Rigden DJ, Melo FR, Grossi-de-Sá MF (2002) Plant α-amylase inhibitors and their interaction with insect α-amylases. FEBS J 269:397–412

Franco OL, Rigden DJ, R Melo F, Bloch C, Silva CP, Grossi de Sá MF (2000) Activity of wheat α-amylase inhibitors towards bruchid α-amylases and structural explanation of observed specificities. FEBS J 267:2166–2173

Friebe A, Vilich V, Hennig L, Kluge M, Sicker D (1998) Detoxification of benzoxazolinone allelochemicals from wheat by *Gaeumannomyces graminis* var. *tritici*, *G. graminis* var. *graminis*, *G. graminis* var. *avenae*, and *Fusarium culmorum*. Appl Environ Microbiol 64:2386–2391

Friesen TL, Stukenbrock EH, Liu Z, Meinhardt S, Ling H, Faris JD, Rasmussen JB, Solomon PS, McDonald BA, Oliver RP (2006) Emergence of a new disease as a result of interspecific virulence gene transfer. Nat Genet 38:953–956

Frye CA, Tang D, Innes RW (2001) Negative regulation of defense responses in plants by a conserved MAPKK kinase. Proc Natl Acad Sci USA 98:373–378

Galiana E, Bonnet P, Conrod S, Keller H, Panabières F, Ponchet M, Poupet A, Ricci P (1997) RNase activity prevents the growth of a fungal pathogen in tobacco leaves and increases upon induction of systemic acquired resistance with elicitin. Plant Physiol 115:1557–1567

Galletti R, Ferrari S, De Lorenzo G (2011) *Arabidopsis* MPK3 and MPK6 play different roles in basal and oligogalacturonide-or flagellin-induced resistance against *Botrytis cinerea*. Plant Physiol 157:804–814

Gao X, Chen X, Lin W, Chen S, Lu D, Niu Y, Li L, Cheng C, McCormack M, Sheen J (2013) Bifurcation of arabidopsis NLR immune signaling via Ca^{2+}-dependent protein kinases. PLoSPathog. 9:e1003127

García-Olmedo F, Molina A, Alamillo JM, Rodríguez-Palenzuéla P (1998) Plant defense peptides. Pept Sci 47:479–491

Gardiner DM, Upadhyaya NM, Stiller J, Ellis JG, Dodds PN, Kazan K, Manners JM (2014) Genomic analysis of *Xanthomonas translucens* pathogenic on wheat and barley reveals cross-kingdom gene transfer events and diverse protein delivery systems. PLoS ONE 9:e84995

Geraats BP, Bakker PA, Lawrence CB, Achuo EA, Höfte M, Van Loon L (2003) Ethylene-insensitive tobacco shows differentially altered susceptibility to different pathogens. Phytopathology 93:813–821

Givovich A, Niemeyer H (1991) Hydroxamic acids affecting barley yellow dwarf virus transmission by the aphid *Rhopalosiphumpadi.* Entomol Exp Appl 59:79–85

Glazebrook J (2005) Contrasting mechanisms of defense against biotrophic and necrotrophic pathogens. Annu Rev Phytopathol 43:205–227

González-Lamothe R, Mitchell G, Gattuso M, Diarra MS, Malouin F, Bouarab K (2009) Plant antimicrobial agents and their effects on plant and human pathogens. Int J Mol Sci 10:3400–3419

Gottwald S, Samans B, Lück S, Friedt W (2012) Jasmonate and ethylene dependent defense gene expression and suppression of fungal virulence factors: two essential mechanisms of *Fusarium* head blight resistance in wheat? BMC Genom 13:369

Graham MA, Silverstein KA, Cannon SB, VandenBosch KA (2004) Computational identification and characterization of novel genes from legumes. Plant Physiol 135:1179–1197

Gururani MA, Venkatesh J, Upadhyaya CP, Nookaraju A, Pandey SK, Park SW (2012) Plant disease resistance genes: current status and future directions. Physiol Mol Plant Pathol 78:51–65

Gust AA, Biswas R, Lenz HD, Rauhut T, Ranf S, Kemmerling B, Götz F, Glawischnig E, Lee J, Felix G (2007) Bacteria-derived peptidoglycans constitute pathogen-associated molecular patterns triggering innate immunity in *Arabidopsis.* J Biol Chem 282:32338–32348

Gust AA, Pruitt R, Nürnberger T (2017) Sensing danger: key to activating plant immunity. Trends Plant Sci 22:779–791

Haas M, Menke J, Chao S, Steffenson BJ (2016) Mapping quantitative trait loci conferring resistance to a widely virulent isolate of *Cochliobolus sativus* in wild barley accession PI 466423. Theor Appl Genet 129:1831–1842

Halterman DA, Wei F, Wise RP (2003) Powdery mildew induced Mla mRNAs are alternatively spliced and contain multiple upstream open reading frames. Plant Physiol 131:558–567

Halterman D, Zhou F, Wei F, Wise RP, Schulze-Lefert P (2001) The MLA6 coiled-coil, NBS-LRR protein confers AvrMla6-dependent resistance specificity to *Blumeriagraminis* f. sp. hordei in barley and wheat. Plant J 25:335–348

Hammond-Kosack KE, Jones J (1996) Resistance gene-dependent plant defense responses. Plant Cell 8:1773–1791

Hammond-Kosack KE, Jones JD (1997) Plant disease resistance genes. Ann Rev Plant Biol 48:575–607

Hejgaard J, Jacobsen S, Svendsen I (1991) Two antifungal thaumatin-like proteins from barley grain. FEBS Lett 291:127–131

Hengartner MO, Horvitz HR (1994) Programmed cell death in *Caenorhabditis elegans.* Curr Opin Genet Dev 4:581–586

Holeski LM, Jander G, Agrawal AA (2012) Transgenerational defense induction and epigenetic inheritance in plants. Trends Ecol Evol 27:618–626

Holt BF, Boyes DC, Ellerström M, Siefers N, Wiig A, Kauffman S, Grant MR, Dangl JL (2002) An evolutionarily conserved mediator of plant disease resistance gene function is required for normal arabidopsis development. Dev Cell 2:807–817

Horinouchi S, Kito M, Nishiyama M, Furuya K, Hong S-K, Miyake K, Beppu T (1990) Primary structure of AfsR, a global regulatory protein for secondary metabolite formation in *Streptomyces coelicolor* A3 (2). Gene 95:49–56

Hugot K, Ponchet M, Marais A, Ricci P, Galiana E (2002) A tobacco S-like RNase inhibits hyphal elongation of plant pathogens. Mol Plant Microbe Interact 15:243–250

Ichimura K, Shinozaki K, Tena G, Sheen J, Henry Y, Champion A, Kreis M, Zhang S, Hirt H, Wilson C (2002) Mitogen-activated protein kinase cascades in plants: a new nomenclature. Trends Plant Sci 7:301–308

Inoue H, Hayashi N, Matsushita A, Xinqiong L, Nakayama A, Sugano S, Jiang C-J, Takatsuji H (2013) Blast resistance of CC-NB-LRR protein Pb1 is mediated by WRKY45 through protein-protein interaction. Proc Natl Acad Sci USA 110:9577–9582

Jacob F, Vernaldi S, Maekawa T (2013) Evolution and conservation of plant NLR functions. Front Immunol 4:297

Jashni MK, Mehrabi R, Collemare J, Mesarich CH, De Wit PJ (2015) The battle in the apoplast: further insights into the roles of proteases and their inhibitors in plant-pathogen interactions. Front Plant Sci 6:584

Jaskiewicz M, Conrath U, Peterhänsel C (2011) Chromatin modification acts as a memory for systemic acquired resistance in the plant stress response. EMBO Rep 12:50–55

Jensen MK, Rung JH, Gregersen PL, Gjetting T, Fuglsang AT, Hansen M, Joehnk N, Lyngkjaer MF, Collinge DB (2007) The HvNAC6 transcription factor: A positive regulator of penetration resistance in barley and *Arabidopsis*. Plant Mol Biol 65:137–150

Jiang Z, Ge S, Xing L, Han D, Kang Z, Zhang G, Wang X, Wang X, Chen P, Cao A (2013) RLP1.1, a novel wheat receptor-like protein gene, is involved in the defense response against *Puccinia striiformis* f. sp. *tritici*. J Exp Bot 64:3735–3746

Jones AM, Assmann SM (2004) Plants: the latest model system for G-protein research. EMBO Rep 5:572–578

Jones JD, Dangl JL (2006) The plant immune system. Nature 444:323–329

Juge N (2006) Plant protein inhibitors of cell wall degrading enzymes. Trends Plant Sci 11:359–367

Kadota Y, Sklenar J, Derbyshire P, Stransfeld L, Asai S, Ntoukakis V, Jones JD, Shirasu K, Menke F, Jones A (2014) Direct regulation of the NADPH oxidase RBOHD by the PRR-associated kinase BIK1 during plant immunity. Mol Cell 54:43–55

Kaku H, Nishizawa Y, Ishii-Minami N, Akimoto-Tomiyama C, Dohmae N, Takio K, Minami E, Shibuya N (2006) Plant cells recognize chitin fragments for defense signaling through a plasma membrane receptor. Proc Natl Acad Sci USA 103:11086–11091

Kang Z, Buchenauer H (2000) Ultrastructural and cytochemical studies on cellulose, xylan, and pectin degradation in wheat spikes infected by *Fusarium culmorum*. J Phytopathol 148:263–275

Kärkönen A, Kuchitsu K (2015) Reactive oxygen species in cell wall metabolism and development in plants. Phytochemistry 112:22–32

Kettles GJ, Drurey C, Schoonbeek HJ, Maule AJ, Hogenhout SA (2013) Resistance of *Arabidopsis thaliana* to the green peach aphid, *Myzuspersicae*, involves camalexin and is regulated by microRNAs. New Phytol 198:1178–1190

Kim K-C, Lai Z, Fan B, Chen Z (2008) *Arabidopsis* WRKY38 and WRKY62 transcription factors interact with histone deacetylase 19 in basal defense. Plant Cell 20:2357–2371

Kim J-G, Stork W, Mudgett MB (2013) *Xanthomonas* type III effector XopDdesumoylates tomato transcription factor SIERF4 to suppress ethylene responses and promote pathogen growth. Cell Host Microbe 13:143–154

Kim SH, Qi D, Ashfield T, Helm M, Innes RW (2016) Using decoys to expand the recognition specificity of a plant disease resistance protein. Science 351:684–687

Knepper C, Savory EA, Day B (2011) The role of NDR1 in pathogen perception and plant defense signaling. Plant Signal Behav 6:1114–1116

Kohorn BD, Kohorn SL, Todorova T, Baptiste G, Stansky K, McCullough M (2012) A dominant allele of *Arabidopsis* pectin-binding wall-associated kinase induces a stress response suppressed by MPK6 but not MPK3 mutations. Mol Plant 5:841–851

Koike M, Okamoto T, Tsuda S, Imai R (2002) A novel plant defensin-like gene of winter wheat is specifically induced during cold acclimation. Biochem Biophys Res Commun 298:46–53

Kong L, Anderson JM, Ohm HW (2005) Induction of wheat defense and stress-related genes in response to *Fusarium graminearum*. Genome 48:29–40

Kroj T, Chanclud E, Michel-Romiti C, Grand X, Morel JB (2016) Integration of decoy domains derived from protein targets of pathogen effectors into plant immune receptors is widespread. New Phytol 210:618–626

Kunze G, Zipfel C, Robatzek S, Niehaus K, Boller T, Felix G (2004) The N terminus of bacterial elongation factor Tu elicits innate immunity in *Arabidopsis* plants. Plant Cell 16:3496–3507

Kutschera U, Hossfeld U (2012) Physiological phytopathology: origin and evolution of a scientific discipline. J Appl Bot Food Qual 85:1–5

Lacerda AF, Vasconcelos ÉA, Pelegrini PB, de Sa MFG (2014) Antifungal defensins and their role in plant defense. Front Microbiol 5:116

Lamb C, Dixon RA (1997) The oxidative burst in plant disease resistance. Ann Rev Plant Biol 48:251–275

Lannoo N, Van Damme EJ (2014) Lectin domains at the frontiers of plant defense. Front Plant Sci 5:397

Le Roux C, Huet G, Jauneau A, Camborde L, Trémousaygue D, Kraut A, Zhou B, Levaillant M, Adachi H, Yoshioka H (2015) A receptor pair with an integrated decoy converts pathogen-disabling transcription factors to immunity. Cell 161:1074–1088

Le T-N, Schumann U, Smith NA, Tiwari S, Au PCK, Zhu Q-H, Taylor JM, Kazan K, Llewellyn DJ, Zhang R (2014) DNA demethylases target promoter transposable elements to positively regulate stress-responsive genes in *Arabidopsis*. Genome Biol 15:458

Lecourieux D, Ranjeva R, Pugin A (2006) Calcium in plant defense-signaling pathways. New Phytol 171:24–269

Leister RT, Ausubel FM, Katagiri F (1996) Molecular recognition of pathogen attack occurs inside of plant cells in plant disease resistance specified by the *Arabidopsis* genes RPS2 and RPM1. Proc Natl Acad Sci USA 93:15497–15502

Leubner-Metzger G, Meins F Jr (1999) 3 Functions, and regulation of plant β-(pr-2). In: Pathogenesis-related proteins in plants. CRC Press, Boca Raton, FL, USA

Levine A, Tenhaken R, Dixon R, Lamb C (1994) H_2O_2 from the oxidative burst orchestrates the plant hypersensitive disease resistance response. Cell 79:583–593

Li G, Meng X, Wang R, Mao G, Han L, Liu Y, Zhang S (2012) Dual-level regulation of ACC synthase activity by MPK3/MPK6 cascade and its downstream WRKY transcription factor during ethylene induction in arabidopsis. PLoS Genet 8:e1002767

Liang H, Maynard CA, Allen RD, Powell WA (2001) Increased Septoria musiva resistance in transgenic hybrid poplar leaves expressing a wheat oxalate oxidase gene. Plant Mol Biol 45:619–629

Lin K-C, Bushnell WR, Szabo LJ, Smith AG (1996) Isolation and expression of a host response gene family encoding thaumatin-like proteins in incompatible oat-stem rust fungus interactions. Mol Plant Microbe Interact 9:511–522

Lind C, Dreyer I, López-Sanjurjo EJ, von Meyer K, Ishizaki K, Kohchi T, Lang D, Zhao Y, Kreuzer I, Al-Rasheid KA (2015) Stomatal guard cells co-opted an ancient ABA-dependent desiccation survival system to regulate stomatal closure. Curr Biol 25:928–935

Liu F, Wang M, Wen J, Yi B, Shen J, Ma C, Tu J, Fu T (2015) Overexpression of barley oxalate oxidase gene induces partial leaf resistance to *Sclerotinia sclerotiorum* in transgenic oilseed rape. Plant Pathol 64:1407–1416

Liu Z, Gao Y, Kim YM, Faris JD, Shelver WL, Wit PJ, Xu SS, Friesen TL (2016) SnTox1, a *Parastagonosporanodorum* necrotrophic effector, is a dual-function protein that facilitates infection while protecting from wheat-produced chitinases. New Phytol 211:1052–1064

Liu Z, Zurn JD, Kariyawasam G, Faris JD, Shi G, Hansen J, Rasmussen JB, Acevedo M (2017) Inverse gene-for-gene interactions contribute additively to tan spot susceptibility in wheat. Theor Appl Genet 130:1267–1276

Lo Presti L, Lanver D, Schweizer G, Tanaka S, Liang L, Tollot M, Zuccaro A, Reissmann S, Kahmann R (2015) Fungal effectors, and plant susceptibility. Ann Rev Plant Biol 66:513–545

López Sánchez A, Stassen JH, Furci L, Smith LM, Ton J (2016) The role of DNA (de)methylation in immune responsiveness of *Arabidopsis*. Plant J 88:361–374

Lorang J, Kidarsa T, Bradford C, Gilbert B, Curtis M, Tzeng S-C, Maier C, Wolpert T (2012) Tricking the guard: exploiting plant defense for disease susceptibility. Science 338:659–662

Lorenzo O, Chico JM, Sánchez-Serrano JJ, Solano R (2004) Jasmonate-insensitive1 encodes an MYC transcription factor essential to discriminate between different jasmonate-regulated defense responses in *Arabidopsis*. Plant Cell 16:1938–1950

Lozano-Durán R, Macho AP, Boutrot F, Segonzac C, Somssich IE, Zipfel C (2013) The transcriptional regulator BZR1 mediates the trade-off between plant innate immunity and growth. Elife 2:e00983

Lozano-Torres JL, Wilbers RH, Gawronski P, Boshoven JC, Finkers-Tomczak A, Cordewener JH, America AH, Overmars HA, Van't Klooster JW, Baranowski L (2012) Dual disease resistance mediated by the immune receptor Cf-2 in tomato requires a common virulence target of a fungus and a nematode. Proc Natl Acad Sci USA 109:10119–10124

Lu J, Ju H, Zhou G, Zhu C, Erb M, Wang X, Wang P, Lou Y (2011) An EAR-motif-containing ERF transcription factor affects herbivore-induced signaling, defense and resistance in rice. Plant J 68:583–596

Manning VA, Chu AL, Steeves JE, Wolpert TJ, Ciuffetti LM (2009) A host-selective toxin of *Pyrenophoratritici-repentis*, PtrtoxA, induces photosystem changes and reactive oxygen species accumulation in sensitive wheat. Mol Plant Microbe Interact 22:665–676

Mao G, Meng X, Liu Y, Zheng Z, Chen Z, Zhang S (2011) Phosphorylation of a WRKY transcription factor by two pathogens responsive MAPKks drives phytoalexin biosynthesis in *Arabidopsis*. Plant Cell 23:1639–1653

Marathe R, Dinesh-Kumar S (2003) Plant defense: one post, multiple guards?! Mol Cell 11:284–286

Marino D, Peeters N, Rivas S (2012) Ubiquitination during plant immune signaling. Plant Physiol 160:15–27

Marone D, Russo MA, Laidò G, De Leonardis AM, Mastrangelo AM (2013) Plant nucleotide binding site–leucine-rich repeat (NBS-LRR) genes: active guardians in host defense responses. Int J Mol Sci 14:7302–7326

Martinez DA, Loening UE, Graham MC (2018) Impacts of glyphosate-based herbicides on disease resistance and health of crops: a review. Environ Sci Eur 30:2

McConn M, Creelman RA, Bell E, Mullet JE (1997) Jasmonate is essential for insect defense in Arabidopsis. Proc Natl Acad Sci USA 94:5473–5477

McKay SAB, Hunter WL, Godard K-A, Wang SX, Martin DM, Bohlmann J, Plant AL (2003) Insect attack and wounding induce traumatic resin duct development and gene expression of (−)-pinene synthase in Sitka spruce. Plant Physiol 133:368–378

McLellan H, Boevink PC, Armstrong MR, Pritchard L, Gomez S, Morales J, Whisson SC, Beynon JL, Birch PR (2013) An RXLR effector from *Phytophthora infestans* prevents re-localization of two plant NAC transcription factors from the endoplasmic reticulum to the nucleus. PLoS Pathog 9:e1003670

Melech-Bonfil S, Sessa G (2010) Tomato MAPKKKε is a positive regulator of cell-death signaling networks associated with plant immunity. Plant J 64:379–391

Melotto M, Underwood W, Koczan J, Nomura K, He SY (2006) Plant stomata function in innate immunity against bacterial invasion. Cell 126:969–980

Melotto M, Underwood W, He SY (2008) Role of stomata in plant innate immunity and foliar bacterial diseases. Annu Rev Phytopathol 46:101–122

Mendez E, Moreno A, Colilla F, Pelaez F, Limas GG, Mendez R, Soriano F, Salinas M, Haro C (1990) Primary structure and inhibition of protein synthesis in eukaryotic cell-free system of a novel thionin, γ-hordothionin, from barley endosperm. FEBS J 194:533–539

Méndez E, Rocher A, Calero M, Girbés T, Citores L, Soriano F (1996) Primary structure of ω-hordothionin, a member of a novel family of thionins from barley endosperm, and its inhibition of protein synthesis in eukaryotic and prokaryotic cell-free systems. FEBS J 239:67–73

Meng X, Zhang S (2013) MAPK cascades in plant disease resistance signaling. Ann Rev Phytopathol 51:245–266

Meng X, Xu J, He Y, Yang K-Y, Mordorski B, Liu Y, Zhang S (2013) Phosphorylation of an ERF transcription factor by *Arabidopsis* MPK3/MPK6 regulates plant defense gene induction and fungal resistance. Plant Cell 25:1126–1142

Mentlak TA, Kombrink A, Shinya T, Ryder LS, Otomo I, Saitoh H, Terauchi R, Nishizawa Y, Shibuya N, Thomma BP (2012) Effector-mediated suppression of chitin-triggered immunity by *Magnaporthe oryzae* is necessary for rice blast disease. Plant Cell 24:322–335

Meyers BC, Kozik A, Griego A, Kuang H, Michelmore RW (2003) Genome-wide analysis of NBS-LRR–encoding genes in *Arabidopsis*. Plant Cell Online 15:809–834

Michelmore RW, Christopoulou M, Caldwell KS (2013) Impacts of resistance gene genetics, function, and evolution on a durable future. Ann Rev Phytopathol 51:291–319

Ming D, Hellekant G (1994) Brazzein, a new high-potency thermostable sweet protein from *Pentadiplandrabrazzeana* B. FEBS Lett 355:106–108

Mitra RM, Long SR (2004) Plant, and bacterial symbiotic mutants define three transcriptionally distinct stages in the development of the *Medicago truncatula/Sinorhizobiummeliloti* symbiosis. Plant Physiol 134:595–604

Mittler R, Herr EH, Orvar BL, Van Camp W, Willekens H, Inzé D, Ellis BE (1999) Transgenic tobacco plants with reduced capability to detoxify reactive oxygen intermediates are hyperresponsive to pathogen infection. Proc Natl Acad Sci USA 96:14165–14170

Moffat CS, Ingle RA, Wathugala DL, Saunders NJ, Knight H, Knight MR (2012) ERF5 and ERF6 play redundant roles as positive regulators of JA/Et-mediated defense against *Botrytis cinerea* in *Arabidopsis*. PLoS ONE 7:e35995

Monaghan J, Zipfel C (2012) Plant pattern recognition receptor complexes at the plasma membrane. Curr Opin Plant Biol 15:349–357

Montillet J-L, Leonhardt N, Mondy S, Tranchimand S, Rumeau D, Boudsocq M, Garcia AV, Douki T, Bigeard J, Laurière C (2013) An abscisic acid-independent oxylipin pathway controls stomatal closure and immune defense in *Arabidopsis*. PLoS Biol 11:e1001513

Morrissey JP, Osbourn AE (1999) Fungal resistance to plant antibiotics as a mechanism of pathogenesis. Microbiol Mol Biol Rev 63:708–724

Mosher RA, Durrant WE, Wang D, Song J, Dong X (2006) A comprehensive structure-function analysis of *Arabidopsis* SNI1 defines essential regions and transcriptional repressor activity. Plant Cell 18:1750–1765

Moutinho-Pereira J, Magalhaes N, Torres de Castro L, Manuela Chaves M, Torres-Pereira J (2001) Physiological responses of grapevine leaves to bordeaux mixture under the light. Vitis 40:117–121

Mundy J, Svendsen I, Hejgaard J (1983) Barley α-amylase/subtilisin inhibitor. I. Isolation and characterization. Carlsberg Res Commun 48:81–90

Nagano N, Orengo CA, Thornton JM (2002) One-fold with many functions: the evolutionary relationships between TIM barrel families based on their sequences, structures, and functions. J Mol Biol 321:741–765

Namba S, Ling K, Gonsalves C, Slightom JL, Gonsalves D (1992) Protection of transgenic plants expressing the coat protein gene of watermelon mosaic virus ii or zucchini yellow mosaic virus against six potyviruses. Phytopathology 82:940–946

Naumann TA, Wicklow DT, Price NP (2011) Identification of a chitinase-modifying protein from *Fusarium verticillioides* truncation of a host resistance protein by a fungalysin metalloprotease. J Biol Chem 286:35358–35366

Navarro L, Jay F, Nomura K, He SY, Voinnet O (2008) Suppression of the microRNA pathway by bacterial effector proteins. Science 321:964–967

Niehl A, Wyrsch I, Boller T, Heinlein M (2016) Double-stranded RNAs induce a pattern-triggered immune signaling pathway in plants. New Phytol 211:1008–1019

Niemeyer HM, Jerez JM (1997) Chromosomal location of genes for hydroxamic acid accumulation in *Triticum aestivum* L. (wheat) using wheat aneuploids and wheat substitution lines. Heredity 79

Nimchuk Z, Marois E, Kjemtrup S, Leister RT, Katagiri F, Dangl JL (2000) Eukaryotic fatty acylation drives plasma membrane targeting and enhances the function of several type III effector proteins from *Pseudomonas syringae*. Cell 101:353–363

Noutoshi Y, Ito T, Seki M, Nakashita H, Yoshida S, Marco Y, Shirasu K, Shinozaki K (2005) A single amino acid insertion in the WRKY domain of the Arabidopsis TIR-NBS-LRR-WRKY -type disease resistance protein SLH1 (sensitive to low humidity 1) causes activation of defense responses and hypersensitive cell death. Plant J 43:873–888

Oh C-S, Martin GB (2011) Effector-triggered immunity mediated by the Pto kinase. Trends Plant Sci 16:132–140

Oh IS, Park AR, Bae MS, Kwon SJ, Kim YS, Lee JE, Kang NY, Lee S, Cheong H, Park OK (2005) Secretome analysis reveals an *Arabidopsis* lipase involved in defense against *Alternaria brassicicola*. Plant Cell 17:2832–2847

Okada A, Okada K, Miyamoto K, Koga J, Shibuya N, Nojiri H, Yamane H (2009) OsTGAP1, a bZIP transcription factor, coordinately regulates the inductive production of diterpenoid phytoalexins in rice. J Biol Chem 284:26510–26518

Osbourn A, Clarke B, Dow J, Daniels M (1991) Partial characterization of avenacinase from *Gaeumannomycesgraminis* var. *avenae*. Physiol Mol Plant Pathol 38:301–312

Osbourn A, Clarke B, Lunness P, Scott P, Daniels M (1994) An oat species lacking avenacin is susceptible to infection by *Gaeumannomycesgraminis* var. *tritici*. Physiol Mol Plant Pathol 45:457–467

Pandey SP, Roccaro M, Schön M, Logemann E, Somssich IE (2010) Transcriptional reprogramming regulated by WRKY18 and WRKY40 facilitates powdery mildew infection of *Arabidopsis*. Plant J 64:912–923

Park C-J, Ronald PC (2012) Cleavage, and nuclear localization of the rice xa21 immune receptor. Nat Commun 3:920

Park HJ, Kim W-Y, Park HC, Lee SY, Bohnert HJ, Yun D-J (2011) Sumo and sumoylation in plants. Mol Cells 32:305–316

Passardi F, Cosio C, Penel C, Dunand C (2005) Peroxidases have more functions than a Swiss army knife. Plant Cell Rep 24:255–265

Pavet V, Olmos E, Kiddle G, Mowla S, Kumar S, Antoniw J, Alvarez ME, Foyer CH (2005) Ascorbic acid deficiency activates cell death and disease resistance responses in *Arabidopsis*. Plant Physiol 139:1291–1303

Pedras MSC, Hossain S, Snitynsky RB (2011) Detoxification of cruciferous phytoalexins in *Botrytis cinerea*: Spontaneous dimerization of a camalexin metabolite. Phytochemistry 72:199–206

Pekkarinen AI, Longstaff C, Jones BL (2007) Kinetics of the inhibition of *Fusarium serine* proteinases by barley (*Hordeum vulgare* L.) inhibitors. J Agric Food Chem 55:2736–2742

Peng Y, Bartley LE, Chen X, Dardick C, Chern M, Ruan R, Canlas PE, Ronald PC (2008) OsWRKY62 is a negative regulator of basal and Xa21-mediated defense against *Xanthomonas oryzae*pv. *oryzae* in rice. Mol Plant 1:446–458

Petrucci T, Rab A, Tomasi M, Silano V (1976) Further characterization studies of the α-amylase protein inhibitor of gel electrophoretic mobility 0.19 from the wheat kernel. Biochim Biophys Acta (BBA)-Protein Struct 420:288–297

Pieterse CM, Leon-Reyes A, Van der Ent S, Van Wees SC (2009) Networking by small-molecule hormones in plant immunity. Nat Chem Biol 5:308–316

Plett JM, Daguerre Y, Wittulsky S, Vayssières A, Deveau A, Melton SJ, Kohler A, Morrell-Falvey JL, Brun A, Veneault-Fourrey C (2014) Effector MiSSP7 of the mutualistic fungus *Laccaria bicolor* stabilizes the Populus JAZ6 protein and represses jasmonic acid (JA) responsive genes. Proc Natl Acad Sci USA 111:8299–8304

Poovaiah B, Du L, Wang H, Yang T (2013) Recent advances in calcium/calmodulin-mediated signaling with an emphasis on plant-microbe interactions. Plant Physiol 163:531–542

Prince DC, Drurey C, Zipfel C, Hogenhout SA (2014) The leucine-rich repeat receptor-like kinase brassinosteroid insensitive1-associated kinase1 and the cytochrome p450 phytoalexin deficient3 contribute to innate immunity to aphids in arabidopsis. Plant Physiol 164:2207–2219

Pumplin N, Voinnet O (2013) RNA silencing suppression by plant pathogens: defence, counter-defence and counter-counter-defence. Nat Rev Microbiol 11:745–760

Qi D, Dubiella U, Kim SH, Sloss DI, Dowen RH, Dixon JE, Innes RW (2014) Recognition of the protein kinase AVRPPHB SUSCEPTIBLE1 by the disease resistance protein resistance to PSEUDOMONAS SYRINGIAE5 is dependent on S-acylation and an exposed loop in AVRPPHB SUSCEPTIBLE1. Plant Physiol 164:340–351

Qiao Y, Liu L, Xiong Q, Flores C, Wong J, Shi J, Wang X, Liu X, Xiang Q, Jiang S (2013) Oomycete pathogens encode RNA silencing suppressors. Nat Genet 45:330–333

Qiao Y, Shi J, Zhai Y, Hou Y, Ma W (2015) Phytophthora effector targets a novel component of small RNA pathway in plants to promote infection. Proc Natl Acad Sci USA 112:5850–5855

Qiu JL, Fiil BK, Petersen K, Nielsen HB, Botanga CJ, Thorgrimsen S, Palma K, Suarez-Rodriguez MC, Sandbech-Clausen S, Lichota J (2008) *Arabidopsis* MAP kinase 4 regulates gene expression through transcription factor release in the nucleus. EMBO J 27:2214–2221

Ramputh A, Arnason J, Cass L, Simmonds J (2002) Reduced herbivory of the European corn borer (*Ostrinianubilalis*) on corn transformed with germin, a wheat oxalate oxidase gene. Plant Sci 162:431–440

Reddy AS, Ali GS, Celesnik H, Day IS (2011) Coping with stresses: Roles of calcium-and calcium/calmodulin-regulated gene expression. Plant Cell 23:2010–2032

Reisige K, Gorzelanny C, Daniels U, Moerschbacher BM (2006) The C28 aldehyde octacosanal is a morphogenetically active component involved in host plant recognition and infection structure differentiation in the wheat stem rust fungus. Physiol Mol Plant Pathol 68:33–40

Rentel MC, Lecourieux D, Ouaked F, Usher SL (2004) Oxi1 kinase is necessary for oxidative burst-mediated signaling in *Arabidopsis*. Nature 427:858–861

Rhoades VH (1935) The location of a gene for disease resistance in maize. Proc Natl Acad Sci USA 21:243–246

Rinerson CI, Rabara RC, Tripathi P, Shen QJ, Rushton PJ (2015) The evolution of WRKY transcription factors. BMC Plant Biol 15:66

Roberts WK, Selitrennikoff CP (1990) Zeamatin, an antifungal protein from maize with membrane-permeabilizing activity. Microbiology 136:1771–1778

Robert-Seilaniantz A, Grant M, Jones JD (2011) Hormone crosstalk in plant disease and defense: more than just jasmonate-salicylate antagonism. Ann Rev Phytopathol 49:317–343

Römer P, Hahn S, Jordan T, Strauß T, Bonas U, Lahaye T (2007) Plant pathogen recognition mediated by promoter activation of the pepper Bs3 resistance gene. Science 318:645–648

Ron M, Avni A (2004) The receptor for the fungal elicitor ethylene-inducing xylanase is a member of a resistance-like gene family in tomato. Plant Cell 16:1604–1615

Rooney HC, van't Klooster JW, van der Hoorn RA, Joosten MH, Jones JD, de Wit PJ (2005) *Cladosporium* Avr2 inhibits tomato Rcr3 protease required for Cf-2-dependent disease resistance. Science 308:1783–1786

Ruther J, Kleier S (2005) Plant–plant signaling: Ethylene synergizes volatile emission in *Zea mays* induced by exposure to (Z)-3-hexen-1-ol. J Chem Ecol 31:2217–2222

Saintenac C, Zhang W, Salcedo A, Rouse MN, Trick HN, Akhunov E, Dubcovsky J (2013) Identification of wheat gene Sr35 that confers resistance to Ug99 stem rust race group. Science 341:783–786

Sarma BK, Singh HB, Fernando D, Silva RN, Gupta VK (2016) Enhancing plant disease resistance without R genes. Trends Biotechnol 34:523–525

Sarris PF, Duxbury Z, Huh SU, Ma Y, Segonzac C, Sklenar J, Derbyshire P, Cevik V, Rallapalli G, Saucet SB (2015) A plant immune receptor detects pathogen effectors that target WRKY transcription factors. Cell 161:1089–1100

Schmelz EA, Alborn HT, Tumlinson JH (2003) Synergistic interactions between volicitin, jasmonic acid, and ethylene mediate insect-induced volatile emission in *Zea mays*. Physiol Plant 117:403–412

Schornack S, Moscou MJ, Ward ER, Horvath DM (2013) Engineering plant disease resistance based on tal effectors. Ann Rev Phytopathol 51:383–406

Seidl MF, Thomma BP (2017) Transposable elements direct the coevolution between plants and microbes. Trends Genet 33:842–851

Sela H, Spiridon LN, Petrescu AJ, Akerman M, Mandel-Gutfreund Y, Nevo E, Loutre C, Keller B, Schulman AH, Fahima T (2012) Ancient diversity of splicing motifs and protein surfaces in the wild emmer wheat (*Triticum dicoccoides*) LR10 coiled-coil (CC) and leucine-rich repeat (LRR) domains. Mol Plant Pathol 13:276–287

Seo PJ, Kim MJ, Park JY, Kim SY, Jeon J, Lee YH, Kim J, Park CM (2010) Cold activation of a plasma membrane-tethered NAC transcription factor induces a pathogen resistance response in *Arabidopsis*. Plant J 61:661–671

Shabab M, Shindo T, Gu C, Kaschani F, Pansuriya T, Chintha R, Harzen A, Colby T, Kamoun S, van der Hoorn RA (2008) Fungal effector protein AVR2 targets diversifying defense related CYS proteases of tomato. Plant Cell 20:1169–1183

Shao F, Golstein C, Ade J, Stoutemyer M, Dixon JE, Innes RW (2003) Cleavage of *Arabidopsis* PBS1 by a bacterial type III effector. Science 301:1230–1233

Shen Q-H, Saijo Y, Mauch S, Biskup C, Bieri S, Keller B, Seki H, Ülker B, Somssich IE, Schulze-Lefert P (2007) Nuclear activity of MLA immune receptors links isolate-specific and basal disease-resistance responses. Science 315:1098–1103

Shi G, Zhang Z, Friesen TL, Raats D, Fahima T, Brueggeman RS, Lu S, Trick HN, Liu Z, Chao W (2016) The hijacking of a receptor kinase–driven pathway by a wheat fungal pathogen leads to disease. Sci Adv 2:e1600822

Silverstein KA, Graham MA, Paape TD, VandenBosch KA (2005) Genome organization of more than 300 defensin-like genes in arabidopsis. Plant Physiol 138:600–610

Singh A, Kirubakaran SI, Sakthivel N (2007) Heterologous expression of new antifungal chitinase from wheat. Protein Express Purif 56:100–109

Singh S, Tripathi RK, Lemaux PG, Buchanan BB, Singh J (2017) Redox-dependent interaction between thaumatin-like protein and β-glucan influences malting quality of barley. Proc Natl Acad Sci USA 201701824

Sinha M, Singh RP, Kushwaha GS, Iqbal N, Singh A, Kaushik S, Kaur P, Sharma S, Singh TP (2014) Current overview of allergens of plant pathogenesis-related protein families. Sci World J

Skadsen R, Sathish P, Kaeppler H (2000) Expression of thaumatin-like permatin PR-5 genes switches from the ovary wall to the aleurone in developing barley and oat seeds. Plant Sci 156:11–22

Slavokhotova AA, Naumann TA, Price NP, Rogozhin EA, Andreev YA, Vassilevski AA, Odintsova TI (2014) Novel mode of action of plant defense peptides–hevein-like antimicrobial peptides from wheat inhibit fungal metalloproteases. FEBS J 281:4754–4764

Song J, Win J, Tian M, Schornack S, Kaschani F, Ilyas M, van der Hoorn RA, Kamoun S (2009) Apoplastic effectors secreted by two unrelated eukaryotic plant pathogens target the tomato defense protease Rcr3. Proc Natl Acad Sci USA 106:1654–1659

Song YY, Cao M, Xie LJ, Liang XT, Zeng RS, Su YJ, Huang JH, Wang RL, Luo SM (2011) Induction of DIMBOA accumulation and systemic defense responses as a mechanism of enhanced resistance of mycorrhizal corn (*Zea mays* L.) to sheath blight. Mycorrhiza 21:721–731

Spelbrink RG, Dilmac N, Allen A, Smith TJ, Shah DM, Hockerman GH (2004) Differential antifungal and calcium channel-blocking activity among structurally related plant defensins. Plant Physiol 135:2055–2067

Steppuhn A, Baldwin IT (2007) Resistance management in a native plant: nicotine prevents herbivores from compensating for plant protease inhibitors. Ecol Lett 10:499–511

Stotz HU, Thomson J, Wang Y (2009) Plant defensins: defense, development, and application. Plant Signal Behav 4:1010–1012

Sugawara T, Trifonova EA, Kochetov AV, Kanayama Y (2016) Expression of an extracellular ribonuclease gene increases resistance to cucumber mosaic virus in tobacco. BMC Plant Biol 16:246

Sun Y, Li L, Macho AP, Han Z, Hu Z, Zipfel C, Zhou J-M, Chai J (2013) Structural basis for flg22-induced activation of the arabidopsis FLS2-BAK1 immune complex. Science 342:624–628

Tada Y, Spoel SH, Pajerowska-Mukhtar K, Mou Z, Song J, Wang C, Zuo J, Dong X (2008) Plant immunity requires conformational charges of NPR1 via S-nitrosylation and thioredoxins. Science 321:952–956
Takahashi F, Mizoguchi T, Yoshida R, Ichimura K, Shinozaki K (2011) Calmodulin-dependent activation of MAP kinase for ROS homeostasis in arabidopsis. Mol Cell 41:649–660
Takken FL, Goverse A (2012) How to build a pathogen detector: Structural basis of NB-LRR function. Curr Opin Plant Biol 15:375–384
Tao Z, Liu H, Qiu D, Zhou Y, Li X, Xu C, Wang S (2009) A pair of allelic WRKY genes play opposite roles in rice-bacteria interactions. Plant Physiol 151:936–948
Temple BR, Jones AM (2007) The plant heterotrimeric G-protein complex. Annu Rev Plant Biol 58:249–266
Temple BR, Jones CD, Jones AM (2010) Evolution of a signaling nexus constrained by protein interfaces and conformational states. PLoSComput Biol 6:e1000962
Terras FRG, Eggermont K, Kovaleva V, Raikhel NV, Osborn RW, Kester A, Rees SB, Torrekens S, Van Leuven F, Vanderleyden J (1995) Small cysteine-rich antifungal proteins from radish: their role in host defense. Plant Cell 7:573–588
Thomas EL, van der Hoorn RA (2018) Ten prominent host proteases in plant-pathogen interactions. Int J Mol Sci 19:639
Thomma BP, Penninckx IA, Cammue BP, Broekaert WF (2001) The complexity of disease signaling in arabidopsis. Curr Opin Immunol 13:63–68
Thomson VP, Cunningham SA, Ball MC, Nicotra AB (2003) Compensation for herbivory by *Cucumis sativus* through increased photosynthetic capacity and efficiency. Oecologia 134:167–175
Tian D, Wang J, Zeng X, Gu K, Qiu C, Yang X, Zhou Z, Goh M, Luo Y, Murata-Hori M (2014) The rice TAL effector–dependent resistance protein XA10 triggers cell death and calcium depletion in the endoplasmic reticulum. Plant Cell 26:497–515
Tian B, Li J, Oakley TR, Todd TC, Trick HN (2016) Host-derived artificial microRNA as an alternative method to improve soybean resistance to soybean cyst nematode. Genes 7:122
Ton J, Flors V, Mauch-Mani B (2009) The multifaceted role of ABA in disease resistance. Trends Plant Sci 14:310–317
Toruño TY, Stergiopoulos I, Coaker G (2016) Plant-pathogen effectors: cellular probes interfering with plant defenses in spatial and temporal manners. Ann Rev Phytopathol 54:419–441
Tran DT, Chung E-H, Habring-Müller A, Demar M, Schwab R, Dangl JL, Weigel D, Chae E (2017) Activation of a plant NLR complex through heteromeric association with an autoimmune risk variant of another NLR. Curr Biol 27:1148–1160
Trümper C, Paffenholz K, Smit I, Kössler P, Karlovsky P, Braun H-P, Pawelzik E (2016) Identification of regulated proteins in naked barley grains (*Hordeum vulgare nudum*) after *Fusarium graminearum* infection at different grain ripening stages. J Proteom 133:86–92
Trusov Y, Botella JR (2016) Plant G-proteins come of age: Breaking the bond with animal models. Front Chem 4:24
Tsuba M, Katagiri C, Takeuchi Y, Takada Y, Yamaoka N (2002) Chemical factors of the leaf surface are involved in the morphogenesis of *Blumeria graminis*. Physiol Mol Plant Pathol 60:51–57
Tsuchiya T, Eulgem T (2013) An alternative polyadenylation mechanism coopted to the *Arabidopsis* RPP7 gene through intronic retrotransposon domestication. Proc Natl Acad Sci USA 110:E3535–E3543
Tsuda K, Somssich IE (2015) Transcriptional networks in plant immunity. New Phytol 206:932–947
Tsuda K, Mine A, Bethke G, Igarashi D, Botanga CJ, Tsuda Y, Glazebrook J, Sato M, Katagiri F (2013) Dual regulation of gene expression mediated by extended MAPK activation and salicylic acid contributes to robust innate immunity in *Arabidopsis thaliana*. PLoS Genet 9:e1004015
Turlings TC, Tumlinson JH, Lewis WJ (1990) Exploitation of herbivore-induced plant odors by host-seeking parasitic wasps. Science 250:1251

Turlings TC, Tumlinson JH, Heath RR, Proveaux AT, Doolittle RE (1991) Isolation, and identification of allelochemicals that attract the larval parasitoid, *Cotesia marginiventris* (cresson), to the microhabitat of one of its hosts. J Chem Ecol 17:2235–2251

Turlings TC, McCall PJ, Alborn HT, Tumlinson JH (1993) An elicitor in caterpillar oral secretions that induces corn seedlings to emit chemical signals attractive to parasitic wasps. J Chem Ecol 19:411–425

Valjavec-Gratian M, Steffenson B (1997) Pathotypes of *Cochliobolus sativus* on barley in North Dakota. Plant Dis 81:1275–1278

van den Burg HA, Harrison SJ, Joosten MH, Vervoort J, de Wit PJ (2006) *Cladosporium fulvum* Avr4 protects fungal cell walls against hydrolysis by plant chitinases accumulating during infection. Mol Plant Microbe Interact 19:1420–1430

Van Der Biezen EA, Jones JD (1998) Plant disease-resistance proteins and the gene-for-gene concept. Trends Biochem Sci 23:454–456

Van Der Weerden NL, Lay FT, Anderson MA (2008) The plant defensin, NaD1, enters the cytoplasm of *Fusarium oxysporum* hyphae. J Biol Chem 283:14445–14452

Van Der Weerden NL, Hancock RE, Anderson MA (2010) Permeabilization of fungal hyphae by the plant defensin NaD1 occurs through a cell wall-dependent process. J Biol Chem 285:37513–37520

van der Hoorn RA, Kamoun S (2008) From guard to decoy: a new model for perception of plant pathogen effectors. Plant Cell 20:2009–2017

Van der Weerden NL, Anderson MA (2013) Plant defensins: common fold, multiple functions. Fungal Biol Rev 26:121–131

van Loon LC, Rep M, Pieterse CM (2006) Significance of inducible defense-related proteins in infected plants. Annu Rev Phytopathol 44:135–162

Van Loon L, Van Strien E (1999) The families of pathogenesis-related proteins, their activities, and comparative analysis of PR-1 type proteins. Physiol Mol Plant Pathol 55:85–97

Vigers AJ, Roberts WK, Selitrennikoff CP (1991) A new family of plant antifungal proteins. Mol Plant Microbe Interact 4:315–323

Vijayan P, Shockey J, Lévesque CA, Cook RJ (1998) A role for jasmonate in pathogen defense of arabidopsis. Proc Natl Acad Sci USA 95:7209–7214

Vleeshouwers VG, Oliver RP (2014) Effectors as tools in disease resistance breeding against biotrophic, hemibiotrophic, and necrotrophic plant pathogens. Mol Plant Microbe Interact 27:196–206

Vlot AC, Dempsey DMA, Klessig DF (2009) Salicylic acid, a multifaceted hormone to combat disease. Ann Rev Phytopathol 47:177–206

Vriens K, Peigneur S, De Coninck B, Tytgat J, Cammue BP, Thevissen K (2016) The antifungal plant defensin AtPDF2.3 from *Arabidopsis thaliana* blocks potassium channels. Sci Rep 6

Walters DR, Paterson L (2012) Parents lend a helping hand to their offspring in plant defense. Biol Lett 8:871–873

Wang S, Rao P, Ye X (2009a) Isolation and biochemical characterization of a novel leguminous defense peptide with antifungal and antiproliferative potency. Appl Microbiol Biotechnol 82:79–86

Wang X, Goregaoker SP, Culver JN (2009b) Interaction of the Tobacco mosaic virus replicase protein with a NAC domain transcription factor is associated with the suppression of systemic host defenses. J Virol 83:9720–9730

Wang M, Weiberg A, Dellota E Jr, Yamane D, Jin H (2017b) Botrytis small RNA Bc-siR37 suppresses plant defense genes by cross-kingdom RNAi. RNA Biol 14:421–428

Wang J, Tao F, Tian W, Guo Z, Chen X, Xu X, Shang H, Hu X (2017a) The wheat WRKY transcription factors TaWRKY49 and TaWRKY62 confer differential high-temperature seedling-plant resistance to *Puccinia striiformis* f. sp. *tritici*. PLoS One 12:e0181963

Wanjiru WM, Zhensheng K, Buchenauer H (2002) Importance of cell wall degrading enzymes produced by *Fusarium graminearum* during infection of wheat heads. Eur J Plant Pathol 108:803–810

War AR, Paulraj MG, Ahmad T, Buhroo AA, Hussain B, Ignacimuthu S, Sharma HC (2012) Mechanisms of plant defense against insect herbivores. Plant Signal Behav 7:1306–1320

Wei Y, Zhang Z, Andersen CH, Schmelzer E, Gregersen PL, Collinge DB, Smedegaard-Petersen V, Thordal-Christensen H (1998) An epidermis/papilla-specific oxalate oxidase-like protein in the defense response of barley attacked by the powdery mildew fungus. Plant Mol Biol 36:101–112

Weiberg A, Wang M, Lin F-M, Zhao H, Zhang Z, Kaloshian I, Huang H-D, Jin H (2013) Fungal small RNAs suppress plant immunity by hijacking host RNA interference pathways. Science 342:118–123

Weibull J, Niemeyer H (1995) Changes in dihydroxymethoxybenzoxazinone glycoside content in wheat plants infected by three plant pathogenic fungi. Physiol Mol Plant Pathol 47:201–212

Wel H, Loeve K (1972) Isolation and characterization of thaumatin I and II, the sweet-tasting proteins from *Thaumatococcusdanielliib*enth. FEBS J 31:221–225

Whetzel HH (1918) An outline of the history of phytopathology. WB Saunders, Philadelphia, PA, USA

Wiermer M, Feys BJ, Parker JE (2005) Plant immunity: the EDS1 regulatory node. Curr Opin Plant Biol 8:383–389

Wijaya R, Neumann GM, Condron R, Hughes AB, Polya GM (2000) Defense proteins from the seed of *Cassia fistula* include a lipid transfer protein homologue and a protease inhibitory plant defensin. Plant Sci 159:243–255

Wilcoxson R, Rasmusson D, Miles M (1990) Development of barley resistant to spot blotch and genetics of resistance. Plant Dis 74:207–210

Woldemariam MG, Dinh ST, Oh Y, Gaquerel E, Baldwin IT, Galis I (2013) NaMYC2 transcription factor regulates a subset of plant defense responses in *Nicotiana attenuata*. BMC Plant Biol 13:73

Woloshuk CP, Kolattukudy P (1986) Mechanism by which contact with plant cuticle triggers cutinase gene expression in the spores of *Fusarium solani* f. sp. *pisi*. Proc Natl Acad Sci USA 83:1704–1708

Wouters FC, Blanchette B, Gershenzon J, Vassão DG (2016) Plant defense, and herbivore counter-defense: benzoxazinoids and insect herbivores. Phytochem Rev 15:1127–1151

Wu H, Pratley J, Lemerle D, Haig T (2001) Allelopathy in wheat (*Triticum aestivum*). Ann Appl Biol 139:1–9

Wu Y, Deng Z, Lai J, Zhang Y, Yang C, Yin B, Zhao Q, Zhang L, Li Y, Yang C (2009) Dual function of *Arabidopsis* ATAF1 in abiotic and biotic stress responses. Cell Res 19:1279–1290

Wu C-H, Krasileva KV, Banfield MJ, Terauchi R, Kamoun S (2015) The, "sensor domains" of plant NLR proteins: more than decoys? Front Plant Sci 6:134

Wu CH, Belhaj K, Bozkurt TO, Birk MS, Kamoun S (2016) Helper NLR proteins NRC2a/b and NRC3 but not NRC1 are required for PTO-mediated cell death and resistance in *Nicotiana benthamiana*. New Phytol 209:1344–1352

Wu CH, Abd-El-Haliem A, Bozkurt TO, Belhaj K, Terauchi R, Vossen JH, Kamoun S (2017) NLR network mediates immunity to diverse plant pathogens. Proc Natl Acad Sci USA 114:8113–8118

Xu J, Li Y, Wang Y, Liu H, Lei L, Yang H, Liu G, Ren D (2008) Activation of MAPK kinase 9 induces ethylene and camalexin biosynthesis and enhances sensitivity to salt stress in *Arabidopsis*. J Biol Chem 283:26996–27006

Yang S, Tang F, Zhu H (2014) Alternative splicing in plant immunity. Int J Mol Sci 15:10424–10445

Zeidler D, Zähringer U, Gerber I, Dubery I, Hartung T, Bors W, Hutzler P, Durner J (2004) Innate immunity in *Arabidopsis thaliana*: Lipopolysaccharides activate nitric oxide synthase (NOS) and induce defense genes. Proc Natl Acad Sci USA 101:15811–15816

Zhai J, Jeong D-H, De Paoli E, Park S, Rosen BD, Li Y, González AJ, Yan Z, Kitto SL, Grusak MA (2011) MicroRNAs as master regulators of the plant NB-LRR defense gene family via the production of phased, trans-acting siRNAs. Genes Dev 25:2540–2553

Zhan J, Thrall PH, Papaïx J, Xie L, Burdon JJ (2015) Playing on a pathogen's weakness: using evolution to guide sustainable plant disease control strategies. Ann Rev Phytopathol 53:19–43

Zhang J, Shao F, Li Y, Cui H, Chen L, Li H, Zou Y, Long C, Lan L, Chai J (2007) A *Pseudomonas syringae* effector inactivates MAPKs to suppress PAMP-induced immunity in plants. Cell Host Microbe 1:175–185

Zhang H, Gao Z, Zheng X, Zhang Z (2012) The role of G-proteins in plant immunity. Plant Signal Behav 7:1284–1288

Zhao J (2015) Phospholipase d and phosphatidic acid in plant defense response: from protein-protein and lipid-protein interactions to hormone signaling. J Exp Bot 66:1721–1736

Zhao Y, Wei T, Yin KQ, Chen Z, Gu H, Qu LJ, Qin G (2012) *Arabidopsis* RAP2.2 plays an important role in plant resistance to *Botrytis cinerea* and ethylene responses. New Phytol 195:450–460

Zhou H, Steffenson B (2013) Genome-wide association mapping reveals genetic architecture of durable spot blotch resistance in US barley breeding germplasm. Mol Breed 32:139–154

Zhou H, Li S, Deng Z, Wang X, Chen T, Zhang J, Chen S, Ling H, Zhang A, Wang D (2007) Molecular analysis of three new receptor-like kinase genes from hexaploid wheat and evidence for their participation in the wheat hypersensitive response to stripe rust fungus infection. Plant J 52:420–434

Zimmermann G, Bäumlein H, Mock H-P, Himmelbach A, Schweizer P (2006) The multigene family encoding germin-like proteins of barley. Regulation and function in basal host resistance. Plant Physiol 142:181–192

Zipfel C (2014) Plant pattern-recognition receptors. Trends Immunol 35:345–351

Zipfel C, Robatzek S, Navarro L, Oakeley EJ (2004) Bacterial disease resistance in *Arabidopsis* through flagellin perception. Nature 428:764–767

Zipfel C, Kunze G, Chinchilla D, Caniard A, Jones JD, Boller T, Felix G (2006) Perception of the bacterial PAMP EF-TU by the receptor EFR restricts *Agrobacterium*-mediated transformation. Cell 125:749–760

Zou H, Henzel WJ, Liu X, Lutschg A, Wang X (1997) Apaf-1, a human protein homologous to *C. elegans* CED-4, participates in cytochrome c-dependent activation of caspase-3. Cell 90:405–413

Genome Editing Technologies for Resistance Against Phytopathogens

Jagmohan Singh, Chiti Aggarwal, Prexha Kapoor, Rahul Kumar Tiwari, Ravinder Kumar, Milan Kumar Lal, Manuj Saini, Manjunath S. Hurakadli, and Vipul Bajwa

Abstract Phytopathogens pose a significant threat to global food security by causing widespread crop diseases. To combat this challenge, genome editing technologies have emerged as powerful tools for developing crops with enhanced resistance to phytopathogens. This chapter explores the latest advancements in genome editing techniques, such as CRISPR/Cas9, TALENs, and ZFNs, and their applications in engineering resistance against various phytopathogens. The book chapter begins by summarizing the molecular mechanisms underlying plant-pathogen interactions and the identification of key genes involved in the host defense response. It then delves into the application of genome editing technologies to precisely modify these genes, enhancing resistance without compromising plant yield. Case studies highlight successful implementations of genome editing in major crops like rice, wheat, and tomatoes, showcasing the potential for broad-spectrum resistance. Additionally, discusses the ethical considerations and regulatory challenges associated with deploying genome-edited crops in agriculture. Future prospects and ongoing

J. Singh (✉) · M. S. Hurakadli · V. Bajwa
Department of Plant Pathology, CCS Haryana Agricultural University, Hisar, Haryana, India
e-mail: dhillonjagmohansingh@gmail.com

C. Aggarwal
Department of Plant Science, University of Maryland, Baltimore, MD, USA

P. Kapoor
Department of Molecular Biology and Biotechnology, CCS Haryana Agricultural University, Hisar, Haryana, India

R. K. Tiwari
Division of Crop Protection, ICAR-Indian Institute of Sugarcane Research, Lucknow, India

R. Kumar
Divison of Plant Pathology, ICAR-Indian Agricultural Research Institute, New Delhi, India

M. K. Lal
ICAR-National Rice Research Institute, Cuttack, India

M. Saini
Department of Genetics & Plant Breeding, CCS Haryana Agricultural University, Hisar, Haryana, India

J. Chen et al. (eds.), *Molecular and Biotechnological Tools for Plant Disease Management*, Interdisciplinary Biotechnological Advances,
https://doi.org/10.1007/978-981-97-7510-1_16

research efforts aimed at refining genome editing technologies for sustainable and environmentally friendly phytopathogen resistance are also addressed. In conclusion, genome editing technologies hold immense promise for developing crops with durable resistance against phytopathogens, offering a viable and targeted approach to address the evolving landscape of agricultural challenges.

Keywords Genome editing · CRISPR-cas9 · Phyto-pathogen · Plant resistance · Food security

1 Introduction

The main aim of agriculture in the twenty-first century is sustainable agriculture, which achieves food security while minimizing negative impacts on the environment. By 2050, the world population is expected to vary between 8 and 10 billion (Cohen 2001). Feeding such a large population is going to be a major concern. So, the production of food must increase while the detrimental effects of agriculture on the land, water, and climate must be reduced (Olesen and Bindi 2002). Crop production is threatened by various biotic as well as abiotic stress. Phytopathogens which cause diseases are one of the major obstacles that poses a risk to agricultural sustainability and global food security (Strange and Scott 2005). They constitute a diverse group of microscopic organisms, like bacteria, fungi, nematodes, and viruses. Diseases have a negative impact on plant growth and development, which ultimately leads to a decrease in crop yields (Strange and Scott 2005). Furthermore, they also have a detrimental effect on the quality of agricultural products, both in the field and during the storage process. Reduced agricultural outputs give rise to financial difficulties for farmers, thereby jeopardizing their means of livelihood (Das and Ansari 2021). Furthermore, poor-quality crop has an impact on the nutritional composition and can lead to food shortages, particularly in regions that are heavily reliant on farming. Postharvest disease losses can be devastating, particularly in cases where farms are remote from markets and where supply chain and infrastructural standards are inadequate (Stathers et al. 2013). Additionally, numerous pathogens also produce toxins that give rise to severe health complications for consumers. Diseases reduce crop production up to 20–40% worldwide (FAO). Furthermore, plant diseases destroy natural ecosystems, increasing environmental issues brought on by habitat loss and careless land management. Severe outbreaks led to entire crop failure, endangering the availability of food and generating financial crises. A few examples include the Irish potato famine in the 1840s in Ireland brought on by *Phytophthora infestans*, the Great Bengal Famine of 1943 brought on by *Cochliobolus miyabeanus* in rice, and the southern corn leaf blight epidemic in the United States between 1970 and 1971 (Raychaudhuri et al. 1972; Ullstrup 1972). Plant diseases have been managed using a variety of techniques and approaches. These techniques include the use of agrochemicals, the development of resistant varieties through plant breeding, genetically engineered plants, physical methods, biological methods, etc.

For sustainable agriculture, plant disease resistance must be genetically improved. Plant disease resistance is one of the most efficient ways to manage plant diseases, as it is inexpensive and environmentally friendly (Darshan et al. 2020; Martinelli et al. 2015). Plant disease resistance is essential for the consistent production of food, and it also provides considerable reductions in agricultural use of land, fuel, water, and other inputs. Plants with disease resistance protect plants from phytopathogens in two different ways: by pre-formed structures and chemicals, and by infection-induced responses of the plant immune system (Yang et al. 2022). Natural breeding, cultural practices, and genetic engineering are strategies for developing resistance in plants against disease.

Crop protection, which is largely accomplished by applying chemical pesticides, is in a transitional phase. To avoid phytopathogens, farmers frequently use chemical treatments, including fungicides and bactericides, which add costs and create environmental issues. However, relying too much on these chemicals might have several negative effects (Jat et al. 2021; van Bruggen et al. 2016). First, these chemical substances have a risk of polluting the environment since they may accumulate in soil and water, endangering ecosystems and non-target organisms (Jutsum et al. 1998). Furthermore, continuous use of these chemicals may lead to the development of resistant strains of pathogens, thereby gradually decreasing the effectiveness of the pesticides. Moreover, the use of chemicals may have unintended consequences, harming beneficial organisms like natural predators and pollinators, thereby disrupting the ecological balance of our agricultural systems (Chagnon et al. 2015; Kashyap et al. 2023). Another non-resistance-based strategy for controlling plant disease is crop rotation, which, while helping in breaking disease cycles, might not be feasible for all crops or in all areas due to particular cropping systems and land limitations. Furthermore, it may not offer complete protection against all kinds of pathogens. These drawbacks emphasize the necessity of a holistic and integrated strategy for controlling phytopathogens, judiciously combining various methods to address the challenges while minimizing negative environmental impact and advancing sustainable agricultural practices (Singh et al. 2022a).

For hundreds of years, traditional breeding techniques have been exploited to develop crops with enhanced resistance against plant pathogens. Numerous traditional breeding approaches namely Marker-Assisted Selection (MAS), hybridization, gene pyramiding, recurrent selection, mutation breeding, pure line, pedigree, and backcross methods, etc. have been effectively developed for building resistance against plant diseases (Hasan et al. 2021; Kumar et al. 2022). The benefits of traditional breeding include consumer acceptance, accessibility, broad-spectrum resistance, and sustainability. Although traditional breeding has been successful over the past few decades, it has several limitations initially, it can be performed only between plants that can reproduce with one another (Tester and Langridge 2010). Secondly, it is heavily reliant on populations of plants that have sufficient genetic diversity. Thirdly, in addition to the trait of interest, i.e., resistance it often introduces a variety of traits, some of which have negative impacts on yield. The introduction of host resistance through traditional breeding is difficult because of the linkage drag and limited genetic diversity within elite germplasm (Tester and Langridge 2010). Fourth, traditional breeding requires genetic crosses and segregant progeny selection which is labor-intensive and requires a lot of effort and time.

These challenges with our current agricultural practices point to the need for new technologies and genome editing emerges as a revolutionary modern and advanced molecular technology that precisely alters the DNA sequence within the genome of living organisms (Kaur et al. 2021). It is also known by some other names namely gene editing, genome engineering, or genome editing with engineered nucleases (GEEN). It is a highly powerful tool that allows scientists to alter the genetic information in cells, tissues, and organisms either by deleting, adding, or replacing genetic material with a high level of precision and accuracy. Successful genome editing requires molecular information about the target gene and known host genome sequences (Ceasar et al. 2016). The main objectives of genome editing are to explore the function of particular genes, fixate genetic mutations causing diseases, or introduce novel traits into an organism by manipulating the genetic code in a regulated way. Genome editing technologies have a vast range of applications, from fundamental research to agricultural and medical enhancements (Gao 2021; Singh et al. 2022b). It possesses significant potential for surpassing the constraints associated with conventional breeding methods.

2 Molecular Targets for Phytopathogen Resistance

Understanding the molecular targets critical for escalating an effective defense is fundamental in harnessing the complete potential of genome editing in plant protection system (Yin and Qiu 2019). At the forefront of molecular targets are the genes governing plant-pathogen interactions. Identifying key components included in pathogen recognition, signal transduction, and defense response pathways permits the strategic deployment of genome editing tools (Zaidi et al. 2018; Barka and Lee 2022). Research efforts are directed toward employing specific genes accountable for conferring resistance, disrupting pathogen recognition mechanisms, or enhancing the expression of defense-related genes. Explicating the intricacies of plant immune systems at the molecular level allows for the development of crops with durable and broad-spectrum resistance, reducing the reliance on traditional chemical pesticides.

2.1 Identification and Characterization of Susceptibility Genes

The approach of enhancing disease resistance by targeting susceptibility genes represents an evolving approach in resistance breeding. This technique exploits the organized functions of certain genes, specifically those associated with both resistance and susceptibility factors within certain plant species (Barka and Lee 2022). With advancements in the understanding of resistance mechanisms, there is an increasing ability to rapidly enhance disease resistance through genome editing based on susceptibility factors. This technique is widely recognized as a prevalent and effective strategy for resistance breeding. The advancement of precision in editing, particularly with the advent of CRISPR, ODM, TALE, and ZF nucleases, has enabled the targeted modification of susceptibility genes. The application of these advanced techniques has established significant promise in various staple crops and vegetables (Zaidi et al. 2018; Yin and Qiu 2019). These methods are being progressively evolved to develop resistance against widespread and challenging pathogens, including viruses, showcasing the potential of susceptibility-based genome editing in fortifying crops against diseases. Table 1 provides a comprehensive summary of diverse susceptibility genes that have been successfully silenced in three solanaceous crops, either through genome editing or RNA interference (RNAi). As elucidated by van Schie and Takken, these susceptibility genes can be grouped into three distinct classes based on their roles in susceptibility (Van Schie and Takken 2014). The first category/class involves genes essential for pathogen infection and penetration, including those involved in spore germination for spore-forming fungal pathogens. Notable examples within this class include conserved membrane-anchored protein-encoding orthologous genes like mildew locus O (Mlo), found in tomato, pepper, and various cereal crops (Barka and Lee 2022).

The second class includes negative regulators of plant immunity, also known as defense suppressors, which employ a negative impact on the expression of cellulose synthase genes such as CeSA3(Van Schie and Takken 2014). Lastly, the third class of susceptibility genes, exemplified by SWEET genes, contributes in sugar biosynthesis and transport, crucial for pathogen nutrition and replication (Gupta et al. 2021). A captivating achievement in the realm of genome editing is the conversion of the Arabidopsis eIF4E1 susceptibility gene into resistance against Clover Yellow Vein Virus (CYVV). This alteration was accomplished through CRISPR/Cas9-cytidine base editor (CBE)-based genome editing, representing its efficacy not only within a single plant species but also across diverse plant species within the solanaceous family (Bastet et al. 2019).

2.2 Understanding Plant-Pathogen Interactions

Explicating plant-pathogen interactions is essential in harnessing the potential of genome editing for developing crops with boosted resistance (Zaidi et al. 2018). Plant-pathogen interactions are intriguing, dynamic mechanism that involve a regulated interplay between the plant's defense mechanisms and the strategies employed by pathogens to invade and colonize the host (Tiwari et al. 2020, 2022, 2023). Genome editing technologies, such as CRISPR-Cas9, provide unprecedented opportunities to dissect and manipulate the key components of these interactions at the molecular level. At the lead of this understanding is the documentation and characterization of genes involved in both plant defense and pathogen virulence. By unraveling the

Table 1 Susceptibility genes targeted for disease resistance in diverse crops

Crop	Susceptibility gene targeted	Disease/pathogen	References
Apple	DIPM-1, DIPM-2, and DIPM-4	Fire blight disease (*Erwinia amylovora*)	Malnoy et al. (2016a, b, c)
Arabidopsis	eIF4E	TuMV	Pyott et al. (2016)
Cucumber	eIF4E	CVYV (ipomovirus), ZYMV and PRSMV (potyvirus)	Chandrasekaran et al. (2016)
Citrus	CsLOB1	Citrus canker	Peng et al. (2017)
Grape	MLO-7	Powdery mildew	Malnoy et al. (2016a, b, c)
Pepper	CaWRKY58	Bacterial wilt	Wang et al. (2013)
	pvr1, pvr2	Potyvirus	Sun et al. (2016b)
Potato	DMR6	*Phytophthora infestnas*	Sun et al. (2016b)
	DND1	Bacterial wilt	Sun et al. (2016a)
	BIK1	Brown spot	Sun et al. (2016b)
Rice	OsSWEET14	Bacterial Blight	Jiang et al. (2013)
	OsMPK5	Fungal (*Magnaporthe grisea*) and bacterial (*Burkholderia glumae*) pathogens	Xie and Yang (2013)
Tomato	DMR6	*Pseudomonas syringae, Phytophthora capsici, and Xanthomonas spp*	de Toledo Thomazella et al. (2016)
	Sitiens/Sit	Gray mold/rot, soft rot	Curvers et al. (2010)
Wheat	TaMLO-A1, TaMLO-B1, TaMLO-D1	Powdery mildew	Wang et al. (2014)

genetic basis of resistance and susceptibility, researchers can pinpoint specific molecular targets for intervention (Jiang et al. 2013; Wang et al. 2020). This knowledge enables the precise modification of plant genes to enhance resistance or disrupt the strategies employed by pathogens to overcome plant defenses. Plant immune systems are governed by intricate signaling pathways, receptor proteins, and defense-related genes. Genome editing allows researchers to target and modify these components to bolster the plant's ability to recognize and respond to pathogen threats effectively. Additionally, exploring the mechanisms of pathogen virulence assists in identifying weaknesses that can be exploited through genome editing to hinder the pathogen's ability to infect plants. The advent of advanced genome editing tools has facilitated the manipulation of plant-pathogen interactions with unparalleled correctness. Techniques like CRISPR-Cas9, CRISPR-Cas12, and base editing allow researchers to edit specific genes or regulatory elements involved in these interactions. Moreover, the use of synthetic biology approaches enables the engineering of plants with customized defense mechanisms, offering a promising avenue for sustainable and resilient agriculture.

2.3 Targeting Key Genes Involved in Plant Defense Mechanisms

Disease resistance breeding through genome editing extends beyond gene disruption or replacement, encompassing precise gene regulation. For example, in rice, japonica rice plants carrying the Xa3 gene exhibit an expanded resistance spectrum and enhanced resistance to *Xanthomonas oryzae* pv. *oryzae* (Xoo) compared to indica rice plants. This increased resistance is because of the elevated expression of Xa3 in japonica rice, highlighting the potential of regulating gene expression to enhance resistance (Liu et al. 2020).

Likewise, in maize, the Hm2 gene provides dosage-dependent resistance, against leaf spot and ear mold disease. The level of resistance is correlated with the transcription level of Hm2, showcasing the importance of gene dosage in conferring resistance (Chintamanani et al. 2008). In contrast, the cotton GhLMMD gene showcases dosage-dependent regulation of programmed cell death and immunity. Down-regulation of GhLMMD in cotton resulted in resistance to infection by *Verticillium dahlia* (Chai et al. 2017). These examples highlight the importance of gene regulation in developing resistance traits. The advancements in the form of CRISPR activation (CRISPRa) and CRISPR interference (CRISPRi) techniques enable precise and multiplex gene regulation. As a result, broad-spectrum resistance in crops can be achieved by modulating genes associated with resistance. In addition, genome editing can fine-tune mRNA translation if target genes possess upstream open reading frames.

Genomics engineering of bacteria has emerged as a promising strategy for combating plant pathogens. Upon detecting *N*-acyl homoserine lactone secreted by the human pathogen *Pseudomonas aeruginosa*, the genetically engineered probiotic *Escherichia coli* can lyse itself. As a result, toxins and antibiofilm enzymes are released, preventing *P. aeruginosa* gut infections (Karlsson et al. 2012). *Pseudomonas fluorescens* expressing the chiA gene also protected bean seedlings from *Rhizoctonia solani* infection. By modifying enzyme-encoding genes in plant-associated bacteria, CRISPR technology can provide tailored resistance to plant pathogens (Downing and Thomson 2000). As a result of these advancements, genome editing is increasingly being used for fine-tuning gene regulation to enhance plant and bacterial resistance, paving the way for innovative strategies in agriculture and plant protection (Table 2).

Table 2 Genome editing for phytopathogen resistance in different crop plants

Sl. no	Target gene	Host plant	Effector protein/ nuclease	Trait improvement	References
Plant genome editing against viruses					
1	Integrated genome of banana streak virus (BSV)	Banana	PcUbi-Cas9 OsU6 PsgRNA	Conferred resistance against Banana streak virus (BSV)	Tripathi et al. (2019)
2	Coilin gene	*S. tuberosum*	SpCas9	Resistance to *poty viruses* PVY	Makhotenko et al. (2019)
3	eIF4E gene	*Arabidopsis thaliana*	Sp-nCas9-cytidine deaminase	Transgene free resistant plants against *potyvirus* i.e. Clover yellow vein virus	Bastet et al. (2019)
4	elF4E isoforms nCBP-1 & nCBP-2	*Manihot esculenta*	SpCas9	Resistance to cassava brown streak virus	Gomez et al. (2019)
5	eIF4G	*Oryza sativa* L. *O. japonica*	SpCas9	Resistance to Rice tungro spherical virus (RTSV)	Macovei et al. (2018)
6	CP (coat protein) or Rep gene	*N. benthamiana*	SpCas9	Enhanced resistance against tomato yellow leaf curl virus (TYLCV)	Tashkandi et al. (2018)
7	GFP, Hc-Pro and CP	Tobacco	CRISPR/ Cas13a	Resistance to TuMV	Aman et al. (2018)
8	Cs-eIF4E	*Cucumis sativus*	SpCas9	Resistance to -potyviruses such as Cucumber vein yellowing virus (CVYV), Zucchini yellow mosaic virus (ZYMV) and Papaya ringspot virus (PRSV); no resistance against CMV or CGMMV	Chandrasekaran et al. (2016)
9	eIF(iso)4E	*Arabidopsis thaliana*	SpCas9	Resistance to Turnip mosaic virus-TuMV-Potyvirus	Pyott et al. (2016)
10	Rep	Tobacco	TALE	Resistance to TbCSV, TYCCNV and TLCYnV	Cheng et al. (2015)
11	Rep	Tobacco	ZFN	Resistance to TYLCCNV and TbCSV	Chen et al. (2014)

(continued)

Table 2 (continued)

Sl. no	Target gene	Host plant	Effector protein/ nuclease	Trait improvement	References
12	Replication origin	Arabidopsis	AZP	Inhibition of virus DNA replication and Resistance to DNA viral disease	Sera et al. (2005)
Plant genome editing against fungi					
1	StDMR6-1 and StDMR6-2	*Solanum tuberosum* L	SpCas9	Editing confers increased late blight resistance	Kieu et al. (2021)
2	VvPR4b	*Vitis vinifera*	SpCas9	Editing decreases downy mildew (*Plasmopara viticola*) resistance in grapevine	Li et al. (2020)
3	PMR4	*Solanum lycopersicum* L	SpCas9	knock-down of SlPMR4 enhances resistance against powdery mildew	Martínez et al. (2020)
4	VvWRKY52	*Vitis vinifera*	SpCas9	Resistance to Graymold (*Botrytis cinerea*)	Leisen et al. (2020)
5	Clpsk1	Watermelon (*Citrullus lanatus*)	SpCas9	Loss-of-function rendered watermelon seedlings more resistant to infection by *Fusarium oxysporum* f.sp. *niveum*	Giacomelli et al. (2018)
6	TcNPR3, a suppressor of the defense response	*Theobroma cacao*	SpCas9	Increased resistance to infection with the cacao pathogen *Phytophthora tropicalis*	Fister et al. (2018)
7	alEPIC8/cds region	Papaya	SpCas9	Increased resistance against *P. palmivora*	Gumtow et al. (2018)
8	TaEDR1	*Triticum aestivum*	SpCas9	Resistance against *Blumeria graminis* f. sp. *tritici* (powdery mildew)	Zhang et al. (2017)
9	SlMlo1/exon	Tomato	CRISPR/ Cas9	Enhanced resistance to powdery mildew of tomato (*Oidium neolycopersici*)	Nekrasov et al. (2017)
10	OsERF922/ exon	Rice	CRISPR/ Cas9	Resistance to *Magnaporthe oryzae* (Rice Blast)	Wang et al. (2016)

(continued)

Table 2 (continued)

Sl. no	Target gene	Host plant	Effector protein/ nuclease	Trait improvement	References
11	TaMLO/exon	Wheat	TALEN	Enhanced resistance against *Blumeria graminis* f. sp. *tritici*	Wang et al. (2014)
Plant genome editing against bacteria					
1	OsSWEET14	*O. sativa*	SpCas9	Disruption of OsSWEET14 in the Zhonghua 11 background confers strong resistance to African Xoo strain AXO1947 and Asian Xoo strain PXO86	Zeng et al. (2020)
2	EBEs of OsSWEET14 gene	Super Basmati Rice	SpCas9	Resistance in Super Basmati rice lines against Xoo strain carrying AvrXa7 TALE	Zafar et al. (2020a, b)
3	EBE of three promoters i.e. SWEET11, SWEET13 and SWEET14	*O. sativa*	SpCas9	Enhanced broad-spectrum disease resistance to *Xanthomonas oryzae* pv. *oryzae*	Oliva et al. (2019)
4	EBE of OsSWEET11 and OsSWEET14	*O. sativa*	SpCas9	Enhanced broad-spectrum disease resistance (*Xanthomonas oryzae* pv. *oryzae*)	Xu et al. (2019)
5	S1JAZ2	*Solanum lycopersicum*	SpCas9	Resistance against bacterial speck caused by *Pseudomonas syringae*, DC3000	Ortigosa et al. (2019)
6	Os8N3	*O. sativa*	CRISPR/ Cas9 mutagenesis	Confer resistance to *Xanthomonas oryzae* pv. *oryzae*	Kim et al. (2019)
7	CsLOB1	*Citrus sinensis*	SpCas9	Resistant to Citrus canker (*Xanthomonas citri* sub spp. *citric*)	Peng et al. (2017)
8	DIPM-1, 2 and 4	Apple	SpCas9	Resistance to fire blast disease (*Erwinia amylovora*)	Malnoy et al. (2016a, b, c)

(continued)

Table 2 (continued)

Sl. no	Target gene	Host plant	Effector protein/ nuclease	Trait improvement	References
9	SlDMR6	*Solanum lycopersicum*	SpCas9	Enhanced resistance against Xanthomonas, Pseudomonas, and *Phytophthora capsici*	Thomazella et al. (2016)
10	effector/TALE PthXo2 of OsSWEET13	*O. sativa*	SpCas9	Resistance to bacterial blight	Zhou et al. (2015)
11	OsSWEET14/ promoter	*O. sativa*	TALEN	Resistance to *Xanthomonas oryzae* pv. *oryzae* (Rice bacterial blight)	Li et al. (2012)

3 Strategies for Genome Editing in Plants

3.1 Gene Knockout and Knockdown Strategies

Gene knockout strategies, including CRISPR-Cas9, TALENs, and ZFNs, have transformed the field of genome editing by bringing targeted double-strand breaks, resulting in the disruption or elimination of specific genes. CRISPR-Cas9, with its effortlessness and multiplexing capabilities, has become a principal choice for researchers (Jiang et al. 2013; de Toledo Thomazella et al. 2016; Zaidi et al. 2018). TALENs and ZFNs, while historically noteworthy, have seen a decline in use due to their complex design compared to the user-friendly CRISPR-Cas9. The emergence of CRISPR-Cas12 and CRISPR-Cas13 further extends the options for precise gene knockout. In contrast, gene knockdown strategies, such as RNA interference (RNAi), antisense oligonucleotides (ASOs), ribozymes, CRISPR interference (CRISPRi), and anti-sense RNA expression, focus on reducing gene expression rather than completely eliminating it. RNAi utilizes small RNA molecules to degrade mRNA, while ASOs and ribozymes interfere with mRNA stability or cleave specific RNA sequences (Ghag and Ganapathi 2019). CRISPRi employs a nuclease-dead Cas9 to interfere with transcription, and anti-sense RNA expression involves expressing RNA molecules complementary to target mRNA. Each strategy offers unique advantages, providing researchers with a diverse toolkit for precise modulation of gene expression (Fig. 1).

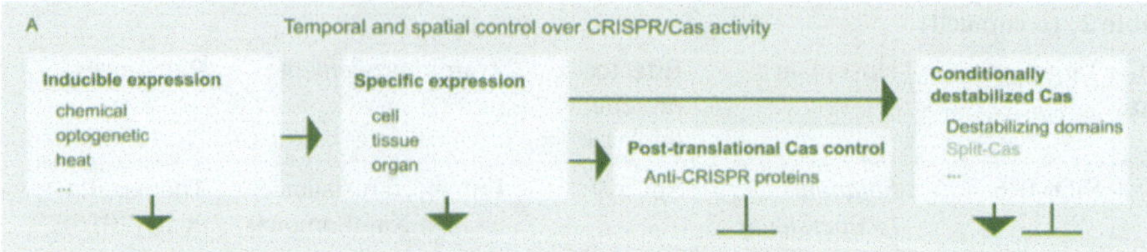

Fig. 1 Conditional knockout (Pfeiffer et al. 2022)

3.2 *Gene Insertion and Gene Replacement Approaches*

Gene insertion and replacement are indispensable genetic engineering approaches that allow the specific modification of an organism's genome to introduce new traits, correct genetic defects, or enhance chosen functions (Fang and Tyler 2016). In gene insertion, foreign genes are introduced into the host organism's genome, typically achieved through viral vectors like retroviruses or adenoviruses in mammalian cells. Bacterial artificial chromosomes (BACs) are also employed to stably integrate large gene inserts into the genomes of plants or animals. The CRISPR-Cas9 system further enables gene insertion by utilizing homology-directed repair (HDR), guiding the insertion of specific genes at target genomic loci. Conversely, gene replacement involves relieving an existing gene with an altered or foreign gene. CRISPR-Cas9, along with homologous recombination, allows for the precise replacement of a targeted gene with an exogenous DNA template. Zinc Finger Nucleases (ZFNs) and Transcription Activator-Like Effector Nucleases (TALENs) can induce double-strand breaks for gene replacement through homologous recombination (Takenaka et al. 2007). Adeno-Associated Viral Vectors (AAVs) are utilized in gene therapy to replace faulty genes with functional ones, addressing genetic disorders. These approaches provide researchers with versatile tools for manipulating genetic material, offering opportunities for advancements in agriculture, biotechnology, and medicine by tailoring genetic content to specific needs (Fig. 2).

3.3 *Multiplex Genome Editing for Enhanced Resistance*

The usefulness of CRISPR/Cas9 technology is demonstrated by its incomparable ability for multiplexing—simultaneously targeting multiple genes using a single molecular construct. This feature holds as a major benefit over other genome editing tools like MN, ZFN, or TALEN. A notable demonstration of this capability is evident in Arabidopsis, where a single construct successfully induced mutations in 14 different genes simultaneously (Peterson et al. 2016). In the realm of crop editing, several early strategies for multiplex genome editing (MGE) have been reported, all sharing a common approach of assembling multiple guide RNAs (gRNAs) under the control of a U3 or U6 promoter into a single construct. For instance, in maize, the

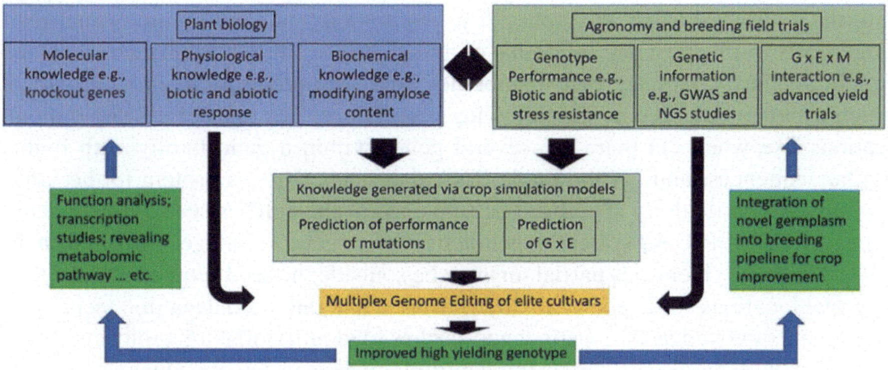

Fig. 2 Role of multiplex genome (Abdelrahman et al. 2021)

ISU Maize CRISPR platform enables the cloning of up to four gRNAs for efficient multiplex gene targeting (Char et al. 2017). Recent progressions in multiplex systems exploit the self-cleavage ability of RNA molecules comprising transfer RNA (tRNA) sequences. Constructs employing alternating single-guide RNA (sgRNA) and tRNA sequences under the control of a single U3 or U6 promoter reduce construct size and alleviate the risk of silencing due to repeated promoter sequences. This strategy, known as the polycistronic tRNA-gRNA (PTG) system, was applied in hexaploid wheat to generate hereditable mutations in TaLpx-1 and TaMLO genes (Wang et al. 2014). The PTG system, evolved from a prior study on gene silencing of TaLpx-1, the wheat 9-lipoxygenase resistance gene to *Fusarium graminearum* (Nalam et al. 2015). The PTG system's effectiveness was confirmed in wheat, affirming its competence in gene editing and providing a powerful tool for rapidly pyramiding traits in breeding programs. This exemplifies the continued evolution of CRISPR/Cas9 methodologies, enhancing precision and efficiency in multiplex genome editing for crop improvement.

4 Challenges and Considerations in Using Genome Editing for Resistance Against Phytopathogens

4.1 Off-Target Effects and Unintended Consequences

Genome editing has applications in plant sciences wherein rendering plants resistant to phytopathogens holds prime focus. Although the technology is simple and robust, it suffers from a plethora of unforeseen effects and bottlenecks. Unintended cleavage and mutations at untargeted genomic regions displaying a similar but distinct sequence in comparison to the target site are known as off-target effects (Modrzejewski et al. 2019). Off-target edits may have negative consequences that interfere with the main goal, such as cytotoxicity, genotoxicity, or even chromosomal rearrangements in plants. They might have immediate harmful consequences that stop protein synthesis, damage cell membranes, and stop plants from growing, all of which could be inherited by the following generations. The odds of off-target modifications rise when (1) there are several genes within a gene family with highly similar sequences, and (2) the guide RNA that leads the Cas protein to the target region is not particularly specific. The sequence of the sgRNA seeds mostly determines the CRISPR/Cas9 system's specificity. Off-target edits are generated when the sgRNA sequence identifies partial mismatches outside the seed sequence rather than on-target locations (Osakabe et al. 2016). Biased and unbiased detection techniques have been employed as two distinct approaches to identify off-target impacts. Many research have examined off-target impacts utilizing biased detection techniques up to this point (Modrzejewski et al. 2019) Zinc Finger Nucleases (ZFNs) have low affinity and low specificity, which results in large frequency of off-target edits during gene editing (Malzahn et al. 2017a, b). Rice paralogs OsBEIIb and OSBEIIa were used in a study on the target precision and efficiency of the CRISPR/Cas9 system (Baysal et al. 2016). The results of the study demonstrate the disparity between gRNA prediction and mutagenesis efficiency, demonstrating that, despite the prediction suggesting different targets with high mutagenesis scores, gRNA with low anticipated efficiency can achieve high mutation frequency (off-target). According to estimates, dimeric ZFN or TALENs systems, which employ two neighboring target sequences to recognize a longer target sequence, are less likely to experience off-target effects than monomeric CRISPR/Cas systems (Lee et al. 2016).

4.2 Regulatory Issues and Public Acceptance

Diverse countries have different approaches to regulating species that have had their genomes altered. While some stakeholder groups, including IFOAM EU Group, TestBiotech, and Confédération Paysanne, have strongly opposed genetically modified organisms (GMOs), they typically refer to genome-edited organisms as new GMOs. In contrast, organizations such as the European Plant Science Organization (EPSO), EU-SAGE, Leopoldina, and other academic institutions tend to distinguish between genome-edited crops and transgenic crops, which do not contain any foreign genetic material (TestBiotech 2021). Argentina was the first to adopt a case-by-case approach to evaluate whether genome-edited crops are subject to GMO legislation, based on the Cartagena Protocol on Biosafety's definition of a 'Living Modified Organism' (LMO) (Lema 2019). This protocol considers a genome-edited organism as an LMO only if its genetic alteration leads to a 'novel combination of genetic material.' In terms of regulation, some countries focus on the process of genome editing, while others consider the characteristics of the final product (Eckerstorfer et al. 2019). Stringent regulatory objectives, as noted by Schmidt et al. (2020), impede the broad application of genome editing, thereby limiting the development of next-generation crops. The decision to regulate genetically engineered (GE) crops and foods depends on the existing GMO regulatory framework in each country. Those with process-based systems, perceiving products derived from regulated processes as inherently different or riskier, are more likely to regulate GE crops and foods under GMO laws. Conversely, countries employing a product-based regulatory approach may not subject GE crops and foods to GMO laws, focusing instead on the final product's traits rather than its production method.

4.3 Delivery Methods and Transformation Efficiency

The success of genome editing critically depends on the delivery method used to introduce editing components into plant cells and the subsequent transformation efficiency, the proportion of edited cells that develop into whole plants. One commonly employed delivery method is *Agrobacterium*-mediated transformation, which utilizes natural DNA transfer ability of *Agrobacterium tumefaciens* to transfer genetic material into plant cells (Zhang et al. 2020). The protein AvrPto and fragments of GFP were delivered by modified strains of Agrobacterium with a Pseudomonas type III secretion system (PT3SS). This highlights the significance of fine-tuning the Agrobacterium-mediated transformation protocol for particular plant species and genotypes, as demonstrated by increased transformation efficiency and precision (Raman et al. 2022). Another prominent method, biolistic particle delivery or the "gene gun" approach, physically propels DNA-coated particles into plant cells using specialized devices (Kocsisova and Coneva 2023; Liang et al. 2018).

Protoplast transformation, which involves the isolation of plant cells with their cell walls removed, offers versatility, and can be applied across various tissues and plant species (Laforest and Nadakuduti 2022; Nicolia et al. 2021). Optimizing the transfection process and culture conditions, making protoplast transformation an increasingly viable option for genome editing. Plasmid DNA, IVTs, or RNPs can be delivered into protoplasts via either electro- or polyethylene glycol (PEG)-mediated transfection for brief expression of CRISPR cassettes. Tissue culture techniques can then be used to regenerate altered plants from transfected protoplasts (Weiss et al. 2020; Klimek-Chodacka et al. 2021). Plant viruses have also served as delivery vectors for genome editing components, enabling systemic spread throughout the plant (Ma et al. 2020; Kujur et al. 2021; Li et al. 2021; Laforest and Nadakuduti 2022). To achieve heritable genome editing, the barley stripe mosaic virus (BMSV) has been modified to transport sgRNAs to wheat.

Transformation efficiency is influenced by several factors. First and foremost, the choice of delivery method significantly impacts transformation efficiency, with each method having unique characteristics that may perform differently depending on the plant species, tissue type, and genotype (Sandhya et al. 2020; Nowak et al. 2016). The quality and purity of CRISPR-Cas9 components, including guide RNAs and Cas9 protein, are crucial determinants of transformation efficiency (Xie et al. 2015). The ability of edited plant cells to regenerate into whole plants is another key factor influencing transformation efficiency. Some plant species and genotypes exhibit better regeneration capabilities, making them more amenable to genome editing. Furthermore, minimizing off-target effects is essential for ensuring precision in genome editing and, subsequently, enhancing transformation efficiency (Azhakanandam and Zhang 2015; LeBlanc et al. 2018; Alok et al. 2020; Sandhya et al. 2020).

4.4 Intellectual Property Rights and Access to Technology

In most cases, the commercialization of genome-edited crops necessitates obtaining a license for CRISPR-related intellectual property (IP) (Jefferson et al. 2021). The landscape of CRISPR-Cas patents and patent applications is intricate, encompassing various components of the CRISPR-Cas machinery, derived tools, and their applications, with over 250 published patent families related to CRISPR-Cas use in plants. A genetically modified organism (GMO) plant variety might consist of a stack of multiple patented traits, some owned by the variety holder and others in-licensed, typically without sublicense rights. The complexity extends beyond the sheer number of patent families to include overlaps between them. Despite Doudna and Charpentier's Nobel Prize for discovering Cas9, multiple entities are vying for their share of the foundational Cas9 patents. Depending on the specific Cas9 enzyme use, four or more licenses may be necessary for a simple plant edit. Significant costs are incurred by parties defending their rights or challenging the rights of others. In the future, farmers may need to engage with the holder of plant breeder's rights (PBR) and multiple patent holders simultaneously, as current laws lack guidance on navigating multiple rights. It's often argued that complex IP situations ("patent thickets") hinder innovation by raising transaction costs and delaying market entry.

5 Future Prospects and Emerging Trends

Phytopathogens, including bacteria, viruses, fungi, and other microorganisms, pose significant challenges to global agriculture by causing substantial crop losses. In recent years, genome editing technologies have emerged as a transformative tool in the fight against phytopathogens, enabling precise and efficient modifications in plant genomes to confer resistance. The ability to edit specific genes responsible for susceptibility has revolutionized the field of plant pathology, offering innovative solutions to combat these detrimental agents and enhance crop resilience.

5.1 Advances in Precision and Efficiency of Genome Editing Techniques

The CRISPR-Cas9 system, renowned for its versatility, has been fine-tuned to achieve higher precision in targeting specific susceptibility genes. Researchers have employed this technology to edit genes involved in disease recognition, signaling, and defense pathways. In an investigation, mutagenesis of OsSWEET13, a susceptibility (S) gene encoding a sucrose transporter involved in plant-pathogen interaction has been performed in rice via CRISPR/Cas9 to achieve resistance to bacterial blight disease caused by γ-proteobacterium *Xanthomonas oryzae* pv. *oryzae* (Zhou et al. 2015). This showcases how precise edits in key genes can bolster the plant's natural defense mechanisms. Efficiency gains in genome editing have been achieved through the development of delivery methods that ensure successful modification of the target genes. Nanoparticle-based delivery systems, viral vectors, and improved transformation protocols have enhanced the delivery of genome editing components to plant cells, resulting in higher editing rates and reduced off-target effects (Liu et al. 2016). Additionally, the optimization of CRISPR-Cas systems and the engineering of high-fidelity variants have further increased the specificity of edits, minimizing unintended alterations. Base editing and prime editing are cutting-edge techniques that have revolutionized the precision of genome editing for phytopathogen resistance. Base editing allows the direct conversion of one DNA base pair into another, while prime editing permits precise insertion, deletion, or substitution of DNA sequences. These techniques are highly specific and offer the advantage of minimizing disruptions to the surrounding genome, increasing the potential for successful pathogen resistance development (Anzalone et al. 2019; Komor et al. 2016). Furthermore, the integration of computational tools and artificial intelligence has facilitated the design of optimal guide RNAs for precise genome editing. These tools predict potential off-target sites and assist in selecting guide RNAs with minimal risk of unintended edits, enhancing the overall precision of the process (Kim et al. 2020).

5.2 Integration of Genome Editing with Other Plant Breeding Methods

The integration of genome editing techniques with traditional plant breeding strategies offers a potent approach to developing robust resistance against phytopathogens. By combining the precision of genome editing with the genetic diversity and selection mechanisms of conventional breeding, researchers are enhancing the effectiveness and sustainability of disease management in crops. One noteworthy integration involves the coupling of genome editing with marker-assisted selection (MAS). This approach streamlines the identification and incorporation of genes associated with disease resistance. For example, researchers have combined CRISPR-Cas9-mediated genome editing with molecular markers to create wheat varieties resistant to powdery mildew, a devastating fungal disease. By targeting and modifying specific susceptibility genes, while concurrently introducing molecular markers linked to resistance genes, the breeding process is accelerated while ensuring the accurate transfer of resistance traits (Liang et al. 2017). Wide hybridization, the introduction of genetic diversity from wild or distantly related species into cultivated crops, is another method integrated with genome editing for enhanced disease resistance. Genome editing can be used to modify the introduced genes in hybrids to improve their resistance while maintaining desirable agronomic traits. In rice, for instance, researchers have used CRISPR-Cas9 to edit a susceptibility gene in hybrid offspring obtained through wide hybridization, resulting in enhanced resistance against bacterial blight (Zafar et al. 2020a, b). Furthermore, quantitative trait loci (QTL) mapping combined with genome editing offers a powerful approach to develop phytopathogen-resistant crops. QTL mapping involves identifying regions of the genome linked to specific traits. By editing genes within these regions, disease resistance can be enhanced while retaining the overall genetic background of the crop. This integrated strategy has been demonstrated in hexaploid wheat, where CRISPR/Cas9-guided knockout of *TaeIF4E* improved Wheat yellow mosaic virus (WYMV) resistance without any yield penalty (Kan et al. 2023). Integrating genome editing with established plant breeding methods addresses the limitations of each approach while capitalizing on their strengths. Traditional breeding methods often lack precision, while genome editing offers targeted modifications. Combining these techniques results in crops that possess both desirable genetic traits and enhanced disease resistance.

5.3 Potential Applications in Non-Model Plant Species

Genome editing technologies have transcended the boundaries of model organisms, opening up exciting avenues for enhancing resistance against phytopathogens in a wide range of non-model plant species. The precision and versatility of genome editing offer novel solutions for addressing the challenges posed by devastating plant diseases, promoting sustainable agriculture, and ensuring global food security. One significant application is the development of disease-resistant varieties in economically important crops. Non-model species such as cassava, a staple food for millions in Africa, are susceptible to various pathogens. Genome editing can be harnessed to modify specific genes that confer resistance. In a pioneering study, researchers used CRISPR-Cas9 to edit the cassava genome and enhance its resistance against cassava mosaic disease, a viral infection that severely impacts crop yields (Tuo et al. 2023). This approach showcases the potential of genome editing to combat phytopathogens in crops critical for food production. Non-model species also encompass plants with unique traits, offering opportunities for enhancing natural defenses. For instance, the wild tobacco plant *Nicotiana benthamiana* is valued for its use in studying plant-pathogen interactions due to its susceptibility to a wide range of pathogens. Genome editing can be employed to bolster its resistance by modifying susceptibility genes or enhancing defense pathways. Recent research demonstrated the use of CRISPR-Cas9 to target and edit immune receptor genes in *N. benthamiana*, resulting in increased resistance against multiple viruses (Borrelli et al. 2018). Additionally, ornamental and horticultural plants, often non-model species, can benefit from genome editing for phytopathogen resistance. A ribonucleoprotein complex (CRISPR/Cas9 RNPs) comprising the Cas9 nuclease and sgRNA was utilized to modify apple protoplasts to produce the DIPM-1, -2, and -4 genes that encode negative regulators of resistance to the bacterial fire blight of fruit crops caused by *Erwinia amilovora.* Fungal infections also contribute to a variety of plant diseases. It is well known that sensitivity genes of plants make it easier for pathogens to invade and infect them. CRISPR/Cas9 technology enabled the production of grapevine plants resistance to powdery mildew, *Erysiphe necator* with the knockout of the MLO-7 gene, which encodes the negative regulators of resistance (Malnoy et al. 2016a, b, c; Rukavtsova et al. 2023).

5.4 Ethical Considerations and Responsible Use of Genome Editing

However, ethical considerations and regulatory frameworks must accompany these advancements. The introduction of edited plants into the environment raises questions about potential ecological impacts and gene flow to wild relatives. Comprehensive risk assessments and thorough consultations with experts are essential to ensure responsible deployment. One of the central ethical concerns revolves around the potential impact of edited crops on ecosystems and biodiversity. Introducing genome-edited plants into the environment could lead to unintended ecological disruptions, such as unintended gene flow to wild relatives or the emergence of new plant-pathogen interactions. The ethical imperative here is to conduct thorough risk assessments to predict and mitigate any potential negative impacts (Podevin et al. 2013). Moreover, responsible use requires careful consideration of unintended off-target effects. While genome editing techniques like CRISPR-Cas9 offer high precision, there is always a risk of unintended genetic changes occurring in non-target regions of the genome. Employing advanced computational methods to predict and minimize off-target effects, coupled with thorough characterization of edited plants, is essential (Hsu et al. 2013). Equity and access are also vital ethical considerations. While genome editing holds the potential to benefit agriculture and food security globally, disparities in access to these technologies could exacerbate existing inequalities between developed and developing regions. Ensuring that the benefits of genome editing for phytopathogen resistance reach marginalized communities is an ethical imperative that requires collaboration, technology transfer, and capacity-building initiatives (Lowder et al. 2015). Transparency and informed consent are integral to responsible use. The public should be informed about the potential benefits and risks of edited crops, and their input should be sought to guide decision-making. Open communication fosters public trust and allows for well-informed discussions regarding the deployment of genome-edited plants in agriculture (Ishii 2017). A precedent for addressing ethical considerations in genetically modified organisms (GMOs) can be found in the Cartagena Protocol on Biosafety, a framework that aims to ensure the safe transfer, handling, and use of GMOs. Genome-edited plants for phytopathogen resistance would fall under such regulatory frameworks, which must be adapted to accommodate new technologies while upholding ethical values (Lassoued et al. 2019; Kawall et al. 2020). While genome editing holds immense potential for enhancing phytopathogen resistance in crops, it also demands rigorous ethical considerations. Responsible use necessitates a holistic approach that includes risk assessment, transparency, equitable access, and open dialogue with stakeholders (Idris et al. 2023). Integrating these ethical principles into the deployment of genome-edited crops, we can harness the benefits of this technology while safeguarding the environment, promoting social justice, and fostering sustainable agriculture.

6 Conclusion

Genome editing techniques have opened up remarkable possibilities for developing precision and efficient resistance against phytopathogens. From CRISPR-Cas9 to RNAi, base editing, and CRISPR-Cas13, researchers are harnessing these tools to target susceptibility genes, enhance plant immunity, and mitigate crop losses. The future of genome editing for phytopathogen resistance is promising and transformative. Advances in precision and efficiency, exemplified by high-fidelity CRISPR-Cas9 variants and innovative editing techniques like base editing and prime editing, are enhancing the precision and efficacy of disease management strategies. The integration of genome editing with traditional breeding methods ensures a comprehensive approach to developing disease-resistant crops. The synergy between genome editing's precision and traditional breeding's genetic diversity leads to crops that are not only resistant to diseases but also possess desired agronomic traits. Additionally, the expansion of genome editing technologies to non-model plant species holds the potential to address global food security concerns and promote sustainable agriculture. However, ethical considerations and responsible deployment must remain at the forefront to navigate potential risks and societal concerns. As these trends continue to shape the landscape of genome editing for phytopathogen resistance, collaboration between scientists, policymakers, ethicists, and the public is essential. This collective effort will guide the development and deployment of genome-edited crops that enhance resistance against phytopathogens while upholding ethical values. As genome editing continues to evolve, its potential to create disease-resistant crops holds significant promise for sustainable agriculture and global food security.

References

Abdelrahman M, Wei Z, Rohila JS, Zhao K (2021) Multiplex genome-editing technologies for revolutionizing plant biology and crop improvement. Front Plant Sci 12:721203

Alok A, Sandhya D, Jogam P, Rodrigues V, Bhati KK, Sharma H, Kumar J (2020) The rise of the CRISPR/Cpf1 system for efficient genome editing in plants. Front Plant Sci 11:264

Aman R, Ali Z, Butt H, Mahas A, Aljedaani F, Khan MZ, Ding S, Mahfouz M (2018) RNA virus interference via CRISPR/Cas13a system in plants. Genome Biol 19:1

Anonymous (2020) Figure from Irvine, C. Update on the CRISPR IP saga and lessons to be learnt. Available online: https://www.hgf.com/media/1570220/Update-on-the-CRISPR-IP-Saga-and-lessons-to-be-learnt-by-Claire-Irvine-and-Cath-Coombes.pdf

Anzalone AV, Randolph PB, Davis JR, Sousa AA, Koblan LW, Levy JM, Chen PJ, Wilson C, Newby GA, Raguram A, Liu DR (2019) Search-and-replace genome editing without double-strand breaks or donor DNA. Nature 576(7785):149–157

Azhakanandam K, Zhang ZJ (2015) Sorghum transformation: achievements, challenges, and perspectives. Recent advancements in gene expression and enabling technologies in crop plants, pp 291–312

Bao A, Burritt DJ, Chen H, Zhou X, Cao D, Tran LP (2019) The CRISPR/Cas9 system and its applications in crop genome editing. Crit Rev Biotechnol 39:1–16. https://doi.org/10.1080/07388551.2018.1554621

Bastet A, Zafirov D, Giovinazzo N, Guyon-Debast A, Nogué F, Robaglia C, Gallois JL (2019) Mimicking natural polymorphism in eIF 4E by CRISPR-Cas9 base editing is associated with resistance to potyviruses. Plant Biotechnol J 17:1736–1750

Baysal C, Bortesi L, Zhu C, Farré G, Schillberg S, Christou P (2016) CRISPR/Cas9 activity in the rice OsBEIIb genedoes not induce off-target effects in the closely related paralog OsBEIIa. Mol Breeding 36:108. https://doi.org/10.1007/s11032-016-0533-4

Boch J, Scholze H, Schornack S, Landgraf A, Hahn S, Kay S (2009) Breaking the code of DNA binding specificity of TAL-type III effectors. Science 326:1509–1512. https://doi.org/10.1126/science.1178811

Borrelli VMG, Brambilla V, Rogowsky P, Marocco A, Lanubile A (2018) The enhancement of plant disease resistance using crispr/cas9 technology. Front Plant Sci 9:1–15

Ceasar SA, Rajan V, Prykhozhij SV, Berman JN, Ignacimuthu S (2016) Insert, remove or replace: a highly advanced genome editing system using CRISPR/Cas9. Biochem Biophys Acta 1863(9):2333–2344. https://doi.org/10.1016/j.bbamcr.2016.06.009

Chagnon M, Kreutzweiser D, Mitchell EAD, Morrissey CA, Noome DA, Van der Sluijs JP (2015) Risks of large-scale use of systemic insecticides to ecosystem functioning and services. Environ Sci Pollut Res Int 22(1):119–134. https://doi.org/10.1007/s11356-014-3277-x

Chandrasekaran J, Brumin M, Wolf D, Leibman D, Klap C, Pearlsman M, Sherman A, Arazi T, Gal-On A (2016) Development of broad virus resistance in non-transgenic cucumber using CRISPR/Cas9 technology. Mol Plant Pathol 17:1140–1153

Chen W, Qian Y, Wu X, Sun Y, Wu X, Cheng X (2014) Inhibiting replication of begomoviruses using artificial zinc finger nucleases that target viral conserved nucleotide motif. Virus Genes 48:494–501

Cheng X, Li F, Cai J, Chen W, Zhao N, Sun Y, Guo Y, Yang X, Wu X (2015) Artificial TALE as a convenient protein platform for engineering broad-spectrum resistance to begomoviruses. Viruses 7:4772–4782

Civáň P, Brown TA (2018) Misconceptions regarding the role of introgression in the origin of *Oryza sativa* subsp. *indica*. Front Plant Sci 9:1750

Cohen JE (2001) World population in 2050: assessing the projections. Conf Ser-Federal Reserve Bank Boston 46:83

Darshan K, Singh J, Yadav S, Venugopala KM, Aggarwal R (2020) Root border cells: a pioneer's of plant defence in rhizosphere. Indian J Agric Sci 90(10):1850–1855. https://doi.org/10.56093/ijas.v90i10.107884

Das U, Ansari MA (2021) The nexus of climate change, sustainable agriculture and farm livelihood: contextualizing climate smart agriculture. Clim Res 84:23–40. https://doi.org/10.3354/cr01648

de Toledo Thomazella DP, Brail Q, Dahlbeck D, Staskawicz BJ (2016) CRISPR-Cas9 mediated mutagenesis of a DMR6 ortholog in tomato confers broad-spectrum disease resistance. BioRxiv. Retrieved from https://doi.org/10.1101/064824

Dong OX, Ronald PC (2021) Targeted DNA insertion in plants. Proc Natl Acad Sci 118:e2004834117

Dong OX, Yu S, Jain R, Zhang N, Duong PQ, Butler C, Li Y, Lipzen A, Martin JA, Barry KW (2020) Marker-free carotenoid-enriched rice generated through targeted gene insertion using CRISPR-Cas9. Nat Commun 11:1178

Downing KJ, Thomson JA (2000) Introduction of the *Serratia marcescens* chiA gene into an endophytic *Pseudomonas fluorescens* for the biocontrol of phytopathogenic fungi. Can J Microbiol 46(4):363–369. https://doi.org/10.1139/w99-147

Eckerstorfer MF, Engelhard M, Heissenberger A, Simon S, Teichmann H (2019) Plants developed by new genetic modification techniques—comparison of existing regulatory frameworks in the EU and non-EU countries. Front Bioeng Biotechnol 7:26. https://doi.org/10.3389/fbioe.2019.00026

Fang Y, Tyler BM (2016) Efficient disruption and replacement of an effector gene in the oomycete *Phytophthora sojae* using CRISPR/Cas9. Mol Plant Pathol 17(1):127–139. https://doi.org/10.1111/mpp.12318

Fister AS, Landherr L, Maximova SN, Guiltinan MJ (2018) Transient expression of CRISPR/Cas9 machinery targeting TcNPR3 enhances defense response in *Theobroma cacao*. Front Plant Sci 9:268

Gao C (2021) Genome engineering for crop improvement and future agriculture. Cell. https://doi.org/10.1016/j.cell.2021.01.005

Ghag SB, Ganapathi TR (2019) RNAi-mediated protection against banana diseases and pests. Biotech 9(3):112. https://doi.org/10.1007/s13205-019-1650-7

Giacomelli L, Zeilmaker T, Malnoy M, Rouppe van der Voort J, Moser C (2018) Generation of mildew-resistant grapevine clones via genome editing. Acta Hortic 1248:195–200

Gumtow R, Wu D, Uchida J, Tian M (2018) A *Phytophthora palmivora* extracellular cystatin-like protease inhibitor targets papain to contribute to virulence on papaya. Mol Plant-Microbe Interact 31:363–373

Gupta PK, Balyan HS, Gautam T (2021) SWEET genes and TAL effectors for disease resistance in plants: present status and future prospects. Mol Plant Pathol 22(8):1014–1026. https://doi.org/10.1111/mpp.13075

Hasan N, Choudhary S, Naaz N, Sharma N, Laskar RA (2021) Recent advancements in molecular marker-assisted selection and applications in plant breeding programmes. J Genet Eng Biotechnol 19(1):128. https://doi.org/10.1186/s43141-021-00231-1

Hsu PD, Scott DA, Weinstein JA, Ran FA, Konermann S, Agarwala V, Li Y, Fine EJ, Wu X, Shalem O, Cradick TJ, Marraffini LA, Bao G, Zhang F (2013) DNA targeting specificity of RNA-guided Cas9 nucleases. Nat Biotechnol 31(9):827–834

Idris SH, Sharleeza N, Jalaluddin MAT, Chang LW (2023) Ethical and legal implications of gene editing in plant breeding: a systematic literature review. J Zhejiang Univ-SCIENCE B 1–13

Ishii T (2017) A future scenario of the global regulatory landscape regarding genome-edited crops. GM Crops Food 8(1):44–56

Jat M, Dohling PNK, Ahuja A, Singh J (2021) Effect of pesticides on soil ecosystem services and processes. Int J Ecosyst 1–10. https://doi.org/10.55446/IJE.2021.5

Jefferson OA, Lang S, Williams K et al (2021) Mapping CRISPR-Cas9 public and commercial innovation using the lens institutional toolkit. Transgenic Res 30:585–599. https://doi.org/10.1007/s11248-021-00237-

Jutsum AR, Heaney SP, Perrin BM, Wege PJ (1998) Pesticide resistance: assessment of risk and the development and implementation of effective management strategies. Pestic Sci 54(4):435–446. https://doi.org/10.1002/(SICI)1096-9063(199812)54:4%3c435::AID-PS844%3e3.0.CO;2-K

Kan J, Cai Y, Cheng C, Chen S, Jiang C, He Z, Yang P (2023) CRISPR/Cas9-guided knockout of eIF4E improves Wheat yellow mosaic virus resistance without yield penalty. Plant Biotechnol J 21(5):893–895

Karlsson T, Turkina MV, Yakymenko O, Magnusson K-E, Vikström E (2012) The *Pseudomonas aeruginosa N*-acylhomoserine lactone quorum sensing molecules target IQGAP1 and modulate epithelial cell migration. PLoS Pathogens 8(10):e1002953. https://doi.org/10.1371/journal.ppat.1002953

Kashyap R, Grover S, Puri H, Kaur S, Singh J, Sandhu KS, Kaur G, Kharva H, Kaur S, Kaur B (2023) Insect and pest management for sustaining crop production under changing climatic patterns of drylands. In: Naorem A, Machiwal D (eds) Enhancing resilience of dryland agriculture under changing climate: interdisciplinary and convergence approaches. Springer Nature Singapore, pp 441–457. https://doi.org/10.1007/978-981-19-9159-2_21

Kaur B, Sandhu KS, Kamal R, Kaur K, Singh J, Röder MS, Muqaddasi QH (2021) Omics for the improvement of abiotic, biotic, and agronomic traits in major cereal crops: applications, challenges, and prospects. Plants 10(10). https://doi.org/10.3390/plants10101989

Kawall K, Cotter J, Then C (2020) Broadening the GMO risk assessment in the EU for genome editing technologies in agriculture. Environ Sci Eur 32(106):1–24

Kieu NP, Lenman M, Wang ES, Petersen BL, Andreasson E (2021) Mutations introduced in susceptibility genes through CRISPR/Cas9 genome editing confer increased late blight resistance in potatoes. Sci Rep 11:1–2

Kim N, Kim HK, Lee S, Seo JH, Choi JW, Park J, Min S, Yoon S, Cho SR, Kim HH (2020) Prediction of the sequence-specific cleavage activity of Cas9 variants. Nat Biotechnol 38(11):1328–1336

Kim YA, Moon H, Park CJ (2019) CRISPR/Cas9-targeted mutagenesis of Os8N3 in rice to confer resistance to *Xanthomonas oryzae* pv. *oryzae*. Rice 12:67

Klimek-Chodacka M, Gieniec M, Baranski R (2021) Multiplex site-directed gene editing using polyethylene glycol-mediated delivery of CRISPR gRNA: Cas9 ribonucleoprotein (RNP) complexes to carrot protoplasts. Int J Mol Sci 22(19):10740

Kocsisova Z, Coneva V (2023) Strategies for delivery of CRISPR/Cas-mediated genome editing to obtain edited plants directly without transgene integration. Front Genome Editing 5:1–17

Komatsuda T, Pourkheirandish M, He C, Azhaguvel P, Kanamori H, Perovic D, Stein N, Graner A, Wicker T, Tagiri A (2007) Six-rowed Barley originated from a mutation in a homeodomain-leucine zipper I-class homeobox gene. Proc Natl Acad Sci 104:1424–1429

Komor AC, Kim YB, Packer MS, Zuris JA, Liu DR (2016) Programmable editing of a target base in genomic DNA without double-stranded DNA cleavage. Nature 533(7603):420–424

Kujur S, Senthil-Kumar M, Kumar R (2021) Plant viral vectors: expanding the possibilities of precise gene editing in plant genomes. Plant Cell Rep 40:931–934

Kumar P, Singh J, Kaur G, Adunola PM, Biswas A, Bazzer S, Kaur H, Kaur I, Kaur H, Sandhu KS, Vemula S, Kaur B, Singh V, Tseng TM (2022) OMICS in fodder crops: applications, challenges, and prospects. Curr Issues Mol Biol 44(11):5440–5473. https://doi.org/10.3390/cimb44110369

Laforest LC, Nadakuduti SS (2022) Advances in delivery mechanisms of CRISPR gene-editing reagents in plants. Front Genome Editing 4:1–10

Lassoued R, Maximiliano D, Smyth SJ, Phillips PWB, Hesseln H (2019) Current research in biotechnology risk and safety considerations of genome edited crops: expert opinion. Curr Res Biotechnol 1(1):11–21

LeBlanc C, Zhang F, Mendez J, Lozano Y, Chatpar K, Irish VF, Jacob Y (2018) Increased efficiency of targeted mutagenesis by CRISPR/Cas9 in plants using heat stress. Plant J 93(2):377–386

Lee J, Chung J-H, Kim HM, Kim D-W, Kim H (2016) Designed nucleases for targeted genome editing. Plant Biotechnol J 14:448–462. https://doi.org/10.1111/pbi.12465

Leisen T, Bietz F, Werner J, Wegner A, Schaffrath U, Scheuring D, Willmund F, Mosbach A, Scalliet G, Hahn M (2020) CRISPR/Cas with ribonucleoprotein complexes and transiently selected telomere vectors allows highly efficient markerfree and multiple genome editing in *Botrytis cinerea*. PLoS Pathog 16:e1008326

Lema MA (2019) Regulatory aspects of gene editing in Argentina. Transgenic Res 28(Suppl 2):147–150. https://doi.org/10.1007/s11248-019-00145-2

Li T, Liu B, Spalding MH, Weeks DP, Yang B (2012) High-efficiency TALEN-based gene editing produces disease-resistant rice. Nat Biotechnol 30:390–392

Li T, Yang X, Yu Y, Si X, Zhai X, Zhang H, Dong W, Gao C, Xu C (2018) Domestication of wild tomato is accelerated by genome editing. Nat Biotechnol 36:1160–1163

Li S, Shen L, Hu P, Liu Q, Zhu X, Qian Q, Wang K, Wang Y (2019) Developing disease-resistant thermosensitive male sterile rice by multiplex gene editing. J Integr Plant Biol 61:1201–1205

Li MY, Jiao YT, Wang YT, Zhang N, Wang BB, Liu RQ, Yin X, Xu Y, Liu GT (2020) CRISPR/Cas9-mediated VvPR4b editing decreases downy mildew resistance in grapevine (*Vitis vinifera* L.). Hortic Res 149:1–11

Li T, Hu J, Sun Y, Li B, Zhang D, Li W, Liu J, Li D, Gao C, Zhang Y et al (2021) Highly efficient heritable genome editing in wheat using an RNA virus and bypassing tissue culture. Mol Plant 14(11):1787–1798

Liang Z, Chen K, Li T, Zhang Y, Wang Y, Zhao Q, Liu J, Zhang H, Liu C, Ran Y, Gao C (2017) Efficient DNA-free genome editing of bread wheat using CRISPR/Cas9 ribonucleoprotein complexes. Nat Commun 8(1):6–10

Liang Z, Chen K, Zhang Y, Liu J, Yin K, Qiu J-L, Gao C (2018) Genome editing of bread wheat using biolistic delivery of CRISPR/Cas9 in vitro transcripts or ribonucleoproteins. Nat Protoc 13(3):413–430

Liu J, Shui S (2016) Delivery methods for site-specific nucleases: achieving the full potential of therapeutic gene editing. J Controlled Release 244:83–97

Liu F, Zhang W, Schwessinger B, Wei T, Ruan D, Ronald P (2020) The rice Xa3 gene confers resistance to *Xanthomonas oryzae* pv. *oryzae* in the model rice kitaake genetic background. Front Plant Sci 11:49. https://doi.org/10.3389/fpls.2020.00049

Lowder LG, Zhang D, Baltes NJ, Iii JWP, Tang X, Zheng X, Voytas DF, Hsieh T, Zhang Y, Qi Y (2015) A CRISPR/Cas9 toolbox for multiplexed plant genome editing and transcriptional regulation. Plant Physiol 169:971–985

Ma X, Zhang X, Liu H, Li Z (2020) Highly efficient DNA-free plant genome editing using virally delivered CRISPR–Cas9. Nat Plants 6(7):773–779

Macovei A, Sevilla NR, Cantos C, Jonson GB, Slamet-Loedin I, Čermák T, Voytas DF, Choi IR, Chadha-Mohanty P (2018) Novel alleles of rice eIF4G generated by CRISPR/Cas9-targeted mutagenesis confer resistance to Rice tungro spherical virus. Plant Biotechnol J 16:1918–1927

Makhotenko AV, Khromov AV, Snigir EA, Makarova SS, Makarov VV, Suprunova TP, Kalinina NO, Taliansky ME (2019) Functional analysis of coilin in virus resistance and stress tolerance of potato *Solanum tuberosum* using CRISPRCas9 editing. Dokl Biochem Biophys 484:88–91

Malnoy M, Viola R, Jung MH, Koo OJ, Kim S, Kim JS, Velasco R, Nagamangala Kanchiswamy C (2016c) DNA-free genetically edited grapevine and apple protoplast using CRISPR/Cas9 ribonucleoproteins. Front Plant Sci 7:1904

Malnoy M, Viola R, Jung M, Koo O, Kim S, Kim J, Velasco R, Kanchiswamy CN (2016a) DNA-free genetically edited grapevine and apple protoplast using CRISPR/Cas9 ribonucleoproteins. 7:1–9

Malnoy M, Viola R, Jung M-H, Koo O-J, Kim S, Kim J-S, Nagamangala Kanchiswamy C (2016) DNA-free genetically edited grapevine and apple protoplast using CRISPR/Cas9 Ribonucleoproteins. Front Plant Sci 7:1904. https://doi.org/10.3389/fpls.2016.01904

Malzahn A, Lowder L, Qi Y (2017a) Plant genome editing with TALEN and CRISPR. Cell Biosci 7:21. https://doi.org/10.1186/s13578-017-0148-4

Malzahn A, Lowder L, Qi Y (2017b) Plant genome editing with TALEN and CRISPR. Cell Biosci 7:21. https://doi.org/10.1186/s13578-017-0148-4

Mao H, Wang H, Liu S, Li Z, Yang X, Yan J, Li J, Tran L-SP, Qin F (2015) A transposable element in a NAC gene is associated with drought tolerance in maize seedlings. Nat Commun 6:8326

Martinelli F, Scalenghe R, Davino S, Panno S, Scuderi G, Ruisi P, Villa P, Stroppiana D, Boschetti M, Goulart LR, Davis CE, Dandekar AM (2015) Advanced methods of plant disease detection. A review. Agron Sustain Develop 35(1):1–25. https://doi.org/10.1007/s13593-014-0246-1

Martinez MIS, Bracuto V, Koseoglou E, Appiano M, Jacobsen E, Visser RG, Wolters AMA, Bai Y (2020) CRISPR/Cas9-targeted mutagenesis of the tomato susceptibility gene PMR4 for resistance against powdery mildew. BMC Plant Biol 20:1–13

Modrzejewski D, Hartung F, Sprink T, Krause D, Kohl C, Wilhelm R (2019) What is the available evidence for the range of applications of genome-editing as a new tool for plant trait modification and the potential occurrence of associated off-target effects: a systematic map. Environ Evidence 8:11. https://doi.org/10.1186/s13750-019-0171-5

Moscou MJ, Bogdanove AJ (2009) A simple cipher governs DNA recognition by TAL effectors. Science 326:1501. https://doi.org/10.1126/science.1178817

Nalam VJ, Alam S, Keereetaweep J, Venables B, Burdan D, Lee H, Shah J (2015) Facilitation of *Fusarium graminearum* infection by 9-lipoxygenases in Arabidopsis and wheat. Mol Plant-Microbe Interact 28(10):1142–1152. https://doi.org/10.1094/MPMI-04-15-0096-R

Nekrasov V, Wang C, Win J, Lanz C, Weigel D, Kamoun S (2017) Rapid generation of a trans-genefree powdery mildew resistant tomato by genome deletion. Sci Rep 7:482. https://doi.org/10.1038/s41598-017-00578-x

Nicolia A, Andersson M, Hofvander P, Festa G, Cardi T (2021) Tomato protoplasts as cell target for ribonucleoprotein (RNP)-mediated multiplexed genome editing. Plant Cell, Tissue Organ Cult (PCTOC) 144(2):463–467

Nowak CM, Lawson S, Zerez M, Bleris L (2016) Guide RNA engineering for versatile Cas9 functionality. Nucleic Acids Res 44(20):9555–9564

Olesen JE, Bindi M (2002) Consequences of climate change for European agricultural productivity, land use and policy. Eur J Agron 16(4):239–262. https://doi.org/10.1016/S1161-0301(02)00004-7

Oliva R, Ji C, Atienza-Grande G, Huguet-Tapia JC, Perez-Quintero A, Li T, Eom JS, Li C, Nguyen H, Liu B, Auguy F (2019) Broad-spectrum resistance to bacterial blight in rice using genome editing. Nat Biotechnol 37:1344–1350

Ortigosa A, Gimenez Ibanez S, Leonhardt N, Solano R (2019) Design of a bacterial speck resistant tomato by CRISPR/Cas9-mediated editing of Sl JAZ 2. Plant Biotechnol J 17:665–673

Osakabe K, Osakabe Y, Toki S (2010) Site-directed mutagenesis in Arabidopsis using custom-designed zinc finger nucleases. Proc Natl Acad Sci USA 107:12034–12039. https://doi.org/10.1073/pnas.1000234107

Osakabe Y, Watanabe T, Sugano SS, Ueta R, Ishihara R, Shinozaki K et al (2016) Optimization of CRISPR/Cas9 genome editing to modify abiotic stress responses in plants. Sci Rep 6:26685. https://doi.org/10.1038/srep26685

Peng A, Chen S, Lei T, Xu L, He Y, Wu L, Zou X (2017) Engineering canker-resistant plants through CRISPR/Cas9-targeted editing of the susceptibility gene CsLOB1 promoter in citrus. Plant Biotechnol J 15(12):1509–1519. https://doi.org/10.1111/pbi.12733

Peterson BA, Haak DC, Nishimura MT, Teixeira PJPL, James SR, Dangl JL, Nimchuk ZL (2016) Genome-wide assessment of efficiency and specificity in CRISPR/Cas9 mediated multiple site targeting in arabidopsis. Plos One 11(9):e0162169. https://doi.org/10.1371/journal.pone.0162169

Pfeiffer ML, Winkler J, Van Damme D, Jacobs TB, Nowack MK (2022) Conditional and tissue-specific approaches to dissect essential mechanisms in plant development. Curr Opin Plant Biol 65:102119

Podevin N, Davies HV, Hartung F, Casacuberta JM, Nogue F (2013) Site-directed nucleases : a paradigm shift in predictable, knowledge-based plant breeding. Trends Biotechnol 31(6):375–383

Puchta H (2002) Gene replacement by homologous recombination in plants. Funct Genomics 173–182

Pyott DE, Sheehan E, Molnar A (2016) Engineering of CRISPR/Cas9-mediated potyvirus resistance in transgene-free Arabidopsis plants. Mol Plant Pathol 17(8):1276–1288. https://doi.org/10.1111/mpp.12417

Raman V, Rojas CM, Vasudevan B, Dunning K, Kolape J, Oh S, Yun J, Yang L, Li G, Pant BD et al (2022) Agrobacterium expressing a type III secretion system delivers Pseudomonas effectors into plant cells to enhance transformation. Nat Commun 13(1):2581

Raychaudhuri SP, Verma JP, Nariani TK, Sen B (1972) The history of plant pathology in india. Annu Rev Phytopathol 10(1):21–36. https://doi.org/10.1146/annurev.py.10.090172.000321

Rukavtsova EB, Zakharchenko NS, Lebedev VG, Shestibratov KA (2023) CRISPR-Cas genome editing for horticultural crops improvement: advantages and prospects. MDPI Horticulturae 9(38):1–17

Sandhya D, Jogam P, Allini VR, Abbagani S, Alok A (2020) The present and potential future methods for delivering CRISPR/Cas9 components in plants. J Genetic Eng Biotechnol 18(1):25

Schmidt SM, Belisle M, Frommer WB (2020) The evolving landscape around genome editing in agriculture: many countries have exempted or move to exempt forms of genome editing from GMO regulation of crop plants. EMBO Rep 21:e50680

Sera T (2005) Inhibition of virus DNA replication by artificial zinc finger proteins. J Virol 79:2614–2619

Shah T, Andleeb T, Lateef S, Noor MA (2018) Genome editing in plants: advancing crop transformation and overview of tools. Plant Physiol Biochem 131:12–21. https://doi.org/10.1016/j.plaphy.2018.05.009

Shukla VK, Doyon Y, Miller JC, Dekelver RC, Moehle EA, Worden SE (2009) Precise genome modification in the crop species *Zea mays* using zinc-finger nucleases. Nature 459:437–441. https://doi.org/10.1038/nature07992

Singh J, Aggarwal R, Bashyal BM, Darshan K, Meena BR, Yadav J, Saharan MS, Hussain Z (2022a) Temporal transcriptome of tomato elucidates the signaling pathways of induced systemic resistance and systemic acquired resistance activated by *Chaetomium globosum*. Front Genet 13:1048578. https://doi.org/10.3389/fgene.2022.1048578

Singh J, Sharma D, Brar GS, Sandhu KS, Wani SH, Kashyap R, Kour A, Singh S (2022b) CRISPR/Cas tool designs for multiplex genome editing and its applications in developing biotic and abiotic stress-resistant crop plants. Mol Biol Rep 49(12):11443–11467. https://doi.org/10.1007/s11033-022-07741-2

Stathers T, Lamboll R, Mvumi BM (2013) Postharvest agriculture in changing climates: its importance to African smallholder farmers. Food Security 5(3):361–392. https://doi.org/10.1007/s12571-013-0262-z

Strange RN, Scott PR (2005) Plant disease: a threat to global food security. Annu Rev Phytopathol 43:83–116. https://doi.org/10.1146/annurev.phyto.43.113004.133839

Sun K, Wolters A-MA, Loonen AEHM, Huibers RP, van der Vlugt R, Goverse A, Bai Y (2016a) Down-regulation of Arabidopsis DND1 orthologs in potato and tomato leads to broad-spectrum resistance to late blight and powdery mildew. Transgenic Res 25(2):123–138. https://doi.org/10.1007/s11248-015-9921-5

Sun K, Wolters A-MA, Vossen JH, Rouwet ME, Loonen AEHM, Jacobsen E, Bai Y (2016b) Silencing of six susceptibility genes results in potato late blight resistance. Transgenic Res 25(5):731–742. https://doi.org/10.1007/s11248-016-9964-2

Takenaka K, Koshino-Kimura Y, Aoyama Y, Sera T (2007) Inhibition of tomato yellow leaf curl virus replication by artificial zinc-finger proteins. Nucl Acids Symp Ser 51:429–430. https://doi.org/10.1093/nass/nrm215

Tashkandi M, Ali Z, Aljedaani F, Shami A, Mahfouz MM (2018) Engineering resistance against tomato yellow leaf curl virus via the CRISPR/Cas9 system in tomato. Plant Signal Behav 13:e1525996

Testbiotech (2021). https://www.testbiotech.org/en/news/what-conventional-gmo

Tester M, Langridge P (2010) Breeding technologies to increase crop production in a changing world. Science 327(5967):818–822. https://doi.org/10.1126/science.1183700

Thomazella DP, Brail Q, Dahlbeck D, Staskawicz B (2016) CRISPR-Cas9 mediated mutagenesis of a DMR6 ortholog in tomato confers broad-spectrum disease resistance. BioRxiv 1:064824
Thorneycroft D, Sherson SM, Smith SM (2001) Using gene knockouts to investigate plant metabolism. J Exp Bot 52:1593–1601
Tiwari RK, Kumar R, Sharma S, Sagar V, Aggarwal R, Naga KC, Kumar M (2020) Potato dry rot disease: current status, pathogenomics and management. Biotech 10(11):503. https://doi.org/10.1007/s13205-020-02496-8
Tiwari RK, Lal MK, Kumar R, Mangal V, Altaf MA, Sharma S, Kumar M (2022) Insight into melatonin-mediated response and signaling in the regulation of plant defense under biotic stress. Plant Mol Biol 109(4–5):385–399. https://doi.org/10.1007/s11103-021-01202-3
Tiwari RK, Kumar R, Lal MK, Kumar A, Altaf MA, Devi R, Aftab T (2023) Melatonin-polyamine interplay in the regulation of stress responses in plants. J Plant Growth Regul 42(8):4834–4850. https://doi.org/10.1007/s00344-022-10717-y
Tripathi J, Ntui V, Ron M, Muiruri S, Britt A, Tripathi L (2019) CRISPR/Cas9 editing of endogenous banana streak virus in the B genome of *Musa* spp. overcomes a major challenge in banana breeding. Commun Biol 2:46. https://doi.org/10.1038/s42003-019-0288-7
Tuo D, Yao Y, Yan P, Chen X, Qu F, Xue W, Liu J, Kong H, Guo J, Cui H, Dai Z, Shen W (2023) Development of cassava common mosaic virus-based vector for protein expression and gene editing in cassava. Plant Methods 19(1):1–12
Ullstrup AJ (1972) The impacts of the southern corn leaf blight epidemics of 1970–1971. Annu Rev Phytopathol 10(1):37–50. https://doi.org/10.1146/annurev.py.10.090172.000345
van Bruggen AHC, Gamliel A, Finckh MR (2016) Plant disease management in organic farming systems. Pest Manag Sci 72(1):30–44. https://doi.org/10.1002/ps.4145
van Schie CCN, Takken FLW (2014) Susceptibility genes 101: how to be a good host. Ann Rev Phytopathol 52(1):551–581. https://doi.org/10.1146/annurev-phyto-102313-045854
Wang F, Wang C, Liu P, Lei C, Hao W, Gao Y, Liu YG, Zhao K (2016) Enhanced rice blast resistance by CRISPR/Cas9-targeted mutagenesis of the ERF transcription factor gene OsERF922. PLoS ONE 11:e0154027
Wang Y, Dang F, Liu Z, Wang X, Eulgem T, Lai Y, He S (2013) CaWRKY58, encoding a group I WRKY transcription factor of *Capsicum annuum*, negatively regulates resistance to *Ralstonia solanacearum* infection. Mol Plant Pathol 14(2):131–144. https://doi.org/10.1111/j.1364-3703.2012.00836.x
Wang Y, Cheng X, Shan Q, Zhang Y, Liu J, Gao C, Qiu J-L (2014) Simultaneous editing of three homoeoalleles in hexaploid bread wheat confers heritable resistance to powdery mildew. Nat Biotechnol 32(9):947–951. https://doi.org/10.1038/nbt.2969
Wang X, Shang X, Huang X (2020) Next-generation pathogen diagnosis with CRISPR/Cas-based detection methods. Emerg Microbes Infect 9(1):1682–1691. https://doi.org/10.1080/22221751.2020.1793689
Weiss T, Wang C, Kang X, Zhao H, Elena Gamo M, Starker CG, Crisp PA, Zhou P, Springer NM, Voytas DF et al (2020) Optimization of multiplexed CRISPR/Cas9 system for highly efficient genome editing in Setaria viridis. Plant J 104(3):828–838
Xie K, Minkenberg B, Yang Y (2015) Boosting CRISPR/Cas9 multiplex editing capability with the endogenous tRNA-processing system. Proc Natl Acad Sci 112(11):3570–3575
Xie K, Yang Y (2013) RNA-guided genome editing in plants using a CRISPR-Cas system. Mol Plant 6(6):1975–1983. https://doi.org/10.1093/mp/sst119
Xu Z, Xu X, Gong Q, Li Z, Li Y, Wang S, Yang Y, Ma W, Liu L, Zhu B, Zou L (2019) Engineering broad-spectrum bacterial blight resistance by simultaneously disrupting variable TALE-binding elements of multiple susceptibility genes in rice. Mol Plant 12:1434–1446
Yang C, Dolatabadian A, Fernando WGD (2022) The wonderful world of intrinsic and intricate immunity responses in plants against pathogens. Can J Plant Path 44(1):1–20. https://doi.org/10.1080/07060661.2021.1960610

Yin K, Qiu J-L (2019) Genome editing for plant disease resistance: applications and perspectives. Philos Trans Roy Soc Lond. Ser B, Biol Sci 374(1767):20180322. https://doi.org/10.1098/rstb.2018.0322

Zafar K, Khan MZ, Amin I, Mukhtar Z, Yasmin S, Arif M, Ejaz K, Mansoor S (2020a) Precise CRISPR-Cas9 mediated genome editing in super basmati rice for resistance against bacterial blight by targeting the major susceptibility gene. Front Plant Sci 11:1–11

Zafar K, Khan MZ, Amin I, Mukhtar Z, Yasmin S, Arif M, Ejaz K, Mansoor S (2020b) Precise CRISPR-Cas9 mediated genome editing in super basmati rice for resistance against bacterial blight by targeting the major susceptibility gene. Front Plant Sci 11:575

Zaidi SS-E-A, Mukhtar MS, Mansoor S (2018) Genome editing: targeting susceptibility genes for plant disease resistance. Trends Biotechnol 36(9):898–906. https://doi.org/10.1016/j.tibtech.2018.04.005

Zeng X, Luo Y, Vu NTQ, Shen S, Xia K, Zhang M (2020) CRISPR/Cas9-mediated mutation of OsSWEET14 in rice cv. Zhonghua11 confers resistance to *Xanthomonas oryzae* pv. *oryzae* without yield penalty. BMC Plant Biol 20:1–11

Zhang K, Chen K, Gao C (2014) Targeted mutagenesis in Zea mays using TALENs and the CRISPR/Cas system. J Genet Genomics 41:63–68

Zhang Y, Bai Y, Wu G, Zou S, Chen Y, Gao C, Tang D (2017) Simultaneous modification of three homoeologs of Ta EDR 1 by genome editing enhances powdery mildew resistance in wheat. Plant J 91:714–724

Zhang Y, Li D, Zhang D, Zhao X, Cao X, Dong L (2018) Analysis of the functions of TaGW2 homoeologs in wheat grain weight and protein content traits. Plant J 94:857–866. https://doi.org/10.1111/tpj.13903

Zhang Y, Zhang Q, Chen QJ (2020) Agrobacterium-mediated delivery of CRISPR/Cas reagents for genome editing in plants enters an era of ternary vector systems. Sci China Life Sci 63(10):1491–1498

Zhou J, Peng Z, Long J, Sosso D, Liu B, Eom JS, Huang S, Liu S, Vera Cruz C, Frommer WB, White FF, Yang B (2015a) Gene targeting by the TAL effector PthXo2 reveals cryptic resistance gene for bacterial blight of rice. Plant J 82(4):632–643

CRISPR/Cas9 System of Crop Improvement: Understanding the Underlying Machinery

Subrata Das, Anupam Das Talukdar, Dipika Das, and Deepa Nath

Abstract The population explosion and climate change in the growing world have become a serious concern with respect to the sustainability and production of crop plants. In this modern era, it has become essential to achieve a balanced production and harvesting of crops by minimizing the impact of pathogens and climate change. The Green Revolution has become successful in achieving the improved production of crops. However, the rising incidence of susceptibility of crop plants to diseases and the high demand for crop yield makes the scientific world think of newer and improved techniques to mitigate the problems associated with the production of high-yield disease-resistant crops. According to a United Nations assessment, to feed the world's population and end hunger by the year 2030, food output must be improved and sustained. The present-day available molecular and genetic methods of crop improvement are widely used in the production of better crop yield in terms of quality and quantity. However, limitations in time and resource utilization with more target-specific genome engineering require the search for advanced techniques of target-specific genome editing. Plant genome editing now makes considerable use of the Clustered regularly interspaced short palindromic repeats-associated protein (CRISPR-Cas) technology. It is one of the most widely accepted and improved tools in the production of disease-free and high-yielding food crop varieties. The use of CRISPR-cas tools will boost the agricultural production of improved crop cultivars compared with conventional breeding.

Keywords CRISPR-Cas · Genome editing · Crop improvement

S. Das (✉)
Department Botany, Karimganj College, Karimganj, Assam 788710, India
e-mail: subrata555das@gmail.com

A. Das Talukdar
Department of Life Science and Bioinformatics, Assam University, Silchar, Assam 788011, India

D. Das
District Institute of Education and Training, Karimganj, Assam 788720, India

D. Nath
Department Botany, Gurucharan College, Silchar, Assam 788004, India

J. Chen et al. (eds.), *Molecular and Biotechnological Tools for Plant Disease Management*, Interdisciplinary Biotechnological Advances,
https://doi.org/10.1007/978-981-97-7510-1_17

1 Introduction

Crop plants have been facing serious issues regarding their physiology of growth, development, and production. Climate change and population explosion are the two major threats responsible for the alteration in the productivity of crop plants. As a result, food scarcity is increasing day by day in line with the global food security index. A report by the United Nations states that by 2030, the global food demand and the growing world population will increase to 11.6 billion tons and 8.5 billion, respectively. By 2030, 840 million people could be afflicted by hunger, if current world patterns continue (Atukunda et al. 2021; Foster, 2023). To prevent this, the UN created Sustainable Development Goal 2 (SDG2), which is to end hunger. To help the UN achieve its zero-hunger goal, global agricultural food production needs to increase 15–20% more than the present-day yields (Ahmad et al. 2021). The sustainable production and utilization of crops is a way forward toward combating the challenges of climate change and the world's population explosion. Changes in environmental factors, both abiotic and biotic, are one of those major hindrances in achieving the sustainable production and utilization of crops. Plant pathogens are a major threat among the several biotic stresses for sustainable crop production.

In this state of the growing human population, the demands for food crops especially, cereals, sugar, oilseed, vegetables, and tubers are increasing. Food production needs to be increased keeping in mind that their quality should remain healthy for human consumption to reach SDG 2 of the UN (Ahmad et al. 2021; Nations 2019). Human efforts in various technological and breeding approaches resulted in high yields of crops in the past. Some of the approaches like during the 1940s to the late 1960s Green Revolution produces improved varieties of crops in terms of yield and production. This improvement was achieved by the growing of disease-resistant, high-yielding, short-stature crop varieties supported by the use of pesticides, chemical fertilizers, fungicides, and insecticides (Tilman 1998). The Green Revolution produces major-yielding crops like rice (*Oryza sativa*), wheat (*Triticum aestivum*), and maize (*Zea mays*) with high yields and better quality crops (Pingali 2012). The resulting increase in global food production improved the socioeconomic status of many geographical areas and accelerated agricultural progress by reducing hunger. (Ahmad et al. 2021). The current trends in agriculture and food industries mainly rely on the production of improved varieties of crops and food products of available agricultural resources. Traditionally, crop improvement has been done by several methods of plant breeding, which include, selection and crossing. These methods are not adequate to obtain improved crop varieties for the current situation. This makes the necessity of modern and improved techniques of crop production to obtain large quantities of crop products. The use of genome editing techniques in agricultural crop improvement allows editing and modifying specific target DNA sequences (Jaganathan et al. 2018). The most common use of genome editing technique is the introduction of mutation in the sequence-specific target DNA. This method involves three prime steps. First, the target DNA sequence is recognized by an exogenously engineered nuclease comprising a nuclease domain and recognition module. Then, the induction of engineered nuclease at the target DNA sequence, a double strand breaks (DSBs) is achieved. The double-stranded fractures are repaired by homology-directed repair (HDR) or nonhomologous end joining (NHEJ). In the repair process, the NHEJ often results in the induction of mutation-like insertion or deletion (Indel). These processes have successful applications in plant propagation for improved varieties (Wada et al. 2020).

2 Genetic Strategies in Crop Improvement

Genome editing or specific gene editing technologies are the most applicable strategies of crop improvement, which impart more precise alteration of the desired DNA. Recent years have seen the development of several significant genome editing methods. The engineered nuclease-like zinc finger nuclease (ZFN) and transcription activator-like effector nuclease (TALEN) have been identified as useful tools for genome editing. Oligonucleotide-directed mutagenesis (ODM) has also been reported as a genome editing tool (Wang et al. 2016). Finally, the most precise and recently developed tool of genome editing is clustered regularly interspaced short palindromic repeat (CRISPR)/CRISPR-associated protein 9 nucleases (Cas9). It has been noted to be a more versatile, focused, and targeted nuclease for gene modification. (Li et al. 2023). TALEN and ZFN comprise a sequence-specific DNA binding element and a domain of FokI nuclease. Dimerization of the FokI nuclease domain is required for it to become an active nuclease. Therefore, two modules must be created to specifically target closely related DNA sequences for the dimerization of DNA sequences with specific FokI sites. The specificity for TALENs and ZFNs is provided by this dimerization. However, designing active nucleases for this dimerization is more challenging and expensive (Jaganathan et al. 2018; Wada et al. 2020). On the contrary, the experimental setup of CRISPR/Cas9 is easier and inexpensive. This genome editing technique of the CRISPR/Cas9 system was introduced in 2012 and now it has become the most suitable and useful tool in target-specific genome alteration for agricultural crop production (Cong et al. 2013). This tool of genome editing is widely applicable for its high efficiency and specificity with ease of use.

3 Components of CRISPR/Cas9

The Cas9 protein and guide RNA (gRNA) are two essential components of the CRISPR/Cas9 system. Together with gRNA, Cas9, an RNA-dependent DNA endonuclease, creates a complex. The Cas9 enzyme must be attracted to the precise target site of the desired sequence using the gRNA, a short RNA with 20 nucleotides that is complementary to the desired sequences. The key differences between CRISPR/Cas9 technology and other genome editing methods are highlighted by the fact that the latter rely on DNA-RNA interaction rather than DNA–protein interaction for target

DNA sequence recognition. When employing TALENs and ZFNs, which act on the specific target sequence by DNA–protein interaction, two distinct DNA binding domains (500–700 amino acids for TALEN) must be designed and expressed per target site. This process takes a lot of time and money. CRISPR/Cas9, on the other hand, requires the development of an 18–20 bp oligonucleotide and a DNA–RNA interaction. Due to its straightforward operating principle, the CRISPR/Cas9 system has been exploited as a platform for genome editing in all genetic changes (Hamdan et al. 2023). The gRNA and Cas9 components of the CRISPR/Cas system must bind to a certain protospacer adjacent motif (PAM) sequence in order for the system to function as a tool for genome editing. The 3' end of the target sequence, such as the 5'-NGG-3' sequence in *Streptococcus pyogenes*, contains the brief nucleotide sequence known as a PAM (Wada et al. 2020; Hamdan et al. 2023).

4 Advantages of CRISPR/Cas9 System

The CRISPR/Cas9 system offers certain advantages over previous genome editing technologies (Mukhtar et al. 2019):

1. A large selection of workable targets: CRISPR/Cas can be used for a range of genomes because it only needs a 20-base pair target sequence before the 5′-NGGPAM.
2. Simple delivery into cells: Compared to ZFN/TALEN, which demands longer sequences, transport into cells is simpler with shorter guide RNA (gRNA).
3. Single-mode engineering: To target a single site, two distinct ZFNs, and TALENs that contain a number of repetitive sequences must be created. However, in the CRISPR/Cas9 system, there is no involvement in the re-engineering of the new target site for Cas9 protein. The final constructions utilizing single-guide RNAs (sgRNAs) are produced when a target site has been selected.
4. Multiplexing: Because the specificity of the target in the CRISPR/Cas system is fully determined by sgRNAs, which are encoded by short sequences of less than 100 base pairs, it is conceivable that multiple sgRNAs can be co-transformed to perform multiplex gene editing of plant loci simultaneously.

The most recent cutting-edge genome editing technique, CRISPR/Cas9, has enormous potential because of its effectiveness, specificity, adaptability, and ease of use when compared to other genome editing methods (Mushtaq et al. 2021; Vats et al. 2019). The CRISPR/Cas9 technology has evolved into an adaptive immune system for bacteria and archaea and is frequently split into classes I and II depending on the type of effector molecules used. They are again split into six distinct categories (I through VI) based on the component of CRISPR-associated loci and the Cas proteins type with crRNA for CRISPR interference. (Makarova et al. 2018). The most widely used tool for genome editing is the CRISPR type II system developed from *Streptococcus pyogenes*, which uses a different DNA endonuclease called Cas9 to recognize and cut the dsDNA substrate with a different nuclease domain called histidine asparagine histidine (HNH) or resolvase enzyme domain called RuvC. Cas9 uses crRNA/gRNA to integrate an effector complex which arranges the Cas9 complex to generate double-stranded breaks by invading the DNA. The processing of crRNA and cleavage of the complex Cas9 nuclease depend on a new RNA dubbed "trans-activating crRNA" (TracrRNA), which has 25 bases. (Marzec et al. 2020). The tracrRNA is complementary to CRISPR repeats. Because of the complementarity, crRNA, and tracrRNA hybridize into a mature product and which is processed by RNaseIII. When RNA reaches maturity, it creates a gRNA complex that directs the nuclease Cas to recognize and cleave at a particular target area exactly next to the PAM recognition sequence, which consists of 2 to 6 base pairs at the 3′ end. (Bisht et al. 2019). The CRISPR/Cas9 system is the most effective method for editing the plant genome and may also be used to create knockins or knockouts of genes using homology-directed repair (HRD). (Makarova et al. 2018; Marzec et al. 2020). The CRISPR/Cas9 system's inability to create off-target cleavage sites has led to the development of variants of its structural design. (Bhat and Kumar 2022; Tyagi et al. 2020) (Table 1).

5 CRISPR-CAS Applications in Plant Breeding

The present plant breeding technique has been revolutionized by the discovery of the CRISPR technology for genome editing. The advancement in the CRISPR-Cas system allowed for more effective use of it in crop modification. There are some of the applications of the CRISPR system mentioned as follows.

5.1 Male Sterility Induction

Production of hybrid plants in the crop improvement process has been an important step for the generation of crops with high yields and improved characteristics. Large-scale hybrid seed generation is a very laborious procedure because it is very difficult to control the female parent's self-pollination to reject homozygous seeds. To overcome this, an important method is the induction of male sterility. In recent times, by knocking out TMS5, thermogenic male sterility was developed in rice. Likely, TaNP1, codes for glucose-methanolcholine oxidoreductase and controls the development of tapetum and formation of pollen exine in wheat. TaNP1 was deleted at three homologous loci using the CRISPR-Cas system, producing male-sterile wheat plants (Ahmad et al. 2021; Li et al. 2020). Table 2 lists further instances of the CRISPR-Cas system being used to induce genic male sterility.

Table 1 Several variations in tools of genome editing and their important features

Genome editing tools	Variants	Alteration	Effectiveness in plants	References
TALENs	Compact TALENs (cTALENs)	• I-TevI with a wild-type N- or C-terminal fused to a scaffold of TALE DNA binding	• In tobacco protoplasts design was authenticated • The neomycin phosphotransferase II gene was chosen to include a TALEN target site (TTS)	Kazama et al. (2019); Tyagi et al. (2020)
		• The catalytic domain of I-TevI's N-terminus, a HE member of the protein family of GIYYIG, contributes to specificity through DNA cleavage selectivity	• Design for cTALEN shown action comparable to that of traditional FokI TALEN	
	mitoTALENs	With mitochondrial signal peptides, TALENs are marked	• Male sterility was restored in rice and canola after the mitochondrial genome's CMS-causing genes were deleted utilizing the mito-TALENS technique	Sauer et al., (2016); Tyagi et al. (2020)
CRISPR/Cas	Lenti-multi guide	Doxycycline-inducible EGFP reporter gene-carrying lentiviral vector (lenti-iCas9-neo), downstream SpCas9, and lentiviral vector with various gRNAs targeting different genes	• A commercially available vector called lenti multi CRISPR can be used to instantly target several genes in plants	Tyagi et al. (2020); Vats et al. (2019)
	Dead cas9	A Cas9 mutation devoid of endonuclease activity	• They can control the number of transcripts expressed in plants • Used in *Arabidopsis thaliana* to suppress *Arabidopsis* Cleavage Stimulating Factor64 (AtCSTF64) and stimulate the *Arabidopsis* Production of Anthocyanin Pigment1 (AtPAP1) gene	Lin et al. (2020); Marzec et al. (2020); Tyagi et al. (2020)

(continued)

Table 1 (continued)

Genome editing tools	Variants	Alteration	Effectiveness in plants	References
	Prime-editing	Prime Editor is a CRISPR/Cas9 nickase/ reverse transcriptase fusion protein that is programmed with pegRNAs that encode the desired edit	Using prime editors, insertions, point mutations, and deletions in the survival gene OsCDC48 have been documented in rice and wheat protoplasts	Lin et al., (2020); Marzec et al. (2020); Tyagi et al. (2020)
	Chimeric Cas9-VirD2	A protein that combines the functions of the relaxase from Agrobacterium VirD2 and the Cas9 enzyme, which produces particular DNA DSBs	• Rice (*Oryza sativa*) has undergone precise modification for a number of features, including plant architecture, herbicide resistance, and the ability to create in-frame fusions with the HA epitope at the OsHDT locus	Ali et al. (2020); Tyagi et al. (2020)

Table 2 CRISPR-Cas system in male sterility induction

Crop	Genome editing effector	Mutant type	Target gene/ region	References
Wheat	Cas9	Indel	TaCalS5 and TaRPG1	Milner et al. (2020)
Wheat	Cas9	Indel	TaNP1	Li et al. (2020)
Wheat	Cas9	Indel	TaMS1	Okada et al. (2019)
Maize	Cas9	Indel	ZmDCL5	Teng et al. (2020)
Tomato	Cas9	Indel	SlMS10	Jung et al. (2020)
Maize	Cas9	Indel	ZmTMS5	Li et al. (2017)
Rice	Cas9	Indel	PTGMS2-2	Lan et al. (2019)
Rice	Cas9	Indel	OsACOS12	Ma et al. (2019)
Rice	Cas9	C to T substitution	Oswbc11	Xuan et al. (2019)
Rice	Cas9	Indel	OsTMS5	Barman et al. (2019)
Rice	Cas9	Indel	OsDGD2β	Basnet et al. (2019)
Rice	Cas9	Indel	OsRMS	Suketomo et al. (2020)
Rice	Cas9	Indel	OsCYP703A3	Song et al. (2021)

5.2 *Fixation of Hybrid Vigor*

After hybridization, the hybrid vigor formed has to be maintained for high-yield crops and apomixis is a very vital process to maintain the hybrid vigor if needed. In this method, the seeds produced are the clones of the maternal parent. The hybrid vigor or heterosis produced was fixed by the process of mitosis instead of meiosis (MiMe) in rice plants. It was done through the simultaneous alteration of four genes, viz, OsPAIR1, OsREC8, OsOSD1, and OsMTL by CRISPR-Cas editing. The hybrid vigor was fixed as a result, and the plants were multiplied using clonal seed (Wang 2020). Another scenario involves the reported acceleration of the embryogenesis process in MiMe rice egg cells via CRISPR-Cas altered expression of Baby Boom1 (Khanday et al. 2019). In an experiment, OsMTL and OsMiMe genes were knocked out in rice, resulting in the production of clonal diploid embryos that exclude the paternal genome after fertilization and fix the hybrid vigor (Ahmad et al. 2021; Wang et al. 2019). A few examples of the hybrid fixation of crops are presented in Table 3.

6 Induction of Haploid

OsMTL a gene encodes for pollen-specific phospholipase in rice, by knocking out this gene through the CRISPR-Cas system, induction of haploid was achieved (Yao et al. 2018). Using the CRISPR-Cas technique, the Centromeric Histone3 ($CENH_3$) gene in maize was altered to produce maternal haploids. The interruption of kinetochore nucleation and spindle formation during meiosis and mitosis caused by $CENH_3$ gene knockout results in the induction of haploidy and the removal of the paternal genome (Kuppu et al. 2020). Several genes have been targeted for haploid induction through CRISPR-Cas editing in maize, rice, wheat, and *Arabidopsis thaliana* as presented in Table 4 as follows.

7 Manipulating Self-Incompatibility

Self-incompatibility is a mechanism that plants have developed to prevent self-fertilization. By inhibiting self-fertilization, this mechanism can be changed to increase the creation of hybrid plant lines. To produce rice at high temperatures, it is possible to modify the rice plant to delete the genes that cause self-incompatibility (Endo et al. 2009). Self-incompatibility was eliminated in potato by knocking off the S-RNase gene (Enciso-Rodriguez et al. 2019). The CRISPR-Cas system's editing capability was used to break the self-incompatibility of oilseed rape and cabbage (*Brassica oleracea* var. capitata) by destroying the S-receptor kinase and M-locus protein kinase (Chen et al. 2019). Some of the examples of alteration of self-incompatibility through the CRISPR-Cas system are illustrated in Table 5.

Table 3 CRISPR-Cas system in fixation of hybrid

Crop	Genome editing effector	Mutant type	Target gene/ region	References
Rice	Cas9	Indel	OsOSD1	Wang (2020)
Rice	Cas9	Indel	OsPAIR1	Wang (2020)
Hieracium	Cas9	Indel	PDS	Henderson et al. (2020)
Rice	Cas9	Indel	OsMTL	Wang (2020)
Rice	Cas9	Indel	OsREC8	Wang (2020)
Poa pratensis	Cas9	Indel	APOSTART_6	Marconi et al. (2020)
Arabidopsis thaliana	Cas9	Indel	AtSPO11-1 + AtREC8 + AtOSD1	Scheben (2020)
Rice	Cas9	Indel	OsSPO11-1 + OsREC8 + OsOSD1 + OsMATL	Scheben (2020)
Rice	Cas9	Indel	AtSPO11-1 + AtREC8 + AtOSD1 + BBM1	Scheben (2020)
Arabidopsis thaliana	Cas9	Indel	AtSPO11-1 + AtREC8 + AtOSD1 + GEM	Scheben (2020)
Rice	Cas9	Indel	OsPAIR1 + OsREC8 + OsOSD1 + OsMATL	Scheben (2020)
Rice	Cas9	Indel	dyad + GEM	Scheben (2020)

8 Scope and Prospect

Evolution of gene editing technologies has gained an important useability in the plant breeding process. In recent times, agricultural crop improvement in terms of high yield is the primary need to mitigate the challenges of shortage of food crops generated through population explosion. In this context, the advanced and more appropriate process of revolution in the agricultural process will provide great relief. The various method of gene editing technologies fits the purpose of the agricultural revolution. The fundamental ideas behind gene editing methods for crop enhancement have been completely transformed by the CRISPR-Cas genome editing system. It has attained the ability to alter the target gene of interest for the improvement of a specific crop in terms of yield and quality. Till today the application of the CRISPR-Cas9 system generated a number of crop varieties like rice, maize, wheat, potato,

Table 4 Application of CRISPR-Cas in haploid induction

Crop	Genome editing effector	Mutant type	Target gene/region	References
Maize	Cas9	Indel	MTL gene	Kelliher et al. (2017)
Maize	Cas9	Indel	$CENH_3$	Kuppu et al. (2020)
Maize	Cas9	Indel	ZmPLA1	Liu et al. (2017)
Arabidopsis thaliana	Cas9	Indel	CENH3	Kuppu et al. (2020)
Maize	Cas9	Indel	ZmDMP	Zhong et al. (2019)
Wheat	Cas9	Indel	TaMTL	Kuppu et al. (2020)
Wheat	Cas9	Indel	TaPLA-A and TaPLA-D	Liu et al. (2020)
Rice	Cas9	Indel	OsMATL	Yao et al. (2018)
Tobacco	Cas9	Indel	PsASGR-BBML	Zhang et al. (2020a)

Table 5 Alteration of self-incompatibility by CRISPR-Cas system

Crop	Genome editing effector	Mutant type	Target gene/region	References
Cabbage	Cas9	Indel	S-receptor kinase	Sehgal and Singh (2018)
Oilseed Rape	Cas9	Indel	M-locus protein kinase	Chen et al. (2019)
Potato	Cas9	Indel	S-RNase	Enciso-Rodriguez et al. (2019)
Brassica napus	Cas9	Indel	BnaA3.MLPK, BnaC4.MLPK, BnaA4.MLPK and BnaC3.MLPK	Chen et al. (2019)
Rye Grass	Cas9	Indel	LpDMC1 and LpCENH3	Zhang et al. (2020)
Cabbage	Cas9	Indel	BoPDS	Ma et al. (2019a)
Brassica oleracea	Cas9	Indel	BoSRK3	Ma et al. (2019b)
Tomato	Cas9	Indel	FPS2	Qin et al. (2018)
Brassica oleracea		Indel	BoMS1	Ma et al. (2019b)

tomato, cabbage, etc. with high nutritional value and improved agronomic traits. Based on the recent advances of the CRISPR-Cas9 system, it may be repurposed for novel applications, like genome editing for de novo speciation, enhancement of photosynthesis through editing of chloroplast and mitochondrial genome, generation of stress and pest-resistant plants. It may also be exploited to alter the season and breeding time of the crops.

References

Ahmad S, Tang L, Shahzad R, Mawia AM, Rao GS, Jamil S, Wei C, Sheng Z, Shao G, Wei X (2021) CRISPR-based crop improvements: a way forward to achieve zero hunger. J Agric Food Chem 69(30):8307–8323

Ali Z, Shami A, Sedeek K, Kamel R, Alhabsi A, Tehseen M, Hassan N, Butt H, Kababji A, Hamdan SM (2020) Fusion of the Cas9 endonuclease and the VirD2 relaxase facilitates homology-directed repair for precise genome engineering in rice. Commun Biol 3(1):44

Atukunda P, Eide WB, Kardel KR, Iversen PO, Westerberg AC (2021) Unlocking the potential for achievement of the UN sustainable development Goal 2–'Zero Hunger'–in Africa: targets, strategies, synergies and challenges. Food Nutri Res 65

Barman HN, Sheng Z, Fiaz S, Zhong M, Wu Y, Cai Y, Wang W, Jiao G, Tang S, Wei X (2019) Generation of a new thermo-sensitive genic male sterile rice line by targeted mutagenesis of TMS5 gene through CRISPR/Cas9 system. BMC Plant Biol 19:1–9

Basnet R, Hussain N, Shu Q (2019) OsDGD2β is the sole digalactosyldiacylglycerol synthase gene highly expressed in anther, and its mutation confers male sterility in rice. Rice 12(1):1–13

Bhat S, Kumar S (2022) Improvement of resistance in plants against insect-pests using genome editing tools, genome editing: current technology advances and applications for crop improvement. Springer, pp 237–249

Bisht DS, Bhatia V, Bhattacharya R (2019) Improving plant-resistance to insect-pests and pathogens: the new opportunities through targeted genome editing, seminars in cell & developmental biology. Elsevier, pp 65–76

Chen F, Yang Y, Li B, Liu Z, Khan F, Zhang T, Zhou G, Tu J, Shen J, Yi B (2019) Functional analysis of M-locus protein kinase revealed a novel regulatory mechanism of self-incompatibility in *Brassica napus* L. Int J Mol Sci 20(13):3303

Cong L, Ran FA, Cox D, Lin S, Barretto R, Habib N, Hsu PD, Wu X, Jiang W, Marraffini LA (2013) Multiplex genome engineering using CRISPR/Cas systems. Science 339(6121):819–823

Enciso-Rodriguez F, Manrique-Carpintero NC, Nadakuduti SS, Buell CR, Zarka D, Douches D (2019) Overcoming self-incompatibility in diploid potato using CRISPR-Cas9. Front Plant Sci 10:376

Endo M, Tsuchiya T, Hamada K, Kawamura S, Yano K, Ohshima M, Higashitani A, Watanabe M, Kawagishi-Kobayashi M (2009) High temperatures cause male sterility in rice plants with transcriptional alterations during pollen development. Plant Cell Physiol 50(11):1911–1922

Foster MM (2023) Creating space for student action and reflection on sustainable development goal 2: zero hunger, educating the sustainability leaders of the future. Springer, pp 47–57

Hamdan MF, Hensel G, Alok A, Tan BC (2023) Genome editing and biotechnological advances for crop improvement and future agriculture. Front Plant Sci 14:1132821

Henderson SW, Henderson ST, Goetz M, Koltunow AMG (2020) Efficient CRISPR/Cas9-mediated knockout of an endogenous phytoene desaturase gene in T1 progeny of apomictic hieracium enables new strategies for apomixis gene identification. Genes 11(9):1064

Jaganathan D, Ramasamy K, Sellamuthu G, Jayabalan S, Venkataraman G (2018) CRISPR for crop improvement: an update review. Front Plant Sci 9:985

Jung YJ, Kim DH, Lee HJ, Nam KH, Bae S, Nou IS, Cho Y-G, Kim MK, Kang KK (2020) Knockout of SlMS10 gene (Solyc02g079810) encoding bHLH transcription factor using CRISPR/Cas9 system confers male sterility phenotype in tomato. Plants 9(9):1189

Kazama T, Okuno M, Watari Y, Yanase S, Koizuka C, Tsuruta Y, Sugaya H, Toyoda A, Itoh T, Tsutsumi N (2019) Curing cytoplasmic male sterility via TALEN-mediated mitochondrial genome editing. Nat Plants 5(7):722–730

Kelliher T, Starr D, Richbourg L, Chintamanani S, Delzer B, Nuccio ML, Green J, Chen Z, McCuiston J, Wang W, Liebler T (2017) MATRILINEAL, a sperm-specific phospholipase, triggers maize haploid induction. Nature 542(7639):105–109

Khanday I, Skinner D, Yang B, Mercier R, Sundaresan V (2019) A male-expressed rice embryogenic trigger redirected for asexual propagation through seeds. Nature 565(7737):91–95

Kuppu S, Ron M, Marimuthu MP, Li G, Huddleson A, Siddeek MH, Terry J, Buchner R, Shabek N, Comai L, Britt AB (2020) A variety of changes, including CRISPR/Cas9-mediated deletions, in $CENH_3$ lead to haploid induction on outcrossing. Plant Biotechnol J 18(10):2068–2080

Lan S, Guojun D, Yu Z, Guocheng H, Qiang Z, Guanglian H, Bo X, Deyong R, Jiang H, Li Z (2019) Rapid creation of new photoperiod-/thermo-sensitive genic male-sterile rice materials by CRISPR/Cas9 system. Rice Sci 26(2):129–132

Li J, Wang Z, He G, Ma L, Deng XW (2020) CRISPR/Cas9-mediated disruption of TaNP1 genes results in complete male sterility in bread wheat. J Genet Genom 47(5):263–272

Li Y, Liang J, Deng B, Jiang Y, Zhu J, Chen L, Li M, Li J (2023) Applications and prospects of CRISPR/Cas9-mediated base editing in plant breeding. Curr Issues Mol Biol 45(2):918–935

Li J, Zhang H, Si X, Tian Y, Chen K, Liu J, Chen H, Gao C (2017) Generation of thermosensitive male-sterile maize by targeted knockout of the ZmTMS5 gene. J Genet Genom = Yi chuan xue bao 44(9):465–468

Lin Q, Zong Y, Xue C, Wang S, Jin S, Zhu Z, Wang Y, Anzalone AV, Raguram A, Doman JL (2020) Prime genome editing in rice and wheat. Nat Biotechnol 38(5):582–585

Liu C, Li X, Meng D, Zhong Y, Chen C, Dong X, Xu X, Chen B, Li W, Li L, Tian X (2017) A 4-bp insertion at ZmPLA1 encoding a putative phospholipase A generates haploid induction in maize. Mol Plant 10(3):520–522

Liu C, Zhong Y, Qi X, Chen M, Liu Z, Chen C, Tian X, Li J, Jiao Y, Wang D, Wang Y (2020) Extension of the in vivo haploid induction system from diploid maize to hexaploid wheat. Plant Biotechnol J 18(2):316

Ma C, Liu M, Li Q, Si J, Ren X, Song H (2019a) Efficient BoPDS gene editing in cabbage by the CRISPR/Cas9 system. Horticult Plant J 5(4):164–169

Ma C, Zhu C, Zheng M, Liu M, Zhang D, Liu B, Li Q, Si J, Ren X, Song H (2019b) CRISPR/Cas9-mediated multiple gene editing in *Brassica olerace*a var. capitata using the endogenous tRNA-processing system. Horticult Res 6

Ma K, Han J, Hao Y, Yang Z, Chen J, Liu Y-G, Zhu Q, Chen L (2019) An effective strategy to establish a male sterility mutant mini-library by CRISPR/Cas9-mediated knockout of anther-specific genes in rice. J Genet and Genom 46(5):273–275

Makarova KS, Wolf YI, Koonin EV (2018) Classification and nomenclature of CRISPR-Cas systems: where from here? CRISPR J 1(5):325–336

Marconi G, Aiello D, Kindiger B, Storchi L, Marrone A, Reale L, Terzaroli N, Albertini E (2020) The role of APOSTART in switching between sexuality and apomixis in *Poa pratensis*. Genes 11(8):941

Marzec M, Brąszewska-Zalewska A, Hensel G (2020) Prime editing: a new way for genome editing. Trends Cell Biol 30(4):257–259

Milner MJ, Craze M, Bowden S, Bates R, Wallington EJ, Keeling A (2020) Identification of genes involved in male sterility in wheat (*Triticum aestivum* L.) which could be used in a genic hybrid breeding system. Plant Direct 4(3):e00201

Mukhtar S, Raina M, Sudan J, Dar AA, Mustafiz A, Kumari S (2019) CRISPR-based tools for crop improvement: understanding the plant–pathogen interaction. Disease Resistance Crop Plants: Mol Genet Genomic Perspect 45–59

Mushtaq M, Ahmad Dar A, Skalicky M, Tyagi A, Bhagat N, Basu U, Bhat BA, Zaid A, Ali S, Dar T-U-H (2021) CRISPR-based genome editing tools: insights into technological breakthroughs and future challenges. Genes 12(6):797

Nations U (2019) World population prospects 2019. Population Division

Okada A, Arndell T, Borisjuk N, Sharma N, Watson-Haigh NS, Tucker EJ, Baumann U, Langridge P, Whitford R (2019) CRISPR/Cas9-mediated knockout of Ms1 enables the rapid generation of male-sterile hexaploid wheat lines for use in hybrid seed production. Plant Biotechnol J 17(10):1905–1913

Pingali PL (2012) Green revolution: impacts, limits, and the path ahead. Proc Natl Acad Sci 109(31):12302–12308

Qin X, Li W, Liu Y, Tan M, Ganal M, Chetelat RT (2018) A farnesyl pyrophosphate synthase gene expressed in pollen functions in S-RN ase-independent unilateral incompatibility. Plant J 93(3):417–430

Sauer NJ, Narváez-Vásquez J, Mozoruk J, Miller RB, Warburg ZJ, Woodward MJ, Mihiret YA, Lincoln TA, Segami RE, Sanders SL (2016) Oligonucleotide-mediated genome editing provides precision and function to engineered nucleases and antibiotics in plants. Plant Physiol 170(4):1917–1928

Scheben AAH (2020) Can we use gene-editing to induce apomixis in sexual plants? Genes 11(7):781

Sehgal N, Singh S (2018) Progress on deciphering the molecular aspects of cell-to-cell communication in Brassica self-incompatibility response. Biotech 8(8):347

Song S, Wang T, Li Y, Hu J, Kan R, Qiu M, Deng Y, Liu P, Zhang L, Dong H (2021) A novel strategy for creating a new system of third-generation hybrid rice technology using a cytoplasmic sterility gene and a genic male-sterile gene. Plant Biotechnol J 19(2):251–260

Suketomo C, Kazama T, Toriyama K (2020) Fertility restoration of Chinese wild rice-type cytoplasmic male sterility by CRISPR/Cas9-mediated genome editing of nuclear-encoded retrograde-regulated male sterility. Plant Biotechnol 37(3):285–292

Teng C, Zhang H, Hammond R, Huang K, Meyers BC, Walbot V (2020) Dicer-like 5 deficiency confers temperature-sensitive male sterility in maize. Nat Commun 11(1):2912

Tilman D (1998) The greening of the green revolution. Nature 396(6708):211–212

Tyagi S, Kesiraju K, Saakre M, Rathinam M, Raman V, Pattanayak D, Sreevathsa R (2020) Genome editing for resistance to insect pests: an emerging tool for crop improvement. ACS Omega 5(33):20674–20683

Vats S, Kumawat S, Kumar V, Patil GB, Joshi T, Sonah H, Sharma TR, Deshmukh R (2019) Genome editing in plants: exploration of technological advancements and challenges. Cells 8(11):1386

Wada N, Ueta R, Osakabe Y, Osakabe K (2020) Precision genome editing in plants: state-of-the-art in CRISPR/Cas9-based genome engineering. BMC Plant Biol 20:1–12

Wang K (2020) Fixation of hybrid vigor in rice: synthetic apomixis generated by genome editing. Abiotech 1(1):15–20

Wang H, La Russa M, Qi LS (2016) CRISPR/Cas9 in genome editing and beyond. Annu Rev Biochem 85:227–264

Wang C, Liu Q, Shen Y, Hua Y, Wang J, Lin J, Wu M, Sun T, Cheng Z, Mercier R, Wang K (2019) Clonal seeds from hybrid rice by simultaneous genome engineering of meiosis and fertilization genes. Nat Biotechnol 37(3):283–286

Xuan C, Liu Y, Liu C, Lusuwi LBT, Fu X, Chen X, Ma B (2019) Identification and gene cloning of a male-sterile mutant Oswbc11 in rice. Plant Breeding 138(3):290–298

Yao L, Zhang Y, Liu C, Liu Y, Wang Y, Liang D, Liu J, Sahoo G, Kelliher T (2018) OsMATL mutation induces haploid seed formation in indica rice. Nature Plants 4(8):530–533

Zhang Z, Conner J, Guo Y, Ozias-Akins P (2020b) Haploidy in tobacco induced by PsASGR-BBML transgenes via parthenogenesis. Genes 11(9):1072

Zhang Y, Ran Y, Nagy I, Lenk I, Qiu JL, Asp T, Jensen CS, Gao C (2020a) Targeted mutagenesis in ryegrass (*Lolium* spp.) using the CRISPR/Cas9 system. Plant Biotechnol J 18(9):1854

Zhong Y, Liu C, Qi X, Jiao Y, Wang D, Wang Y, Liu Z, Chen C, Chen B, Tian X, Li J (2019) Mutation of ZmDMP enhances haploid induction in maize. Nature Plants 5(6):575–580

CRISPR-Edited Plants for Plant-Disease Management

Alka Devi, Krutika Patil, Arti Kumari, and Rahul Goyal

Abstract Omics approaches, especially targeting genome editing, have generated considerable interest in crop improvement in agriculture. Plant-pathogen interaction is crucial for inducing a defense response and disease development. Plants use signaling molecules called pathogen-associated molecular patterns (PAMP)-triggered immunity (PTI) and effector-triggered immunity (ETI) to recognize and identify invading pathogens. CRISPR, or clustered regularly interspaced short palindromic repeats, is a popular genomic editing technology utilized in both prokaryotes and eukaryotes. CRISPR/Cas9 is a synthetic endonuclease with palindromic repeats that can be reprogrammed into various genome editing tools. CRISPR is simple, versatile, and highly efficient for gene knockout, targeting major susceptibility genes and negative regulators of host defense pathways for gaining resistance to bacterial, fungal, and viral pathogens in plants. This chapter explains the fundamentals of developing CRISPR-based tools and how they may be used to improve crops. It also covers tests that are aimed at plant pathogens and other possible uses in pest management and plant-pathogen interactions that will enhance plant breeding for crop protection.

A. Devi (✉)
National Institute of Plant Genomic Research, New Delhi 110067, India
e-mail: alkamondela@gmail.com

Dhiti Lifesciences, Pvt. Ltd., New Delhi 110020, India

K. Patil
Division of Microbiology, Indian Agricultural Research Institute, New Delhi 110012, India

A. Kumari
Nalanda college of Horticulture, Noorsarai, Nalanda 803113, Bihar, India

R. Goyal
Centre for Nanotechnology, Indian Institute of Technology Roorkee, Roorkee, Uttarakhand 247667, India

Department of Veterinary Public Health and Epidemiology, College of Veterinary Sciences, Lala Lajpat Rai University of Veterinary and Animal Sciences, Hisar, Haryana 125004, India

J. Chen et al. (eds.), *Molecular and Biotechnological Tools for Plant Disease Management*, Interdisciplinary Biotechnological Advances,
https://doi.org/10.1007/978-981-97-7510-1_18

Keywords CRISPR/Cas9 · Genome editing · Plant–pathogen interactions · Crop improvement

1 Introduction

Increasing the world's population, food demands, climate changes, and reduced free land have challenged agricultural science to develop modified crops to fulfill the food demand. The United Nations predicts that by 2050, there will be up to 10 billion people on the planet. In the present scenario, one of the most significant challenges is to ensure food security for an increasingly growing population. As the agricultural lands are constantly degrading, it is crucial to revise the farming practices and improve the varieties to fulfill the present needs. To achieve this, various studies have been implemented, such as diversity studies, genotypic, phenotypic, and metabolic variations across multiple crop species for improvement (Rao and Wang 2021). For the past three decades, genetic manipulation techniques have been widely used to develop transgenic species for crop improvement. Due to climate changes, it was observed that crop losses account for 20–40% of the total agricultural productivity, which costs about $290 billion annually (Sahu et al. 2023). Plants provide essential nutritional requirements to us, yet all are subjected to diseases in the field and post-harvest conditions. Therefore, it is a significant challenge to develop disease-resistant crops in changing environmental conditions (Singh et al. 2023a, b). These biotic stresses affect crop yield and raise the question of food security. History has many examples of when the pathogens resulted in calamity to humankind.

A famous example is the great Irish famine, which affected Ireland's population due to the infestation of potato crops by *Phytophthora infestans* in the 1840s (Waheed et al. 2023; Birch and Whisson 2001). The Great Bengal Famine was due to the reduction in rice production when the rice crop was infected with *Cochliobolus miyabeanus*, formerly known as *Helminthosporium oryzae*, causing brown spots in rice. Such threats can be minimized by reducing inoculum, proper quarantine facilities for imported materials, developing disease-resistant varieties and genetically modified crops, etc. (Shamim et al. 2024).

Plant diseases are caused by viral, fungal, and bacterial pathogens, etc. Among the bacterial diseases, 350 different plant diseases are caused by *Xanthomonas* species. *Xanthomonas oryzae* pv. *oryzae* is causing bacterial blight of rice (Niño-Liu et al. 2006). All the basmati rice varieties are highly susceptible, and this disease severely affects production. Another important bacterial pathogen, *Ralstonia solanacearum* infects more than 200 host species, including potato, tomato, eggplant, peppers, and banana (Balamurugan et al. 2020). Barley yellow dwarf viruses (BYDV) infect about 150 species of *Poaceae* globally, including barley, wheat, oat, maize, and rice (Choudhury et al. 2017).

The focus should shift from managing plant pathogens to host plants to achieve a sustainable outcome. The understanding of plant-pathogen-environment interaction and plant-pathogen-abiotic factors, along with the knowledge of the principles viz. soil, water, nutrition, seed, light, air, farm machinery, population density, farm management, and plant protection, is of paramount importance (Geffersa et al. 2023). Plant disease management faces challenges as demands for food are increasing, competition for fertile land is increasing, natural resources are shrinking, and the risk of disease epidemics is increasing due to monoculture (He et al. 2016). For decades, biological and agricultural technologies interventions have been developed to understand host–pathogen interactions, detect and diagnose plant pathogens, and breed disease-resistant varieties. The conventional approaches to breeding disease resistance plants are time-consuming and are occasionally associated with genes affecting plant development (Gupta et al. 2023). According to Khanna et al. (2023), the use of Cluster regularly interspersed short palindromic repeats/CRISPR associated (CRISPR/Cas) technology has revolutionized the field of genome editing and greatly benefited biomedicine, biotechnology, and agriculture. Furthermore, a wide range of applications for CRISPR/Cas technology exist, including chromatin imaging, base editing, epigenome editing, gene repression, and activation of genes (Lučanský et al. 2023). This technology has immense application in plant pathology and disease management. Through the cleavage of particular nucleic acid sequences, the CRISPR/Cas system, an adaptive immune system found in bacteria and archaea, offers defense against viruses and foreign genetic material (Bhatia and Yadav 2023). At the CRISPR array locus, the bacteria preserve sections of foreign genetic material in their genomic DNAs (Santiago-Frangos et al. 2023). CRISPR RNAs (crRNAs), which direct Cas nucleases for target-specific cleavage of nucleic acid sequences, are created via transcription and processing of CRISPR arrays (Pandya et al. 2023). When compared to other genome editing tools like zinc finger nuclease (ZFN) and transcription activator-like effector nuclease (TALEN), CRISPR/Cas9 has emerged as the most popular option for genome editing due to its simplicity, efficacy, multiplex editing capability, and programmability (Bajpai et al. 2023).

Three crucial conditions for disease development in plants are described as the presence of a pathogen, a susceptible host (plant), and favorable environmental conditions (Andrivon et al. 2022). Different strategies have been used to tailor plant genomes (host) for disease management, viz. modification of resistance genes (R genes), modification of susceptibility genes (S gene), and targeted degradation of plant pathogen genomes (Ngou et al. 2022). R genes play a role in pathogen recognition. Therefore, resistance can be developed by increasing the expression of R genes and editing their recognition-related regions. The S gene facilitates the infection process and supports compatibility with a pathogen. Modifying the S gene by inhibiting S gene expression and mutation in its effector recognition site confers resistance to plant diseases. CRISPR plants are produced by incorporating CRISPR/Cas system into the plants for targeted cleavage of plant pathogen genome (Karmakar et al. 2022). This chapter provides an overview of how CRISPR/Cas9 works as a genome editing tool and discusses how it can be used to manage diseases by fighting bacterial, fungal, and viral pathogens. It also provides examples of several plants that are vital for agriculture that have been generated using CRISPR/Cas9 technology (Fig. 1).

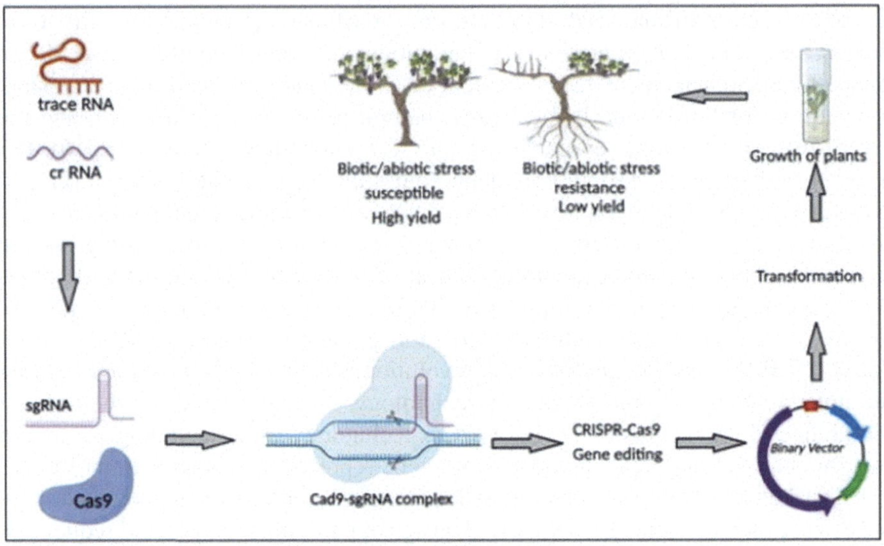

Fig. 1 Schematic representation of the role of CRISPR/Cas9 in crop improvement

2 Host–Pathogen Interaction Mechanisms

Regularly, plants tend to get exposed to various pathogens such as viruses, fungi, and bacteria but can develop resistance toward all microorganisms. Plants secrete phytoalexin as a defense system for the invasion of the pathogen, inhibiting the growth of microorganisms. Plant-pathogen interaction is a constantly evolving process. Plant pathogenicity or plant pathogens predominantly focus on the interaction of the host with a single pathogen, while plants interact with multiple pathogens.

Plant–pathogen interaction works in two steps:

1. **Recognition**: Recognition generally depends upon elicitors' production, which decodes whether the pathogen is disease-causing or not.
2. **Transduction**: It is a cellular-level mechanism in which signaling takes place that comprises (a) Perception, a surface phenomenon in which the host cell receptors recognize elicitors from pathogens (b) Signaling: this includes hypersensitive response (HR) where the plant undergoes rapid cell death to contain the pathogen spread.

Sometimes, the pathogen produces effector molecules to suppress the plant's defense system. In response, plants have the resistance (R) protein, which triggers the effector-trigged immunity (ETI).

CRISPR-Cas was identified in 1987 as an autoimmune defense mechanism against viral infections in bacterial systems (Ishino et al. 1987). CRISPR functions by introducing tiny pieces of pathogen DNA into the host chromosome's CRISPR locus during the pathogen attack. Subsequently, the host transcribes this locus into short

RNA molecules known as CRISPR RNA (crRNA), which, in association with another RNA molecule called transactivating CRISPR RNA (tracrRNA), forms a complex referred to as guide RNA (gRNA) (Deltcheva et al. 2011; Jinek et al. 2012). This gRNA complex aids in properly guiding the Cas9 enzyme to the target region, which is usually the pathogen DNA, as the name suggests. By looking for the Protospacer Adjacent Motif (PAM) sequence in the genome of the invasive pathogen, the endonuclease enzyme Cas9 produces double-strand breaks (DSBs) at the designated location. In bacteria, this mechanism serves as a memory unit that helps them defend against possible viral attacks (Bolotin et al. 2005; Barrangou et al. 2007; Mojica et al. 2005). For Cas9, the canonical PAM sequence is typically 5′-NGG-3′. However, variations such as 5′-NGA-3′ and 5′-YTN-3′ have also been documented (Esvelt et al. 2013; Kleinstiver et al. 2015). The gRNA bears the complementary sequence for the PAM, facilitating its binding and subsequently guiding Cas9 to cleave the pathogenic DNA, effectively countering the viral assault. The resulting double-stranded breaks (DSBs) are subsequently mended through either Non-Homologous End Joining (NHEJ), which introduces minor insertions or deletions (indels) at the site, or Homology Directed Repair (HDR), a process necessitating a homologous donor fragment to serve as a template for the repair.

CRISPR represents a compelling departure from methods like ZFNs and TALENs for gene editing and regulation. This is due to the consistent use of CRISPR-associated (Cas) proteins across different gene targets, with only the guide RNA's short sequence modified to redirect the site-specific cleavage (Jiang et al. 2020).

3 An Overview of CRISPR-Based Tools

CRISPR/Cas systems are divided into two categories, taking into account their phylogenetic, structural, and functional attributes. These classes have unique mechanisms for the biosynthesis of guide RNA (gRNA) and target interference. Class 1 contains three types (I, III, and IV) with 16 subtypes, and their effector molecules consist of multiple Cas proteins. On the other hand, Class 2 comprises three types (II, V, and VI) with 17 subtypes with a single large protein as an effector module (Makarova et al. 2018; Gosavi et al. 2020). Table 1 specifies a concise overview of the features of various types of Class 2 CRISPR/Cas systems participating in interactions between plants and pathogens.

In the last ten years, the CRISPR-Cas system has been adapted into various editing tools. However, fewer gene editing tools have been created due to the intricate nature of class 1 systems (Hu and Li 2022). Extensive efforts are being made to harness the gene editing potential of class 2 systems, leveraging their ability to bind and cleave nucleic acids. Class 2 systems are further categorized into DNA-targeting effectors, like Cas9 and Cas12, and RNA-targeting effectors, like Cas13. Among these, the type II CRISPR/SpCas9 system, derived from *S. pyogenes*, is the most widely studied and extensively employed for targeted genome editing in plants. Originally consisting of three elements—SpCas9 nuclease, trans-activating small

Table 1 Distinctive attributes of various CRISPR/Cas systems participating in plant-pathogen interactions

CRISPR system	Cas9	Cas12a	Cas12b	Cas13	Cas14
Type	II	V	V	VI	V
Substrate	dsDNA	dsDNA	dsDNA	RNA	ssDNA
Nuclease domain	RuvC, HNH	RuvC	RuvC	HPEN	RuvC
Pre-crRNA processing	×	✓	×	✓	✓
tracrRNA	✓	×	✓	×	✓
PAM/PFS	3′, G-rich	5′, T-rich	5′, T-rich	3′, non-G PFS	PAM independently

RNA (tracrRNA), and small mature CRISPR RNA (crRNA) (Bhaya et al. 2011)—this system has been optimized to include just two elements: SpCas9 and guide RNA (gRNA), a fusion RNA comprising crRNA and the scaffold tracrRNA (Cong et al. 2013; Jinek et al. 2012). Another well-documented CRISPR system is CRISPR/Cas12, which is categorized as type V and has a different structural architecture and evolutionary origin than Cas9. Several RuvC-like domain-comprising Cas12 systems, such as CRISPR/Cas12b (C2c1) and CRISPR/Cas12a (Cpf1), have been effectively utilized for genome editing in plants (Zaidi et al. 2017; Ming et al. 2020). Cas12a functions with a single mature crRNA for DNA targeting, while Cas12b requires both tracrRNA and crRNA to operate (Teng et al. 2018). Cas14, a distinct type V protein identified in nonculturable archaea, has been uncovered, demonstrating the ability to cleave single-stranded DNA without sequence specificity. This makes Cas14 an excellent means for enhancing resistance to ssDNA plant viruses and conducting highly reliable SNP genotyping due to its ability to cleave DNA independently of its sequence (Khan et al. 2019). Type VI CRISPR/Cas13 systems behave as ribonucleases and are categorized into the following subtypes: Cas13a, Cas13b, Cas13c, and Cas13d (Cox et al. 2017; Shmakov et al. 2017). These Cas13 proteins focus on RNA instead of DNA molecules and are distinguished by the presence of two separate HEPN RNase domains. They form complexes with single crRNA to guide their RNA-targeting activities and are being employed for the localization, detection, and tracking of various RNA molecules.

The practical implementation of control measures relies largely on the accurate identification of diseases and the organisms causing them. Therefore, the initial step in disease management is the crucial process of disease diagnosis. When compared to traditional nucleic acid detection methods of PCR, Cas-based detection offers the advantage of requiring relatively lower inputs while maintaining consistent detection capabilities at femtomolar to zeptomolar sensitivity, based on the specific methods employed (Wheatley and Yang 2021). Recent breakthroughs and an improved understanding of diverse Cas proteins have opened avenues for creating portable, highly sensitive, and cost-effective point-of-care (POC) testing tools based on nucleic acids.

While durable and reliable platforms for nucleic acid detection have been established through the utilization of Cas9 proteins (Quan et al. 2019; Jiao et al. 2021), the finding of parallel cleavage activities in Cas12a and Cas13a has brought about a revolutionary change in the field of detection of nucleic acids (Gootenberg et al. 2018). Cas13a and Cas12a have become the foundation for numerous CRISPR-dependent detection techniques (Fig. 2). The following text briefly explains the mechanisms by which these tools operate.

(a) **NASBACC** (Nucleic Acid Sequenced Based Amplification-CRISPR Cleavage): It relies on the specific target DNA cleavage by Cas9 and operates on the underlying basis of a "toehold switch". It uses PAM-dependent target detection, sequence-based amplification, and Cas9-mediated cleavage, along with the toehold sensor, to generate the final output. The toehold sensor is an RNA switch that encodes a reporter protein. Within this sensor, a hairpin structure conceals the ribosome binding site (RBS). When the toehold sensor attaches to a trigger RNA, it exposes the RBS, allowing for the translation of the reporter protein (Ching-Yuet To et al. 2018). Once the sgRNA-Cas9 complex locates the PAM sequence and, subsequently, the target site, it initiates DNA cleavage, producing truncated RNA via in vitro transcription. This truncated RNA is unable to activate the toehold switch. Meanwhile, a different genotype lacking the PAM sequence remains uncut by Cas9, generating full-length RNA that triggers the toehold reaction and a visible change in color. As a result,

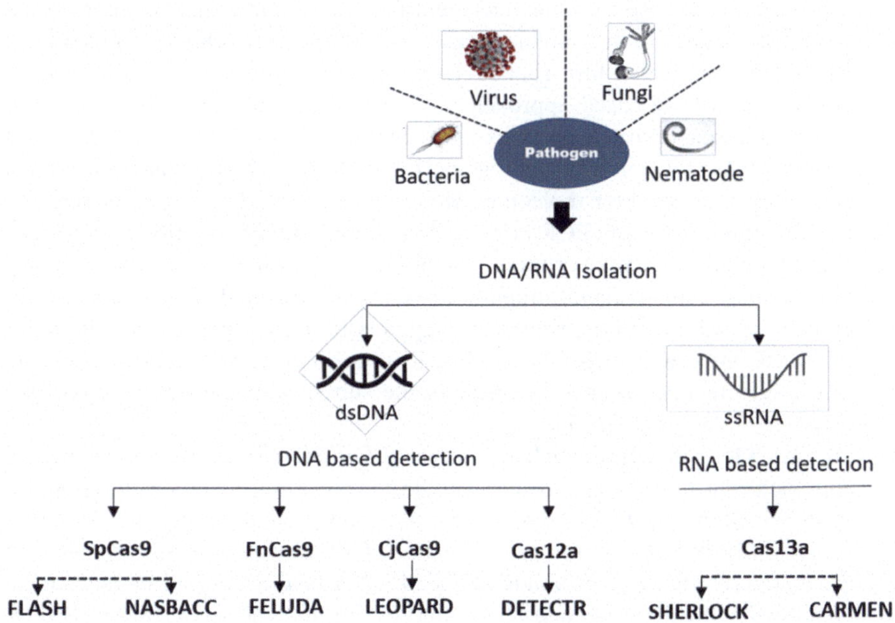

Fig. 2 A broad overview of different disease diagnostic methods based on CRISPR-Cas technology

this technique can undeniably be employed for discriminating between closely related viral strains by utilizing strain-specific PAM sites and strategically designed sgRNA.

(b) **LEOPARD** (Leveraging Engineered tracr RNAs and On-target DNAs for PArallel RNA Detection): This method is based on CjCas9 (Cas9 from *Campylobacter jejuni*) has been developed, enabling the simultaneous identification of multiple disease-related biomarkers with single-nucleotide precision, during their investigation of CjCas9 within its host. Jiao et al. (2021) observed that the tracrRNA of the CjCas9 system formed unconventional crRNAs (ncrRNAs) by hybridizing with cellular RNAs. Realizing the potential for programming the tracrRNA to target a specific RNA molecule, they engineered a reprogrammed tracrRNA (Rptr) to bind and cleave a required RNA. This cleavage process forms a functional guide RNA from a segment of the target RNA. The final detection step, through gel-based analysis, occurs when a ncrRNA and Rptr collectively cleave a fluorescent DNA sensor, producing a measurable readout.

(c) **FLASH** (Finding Low-Abundance Sequences by Hybridization): This method involves the precise cleavage of the desired sequence utilizing a specifically designed Cas9 guide RNAs (gRNAs). A phosphate treatment is applied to block the genomic DNA input in the initial step. Subsequently, the complex formed by Cas9 and the designed gRNAs effectively digests the input DNA, resulting in cleaved products that can be readily joined with universal sequencing adaptors. Following amplification, the target sequences are enhanced onto the background and are prepared for attachment to the sequencing flow cell. Notably, this approach enables the simultaneous identification of numerous targets within a single test, thanks to the implementation of multiplexing (Quan et al. 2019).

(d) **FELUDA** (FnCas9 Editor-Linked Uniform Detection Assay): This method utilizes an affinity-based approach to detect pathogenic DNA or RNA with remarkable sensitivity. It involves using FnCas9 (*Franscisella novicida* Cas9) for enzymatic detection of pathogenic biomarkers without causing any trans cleavage of the reporter molecule (Mohd et al. 2021). The target sequence is initially amplified using PCR or recombinase polymerase amplification (RPA), employing biotin-labeled primers. The biotin-tagged products are subsequently immobilized onto magnetic streptavidin beads. Following this, a ribonucleoprotein complex binds to the substrate, tagged with a fluorophore and consisting of dCas9 paired with a specific sgRNA. This enzymatic binding process results in changes in fluorescence intensity in the supernatant, indicating a positive reaction (Azhar et al. 2021).

(e) **DETECTR** (DNA Endonuclease-Targeted CRISPR Trans Reporter): In this approach, the initial step involves guiding Cas12a toward a specific double-stranded DNA (dsDNA) target using a complementary guide RNA. Upon successfully binding to the correct target, Cas12a subsequently engages in the collateral cleavage of single-stranded DNA (ssDNA) reporter molecules, combined with a quencher and a fluorophore. The non-selective cleavage activity is then identified by the fluorescence signal produced when the quencher and fluorophore separate (Chen et al. 2018).

(f) **SHERLOCK** (Specific High-Sensitivity Enzymatic Reporter unLOCKing): In this procedure, the initial step involves the isothermal amplification of collected samples. Subsequently, the sample is supplemented with an engineered Cas13, a guide RNA, and an RNA reporter. Cas13 is programmed explicitly with the guide RNA to target viral or bacterial DNA or RNA for detection. A reporter probe, consisting of a short oligonucleotide that separates a fluorophore and a quencher, is also introduced. Once Cas13 successfully binds to its designated target, the initiation of unbiased cleavage of nearby RNA molecules, along with the reporter starts. This action separates the reporter and the quencher, producing distinct signals within the sample (Myhrvold et al. 2018).

(g) **CARMEN** (Combinatorial Arrayed Reactions for Multiplexed Evaluation of Nucleic acids): This approach involves amplifying nucleotide targets through PCR or RPA. Following the amplification step, a unique color code is introduced along with the sample and Cas13, which are then emulsified. Subsequently, droplets from all the samples and the detection mix (consisting of Cas13, a cleavage reporter, and a sequence-specific CRISPR RNA) are pooled together. These mixed droplets are introduced into a microfabricated chip, where each microwell accommodates one pair of sample and detection droplets. An external electric field is applied to merge these droplets, initiating the reaction. The fluorescence indicates a positive sample reaction, signifying that Cas13a has cleaved the reporter after identifying the target (Ackerman et al. 2020).

4 Genome Editing in Plant Pathogens

The advent of effective genome-editing tools, notably CRISPR/SpCas9, has opened up a wide range of potential applications in plant pathogens. The capability to alter the genomes of plant pathogens holds the promise of imparting desirable traits for various purposes. In contrast to conventional methods of manipulating microbial genomes, which are often hindered by ineffective homologous recombination, CRISPR/SpCas9 tools are markedly capable and, in some instances, considerably more straightforward. Additionally, they offer a high-throughput experimental place for investigating gene functions at the whole genome scale in plant pathogens. These emerging CRISPR-based genome-editing technologies are crucial in developing resistance to diseases caused by bacteria, fungi, viruses, and other plant pathogens.

4.1 Genome Editing of Viral Pathogens Through CRISPR/Cas9

The inherent obligate parasitic, fast-evolving genome, and high infectivity of plant viruses represent a significant menace to crop plants, leading to a substantial decrease in crop yields (Das et al. 2019). In recent times, various CRISPR-Cas9-based techniques utilizing Cas9 nucleases have been employed for genetic engineering in plants, aiming to increase resistance to DNA viruses that affect plants. In 2015, there were pioneering reports from three separate research groups on utilizing CRISPR-Cas9-dependent techniques to establish resistance against plant DNA viruses. These groups successfully generated resistance against geminiviruses, like *Tomato leaf curl virus* (TYLCV), *Beet severe curly top virus* (BSCTV), and *Bean yellow dwarf virus* (BeYDV), in model crops like *Nicotiana benthamiana* and *Arabidopsis thaliana* (Ali et al. 2015; Baltes et al. 2015; Hirano et al. 2016). Specifically, the viral genome's coding and non-coding regions, including elements like the capsid protein and replication protein, were aimed using synthetically designed guide RNA (gRNA) and Cas9 nuclease proteins. This approach resulted in reduced symptoms and enhanced viral resistance.

Ali et al. (2016) induced mutations into the non-coding regions of viral genomes, such as *Tomato yellow leaf curl virus* (TYLCV), *Beet curly top virus* (BCTV), *Merremia mosaic virus* (MeMV), and *Cotton leaf curl Kokhran virus* (CLCuKoV). Their study confirmed that the CRISPR tool effectively induced targeted mutations, rendering the viruses less capable of replication and bolstering resistance in plants. Notably, the research demonstrated that the non-coding regions of the viral genome conferred increased resistance compared to the coding sequences. This finding stemmed from the rapid adaptability of the virus, which could quickly modify coding sequences to overcome challenges. In contrast, non-coding regions were less mutable, leading to longer-lasting resistance. It was revealed that viral genomes could develop mutations in the coding region to evade Cas9 cleavage and immune responses.

Whereas, the intergenic region exhibited lower mutagenicity, resulting in a higher resistance level. In addition, Iqbal et al. (2016) harnessed the CRISPR-Cas9 system to combat begomoviruses, a diverse group of viruses known to cause cotton leaf curl disease in cotton plants, posing a significant threat to crop yields. Their study proposed using multiplex-type single guide RNA (sgRNA) sequences that could simultaneously target the genomes of the viral complex and their DNA satellites, thereby generating resistant cotton plants. Similarly, the *endogenous banana streak virus* (eBSV), responsible for infecting banana plants, was deactivated by applying the CRISPR system, explicitly targeting viral sequences (Tripathi et al. 2019). The mutations introduced into the viral genome led to the deactivation of the transcriptional and translational machinery, thereby inhibiting the replication of functional viral proteins and offering resistance to banana streak disease.

Furthermore, a ground-breaking achievement of CRISPR/Cas9-engineered plant immunity against RNA viruses was documented. This research targeted *cucumber mosaic virus* (CMV) and *tobacco mosaic virus* (TMV) utilizing FnCas9 and sgRNA. The results revealed a notable decrease in viral load in transgenic Arabidopsis and tobacco plants, attributable to the RNA cleavage action of the FnCas9 enzyme (Zhang et al. 2018). The CRISPR/Cas9 technique was employed to deactivate the crucial gene *Dicer-like 2 (DCL2)* in tomatoes. The dcl2 mutants displayed susceptibility to *tobacco mosaic virus*, *potato virus X*, and *tomato mosaic virus*, highlighting the pivotal function of *DCL2* in providing resistance against RNA viruses (Wang et al. 2018). These findings hold significant implications for future research, highlighting the potential to develop plant resistance by either activating or expressing genes that identify and target viral genomes, ultimately leading to enhanced viral resistance.

4.2 Resistance Against Fungal Pathogens Using CRISPR/Cas9-Mediated Tools

Fungal pathogens significantly threaten plant health, inducing many diseases like rust, smut, powdery mildew, early blight, late blight, rot, wilt, and many more. Beyond the direct harm they inflict through these diseases, fungal pathogens also contribute to substantial losses by producing mycotoxins. This group constitutes the most extensive category of plant pathogens, presenting a universal menace to agricultural production and a tremendous burden on plant protection efforts.

The CRISPR/Cas9 system has proven a precise and efficient tool for inducing targeted mutagenesis in plants, bolstering their resistance to fungal diseases. Considerable focus has been directed toward identifying susceptibility genes and negative regulators that play pivotal roles in the plant's defense mechanisms against disease. CRISPR-Cas9 system was applied to target a negative defense regulator known as *TaEDR1*, resulting in mutant wheat lines that showed heightened resistance to the powdery mildew-causing fungus *Blumeria graminis* f. sp. *tritici* (Zhang et al. 2017). Another noteworthy example is the Microrchidia (MORC) protein, which is significant in both animal and plant DNA mismatch repair pathways and structural rearrangement. In barley, CRISPR-Cas9 created single and double knockout (KO) mutants of *HvMORC1* and *HvMORC6a*. These mutant lines displayed enhanced resistance to both biotrophic (*Bipolaris sorokiniana*) as well as necrotrophic (*Fusarium graminearum*) pathogens compared to single mutants (Galli et al. 2022). The CRISPR-Cas9 system was effectively utilized to induce indels in rice transcription factor genes, specifically *OsERF922*, resulting in enhanced resistance to the rice blast fungus *Magnaporthe oryzae* (Wang et al. 2016). Disruption of *OsSEC3A* through CRISPR-Cas9 led to increased salicylic acid content and improved tolerance against the rice blast fungus, albeit with some adverse phenotypic effects. Furthermore, the knockout of *OsBSR-K* using CRISPR-Cas9 in transgenic rice significantly enhanced resistance to *M. oryzae* (Zhou et al. 2018). In wheat, the knockout of its homologs (*TaMLOs*) via CRISPR/SpCas9 conferred resistance to *Blumeria graminis* f. sp. *tritici* (Wang et al. 2014). A similar strategy was employed in tomato (*SlMLO1*), resulting in resistance against the powdery mildew-causing fungus *Oidium neolycopersici* (Nekrasov et al. 2017). In addition to knockout strategies, precise base editing through CRISPR/Cas9-mediated correction of defective resistant (R) genes has emerged as an efficient and practical method to enhance crop disease resistance. For example, researchers utilized a cytidine base editor (*rBE5*) to swiftly adjust the recessive *pi-d2* gene in the rice variety Kitaake by making a

G > A substitution (M411I) in the genome. This rapid and precise genome engineering approach substantially accelerates resistance breeding programs in comparison to traditional marker-assisted selection and hybrid breeding methods (Ren et al. 2018). In the modern era of agriculture, innovative disease management strategies like the CRISPR/Cas9 tool have become crucial. The editing of susceptible (S) genes using CRISPR/Cas9 technology has been a game-changer, providing broad-spectrum resistance against fungal diseases. These S genes are pivotal in facilitating infection and aiding pathogens in forming compatible communication with plants. Resistance based on S genes is typically inherited in a recessive manner, and mutations in these S alleles can confer even stronger resistance than R genes (Schie and Takken 2014). For instance, the CRISPR/Cas9-mediated knockout of the *Solyc08g075770* gene, which is associated with decreased mycorrhizal colonization (rmc) in tomatoes, resulted in immunity to *Fusarium* (Prihatna et al. 2018). Another study by Gao et al. (2021) demonstrated that CRISPR/Cas9-based knockout mutants without *SlymiR482e-3p* in tomatoes displayed increased resistance against tomato wilt-causing fungus *F. oxysporum* f. sp *lycopersici*. This miRNA targeted a specific protein, enhancing the expression of nucleotide-binding site leucine-rich repeat (NBS-LRR) proteins. Therefore, the targeted modification of host S genes through the CRISPR/Cas9 tool can establish robust and long-lasting plant resistance to various types of fungi (Talakayala et al. 2022).

4.3 Genome Editing in Bacteria and Nematodes Using CRISPR/Cas9–mediated Tools

Bacteria are a significant group of phytopathogens, joining fungi, oomycetes, and viruses in their worldwide impact on crop production and food security. Bacterial pathogens continually disrupt crop yields, threatening global food supplies. During bacterial infections, plants typically secrete type III effectors, disrupting host cell defense systems and activating susceptibility (S) genes, leading to disease development. Utilizing CRISPR-Cas technology to edit S genes can reduce plant susceptibility to bacterial diseases without affecting plant growth. For instance, CRISPR/Cas9 was used to edit susceptible sucrose transporter genes like *OsSWEET13* and *OsSWEET14* in rice to achieve resistance to bacterial blight (Zeng et al. 2020). Knocking out genes such as *Pi21* (*Pyricularia oryzae* resistance protein 21), *Xa13* (*Xanthomonas oryzae* 13), and *TMS5* (thermosensitive male sterility) utilizing CRISPR/Cas9 resulted in thermo-sensitive male sterility and enhanced resistance against bacterial blight and rice blast (Li et al. 2019). Furthermore, a small deletion in the *SlDMR6-1* gene via CRISPR/Cas9 in tomato led to the premature truncation of the protein. Subsequently, it conferred disease resistance to various pathogens, including *Xanthomonas* spp., *P. syringae*, and *P. capsici* (Thomazella et al. 2016). The CRISPR-Cas9 technique was effectively applied in tomatoes to deactivate the *downy mildew resistant 6* (*DMR6*) gene, resulting in mutants that exhibit resistance against various plant pathogens, such as *Pseudomonas syringae*, *Phytophthora capsicum*, and *Xanthomonas* spp. (de Toledo et al. 2016). Furthermore, CRISPR-Cas9 technology was harnessed to create tomato plants featuring dominant Jasmonatezim domain protein 2 (JAZ2) repressors without jasmonate-associated (Jas) domain (JAZ2_jas). These modified repressors conferred resistance against *Pseudomonas syringae*, under natural conditions as well (Ortigosa et al. 2019). CRISPR-based antibacterials represent a unique approach to target specific pathogenic bacteria. Due to its high sequence specificity, CRISPR/Cas can effectively discriminate between pathogenic and commensal bacterial species. To adapt the CRISPR/Cas mechanism for defensive action against bacteria, CRISPR-guided RNAs are developed to target virulent or essential genes unique to pathogens (Greene 2018). For example, Peng et al. (2017) recently engineered citrus plants resistant to canker disease by employing CRISPR/Cas9 to target the susceptibility gene *CsLOB1* promoter. Similarly, genome-editing tools were employed to obtain plants resistant to *Xanthomonas oryzae* pv. *oryzae* by targeting the transcriptional regulation of S genes through effector manipulation (Li et al. 2015).

Plant pathogenic nematodes (PPN) inflict severe damage on their host while promoting secondary infections by other phytopathogens like bacteria and fungi. Various genome-editing techniques utilizing CRISPR/Cas9 have been extensively documented in *Caenorhabditis elegans*, including co-conversion (Arribere et al. 2014), co-CRISPR (Kim et al. 2014), self-excising cassettes (Dickinson et al. 2015), bacterial feeding (Liu et al. 2014), and SapTrap (Schwartz and Jorgensen 2016). The successful translation of CRISPR/Cas9 technology from the model nematode *Caenorhabditis elegans* to various pathogenic nematodes such as *Ascaris suum*, *Haemonchus contortus*, *Strongyloides* spp., and *Brugia malayi* has opened up new avenues for addressing nematode pathogens in plants (Zamanian and Andersen 2016). For instance, Huang et al. (2023) applied the CRISPR/Cas9 system to prompt targeted mutagenesis in the S gene of the rice copper metallochaperone heavy metal-associated plant protein 04 (OsHPP04). They managed to produce genetically stable homozygous rice mutants, with some devoid of transgenic elements. The mutants exhibited heightened resistance to the rice root-knot nematode (*Meloidogyne graminicola*), a notable disease-causing nematode in rice. This marks the initial application of CRISPR/Cas9-mediated knockout of plant S genes for the creation of plant material resistant to PPN (Plant-Parasitic Nematodes). In conclusion, the CRISPR tool offers an effective means of conferring resistance against several devastating nematode and bacterial diseases in plants.

5 CRISPR Plants

CRISPR/Cas9 technique is being used effectively to develop plants resistant to diseases. CRISPR plants are developed using gene disruption via indels in the coding sequence. This method has been applied to several crops for disease resistance. These indels can be created in coding sequences or promoter regions. Gene deletion

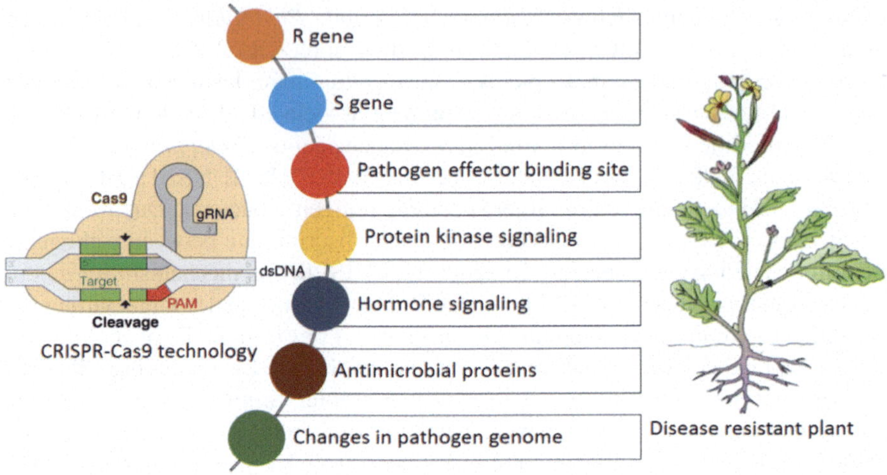

Fig. 3 Strategies used by the CRISPR-Cas9 technology for the development of disease-resistant plants

or gene insertion is achieved by multiplex sgRNAs and homology-directed repair, respectively. For example, in *Arabidopsis* and cucumber (*Cucumis sativa*), resistance against multiple RNA viruses has been introduced by indels in the eukaryotic initiation factor 4E protein (eIF4Es) by using CRISPR/Cas9 technique (Noureen et al. 2022; Kuroiwa et al. 2022). In another approach, the rice sugar transporter gene promoter was modified by disrupting the effector binding site for bacterial blight pathogen, which resulted in blight resistance (Jiang et al. 2020). CRISPR-Cas9-based gene insertion utilizes HDR rather than NHEJ. However, HDR efficiency in plants is actually poor. CRISPR-Cas9 technique was used to incorporate green fluorescent protein in rice, which has the potential to study function, localization, and spatiotemporal expression of the *S* gene (Wang et al. 2020). The CRISPR/Cas9 technology has already been employed in a wide range of plants (Fig. 3), as described in the following subsections.

5.1 Rice

Rice (*Oryza Sativa* L.) is one of the most important cereal crops, providing carbohydrates for 50% of the population worldwide. Rice productivity has been impacted by increased population, changing climatic conditions, and the emergence of novel phytopathogens. Traditional breeding methods, including molecular and mutation breeding, have been practiced for decades to raise disease-resistant rice plants. However, these practices are laborious and time-demanding. To address these issues, precise genome editing methods were utilized to create disease-free rice cultivars

(Li et al. 2018). Blast disease, bacterial leaf blight (BLB) disease, and rice tungro disease are among the major rice diseases causing production constraints. The BLB pathogen (*Xanthomonas oryzae* pv. *oryzae*) targets the promoter of the S gene by introducing DNA-binding proteins known as TALEs (transcription activator-like effectors) into the plant. These proteins target S genes' effector binding domain elements (EBEs). TALEs aim to the sugar-transporting SWEET gene family, which helps in the release of sugar into the apoplast that is utilized as a source of nutrition for the BLB pathogens (Cohn et al. 2014). TALEs act as pathogenic factors of BLB, and SWEET gene family members act as susceptibility factors of rice plants. Therefore, these two are major targets for CRISPR/ Cas-9 in rice. *OsSWEET11* and *OsSWEET14* are two susceptibility genes in the rice variety Kitaake. BLB resistance in rice was developed through the disruption of their EBEs via CRISPR/Cas9 (Xu et al. 2019). Indels in the EBEs of the *OsSWEET* promoter and *OsSWEET13* promoter in a novel rice line and *MS14K* line respectively resulted in broad-spectrum resistance against BLB pathogens. CRISPR/Cas9-targeted deletion of the susceptibility gene *Os8N3* (*OsSWEET11*) improved resistance to the BLB pathogen without compromising agronomic attributes or pollen viability in mutant rice lines (Kim et al. 2019).

In rice, 46 resistance genes for BLB resistance have so far been discovered. The *Xa4, Xa5, Xa7, Xa13, Xa21, Xa33*, and *Xa38* genes have been extensively employed in the development of BLB resistance cultivars (Hsu et al. 2020). *Xa13* resistance gene (fully susceptible) was mutated using CRISPR/Cas12a technology, resulting in improved resistance to BLB disease in rice (Yu et al. 2021). Blast disease (caused by *Magnaporthe oryzae*) accounts for a loss of 30% of global rice production (Nalley et al. 2016). To present, over 100 major blast R genes have been discovered, with *Pi-ta/Pi-ta2, Pi-z, Pi-b*, and *Pi-k/h/m/s* being used to build disease resistance genotypes (Li Y et al., 2018). Ethylene-responsive factors (ERF) are involved in modulating biotic or abiotic stresses. For example, in rice, *OsERF922* functions as a negative regulator of blast resistance. Improved resistance to rice blast has been achieved by modification in *OsERF922* transcription factor using CRISPR/Cas9 in the japonica variety of rice (Wang et al. 2016). Rice *OSEC3A* gene is involved in forming exocyst subunits, and CRISPR/Cas9 targeted editing in *OSEC3A* resulted in enhanced resistance to *Magnaporthe oryzae* (Ma et al. 2018). The combination of two viruses causes rice tungro disease, namely rice tungro bacilliform virus containing a double-stranded DNA genome and rice tungro spherical virus containing a single-stranded RNA genome respectively (Hull 1996). It has been revealed that the susceptibility gene *eIF4G* (*translation initiation factor 4 gamma*) is required for the replication of viruses within the host plant. CRISPR/Cas9-targeted alteration of the eIF4G gene resulted in resistance to rice tungro disease (Kumam et al. 2022). *Bsr-d1* is a broad-spectrum blast-resistant C_2H_2-domain transcription factor from the Chinese rice variety Digu. *Bsr-d1* regulates peroxidase expression, which affects H_2O_2 levels and suppresses immune responses to *M. oryzae*. CRISPR/Cas9 mediated knockout of disease resistance *Bsr-d1* gene resulted in a wide range of resistance to pathogens causing blast disease in rice (Nasir et al. 2018).

5.2 *Citrus Plants*

Citrus is a high-value crop that provides benefits to humans while facing many biotic challenges at the same time. Among the bacterial pathogens, citrus canker (*Xanthomonas citri* pv. *citri*) is causing significant damage to the crop. *CsLOB1* has been discovered as a citrus canker susceptibility gene. CRISPR/Cas9-targeted knockout of the *CsLOB1* gene resulted in resistance to citrus canker (Jia et al. 2017).

5.3 *Tomato*

Tomatoes are one of the most widely consumed vegetables worldwide. However, this crop's productivity is severely harmed by viral infections such as Tomato yellow leaf curl virus (TYLCV) and fungal pathogens such as powdery mildew. In commercial tomato varieties, *SlPelo* carries susceptibility factor (S-gene) for TYLCV, and *Mildew resistance locus o 1* (*SlMlo1*) carries susceptibility factor (S-gene) for powdery mildew resistance in tomatoes. CRISPR/Cas9 mediated knockout of *SlPelo* and *SlMlo1* resulted in TYLCV and powdery mildew resistance in the tomato (Pramanik et al. 2021). Another interesting example is the *Downy mildew resistance-6* (*DMR-6*) gene, whose expression is enhanced during pathogen infection. Mutation in this gene increased the production of defense-related genes and salicylic acid production. A wide range of disease resistance to the pathogens, viz., *Xanthomonas* sp., *P. syringae*, and *P. capsici,* had been achieved by deletion in the *SlDMR-6* gene using CRISPR/Cas9 (Paula et al. 2016). Another susceptibility gene *Powdery mildew resistance 4* (*Pmr4*) was altered by using CRISPR/Cas9, resulting in resistance to powdery mildew in tomato (Santillán Martínez et al. 2020). Mitogen-activated protein kinases (MAPKs) are essential in defense response and upregulation of salicylic acid metabolism. The role of S*lMAPK3* in resistance against *Botrytis cinerea* has been investigated in tomatoes. Knockout mutants of S*lMAPK3* are produced using CRISPR/Cas9 resulting in reduced susceptibility to *Botrytis cinerea* pathogen, causing grey mold in tomatoes (Yu et al. 2019). Hormonal signaling plays a vital role in pathogen resistance, e.g. jasmonic acid (JA) and salicylic acid (SA) signaling. JA inhibits the SA-dependent defense pathway and stimulates stomatal opening, leading to bacterial pathogen invasion and colonization. Tomato bacterial speck disease is triggered by *Pseudomonas syringae* pv. *tomato (pto)*, a bacterial pathogen that produces coronatine (COR, JA mimic). The stomatal response of COR requires a co-receptor called AtJAZ2 (reported in Arabidopsis). SlJAZ2, a functional ortholog in tomato, was edited by CRISPR/Cas9, which prevented stomatal reopening and reduced bacterial entry, increasing resistance to *pto* infection (Ortigosa et al. 2019).

5.4 Watermelon

Watermelon (*Citrullus lanatus*) is a commercially important fruit crop globally. However, this crop suffers many biotic and abiotic stresses. *Fusarium* wilt is one of the most devastating watermelon diseases, caused by the pathogen *Fusarium oxysporum* f. sp. *niveum*is. Phytosulfokine (PSK) signaling is associated with the modulation of plant immunity during pathogen infection. *PSK* genes were successfully used in several plants to develop disease resistance. Mutagenesis in the *psk1* gene of watermelon using CRISPR/Cas9 resulted in resistance to *Fusarium* wilt (Zhang et al. 2020).

5.5 Soybean

Soybean (*Glycin max*) is an economic oilseed crop worldwide. Many diseases affect soybean yield. Soybean root rot (*Phytophthora sojae*) is causing significant yield loss throughout the growth period. *P. sojae* effector recruits an acetyltransferase G*mTAP1* into the nucleus that acetylates histones (H2A and H3) and increases the expression of susceptibility genes. A recent study attempted to CRISPR/Cas9-mediated knockout of *GmTAP1* which resulted in improved resistance to soybean root rot (Liu et al. 2023).

5.6 Wheat

Wheat is the prominent staple food grain, which provides a high-calorie source. Major diseases like stem rust, loose smut, brown rust, yellow rust, black rust, and powdery mildew are causing damage to the crop. Among the fungal diseases, powdery mildew is the most significant, limiting wheat production. Hexaploid nature, big genome size (17 Gb), and high amount of repetitive sequence (80–90%) limit the application of gene modification methods. However, genome editing tools are precisely and widely applicable. Disruption of susceptibility genes is one of the strategies for creating disease resistance. A susceptibility gene namely *Mildew resistance locus O* (MLO) gives wide-range resistance to powdery mildew disease in many plant species. Recently, all three *TaMLO1* homologs were edited using CRISPR/Cas9 technology resulting in powdery mildew resistance in wheat with no yield penalty (Li et al. 2022). Another vital gene, enhanced disease resistance *EDR1* is detrimental to powdery mildew resistance. It affects protein levels and mitogen-activated protein kinase activity (e.g. MPK3 and MPK6) to modulate the plant's innate immunity. The knock-down of three *TaEDR1* homeologs using the CRISPR/Cas9 system resulted in enhanced powdery mildew resistance in wheat (Zhang et al. 2017).

5.7 *Barley*

Wheat dwarf virus (WDV) belongs to *Geminiviridae* family that attacks barley and wheat, causing yield losses. CRISPR/Cas9-mediated WDV targeting created a barley variety that is resistant to this virus (Kis et al. 2019). They used four different sgRNA complementary to the other genomic regions of WDV, and the edited lines were evaluated using the infection process. Microrchidia (MORC) proteins are regulatory proteins involved in chromatin remodeling and gene expression in plants and animals. In barley, seven MORC family proteins were recently discovered, out of which *Hv*MORC1 and *Hv*MORC6a were linked with the RNA-directed DNA methylation pathway. CRISPR/Cas9-mediated single and double knock-out mutants of these genes helped to fight against *Bipolaris* spot blotch and *Fusarium* root rot in Barley (Galli et al. 2022).

5.8 *Banana*

Bananas are an important agricultural crop and an important means of revenue for tropical and subtropical farmers (FAOSTAT 2018). Commercial banana varieties are polyploid clones. These are derived from *Musa accuminata* (A genome) and *Musa balbisiana* (B genome). The main constraint for banana breeding programs is a double-stranded DNA virus known as banana streak virus (BSV). Multiple copies of this viral genome get integrated into the B genome of the host plant during the infection, and these viruses are called endogenous BSV (eBSV). Various stress conditions, *in-vitro* propagation and hybridization activate the multiplication of eBSV. The banana cultivars containing eBSV cannot be used as a parent in the breeding program. Therefore, inactivation of eBSV can be an effective way to control the dissemination of these viruses. Recently, the CRISPR/Cas9 technique was used to manipulate eBSV in the B genome of *Musa* spp., which resulted in the inactivation of eBSV (Tripathi et al. 2019). Another most destructive disease is Banana wilt, which is caused by the bacterial pathogen *Xanthomonas compestris* pv. *musacearum*. The *DMR6* gene is a susceptibility factor for different pathogens which encodes an oxygenase enzyme namely 2-oxoglutarate Fe(II)-dependent oxygenase. When a pathogen infects the plant, this gene becomes active. CRISPR/Cas9 mediated mutation in *MusaDMR6* orthologues resulted in resistance against Banana *Xanthomonas* wilt (BXW) (Tripathi et al. 2021).

6 Advantages and Challenges of CRISPR in Genome Editing

CRISPR is a revolutionary genome-editing tool with great promise, especially in plant disease management. CRISPR offers several advantages in the following areas:

a. **Accuracy and specificity**: CRISPR enables the exact targeting of plant genes responsible for disease resistance. Because of CRISPR accuracy, only the required genetic alterations are performed, minimizing the impacts of off-target effects.
b. **Efficiency**: Generating disease-resistant plants using traditional breeding techniques might take a while. CRISPR quickens the process and hastens the emergence of resistant plant variants by specifically aiming at and altering the pertinent genes.
c. **Broad applicability**: A wide variety of plant species can benefit from the versatility of CRISPR technology. Because of its versatility may be used as a valuable tool to combat a wide range of crop diseases.
d. **Mitigation of ecological consequences**: With CRISPR, the demand for chemical pesticides is decreased by selectively targeting the plant's genes that make them susceptible to disease. This may result in a more sustainable and ecologically friendly method of managing plant diseases.
e. **Higher crop quality and output**: Crop yields can be boosted by breeding plants with greater disease resistance. Plants that can withstand disease are more likely to flourish in harsh environments, which increases agricultural production.
f. **Combination of multiple characteristics**: CRISPR facilitates the simultaneous alteration of numerous genes, leading to the generation of disease-resistant crops. Traditional breeding techniques make this trait stacking difficult, but CRISPR's precision makes it possible.
g. **Remedy for emerging threats**: Rapid response to newly discovered plant diseases is made possible by CRISPR. Scientists can swiftly identify the genes linked to susceptibility and create resistant types to lessen the effects of newly discovered and developing plant diseases.
h. **Reduction of financial burden**: Significant financial losses in agriculture can be attributed to plant diseases. By fostering the growth of more disease-resistant crops, CRISPR helps farmers avoid financial losses due to crop failure.
i. **Biodiversity protection**: There is less need for chemical treatments that can affect the environment and non-target commodities when crops with increased disease resistance are developed. This has the potential to support agricultural environments' biodiversity protection.
j. **Development of regulatory frameworks**: Regulatory frameworks are changing to accommodate the uses of CRISPR technology as it advances. Simplified regulatory procedures have the potential to accelerate the commercialization of CRISPR-modified crops by enabling their quicker deployment.

7 Challenges of CRISPR in Genome Editing for Plant Disease Management

In the scientific community worldwide, CRISPR is regarded as a game-changing technology. Notwithstanding the many advantages that CRISPR technology brings to the genome editing industry, a number of legitimate problems and worries need to be addressed worldwide. The scientific community is now facing the following difficulties:

a. **Intellectual property issues**: Several patents pertaining to CRISPR technology give rise to intellectual property disputes. Legal challenges can impede research and development endeavors and induce unpredictability for research institutions and investigators.
b. **Public opinion**: The adoption of genetically engineered organisms, especially those altered with CRISPR, can be influenced by public opinion. Even when CRISPR-edited crops have advantages, their reception may be hampered by a lack of general knowledge, comprehension, and involvement.
c. **International trade and consumer acceptability**: There are differences in consumer acceptability of genetically modified crops, and importation and cultivation of genetically modified products are prohibited in certain nations or subject to stringent controls. This may affect the international trade in crops modified with CRISPR.
d. **Unexpected long-term consequences**: It is unclear what the long-term implications of CRISPR-mediated genetic alterations will be in plants. Further study is required to evaluate the possible ecological impact over extended periods, as modifications to the transformed plants may have unanticipated effects on environmental relationships.
e. **Ethical Issues**: Concerns regarding unforeseen ecological repercussions, biodiversity loss, and the possibility of technology misuse are among the ethical issues that arise when using CRISPR in agriculture. Gaining the approval and support of the general population requires addressing these issues.
f. **Regulatory issues**: Genetically modified products are subject to different regulatory frameworks worldwide. It can take a long time and be difficult to get regulatory permission for CRISPR-edited crops, which makes it more challenging to implement disease-resistant cultivars on time.
g. **Plant genome complexity obstacles**: Plant genomes can include many redundant genes and functional overlap. Because of this intricacy, it may be difficult to pinpoint and target specific genes linked to disease resistance without also impairing other vital plant activities.

8 Conclusions and Future Aspects

The advent of CRISPR-Cas technology has significantly transformed both fundamental and practical aspects of plant research. Beyond generating indel mutations, CRISPR-Cas-based editors have been engineered to enable precise modifications in the genome. These powerful tools possess unparalleled gene-editing abilities, contributing to the development of numerous crop varieties exhibiting enhanced agricultural traits. Their impact has revolutionized breeding methodologies, facilitating the creation of hundreds of improved crop varieties with superior agronomic performance.

Apart from the mentioned uses, CRISPR-Cas technology holds potential for novel applications. It can be adapted for editing mitochondrial and chloroplast genomes, tracing cell lineages to unveil developmental patterns in plants, constructing genetic circuits to process and transmit signals, creating plant biosensors for detecting internal and external cues, and exploring other facets of plant synthetic biology. Overall, CRISPR-Cas technology stands as a revolutionary force, poised to transform both agriculture and plant biotechnology continually.

References

Ackerman CM, Myhrvold C, Thakku SG et al (2020) Massively multiplexed nucleic acid detection with Cas13. Nature 582:277–282. https://doi.org/10.1038/s41586-020-2279-8

Ali Z, Abulfaraj A, Idris A et al (2015) CRISPR/Cas9-mediated viral interference in plants. Genome Biol 16:238. https://doi.org/10.1186/s13059-015-0799-6

Ali Z, Ali S, Tashkandi M et al (2016) CRISPR/Cas9-mediated immunity to geminiviruses: differential interference and evasion. Sci Rep 6:26912. https://doi.org/10.1038/srep26912

Andrivon D, Montarry J, Fournet S (2022) Plant health in a one health world: missing links and hidden treasures. Plant Pathol 23–29. https://doi.org/10.1111/ppa.13463

Arribere JA, Bell RT, Fu BXH et al (2014) Efficient marker-free recovery of custom genetic modifications with CRISPR/Cas9 in *Caenorhabditis elegans*. Genetics 198:837–846. https://doi.org/10.1534/genetics.114.169730

Azhar M, Phutela R, Kumar M, et al (2021) Rapid and accurate nucleobase detection using FnCas9 and its application in COVID-19 diagnosis. Biosens Bioelectron 183:113207. https://doi.org/10.1016/j.bios.2021.113207

Bajpai AB, Solanki A, Srivastava N, Semwal PP, Thapliyal P, Rayal R, Sahni S (2023) Crispr-Cas system: a revolutionizing tool for genome editing. Biochem Cell Arch 23(1). https://doi.org/10.51470/bca.2023.23.1.623

Balamurugan A, Sakthivel K, Gautam RK, Sharma SK, Kumar A (2020) *Ralstonia solanacearum*: biology and its management in solanaceous vegetable crops. Rhizosphere Microbes: Soil Plant Funct 259–289

Baltes NJ, Hummel AW, Konecna E et al (2015) Conferring resistance to geminiviruses with the CRISPR–Cas prokaryotic immune system. Nat Plants 1:15145. https://doi.org/10.1038/nplants.2015.145

Barrangou R, Fremaux C, Deveau H, Richards M, Boyaval P, Moineau S, Romero DA, Horvath P (2007) CRISPR provides acquired resistance against viruses in prokaryotes. Science 315:1709–1712. https://doi.org/10.1126/science.1138140

Bhatia S, and Yadav SK (2023) CRISPR-Cas for genome editing: classification, mechanism, designing and applications. Int J Biol Macromol 124054. https://doi.org/10.1016/j.ijbiomac.2023.124054

Bhaya D, Davison M, Barrangou R (2011) CRISPR-Cas systems in bacteria and archaea: versatile small RNAs for adaptive defense and regulation. Annu Rev Genet 45:273–297. https://doi.org/10.1146/annurev-genet-110410-132430

Birch PR, Whisson SC (2001) Phytophthora infestans enters the genomics era. Mol Plant Pathol 2(5):257–263. https://doi.org/10.1046/j.1464-6722.2001.00073.x

Bolotin A, Quinquis B, Sorokin A, Ehrlich SD (2005) Clustered regularly interspaced short palindrome repeats (CRISPRs) have spacers of extrachromosomal origin. Microbiology 151:2551–2561. https://doi.org/10.1099/mic.0.28048-0

Chen JS, Ma E, Harrington LB et al (2018) CRISPR-Cas12a target binding unleashes indiscriminate single-stranded DNase activity. Science 360:436–439. https://doi.org/10.1126/science.aar6245

Ching-Yuet To A, Ho-Ting Chu D, Wang AR et al (2018) A comprehensive web tool for toehold switch design. Bioinformatics 34:2862–2864. https://doi.org/10.1093/bioinformatics/bty216

Choudhury S, Hu H, Meinke H, Shabala S, Westmore G, Larkin P, Zhou M (2017) Barley yellow dwarf viruses: infection mechanisms and breeding strategies. Euphytica 213:1–22. https://doi.org/10.1007/s10681-017-1955-8

Cohn M, Bart RS, Shybut M, Dahlbeck D, Gomez M, Morbitzer R, Staskawicz BJ (2014) Xanthomonas axonopodis virulence is promoted by a transcription activator-like effector–mediated induction of a SWEET sugar transporter in cassava. Mol Plant Microbe Interact 27(11):1186–1198. https://doi.org/10.1094/MPMI-06-14-0161-R

Cong L, Ran FA, Cox D et al (2013) Multiplex genome engineering using CRISPR/Cas systems. Science 339:819–823. https://doi.org/10.1126/science.1231143

Cox DBT, Gootenberg JS, Abudayyeh OO et al (2017) RNA editing with CRISPR-Cas13. Science 358:1019–1027. https://doi.org/10.1126/science.aaq0180

Das A, Sharma N, Prasad M (2019) CRISPR/Cas9: a novel weapon in the arsenal to combat plant diseases. Front Plant Sci 9:2008. https://doi.org/10.3389/fpls.2018.02008

de Toledo P, Thomazella D, Brail Q, Dahlbeck D, Staskawicz B (2016) CRISPR-Cas9 mediated mutagenesis of a DMR6 ortholog in tomato confers broad-spectrum disease resistance. Plant Biol. https://doi.org/10.1101/064824

Dickinson DJ, Pani AM, Heppert JK et al (2015) Streamlined genome engineering with a self-excising drug selection cassette. Genetics 200:1035–1049. https://doi.org/10.1534/genetics.115.178335

Esvelt KM, Mali P, Braff JL, et al (2013) Orthogonal Cas9 proteins for RNA-guided gene regulation and editing. Nat Methods 10:1116. https://doi.org/10.1038/nmeth.2681

FAOSTAT (2018). http://www.fao.org/FAOSTAT./en/#data/QC. Accessed 26 Oct 20

Galli M, Martiny E, Imani J et al (2022) CRISPR/ *Sp* Cas9-mediated double knockout of barley Microrchidia MORC1 and MORC6a reveals their strong involvement in plant immunity, transcriptional gene silencing and plant growth. Plant Biotechnol J 20:89–102. https://doi.org/10.1111/pbi.13697

Gao Y, Li S, Zhang S et al (2021) *Sly* miR482e-3p mediates tomato wilt disease by modulating ethylene response pathway. Plant Biotechnol J 19:17–19. https://doi.org/10.1111/pbi.13439

Geffersa AG, Burdon JJ, Macfadyen S, Thrall PH, Sprague SJ, Barrett LG (2023) The socio-economic challenges of managing pathogen evolution in agriculture. Philos Trans R Soc B 378(1873):20220012. https://doi.org/10.1098/rstb.2022.0012

Gootenberg JS, Abudayyeh OO, Kellner MJ et al (2018) Multiplexed and portable nucleic acid detection platform with Cas13, Cas12a, and Csm6. Science 360:439–444. https://doi.org/10.1126/science.aaq0179

Gosavi G, Yan F, Ren B et al (2020) Applications of CRISPR technology in studying plant-pathogen interactions: overview and perspective. Phytopathol Res 2:21. https://doi.org/10.1186/s42483-020-00060-z

Greene AC (2018) CRISPR-based antibacterials: transforming bacterial defense into offense. Trends Biotechnol 36:127–130. https://doi.org/10.1016/j.tibtech.2017.10.021

Gupta AK, Mishra G, Yadav H, Gupta R, Singh A, Singh J, Singh P (2023) Enhancing crop resilience through CRISPR/Cas9-mediated development of disease-resistant cultivars. Int J Environ Clim Change 13(10):2773–2783. https://doi.org/10.9734/ijecc/2023/v13i102942

He DC, Zhan JS, Xie LH (2016) Problems, challenges and future of plant disease management: from an ecological point of view. J Integr Agric 15(4):705–715. https://doi.org/10.1016/S2095-3119(15)61300-4

Hirano H, Gootenberg JS, Horii T et al (2016) Structure and engineering of *Francisella novicida* Cas9. Cell 164:950–961. https://doi.org/10.1016/j.cell.2016.01.039

Hsu YC, Chiu CH, Yap R, Tseng YC, Wu YP (2020) Pyramiding bacterial blight resistance genes in Tainung82 for broad-spectrum resistance using marker-assisted selection. Int J Mol Sci 21(4):1281. https://doi.org/10.3390/ijms21041281

Hu Y, Li W (2022) Development and application of CRISPR-Cas based tools. Front Cell Dev Biol 10:834646. https://doi.org/10.3389/fcell.2022.834646

Huang Q, Lin B, Cao Y et al (2023) CRISPR/Cas9-mediated mutagenesis of the susceptibility gene OsHPP04 in rice confers enhanced resistance to rice root-knot nematode. Front Plant Sci 14:1134653. https://doi.org/10.3389/fpls.2023.1134653

Hull R (1996) Molecular biology of rice tungro viruses. Annu Rev Phytopathol 34(1):275–297. https://doi.org/10.1146/annurev.phyto.34.1.275

Ishino Y, Shinagawa H, Makino K, Amemura M, Nakata A (1987) Nucleotide sequence of the iap gene, responsible for alkaline phosphatase isozyme conversion in Escherichia coli, and identification of the gene product. J Bacteriol 169:5429– 5433. https://doi.org/10.1128/jb.169.12.5429-5433.1987

Iqbal Z, Sattar MN, Shafiq M (2016) CRISPR/Cas9: a tool to circumscribe cotton leaf curl disease. Front Plant Sci. https://doi.org/10.3389/fpls.2016.00475

Jia H, Zhang Y, Orbović V, Xu J, White FF, Jones JB, Wang N (2017) Genome editing of the disease susceptibility gene Cs LOB 1 in citrus confers resistance to citrus canker. Plant Biotechnol J 15(7):817–823. https://doi.org/10.1111/pbi.12677

Jiang N, Yan J, Liang Y, Shi Y, He Z, Wu Y, Peng J (2020) Resistance genes and their interactions with bacterial blight/leaf streak pathogens (*Xanthomonas oryzae*) in rice (*Oryza sativa* L.)—an updated review. Rice 13(1):1–12. https://doi.org/10.1186/s12284-019-0358-y

Jiao C, Sharma S, Dugar G et al (2021) Noncanonical crRNAs derived from host transcripts enable multiplexable RNA detection by Cas9. Science 372:941–948. https://doi.org/10.1126/science.abe7106

Jinek M, Chylinski K, Fonfara I et al (2012) A programmable dual-RNA–guided DNA endonuclease in adaptive bacterial immunity. Science 337:816–821. https://doi.org/10.1126/science.1225829

Karmakar S, Das P, Panda D, Xie K, Baig MJ, Molla KA (2022) A detailed landscape of CRISPR-Cas-mediated plant disease and pest management. Plant Sci 323:111376. https://doi.org/10.1016/j.plantsci.2022.111376

Khan MZ, Haider S, Mansoor S, Amin I (2019) Targeting plant ssDNA viruses with engineered miniature CRISPR-Cas14a. Trends Biotechnol 37:800–804. https://doi.org/10.1016/j.tibtech.2019.03.015

Khanna K, Ohri P, Bhardwaj R (2023) Nanotechnology and CRISPR/Cas9 system for sustainable agriculture. Environ Sci Pollut Res. https://doi.org/10.1007/s11356-023-26482-8

Kim H, Ishidate T, Ghanta KS et al (2014) A co-CRISPR strategy for efficient genome editing in *Caenorhabditis elegans*. Genetics 197:1069–1080. https://doi.org/10.1534/genetics.114.166389

Kim YA, Moon H, Park CJ (2019) CRISPR/Cas9-targeted mutagenesis of Os8N3 in rice to confer resistance to *Xanthomonas oryzae* pv. *Oryzae*. Rice 12(1):1–13. https://doi.org/10.1186/s12284-019-0325-7

Kis A, Hamar É, Tholt G, Bán R, Havelda Z (2019) Creating highly efficient resistance against wheat dwarf virus in barley by employing CRISPR/Cas9 system. Plant Biotechnol J 17(6):1004. https://doi.org/10.1111/pbi.13077

Kleinstiver BP, Pattanayak V, Prew MS, et al (2016) High-fidelity CRISPR–Cas9 nucleases with no detectable genome-wide off-target effects. Nature 529:490–495. https://doi.org/10.1038/nature16526

Kumam Y, Rajadurai G, Kumar KK, Varanavasiappan S, Reddy MK, Krishnaveni D, Sudhakar D (2022) Genome editing of indica rice ASD16 for imparting resistance against rice tungro disease. J Plant Biochem Biotechnol. https://doi.org/10.1007/s13562-021-00765-y

Kuroiwa K, Thenault C, Nogué F, Perrot L, Mazier M, Gallois JL (2022) CRISPR-based knock-out of eIF4E2 in a cherry tomato background successfully recapitulates resistance to pepper veinal mottle virus. Plant Sci 316:111160. https://doi.org/10.1016/j.plantsci.2021.111160

Li L, He Z-Y, Wei X-W et al (2015) Challenges in CRISPR/CAS9 delivery: potential roles of nonviral vectors. Hum Gene Ther 26:452–462. https://doi.org/10.1089/hum.2015.069

Li Y, Xiao J, Chen L, Huang X, Cheng Z, Han B, Wu C (2018) Rice functional genomics research: past decade and future. Mol Plant 11(3):359–380. https://doi.org/10.1016/j.molp.2018.01.007

Li S, Shen L, Hu P et al (2019) Developing disease-resistant thermosensitive male sterile rice by multiplex gene editing. JIPB 61:1201–1205. https://doi.org/10.1111/jipb.12774

Li S, Lin D, Zhang Y, Deng M, Chen Y, Lv B, Gao C (2022) Genome-edited powdery mildew resistance in wheat without growth penalties. Nature 602(7897):455–460. https://doi.org/10.1038/s41586-022-04395-9

Liu P, Long L, Xiong K et al (2014) Heritable/conditional genome editing in *C. elegans* using a CRISPR-Cas9 feeding system. Cell Res 24:886–889. https://doi.org/10.1038/cr.2014.73

Liu T, Ji J, Cheng Y, Zhang S, Wang Z, Duan K, Wang Y (2023) CRISPR/Cas9-mediated editing of GmTAP1 confers enhanced resistance to *Phytophthora sojae* in soybean. J Integr Plant Biol. https://doi.org/10.1111/jipb.13476

Lučanský V, Holubeková V, Kolková Z, Halašová E, Samec M, Golubnitschaja O (2023) Multi-faceted CRISPR/Cas technological innovation aspects in the framework of 3P medicine. EPMA J. https://doi.org/10.1007/s13167-023-00324-6

Ma J, Chen J, Wang M, Ren Y, Wang S, Lei C, Sodmergen (2018) Disruption of OsSEC3A increases the content of salicylic acid and induces plant defense responses in rice. J Exp Bot 69(5):1051–1064. https://doi.org/10.1093/jxb/erx458

Makarova KS, Wolf YI, Koonin EV (2018) Classification and nomenclature of CRISPR-cas systems: where from here? CRISPR J 1:325–336. https://doi.org/10.1089/crispr.2018.0033

Ming M, Ren Q, Pan C et al (2020) CRISPR–Cas12b enables efficient plant genome engineering. Nat Plants 6:202–208. https://doi.org/10.1038/s41477-020-0614-6

Mohd A, Phutela R, Kumar M et al (2021) Rapid and accurate nucleobase detection using FnCas9 and its application in COVID-19 diagnosis. Biosens Bioelectron 183:113207. https://doi.org/10.1016/j.bios.2021.113207

Mojica FJ, Diez-Villaseñor C, Garcia-Martínez J, Soria E (2005) Intervening sequences of regularly spaced prokaryotic repeats derive from foreign genetic elements. J Mol Evol 60:174–182

Myhrvold C, Freije CA, Gootenberg JS et al (2018) Field-deployable viral diagnostics using CRISPR-Cas13. Science 360:444–448. https://doi.org/10.1126/science.aas8836

Nalley L, Tsiboe F, Durand-Morat A, Shew A, Thoma G (2016) Economic and environmental impact of rice blast pathogen (*Magnaporthe oryzae*) alleviation in the United States. PLoS ONE 11(12):e0167295. https://doi.org/10.1371/journal.pone.0167295

Nasir F, Tian L, Chang C, Li X, Gao Y, Tran LSP, Tian C (2018) Current understanding of pattern-triggered immunity and hormone-mediated defense in rice (*Oryza sativa*) in response to *Magnaporthe oryzae* infection. Semin Cell Develop Biol 83:95–105. https://doi.org/10.1016/j.semcdb.2017.10.020

Nekrasov V, Wang C, Win J et al (2017) Rapid generation of a transgene-free powdery mildew resistant tomato by genome deletion. Sci Rep 7:482. https://doi.org/10.1038/s41598-017-00578-x

Ngou BPM, Jones JD, Ding P (2022) Plant immune networks. Trends Plant Sci 27(3):255–273. https://doi.org/10.1016/j.tplants.2021.08.012

Niño-Liu DO, Ronald PC, Bogdanove AJ (2006) *Xanthomonas oryzae* pathovars: model pathogens of a model crop. Mol Plant Pathol 7(5):303–324. https://doi.org/10.1111/j.1364-3703.2006.00344.x

Noureen A, Khan MZ, AminI ZT, Mansoor S (2022) CRISPR/Cas9-mediated targeting of susceptibility factor eIF4E-enhanced resistance against potato virus Y. Front Genet 13:922019. https://doi.org/10.3389/fgene.2022.922019

Ortigosa A, Gimenez-Ibanez S, Leonhardt N, Solano R (2019) Design of a bacterial speck resistant tomato by CRISPR/Cas9-mediated editing of Sl JAZ 2. Plant Biotechnol J 17(3):665–673. https://doi.org/10.1111/pbi.13006

Pandya K, Jagani D, Singh N (2023) CRISPR-Cas systems: programmable nuclease revolutionizing the molecular diagnosis. Mol Biotechnol. https://doi.org/10.1007/s12033-023-00819-7

Paula de Toledo Thomazella D, Brail Q, Dahlbeck D, Staskawicz B (2016) CRISPR-Cas9 mediated mutagenesis of a DMR6 ortholog in tomato confers broad-spectrum disease resistance. BioRxiv 064824. https://doi.org/10.1101/064824

Peng A, Chen S, Lei T et al (2017) Engineering canker-resistant plants through CRISPR /Cas9-targeted editing of the susceptibility gene *Cs LOB 1* promoter in citrus. Plant Biotechnol J 15:1509–1519. https://doi.org/10.1111/pbi.12733

Pramanik D, Shelake RM, Park J, Kim MJ, Hwang I, Park Y, Kim JY (2021) CRISPR/Cas9-mediated generation of pathogen-resistant tomato against tomato yellow leaf curl virus and powdery mildew. Int J Mol Sci 22(4):1878. https://doi.org/10.3390/ijms22041878

Prihatna C, Larkan NJ, Barbetti MJ, Barker SJ (2018) Tomato CYCLOPS/IPD3 is required for mycorrhizal symbiosis but not tolerance to *Fusarium wilt* in mycorrhiza-deficient tomato mutant rmc. Mycorrhiza 28:495–507. https://doi.org/10.1007/s00572-018-0842-z

Quan J, Langelier C, Kuchta A et al (2019) FLASH: a next-generation CRISPR diagnostic for multiplexed detection of antimicrobial resistance sequences. Nucleic Acids Res 47:e83–e83. https://doi.org/10.1093/nar/gkz418

Rao MJ, Wang L (2021) CRISPR/Cas9 technology for improving agronomic traits and future prospective in agriculture. Planta 254:1–16

Ren B, Yan F, Kuang Y et al (2018) Improved base editor for efficiently inducing genetic variations in rice with CRISPR/Cas9-guided hyperactive hAID mutant. Mol Plant 11:623–626. https://doi.org/10.1016/j.molp.2018.01.005

Sahu B, Choudhary VK, Sahu MP, Kumar KK, Sujayanand GK, Gopi R, Ghosh PK (2023) Biotic stress management. In: Trajectory of 75 years of Indian agriculture after independence. Springer Nature Singapore, Singapore, pp 619–653. https://doi.org/10.1007/978-981-19-7997-2_23

Santiago-Frangos A, Henriques WS, Wiegand T, Gauvin CC, Buyukyoruk M, Graham AB, Wiedenheft B (2023) Structure reveals why genome folding is necessary for site-specific integration of foreign DNA into CRISPR arrays. Nat Struct Mol Biol. https://doi.org/10.1038/s41594-023-01097-2

Santillán Martínez MI, Bracuto V, Koseoglou E, Appiano M, Jacobsen E, Visser RG, Bai Y (2020) CRISPR/Cas9-targeted mutagenesis of the tomato susceptibility gene PMR4 for resistance against powdery mildew. BMC Plant Biol 20(1):1–13. https://doi.org/10.1186/s12870-020-02497-y

Schie CCN, Takken FLW (2014) Susceptibility genes 101: how to be a good host. Annu Rev Phytopathol 52:551–581. https://doi.org/10.1146/annurev-phyto-102313-045854

Schwartz ML, Jorgensen EM (2016) SapTrap, a toolkit for high-throughput CRISPR/Cas9 gene modification in *Caenorhabditis elegans*. Genetics 202:1277–1288. https://doi.org/10.1534/genetics.115.184275

Shamim M, Kumar M, Kumar S, Srivastava D, Ranjan T, Kesari R, Jha V B (2024) Economical and environmental impact of rice fungal diseases on global food security. In: Fungal diseases of rice and their management. Apple Academic Press, pp 1–30

Shmakov S, Smargon A, Scott D et al (2017) Diversity and evolution of class 2 CRISPR–Cas systems. Nat Rev Microbiol 15:169–182. https://doi.org/10.1038/nrmicro.2016.184

Singh BK, Delgado-Baquerizo M, Egidi E, Guirado E, Leach JE, Liu H, Trivedi P (2023a) Climate change impacts on plant pathogens, food security and paths forward. Nat Rev Microbiol. https://doi.org/10.1186/s12870-020-02497-y

Singh S, Chaudhary R, Deshmukh R, Tiwari (2023b) Opportunities and challenges with CRISPR-Cas mediated homologous recombination based precise editing in plants and animals. Plant Mol Biol 111(1–2):1–20. https://doi.org/10.1007/s11103-022-01321-5

Talakayala A, Ankanagari S, Garladinne M (2022) CRISPR-Cas genome editing system: a versatile tool for developing disease resistant crops. Plant Stress 3:100056. https://doi.org/10.1016/j.stress.2022.100056

Teng F, Cui T, Feng G et al (2018) Repurposing CRISPR-Cas12b for mammalian genome engineering. Cell Discov 4:63. https://doi.org/10.1038/s41421-018-0069-3

Thomazella DP, Brail Q, Dahlbeck D, Staskawicz B (2016) CRISPR-Cas9 mediated mutagenesis of a *DMR6* ortholog in tomato confers broad-spectrum disease resistance. Plant Biol

Tripathi J N, Ntui VO, Ron M, Muiruri SK, Britt A, and Tripathi L (2019) CRISPR/Cas9 editing of endogenous banana streak virus in the B genome of Musa spp. overcomes a major challenge in banana breeding. Commun Biol 2(1):46. https://doi.org/10.1038/s42003-019-0288-7

Tripathi JN, Ntui VO, Shah T, Tripathi L (2021) CRISPR/Cas9-mediated editing of DMR6 orthologue in banana (Musa spp.) confers enhanced resistance to bacterial disease. Plant Biotechnol J 19(7):1291. https://doi.org/10.1111/pbi.13614

Waheed A, Haxim Y, Islam W, Ahmad M, Muhammad M, Alqahtani FM, Zhang D (2023) Climate change reshaping plant-fungal interaction. Environ Res. https://doi.org/10.1016/j.envres.2023.117282

Wang Y, Cheng X, Shan Q et al (2014) Simultaneous editing of three homoeoalleles in hexaploid bread wheat confers heritable resistance to powdery mildew. Nat Biotechnol 32:947–951. https://doi.org/10.1038/nbt.2969

Wang Z, Hardcastle TJ, Canto Pastor A et al (2018) A novel DCL2-dependent miRNA pathway in tomato affects susceptibility to RNA viruses. Genes Dev 32:1155–1160. https://doi.org/10.1101/gad.313601.118

Wang G, Wang C, Lu G, Wang W, Mao G, Habben JE, Greene TW (2020) Knockouts of a late flowering gene via CRISPR–Cas9 confer early maturity in rice at multiple field locations. Plant Mol Biol 104(1–2):137–150. https://doi.org/10.1007/s11103-020-01031-w

Wang F, Wang C, Liu P et al (2016) Enhanced rice blast resistance by CRISPR/Cas9-targeted mutagenesis of the ERF transcription factor gene OsERF922. PLoS ONE 11:e0154027. https://doi.org/10.1371/journal.pone.0154027

Wheatley MS, Yang Y (2021) Versatile applications of the CRISPR/Cas toolkit in plant pathology and disease management. Phytopathology® 111:1080–1090. https://doi.org/10.1094/PHYTO-08-20-0322-IA

Xu Z, Xu X, Gong Q, Li Y, Wang S, Chen G (2019) Engineering broad-spectrum bacterial blight resistance by simultaneously disrupting variable TALE-binding elements of multiple susceptibility genes in rice. Mol Plant 12(11):1434–1446. https://doi.org/10.1016/j.molp.2019.08.006

Yu W, Wang L, Zhao R, Sheng J, Zhang S, Li R, Shen L (2019) Knockout of SlMAPK3 enhances tolerance to heat stress involving ROS homeostasis in tomato plants. BMC Plant Biol 19(1):1–13. https://doi.org/10.1186/s12870-019-1939-z

Yu K, Liu Z, Gui H, Geng L, Wei J, Liang D, Chen X (2021) Highly efficient generation of bacterial leaf blight-resistant and transgene-free rice using a genome editing and multiplexed selection system. BMC Plant Biol 21(1):1–10. https://doi.org/10.1186/s12870-021-02979-7

Zaidi SS-A, Mahfouz MM, Mansoor S (2017) CRISPR-Cpf1: a new tool for plant genome editing. Trends Plant Sci 22:550–553. https://doi.org/10.1016/j.tplants.2017.05.001

Zamanian M, Andersen EC (2016) Prospects and challenges of CRISPR/Cas genome editing for the study and control of neglected vector-borne nematode diseases. FEBS J 283:3204–3221. https://doi.org/10.1111/febs.13781

Zeng X, Luo Y, Vu NTQ et al (2020) CRISPR/Cas9-mediated mutation of OsSWEET14 in rice cv. Zhonghua11 confers resistance to *Xanthomonas oryzae* pv. *oryzae* without yield penalty. BMC Plant Biol 20:313. https://doi.org/10.1186/s12870-020-02524-y
Zhang Y, Bai Y, Wu G et al (2017a) Simultaneous modification of three homoeologs of *Ta* EDR *1* by genome editing enhances powdery mildew resistance in wheat. Plant J 91:714–724. https://doi.org/10.1111/tpj.13599
Zhang T, Zheng Q, Yi X et al (2018) Establishing RNA virus resistance in plants by harnessing CRISPR immune system. Plant Biotechnol J 16:1415–1423. https://doi.org/10.1111/pbi.12881
Zhang M, Liu Q, Yan X, Xu J, Liu G, Yao X, Lou L (2020) CRISPR/Cas9-mediated mutagenesis of Clpsk1 in watermelon to confer resistance to *Fusarium oxysporum* f. sp. *niveum*. Plant Cell Rep 39:589–595. https://doi.org/10.1007/s00299-020-02516-0
Zhou X, Liao H, Chern M et al (2018) Loss of function of a rice TPR-domain RNA-binding protein confers broad-spectrum disease resistance. Proc Natl Acad Sci USA 115:3174–3179. https://doi.org/10.1073/pnas.1705927115

The manufacturer's authorised representative in the EU is Springer Nature Customer Service Centre GmbH, Europaplatz 3, 69115 Heidelberg, Germany. If you have any concerns regarding our products, please contact ProductSafety@springernature.com

Printed and bound by CPI Group (UK) Ltd, Croydon, CR0 4YY
23/09/2026
02221538-0003